MATHEMATICS FOR THE ENVIRONMENT

MATHEMATICS FOR THE ENVIRONMENT

MARTIN WALTER
University of Colorado
Boulder, Colorado, USA

CRC Press
Taylor & Francis Group
Boca Raton London New York
CRC Press is an imprint of the
Taylor & Francis Group an **informa** business
A CHAPMAN & HALL BOOK

Chapman & Hall/CRC
Taylor & Francis Group
6000 Broken Sound Parkway NW, Suite 300
Boca Raton, FL 33487-2742

International Standard Book Number: 978-1-4398-3472-5 (Hardback)

Visit the Taylor & Francis Web site at
http://www.taylorandfrancis.com

and the CRC Press Web site at
http://www.crcpress.com

Dedication

I dedicate this book to the memory of Judson "Sandy" Sanderson, to Baby Lorax, and to the Joy of my life.

Contents

List of Tables

List of Figures

Why Did I Write This Book?

We are not going to solve the big problems we face by doing things just as we have been. This book/course is my very modest attempt at positive change. I propose that for teachers and students so inclined, a course on mathematics and environmental issues be introduced as a possible alternative to "college algebra," "liberal arts math," and possibly "math for prospective teachers," or as an entry level course in mathematical modeling. More generally, in the ideal democracy citizens should know enough useful, basic mathematics to understand and effectively deal—on their own terms—with economic, environmental, and social crises, about which there is abundant conflicting information. Thus this book could have been titled: "Citizen Math." It is a beginning, perhaps a middle, but not an end.

I am curious about many things. How did the food I ate today get to me? What happened to it along the way? Will the system that provides my food be around and working in a ten years? Is the food safe, does it promote health? Are there any simple lessons to be learned from the economic meltdown of 2008–2009? (Yes, and "we" have not learned them!) Is global climate change really happening? If so, are there any parts of this change that are easily understood? Does it matter? Is there enough solar and wind power to run the U.S. economy? The world economy? Were some humans really doing some serious mathematical thinking 50,000 years ago? What does the second law of thermodynamics have to do with economics? Is it easy to understand? Does history help us solve problems today? Is the "megamedia" providing me with all the information I need? (I don't think so.) How does one go about finding out what one needs to know? (An ongoing exercise, with hints.) Should I encrypt my email? Can somebody steal my identity, get credit cards in my name, and run up a huge debt? Can I prevent that from happening? What is a mathematical proof? What does a proof prove? Why am I not allowed to divide by zero? These and many, many other topics are all connected and covered in this book, and elementary mathematics is ever present. The "good news" is that one can get quite far in understanding these things with no more than one or two years of (American level) high school mathematics; often encountered, less often mastered. I wrote this book for people who do not consider themselves "math whizzes" but are curious about how things work and what mathematics has to do with it. I also wrote this book in the hope of helping sufficiently many of us understand enough about "how things work" so that "things keep working." Allow me to explain a bit.

In many ways the logic of most of human civilization is at odds with the logic of Nature, and there are so many of us that it is beginning to show. Just one example: most human economic systems are wedded to material growth without bounds and the idea that whatever is desired to maintain this growth can be freely taken. I have never seen a serious politician run on a platform of a stable, as opposed to a materially growing, economy. There are several futures for us that are mathematically, scientifically, socially, economically, historically and politically possible. One reason I wrote this book is to avoid a future "Age of Cannibalism." The Alferd Packer Grill, at my university, has the slogan: "Have a friend for lunch." Al Packer (1842–1907) was convicted in Colorado of manslaughter, for allegedly[1] killing and eating his companions after they became snowbound one winter in the Rocky Mountains without (other) sources of food. Human cannibalism has been documented in the prehistoric American Southwest [689],[2] not far from where I am writing, as well as in places such as Europe in historic times [546, 282]. When food is scarce, and even when it is not, eating fellow humans is not without precedent. Thus it remains a possibility in the future—that I would like to avoid. I would even like to avoid "financial cannibalism," a few people consuming the life savings of others. I would like to avoid "slavery-lite" where the few rule and greatly limit society's possibilities. Given the self-organizing generosity that whole populations exhibit from time to time, removing rule by a few would likely lead to happier communities.

At the other end of a spectrum of possible futures is one where everyone has enough to eat without resorting to cannibalism. In fact, it seems possible that all, or at least most, of us could have a reasonably enjoyable, peaceful existence, sustainable for many generations. In between the extremes we humans may muddle through to a mixed future with a few, or possibly massive, casualties. My contention is that what we humans do today will largely determine our possible tomorrows, and that elementary mathematics can help us make informed choices now that will lead us to at least a tolerable future. Parts of this book deal with things that elementary mathematics strongly suggests we do in order to have a future we will enjoy. At the very least a compelling, persuasive argument is required before we ignore what mathematics tells us. For example, the laws of mathematics and physics predict that jumping out of an airplane at 10,000 meters (about 32,808.3 feet) unassisted by some device, leads to a grim future—rather soon. Whereas, a jump with a parachute can be enjoyable, at least for some.

While gravity and its implications are obvious without doing many calculations, mathematics has less obvious but similarly fundamental things to say: about conditions that are necessary for a financial system to be stable and not collapse; about pollution and its effects; about the growth of populations

[1]Packer claimed that Shannon Wilson Bell was the true cannibal, whom he shot in self defense.

[2]This number in brackets refers to a reference in the list of references near the back of the book.

or their decline and extinction; the progression of diseases such as the flu or HIV; about our food supply; our ecological support systems and many other subjects.

There is another reason I wrote this book. From time to time I have taught "College Algebra" and "Math for Liberal Arts," and while such courses have much merit, for me something was missing. Most of the students I have had in these courses were interested in and talented in some area: history, dance, art, music, car mechanics, journalism, literature, poetry, economics, business, anthropology, psychology, philosophy, politics, social science, science, even mathematics. Most, however, were trying to avoid mathematics entirely; and they had been successful until a dreaded math requirement caught up with them. In all of the abovementioned areas of interest that held my students' attention I could/can see some mathematics—real math, not made up math. Couple this with the fact that a mathematical perspective can help us solve some of the major problems that we all face or may face in the near future, and I ended up designing the course: "Mathematics for the Environment." Over a period of ten to fifteen years I wrote various versions of this book for that course. This book is for those who may consider themselves not very mathematically talented, but are at least curious or concerned about some of the topics already mentioned above—and would like to know a bit about how mathematics could possibly be relevant—or even be of help. I have had students of just about every political persuasion. If a student uses what they have learned to help them act in their own self-interest the course(book) has been a success, since a great many people are persuaded by propaganda to act against their own self-interest. Of course, the course(book) is an even greater success to the extent that it helps us all work together to accomplish things we could not accomplish individually.

Finally, I would have been somewhat surprised over a decade ago when I started this project to see how it turned out. It is clear that this book must be interdisciplinary. However, over the years my interaction with other disciplines led me to realize the importance of certain principles/concepts with mathematical content which would not have so starkly stood out had I remained completely within pure mathematics. Among these are: fuzzy logic; the Bio-Copernican Axiom (humans are not the center of the biosphere); the Connection Axiom; cycles and feedback; multiplier effects; the Dunbar Number, 147.8 (a biological constraint on the number of people to whom one person can be "strongly connected"); and others I will cover in this book. Of course, such topics as fractions and logarithms come up, but this is no surprise!

Reading, Learning and/or Teaching from This Book

If you want to learn a language you need to practice, get involved. It is the same for mathematics, and one of the main ways to get involved is to do the exercises. Especially for the student who is afraid of, allergic to, or otherwise distressed by mathematics, I have tried to provide at least one exercise in each group that can be done after a careful reading, perhaps more than once, of the corresponding section. The exercises are numbered so that, for example, Exercise 2.1 is the first exercise in Chapter 2. Each exercise usually has "subparts" numbered with lowercase roman numerals, such as: (i) = 1, (ii) = 2, (iii) = 3, (iv) = 4, (v) = 5 , (vi) = 6, and so on. Some of the exercises are more difficult, some require exploring outside this book. Some exercises are "open ended" in the sense that our answers to them can always be expanded. Occasionally I include an exercise for which I do not know the (complete) answer. The majority of exercises are interdisciplinary, a mixture of math with some *other* part of real life. (Yes, mathematical principles are a part of each of our lives—whether anyone realizes it or not!) The goal is to deepen understanding of how things really work. Even the occasional "pure math" exercises are designed to show how mathematics works. I might add that if you see a word or abbreviation that you do not understand—consult a dictionary! As an exercise you can look up the Latin words that have the following abbreviations; thus, e.g., means "for example," i.e., means "that is," and viz., means "namely" or "that is to say," and cf., means "compare." (I'll do the first one for you; hence e.g. is the abbreviation for the Latin *exempli gratia*, which means as I just said "for example.")

Anytime you have problems reading the text or doing an exercise it is always a useful strategy to ask a friend or classmate. If two or more of you cannot figure out the answer, then your group can approach the teacher, if you have one. It is my intention that this book be useful for self study, but it is advantageous to have an instructor. The chapters were written to be read in the order presented, for the most part; however, it is quite possible to dive into any chapter and start reading without much difficulty. Whenever you see a number in a form such as [28], I am citing a reference in the list of references towards the end of the book just before the index. In this case [28] refers to a book by Albert A. Bartlett, *The Essential Exponential (for the future of our planet).*

I need to emphasize that this is not a pure mathematics text, it is truly interdisciplinary. No one I know is simultaneously an expert in all of the areas visited, including me. This might cause a bit of apprehension in some instructors, it might excite a sense of adventure in others. Fortunately in most of the areas discussed there are simple "necessary observations" that mathematics can help illuminate. (For example, a stable system of finance requires an honest assessment of risk.) It is true, however, that none of us are experts in everything. Such is life, for we as citizens are often confronted with new things, forcing us to make decisions with incomplete information—requiring us to research the best we can in the time we have. In any event, with regard to any of the many topics discussed I do not claim to be offering the "last word" on any of them. Rather consider each chapter an introduction that highlights some (hopefully) interesting information worthy of further investigation. Occasionally it turns out that an "ordinary" person discovers answers that were missed by experts.

I have heard variants of the following comments in the same day: "This book does not have enough mathematics." "The book is very interesting, too bad there is so much mathematics." I find it somewhat pointless to dispute the first comment, since for this book my working definition of mathematics is: *the search for and study of patterns.* Patterns are everywhere, thus the entire book is either mathematics or proto-mathematics. But if you accept as a worthy goal showing how traditional mathematics arises naturally, and is useful, in other subjects, then you have to spend some time talking about the other subjects while introducing the mathematics. On the other hand, the entire book is written from the *perspective* of mathematics. Mathematically averse students have told me that the mathematics is easier to take in the concrete contexts developed herein.

I believe that at least some of the people with mathematical knowledge or talent should devote some of their time seeking solutions to the major problems of our time, or the not too distant future: climate change; overshoot of carrying capacity and/or ecological collapse (local or global); problems associated with food, energy, pollution, environmentally induced disease; economic collapse; disappearing justice—to name a few. It follows that any insights gained should be communicated to others, via our educational institutions and other avenues. If this book is found to be helpful in this regard, feel free to use it, expand and improve it. Adapt it to any circumstances special to you.

Human history becomes more and more a race
between education and catastrophe.

H.G. Wells

In this book I often find it necessary to be critical, a position that is sometimes not appreciated by all. On the other hand, I hope that I have been sufficiently critical where necessary. To remind me of this I have a picture of a landscape brutalized by the hand of man hanging above my desk emblazoned with the following quote from Shakespeare: *"Forgive me thou bleeding piece of earth that I am meek and gentle with these butchers."*

Some Details on How to Read This Book. If you have a command of fractions and are somewhat mature (this is often independent of age), you can dive into just about any chapter that captures your interest. If you want to start by studying "pure" mathematics, you can start with Part II, often abbreviated simply as, II. If you want to warm up to II, we recommend Chapters 1 (on climate change), 2 (financial collapse), and 3 (some basics). If you are uncomfortable with powers of 10, e.g., 10^{12} is a trillion, a 1 followed by 12 zeros, read Section 3.4 and do Exercise 3.11 at the same time you are reading Chapters 1 and 2. One more technical detail: in order to read Chapter 20, and later parts of Part VII, you need to master the Σ notation of Chapter 10.

Part I, or just I, is rather long. It is not necessary to read all of Part I before reading other parts of the book. One approach would be to read, say, the first three chapters of I, and then go on to II; returning to I as the mood strikes.

There is enough material to easily occupy a one semester course, and hopefully a sufficient diversity of topics to catch 'most anyone's attention.

Someday I hope to communicate with you via the Web sites, MartyWalterMath.com or MartyWalterMath.org. For example, when I get the time I plan to post therein additional helpful hints (maybe yours!) on how to teach and learn from and otherwise use the material in this book. By the way, the equation on the cover is the Greenhouse Law for CO_2 and is discussed in Section 1.9.

I (penultimately) close these introductory comments with a bit of humor[3] directed at everyone, including myself.

I finally close with comments on a couple of things to expect and not to expect from this book. What I have tried to do as often as possible is "give

[3]Cartoon reproduced with the permission of Gahan Wilson.

a reason to learn a given topic." As one of my students said recently: "Your book, for the first time, gives me real reasons to learn some mathematics." This, I believe, is one of the main attractions of this book. Once a topic is motivated and introduced, I give only a few exercises. This is because in my many years of teaching I have found that finding a real, attention-getting example that captures, inspires, and motivates the "general student" is the hardest part. If a student truly understands one or two examples of the math involved, he or she does not need endless repetition. Much repetition is useful in gaining proficiency in implementing a given algorithmic solution, but students who "need" this are often unable to perform if the algorithm is changed slightly. Repetition bores the gifted students and most often bores, does not really help, and is unappreciated by the math averse. Some concepts are repeated in this book, for example, use of geometric sums; but each time the mathematics is motivated by a real situation.

Consider the following which is typical: I introduce "percent change" in the second exercise in Chapter 1. If either the student or teacher wishes at that time to do many more exercises on "percent change," such exercises are available on the Web or in easily available library books. (I would be happy to explicitly list some of these resources on MartyWalterMath.org, or even post some additional exercises in the spirit of this book, if I get sufficient feedback from readers desiring such.)

I have chosen to use precious printed space to introduce new topics, contexts, and pressing environmental information and issues, rather than "repeat exercises." The Web is ideal for providing repetition of a topic once discovered; this book is useful for singling out and emphasizing topics not necessarily easily found by someone working individually.

In Part II we study some elementary mathematical structures from a mathematically mature perspective. One of the goals of Part II is to give the "general student" the ability to manipulate mathematical expressions with "letters" or symbols in them that represent numbers. In my interaction with professors in a variety of disciplines, one common request came up again and again; namely, please teach students how to deal with mathematical expressions and equations which contain letters that represent numbers, but do not contain specific numbers. In Part II we learn that numbers are mathematical objects that have certain properties. Thus if a letter represents a number, the student has "handles" or tools that can be used to manipulate it—regardless of the discipline.

Without being encyclopedic, throughout the book we focus on fundamentals: both mathematical and environmental.

Acknowledgments

My debt to others is unbounded, and what I fear most is that I will not mention all of the people to whom I owe this debt. My late parents supported my education, not to mention everything else, when I was young. My math teachers, in particular, the late Judson "Sandy" Sanderson at the University of Redlands and Richard Pawley at San Jacinto High School taught me mathematics and that it was OK to be interested in everything else. Thanks also to my teacher Lawrence Harvill. All of the students who have taken my class, "Mathematics for the Environment," at the University of Colorado since 1992 have taught me plenty. Thanks to all of you for being part of this adventure. Thanks to Professors David Grant and Karl Gustafson, who have taught this course.

Thanks to the several graduate students who have taught this course; but Bob Cohen, now a professor, merits special mention due to his keen interest and valuable input over a period of four years. Dara Parsavand and Haakon Waadeland read early versions of the text and provided valuable comments. Thanks to Topaz Dent, Ben Fusaro, Bill Stone, Pat Kenschaft, Karen Bolinger, Dave Trunnell, Clayton Lewis, Bob McKelvey, John Cogswell Meyer, David Crumpacker, Michael Olinick, and Alexander J. Hahn. Also thanks to Adrienne Anderson, Arlan Ramsay, Larry Baggett, Sasha Gorokhovsky (thank you several times, Sasha), Nat Thiem, Dave Anderson and Dick Holley, Ellen Zweibel, Shelly Miller, Dave Mastronarde, Don Glen, Chris Braider, Rebekka Struik, John H. Conway, (the late) David Brower, and (the late) David Hawkins. Thanks also to economists Michael Luzius and Alexander Tsoucatos. Thanks to Zachary Strider McGregor-Dorsey for teaching the course and providing crucial feedback. Thanks to Betty Ball and Russ Croop.

Abundant thanks to research librarians everywhere, especially at Norlin Library in Boulder. On the technical side, Liz Stimmel helped me with TEX questions in the early stages. Shashi Kumar of CRC Press has helped me with TEX questions for this edition. Don Monk has helped me with Linux questions for many years. (Thanks to all the open source developers out there, an essential, self-organizing community to whom we all owe a great debt.) I want to thank "Unixops," as well as Markus Pflaum, my math department's computer committee chair, Bart Kastermans, and Bruce Fast for tech support. My undying gratitude to Stephen Preston and Anca Rădulescu (all "three" of them) for a last minute tech-rescue.[4]

[4]The tables in this book were generated with *OpenOffice.org3* and/or LATEX. Population

I would like to extend my special appreciation to www.kgnu.org (88.5 FM Boulder, 1390 AM Denver) my community radio station and to everybody who works for or volunteers for it.

Special thanks to hard-working Bob Stern, executive editor for CRC Press; his rare prescient vision has made publication of this book possible. Also many thanks to the production staff at CRC Press, specially the copyreader(s) and cover artists, for greatly improving the presentation. Last, but not least, my family and their love make it all immeasurably more worthwhile.

I do not claim any particular originality in the material presented in this book. Numbers, fractions, algebra, logarithms, etc., have been around for a long time; Schwartz charts, spreadsheets, not as long, but they are not original with me. All of the facts presented herein would not have seen the light of day had it not been for the perseverance of others. My goals include providing basic mathematical information, being interesting and relevant, with, what is a *sine qua non* for me—honesty—where I define honest as "being one with what is." Of course, though I have worked very hard for a very long time, no amount of effort justifies mistakes (of which I hope there are few). I am responsible for the content and any errors or irritants you might find.

Martin Walter
Martin.Walter@Colorado.Edu
Boulder, Colorado

---

data is from U.S. government Web sites and publicly available. Unless otherwise stated either here or in the text, all figures were generated with *inkscape* and/or *The Gimp* and/or LaTeX and/or OpenOffice.org3. Figure 1.4 is from Wikipedia, using publicly available data. Figure 11.1, the Periodic Table, is from the 1967 Canadian edition of *Webster's Seventh New Collegiate Dictionary*; more detailed and newer versions are widely available on the Web! Figure 13.2 was handed to me by Prof. John Horton Conway, Princeton University. Figures 30.1, 30.2, 30.3, 29.5 used *Mathematica*, *inkscape*, and embedding of open-source fonts. Figure 25.3 is reproduced with the very kind permission of Randy Regier. The poem in Section 8.5 was provided to me originally by John Cogswell Meyer from a very old original work; but it is available publicly on Wikipedia.

Part I

Mathematics Is Connected to Everything

Chapter 1

Earth's Climate and Some Basic Principles

1.1 One of the Greatest Crimes of the 20th Century

What was this crime? "...In 1949 a federal jury convicted GM (General Motors), Standard Oil of California, Firestone Tire, and others of conspiring to dismantle trolley lines throughout the country (U.S.A.)."[1]

My father once told me when I was very young, as we sat in a traffic jam, that there used to be electric-powered trolleys (also called streetcars, or light rail) in Los Angeles, before General Motors came in and bought up the trains, tracks and rights of way and did away with them. That comment sat deep in my unaccessed memory until decades later I went to a showing of the movie, "*Taken for a Ride*," by Jim Klein and Martha Olson,[2] at my public library.

Discussion of this crime is not yet part of the educational experience of most Americans, while the effects of this crime have been global. That is one of the reasons I bring it to your attention now. The fact itself is important for understanding current transportation and energy troubles, as we shall see; but the lesson in *media literacy* that it begins (more later) is equally important. This crime is also connected to many other important and timely topics to be discussed momentarily, some with strong mathematical content. Thus this section is an illustration of a pattern about which famous conservationist John Muir (1838–1914) said: "When we try to pick out anything by itself, we find that it is bound fast by a thousand invisible cords that cannot be broken to everything in the universe."

Less poetically I enshrine this pattern in the following *assumption*:

Connection Axiom: *Everything is connected to everything else.*

---

[1]See [416, p. 13]. Although this crime was committed in the U.S.A., it is relevant to the reader for several reasons, irrespective of the country in which the reader resides. This little known fact might have been lost to history had it not been for the work of Bradford Snell, [646]. The verdict was upheld by the United States Court of Appeals for the Seventh Circuit, January 3, 1951, *United States v. National City Lines, 1951* (186 F.2d 562).

[2]For those interested go to http://www.newday.com/films/Taken_for_a_Ride.html where this scholarly, yet entertaining and accessible, video account of "the crime" is available from NEW DAY FILMS.

1.2 Feedback

There are two examples of *feedback loops* associated with the elimination of intraurban electric rail transportation systems in the 20th century: one with *positive* growth and one with *negative* growth.

General Motors established the National Highway Users Conference, consisting of over 3,000 businesses associated with cars, to lobby the federal government for highways. One of their legislative triumphs was to get a federal tax on gasoline dedicated exclusively to road construction, without annual review, [416, p. 12]. Thus gas taxes lead to more roads which lead to more driving (as long as oil is sufficiently cheap) which leads to more gas taxes and repeat. In this next exercise we play with some (made-up to be simple) numbers to see how the mathematics works in a feedback loop with *positive* growth.

Exercise 1.1 A Feedback Loop with Positive Growth

(i) Let's simplify the numbers as follows, until we understand the ideas involved. Suppose that for each 1 mile of road $100,000 is generated each year in gas and other user taxes. Suppose that it costs $1,000,000 to build 1 mile of road. In year 0 start with 1 mile of road. After 1 year, this mile has generated $100,000. Our 1 mile of road can then be extended by how many miles using the $100,000 generated that first year? (Hint: the answer is a fraction of a mile.)

(ii) How much gas tax will 1.1 miles of road generate in year 2? (For simplicity, we assume that the 1 mile of road instantly becomes 1.1 miles long at the end of the first year. Make a similar assumption at the end of the second year and so on.)

(iii) How much additional road can be built for $110,000?

(iv) At the end of 10 years how much road will there be and how much gas tax will it generate in the following year?

(v) After 10 years will any money be required to go back and renovate "potholes" or other degradation of the existing roads?

Over $220 billion was spent on road construction by the U.S. federal government in the last three decades of the last century; with state and local governments spending far more than that on roads. From 1956 to the 70s for every dollar the federal government spent on rail transit it spent over $88 on highways. This ratio improved in the direction of mass transit in the 70s, never reaching parity, but in the 80s federal highway spending increased by 85% while mass transit spending lost 50%, a trend continuing into the 90s, cf., [416, p. 13]. Consider the following exercises about "percents," cf., Exercise 13.2 (viii).

Exercise 1.2 Percent Change

(i) During the 1980s federal highway spending increased by 85%. Now 85% means $\frac{85}{100}$, a fraction, which can also be written .85; thus given $1, the increase is $1 * .85, or $.85, i.e., 85 cents. What is the sum of the base amount, $1, and 85% of the base amount? Note that * means multiplication.

(ii) During the 1980s federal spending on mass transit decreased by 50%. So each dollar spent before 1980 is reduced to how much after 1980?

(iii) Suppose the following: in 1980 you spent \$1. In 1981 you decreased your spending by 50% over the year 1980. Then in 1982 you increased your spending by 50% over the year 1981. How much did you spend in 1982? Exactly one dollar? More than a dollar? Less than a dollar? How much exactly?

The second feedback loop employed by the GM-led conspiracy was to use a bus company, like National City Lines, which the general public did not know was associated with GM, to go into cities. They would buy up the trolley service, then decrease service thereby diminishing demand, leading to further cuts in service—and then repeat until the trolleys ceased to exist. By 1949 more than 100 electric transit systems in more than 45 cities (90% of the trolley network) had been destroyed, replaced first by buses that were slower and less popular, and then eventually by cars, cf., [416, 646]. Let's look at how the math works in such a feedback loop with *negative* growth.

Exercise 1.3 A Feedback Loop with Negative Growth

(i) Suppose you start with 1 unit of demand for an existing 1 unit of trolley service in some city. Suppose National City Lines buys the trolleys and decreases service by 20%. Do you see that now there is .8 units of trolley service?

(ii) The decrease in service makes it more difficult or impossible for some people to use the service so suppose demand drops to .8 units, to meet the amount of available service. Citing the decrease in demand, National City Lines cuts service by 20% again. Do you see that now there are $.8 - .8 * .2 = .8 * (1 - .2) = .8 * .8 = .64$ units of trolley service?

(iii) Suppose that demand drops to meet available service once more, and that service is once more cut by 20%. How many units of trolley service exist now?

(iv) When service drops to less than $\frac{1}{3}$ of a unit, say, the trolleys are discontinued. How many iterations, i.e., repetitions, of the above process does it take for this to happen?

1.3 Edison's Algorithm: Listening to Nature's Feedback

An *algorithm*, when defined generally, is a step-by-step problem-solving procedure. Thomas Alva Edison (1876–1933) was a famous American inventor with more than a thousand patented inventions, including the incandescent light. Edison reportedly replied to those who commented on the enormous number of "failures" he encountered as he tried one material after another in the search for a functioning filament for his "light bulb:" Those were not 100 failures, for I now know 100 things that do not work.

When you find yourself in a situation with insufficient information to solve some problem (and that's all of us at one time or another), you may be left with no other alternative than to try various things and see what happens. Of course, "insufficient information" rarely equates to "no information at all," if for no other reason than other folks have encountered similar problems before us—tried this or that out—and records of what works and what doesn't were kept.

Humans have been doing lots of things for a long time, and *learning from previous mistakes* might be considered a hallmark of cultural progress. I make explicit this folk wisdom in the following.

Edison's Algorithm: *Given a problem, research what is known about what might work and what might not work, i.e., what might lead to a solution. Using this information attempt solutions (**experiment: observe Nature**), and record the results—to the best of your ability.*

I have found this algorithm surfacing in my own life in organic farming, house building, teaching, and especially in mathematics! Keep it in mind while tackling the homework problems.

A corollary of Edison's Algorithm, or a product thereof, is a *list of rules.* (A corollary of a statement 1 is another statement 2 that logically and easily follows from statement 1.) As mistakes (a mistake by definition leads to undesirable results) are made, investigations are (should be) carried out which isolate the cause(s), and rules are (should be) adopted to avoid making those same mistakes anew.

Corollary: *Lists of rules for solving a particular problem and avoiding previously made mistakes arise from successful implementation of Edison's Algorithm.*

Now Edison's Algorithm and its corollary are so simple and obvious that you might wonder why I take the time to write them down. True, following a set of rules does not guarantee success, but not following "the rules" usually leads to the opposite. Unfortunately, powerful decision-makers all too often ignore the simple wisdom of following rules proven to work. Disasters result. Let's look at some specific examples.

Exercise 1.4 Edison's Algorithm and Lists of Rules

(i) On April 5, 2010 there was an explosion at Massey Energy's Upper Big Branch (Coal) Mine killing 29 miners. According to the Mine Safety and Health Administration (MSHA), see http://www.msha.gov, this mine had 515 citations and orders in 2009, and 124 in 2010 as of May 24, 2010. (See also page 29.) A citation is given for violation of a rule. For example, there are standards for mine ventilation which when followed prevent the buildup of methane gas to dangerous (explosive) levels. Investigate at least two of the violations which were likely proximate causes of the explosion. Discuss the extent to which this is an example of not following Edison's Algorithm and Corollary—even a pattern of such. How much authority does MSHA have? How big are the fines associated with the violations? Is there too much regulation of coal mines, too little? Why? Are whistleblowers fired, see part (viii) below? List all of the causes of system failure you can find in this case and rank them in order of importance.

(ii) On April 20, 2010, the Macondo oil well of BP-Deepwater Horizon in the Gulf of Mexico blew out, 11 workers killed. It took about 100 days to bring the well under control. Find 2 rules (at least) BP violated. (See, for example, *CBS 60 Minutes*: http://www.cbsnews.com/stories/2010/05/16/60minutes/main6490197.shtml, "Blowout: The Deepwater Horizon Disaster." See also "BP's Deep Secrets," by Julia Whitty, *Mother Jones*, Sept/Oct 2010.) Just the year before, the Montera oil rig, operated by PTTEP Australasia in the Timor Sea, blew out on August 21, 2009 and was not capped until November 1, 2009. Previously in the Gulf of Mexico, the Ixtoc 1 (1924' N, 9212' W) oil well being drilled by Pemex blew out on June 3, 1979. It was not capped for 297 days. On January 28, 1969, Union Oil's Platform A off the coast of Santa Barbara, California blew out and went uncontrolled for about 8 to 10 days. Oil disasters are too numerous to list in this book. For example, research the fairly well-known oil pollution in Nigeria and Ecuador, and the not-so-well-known oil disaster in Bolivia, cf., [616, Chapter 2]. Then there was the Exxon-Valdez, cf., page 25, and Exercise 13.12.

Two decades after the Exxon-Valdez serious environmental impacts remain, contrary to some prognostications in 1989–90. (On June 25, 2008, after a long wait, plaintiff's original $5 billion dollar punitive damage claim, i.e., equal to about one year's Exxon profit at the time of the spill, was reduced by 90%, to about $\$\frac{1}{2}$ billion by the U.S. Supreme Court.) Evaluate each of the above oil-examples for their ecological and economic impacts (so far). For example, regarding the BP blowout in the gulf: How are the Atlantic bluefin tuna, sperm whales, Kemp's Ridley sea turtles, bottlenose dolphins, the Atakapa-Ishak indigenous people, and all others who make their living from the Gulf waters doing when you read this? The Union Oil blowout near Santa Barbara in 1969 is credited by some with inspiring the environmental movement of the 70s. Did the BP Deep Horizon blowout in the Gulf of Mexico inspire American environmentalism to the extent that renewable fuels/energy are taken seriously? How effective was the legal and public relations efforts of BP in blunting response?

While the MSHA inspectors were handing out citations, see part (i), the MMS, Minerals Management Service (agency of the U.S. government managing oil, gas, and other mineral resources on the outer continental shelf, www.mms.gov., which in June 2010 changed its name from MMS to the Bureau of Ocean Energy Management, Regulation and Enforcement.) was giving exemptions from comprehensive environmental review to projects such as BP's Deepwater Horizon. Find 2 rules, at least, MMS violated. In the first decade of the 20th century the MMS was saturated with its "sex, drugs, and paintball" scandal. Industry representatives were giving "perks" to, sharing drugs with, and having sex with agents from the MMS. (There are a number of other operations the MMS has given exemptions from review, some operating in the Gulf of Mexico.)

Research each of these oil-examples, find as many instances as you can where those involved did not follow known rules for safety and reliability. Investigate the extent to which censorship of information on the part of participating corporations (and government) was/is practiced. In particular, investigate BP operations, such as Deepwater Horizon, Atlantis, and those in Alaska and Texas, decide if you have found a pattern of BP ignoring known rules. Pick some large ecological disasters caused by humans and compare their impacts to the impacts to the U.S. caused by the terrorists of 9/11/2001.

List all of the causes of system failure you can find in each of these cases above and rank them as best you can in order of importance.

(iii) Research the technique of *hydraulic fracturing* for oil and gas recovery. In particular, study the safety and environmental record of this process. Consider the proposal to drill in the Marcellus Shale, cf., page 466, in the geologic formations that provide water for many New Yorkers. Have the heretofor secret names and amounts of chemicals injected into such wells been released by the industry at the time you read this? Given the rate of known accidents, what do you estimate the chances are of conducting natural gas drilling in the Marcellus Shale without greatly, negatively impacting the drinking water resource?

(iv) Can we move to a fossil fuel free economy, and thus eventually avoid most of the problems associated with fossil-fuel extraction? See Chapter 6.

(v) Read about the January 28, 1986, Challenger disaster, cf., Section 8.4. Find several known essential rules that were not followed by those in charge of making decisions. (Richard Feynman discovered several, see [188].)

(vi) Certain rules for governing the financial industry (rules for avoiding collapse, for example) were learned during the first Great Depression of the 1920s and 30s, see Chapter 2. For example, the Glass-Steagall Act of 1933 was passed, and reasonably enforced until it was repealed in 1999. The financial collapse of 2008 and following years led to a re-examination of "the rules." The financial reforms of 2010 notwithstanding, we have not yet been able to fully return to the "rules" that were learned the hard way in the first Great Depression. Why is this?

(vii) Does a successful following of rules inevitably lead to complacency? Is this an intrinsic human problem? Can complacency be avoided?

(viii) Those who are in the best position to discover new rules, or observe enforcement (or non-enforcement) of known rules are often not in a position of authority. Those who dare speak up in such situations are called *whistleblowers.* Allowing whistleblowers to make their information available in a forum that leads to progress, positive change, would seem advisable. I was told that the safety, such as it was (is?), of the U.S. airline transportation industry, owed its success to a program wherein whistleblowers could reveal information anonymously to the FAA, Federal Aviation Administration, www.faa.gov. Research the fate and impact of at least 3 whistleblowers in any areas of your choice.

(ix) When should a list of rules be revisited for possible revision? Are there rules for this?

(x) Since there are often several causes for a given effect, there arises ample opportunity for "spin" or to play the "blame game." Can you think of a method of listing causes of a given effect and ranking them in order of importance. When arguments arise among parties, for example, one should at least create a list of causes that includes all causes mentioned up to that point. Then one can argue a ranking, and give explicit details of one's reasoning.

(xi) Examples above discuss multiple causes for a given effect. Can you think of examples of circular feedback where A causes B, B causes A, and so on?

(xii) If it is known that violating certain rules increases the accident or death rate in a given situation, then decision-makers so empowered who violate those rules are at least statistically linked to the increased number of accidents or deaths. Said another way, the decision-makers who knowingly violate safety and reliability rules do cause accidents and deaths. It is just not always clear which accidents or deaths. Why is it that such decision-makers are not held accountable in our society, as are, say, murderers in the classical sense?

(xiii) Deciding to intentionally ignore known rules (of safety/reliability and so on) for whatever reason (one is in a hurry, one can make more money if one skips a few safety procedures ...) is grounds for isolating that decision-maker from the rest of us—to a degree and for a duration—commensurate with the extent that person's actions have been an unnecessary contributory cause of pain, death, dismemberment, or discombobulation (ecological, economic, social, or otherwise). Comment.

(xiv) Is it a rule (or observable pattern) that in any system run by humans that known rules set in place to avoid mistakes and achieve success will eventually be violated due to incompetence, complaceny, impatience, greed, laziness, or stupidity? What system(s) do you envision can be put in place to avoid this type of meta-mistake? Besides making sure that our educational system includes teaching a respect for the rules that Nature/reality has taught humans over time (with examples), consider Part VIII.

I will now return to our discussion of one of the greatest crimes/mistakes of the last century.

1.4 Fuzzy Logic, Filters, the Bigger Picture Principle

My experience has been that the following claim is generally accepted as true by most people, including experts in history with whom I have communicated. None of the folks with whom I have discussed this subject and who believed the following claim were aware of "the crime" mentioned above.

The Claim: Buses replaced trolleys, and were in turn replaced by cars because the public wanted it that way.

Now it is possible that there was at least one person in 1920, say, that had a car and wished that the trolleys would get out of the way. (For example, in 1918 only 1 family in 13 had a car, by 1929 4 out of 5 families had one.) So "The Claim" is probably not completely false. On the other hand, trolleys were quite popular. In fact, we have the following observation from [430, p. 26]: *"Electric trolleys were born in the USA. They chalked up their first commercial successes in America and they gave Americans a first-class system at a time when public transport in London, Berlin and Tokyo was still mired in horse manure. Trolleys are as indigenous to America as jazz, baseball and abstract expressionism. Why then were they killed off in the United States while they and their latter-day incarnations were viewed throughout most of the world as both winsome and essential?"*

It is quite likely, then, that "The Claim" is not completely true; recall the findings of federal court that begin this chapter.

Mathematicians can handle a situation like this with *fuzzy*,[3] or more precisely, *measured logic*. Thus a truth value of 1 for our claim corresponds to the claim being completely, i.e., 100%, true. A truth value of 0 corresponds to the claim being completely false, i.e., 0% true. Any truth value between 0 and 1, e.g., $\frac{1}{2}$ or $\frac{3}{11}$, is possible.

I have not seen a great many well-documented writings on the claim. However, the following references suggest that the truth value of the claim is near 0, [39, 430, 142, 416, 680].[4]

What I have just done is *filter* the information presented to you. See Figure 32.1 in the final chapter (on Media Literacy) and duplicated below.

This happens to all of us all of the time. I have not included all references to the claim that exist, in fact this is not possible. There are probably references I do not know of. I did not include references without documentation. Some references might pop up after this book is published. Statements have been made in support of the claim—or else the claim likely would not even exist.[5] I have just indirectly referenced some of these statements, but have not found supporting evidence for the claim that holds up under thorough examination. In fact, I claim that we would have both a robust electric rail system and a system of roads and cars had "the crime" not been committed. Thus there are at least two competing views of the claim, quite a common situation.

What is one to do? In a word "triangulate." If an issue is important to you it is necessary to expend the time and energy to find as many references as you can, at least two or three which are as mutually independent as possible. Direct experience certainly counts as "a reference," but more often than not we must rely on the experiences and reporting of others. Do not fall into the trap

[3]Lotfi Zadeh introduced the term "fuzzy sets" in 1965 in order to *more precisely* model certain phenomena. Fuzzy does not mean muddleheaded in this context.

[4]Other references are *Car Trouble* by Steve Nadis, James J. MacKenzie; an article in the February 1981 *Harper's* by Jonathan Kwitny; an article by Morton Mintz in the May 1974 *Washington Post*; and *Corporate Crime and Violence* by Russel Mokhiber, 1988.

[5]For example, see the movie: "Taken for a Ride," cf., page 3.

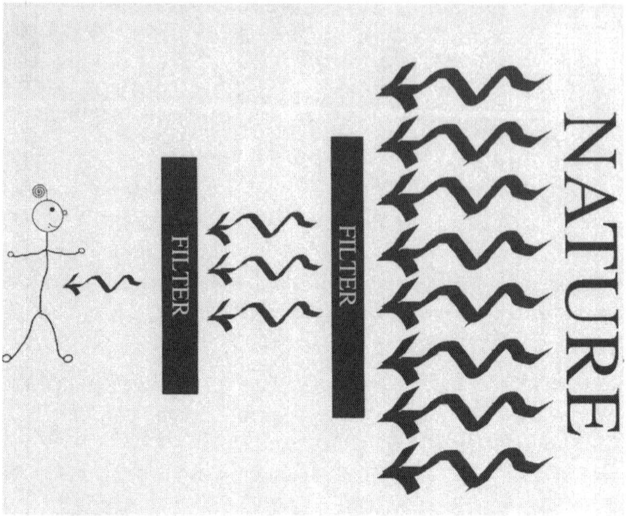

FIGURE 1.1: Filters between You and Nature

of only finding references you agree with. Find contrary sources, arguments; compare them with yours. Which argument is most rigorous? Most convincing? To whom? *Can you find a "Picture" big enough to explain all arguments in contention?* Always document, but realize you can "document" anything; how many references, e.g., Web sites, can you find asserting ("proving") that the earth is flat? Do not fall into the trap of always relying on information that is most easily available. Be *critical*! To paraphrase Upton Sinclair: *it is difficult to get a man to understand* (or be *honest* about)[6] *something when his job* (finances, friendships, or social status) *depends on* **not** *understanding* (or being honest about) *it.* Always check sources for a *conflict of interest*! For example, ask: Who pays the source? How much and why? If possible, repeatedly return to an issue for as long an interval of time as possible. Does new information surface? Do new connections, new patterns emerge? While searching for information keep in mind that as early as 1995 the number of professional propagandists, known as PR (public relations) personnel, 150,000, was larger than the number of (news) reporters, 130,000; by 2008 there were almost 4 PR persons for every 1 employed as a journalist, editor, reporter, or announcer in the newspaper, radio, or TV industries. Often what is presented as "objective news" is just PR disguised as news, [561, p. 2], [181], [424, p. 49].

If you go through the process of the previous paragraph, I say that you have applied the *Bigger Picture Principle.* Is your picture big enough to

[6]Note that I define *honest* to mean *being one with what is.* Nature is what "is," and I suggest that Nature is the ultimate arbiter.

contain, explain, understand all of the information, arguments on the subject under scrutiny? There are whole industries, cf., advertising, public relations/propaganda, lobbyists, and more, devoted to presenting smaller pictures leading to preordained conclusions. It is somewhat surprising to me that such industries can sometimes lead people (sometimes yours truly) into clearly behaving against their own self interest. But humans fooling other humans is quite common. When humans are dealing with Nature, however, honesty is imperative; for *Nature cannot be fooled.* By the way, I am making the tacit assumption that there is one all encompassing reality, which includes all of us;[7] and I call it Nature.

Exercise 1.5 Bigger Pictures are Models

(i) Why do I talk about Bigger Pictures but not the Biggest Picture?

(ii) Exercises 1.1 and 1.3 are examples of simple *models* (of small parts of Nature). I refer to the (entire) picture of Nature that you have in your mind as your *mental model.* Estimate how much of your mental model of Nature, i.e., reality, has been put there by others, and how much you have had a direct, dominant role in creating. How much of your mental model comes from your direct observations of Nature, without intervening (human) filters? How much of your mental model is the product of your imagination? Pick someone else and ask (and try to answer) these same questions about their mental model.

(iii) I have heard the following claim: Anything I really need to know will be presented at or before the time I need to know it in the "mainstream news," e.g., television, newspapers, news magazines, or will have been presented in textbooks and such materials during my educational experience. I assert that "the crime" of section 1.1 is a *counterexample* to this claim. Thus the truth value of this claim is not 1. What is your estimate of the truth value of this claim? Can you think of very important examples, e.g., involving life and death, nations going to war, where the claim is mostly false? Can you think of cases where the claim is mostly true?

(iv) You probably have heard of *spin.* Is spin, to the extent that it does not contain outright falsehoods, just the process of presenting a smaller picture favorable to pre-ordained conclusions?

Many generations ago people believed that eating liver cured night blindness and eating citrus cured scurvy. People were led to this model by direct observation of Nature, and this model *worked* long before the discovery of vitamins A in liver and C in citrus.[8]

Since ancient times the Warlpiri Aborigines of Australia, cf., III, have believed that when harvesting yams, only a fraction (say one-third) of any particular yam should be taken at a time. To take more would anger the Yam God, and the plant would die. Although their mental model of the situation may be considered a *myth*, this mythic mental model passes an important test: it works for them—it contributed/contributes positively to their survival and enjoyment of life. I propose that their model works because it is based on direct, *honest* observations of Nature.

[7] As in *Not Man Apart; Lines from Robinson Jeffers*, author David Brower, Sierra Club Book, January 1, 1965.

[8] See [53] for the history of finding the cure for scurvy, and, much later, vitamin C.

Thus I suggest that one simple way to judge a model is to ask: "Does the model *work*?" When something stops working for you, it is time to ask "why?" and make the necessary changes. One of the results of "an education" should be the ability to analyze a model and make changes to prevent complete collapse *before* it happens. You are *responsible* for your mental model and its consequences. The choices you make of what to read, watch, listen to; of what experiences you have, what you chose to think about—all these and more go into shaping your mental model. Your mental model becomes a fundamental part of you and largely determines with whom and with what you are in harmony or conflict, how you interact with Nature and what you believe is true. Reality acts on you, and you act on reality. This is another example of a feedback loop. Nature will likely, eventually, take care of models that are not honest—maybe sooner than later. Incidentally, grizzly bears learn adaptations to their environment via play, [524]—so perhaps our mental models might benefit from play as well!

1.5 Consequences of the Crime: Suburbia's Topology

There is an axiom of city planners, [496]:

Land Use Axiom: *Transportation determines land use.*

From [496, p. xvii]: *"Until about 1945, the focus of design for American communities was people—pedestrians, bicyclists, kids. Neighborhood streets were places where people walked, socialized, greeted neighbors sitting on their front porches, places where kids played kick the can and rode their bikes. Compared to most of our neighborhoods today, streets were narrower and connected to each other at almost every block, blocks were shorter, street surfaces were often rougher. Cars parked along the roadside, trees shaded streets and sidewalks—and there* were *sidewalks. In city streets, people ran into friends, exchanged greetings with strangers, conducted business, window shopped, waved to acquaintances through shop and restaurant windows."*

If you have ever visited a city that was designed and built for people (before the automobile), such as an old part of a European city or Boston, U.S.A., you see it right away, it is tough to get around in a car. Today's cities and suburbs are designed for the automobile. If you live in the suburbs, since everything you need in life except your house is too far to walk to, you drive—a lot. You want to get there quickly, so streets are wider, traffic faster. To keep children safe, to keep traffic noise away, you don't live on a main road, you live on a cul-de-sac, which connects to a network of a few main arteries which become jammed with traffic periodically or constantly. Over half of the urban space in America, on average, is devoted to cars; in Los Angeles the figure is two-thirds. Some 40,000 to 50,000 people die in traffic accidents every year in the U.S., many of these deaths could be prevented, cf., [525]; and a comparable number die from air pollution, largely because of cars, cf., section 1.6. Parking downtown, if it exists, is expensive, so you shop at a mall surrounded by an ocean of asphalt, i.e., free parking. Life is busy, there is no time to cook, so many of us pick up a quick meal[9] at a drive-through. Less walking means more obesity, less meaningful interaction with others. Many Americans do not know or socialize with their neighbors, driving from home to work, to home with a TV. This model does not apply to all of us, but it applies to many of us.

In the following exercise we look at some mathematical differences between an "Old Town" built on a grid of roads vs. a typical suburban development. (Recall that vs. means versus.)

Exercise 1.6 The Topology of Suburbia vs. Old Town

The following helps explain the relatively higher rates of traffic congestion and lower rates of social interaction in suburbia as opposed to an old town with a grid road system.[10]

(i) Imagine a square with lower left vertex labeled A and the diagonally opposite upper right vertex labeled B. You can draw the square so that its top and bottom are horizontal and its sides are vertical, cf., Figure 1.2 below. How many ways, i.e., paths, are there from A to B, subject to the following rule: From any vertex you can only move to the next vertex to the right or the next vertex above (straight up).

(ii) Add a line segment to the square in (i) that connects the midpoint of one side of the square to the midpoint of the opposite side of the square. Now how many paths are there from A to B, subject to the same rule as in (i)? See Figure 1.2 below.

(iii) Add another line segment to the figure in (ii), perpendicular to the segment you just added, which joins the midpoint of one side of the square to the midpoint of the opposite side. Your original square should now be subdivided into 4 smaller squares. How many paths are there from A to B now, still subject to same rule as in (i)? See Figure 1.2 below.

(iv) Keep this process up until you have subdivided your original square into $9^2 = 81$ equally sized subsquares with an array that has 10 "roads" connecting the top of your square to the bottom and 10 more "roads" connecting the one side of the square to the other. How many paths are there from A to B now, subject to the rule in (i)? See Figure 1.2 below. Hint: Each path can be uniquely labeled with a sequence of 18 letters, 9 of which are R, for move to right, and 9 of which are U, for move up. Each such word uniquely represents a path (that satisfies the rule). Thus from Chapter 17, apply the formula $\frac{18!}{9!9!}$.

(v) Now assume that three sides of the original square are declared "main roads." Erase parts of line segments within the grid of part (iv) until you have everyone inside the square "living on a cul-de-sac, i.e., a dead-end road." How many paths there are from A to B now.

(vi) Suppose two houses are very close to each other but on different cul-de-sacs in suburbia in (v). What is the distance along roads between them? What was the distance along roads, in (iv), before you erased some?

(vii) Describe the differences in the traffic patterns and modes of transportation you would expect in parts (iv) and (v).

(viii) If walking and streetcars were the main mode of transportation, which design do you think would be most popular, (iv) or (v)?

(ix) In general, which road system, (iv) or (v), offers the quickest response time for an emergency vehicle?

---

[9]Possible health effects of an exclusive "fast-food" diet are explored in the DVD, *Super Size Me* (2003), starring John Banzhaf and Bridget Bennet (II), et al.

[10]This exercise is adapted from [496, pp. 30–4]. If you look up the formula therein, be careful in how you use it, there is a small misprint. In Chapter 17 we will take a more systematic look at how to solve problems like this one.

(x) From the Dec. 13, 2009 issue of *The New York Times Magazine*, p. 34, "Cul-de-Sac Ban, The," we read that in the fall of 2009 the state of Virginia became the first state to severely limit cul-de-sacs from future housing developments. All new subdivisions need to attain a level of "connectivity" with sufficiently many through-streets connecting to other neighborhoods and commercial areas. At the time you read this has any other city, county, or state adopted similar regulations?

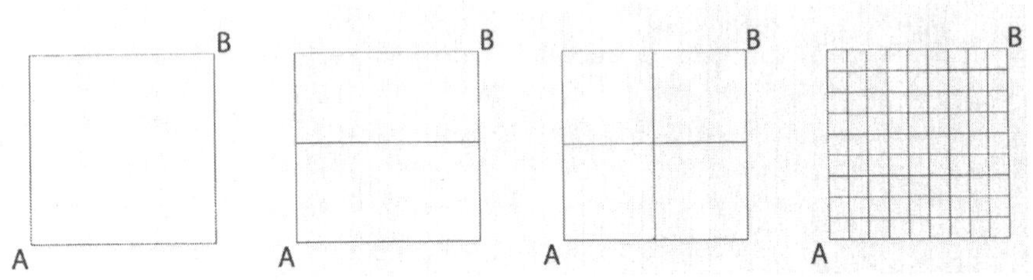

FIGURE 1.2: Street Grids

1.6 A Toxic Consequence of the Crime

In [680] the former head of the California Environmental Protection Agency draws a close parallel between the behavior of the tobacco industry and the behavior of the automobile and fossil fuel energy industries, including their (toxic, according to [680],) contribution to the synthetic soup we all live in.

The research paper [140] reports[11] on one of longest most in-depth studies of the correlation between air pollution and disease/mortality. Of particular concern are pollution air particles, pm 10 or smaller, i.e., particles 10 microns[12] or less in diameter. Of these those of 2.5 microns or smaller were most tightly correlated to increased mortality and disease—the more air pollution particles the more disease and mortality. This study is not the only one which correlates air pollution and disease/mortality,[13] and these studies have

[11] Note http://content.nejm.org/cgi/content/abstract/329/24/1753 is the Web address of the original paper. Note http://www.healtheffects.org/Pubs/Rean-ExecSumm.pdf is the Web address of a later reanalysis of the original data.

[12] A micron is one-millionth of a meter.

[13] A National Academy of Sciences report released Oct. 19, 2009, claimed that nearly 20,000 people die prematurely each year as a result of air pollutants from power plants and vehicles (that is the burning of coal and oil products), cf., "Fossil Fuel's Hidden Cost Is in Billions, Study Says," by Matthew L. Wald, *The New York Times*, Oct. 20, 2009.

survived criticism as have many scientific results with economic implications. This body of science is analogous to the science establishing links between illness and smoking, both in its implications and its being a focus of attack. I might add that before the mandatory introduction of unleaded gasoline, lead and its negative health effects were part of the toxic legacy of automobile exhaust cf., Section 4.5 . Is any of that lead still around? Note that the Clean Air Act of 1970 gave the first administrator of the Environmental Protection Agency the power to remove lead from gasoline in 1973. The auto industry and its allies vigorously opposed the Clean Air Act, [142, p. 68].

I must confess that I learned very early that I am often not sufficiently critical. On many days when I walked home from Dahlia Heights elementary school in east Los Angeles my lungs/chest hurt when I took a deep breath. I asked my dad why. He said it was the smog. In one of the first television programs I saw in my life some official people explained that L.A. smog was produced by dirty industrial plants. No autos were mentioned. Later, while breathing air far away from L.A., I read that it had been discovered that L.A. smog was primarily due to auto exhaust. I never regained the trust in television (or officials) that I had in elementary school.

Exercise 1.7 Meters and Correlations

(i) We will deal with the metric system and conversions to and from the American/English system in section 3.4. For the moment try to understand how big an air pollution particle which is 2.5 microns in diameter is by comparing it with the size of something you know, say various diameters that a human hair might have. I have seen numbers ranging from 181 microns to 17 microns given for the diameter of various human hairs, with 100 microns as "typical." Could you see a pollution particle which was 2.5 microns in diameter?

(ii) The thinnest paper I have seen is $.001''$, i.e., one one-thousandth of an inch, thick. How many microns is this? Hint: Use 1 meter $= 39.37$ inches, or cf., section 3.4.

(iii) It has been stated that a piece of paper, say the one in part (ii) above, cannot be folded in half 12 times by hand. Mathematics can help you understand possible difficulties, for example, how thick would the resulting 12-times-folded-in-half paper be? How thick would be the result if you did this 15 times? 20 times?

(iv) We will discuss correlations very lightly in Section 5.11. For now, a *correlation coefficient* is a number between 1 and -1. If two phenomena have a correlation of $+1$, they are observed to always happen together (if -1, they never happen together). For a small project investigate what is meant by "pollution particles of 2.5 microns or smaller are most tightly correlated to mortality and disease."

(v) Some air pollution particles are small enough to pass from the air through lung tissue into the blood stream, affecting other organs. What are the sizes of such particles?

1.7 Hubbert's Peak and the End of Cheap Oil

In 1956 Marion King Hubbert, an American geophysicist,[14] predicted that in the early 1970s oil production in the United States would reach a peak, i.e., oil production (actually extraction) would rise each year until reaching a high point and then decline each year thereafter. His prediction came true between 1970 and the Spring of 1971. In the 1950s almost everyone in and out of the oil industry rejected Hubbert's work; just as those who use the same methods to predict a peak in world oil production have been, if not rejected, ignored today. Admittedly, there is a bit of magic involved, in addition to some brilliant mathematics.

Hubbert's Mathematics. Briefly, Hubbert hypothesized that production data for the unconstrained exploitation of a nonrenewable resource, like oil, would follow a "bell-shaped curve." The peak of the curve would be attained when half of the resource was exhausted. He also assumed that the *discovery curve* would look very much like the *production/exploitation curve*, except that the discovery curve would occur some fixed number of years earlier than the production curve. Since, according to Hubbert's hypothesis, the peak occurs when half the oil is gone, one key to a successful prediction is knowing/guessing/estimating what the total petroleum reserves are. Using the best estimates available, in the 1950s Hubbert fit the known data for the United States with a curve, a logistic curve, using some of the same math used to analyze population growth.[15] What was Hubbert's prediction long ago is (for the U.S.) now an actual set of data, cf., Figure 1.3 from http://en.wikipedia.or/wiki/Peak_oil, which is a graphical illustration of data from the Energy Information Administration of the U.S. Department of Energy. With 50 years of hindsight, another curve, a *Gaussian*, is used because it apparently gives a better fit to historical data; and presumably it will do a more accurate job of predicting Hubbert's peak for world oil production. As for discovery curves, discovery of oil in the lower 48 states of the U.S. peaked in 1930, cf. [727, p. 312]. Also see [131, pp. 137–3], where it is stated that "More oil was found in the United States during the 1930s than in any decade before or since." Graphs of oil production that include Alaska and deepwater sources can be found in [727, p. 314]. These new sources did not dramatically alter general trends. As for the world, global oil discovery peaked in about 1964, [70] and [727, Appendix 9]. I encourage you to do your own literature

[14] M. King Hubbert (1903–89) was working at the Shell Oil research lab in Houston when he made his original predictions/estimates of future U.S. oil production. He later worked for the USGS, United States Geological Survey.

[15] If you want to look at some of the math, including rate plot techniques, see [130, Chapter 8]. The logistic curve usually encountered in population growth, of Pierre Verhulst, is "S" shaped. But the graph of the rate of growth, i.e., the derivative, rises to a peak then declines. Note we study populations in V.

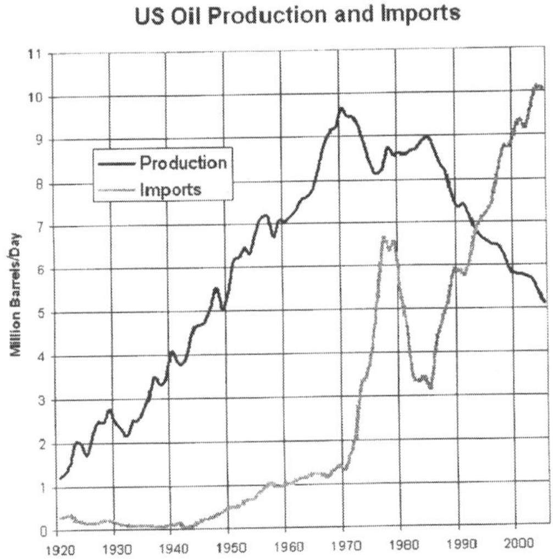

FIGURE 1.3: Hubbert's Peak for U.S.A.

search and see what numbers you come up with; beware, however, that not everyone uses the same definitions of "discovery," "reserves" and "hits."

So When Will the World's Oil Production Peak? The biggest argument comes in estimating total world oil which equals all the oil ever extracted plus what remains in the ground, i.e., estimated reserves. One estimate of $1.8(10^{12})$ *barrels*,[16] when using a Gaussian curve to fit known data, yields 2003 as the peak year. A colleague of mine in the physics department got 2004. Other published estimates range from 2004 to 2009, assuming a little over $2(10^{12})$ *barrels* total world oil. The USGS estimates world oil reserves to be $3.012(10^{12})$ *barrels*; which requires the discovery of an additional amount of oil equivalent to the entire Middle East. Some will say that the peak in world oil production is decades away. The chances of this happening are not zero, but low. Estimating remaining oil reserves, and to a much lesser extent researching past world oil production, gets you involved in guessing which country is fudging its numbers for political purposes, filling in numbers that are held secret by some governments, and guessing what has not been discovered yet. One sobering exercise is to look at a picture of the world with pins stuck wherever a hole has been drilled for oil. There have been a tremendous number of such holes drilled, and the more holes the less chance that future holes will come up with significant oil. The mathematics and data

[16]Note that 10^{12} is 1 trillion, i.e., 1 followed by 12 zeroes. For a complete discussion of this notation see Section 3.4, and Exercise 3.11 in particular.

assign probabilities to each estimate of Hubbert's peak for the world, none are known with absolute precision—as should be clear; and we will not be sure of the peak's date until about 10 years after the fact. What is clear, at least to me, is that Hubbert's peak for world oil will occur within the lifetime of a typical 18 year old reader of this book, and it is likely to occur well before such a reader reaches middle age, and quite likely much sooner than that—in fact, it may have already occurred.[17] Other references are [69, 70, 221, 243, 293, 294, 221, 218, 472] and www.dieoff.org; there are many more.

In the following exercise we will introduce some topics that we do not have room to pursue in this book.

Exercise 1.8 Energy Return on Investment (E.R.O.I.), Self-Organizing Systems, Electric Cars and Trains, How Might It Have Been?

(i) Why are the graphs in Figure 1.3 not smoother?

(ii) Can you sketch the discovery curve of oil in the U.S. based on Figure 1.3 and Hubbert's analysis discussed in the text?

(iii) Did Standard Oil of California simply disappear, or is it still around under a different name? If so what is that name?

(iv) Hubbert's Peak analysis applies to *all* non-renewable energy resources: oil, coal, gas, uranium; and to other non-renewable resources that are not recycled. This is "almost" clear. If you have piles of stuff of variable quality, buried in lots of different places, at various depths, with variable accessibility, one will probably extract the easiest, most valuable piles first, proceeding on to extract piles involving increasing effort/cost until forced to stop due to cost or lack of resource. Trying to predict how long the piles will last requires more sophistication, like Hubbert's analysis. Figuring out how cheap/expensive a resource is at any given time also requires more sophistication. A fundamental way to understand the cost of a resource, like oil for example, is to calculate the Energy Return on Investment—the number of units of energy you get in return for each unit of energy expended in getting said energy. (See VII for a discussion of units of energy.) Why is this method more meaningful/fundamental than calculating the cost of extraction of a resource in dollars, euros, yuan or yen? For a discussion of E.R.O.I. in several contexts see [270]. As an example, Cutler Cleveland, one of the authors of [270], an energy scientist at Boston University who helped develop the concept of E.R.O.I., calculates that from the early 1970s to the early 1990s the E.R.O.I. for oil and natural gas in the United States fell from 25 to 1 to about 15 to 1. Do you think this result is reasonable? Why has this happened? What is oil's E.R.O.I. the year you read this?

(v) Research and/or estimate the E.R.O.I. of every energy source you can think of: coal, tar sands (E.R.O.I estimate of 4 to 1), renewable energy sources, ... For example, is the following statement true? "In 1950, spending the energy equivalent of one barrel of oil in searching for more oil yielded 100 barrels in discovered oil. In 2004, the world's five largest energy companies found less oil energy than they expended in looking for that energy." (This is a quote from Chris Stolz's review on Amazon of [678].)

[17] A summary of all the predictions of Hubbert's peak for global oil production that I know can be found in [727, Appendix 9]. Reference [130] is written by a colleague of M. King Hubbert, is readable and gives a great deal more detailed information than can be provided here. Natural gas will soon have a peak of its own, cf., [127]. For technical reasons having to do with restrictions on output of gas wells created by gas pipeline capacity, the Hubbert peak for natural gas is "flattened" and longer in duration, followed by a more precipitous decline, sometimes called the "gas cliff," cf., [127, pp. 94–6].

(vi) What is the relationship of the E.R.O.I. of a fossil fuel resource to climate change? Hint: See Section 1.9.

(vii) Another method of evaluating a resource is to look at the environmental and social impacts/damages due to the extraction of the resource and due to the disposal of waste after the resource is spent. Investigate such impacts for all non-renewable sources of energy, e.g. coal ash disaster, Kingston, Tennessee. Will renewable energy sources have environmental and social impacts? How do you think they (will?) compare to the impacts of non-renewable sources?

(viii) The fine that was imposed for the crime of dismantling (90% of) the trolley system was $5,000 in 1949. If that money were put in the bank and earned 7% simple interest per year for 60 years, how much money would there be?[18] How much light rail could you build for that amount? A lot? Almost nothing? We will deal with the mathematics of this calculation in more detail in V and VII.

(ix) I propose the following "*democratic, decision-making principle:*" Except in the rarest and most urgent of circumstances,[19] decisions that effect a majority or most of the people in a given unit should be put to a direct vote of the people in that unit.[20] Was "the crime" an example of too few people making a decision for all of us, i.e., a violation of this principle? Was the feedback experienced by the conspirators very strong?

(x) If electric-powered rail transportation had been allowed to evolve, both intercity and intracity, along with an evolving automobile mode of transportation, what are the possible outcomes you imagine could exist today? What would the differences in outcomes be? Mathematically speaking this would have been an example of a (freely) *self-organizing* social system. This is a somewhat new branch of mathematics: self-organizing systems.

(xi) Research the history of the electric car. It was killed twice, once at the beginning of the 20th century and a second time at the beginning of the 21st century. In response to California's passage of a law requiring certain fractions of auto sales to be zero emissions cars by certain dates, General Motors[21] produced the EV1 electric car with Stanford R. Ovshinsky batteries. There was an eager waiting list for these well performing and very popular cars, but under pressure from auto/oil/federal government the California Air Resources Board voted on April 24, 2003 to kill the zero emissions requirements. Within a year GM had reclaimed all EV1s and had them destroyed (only one is in a museum). See the movie (DVD) "Who Killed the Electric Car?" (released 2006), then discuss if General Motors would have been better prepared for the future if California had followed the above democratic decision-making principle?

(xii) One of the technical problems with solar and wind power is storage. If a majority of cars were electric and there were an electric rail system both within and between cities, would this provide a large reservoir into which solar and wind energy could be dumped in real time? (If all the electricity for this alternative transportation scenario were provided by burning coal, would the environment be better/worse or the same as it is today with oil-powered transportation?)

(xiii) The people of Iceland are making a serious effort to create a hydrogen economy for themselves. Research the problems and promise of a hydrogen economy, including fuel cell technology, cf., [155, 576, 680].

(xiv) How much of your life is built upon the assumption that petroleum products will remain cheap or at least affordable—or at least available?

---

[18] Do not try this, since bank accounts are declared inactive if left unattended for 5 years.

[19] Such as an invasion by aliens, perhaps certain epidemics—a list of exceptions decided on by use of the democratic decision-making principle!

[20] As we will discuss a bit more in VI there is no mathematically "perfect" method of "voting" that always reflects the "will of the population." Nevertheless, if one believes in democracy as a form of government, one must strive to implement voting processes that reflect the bottom up self-organizing wisdom of the population, even if it is not perfect.

[21] Other auto corporations produced electric cars as well, but GM was first. GM also simultaneously fought California's zero emissions standards.

(xv) When will the peak of coal production occur? How many years of coal used for energy, assuming no shift to renewables, remain? (The answer may not be nearly as far off in the future as you have been led to believe!)

(xvi) Discuss additional impacts on America of the killing of trolleys. Discuss solutions as we go over Hubbert's peak. Where do we go from here? A few among many references include [675, 341, 492, 266, 84, 684, 377, 470, 193].

1.8 Resource Wars: Oil and Water

We will delve more deeply into conflict over resources in VII; however, due to its importance I will briefly mention it here, together with its connection to "the crime." In Figure 1.3 is reflected the fact that the U.S. went from being an oil exporter nation, to one that was barely self-sufficient, to a major oil importer nation during the 20th century. Since our military, transportation network, chemical industry—even our food supply, cf., Chapter 5, are seriously dependent on reliable oil supplies, maintaining access to such has been a national priority.

In Section 1.6 I very briefly mentioned the impacts of the crime on our air, but there are impacts on water, both salty and fresh, as well. Potable water is becoming an ever more valuable and contentious resource, cf., Chapter 5. Mention oil spill pollution of our oceans and many major accidents around the globe can be cited: Santa Barbara (Union Oil Company) and Prince William Sound (Exxon Valdez, cf., Exercise 13.12), cf., II, and the BP gusher of 2010 immediately come to mind for me. However, an immense amount of oil is released (often deliberately dumped) into the U.S. environment every year, polluting among other things, fresh water.[22]

One-half to one-third of our urban areas are paved. Although there exist paving materials that are partially porous to water, most paving is impenetrable, leading to water runoff, which contains an assortment of pollutants. One example: pesticides/herbicides and fertilizers stacked up in parking lots at various big-box stores contribute to the toxic mix of pollutants deposited on these surfaces by cars and trucks. Rain then quickly delivers this brew to our waterways, often violating the Clean Water Act, [461, p. 118]. In heavy rains paving increases the chances of flooding. Paving interferes with the natural cycle during which water would percolate through the earth, replenishing ground water, being partially filtered and cleansed in the process.

[22] Pick a year, like 1997; it was estimated that 240 million gallons of the 10.8 billion barrels of oil used in U.S. transportation that year were released, cf., [341, p. 95]. Much of this oil ends up in storm sewers, waterways and ground water.

Exercise 1.9 What Have We Done for Oil, What Have We Done to Water?

(i) Assess the truth value of the following statements: America invaded and has occupied Iraq (2003–?) because: (A) Saddam Hussein had links to the 9–11 terrorists and/or was a malevolent dictator. (B) Iraq had weapons of mass destruction. (C) Iraq has/had the second largest oil reserves on earth. (Hint: Your picture should be big enough to explain/understand the National Security Council documents aired during the January 11, 2004 *60 Minutes* interview with former Treasury Secretary Paul O'Neill, cf., http://www.cbsnews.com/stories/2004/01/09/60minutes/main592330.shtml and page 96 of [676].) Note that in 2010 the end of combat/war by American's in Iraq was declared by the U.S. President. How many U.S. troops remained at that time (and when you read this)? How many employees of private military contractors remained at that time (and when you read this)? For example, see the 9/3/10 interview with Phyllis Bennis, www.fair.org.

(ii) What changes would you expect in the past 100 years of Iraq's political history if its principal resource were not oil but, say, broccoli?

(iii) Do your answers to the above questions depend on how carefully you checked the documentation/accuracy of the pronouncements of government officials and the mainstream media during the invasion of Iraq and the years afterward? Discuss with your friends and/or classmates, especially if they do not agree with you. Have any of those government officials or mainstream media changed their story in the interim?

(iv) Estimate how much less dependent America could have been on oil, had electric rail and car transportation been allowed to co-evolve with fossil-fuel powered transport.

(v) Big-box stores and shopping centers alone are responsible for about 900 million cubic feet of runoff each year in the Cleveland metropolitan area, [461, p. 106–7]. Assume Cleveland is roughly 80 square miles (the metro area is bigger, but we ignore this for simplicity) with a shape of a semicircle bounded on its diameter by Lake Erie. Consider two scenarios: the 900 million cubic feet of water ran off into the lake, or the 900 million cubic feet of water went into the ground and is sitting a few feet from the surface, at roughly the same elevation as the surface of the lake. (This latter scenario is possible if porous paving materials are used.) Estimate the difference in energy it would take for 100,000 uniformly distributed wells to pump this water to the surface in one year vs. the energy to transport this water from the lake to the 100,000 pump sites in a year. You can give a qualitative answer in English, or a more quantitative answer using VII.

(vi) Measured in terms of the amount of oil released as a pollutant into the environment, how does the year in which you read this compare to 1997?

(vii) Can you identify at least two indigenous populations in the world who have suffered greatly in terms of health, social, and environmental degradation due to the extraction of oil from their lands? Are there more than two such cases? Do the same exercise for coal, natural gas, and again for uranium, if you have time.

(viii) To what extent is there/has there been armed conflict over resources such as oil, cf., [354, 355]? Have the costs of these conflicts been directly reflected in the price of such resources?

1.9 The CO_2 Greenhouse Law of Svante Arrhenius

In 1824 the French mathematician, Jean Baptiste Joseph Fourier (1768–1830), in order to describe certain observations, created the term "greenhouse effect," [199, 200]. Thus the term greenhouse effect has a long history! In modern language this effect occurs when visible spectrum sunlight passes through an enclosure-creating barrier, like glass or an atmosphere, and the enclosure heats up because the barrier absorbs/emits infrared spectrum radiation or otherwise traps heat.

In 1860, scientist John Tyndall experimentally determined that gases such as CO_2, carbon dioxide, and water vapor were major contributors to the greenhouse effect.[23] He observed that these gases, not the massively more abundant oxygen and nitrogen, were the most effective in trapping thermal energy.

In 1896, Swedish scientist, Svante August Arrhenius (February 19, 1859–October 2, 1927), 1903 Nobel Prize winner in chemistry, studied effects of atmospheric concentrations of CO_2 on ground level temperatures. Motivating Arrhenius was his desire to explain ice ages. His careful calculations were based in part on certain experimental data available at the time. Details in the work of Arrhenius, [16], may easily be criticized from the vantage point of the 21st century. For example, some of the numerical values of constants carefully calculated by Arrhenius are not in agreement with numerical values used now. Also, he likely believed that global warming would be an unalloyed blessing, averting ice ages, an understandable position in 1896. Finally he thought, based on anthropogenic CO_2 emissions at the time he was working, that it would take humans about 3000 years to double atmospheric CO_2 concentrations.

These days the role of atmospheric CO_2 concentrations in the science of ice ages is still considered nontrivial, but other mechanisms are believed to precipitate "ice age events." Also, the time it is taking humans to double the CO_2 concentration in the air is of the order of 100 years, not 1000. But as for his major thesis, which he was the first to articulate, namely: *increasing emissions of CO_2 leads to global warming*—Arrhenius's work remains intact. Although some of the numerical values used or calculated by Arrhenius can be called into question, the basic *form* of his "greenhouse law" remains the same; and he derived this law from basic scientific and mathematical principles. Using this law and the numerical values he calculated for various constants, Arrhenius predicted that doubling CO_2 concentrations would result in a global average temperature[24] rise of 5 to 6 deg C. The Intergovernmental Panel on Climate Change (IPCC) calculated in 2007 a 2 to 4.5 deg C rise; fairly good agreement given that over a century separates the two sets of numbers—which rely on the accuracy of certain experimentally determined constants.[25]

---

[23]Some have argued that increased water vapor in the atmosphere leads to more clouds which reflect some sunlight, hence global warming may not be as big a problem as once thought. It is not that simple. A more detailed analysis than we can go into here indicates that the global warming effect of water vapor is greater than the cooling effects of clouds (besides cloud cover at night slows the escape of heat at ground level, for example). I invite the interested reader to research this and many other related topics, for example, cf., http://www.realclimate.org.

[24]Note that C denotes degrees Centigrade, where $F = \frac{9}{5}$ C + 32, is the formula connecting Fahrenheit temperature, i.e., F, to Centigrade temperature, C.

[25]It should be emphasized that these numbers are global averages. The temperature in the

From a reference published about 102 years after [16], namely, page 2718 of [476], we see Arrhenius's greenhouse law for CO_2 stated as:[26]

$$\Delta F = \alpha \, ln(C/C_0), \qquad \text{(Greenhouse Law for } CO_2\text{)}$$

where C is CO_2 concentration measured in parts per million by volume (ppmv); C_0 denotes a baseline or "unperturbed concentration" of CO_2; α is a constant (IPCC gives $\alpha = 6.3$, [476] gives 5.35); and ΔF is the radiative forcing, measured in Watts per square meter, $\frac{W}{m^2}$, due to the increased (or decreased) value for C, the independent variable. Radiative forcing is directly related to a corresponding (global average) temperature; because by definition radiative forcing is the change in the balance between radiation coming into the atmosphere and radiation going out. A positive radiative forcing tends on average to warm the surface of the Earth, and negative forcing tends on average to cool the surface. (We will not go into the details of the quantitative relationship between radiative forcing and global average temperature.)

For the record, cf., [15], preindustrial concentrations of CO_2 are estimated to have been about 275 ppmv. Also see Figure 1.4, where CO_2 concentrations of the past, viz., 1744 to just after 1950, are measured in ice cores taken at Siple station in Antarctica. We could take $C_0 = 275$ ppmv in Arrhenius's law. From Figure 1.4 (or from [15], p. 43) we see a graph (or table) of global, average annual CO_2 concentrations measured at Mauna Loa, Hawaii from 1960 when it was 316.91 ppmv to 2006 when it was approximately 381.84 ppmv. In this graph/table the function of CO_2 concentration vs. time is essentially an increasing function from 1960 to the present (neglecting the annual fluctuation). We note that in 2008 CO_2 concentrations of 387 ppmv were measured in Svalbard, Norway.

So What's the Point? Arrhenius's CO_2 Greenhouse Law, like the law of gravity, has never been repealed since discovered. Though we may not be able to discuss every detail of the math/physics/chemistry involved here, the approximate quantitative increase in global temperature experienced on the ground corresponding to any particular rise in CO_2 concentrations is understood. Now some comments. Do you think humans burning carbon-based fuels like crazy since the beginning of the Industrial Revolution has had anything to do with the fact that CO_2 concentrations are rising? (See Exercise 1.10 part (xix).) Also, CO_2 is not the only greenhouse gas whose concentrations are going up due to human activities, cf., methane, for example.[27] In

Arctic, for example, is increasing much faster than the global average. From the Toolik field station in Alaska I learned that in parts of the Alaskan arctic temperatures are up 4 degrees Fahrenheit or more since the 1950s, compared to a global average increase of about 1 degree over the last century.

[26] Do not be alarmed just yet if you do not understand every detail of Arrhenius's law. It will be easier after we study ln, the natural logarithm function, in V.

[27] As I write there are news stories about methane being released from formerly stable regions—sea beds, frozen tundra, and the like. Methane is more than 20 times more potent

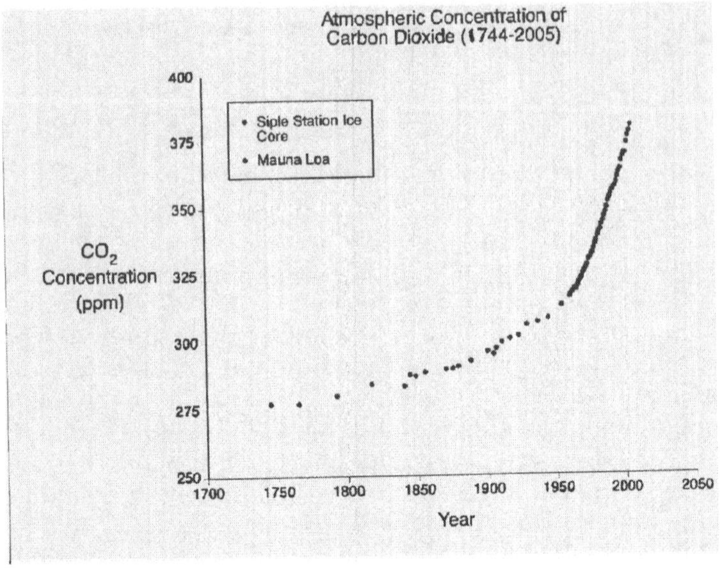

FIGURE 1.4: Carbon Dioxide Concentration in the Atmosphere (1744–2005)

[476] a few other greenhouse gases are discussed along with the corresponding laws from science telling us how much warming we can expect from an increase in their atmospheric concentrations.

Thus those who would deny that global warming is occurring have a prodigious task. They have to (quantitatively) show that some other factors compensate for, i.e., cancel out, the warming predicted by Arrehenius's law. (This warming is significant, cf., Exercise 1.10 part (xvii).) They have to explain why mounting data indicating that warming is going on, [217, 218, 51], to name only three of many references, (not to mention what polar bears and some indigenous arctic folks have to say) is not persuasive. Doubters will no doubt always be with us, as might be expected in any diverse self-organizing system. For example, in late 2009 a number of pundits concluded after "eyeballing" temperature data that the globe was cooling. The Associated Press responded by doing the following experiment. They gave global temperature data to four independent statisticians without telling them what the numbers represented. (This is an example of a "blind test"). The mathematicians found no actual temperature declines over time, despite the fact that 1998 was a record temperature year. So why do I have to even discuss these things? For quite a while, according to [217, 218, 317], the fossil fuel industries and some

a greenhouse gas than is CO_2. What is the status of methane release at the time you read this?

of their associates have financed the disinforming of the public about this science. This disinformation campaign finally, rather late, made the front page of *The New York Times* on Friday, April 24, 2009.[28] As noted in the article, a group, the Global Climate Coalition, ignored its own scientists' findings (that global warming is a fact) as it sowed seeds of doubt in the media, legislative bodies, and among the public about the science of climate change (much of which was understood by Arrhenius in 1896!) and the need to reduce carbon dioxide emissions. The mainstream media gave this effort "equal time" to show "balance," without making it clear who exactly was arguing on each side of the issue. When it comes to science and mathematics, all a PR effort has to do is confuse the public to induce inaction, cf., [317] for a book length critique by a PR professional of the PR effort to challenge climate science on the subject of global warming. Confusion was well funded as we read, again a bit after the fact, in 2009 in [247, p. 366] that Exxon-Mobil was giving millions of dollars to 39 groups "which misinformed the public about climate change . . . ." The view that climate change is not a (human caused) phenomenon apparently still has strong financial support.[29]

On April 1, 2008, during a congressional hearing, Exxon-Mobil Senior Vice President, Stephen Simon, had this to say in reply to Representative Jay Inslee's discussing America's energy portfolio in 2050 and questioning Exxon-Mobil's investing less than one percent of their profits in renewable energy: "..., but the fact is that we are going to have oil and gas and coal, and its going to constitute about 80 percent of the energy equation. With that as a given, ..."[30]

Future generations might be able to use the complex organic molecules in what fossil fuels might remain for them in fantastic ways we have yet to imagine. Burning oil for energy is like burning the Mona Lisa for heat. If the current world's human population succeeds, however, in burning most of the remaining coal, oil, natural gas, tar sands and oil shale, following the vision put forth just above, the journey begun with the destruction of the trolleys in America will likely end in distress. The climate science seems quite clear: stop burning fossil (carbon) fuels.

Exercise 1.10 Global Warming Exercises

(i) A 3 degree rise in temperature centigrade corresponds to how many degrees rise Fahrenheit?

(ii) Can you find a ln button on your calculator? If so what is $ln\,2$?

---

[28]Andrew C. Revkin, "On Climate Issue, Industry Ignored Its Scientists," *The New York Times*, April 24, 2009.

[29]Research the role with (regard to the "flow of information" about climate change) of Koch industries and the various groups it funds, such as AFP, Americans for Prosperity. Charles and David Koch are owners of the largest privately held oil corporation. See, for example, "Covert Operations: The billionaire brothers who are waging a war against Obama," by Jane Mayer, *The New Yorker*, Aug. 30, 2010, pp. 44–55.

[30]See http://www.democracynow.org, April 2, 2008.

(iii) Would you estimate that relatively very few people were able to confuse the global warming debate? Is this number more, less, or about the same as the number of people it took to destroy the trolley systems in America?

(iv) Take a close look at Figure 1.4. Extrapolate, i.e., extend into the future, where you think the graph is going, given present trends. In what year do you predict we hit 400 ppmv? 450 ppmv? 500 ppmv? Are there any circumstances you are aware of that might speed up (or slow down) your predicted arrival dates at each of the above? How might other gases like methane behave as the planet warms?

(v) I claim that James Hansen is one of the most highly regarded climate scientists in the world, cf., [52]; investigate the truth value of my claim. I also claim that Freeman Dyson is one of the most highly regarded mathematicians/physicists in the world; investigate this claim as well. Dr. Hansen has said that we must get the global CO_2 concentrations below 350 ppmv as soon as possible to avoid a "tipping point" in world climate that humans might likely find highly undesirable, cf., www.350.org. Dr. Freeman Dyson does not agree, cf., Nicholas Dawidoff, "The Global-Warming Heretic: How did Freeman Dyson—REVERED SCIENTIST, LIBERAL INTELLECTUAL, PROBLEM-SOLVER —wind up infuriating the environmentalists?", *New York Times Magazine*, March 29, 2009. (I mention Dyson because I have found no evidence that his views were "purchased" by the disinformation campaign discussed in this section.) Whose mental model do you think is closer to reality and why? If we follow the advice of Dr. Hansen and he is wrong, what are the consequences, to whom? If we follow Dr. Dyson and he is wrong, what are the consequences, to whom?

(vi) Investigate the truth value of the following claims. The extraction process associated to any non-renewable fuel has negative environmental impacts that get larger the longer the extraction process continues. The disposal of waste products associated to every non-renewable fuel has environmental impacts that also increase with time. Every non-renewable fuel "runs out" eventually, in the sense that it takes more energy to extract it than what you get, i.e., its E.R.O.I. (see Exercise 1.8) is less than 1; and/or it becomes too expensive to extract. Conclusion: we must eventually rely on renewable energy sources, the only question is when do we start, and why?

(vii) I once heard a media pundit mock climate change scientists with approximately the following argument: "2 degree temperature rise, 4 degree temperature rise, who cares, that can hardly be really noticeable, at least in my lifetime." I wrote a paper, [698], that quite possibly calls this argument into question. Under a very tenable assumption, it follows that a modest increase in average global temperature leads to an immodest increase in extreme weather events, such as tornados, hurricanes, floods, blizzards and so on. What is this assumption? Are we already experiencing this phenomena? Is the data conclusive? Is there a qualitative, intuitive argument for this phenomena? Should we take no action until there is no doubt?

(viii) For at least approximately 10,000 years humans have enjoyed a global climate that has been relatively stable. Research where climate scientists think precipitation will increase/decrease, and what instabilities we can expect with global warming. Make an estimate or at least discuss qualitatively what effects these changes will have on human agriculture.

(ix) Not everyone agrees with my last sentence before Exercise 1.10. They say that we can continue to burn fossil fuels, we just need to sequester the carbon dioxide thus produced. Investigate proposed methods of sequestration such as pumping the CO_2 into underground formations, or liquefying it in pools in the deep ocean and so on. At the time you read this are there any successful/commercial CO_2 sequestration operations? What are the negative environmental consequences of CO_2 sequestration, for example, on groundwater? One interesting fact is that for every 12 tons of pure carbon that is completely burned, about $12 + 16 + 16 = 44$ tons of CO_2, cf., Figure 11.1, is produced and needs to be sequestered.

(x) **Global Warming: A National Security and Economic Threat?** Trillions of dollars have been spent on military activities. Trillions have been spent in attempts to "fix" the financial crisis of 2008 and beyond. How much money has been spent in response to the quite possibly devastating effects on humans of global warming? What does this say, if anything, about humans as a species?

(xi) **Geoengineering?** Suppose some massive engineering project, like putting mirrors in space to reflect sunlight, managed to compensate for the greenhouse effect of CO_2. If humans could then continue with increased CO_2 emissions without worrying about warming, what other catastrophe, possibly greater than global warming, would take place? Hint: see Chapter 5. Projects such as these are called *geoengineering*. Another proposes to put sulfur dioxide into the atmosphere, year after year, mimicking volcanic eruptions. Discuss.

(xii) **Other Reasons for Not Burning Fossil Fuels.** Assume for the sake of argument that burning coal and other fossil fuels did not cause global warming or ocean acidification. Give at least two additional important reasons for switching to renewable energy sources.

(xiii) **Probabilities of Adaptation or Extinction.** It is quite possible (likely?) that humans will not be able to control their greenhouse gas emissions in time to prevent the global climate from passing through a tipping point, which will bring irreversible dramatic changes lasting thousands, perhaps millions, of years. Do you think it will be possible for humans to adapt, or will they go extinct? What sort of adaptations do you envision?

(xiv) **Tax Policy and Global Warming.** Discuss the relative merits of a (revenue neutral) *carbon tax* and a *cap and trade* program of carbon emissions with regard to curbing CO_2 emissions. Briefly, cap and trade is analogous to the system that (partially) mitigated the acid rain effects from SO_2 emissions. It involves putting "caps" on emissions by law, hopefully *declining over time*, and allowing those whose emissions are below their cap a "credit" or "allowance" which becomes a financial instrument that can be sold or traded. If one exceeds their cap, they must buy sufficiently many "credits" to compensate or be fined. (One method is to buy "offsets" in the developing world; but enforcement is difficult and loop holes exist so that net carbon emissions do not necessarily decrease.) A revenue neutral carbon tax is a tax on every source (mine or port of entry), say $x per ton of C. The tax is revenue neutral if, say, income taxes are reduced by the amount the carbon tax brings in. Cap and trade is complicated, possibly hard to enforce, especially internationally, but it has the political advantage of creating new securities, including likely unregulated derivatives that Wall Street can profit from. (See Chapter 2. Is a "carbon bubble" possible? What would be its consequences when it "pops?") The carbon tax is simpler, enforceable, and would reduce emissions, but likely difficult politically. See [247, pp. 342–5], which advocates using both. See James Hansen, "Cap and Fade," *The New York Times*, Dec. 7, 2009, p. A27, who argues for a revenue neutral carbon tax. Hansen also has a book, [272], urgently calling for action to stem global warming. (Note Hansen believes nuclear power is an essential part of the energy mix of the future and that "renewables" cannot completely fill the bill. See our Chapter 6 and www.rmi.org for arguments that "renewables" can indeed fulfill humanities energy requirements.) See also http://www.storyofstuff.com/capandtrade/. In addition to putting a price on carbon, by whatever mechanism, what needs to be done to actually start reducing carbon emissions?

(xv) **Agriculture and Global Warming.** The figures at the climate change Web site, www.ipcc.ch, state that at least 60 percent of all nitrous oxide, N_2O emissions, the most potent greenhouse gas, are caused by industrial agriculture, primarily from the use of synthetic nitrogen fertilizer. Nearly 50 percent of methane, CH_4, the second strongest greenhouse gas, is due to industrial farming practices, much of this from intensive livestock operations. The IPCC then estimates that industrial agriculture contributes at least 14% of greenhouse gas emissions. The Center for Food Safety (an environmental group) estimates that if one looks at a "Big Picture" which includes: all energy inputs into industrial agriculture (such as production and use of pesticides, herbicides, fertilizers); food transport; factory farming impacts; water useage; and displacement of carbon sequestering biodiverse ecosystems; then the IPCC estimate could be revised upward to 25% to 30%. If this is true then industrial agricultural practices would need to be addressed for a complete solution to our climate change challenge. Do your own estimates of the Center for Food Safety estimates and see if you agree.

The Center for Food Safety offers the following solution: *"The potential for rapid change is exciting. For example, studies by the Rodale Institute project [coolfoodscountdown.org] (Scroll down to Regenerative Organic Farming: A Solution to Global Warming) that the planet's 3.5 billion tillable acres could sequester nearly 40 percent of current* CO_2 *emissions if converted to 'regenerative' organic agriculture practices. The same 10-year research project concluded that if U.S. cropland (based on 434 million acres) were converted to organic farming methods, we could reduce nearly 25 percent of our total GHG emissions.*

"Many other studies have drawn similar conclusions. In India, research shows that organic farming practices increase carbon absorption in soils by up to 55 percent (even higher when agro-forestry is added into the mix), and water holding capacity is increased by 10 percent. A study [www.cnr.berkeley.edu] of 20 commercial farms in California found that organic fields had 28 percent more carbon in the soil than industrial farms."

The Center for Food Safety also compares productivity of organic vs. industrial agriculture: *"A comprehensive study [www.ns.umich.edu] of 293 crop comparisons of industrial and organic agriculture demonstrates that organic farm yields are roughly comparable to industrial farm yields in developed countries and result in much higher yields in developing countries (the full study is available for purchase [journals.cambridge.org]).*

"The World Bank and United Nations International Assessment of Agricultural Knowledge, Science and Technology for Development [coolfoodscountdown.org] concluded that a fundamental overhaul of the current food and farming system is needed to get us out of the growing food (and fuel) crisis. They recommend that small-scale farmers and agro-ecological methods 'not industrialization' are the keys to a viable food security. Additionally, numerous studies [coolfoodscountdown.org] unequivocally state that our survival depends on the resiliency and biodiversity of organic farm systems free of fossil fuels and chemical dependency."

For a project research the IPCC and Center for Food Safety statements. We discuss some of these issues in Chapter 5.

(xvi) The United States has come to rely ever increasingly on the Alberta tar sands of Canada for oil. What are the impacts on global CO_2 emissions, on water, on the health of indigenous peoples living downstream? See [494] for an up-close perspective.

(xvii) **The Law of Arrhenius.** Given the numbers relevant to the Greenhouse gas law of Arrhenius, viz., $\frac{390}{275}$ as the ratio $\frac{C}{C_0}$ in 2010, calculate the Watts per square meter increase due to carbon dioxide increases in the atmosphere. Compare this to the natural solar radiation, also measured in Watts per square meter, from the sun, see Exercise 6.6.

(xviii) The U.S. military has considered climate change seriously as a matter of national security. If Americans actually took the potential impacts of climate change seriously how do you think their behavior and, say, the U.S. national budget, would change? For example, would a revenue-neutral carbon tax be possible? (A tax would be levied on carbon fuels at their source, either mine, well, or port of entry, for example; and the money would then be distributed to the population at large in, say, the form of a tax credit.) Would we be spending some defense money in the form of solar energy research and installation of existing technologies? Would we be treating climate change as seriously as, say, the danger of an explosive device or weapon being carried on an airplane?

(xix) The late John Firor, an environmental scholar and former director of the National Center of Atmospheric Research (NCAR), in a lecture given on November 18, 1998, said that there are about 6 billion tonnes of C, i.e., carbon, put into the atmosphere in the form of CO_2 every year. At that time there were about 6 billion people on earth. So the human contribution of C to the atmosphere is easy to remember, about 1 tonne per person per year—on average. However, Firor estimated that the typical American's contribution is 6 to 10 tonnes per person per year while, say, the typical Indian contributes $\frac{1}{5}$ tonne. The exercise is this: verify the accuracy of these statements—either now, or later when you have read more of the book and have hopefully picked up sufficiently many skills to do this verification. In fact, in the intervening years (decade), the amount of C put into the atmosphere per person per year has been increasing somewhat. Is it less than 1 tonne per person per year on average at the time you read this?

(xx) I guess that the fossil fuel industry might disagree with most if not all of this chapter. On page 25 I gave a quote from an Exxon representative. Look up quotes of other representatives of the fossil fuel industry, especially in regard to their position on global warming and/or climate change. Check to the best of your ability the references from which they get the information on which they base their disbelief. For example, the following can be found at www.democracynow.org, April 7, 2010, regarding Massey Energy, the fourth largest coal company in the U.S. (The video clip of the Massey Energy CEO runs from the 32:25 to the 34:25 minute marks of the 41:15 minute piece.) Note that on June 1, 2011, Alpha Natural Resources, Inc. (NYSE: ANR) announced it had completed its acquisition of Massey Energy Company. With this acquisition Alpha Natural Resources controls the second largest coal reserve base in the United States.

AMY GOODMAN: I want to play some of the past comments of Massey Energy CEO Don Blankenship. He's the director of the US Chamber of Commerce. He's strongly opposed any legislation around climate change. These are highlights from the speech Don Blankenship delivered at the Tug Valley Mining Institute in West Virginia in November of 2008.

DON BLANKENSHIP: I don't believe climate change is real. I do believe that the Arctic is melting and the Antarctic is getting colder. I believe it's a normal cycle. This is the first speech in twenty-two years at the Tug Valley Institute that I've made in November while it was snowing outside. So it's not my greatest concern.

Let me be clear about it: Al Gore, Nancy Pelosi, Harry Reid, they don't know what they're talking about. They're totally wrong. What they do is nonsense. And until we begin to call it what it is, people are going to misunderstand, because when we talk about it in more articulate, educated ways, the American public doesn't get it. Pretty simple, they're all crazy. I mean, it is absolutely crazy. How can anybody run for office and say they're going to bankrupt the coal companies and be energy-independent and get elected? I mean, how do you do that? How do you stop us from mining coal while we look for Indiana bats and put up windmills to kill them all? I mean, if they go ahead with the windmills, we wouldn't have a problem. You know, it is absolutely crazy.

It is a great—it is as great a pleasure to me to be criticized by the communists and the atheists of the Gazette as it is to be applauded by my best friends, because I know that they're wrong. I mean, when you have an editor that's, you know, an admitted atheist and when you have people who are clearly of the far-left communist persuasion, would you want them to speak highly of you? You know, it's really crazy when you look at it—and I reuse that word over and over, because what we've got is people cowering away from being criticized by people that are our enemies. I mean, are we going—would we be upset if Osama bin Laden were to be critical of us? I don't think so.

(xxi) Research the global flow of dust. For example, dust from the Sahara in Africa has landed in the Western Hemisphere, as has dust from Asian deserts. Dust from Arizona and Utah have ended up at my home. Why is this becoming a more frequent occurrence? Are there both positive and negative effects of global dust transport?

(xxii) **Ocean Acidification.** When one calculates how much carbon humans have emitted globally into the atmosphere over long periods of time, one finds that Figure 1.4 does not account for it all. (Before the course is over, try to do some of these calculations yourself.) In fact, there are enormous amounts of CO_2 that were emitted over the last century or so that did not end up in the atmosphere. Where to you think it went? Hint: look up *ocean acidification* in the index.

Chapter 2

Economic Instability: Ongoing Causes

But what do they know (referring to big bankers and Wall Street executives)? *The answer, as far as I can tell, is: not much.*

Paul Krugman, economics *Nobel laureate*, January 15, 2010

I am writing this chapter[1] a year to two years after the "great economic meltdown of 2008." In 2008 I would not have thought it necessary to write this chapter, especially in the elementary form it has taken. However, over two years have passed, the "recession has been declared over[2]"; federal legislation has become law which claims to reform the U.S. financial system;[3] yet basic flaws at the heart of the crisis have not been honestly addressed by the people entrusted by society with the power to do so, e.g., our government and our financial industry. Some of these flaws are most easily described using mathematics, and it is on these aspects of the crisis I will concentrate. The good news is that given the immensity of the problems being discussed, the mathematical structures involved are very simple! The bad news is that this chapter is likely to remain timely for a long time to come. In any event, the mathematical structures I am about to discuss which have economic importance are timeless.

[1]A preliminary version of this chapter was delivered as a paper at a Cambridge University, U.K., conference in August 2009. This paper was published as part of the conference proceedings; see "One Mathematical Perspective of Economics, Ecology and Society: Some Natural Necessities in Elementary Interdisciplinary Mathematical Education," *The International Journal of Science in Society*, Volume 1, Issue 2, November 2009, pp.111–120; http://ijy.cgpublisher.com/product/pub.187/prod.25.

[2]For example, it was widely reported in September 2010 that the National Bureau of Economic Research declared the 18 month recession had ended in June 2009.

[3]For example, see the "Restoring American Financial Stability Act of 2010."

2.1 Necessary Conditions for Economic Success

We will be tossing some very large numbers of dollars around, so let's begin by getting an intuitive grasp of how large these numbers are. If you need a refresher on how to write 1 million and so on mathematically, see page 59.

Exercise 2.1 Millions, Billions, Trillions

(i) Is a million seconds longer or shorter than one year? If shorter, what fraction of a year is it?

(ii) How many years is a billion seconds? If you expressed your answer to part (i) in decimal form, what is the easiest way to do this exercise using part (i)? Hint: A billion is one thousand times a million.

(iii) How many years is a trillion seconds? Again, what is the easiest way to do this exercise using part (ii)?

(iv) To simplify the math, suppose a large number of people earn $10 per hour. How long would it take one person at that wage rate to earn a trillion dollars? How many people would have to work for one year at that wage rate to earn a trillion dollars? Assume an eight hour day, five day work week, with two weeks off for vacation (hopefully). (Approximately what fraction of the adult working population of the United States is this number of people?) At the time you read this is the U.S. federal minimum hourly wage more or less than $10? On the day I wrote this, my state of Colorado just *reduced* its minimum wage to $7.24 per hour.

There are two words in mathematics with very special and important uses: *necessary* and *sufficient*. Let's take a look at how mathematicians use these terms; for that's how they will be used in this book.

Exercise 2.2 The Concepts of Necessary and Sufficient.

(i) These two concepts are in some sense part of the logical notion of *implication* used in mathematics. "A implies B" is often written "A $\Longrightarrow$ B." This means that *if* A is true (if A "happens"), *then* B is true (then B "happens"). If it is true that A $\Longrightarrow$ B, then we say that the truth of B is a necessary condition for the truth of A. In order for A to "happen" it is necessary that B "happen." Think about the following way of saying this which it turns out is logically equivalent: If B is not true (if B does not "happen"), then A is not true (then A does not "happen"). The following is a true statement in the real world (so far anyway): If a chicken lays an egg, then the chicken is a hen, i.e., a female chicken. Thus it is necessary for a chicken to be female in order for it to lay an egg. Can you think of other such examples of "B is necessary for A"?

(ii) If it is true that A $\Longrightarrow$ B, then we say that A is sufficient for B, i.e., in order for B to be true (for B to "happen"), it is sufficient for A to be true (for A to "happen"). Said another way, if A "happens" then B will definitely "happen." Thus, if a chicken lays an egg, then the chicken is a hen; we have that for a chicken to be a hen it is sufficient to know that the chicken has laid an egg! Can you think of other examples? Hint: look at the examples you came up with in part (i).

Here are four conditions I propose are necessary for a stable economy, i.e., one that reliably works for us for extended periods of time, say, for several generations. They all involve *feedback*, or said another way: mechanisms of *accountability*.

(1) There must be a *transparent* and *honest* evaluation of the *risk*[4] of any financial transaction. All parties to a transaction should understand it. All parties affected by a transaction should have a representative involved who understands it and can communicate that knowledge to those potentially affected. A regulatory framework (such as existed for about 70 years, characterized by laws such as the Glass-Steagall Act—see Section 2.3.[5]) should always be in place and enforced! This regulatory framework needs to be updated to deal with all financial institutions, including "shadow banks," hedge funds and the like. Gambling (see Section 2.3), whether in the form of "credit default swaps," "derivatives," or any other form, when not outlawed outright, needs to be clearly defined, regulated, and isolated from non-gambling financial activities, also carefully defined. A Tobin tax on financial transactions should be considered, as was originally proposed for currency markets by Nobel laureate, James Tobin. A mechanism should be developed which favors the economically and socially productive functions of financial institutions (such as raising capital to finance innovation, job-creating businesses, loans for shelter, and so on) over financial activities which are, mathematically speaking, games of chance.

(2) We must reinstate laws against *usury*, i.e., there must be legal caps on the amount of interest that can be charged for a monetary loan.

(3) To the extent possible and reasonable, *decision making* processes should be *distributed*.[6] In any event, we should not easily allow major decisions that affect us all to be made by very few, especially when (1) and (2) are not operating. Corporations "too big to fail," which concentrate decision-making and political power of global impact in the hands of a few, should be declared economically and socially too potentially dangerously disruptive to exist. Rules for redistributing them into more manageable entities should be enforced.

(4) Any financial operation or collection of operations that is mathematically similar to a Ponzi scheme should be illegal—just as actual Ponzi schemes are. (See the next section.) Barring a legal ban, when inevitable failure ensues, Ponzi-like operations shall not be "bailed out" at the public's expense, i.e., taxpayers' expense.

At this moment, say, some expert can declare this whole discussion nonsense by saying: Our economy works fine right now and none of the four so-called

[4]The concept of risk is essentially a mathematical one related to the probability or chance of failure. We will discuss probabilities briefly in VI.

[5]Briefly, the Glass-Steagall Act legally separated low-risk financial activities from higher risk ones. For example, money put into a simple interest bearing savings account at a bank could not be comingled with activities in the stock market or other speculation. Similarly, the insurance industry is based on very solid actuarial mathematics, and money in this sector was insulated from banking and stock market sectors.

[6]True democracies usually exhibit, perhaps indirectly, this form of decision making via a voting process, cf., Chapter 24.

necessary conditions hold! To such an expert I have only two words: Just wait.

Exercise 2.3 Financial Reform: How Much?

In mid 2010 a legislative package was passed by the U.S. Congress which was touted as the most sweeping financial reform in the United States since the 1930s, cf., for example, the Glass-Steagall Act of 1933.

(i) Compare the financial sector of the U.S. economy *before* the meltdown of 2008 and *after* the 2010 financial reform. Did the financial sector emerge from 2010 more or less concentrated than it was prior to 2008?

(ii) To what extent does the reform of 2010 address the four necessary conditions above? In particular, what principles of the Glass-Steagall Act were reinstated, which were not?

(iii) What aspects of 2010 reform are considered "pro-consumer"? What aspects are considered "pro-Wall Street"?

(iv) To what extent are whistleblowers within the financial industry (or elsewhere for that matter) protected against retaliation if their claims are true? See Exercise 2.6.

The above list of four conditions is clearly not sufficient to guarantee a stable economy. We need to work together in some sort of self-organizing system to come up with such a list. The underlying mathematics of these four "principles" acts inexorably, whether we take the time to understand or not. So let's study them at least briefly. I believe that the greater the number of citizens who understand the simple mathematics involved in "what is going on" with the financial sector of the economy the greater our chances for financial stability.

2.2 The Mathematical Structure of Ponzi Schemes

If you can multiply by 1.1 or 1.2 or 1.5, for example, you can understand Ponzi schemes, named after Charles Ponzi (1882–1949) who was not the first but one of the most famous practitioners of this particular form of fraud. In the early 20th century he offered his clients a 50% profit (or return) on certain "investments" within 45 days, or 100% within 90 days.

Exercise 2.4 Geometric Ponzi Math

(i) If you invested \$100 with Ponzi, how much total money does he claim you will have in 45 days? The answer is \$100 plus 50% of \$100. How much is that?

(ii) Is your answer to (i) the same as $1.5 * \$100$?

(iii) If you invest \$100 how much does Ponzi promise the total value of your investment will be in 90 days?

(iv) Suppose you invest \$100 and in 45 days you actually have $1.5 * \$100$, which you then reinvest with Ponzi for 45 days more. How much does Ponzi claim your total investment is worth now? Is your answer to (iv) greater, less than, or equal to your answer in (iii)?

The next crucial step in a Ponzi scheme is for you to "tell your friends" what a great deal you just got. In other words, by some mechanism a new group of investors needs to be recruited—and then another new group is recruited after that, and so on. This is very much like a "chain letter" which arrives in your e-mail, convincing you to send in $1 and asking you to resend the letter to 10 of your friends. For each friend that sends in $1, you will get 15 cents. If all goes according to plan, you will have $1.50 after this round of the Ponzi scheme—and the person who started it will have your $1 plus $10 less the $1.50 rebated to you—if indeed the rebate is sent. Remember, you are dealing with Ponzi the crook. For this scheme to work, you need lots of gullible investors brought in one "layer" at a time, each layer larger than the layer that went before. In our chain letter example, each layer is 10 times larger than the previous layer, building a *pyramid* of investors. For this reason, Ponzi schemes are often called pyramid schemes. For the scheme to work, Ponzi must strike a delicate balance among: (1) paying himself or herself, and (2) paying out enough to investors to keep them convinced the system is legit and keep them recruiting new investors, and (3) getting caught by authorities—or the inexorable march of mathematics, which we look at in the following exercise.

Exercise 2.5 An Ideal Pyramid Scheme

(i) Ponzi is level 0 and recruits 10 investors in level 1 to send him $1. At this point how much money lies in level 0? level 1? Assume that for each investor in level 1 $1 is their entire life savings/worth.

(ii) For each investor in level 1, 10 investors in level 2 are recruited, by some means. Each level 2 investor sends in $1 (still assume here and throughout that this is their life savings) to Ponzi, who rebates $.15 of each dollar to the corresponding recruiter from level 1. Of course, if a level 2 investor is recruited by Ponzi himself, or hears about the great deal without being recruited, Ponzi can keep the $.15 as well. How much money (minimum) does Ponzi have at this point, and how many investors are there?

(iii) Repeat, with each investor at level 2 somehow recruiting 10 investors at level 3, each of which sends in $1 to Ponzi, who rebates $.15 of each such dollar to a level 2 investor. How much money does Ponzi have now? How many investors are there total?

(iv) How many levels have to be introduced before the total number of investors is greater than the entire human population of the earth? Just before that happens, how much money does Ponzi have?

Of course, in the real world the ideal Ponzi pyramid will have variations; but no matter how complex or convoluted, every Ponzi scheme exhibits *growth* of prodigious proportions.

American financier, Bernard Madoff, one-time chair of the NASDAQ stock exchange, was able to run a Ponzi scheme for 15 possibly 20 or more years without interruption. He was eventually caught and sentenced to 150 years in prison on June 29, 2009. Upon his arrest $65 billion dollars was missing in his clients' accounts, and at least $18 billion of investors' initial capital was missing. The Madoff example is instructive in many ways.

Madoff was able to get away with this fraud because of his social position and connections, his impeccable reputation, investors who believed, and the extremely lax oversight of such bodies as the SEC, Securities and Exchange Commission, which was created in 1934 to enforce federal securities laws and regulate the securities industry. In addition, Madoff's activities, being totally illegal, also fell under the jurisdiction of the FBI, the Federal Bureau of Investigation. But in 2001, the division of the FBI that deals with white-collar crime was weakened by the transfer of (and never replaced as of 2009) 500 agents to "anti-terrorism," i.e., Homeland Security. Note that even in prosperous times annual losses due to ordinary white-collar crimes are far, far greater than losses due to ordinary property crimes, [585, 584].

It turns out that mathematics can play a very interesting role in uncovering securities fraud. For example, the stock market exhibits certain mathematical behaviors that have been studied a great deal. The market does not behave like an interest bearing bank account, especially over a period of years. Yet Madoff offered nearly constant rates of return. He did not even find it necessary to introduce a little "randomness" as a disguise. Hundreds probably knew, e.g., no large Wall Street institution "invested" with Madoff; but financial analyst, Harry Markopolos, tried to act. He said: "It took me 5 minutes to know that it was a fraud, it took me another almost 4 hours of mathematical modeling to prove that it was fraud." He went to the Security and Exchange Commission in: May 2000; Oct. 2001; Oct., Nov., Dec. 2005; June 2007 and April 2008, without result.[7] As this and the surfacing of the many "Mini-Madoffs" reveals,[8] the watchdogs were asleep—mathematically and otherwise. Madoff might have gone on for a few more years if the downturn in the economy had not led a number of his clients to ask for their money simultaneously—with no new "investors" to cover the scheme. Every Ponzi scheme ends badly for some, it is mathematically guaranteed.

Exercise 2.6 Whistleblowing and Other Forms of Feedback.

(i) I define a *whistleblower* as one who speaks truth (about and) to the powerful. (As opposed to a *snitch* who speaks truth or falsehoods to the powerful about those less powerful.) Whistleblowers often provide an essential form of feedback about system function, in this case, about the functioning of our financial system. Research what happened to whistleblowers in the financial industry during the lead up to and during the meltdown in the first decade of the 21st century. You can start with [320], or Michael Hudson's articles on "Silencing the Whistleblowers," posted on May 9 and May 13, 2010 at www.thebigmoney.com. See also interviews with Michael Hudson and Ed Parker, a former mortgage fraud investigator, www.democracynow.org (May 20, 2010).

(ii) Ratings agencies, such as Standard & Poor's, Moody's and Fitch are supposed to give accurate assessments of the risk associated with financial instruments that they rate. How did they do in the first decade of the 21st century?

(ii) The Security and Exchange Commission (SEC) is supposed to regulate Wall Street and financial activities therein. How did they do in the first decade of the 21st century? For example, if they did not act on Markopolos's mathematical proof of impropriety, did they act on anything less obvious?

(iv) What is the role of the Federal Reserve regarding America's financial industry? Though it is really a private entity, despite the name, which operates in secret, looking back at the first decade of the 21st century, how was their performance?

---

[7]Harry Markopolos, *interview*, "The Man Who Figured Out Madoff's Scheme," CBS 60 Minutes, http://www.cbsnews.com, March 1, 2009.

[8]See pages B1 and B4, Leslie Wayne, "The Mini-Madoffs," *The New York Times*, January 28, 2009

2.3 Dishonest Assessment of Risk

A little history goes a long way in understanding the economic instability of recent times, which is a recurring pattern! I claim that a way to stop this recurring pattern, which I am about to discuss, is for a politically effective number of people to educate themselves just a little bit about the mathematical notion of *risk*, and then act when necessary. Thus, every citizen should be able to detect whether a financial transaction is: very risky (a Ponzi scheme, gambling, a combination of a Ponzi scheme and gambling in disguise); an investment with reasonable risk; or a more stable financial activity such as an interest bearing savings account or an insurance policy based on solid actuarial mathematical models. This task has recently become more difficult, so I suggest adopting the principle of *caveat emptor*. Assume you are being lied to until proven otherwise. Your life savings may be in the balance.

In April of 1998 Citicorp, i.e., Citibank, merged with Travelers Insurance Company, and I had a modest retirement account with Travelers at the time. This merger violated the Glass-Steagall Act of 1933, but regulators gave them a technical exemption for two years. During the Great Depression of the 1930s firewalls, such as Glass-Steagall, were erected between "gambling," e.g., the stock market/speculation, and banking activities like checking and savings. Actuarial-based insurance activities were similarly isolated from speculation. An expensive campaign (which continues, for example, $5.1 billion was spent by Wall Street in Washington D.C. on lobbying and campaign contributions from 1998–2008) was launched to make the Citicorp-Travelers merger legal, culminating in the passage of the Gramm-Leach-Bliley Act (November 1999), which set off a wave of mergers among banks, securities and insurance companies. Then Senator Phil Gramm from Texas was instrumental in the repeal of Glass-Steagall and other regulatory legislation. (See [210] for the preceding.) Financial collapse was predictable; in fact, at least as far back as 1966,[9] economist John Kenneth Galbraith said:

The financial bubbles that led to the Depression will return as soon as America is led by politicians who didn't live through it. Capitalism has no permanent memories, but one of its permanent vices is gambling with other people's money. So a post-Depression generation politics will dismantle the safeguards the New Deal put in place, and another financial collapse will then be inevitable.

I moved my retirement account. For the above and much more, cf., [210, 716, 717].

Exercise 2.7 Are Financial Bubbles Created?

(i) I will define an economic or financial bubble to be a period of economic trading in ever increasing volume of an entity at ever increasing prices not based on the intrinsic value

[9]Carl Pope, "Some Questions on the Bailout," www.huffingtonpost.com, September 22, 2008.

of that entity. I want you to look up, at least briefly, at least 10 financial bubbles, cf., [350, 48, 79]. I will start you out by listing a few: Tulip mania (about 1637); South Sea Company bubble (1720); Railway mania (1840s); American economic bubble of the 1920s; dot-com bubble (1995–2000); U.S. housing bubble (ending in 2008). Bubbles, it turns out, are very common. Why do you think this is?

(ii) Do all of these, and other bubbles, manifest relatively rapid growth?

(iii) Are bubbles just examples of self-organizing systems? Are bubbles created? Is their formation knowingly facilitated? Pick a bubble and try to decide if it was created, facilitated or just self-organized (or a mixture)?

(iv) Do bubbles behave mathematically like Ponzi schemes?

(v) Define *leverage* as making (often risky) investments with borrowed (that is, somebody else's) money. What role did increasing leverage have in the latest financial collapse? (Leverage went from 12:1 up to 30:1 and 40:1 before the collapse of 2008.) In earlier financial collapses? To understand leverage see the section on fractional reserve banking and the exercises therein, cf., page 503.

(vi) Why do bubbles always burst, i.e., come to an end?

(vii) Author Edward Abbey said that growth for the sake of growth is the ideology of the cancer cell. A mathematician might say that planning for perpetual (prodigious) growth is one of the mathematical defining properties of a Ponzi scheme. What relevance, if any, do such comments have about "mainstream" economic thinking?

(viii) In [669], Joseph Stiglitz gives a readable and detailed analysis of the causes (and possible remedies) of the 2008 and beyond meltdown. At the time you read this has there been a credible "counteranalysis" given by those who disagree with Stiglitz?

(ix) On page 488 there is a footnote that discusses the total value of derivatives (an invented type of financial instrument) in 2001. To wit: "... in 2001 the total value of derivatives contracts traded approached one hundred trillion dollars, which is approximately the value of the total global manufacturing production for the last millennium." Is this an indication of a "Ponzi-like" financial venture?

(x) Investigate the "food bubble of 2008." You might start with Frederick Kaufman's article in the July 2010 issue of *Harper's Magazine*, "The Food Bubble: How Goldman Sachs and Wall Street Starved Millions and Got Away with It," pp. 27–34. These activities of Wall Street have direct consequences in the lives of people throughout the world, cf., the food riots of 2008. See also Chapter 5 and Chapter 6, and Exercise 6.7, page 168.

Let's look at the "housing bubble" which burst in 2008, followed by further deflation. How one refers to this bubble provides an interesting lesson in the art of "determining the frame of debate." In a *Columbia Journalism Review* article[10] the use of the two terms "predatory lending" and "subprime lending" are studied in detail. The number of times each term is used each year by "mainstream media" is graphed. Between 2000 and 2004 they are used somewhat equally, neither being used more than 1000 times a year. From 2004 to 2006 usage of subprime gains on usage of predatory, but in 2007 usage of predatory peaked at a little over 1,000 then proceeds to fall, while the usage of subprime exploded to over 75,000 in 2007, falling to 55,800 in 2008. The term predatory conjures up a certain image of the lender, while subprime tends to emphasize the borrower (as being less qualified). I will leave it to the reader to decide which term is most honest. In any event, some lenders have been criminally prosecuted, while most have not. Fooling someone who owns their

[10]Elinore Longobardi, How 'Subprime' Killed 'Predatory' and what it tells us about language, business journalism, and the way we think about the economic crisis, *Columbia Journalism Review*, September/October 2009, pp. 45–9.

home into taking out a loan on their house which they cannot repay is not necessarily illegal, but it is a tactic worthy of a Ponzi scheme artist who needs to recruit more "players."

I recommend the reader study this subject in a great deal more detail than I have room for here, but I would like to quote[11] former bank regulator and author, [40], William Black. Black starts out recounting the events/mathematical structure of the Savings and Loan scandal of the 1980s, cf., VII, and then states that the 2008 and on crisis has the same structure:

"What happened then was an epidemic of what we call in criminology 'control fraud.' And that means what happens when the fraud is led by the person who controls a seemingly legitimate corporation or government agency. In this case, they were savings and loans. And these frauds were growing at an annual rate of over 50 percent. Their weapon of choice is accounting fraud. So it's real easy. It's a three-part optimization. First thing you do is grow like crazy, Ponzi-like scheme. Second thing you do is deliberately make really bad loans, because they have a higher interest and higher expenses associated with them, so you report more profits. And the third thing you do is have extraordinary leverage. Leverage is simply lots of debt compared to your equity. And the point of this is, if you do those three things, you are mathematically guaranteed to report not just profits, but record profits. ... At that kind of growth rate, with people concentrating on whatever the optimal area is for the fraud, you produce financial bubbles. In the case of the savings and loan crisis, we re-regulated the industry in the face of opposition from the Reagan administration, the House of Representatives and the Senate. And we looked for the Achilles heel for this kind of scheme, which is growth. And so, we restricted growth. And this kind of fraud also creates a distinctive pattern of operations, and we used that to triage and to go after these institutions while they were still reporting they were the most profitable savings and loans in America. People thought we were crazy, contemporaneously, who were conservative economists. But it turned out we were right about every single one of these institutions. What does it mean for today? The same thing. We have another epidemic of accounting fraud. In this case, it's not commercial real estate, which it was in the savings and loan crisis. It started out with, in the United States' context, with home mortgages."

Further on in the same interview Black says: "To add to your point about appraisers, the only reason you inflate an appraisal is for fraud. There's no other purpose in the world. And we have survey information that's quite good on appraisers. In 2003, 70 percent reported that they had been the subject of an attempt to intimidate them to inflate appraisal values in that year alone. When we did the same survey in 2007, that percentage was up to 90 percent. So we have horrific, endemic fraud, and it's coming out of the lenders, not the poor people who can't pay the mortgages. And that is what brought this crisis."

Derivatives and Gambling. Though not as simple as an ideal Ponzi-pyramid, the housing bubble shares its essential mathematical structure of constant recruiting of players with the infusion of their money. Things broke down

[11]The quote is taken from an interview on www.democracynow.org, Oct. 15, 2009

when the price of housing started to decline and infusions of fresh cash slowed/stopped.

There is an additional elementary mathematical aspect to the crisis of 2008 and on: dishonest evaluation of risk. A number of innovative financial instruments were created such as credit default swaps, derivatives, etc., which in retrospect a lot of people did not understand. True innovation is usually welcome, but a "new" financial instrument that was not an actual investment in, say X, but rather a bet as to whether the price of X would go up or down, is clearly a form of gambling. The gamblers must have known what they were doing, however, since they lobbied the passage of the Commodity Futures Modernization Act (of 2000) which exempted derivatives from being regulated as gambling—by state or federal agencies.[12] The involvement of insurance giant AIG in the "financial collapse of 2008" almost surely was not based on rigorous actuarial mathematics and data. The credit rating agencies of up-til-then good repute, such as Standard & Poor's, Moody's and Fitch,[13] that gave AAA-ratings to gambling infused securities (the owners of which paid the rating agencies for doing the rating!) completed the near perfect storm (making profits along the way).[14] Perfection was realized when "educated guesses" were substituted for required data in decent risk models.

So predatory lenders collect commissions on mortgages of questionable value, which are then sliced, diced and bundled and, along with their risk, passed up the financial chain[15]—creating paper profits and real executive bonuses. This behavior dissociates risks and rewards and is destined to fail, and it did!

The response to collapse continued to show willful mathematical ignorance. No less an authority than economics *Nobel laureate*, Joseph Stiglitz points out that Lehman Brothers, the first big casualty, should not have been let go *in toto*. The U.S. treasury secretary should have distinguished between the solid

[12]Ms. Brooksley Born was the head of the Commodity Futures Trading Commission (CFTC) in the late 1990s. When the rather secretive business in over the counter derivatives came to her attention, she studied them and came to the conclusion that they needed to be regulated. The then Federal Reserve Chairman, Alan Greenspan, Treasury Secretary Robert Rubin, Assistant Treasury Secretary Larry Summers, and colleagues stopped her efforts and ultimately prevailed on Congress to limit regulation of derivatives, cf., www.pbs.org/frontline/warning.

[13]James Surowiecki, "Ratings Downgrade," *The New Yorker*, Sept. 28, 2009, p.25

[14]Maria Hinojosa, David Brancaccio, *Credit and Credibility*, NOW on PBS, www.pbs.org, week of 12.26.08

[15]Interestingly some folks have managed to save their homes from foreclosure by demanding that someone document ownership of the mortgage in question. When such documentation can not be produced, people have sometimes kept their homes, cf., Judge Robert D. Drain, federal bankruptcy court in the Southern District of New York; Judge Arthur Schack of the New York Supreme Court; Judge Randolph Haines, based in Phoenix, a U.S. bankruptcy judge for the District of Arizona. It came to light in Sept. 2010, see "Bank's Flawed Paperwork Throws Some Foreclosures Into Chaos," by Gretchen Morgenson, October 4, 2010, The New York Times, p. A1, that the foreclosure processes of the banks were (are?) not only questionable, but in some cases illegal.

systemically important parts of Lehman and its gambling.[16] The resulting widespread fear following the collapse of Lehman, possibly useful to some, was historically predictable. The bailout came.

The number most often associated with the bailout is $700 billion plus of TARP, Toxic Asset Relief Program, money. Besides the fact that this could just as easily be called the Toxic Liability Relief Program, there are many **trillions** of dollars left out of this picture. Journalist and former managing director at Goldman Sachs, Nomi Prins, has shown that the bailout is at least $17.5 trillion! (Note: The Special Inspector General set up to oversee the bailout estimates that government agencies, including the Federal Reserve, will ultimately put out more than $23 trillion in various programs and supports related to the financial crisis. In adjusted dollars, this total is almost three times what was spent on World War II.[17]) And it is not clear where a lot of the bailout money is at the moment. If this were widely appreciated, the political consequences would be interesting. Details can be found in [555, 556]. This is the largest transfer of wealth from the many to the few in world history (so far). By the way, and as an example, over the past 80 years how many times has the United States government engineered rescues of the institution now known as Citigroup?[18]

There is likely something powerful going on here with a basis in anthropology and biology, cf., Dunbar Number, Chapter 9. The lessons of the first Great Depression are being ignored. While unemployment remains high, foreclosures mount, as do enormous profits in the financial industry. A political-economic system that does not adequately, honestly regulate (give feedback to) risk and fraud is mathematically questionable. The system must make clear legal distinctions among such activities as fraud, gambling, actuarial-based insurance, and simple savings accounts. Gamblers (which can be mathematically defined) cannot expect and must not be allowed protections from society even remotely similar to those afforded a person with a simple interest-bearing savings account. Bailouts for gamblers will eventually bankrupt the entire system. We must return to the days when it was illegal to gamble with other people's money; otherwise future collapse, to quote John Kenneth Galbraith, is inevitable.

---

[16] Joseph Stiglitz, *interview*, "Nobel Prize-Winning Economist Joseph Stiglitz: Obama Has Confused Saving the Banks with Saving the Bankers," audio archives at http://www.democracynow.org, February 25, 2009.

[17] Robert Weissman, "The More Things Change, the More They Stay the Same: One Year Into the Meltdown," *The Progressive Populist*, Vol. 15, No. 18, October 15, 2009.

[18] The answer is 4 as of Nov. 1, 2009, cf., Andrew Martin, Gretchen Morgenson, "Can Citigroup Carry Its Own Weight?," *The New York Times*, Sunday Business, November 1, 2009, pp. 1 and 6.

Exercise 2.8 What is Your Family's Share of the Bailout?
If instead of having to pay the bailout, assuming the $17.5 trillion were equally divided among roughly 300 million Americans, what would be your family's share? What would be that share for $23 trillion?

2.4 One Reason Why Usury Should Again Be Illegal

One more mathematically necessary change to heal our economy is this: bring back the laws against usury. In 1950 29.3% of U.S. Gross Domestic Product was due to manufacturing and 10.9% was from financial services. From 1950 to 2005 manufacturing's share gradually dropped to 12.0% while the financial services share rose to 20.4%, [532]. During this same period effective caps on allowable interest rates were removed, allowing financial services to produce—through the miracle of compound interest—much larger percentage profits from other people's debts than say, profits from the manufacture of cars, or just about anything else [219]. Money seeking the highest return on investment thus will tend to expand the financial sector over manufacturing. Capping interest rates is perhaps not sufficient to make most manufacturing competitive with finance again, but it is certainly necessary. (Note that many of the states in the U.S. have laws that would control interest rates, but the financial industry sought and obtained federal legislation that preempts these laws, i.e., the states can no longer regulate the interest rate charged by any institution that has any kind of federal banking connection, which means most of them.)

While not usury, a great deal of innovation went into creating financial instruments, e.g., credit default swaps, derivatives, and others, which were complex enough in details to evade much understanding and regulation, but simple enough in that they involved very high returns while doing little in the way of promoting productive enterprise in the "real economy," i.e., on "main street." As an example I cite the March 21, 2010 article in *The New York Times Magazine* by Roger Lowenstein, to wit:

> *For much of Wall Street, capital-raising is now a side show. At Goldman, trading and investing for the firm's account produced 76 percent of revenue last year. Investment banking, which raises capital for productive enterprise, accounted for a mere 11 percent. Other than that, it could have been a hedge fund.*

Amazingly to me, but not to "Wall Street insiders," the following is an example of what can happen and has happened. While a financial institution is brokering a deal with a client, at the same time the same institution *bets against the deal* using, say, derivatives. The short-term profits made possible by the bet can exceed profits from promoting deals productive from the standpoint of creating jobs and real "stuff." Thus, for example, money that could have been invested by the financial institution in high speed rail or renewable energy, creating "green" jobs and product, is more profitably employed in making bets. Finally, I was shocked to learn for the first time from a Michael Moore documentary movie that many corporations "hedge bets" on the lives of their employees via *dead peasant* life insurance policies, cf., page 215.

Exercise 2.9 Financial Fallout from Usury and Related High Return Activities
(i) Suppose you have a billion dollars to invest. The automobile company in Detroit will provide 6% return annually if you invest with them. There are limits on how profitable a manufacturing business can be, since they have to deal with many constraints in the real world, not the least of which is the ability of customers to not buy your product. How much money will you have at the end of one year if you invest in the auto company?

(ii) People often find themselves in a position where they must borrow money. In the United States this could happen when a medical emergency occurs and large medical bills must be paid. Also, few people have enough money to pay cash for a house, or a car for that matter. When there are no caps on the amount of interest that can be charged on a loan, it is well possible that you will have to pay 12% annual interest. (In fact, interest rates considerably higher, e.g., 29.99% on credit cards and over 400% on pay day loans, are common as I write.) Suppose the billionaire in part (i) decides to invest his one billion dollars with the financial sector and is promised 12% return annually. How much money does the billionaire make?

(iii) Compare your answers to (i) and (ii) after two consecutive years, then three, then four, finally five consecutive years? Are the answers for part (ii) twice or more than twice the corresponding answers for part (i)?

(iv) Follow up on Lowenstein's observation in his quote immediately before this exercise, cf., for example, [415]. Simply, might it be said that financial institutions have found a way to use their access to money to make more money more rapidly (for themselves) using "innovations" that have had little positive effect on the real economy and the production of jobs? In fact, more money can be made by financial institutions using these "innovations" than can be made performing the traditional functions of raising capital for productive enterprise. Is this prototypical WACU behavior? (WACU refers to the "We Are the Center of the Universe" pattern of behavior, cf., see page 85.) What are the implications for society at large of this behavior? For you in particular?

Exercise 2.10 Everything is Connected
(i) Consider programs such as the WPA, the Work Projects Administration, and the CCC, Civilian Conservation Corps, during the 1930s and more recently the Peace Corps, AmeriCorps. Can you think of a way of "bailing out America" that would save the financial industry, and lower unemployment, i.e., save working people, at the same time? Consider [194], are there many jobs that need to be done to rebuild America's infrastructure? Could we rebuild an electric rail system throughout the U.S.? Could we build an energy infrastructure based on sustainable/renewable energy? Would this create jobs? How much would this cost, compared to, say just a few trillion dollars, cf., $23 trillion bailout, page 41? What would be the social and economic "return" on such investments?

(ii) At the start of the 21st century China is serving as "the banker" of the U.S. in that it holds over $800 billion in U.S. debt (as of 2009, and it does this by buying U.S. treasury bills—thus financing the U. S. government). This gives the Chinese considerable political leverage in the world economy, and the U.S. economy in particular. If global warming plays out as expected, over two billion people who are dependent on the glaciers of the Himalayas and the Tibetan plateau for dependable agricultural water will find themselves on the world market buying grains they would have produced themselves had those glaciers not melted. (In 2009 China was the world's leading producer of wheat, India was number two, the U.S. number three.) Thus the Chinese will likely be buying American grain, and the world price for grain will rise. Does this scenario connect Chapter 1 with the this chapter? See [60].

(iii) The following is a quote from George Lakoff: "Right now we have an economic disaster and an ecological disaster upon us, and those two disasters have the same cause:

namely, short-term greed combined with a lack of an understanding of systemic risk, of how systems work both in ecology and in the economy. You put those together, and you have the same cause for twin disasters." (www.democracynow.org, Nov. 18, 2009) Discuss.

(iv) From a special report by Greg Palast covering the World Trade Organization (WTO) meeting in Geneva, November 30, 2009, the 10th anniversary of the first global protest against the WTO in Seattle, 1999, we read: "But we got our hands on a document you certainly won't find on the WTO website; something very confidential: a secret demand of the European Union and the USA, leaning on emerging nations to open their borders to trade in financial derivatives and exotic, even toxic, financial products."

"These unregulated products are precisely at the seismic center of the financial meltdown of 2008 and beyond! It seems to me that it would be wise for the USA and Europe to have never gone down this road, now they are trying to force developing countries into this mess." Further on in Palast's report we read about what happens if a country decides that they do not want to get involved with derivatives, toxic products etc., viz., they end up paying hefty "fines."

"The President of Ecuador told me he'd like to get out of the globalization jungle. But if Ecuador dares to bar US banks, the WTO will let the USA stick a tariff onto every banana imported from Ecuador. If the US tallies those bananas, Ecuador's economy will go splat!" (See www.gregpalast.com.)

Who do you think wrote those rules for the WTO? Why? (The Palast report is also available at www.democracynow.org, Nov. 30, 2009.) You might find two books of interest in regard to the 1999 WTO protest: *The Battle of the Story of the Battle of Seattle*, by Rebecca Solnit and David Solnit; and the book by Norm Stamper, the Seattle Police Chief during the protests, *Breaking Rank: A Top Cop's Exposé of the Dark Side of American Policing.*

(v) Research the role of the *Federal Reserve* in the U.S. economy in general and in the bailouts of 2008–9, cf., [258, 555]. Does this institution have immense economic power with minimal accountability? What are its priorities? Did it execute its duties responsibly leading up to the financial crash of 2008?[19] If Congress creates a financial consumer protection agency, why should it be independent as such an agency is in Canada, cf., page 211, and not housed in the the Federal Reserve as proposed in the financial reforms of 2010?

(vi) What are the G8 and the G20? Who created the G20, why, and when?[20] Are meetings of the G8/G20 reminiscent of the 1999 meeting of the WTO in Seattle (see part (iv) of this Exercise above)?

[19]For example, did the head of the New York branch of the Federal Reserve know that Lehman Brothers was "cooking its books" well before 2008?

[20]Hint: see the inteview with Naomi Klein, www.democracynow.org, June 28, 2010.

Chapter 3

What Is Mathematics? More Basics

Numbers are a part of mathematics; but unfortunately, when asked, too many people think that numbers are all there is to mathematics. When pressed, some say that geometry and algebra are part of mathematics; some even say that mathematics is a language—the language of science. I claim that mathematics includes all of these things but much more. Mathematics can arise in various forms in any human endeavor, some of which have nothing to do with numbers! I will partially demonstrate this claim in this book.

3.1 The Definition of Mathematics Used in This Book

Definitions. In formal mathematics one of the most important concepts is that of *definition*. Definitions, i.e., the meanings of words/symbols, stated as precisely as is humanly possible, are indispensable stones in the foundation upon which mathematics is built. Even though this book is not completely formal, I owe you the best definition of mathematics that I know. The following is the beginning not the end of a discussion; here is the definition of *mathematics* that I use in this book.

Definition: *Mathematics* is the search for and study of patterns.

For example, patterns in counting give rise to numbers, arithmetic and eventually number theory and algebra. Patterns in space give rise to geometry. Patterns in human thought give rise to mathematical logic. Patterns in motion, in part, give rise to calculus, cf., [136]. Patterns of patterns can be very interesting also. For me there are no limits, any pattern is worth a look—especially if it captures your curiosity.

Exercise 3.1 What is Mathematics?

(i) Consider the following quote: *"Thus (pure) mathematics may be defined as the subject in which we never know what we are talking about, nor whether what we are saying is true."*(Bertrand Russell) What do you think this means? Comment.

(ii) Again, consider: *"As far as the laws of mathematics refer to reality, they are not certain; and as far as they are certain, they do not refer to reality."* (Albert Einstein, 1921). Is there any relationship of this quote of Einstein to the previous one due to Russell?

(iii) An amazing thing is that mathematics is so useful. I attribute this miracle to the fact that ultimately mathematics rests on patterns observed in Nature. A poetic statement of this miracle is: *"The mathematician may be compared to a designer of garments, who is utterly oblivious of the creatures whom his garments may fit. To be sure, his art originated in the necessity for clothing such creatures, but this was long ago; to this day a shape will occasionally appear which will fit into the garment as if the garment has been made for it. Then there is no end of surprise and delight."* See [126]. Comment.

(iv) A pattern is an example of a "simplification," an extraction of information common to a number of different situations, events, or processes. One hallmark of genius is the ability to sift through available data and extract what is "truly important," to simplify appropriately. Compare the following quote from [358], *"A good simplification should minimize the loss of information relevant to the problem of concern."* Does this indicate that starting with a problem can lead to patterns?

(v) Do any (or all) of the above quotes have anything to do with the search for and study of patterns?

When using pure mathematics or studying applications, always be as clear as possible in your own mind about the definitions of everything!

Axioms and Assumptions: Hidden or Not. If definitions are one type of foundation stone for mathematics, *axioms*, sometimes called *assumptions* are another. Axioms are statements that are assumed to be true; they are often statements of patterns. In pure mathematics the axioms are usually clearly stated, cf., II. When applying mathematics to a "real life" problem, there are often *hidden assumptions.* If hidden assumptions are not rooted out and made explicit, severe and unnecessary difficulties may interfere with the process of finding solutions. One of my favorite examples of hidden assumptions giving rise to difficulties is the following *actual* log of a radio transmissions between between a U.S. naval ship and the Canadian authorities off the coast of Newfoundland in October, 1995 (as released by the Chief of Naval Operations 10, 1995 and quoted in *The New York Times*, Sunday, July 5, 1998):

Canadians: Please divert your course 15 degrees to the south to avoid a collision.

Americans: Recommend you divert your course 15 degrees to the north to avoid a collision.

Canadians: Negative. You will have to divert your course 15 degrees to the south to avoid a collision.

Americans: This is the Captain of a U.S. Navy ship. I say again, divert YOUR course.

Canadians: No. I say again, you divert YOUR course.

Americans: THIS IS THE AIRCRAFT CARRIER U.S.S. LINCOLN, THE SECOND-LARGEST SHIP IN THE UNITED STATES' ATLANTIC FLEET. WE ARE ACCOMPANIED BY THREE DESTROYERS, THREE CRUISERS AND NUMEROUS SUPPORT VESSELS. I DEMAND THAT YOU CHANGE YOUR COURSE 15 DEGREES NORTH. I SAY AGAIN, THAT'S ONE FIVE DEGREES NORTH, OR COUNTER-MEASURES WILL BE UNDERTAKEN TO ENSURE THE SAFETY OF THIS SHIP.

Canadians: This is a lighthouse. Your call.

Another quite common hidden assumption that can shut down an urgently needed discussion/debate is the assumption that Aristotelian (yes-no, true-false, black-white) logic applies to a situation when actually fuzzy (or measured) logic which allows any numerical truth value between 0 (false) and 1 (true) is far more applicable, cf., Sections 1.4 and 8.2.

Exercise 3.2 All or Nothing Arguments are Usually Needless

(i) Person A insists that Mr. X, our local congressional representative, is an environmentalist because he has introduced many pro-environmental bills in the legislature, and because he loves the outdoors. Person B vehemently disagrees, saying no one who drives a gas guzzling vehicle as Mr. X does could be an environmentalist.

Use fuzzy/measured logic to more deeply analyze the above discussion and hopefully make it more productive.

(ii) Discuss: "You are either with me or against me." (Said by Roman Tribune Messala, representing the Roman Emperor, to Judah Ben-Hur, Charlton Heston, in the 1959 movie, *Ben-Hur*. Also uttered by some contemporary politicians.)

(iii) Suppose on a 0 to 1 scale of "objectivity values," if a statement is a "pure fact" or is "purely objective" it is assigned a 1. If it is "pure opinion" or "purely subjective" it is assigned a 0. Can you find a reasonably complicated sentence that is pure fact, i.e., purely objective? Who agrees/disagrees with you?

(iv) Suppose in a conversation/debate you are being critical of American (or fill in the blank with some other country) foreign and environmental policies. The other party turns hostile and says: "Can't you say anything good about America?" Using careful definitions and fuzzy logic how might you productively respond?

3.2 The Logic of Nature and the Logic of Civilization

A case can be made, and we do so in part in this book, that humans face some daunting problems: global warming with accompanying disruptions to climate and agriculture, collapse of most of the earth's wild fish populations, vanishing native forests, fresh water shortages, presence of persistent pollutants nearly everywhere, species extinctions, human populations beyond the carrying capacity of ecosystems that support them, and so on.

There are various possible futures that are mathematically possible. Many are grim, and were I writing a polemic I would now recount in detail the most ferocious possibilities to encourage action on the reader's part. However, with limited space and time I will instead focus on solutions. Human extinction, being one of the near term possible futures, I will make the assumption that most humans would like to avoid such.

Technology Will Rescue Us? There is a view with many adherents, some with best-selling books, that most if not all of our problems (will) have technological solutions. But consider that chlorofluorocarbons, CFCs, DDT, leaded gasoline were all widely heralded and appreciated technological solutions to certain pressing problems, cf., Section 4.5. Each of these solutions created problems, which in some cases were far greater than the initial problem that was "solved." If you would like to read an entire book of similar examples, see [682]. Thus we are dealing with a *pattern*! I enshrine this pattern as an axiom, which can also be considered a corollary[1] of the Connection Axiom:

[1]A corollary of a mathematical statement, like a theorem which has been proven or an axiom that has been assumed, is a proposition which logically follows from said statement, often with little proof required.

The Principle of Unintended Consequences: *In a complex system, it is not possible to do just one thing.*

Now from many perspectives it would not be appropriate to ban technology; besides, many of us find technology a lot of fun. In fact, there are examples of technology that have had immense positive consequences, like the polio vaccine. Thus an approach to this situation employed in many societies is the *Precautionary Principle*, which I invite you to investigate. This principle states, briefly, that the burden of proof of safety for the (commercial or widespread as opposed to experimental) introduction of a technology falls on those who would introduce said technology. Since every solution to a problem will have additional consequences, some unimaginable at the time, it is wise to take the time to investigate what those unintended consequences might be. A major difficulty is that very often fortunes are to be made or the benefits are so "obvious" with a new technology, that there is a rush to commercialization without adequate caution. Now, reliance on the precautionary principle may not avoid all technological disasters; no system is perfect; most species (probably ours included) will eventually go extinct. But using the Precautionary Principle greatly increases the chances that we will avoid technological disaster and might even postpone human extinction.

A Completely Different Approach—The Logic of Civilization Must be Compatible with the Logic of Nature. While pondering technological solutions, I suggest that there are problems whose very nature require a completely different approach. For example, suppose you are driving a bus down a long road greatly enjoying the scenery, the sense of motion, the wind blowing in the open windows. However, there is a brick wall at the end of this road that for some reason you cannot see, or refuse to see, even though a few (but only a few) of the many folks on board are shouting: "Slow down, turn off this road, we are going to hit that brick wall." Any improvements to your bus's efficiency, innovations that allow you to power the bus with wind and sun, and so on, do nothing to alter the fact that your trip will not end well—unless you stop and re-evaluate why it is that you are traveling down that particular road, in that particular direction.

I propose that we take the time to examine the fundamental *logic* of our civilization. What are our collective hidden assumptions? Do we need to change the way we think? Will this change in our thinking lead to changes in our behavior that Nature requires for our long term survival? I contend that from the point of view of humans, *Nature has its own logic.* There are laws of Nature which humans may or may not perceive—but nevertheless operate. The law of gravity is an example of a law of Nature that can be perceived at many different levels. The Spanish architect Gaudi built a multi-story apartment complex with a narrow, railingless, external staircase which Gaudi said brought the residents in touch with their own mortality as they climbed. This is a visceral level at which we all understand gravity. Newton created a mathematical level of the understanding of gravity with his inverse square law, and there is yet a more detailed level of understanding with relativistic refinements due to Einstein. There are laws of physics, chemistry, biology, economics, sociology, "human nature," mathematics and so on, which I assert are human efforts to understand the *logic of Nature.* I will discuss the role of the laws of thermodynamics in economics in VII, for example. Clearly an entire library of books could be devoted to this topic; however, I will restrict attention here to just one axiom that I see as essential to the long-term survival of humans. If adopted as a fundamental logical principle this axiom would completely change the way most humans now behave on earth. The consequences of this rather innocent axiom are truly revolutionary.

The Bio-Copernican Axiom: *Humans are not the "center" of the biosphere,*[2] *i.e., earth's system of life. Humans are a part of Nature, not apart from Nature. Humans are subject to the laws of Nature.*

Notions equivalent to this axiom probably were (are?) part of the fabric of some indigenous peoples' cultures before contact and dissolution by "modern man." Before briefly discussing this axiom and some of its implications, I would like to revisit the history of the original Copernican Axiom, a history that provides contemporary insights that go deeper than a simple analogy.

Aristarchus of Samos (310 B.C.– ca. 230 B.C.) was a Greek astronomer and mathematician and the first person of which I am aware to present an explicit heliocentric, i.e., sun-centered, model of our solar system. His model was not adopted; instead, the geocentric, i.e., earth-centered, model of Aristotle and Ptolemy was accepted as the obviously correct model for millennia. Nicolaus Copernicus (1473–1543) redeveloped the theory of the heliocentric solar system and wrote a book in Latin on the subject, *On the Revolution of Heavenly Spheres*, published upon his death, since he apparently did not want to deal with the inevitable backlash from the powerful of his day who held steadfast to the geocentric model. His book provides many mathematical tables for doing astronomical calculations. One of the justifications for Copernicus's work—as he mentioned in a letter to Pope Paul III—was that the ecclesiastical calendar could be improved in accuracy and ease of calculation. For example, using Copernicus's sun-centered model, as opposed to the earth-centered model of Ptolemy, it was easier to calculate when Easter would occur. However, such mathematical arguments did not influence deeply held convictions about the centrality of the Earth in the universe, as the yet later experience of Galileo demonstrates.

Galileo (1564–1642) challenged the still dominant assumption that the earth is at the center of the universe; and it is well known that a long series of popes took a dim view of Galileo's theories. The Inquisition condemned Galileo in 1633 because his teachings were believed (by those in power) to clash with the *Bible.* The thinking of the time is reflected in the following two quotes from experts of the day. First, Scipio Chiaramonti, Professor of Philosophy and

[2] I define the biosphere as does V.I. Vernadsky in [693], a reprint of his 1926 edition.

Mathematics at the University of Pisa, 1633, [77, p. 6]: *Animals, which move, have limbs and muscles; the earth has no limbs and muscles, hence it does not move.*

Earlier, Martin Luther, leader of the Protestant Reformation, said (about 1543): *People give ear to an upstart astrologer [Copernicus] who strove to show that the earth revolves, not the heavens or the firmament, the sun and the moon. Whoever wishes to appear clever must devise some new system, which of all systems is of course the very best. This fool wishes to reverse the entire science of astronomy.*

Galileo was, in fact, only recently pardoned in 1992 by the Pope for his heresies 359 years before.[3]

What is not well understood is the fact that from the point of view of pure mathematics any point or celestial body can be taken to be the "center of the universe" or "the center of the solar system." What distinguishes the mathematical choice of taking the sun to be the center of the solar system is *simplicity.* Taking the earth as the center of the solar system leads to the math of Ptolemy. While the orbits of the planets are simple ellipses ("slightly flattened circles") in Copernicus's model, the planets in Ptolemy's model follow paths called (curtate and/or prolate) epitrochoids and/or hypotrochoids.[4] In order to reconcile the motion of the planets as viewed by Ptolemy with the motion of rocks and apples on earth, little "angels" were introduced at appropriate times to act on the planets.

It certainly was easier for Newton to discover his inverse square law for gravitation while studying the data accompanying the heliocentric model, as opposed to trying to extract that pattern from the trochoids of Ptolemy. In fact, it is not clear that the latter would have been possible then.

Exercise 3.3 The Logic of Civilization

(i) Given the geocentric and heliocentric models of the solar system, speaking as a pure mathematician, is one true and the other false? Is one far simpler and more useful than the other? Why do you think we use the heliocentric model now, whereas we did not for millennia?

(ii) In switching from the geocentric model to the heliocentric model were there political winners and losers? Did the social system evolve (with some lag-time) along with the science?

(iii) Switching between the heliocentric and geocentric models of the solar system is a simple process of "changing co-ordinates" for today's mathematical physicist. Asserting that humans are not the central/dominant/most important species of life on earth is not so easily viewed as a "change of co-ordinates." The Bio-Copernican Axiom involves biology, ecology—among other subjects. Is this Axiom true or false (or a fuzzy truth value between 0 and 1)? Is this Axiom a useful myth, like the Yam God of the Warlpiri, cf. page 11? Is this Axiom a destructive myth? Is this Axiom a mixture of morality and science?

(iv) What behavioral changes would you anticipate if we switched from an anthropocentric, i.e., human-centered world view, to an all-life-centered world view? What might

[3]Somewhat quicker, according to *The New York Times*, November 2, 2001, on Halloween, October 31, 2001, the state of Massachusetts officially exonerated five women who were tried and hanged as unrepentant witches on Gallows Hill in Salem, Massachusetts over three hundred years before in 1692.

[4]My mathematical analysis of these beautiful objects was deemed sufficiently complicated to be a high school science fair project!

prompt some subcultures to make this change? Would there be political winners and losers? Does your answer to this last question depend on how the change takes place? If this Axiom were adopted, what simplifications might it lead to (in analogy with the simplifications yielded by the heliocentric model)?

(v) Are there any problems we currently face that would become more amenable if all humans adopted the Bio-Copernican Axiom? Pick at least one such problem and its would-be solution in the Bio-Copernican society.

(vi) What cultural manifestations are there which indicate that humans think of themselves as the center of the biosphere, e.g., the "highest ranked, most important species?" Children (should?) learn early on that society does not "revolve" around them, i.e., to not be (totally) self-centered. Is the Bio-Copernican Axiom just an application of this process at the species level?

(vii) If all humans adopted the Bio-Copernican Axiom, what do you think the total human population on earth would be? Would it be greater, less than or equal to what it is at the moment?

(viii) If all humans assumed the Bio-Copernican Axiom, what cultural feedback mechanisms would there be to regulate the total human population? Why?

(ix) Pick a popular religion. Can that religion adapt and become compatible with the Bio-Copernican Axiom?

(x) Discuss the logical implications of *monoculture* in agriculture vs. crop diversity. Which is more likely sustainable for long periods of time?

(xi) In what countries is the Precautionary Principle taken seriously? What have been its effects/consequences?

(xii) Find a country that does not take the Precautionary Principle seriously. What have been the effects/consequences?

(xiii) Geoengineering has been proposed to solve the global warming crisis. What are some examples of geoengineering? Are there (might there be) any unintended consequences?

(xiv) How many humans share the assumption that civilization must continue to burn fossil fuels in order to function? Are there widely held assumptions that are incompatible with the long-term survival of human civilization?

(xv) What do you think the implications of the Bio-Copernican Axiom are for the academic discipline of economics, cf., VII? Is material economic growth without bounds consistent with the Bio-Copernican Axiom? If humans accepted the Bio-Copernican Axiom would we "ask" before "taking," i.e, would we at least consider the effects of human actions on other life forms before acting?

(xvi) For a project, come up with a mathematical definition of a complex system and some axioms for a complex system, then try to prove, or at least give a convincing argument, that the Principle of Unintended Consequences is actually a theorem. (A theorem is a statement with a proof that relies on axioms, definitions, traditional logic and other theorems. "Little theorems" that help prove bigger theorems are called *lemmas*, in case you run into that term.)

3.3 Box-Flow Models

One-Box Models in Steady State. One of the simplest and most versatile tools for making models of certain parts of Nature is the *box-flow* model. Boxes, sometimes called *compartments*, are imaginary containers that contain some stuff, usually some type of matter, that is flowing in and flowing out of the container. There are many types of boxes. A few examples are the earth's atmosphere, the airshed over an ecosystem, the stratosphere, a university (with students flowing in and out), a single person, a lake, an anteater, a tree, a house.

Bathtubs. A simple box-flow model is a bathtub with one faucet, through which water flows in, and one drain, through which water flows out. If the rate at which water flows in is equal to the rate at which water flows out of the tub, I will say that our box-flow model is in *steady state.* The amount of stuff in the box, in this case the water in the tub, is called the *stock* of matter in the box.

Steady-State Residence Times. Thus far we have a box with a stock of matter in it and a single rate of flow of matter in/out of the box. There is a third concept associated with this model: *residence time* of the matter in the box. In the case of our bathtub with water flowing in and out at the same rate, I would like to have some idea of how long some bit of water remains in the tub before flowing out. Some bits of water might flow out very soon after having entered the tub, while others might hang around for a very long time before getting close to and then sucked down the drain. Now keeping track of all the individual bits of water is an exercise that is far more challenging than I want to attempt.[5] There is something much easier, but still informative, that we can do. Let's divide the number which represents the amount of water in the tub by the number that represents the rate of flow of the water, either in or out. The result will be a number which represents an amount of time. We can think of this as the average amount of time one might expect a bit of water to remain in the tub before going down the drain.

For a concrete example, suppose our tub has 50 *liters*[6] of water in it and water is flowing into, and out of, the tub at the rate of 5 *liters per minute*, i.e., $\frac{5\ liters}{minute}$. If I divide 50 *liters* by $\frac{5\ liters}{minute}$, viz., $\frac{50\ liters}{\frac{5\ liters}{minute}}$, I get 10 *minutes* for the answer. Note you can perform the last operation mechanically as follows: $\frac{50\ liters}{\frac{5\ liters}{minute}} = \frac{50\ liters}{1}\frac{minute}{5\ liters} = 10\ minutes$, where the *liters* cancel out and 50 is divided by 5.

I can summarize our discussion of this one-box model quite succinctly, if you allow me the use of a few symbols. Suppose I have one box, let M represent the amount of stuff or matter, i.e., the stock, that is in the box. This will be a number with some *units* attached, such as *liters* above. Let F_{in} be the rate of flow of matter into the box; let F_{out} be the rate of flow of matter out of the box. Using these symbols I can make two definitions:

Definition: A single box into which matter flows at rate F_{in} and out of which matter flows at rate F_{out} is in *steady state* if $F_{in} = F_{out}$.

---

[5]To do this we would have to know a lot about the tub. For example, What is the shape of the tub? Where are the drain and faucet located physically? Is anything stirring the water in the tub? How is the stirring being performed?

[6]A liter is a unit of volume equal to 1000 cubic centimeters, cf., Exercise 3.11 (ii). A liter is also equal to 1.057 U.S. quarts

Definition: If a box is in steady state with a stock M of matter in it and flows $F = F_{in} = F_{out}$, then the *residence time*, T, of matter in the box is $T = \frac{M}{F}$.
Thus for a box in steady state: $M = FT$, where the juxtaposition FT means F multiplied times T.

Boxes are Everywhere. The discussion of one-box models so far might be either so easy it is boring you to tears, or it might be a bit too brisk—or something in between.

If you find what we have done so far to be easy, I recommend you look at [283], [73], and the rest of this book as well. If the box is your house and the flowing matter is radon gas, you are looking at a problem of importance to your health. If the box is the airshed over the northeastern United States and the flowing matter is a certain type of pollution from coal-fired power plants, you are investigating acid rain. These are some of the box models in [283], there are others. Once you become aware of this tool you can find dozens of other box models. We will look at some later.

If you find what we have done so far is a bit overwhelming, using letters to stand for numbers, using units like *liters*, or if you are a bit rusty when it comes to multiplying and dividing, all is not yet lost. In the next section I will begin introducing units and such; and in Part II I will systematically study all these things and more; hopefully this will work for you. If not, don't hesitate to study with friends. When friends spend some of their time studying in small groups, everyone benefits. If you are still lost, there should be a hungry, mathematically talented person nearby who will help you for a fee.

Exercise 3.4 Some One-Box Models: Steady State and Not

(i) Suppose you have a hot tub which holds 1000 liters of water. Suppose water is flowing in, and out, of the tub at the rate of 25 liters per minute. What is the residence time of water in the tub?

(ii) Suppose you are in the situation described in part (i) except that the water in the tub is contaminated with a nasty bacteria floating around in the tub, none sticking to the tub. Suppose that the answer to part (i) is T. If you wait an amount of time T, will the tub be free of bacteria?

(iii) Suppose your hot tub has a capacity of 2000 *liters*, but at noon the tub is half full. Suppose also that water is flowing into the tub at $\frac{25\ liters}{minute}$ and that water is flowing out of the tub at $\frac{3\ liters}{minute}$. When will the tub be full to capacity?

(iv) Suppose a one-box model is in steady state with a given flow rate. If you increase the rate of flow while keeping the stock fixed, is there anything you can say about the residence time?

(v) If a national park biologist estimates that a typical bear in a certain park lives 25 years and that no bears leave the park permanently except by death of which there are 6 deaths per year, how many bears live in the park? Assume that the bear population is in steady state. This might not be true, but this assumption makes the problem easy.

(vi) Suppose your box is the earth's human population. The flow in is the birth rate, the flow out is the death rate. (Note: The number of people leaving the earth at this time by any other means than death is negligible. The number of people arriving on earth by any other means than birth on earth—well if you find such a person let me know immediately.) At the time you read this, is this system in steady state or not? See V.

Multi-Box Models in Steady State. To start with, suppose our university is one box with 25,555 students. If we assume a one-box, steady-state model, i.e., the stock is 25,555 students all of the same type (they all graduate), then I can calculate the residence time of the students (the typical length of time from entering to graduation) if I know the graduation rate. If we assume the graduation rate is 5,555 students per year, then you should be able to check that the residence time for this model is approximately 4.600360036 years, which I round off to 4.6 years.

Now a one-box model of a university, with a fixed population, students entering at a fixed rate and leaving at the same rate is not realistic. I could make this model a little bit more real by assuming that some first-year students flunk out at the end of their first year, but that everyone else graduates—and no one else flunks out at any other time. This leads us to the following:

Exercise 3.5 A Two-Box Model of a University

(i) Suppose our university has 25,555 students and that students leave only by graduating or by flunking out at the end of the first year—no sooner or later. Suppose that 555 students flunk out each year, and that the residence time for all students taken together is 4.6 years. What is the residence time of the students who graduate? Hint: Think of two boxes, a box that contains the students who flunk out and a box that contains the students who eventually graduate. A third box containing all of the students together is also of use.

(ii) In the model of part (i) there are 5,555 students entering each year. (Note: Actually the number is closer to 5555.434783, but fractional people do not exist.) In the first year then we have 5,555 students entering and 555 students leaving. What can you say about the number of students in years 2, 3, 4, and so on? How can you get a total of 25,555 students?

(iii) By using a tool called a spreadsheet, we can do far more complicated models of our university. For example, we could allow students to flunk out at the end of any year or enter at the beginning of any year (by transfer, say). By using a computer we could let each student be in his/her individual box! If you want to start learning how to do some of this immediately see V or [73].

Exercise 3.6 A Cycle with Two or More Boxes

(i) Imagine the following system: Two boxes, A and B, with matter, say water, flowing from box A to box B and vice versa. Suppose both boxes hold the same amount of water, M. If the system is in steady state in the long term, i.e., both boxes continue to hold the same amount of water, M, then what can you say about the flows between the boxes? Can you say anything about the residence times of water in box A relative to box B? This clearly is an example of a *cycle*.

(ii) Again imagine a two box system as in (i) except that Box A holds an amount of water M, and box B holds an amount of water 1000 M. If the system is in steady state, what can you say about the relative rates of flow from one box to another and the relative residence times of water in each box? Related to this exercise is Exercise 5.11.

(iii) Generalize this exercise by having a string of boxes in a circle, each box holding a different amount of water, with a circular flow of water that goes from one box to the next around the circle. If you assume the system is in steady state, i.e., the amount of water in each box does not change in time, can you say anything about the various rates of flow and the various residence times?

(iv) This example is a bit more complex. Suppose two pipes come out of box A, one to box B and another to box C. Then suppose that pipes come out of box B and box C both going to box D. Now assume that a pipe goes from box D to box A. If each box contains some fixed amount of water over time, is there anything you can say about the rates of flow and residence times? Can you make this problem more complicated?

3.4 Cycles and Scales in Nature and Mathematics

Nature's Cycles. Cycles are among the most basic patterns in Nature. Cycles inspired our ancestors to create a good deal of mathematics, some of the earliest mathematics. One of the first cycles of Nature that humans must have observed was that of night and day. Clearly associated with this cycle were the sun, moon, stars and planets. Depending on location, cycles of the seasons with associated weather cycles of rain/drought, heat/cold must have also become obvious if for no other reason than the availability of various foods often correlates with the aforementioned cycles—and humans must eat.[7]

If hunter-gatherers were aware of cycles, agriculturalists[8] were likely even more so. The knowledge of when to plant crops flowed from careful observations of the cycles of heavenly bodies and from concomitant calculations involving rather sophisticated symbols and mathematics. In some early civilizations a lot of social power came to those who were able to precisely deal with astronomical data. Imagine a priesthood that could predict eclipses. Such individuals might convince others that they were closer than most to the gods.[9]

Exercise 3.7 Ancient Mathematics

If you are interested in history, philosophy or cultural anthropology, the stories of how mathematics originated in various places around the world is fascinating. To name a few such cultures consider: the Aztecs and Mayans, in what is now Central America and parts of Mexico; the Incas, in what is now South America, Peru in particular; the Chinese; the Indians; the Greeks; the Sumerians, Babylonians to name only two peoples of the Middle East; the Egyptians, the people of the Great Zimbabwe, to name only two African peoples. Do not limit yourself to this list. For example, likely the oldest of all mathematics with more sophistication than simple counting was created by the Australian Aborigines. I discuss this in some depth in III. See [643], [485], [72], [17].

(i) Investigate the mathematics in two or more ancient cultures. Were there any common natural patterns that inspired the mathematics that you find?

(ii) Does mathematics depend on the culture in which it is developed?

(iii) Is there anything universal about mathematics that transcends the various symbolic systems invented in the various cultures?

(iv) Do you think that universal mathematical properties reflect aspects of the human mind, Nature, both, neither, something else?

(v) If you can find out much information like when to plant crops, when eclipses will occur, by careful observations of the heavens, why not get additional information about

[7] There once was a group of people, self-named "'Breathairians," in my hometown that claimed they lived on air and water only—until one of the group was seen eating junk food at 2 a.m. at a gas station.

[8] The invention of agriculture, about 12,000 B.C. to 8000 B.C., had both positive and negative effects. *"There is some evidence that human health generally declined with the onset of agriculture."* See [169, p. 237]. In [169, p. 236] it also says: *"However it originated, agriculture started a positive feedback system that put humanity on the road to sociopolitical complexity."*

[9] Can you think of any contemporary "priesthoods"?

your personal life from the same place? This is my (untested) hypothesis of how astrology got started. Does careful observation of Nature confirm or contradict the accuracy of astrological predictions, such as occur in the daily horoscopes found in many newspapers?

The Biosphere and Matter Cycles. There are many other natural cycles of importance, and I will discuss some of them in this book. The *hydrologic cycle*, i.e., *water cycle*, consists of the flow of water among three main boxes of water and their subboxes, viz., the atmosphere (gas) box containing water vapor, the ice box containing water in its solid state, and the liquid box consisting of subboxes such as oceans, lakes, groundwater aquifers.

Though he was not the first to use the term, V.I. Vernadsky in the 1926 edition of [693] defined and described the "life box" or *biosphere* as that term is used today. Briefly, the biosphere is the box that contains all of earth's life forms and the parts of the earth that interact with life forms. Here the term earth includes its atmosphere. Vernadsky's book, available in Russian, French and now English is of current as well as historical interest. Realizing that there are some exceptions[10] to the idea that the biosphere is totally self-contained, I present an axiom:

Axiom on Matter Cycles: *Matter in the biosphere tends to flow in cycles that stay within the biosphere.*

Water cycles within the biosphere. Ecologists study how matter such as carbon cycles in the biosphere, and it is naturally called the *carbon cycle*. I will study the concept of *recycling*, which involves matter in forms referred to as resources and garbage, cf., V, [581], [568], [249]. Mathematics also deals with more abstract cycles that occur in the economy, history, electronics, sociology, politics and more. For example, is there any mathematical structure in [605]?

Cycles and Time. Cycles lead us to different ways of measuring time. First in history was *astronomical time*, which was the standard before 1971. The time it takes for one rotation of the earth about its axis is *one day*, the time for the earth to make one trip around the sun is *one year*. It is interesting to note that the convention of dividing hours into 60 minutes and a minute into 60 seconds goes back to ancient times, and the choice of the number 60 was mathematically motivated. The Babylonians, for example, probably chose the number 60 since they did many astronomical calculations; and the number 60 admits many divisors, minimizing the appearance of fractions.

Exercise 3.8 Astronomical Time

(i) Can you list all of the whole numbers bigger than zero that divide 60 without leaving a remainder? For example, 2 is such a number, since 60 divided by 2 is 30 without remainder.

(ii) Can you think of any real-life situations where astronomical time, i.e., dividing a year into 365 days, a day into 24 hours and so on, is not adequate?

(iii) An anthropologist once told me that the Innuit people, the indigenous people of Alaska and northern Canada, do not have a word for time. In Ethiopia the native time is

[10]There are exceptions such as: astronauts leaving things on the moon; space probes which never return to earth; asteroids which come from outer space and collide with earth (some believe that life may have been initialized on earth in this way).

found by dividing the time from dawn to dusk into 12 equal parts, which works since they live near the equator. Can you find any other examples of cultures which treat or measure time differently from the way Americans do now?

Since 1971 there has been a new standard of time measurement, called *atomic time*.[11] In the early 1990s the standard became based on cycles or oscillations associated with the substance cesium. The smallest unit of cesium is called an atom of cesium; and the the fundamental unit of time measurement, *the international second*, was defined to be 9,192,631,770 periods of the radiation corresponding to the transition between the two hyperfine levels of the ground state of the cesium 133 atom. What this means in practice is that you can actually buy a cesium clock that counts the oscillations (periods) of the cesium-atom "pendulum," and it ticks off one second for each 9,192,631,770 oscillations.

In 1997 scientists at N.I.S.T. announced a new design for the atomic clock called the fountain clock, which is about 10 to 100 times more accurate than the cesium clock. In the journal *Science*, on July 12, 2001, a team of researchers from N.I.S.T. and Germany announced a mercury clock which is 1,000 times more accurate than the cesium clock. This clock makes use of laser technology to trap and cool a single mercury ion and count its 1,000 trillion oscillations per second. It is estimated that it will take until 2015 for this new clock to replace the cesium clock as the accepted standard.

Exercise 3.9 Atomic Clocks

(i) How much more accurate is the cesium clock than the quartz-crystal[12] clock ?

(ii) How much more accurate is the mercury clock than the quartz-crystal clock?

Measuring Length Using Time. The definition of *to measure* is "to compare to a standard." So, for example, not long ago the *meter*, a unit of length, was defined to be one ten-millionth part of the distance along a meridian of the distance between the equator to the pole.[13] For the record a $meter = 39.37\ inches$.[14] Today, the meter is defined to be the distance light travels in a vacuum in $\frac{1}{299{,}792{,}458}$ *seconds*. This definition was adopted in October 1983, it is implemented in practice using laser technology.

The meter and the second are on what might be called a human scale, but their modern hyperaccurate definitions have left the realm of experience

[11]If you ever want to know the latest regarding measurements of time (or length) contact the National Institute of Standards and Technology, N.I.S.T.,(formerly the National Bureau of Standards) in Boulder, Colorado, cf., http://www.nist.gov.

[12]Note that the typical quartz-crystal wristwatch ticks off one second for about each 32,000 oscillations that it counts.

[13] *Webster's New Twentieth Century Dictionary*, Simon & Schuster, a Division of Gulf & Western Corp., 1979.

[14]The "standard" inch at one point was the length of the king's thumb, below the knuckle; the "standard" yard was the distance from the king's nose to the tip of the longest finger on his horizontally outstretched–to the side–arm and hand. Today, $1\ yard = 36\ inches$, and the inch has just been defined in terms of a meter.

for most of us. None of this would make sense were it not for the theory of relativity introduced by Albert Einstein in 1905. This theory postulates, and it is experimentally verifiable, that the speed of light in a vacuum is a universal constant. The speed of light in a vacuum is also independent of the frequency (color) of the light, as has been verified by astronomers who looked at the red and blue light coming from a distant binary star.

Starting in 1977 satellites were launched that became part of the *geographical information system*, i.e., GIS. A colleague of mine in the Physics Department, Prof. Neil Ashby, was a consultant to that project and had a difficult time convincing the government and private leaders that the system should be designed with relativity theory built in—they did not believe relativity was relevant to GIS. Fortunately they agreed to launch the satellites with a "switch" so that relativity could be turned on if necessary. The satellites were launched without relativity. The system did not work. Relativity was switched on and the system worked with amazing accuracy.

An example of the capabilities of the satellite systems is the work of another colleague of mine, Professor Kristine Larson, of Aerospace Engineering. She determined in collaboration with Stanford University Professor Jerry Freymueller, that the continents of Antarctica and Australia are moving apart 2 to 3 inches a year.[15]

I conclude this subsection on a personal note. One autumn I was climbing a small mountain in a national forest not far from where I am writing this. On top was a not-yet-completed communications facility, to be used for either cell phones, TV or GPS. Next to the modern facility was a taller structure, an old fire lookout tower, the pieces of which had been carried in on a narrow, minimally invasive trail built by the Civilian Conservation Corps many years ago. The pieces were then bolted together on site like a large version of a toy erector set. Unfortunately, the new building and our current culture caused a wide road to be bulldozed to the top of the peak to facilitate construction. Technology can be great, but we must be careful not to implement an electronic version of the following words of poet, Kahlil Gibran:

Trees are a poem the Earth writes across the Sky. Humanity cuts them down for paper so we may record our emptiness.

Scales, Units and Powers of 10. The meter and the minute are on a human scale. Soon we will develop other words and symbols to deal with very large and very small scales. But first I want to mention some facts about what I will call the American/English system of measurement (used only in the United States).

The American/English System of Units and Measurement. This system was developed with human dimensions in mind, as noted in the approximate

[15]The U.S. Department of Defense and the U.S. Department of Transportation operate what is known as the Global Positioning System, GPS, a system of 25 satellites coupled with thousands of receivers around the world. This system is used by a number of groups to better understand the earth, including, for example, earthquakes.

definition of the inch in terms of a thumb. The next scale up is that of the foot; it takes a counting cycle of 12 inches to get a foot, i.e., $1\ foot = 12\ inches$. It takes a counting cycle of 3 to pass from the scale of one foot to that of a yard, i.e., $1\ yard = 3\ feet$. There are 5280 *feet* in a *mile*. Scales below that of the inch have a counting cycle of 2, i.e., *half-inch*, *quarter-inch*, *eighth-inch*, *sixteenth-inch*, and so on.

Human stomach size presumably led to the unit of volume called a *pint*. The next smaller scale is the *cup*, $2\ cups = 1\ pint$. The next smaller scale is that of an *ounce*, i.e., $8\ ounces = 1\ cup$, $16\ ounces = 1\ pint$. Going up in scale, $2\ pints = 1\ quart$, $4\ quarts = 1\ gallon$.

Powers of 10. Before discussing the *metric system* of measurement, based on one counting cycle, namely 10, and used throughout the world, I need to give names to the various powers of 10. I assume you know what the number 10 means; and the fact that we are going to base our discussions in this section on the number 10 is due, perhaps, to the fact that most people have 10 fingers (and 10 toes). If we agree to use counting cycles of length 10 in going from one scale to the next, then the next scale up would be 10 *times* 10, or 100. This is written as 10^2 in mathematics. The 2 in 10^2 is called an *exponent*, and we say[16] that we "raised 10 to the power 2" to get 10^2. The exponent 2 in 10^2 tells us how many 10s we need to multiply together to get $10^2 = 100$, i.e., $10^2 = (10)(10) = ten\ times\ ten = 100 = one\ hundred$.

The next scale up would be $(10)(100)$. Other ways to write this are: $(10)(100) = (10)(10)(10) = 10^3 = one\ thousand = 10\ raised\ to\ the\ power\ 3$. We multiply by 10 again to get the next scale up: $(10)(1000) = (10)(10)(10)(10) = 10^4 = ten\ thousand = 10\ raised\ to\ the\ power\ 4$. We could keep going, and we will, which leads mathematicians to write 10^n, where $n = 1, 2, 3, 4, 5, 6, ...$[17] This last step should not bother you if you realize that 10^n is just a shorthand way to talk about the various "powers of 10" without telling you which one in particular we are talking about. At first this may seem silly, but it turns out to be quite necessary. I will have a lot more to say about this notation in II. For now, please accept for a while that this way of writing/talking is essential. By the way, $10^1 = 10$. Thus the notation 10^n means "n tens multiplied together." I can also think of 10^n in another way. I write "1." to mean 1 followed by a decimal point. Then 10^n means "1. with the decimal point moved n places to the right, filling in with n zeros."

Now just as medical doctors use Greek and Latin words to refer to body parts and diseases, scientists (including mathematicians) use Greek words, actually Greek prefixes, to refer to the powers of 10—not English. Thus,

---

[16]Another way this operation is referred to in English is this: "we *exponentiated* 2 to base 10" to get 10^2. Thus "to exponentiate" means "to make an exponent out of."

[17]Note that 1,2,3,4,5,6, ... means 1,2,3,4,5,6, and so on. Three of the dots following the 6 means "and so on" or "keep going" or "keep counting with whole numbers." The last dot is the period for the sentence.

we will not refer to one thousand meters, but rather to *one kilometer*, i.e., 10^3 *meters*. See Table Greek Prefixes below.

10^1	deka (da)	10^{-1}	deci (d)
10^2	hecto (h)	10^{-2}	centi (c)
10^3	kilo (k)	10^{-3}	milli (m)
10^6	mega (M)	10^{-6}	micro (μ)
10^9	giga (G)	10^{-9}	nano (n)
10^{12}	tera (T)	10^{-12}	pico (p)
10^{15}	peta (P)	10^{-15}	femto (f)
10^{18}	exa (E)	10^{-18}	atto (a)

Table Greek Prefixes

Notice in Table Greek Prefixes that negative powers of 10 appear, for example, 10^{-1}. The number 10^{-1} means "one tenth" or "one divided by 10." Thus we can write 10^{-1} in several ways: $10^{-1} = .1 = \frac{1}{10} = one\ tenth$. What is 10^{-2}? I will explain this in more detail in II, but the following are all equal: $10^{-2} = 1\ divided\ by\ 10^2 = 1\ divided\ by\ 100 = one\ one-hundredth = \frac{1}{100} = (\frac{1}{10})(\frac{1}{10}) = (.1)(.1) = (.1)^2 = .01$.

For 10^{-3} we have: $10^{-3} = 1\ divided\ by\ 10^3 = 1\ divided\ by\ 1000 = one\ one\text{-}thousandth = \frac{1}{1000} = (\frac{1}{10})(\frac{1}{10})(\frac{1}{10}) = (.1)(.1)(.1) = (.1)^3 = .001$.

The meaning of 10^{-n} is similar to the meaning of 10^n: 10^{-n} means "1 divided by 10^n." It also means n $\frac{1}{10}s$ multiplied together, i.e., $(\frac{1}{10})^n = (.1)^n$. Regarding decimal representations, 10^{-n} is "1. with the decimal moved n places to the left, filling in with $n-1$ zeros." (Question: How many zeros are in the decimal representation of 10^7 vs. 10^{-7}?)

Finally, $10^0 = 1$. This is natural if you give it some thought. I will discuss this a bit more in II.

The Metric System. To discuss units and measurement in the metric system we must be familiar with the names of the powers of 10, both positive and negative. Thus, for example, 1 *centimeter* = 1 cm = 10^{-2} *meters*. One *meter* = 1 *m*. One *millimeter* = 1 *mm* = 10^{-3} *m*. One *kilometer* = 1 *km* = 10^3 *m*. One *gigabyte* = 10^9 *bytes*. One *terawatt* = 10^{12} *watts*.

If you know how to measure lengths, then you should be able to measure areas and volumes. For example, a square that is one *meter* on a side has an area of one m^2, i.e., "one square meter" or "one meter squared." This same square is 100 *cm* on a side, so 1 m^2 = (100 *cm*) *times* (100 *cm*) = 10,000 *square cm* = 10^4 cm^2. A cube each edge of which is one *meter* long has a volume of one *cubic meter* = 1 m^3. Each edge of this cube is 100 *cm* long, so the volume of this cube in cubic centimeters is (100 *cm*) *times* (100 *cm*) *times* (100 *cm*) =

$1,000,000\ cm^3 = 10^6\ cm^3$. Actually drawing pictures of the relevant squares and cubes might be helpful in visualizing what I have just done.

Measuring Mass in the Metric and American/English Systems. After measuring time, lengths, areas and volumes, there is one basic type of measurement left, viz., that of mass. In the metric system the basic unit of mass is the *gram*$= 1\ g$. In the American/English system the basic units of mass are the *once* $= oz$ or the *pound* $= 1\ lb$. There are 16 oz in 1 lb. One *kilogram* $= 10^3\ g = 1\ kg = 2.2\ lb$.

There is one fact from basic science that we will use now and then. A cube of pure water 1 cm on a side, i.e., 1 cm^3 of water, has a mass of 1 *gram* $= 1\ g$. In the American/English system this fact becomes 1 *pint* of water has a mass of 1 *pound*. The rhyme: "a pint's a pound the world around" is one way to remember this. My students usually find the American/English system more confusing than the metric system for the purposes of doing mathematics, so I will minimize my use of the former.

Exercise 3.10 How Much Water Can a Pickup Truck Hold?

(i) If a metric tonne is 10^3 kilograms, how much volume is occupied by 1 metric tonne of pure water? Can you express your answer in terms of cubic meters, i.e., m^3? Can you express your answer in terms of pounds?

(ii) Based on your answer to (i) do you think that the back of a full-sized pickup truck could hold a tonne of watermelons?

Exercise 3.11 Units and Powers of 10

(o) Draw a picture as accurately as you can of one square centimeter, that is 1 cm^2. Make a model with paper and sticky tape of one cubic centimeter, i.e., 1 cm^3.

Draw a picture as accurately as you can of a square which is 3 cm on a side; count how many square centimeters are inside of this square.

Draw a cube that is 3 cm on a side; find the number of cubic centimeters inside of this cube.

(i) How many centimeters are in a meter? meters in a kilometer? grams in a kilogram? cm^2 in a m^2 (square centimeters in a square meter)? cm^3 in a m^3 (cubic centimeters in a cubic meter)?

(ii) One liter, denoted l, is $10^3\ cm^3$. How many l are in a m^3? One tonne, also called a metric ton (MT), is 10^3 kilograms. How many grams in one MT?

(iii) Convert the following American English words into powers of 10: one, ten, one hundred, one thousand, one million, one billion, one trillion.

(iv) Convert the following American English words into powers of 10: one tenth, one hundredth, one thousandth, one millionth, one billionth, one trillionth.

(v) Note that an nm is a nanometer (or one billionth of a meter), and a μm is a micrometer, or micron. Find the American English word equivalent to nm, cm, Gm, Tm, mm, μm, pm, dm, km, hm, dam, Em, am.

(vi) Express in decimal form, and in at least one other form, the following: $\frac{1}{10}$, $\frac{1}{10^{-1}}$, $\frac{1}{10^{-5}}$, $\frac{1}{10^8}$, 10^{-9}, 10^7. Hint: $\frac{1}{10^{-n}} = \frac{1}{(\frac{1}{10^n})} = (1)(\frac{10^n}{1}) = 10^n$.

(vii) Express in exponential form, i.e., 10 to some power, the following:

$$(a)\ .0000000000000000001,\ (b)\ 100,000,000,000,000,000,000.$$

and (c)

$$(1,000,000,000,000,000,000,000,000,000.)(.000000000000000000000000000001)$$

Do you see how exponential notation saves time?

(viii) Express as a single power of 10: $(10^2)(10^{-7})$, $\frac{10^{56}}{10^{60}}$, $(10)(10^7)(10^{-3})(\frac{1}{10^{-6}})$.

(ix) What is $(10^{-1})^2$? Review the two paragraphs just before this exercise.

(x) Write as a single power of 10: $(10^3 10^4)^2$, $(\frac{10^4}{10^{-3}})^3$, 1,

$$\frac{(\frac{10^{-2}}{10^3})}{(\frac{10^2}{10^{-90}})}.$$

(xi) What are the following: $\frac{100}{100}$, $\frac{10^5}{10^5}$, $\frac{10^{-7}}{10^{-7}}$, $(10^3)(10^{-3})$, $(10^n)(10^{-n})$, 10^0, $(10^n)^0$?

(xii) Since too many of my students have missed the following in the past I will ask again. How many cm^2 are there in $1\,m^2$? How many cm^3 are there in $1\,m^3$? Draw a picture of a square meter and a cubic meter and really understand what you are doing. Hint: Can you visualize a square meter as an array of little square centimeters, arranged in rows with 100 rows, and 100 little squares in each row? So the total number of little square centimeters would be 100 in the first row plus 100 in the second row plus ... plus 100 in the last, that is, one-hundredth row. Can you similarly visualize a cubic meter as an array of little cubic centimeters arranged in 100 "slices" with 100 times 100 little cubic centimeters in each slice?

Exercise 3.12 Accuracy

I once read on a cereal box an "educational"comment which said that the metic system was superior to the English system of measurement because the metric system was more accurate. Do you agree?

Exercise 3.13 Unit Conversions in America

Note that if you forget some fact, such as there are 12 *inches* in a *foot*, you can find the information you need in a dictionary or on the Web (or you can look in the index of this book to find the relevant page).

(i) How many cubic inches are there in a quart? Hint: 264.2 *U.S. gallons* = 35.31 *cubic feet*.

(ii) How many square inches are in a square foot? an acre? a square mile? Hint: 1 *acre* = 43,560 *square feet*. Also, there are 5280 *feet* in a *mile*.

(iii) How many *grams* of water are there in a *pint* of water?

Exercise 3.14 The Concept of Scale is Important in Science

(i) Is there a natural scale in the design of animals? If you scale up a person's height, thickness and width by a factor of 2, i.e., you double height, thickness, and width, by what factor is that person's mass scaled up? Are there any implications for the relative size of the bones in our person, before and after scaling up by a factor of 2?

(ii) Some mathematical laws of science are unchanged (at least approximately) by a change in scale, others are. Find an example of each, cf., [614, p. 63]. Hint: Look for power laws.

Orders of Magnitude, Numbers in Standard Form, Significant Digits. Before we look at some examples of Nature at various scales, I want to remind you about how scientists write numbers. In this book we will very often but not always write numbers in *standard form*. To write a number in standard form you do the following: move the decimal so that one (nonzero) digit stands to the left of the decimal and then multiply the number by the appropriate power of 10—so that the resulting number you end up with is the same as the number you started with. In general, if you have a number and you create a new, smaller number by moving the decimal n places to the left, you can recover the original number by multiplying the new number by 10^n. If you have a number and you make a new, bigger number by moving the decimal n places to the right, you can recover the original number by multiplying the new number by 10^{-n}.

Thus given the number 31415.667, we write this in standard form[18] as follows: $3.1415667(10^4)$. Another example: $.0067 = 6.7(10^{-3})$. Finally, $.00093(10^2) = 9.3(10^{-4})(10^2) = 9.3(10^{-2})$. When a number is written in standard form you can immediately see two things. First, you can see the *order of magnitude* of the number; this is *the exponent of the power of ten closest to our number.* This will be the exponent of the power of ten of the number in standard form (plus 1 if the number to the left of the power of 10 is 5.0 or more).[19] For example, $3.9(10^6)$ has order of magnitude 6; $5.0(10^6)$ has order of magnitude 7; and $5.9(10^{-3})$ has order of magnitude -2.

Second, you can see the *significant digits* of the number, i.e., the digits used to the left of the power of ten. Thus $3.1415667(10^4)$ has 8 significant digits and is of order of magnitude 4, $6.7(10^{-3})$ has 2 significant digits and has order of magnitude -2, and $9.3(10^{-2})$ has 2 significant digits and order of magnitude -1. Our definitions of "significant digits" and "order of magnitude" are not complete or perfect; for example, see part (iii) of exercise below. The idea that we are trying to communicate is this: when a scientist writes a number, which comes from an actual measurement, in standard form he/she is not supposed to write a digit if it is not meaningful or significant, i.e., if it is not actually measured. There is an additional complication: all measurements involve error. So a number is not complete unless you are told how much error there might be. For the most part we are going to ignore the complications caused by error in measurements.

Exercise 3.15 Significant Digits and Orders of Magnitude

(i) Write the number π in standard form with 8 significant digits. (Feel free to use a calculator or a book to look up the number.)

(ii) Write the number .0000000985 in standard form. How many significant digits are there? What is the order of magnitude?

(iii) Write the number 102,000 in standard form. How many significant digits are there? Can there be more than one possible correct answer to this question? What additional information do you need in order to know for sure how many digits are really significant? What is the order of magnitude of this number? Does the order of magnitude depend on any of the things you had to consider to determine the correct number of significant digits?

Nature's Smaller Scales. Let's look at some examples of Nature at different scales. First, the human scale is roughly of order of magnitude 0. Most

---

[18] Do not forget that multiplication of 3.1415667 by 10^4 can be written in at least three equivalent ways as $3.1415667 * 10^4$, as $3.1415667\ 10^4$ or as $3.1415667(10^4)$.

[19] This is a simple and coarse way to "round off." Many others prefer the more delicate approach. For example, a number between 0.3 and 3 is "rounded" to $10^0 = 1$, a number between 3 and 30 is rounded to 10^1 and so on.

adult humans are from $1(10^0)$ to $2(10^0)$ *meters* tall. Recall that $10^0 = 1$, see Exercise 3.11 (xi). The limit of resolution[20] of the unaided eye is about .1 *mm* or 10^{-4} *meters*. The compound light microscope can resolve points about 400 times closer than can the naked eye, i.e., the light microscope has a limit of resolution of about $2.5(10^{-7})$ *m*. Now .1 *mm* is 100 microns (μm), and 10^{-7} *m* is .1 microns. At this scale of Nature we find the human egg and the amoeba (size 100 μm), liver cell (20 μm), red blood cell (7 μm), and typhoid bacillus (.2 to .5 μm).[21]

I have a friend, David, who is a cell biologist. He works in the world of the living cell, and he talks about motor molecules and microtubules. He uses an electron microscope with a theoretical limit of resolution of about 2.5 (10^{-12}) *m*. In practice, however, he works with cell parts from 1 micron to 1 nanometer (10^{-9} *m*) in size. On this scale of nature we find the typical bacterial virus (80 millimicrons) and the haemoglobin molecule (7 millimicrons).

Exercise 3.16 Millimicrons and Nanometers

(i) What is the relationship between a millimicron and a nanometer?

(ii) What is nanotechnology and what effects might it have on your life? Nanoparticles are likely added to some commercial products, which ones? Is nanotechnology regulated? Might there be (are there) any unintended consequences?[22]

(iii) The smallest known life form is *Mycoplasma*, diameter 200 *nm*. What is the status of research on *nanobacteria*, (10 – 200 *nm*), at the time you read this? See the Jan. 2010 *Scientific American*, for example. What is the status of research on *prions*, (13 *nm*), when you read this? While not considered "living" are nanobacteria and prions "on the border" between animate and inamimate? Might there be a fuzzy boundary between animate and inanimate that has made it difficult thus far to give a precise definition of *life*?

At 10^{-9} meters we are getting down to the scale of molecules. A *molecule* of a chemical is the smallest amount of the chemical that has the defining properties of the chemical. As an example there is an experiment that you can actually do to measure the size of an oil molecule, see Exercise 13.12. One order of magnitude less, 10^{-10} *m*, and we are on the scale of an atom, the building blocks from which molecules are made. The most elementary chemicals are called *elements*. All chemicals are made of elements, but the elements cannot be broken down into simpler constituents by any chemical process. An atom of an element is the smallest amount of that element that can exist. For

[20]The "limit of resolution" is defined to be the minimum distance between two points which can be discerned as separate entities.

[21]The most common waterborne disease in the United States today is Giardiasis, caused by a protozoan *Giardia lamblia*. Giardiasis is common in hikers and backpackers who do not treat their water. This little protozoan is just a "few" microns in diameter. Some water filters can remove all water borne objects larger than a "few" microns. If you rely on such a water filter, it is important to know the size (in microns) of the smallest particle that can pass through it. Just how big is *Giardia lamblia* anyway?

[22]The following paper indicates that nanoparticles already in common household items, e.g., cosmetics, sunscreen, vitamins, toothpaste, and hundreds of other products, have caused genetic damage in mice: http://www.eurekalert.org/pub_releases/2009-11/uoc–nui111609.php

example, the hydrogen atom, denoted H, the carbon atom, denoted C, and the oxygen atom, denoted O, are all about 10^{-10} m in diameter.[23] At the atomic and molecular scale 10^{-10} m is such a convenient unit that it is given its own name. We define 10^{-10} m to be 1 *angstrom*, which is denoted by the symbol 1 Å. The angstrom is named after Anders Jöns Angström, a Swedish physicist at the University of Uppsala in the 19th century who studied light. The Å unit is used to measure the wavelength of light. It is interesting to note that at the atomic and subatomic scale the distinction between "wave-like" phenomena and "particle-like" phenomena is not as hard and fast as it appears to us at the human scale. In a branch of physics called quantum mechanics one learns that very small particles often act as though they are waves. In fact the electron microscope mentioned above works on the principle that very fast (high energy) electrons act like waves with a very short wavelength (shorter than the wavelengths of visible light for example) and hence this beam of electrons can be used to look at very small things that can not be seen with an ordinary microscope that sees things using light.

Quantum mechanics also tells us that waves, such as light waves, can exhibit particle behavior. Have you ever heard of photons ("particles of light")? In this world of the very small we have to develop an entirely new intuition about how things behave. Mathematics is an invaluable tool for building this intuition.

Exercise 3.17 What are the Smallest Things that Affect Your Life?

(i) What are the smallest measurements of length, area, volume, time and mass that you have ever heard of, seen or used—outside of this book?

(ii) What are the smallest things that have ever had a measurable impact on your life? For example, see Exercise 3.16 and Section 1.6.

(iii) The scale between the size of an atom and the size of a bacteria is called the *mesoscale*. It is a transition zone between classical physics and quantum mechanics, and is the zone of "programmable atoms," which is somewhere between serious physics and science fiction. What are programmable atoms? See [419].

Even in countries that have adopted the metric system[24] the people who build furniture and houses often use systems of measurement that were used before the metric system was introduced.[25] Thus it should not come as a surprise that when Rich and Chris built some cabinets for me they used American/English units of measurement. I learned that fine cabinetry leaves gaps no larger than $\frac{1}{128}th$ of an inch. Rich will say things like: "You can sometimes get away with $\frac{1}{64}th$ of an inch, but I can always see gaps that big and I fill 'em."

[23]See the periodic table of the elements in on page 253 for the various kinds of atoms/elements.

[24]Newspapers reported on October 1, 1995, that England has gone nearly all metric. Pints of beer and milk will still be available. The change over to metric measurement has not been met with universal approval.

[25]Such is the case in Norway, for example, where units of measurement as old as the Vikings still exist.

When I built my own house with Richard, Joe, Lee and Fred, getting the rough framing within $\frac{1}{16}th$ of an inch was good, within $\frac{1}{32}nd$ of an inch was overdoing it; and at the end of a long hard day with the wind howling, a thunderstorm pounding down, and almost in the dark with a warped piece of wood, $\frac{1}{8}th$ of an inch, plus or minus, was looking mighty fine.

Exercise 3.18 Nonmetric Units Persist

(i) Convert the English unit measurements used in cabinetry and carpentry, as mentioned above, to the metric system. Recall that $1\ cm = 10^{-2}\ m$ and that $1\ in = 2.54\ cm$.

(ii) Why did I use American/English units to build my house?

(iii) Why do you think that pubs in England still serve pints of ale, even after officially converting to the metric system in 1995?

(iv) Why do you think that nonmetric units persist in some realms but not in others? Why do we still use 24 hours in a day (60 minutes in an hour) and not 10 (or 100)?

Nature's Larger Scales. The area of the earth is about $5.10(10^{14})\ m^2$. The mass of the earth is about $5.98(10^{24})\ kg$. The mass of the atmosphere is about $5.14(10^{18})\ kg$.[26] The volume of water in the oceans is about $1.35(10^{18})\ m^3$. These are the kinds of numbers we need in order to talk about things on the global scale.

Amory Lovins, at the Rocky Mountain Institute in Snowmass, Colorado, often talks about power in terawatts (a terawatt is $10^{12}\ W = 1\ TW$) and energy in gigajoules (a gigajoule is $10^9\ J = 1\ GJ$) when he talks about energy consumption and the economy of the United States.

Between the global and human scales are Nature's ecosystems. The terms ecosystem, ecological system and environmental system will be synonymous for us. These words are not used consistently by everyone. For the sake of at least the appearance of precision I will take Howard T. Odum's, [503], definition of ecosystem: "An organized system of land, water, mineral cycles, living organisms, and their programmatic behavioral control mechanisms."

Forests, seas and the earth's entire biosphere are examples of large ecosystems. A pond, coral head and an aquarium are all examples of small ecosystems. Before passing in the next paragraph to outer space I would like to remark that one of my motivations for writing this book is clearly stated in the following quote from H.T. Odum's book: "..., and there is growing recognition that humans are incomplete without the life support of self-maintaining natural ecosystems."

An astrophysicist friend of mine, Professor Ellen Zweibel, University of Wisconsin, studies cosmology, the sun and the stars among other things. She routinely talks of distances in light-years. A light-year is the distance light travels in one year. The speed of light in metric units is about $3(10^8)\ m/sec$.[27] In fact, she even speaks of *parsecs* and *megaparsecs* (a *megaparsec* is 10^6 *parsecs*).

[26] Recall that kg stands for kilograms. How many grams are in a kilogram? Don't forget Table Greek Prefixes as you read this subsection.

[27] More precisely, the speed of light in a vacuum is 299,792,458 m/sec. Where in this book have you see this number before?

The star nearest to our sun is 1.2 *parsecs* away, and 1 $parsec = 3.26$ *light-years*.[28] She tells me that the sun's mass is $1.989(10^{33})$ g. She also studies magnetic fields in space, which are measured in terms of a unit called a *gauss*. Between galaxies the magnetic field is very weak, 10^{-9} *gauss*; and on a neutron star the magnetic field is a *teragauss*.

Exercise 3.19 What are the Largest Things that Affect Your Life?
(i) What is the largest length, area, volume or mass that you have ever heard of, seen or used—outside of this book?
(ii) What is the largest object or process that has ever affected you personally?
(iii) What Greek prefix expresses 10^{-9} *gauss*? How many *gauss* is a *teragauss*?
(iv) How many *kilometers* are in a *light-year*?
(v) Express the speed of light in *miles/sec*.
(vi) How many *miles* are in a *light-year*?
(vii) How many *tonnes* of mass does the sun have? A $tonne = 10^3$ kg. Can you express the sun's mass using some Greek prefixes?
(viii) This exercise compares the global scale to the atomic scale. Calculate the surface area of the earth in square *angstroms*, $Å^2$. Calculate the volume of the earth in cubic *angstroms*, i.e., $Å^3$, and in cm^3. Hint: The earth is (not quite) a sphere, with a radius at the equator of $6.38(10^6)$ m and a radius at the poles of $6.36(10^6)$ m. The volume of a sphere of radius R is $\frac{4}{3}\pi R^3$. What is the area of a sphere of radius R?

3.5 The Art of Estimating

Sometimes you would like to count or otherwise measure something which at first seems "unknowable," or at least difficult or impossible to look up. If you stop and think about what you do know, perhaps even do a little data gathering and/or make some educated guesses, you often can give a rather solid justification for an estimate of the number or measure you are looking for. The following exercise was once suggested to me by the late David Brower.

Our First Estimate: Something to Chew on. Suppose you wanted to know how many times your tongue "gets out of the way of your teeth" while you are eating in, say, one day. This may not be the most pressing issue in your (my) life, but it is certainly a number that is not easily looked up. And I want to know it.

Let's begin. I will suppose that today has been a "serious" eating day, with three full meals reasonably leisurely eaten. For breakfast I had three bowls of cereal and some fruit. I spent 15 minutes at breakfast, with 10 minutes devoted to chewing at the rate of about 50 chews per minute. That gives me 500 chews for breakfast. Lunch was moderate, lasting about 20 minutes, with about 15 minutes devoted to chewing. That gives 750 chews for lunch. Dinner lasted about 35 minutes with four courses of Chinese mixed vegetables, rice

[28] A *light-year* is the distance light travels in one year.

and dessert. About 25 minutes were devoted to chewing. This gives 1250 chews for a total of $1250 + 750 + 500 = 2500$ chews. I did not bite my tongue at all today, so my tongue successfully got out of the way of my chewing teeth 2500 times today.

Now suppose I missed a meal, or one meal became an extended, ravenous feast. It would be easy to adjust the estimates. The main point is that these estimates can be made, and they are not devoid of meaning. In fact, a series of estimates based on slightly altered assumptions and educated guesses will yield *an interval of numbers.* Such a set of numbers is probably more realistic than any one number might be. Finally, you do not want to ascribe more meaning to these estimates than is really there. For example, in the above example, at most two digits are significant, possibly only one is.

Estimating is a Mixture of Art and Science. The process of estimating is not carefully defined and the meaning of your estimate depends entirely on the amount and quality of thought you put into making the estimate. The least meaningful estimate is a wild guess, with nothing to substantiate it. An estimate is not just the number, or interval of numbers, you end up with. It is that, plus the reasoning process you have to back up your estimate(s). Use your critical thinking skills to make the following estimates.

Exercise 3.20 Making Estimates

(i) Estimate the number of people that are within an hour's walk from where you are right now.

(ii) Estimate the number of people who were within an hour's walk of your present location 25 years ago and 100 years ago.

(iii) Estimate the number of medical doctors now in the United States. (Try to do this estimate without directly consulting the Web or the American Medical Association. After you have your own estimate, check it with these two and/or other sources.)

(iv) Estimate the number of television sets in the United States today.

(v) (a) Estimate the number of four-year colleges in the United States today.

(b) Estimate the fraction of the population in the United States that has the opportunity to attend a four-year college—for four years (whether or not they take advantage of it).

(c) Estimate the fraction of the world's (human) population that has the opportunity to attend a four-year college. You may want to look at V for data you can use to estimate the world's human population.

(vi) If you have decided "what you want to be when you grow up," estimate the number of job openings of the type you desire in the year you will be looking for employment. Also estimate the number of people that will be looking for that same type of job.

(vii) Estimate the number of revolutions of your bicycle (or car) tire it takes to wear your tread down by a thickness of 100 Å.

(viii) In a popular book, [552], some frightening afflictions, like the Ebola virus disease, endemic to tropical rain forests are discussed. Estimate the minimum number of hours it would take for an Ebola virus to travel from its native habitat deep in the jungles of Africa to your home. Make two estimates, one before and one after a road is built into the jungle where the Ebola virus lives. I call these estimates *connection times.*

(ix) Estimate the number of people in the world whose birthday falls on the day you happen to be reading this.

(x) Estimate the number of hours Americans spent preparing their income tax forms this year.

(xi) Estimate the number of billionaires in the world, in the United States, in Mexico and India at the time you read this.

(xii) Estimate how much you are paying for each class period where you are studying.

Using Estimation Techniques to Check "Facts." Sometimes you are presented with a "fact" or argument that seems suspicious. Using estimating techniques and/or a little mathematics you can often verify, debunk or otherwise check the reasonableness of the information being presented. The following is a somewhat famous quote from an article titled: "On the American Pet," which appeared in the December 23, 1974 edition of *Time Magazine.*

"[There are] one hundred million dogs and cats in the U.S. ... Each day across the nation, dogs deposit an estimated four million tons of feces."

You should get in the habit of not taking everything you hear or read for granted, even—or especially—when numbers are involved. Thus it is with the above quote from a major news source. Are the numbers reasonable? Off the top of my head, 10^8 dogs and cats is a reasonable number compared to the U.S. population in 1974 of roughly $2(10^8)$ people. The $4(10^6)$ tons of feces, however, seems a bit high. Is it?

Well, suppose for the moment that all of the 10^8 dogs and cats are dogs. A *ton* is 2000 *pounds* in the American/English system of units, so our 10^8 dogs are (according to the quote) putting out $4(10^6) * 2000 = 8(10^9)$ *pounds* of feces each day. Thus, on average, a single dog is producing $\frac{8(10^9)}{10^8} = 80$ *pounds* of feces per day! Actually, the article is asserting a greater production rate per dog, since not all of the 10^8 pets are dogs.

Exercise 3.21 A Mistake in Major Media Involving Dogs and Cats

(i) If half of the pets in question are dogs and half are cats, what is the average daily production of dog feces per dog according to the above quote from *Time Magazine*?

(ii) Assume that the article had a misprint. Suppose that the last line of the article says: *"Each day across the nation, dogs and cats deposit an estimated four million tons of feces."* Suppose that there are an equal number of cats and dogs in the U.S. and that dogs produce on average three times the feces that cats do on average. What then is the daily feces production per dog?

Well, you get my point. No matter how you look at them, the numbers are ridiculous. Not every mistake or use of spin is so obvious. Accuracy, being one with what is, these are the hallmarks of scholarly journals, which often lose entertainment value in the drive for precision. Most of our media fall short of such scholarly standards, as do even some scholarly works.

Sometimes misleading mathematics is more subtle. Consider the following material, brought to my attention by www.fair.org, that comes from an article titled: "What Price for Good Coffee?" which appeared in the October 5, 2009 edition of *Time.*

The article's lead sentence is: *"Fair Trade practices were created to help small farmers. But they may have hit their limits."* The article then goes on to provide the following information which for the moment I accept as accurate. Fair Trade pays $1.55 per lb. for coffee from small farmers, almost 10% more than the market price. Then Fair Trade researcher, Christopher Bacon of the University of California, Berkeley, says that (for various reasons) at least $2 per lb. is needed if small farmers are to rise above subsistence level. It is also noted that if the farmer gets $1.55 per lb., the retail customer pays $10 per lb.

Later, the article interviews a coffee drinker and says: ... *"The company declined to comment on whether Fair Trade's benefits fall short of its vision or how much it would need to raise prices if* **coffee** *were to climb to $2 per lb. Fair Trade 'isn't the only reason I drink Starbucks, but it's a big one,' says Connie Silver, a nurse, sipping a large, $4.15 Frappuccino outside a Miami store. Asked if she'd pay, say $4.50 or even $5 to help absorb higher Fair Trade prices, Silver raises her eyebrows and says, 'Wow, these days, that's a tough one.' "*

My conclusion from a cursory reading of this article is that if the Fair Trade price paid the small farmer goes up $.45 per lb, the price of 1 Frappuccino will go up $.35 to $.85, and this is untenable, especially for the nonrich. But let's take a closer look at the math here. The price paid for a cup of coffee includes many things, only one of which is coffee: for example, milk, sugar or any other ingredients, the paper cup, the rent for the coffee shop, the wages of various people involved in bringing the coffee from the farmer to the shop, and so on. A quick estimate (other details are discussed in the following exercise) goes as follows. Suppose you can get 20 Frappuccinos out of 1 lb. of coffee. Then the intrinsic cost increase in the price for *one* Frappuccino due to the coffee would start at around $\frac{\$.45}{20} \approx \$.02$ *or* $.03. So depending on the mathematics we get two opposite conclusions!

Exercise 3.22 Is Fair Trade Affordable?

(i) Using the same approach as above, what is the increase in cost of one Frappuccino if you only get 10 from 1 lb. of coffee? Same problem, if you get 25 Frappuccinos from a lb. of coffee? Same problem, if you get 30 Frappuccinos from a lb.?

(ii) Suppose retailers look at the problem as follows. If $1.55 is paid per lb. wholesale, the retailers charge $10 per lb. If the same percentage markup is passed on at $2 per lb. paid to the small farmer, what is the new retail price of 1 lb. of coffee?

(iii) If you can get 20 Frappuccinos from 1 lb. of coffee, and the total price of 20 Frappuccinos is $83, what percentage of this $83 is for the coffee, if the coffee retails for $10 per lb.?

(iv) If 12% of the price of a cup of coffee is for the coffee and the price of coffee goes up 30%, what would be the increase in the $4.15 price for a cup of coffee? This is the highest estimate, assuming all costs are passed on with the same retail markup rate. Why is this unlikely if most consumers are truly unable to afford such an increase? Do large chain coffee shops pay retail or wholesale prices for coffee? What is the answer to this exercise if only 1.9% of the price of a cup of coffee is for the coffee.

(v) Investigate how much of a pound of coffee is used by your local coffee shop to make a large Frappuccino. Incidentally [521, pp. 8–11] gives a set of numbers concerning coffee independent of the above discussion.

(vi) Is there a negative effect of coffee production on endangered species, e.g., the jaguar, that is partially mitigated by Fair Trade practices?

Try your estimation skills on the following.

Exercise 3.23 Do Americans Really Watch That Much TV?

(i) In his book, [441], Jerry Mander states that the typical American sees 21,000 television advertisements a year. Estimates usually are more meaningful when stated as intervals of numbers, say 17,000 to 25,000 advertisements, for example, in this case. Adjusting for the fact that Mander is writing in 1992, do you think his estimate is accurate? What interval of numbers do you get in estimating the number of TV ads seen by a "typical" American in 1 year?

(ii) How many TV ads do you see in 1 year?

(iii) How many Americans do you estimate see between 0 and 100 T.V. ads in 1 year?

The above exercise was relatively easy. The following exercise is more interesting and a bit more difficult. References include [63, 499, 104, 327, 14, 83, 516, 158].

Exercise 3.24 Prisoners and Farmers in America

(i) Estimate the number of Americans incarcerated at various times during the last century up to the present. Be careful to distinguish among inmates of federal prisons, state prisons, jails (city and county), and folks on probation or parole.

(ii) Estimate the number of American farmers at various times from 1900 to the present. Define farmer to be a person who declares farming to be their principal occupation.

(iii) At one point in the last century the number of American farmers was far, far higher than the number of American prisoners, no matter how the latter might have been counted. For some time the number of American farmers has been less than the number of Americans incarcerated. Does this mean that at some time these two numbers were approximately equal? Why? If so, at what time(s) were these two numbers equal?

(iv) Pick a state, estimate the cost per year for that state to maintain one prisoner.

(v) For the same state, estimate the cost per year for one student to attend a state university.

(vi) Estimate the number of people in the United States on death row (executed or not) who were or are innocent. Hint: see "innocence project."

(vii) Compare incarceration rates for various countries around the world, i.e., the number incarcerated for every 100,000 people. Which country has the highest incarceration rate? Why?

Chapter 4

We All Soak in a Synthetic Chemical Soup

I begin with a horror story that could happen to any one of us at any time, hopefully with low probability—but it is likely that no one knows what those chances are.[1]

4.1 Thomas Latimer's Unfortunate Experience

On July 20, 1985 Thomas Latimer, a petroleum engineer living in Dallas, went out to mow his lawn. Twice the previous month, Latimer had sprayed his lawn with a pesticide made by Chevron-Ortho that contained diazinon. (Diazinon was banned for residential use on December 31, 2004, but it is still allowed for agricultural uses. It is, however, legal for consumers to use products containing diazinon purchased before 12/31/04. It was found in hundreds of products, including such brand names as Spectracide, Bug-B-Gon and GardenTox.) Halfway through that July day, Latimer got tired, his head began to hurt, and he felt dizzy and nauseated. He rested for the remainder of the day. The next morning he tried to finish the job, but the symptoms returned. This time, he experienced impaired vision.

His symptoms persisted for a week, he went to the doctor and had several tests. The doctor suggested that Latimer stop taking Tagamet, an antacid he had been prescribed. Latimer's test came back: he had been poisoned. A specialist later identified the toxin: diazinon (which Latimer picked up via the air and his skin while mowing and handling grass clippings.) It is believed by some that the Tagament was interfering with his liver function, which should have filtered the poison from his blood. (His wife worked along with him, was not taking antacids, and did not have her husband's reaction.)

In testimony before the Senate Environment and Public Works Subcommittee on Toxic Substances, Environmental Oversight, Research, and Development, May 1991, Latimer said that when he first became ill he called Chevron-Ortho's emergency toll-free number and

[1]The account in this section of Thomas Latimer's unfortunate experience is mostly taken from the newsletter of the Center for Public Integrity, Vol. 4, No. 3, August 1998, Web site www.publicintegrity.org. One can find other accounts. The manufacturers of Diazinon and Tagament do not agree that there is a causal relation between their products and Mr. Latimer's experience; and, in fact, if one does a Google search on the subject "Thomas Latimer+Tagament+Diazinon" the first item to appear is a ruling, Dec. 14, 1990, from the U.S. Court of Appeals, Fifth Circuit, where the court finds in favor of Ortho Consumer Products and SmithKline & French Laboratories. The court agreed with these companies that Mr. Latimer did not prove a causal link. The citation for the case is 919 F.2d 301.

asked if the symptoms he was experiencing could be related in any way to the pesticide he sprayed on his lawn. "I told the representative on the phone that I was taking the medication Tagament and asked, 'Could this have resulted in an interaction poisoning?"'

"The Chevron-Ortho representative said it was not possible to have a problem with diazinon if I was on Tagament. The representative claimed to me that diazinon was so safe that I could drink an entire bottle and the only problem I would have is that I would be nauseated for a few days."

Latimer's professional training gave him enough knowledge to handle his lawn chemicals carefully. Latimer suffered permanent damage: "I live every waking moment in constant, unrelenting head pain," he explained to the Senate committee. "My eyesight damage has been verified by three neuro-ophthalmologists. My ability to read is limited to ten minutes at a time ... I suffer from brain seizures, panic and fear attacks both day and night, and nightmares ...I suffer a degree of physical retardation and motor-skill damage. I cannot run or swim. I have also suffered from viral growths on my vocal cords, which have required laser surgery three times. It is likely I will need vocal-chord surgery every year for the rest of my life ... I cannot yell or talk loudly. I must talk softly and on a limited basis. Many days, I have to be virtually silent. The frustration and anger level due to my voice being restricted is very high ..."

One explanation for the above is that Mr. Latimer was hit by a *synergistic* interaction of two readily available *synthetic chemicals.*[2] We will learn in Exercise 17.19 that given the many tens of thousands of such chemicals registered for use in the U.S., it is difficult to pretest for all such interactions. Thus for those who care about such things, caution is advised. It is often tough to prove "cause and effect" beyond a reasonable doubt either scientifically or in a court of law. Witness the decades long relationship between smoking and health as it evolved in the media, culture, and the courts.

Who Cares? I once took a trip to a country rather infamous for muggings and theft. In fact, one of my guide books was blunt: "Assume you will be robbed." While there I did an informal survey of several of the people I met, asking them if theft/mugging was a problem. About half the people said there was no problem. The other half said I should take great care, mugging was a serious problem—perhaps my life could be in danger.

For me the most interesting part of this informal survey was this: Those who saw a problem had been mugged; those who had not been mugged saw no problem. This type of response is primitive, devoid of mathematical perspective, and it is inadequate for dealing not only with mugging but with more subtle threats to well being posed by, for example, toxins and climate change. The position: "If it kills me I will take note, if it doesn't I won't;" needs to be replaced by a bigger picture containing a more nuanced understanding. To appreciate and understand a looming threat before it is obvious requires abstract thought, and this thought invariably contains some form of

[2] *Synergy*, a phenomenon wherein the combined effect of two or more entities is more than the aggregation of the effects of each entity taken separately. Said another way: The whole is more than, perhaps quite different from, the "sum" of its parts. This can be a very positive experience, as when several people work together to accomplish much more than they ever could have working individually. It can also be a negative experience. We define *synthetic chemicals* to be those which did not exist until created by humans.

mathematics. This abstract understanding has to be deep enough to induce sufficient action to neutralize the threat.

In this section I will be dealing primarily with the flow of known and suspected toxins from our surroundings into our bodies. I will give evidence that such flows actually exist and that some harm has resulted. Whether a given person cares or not, or how deeply that concern goes, seems to depend in part on whether that person thinks he/she has been assaulted by toxins or not—and if that person believes he/she or a loved one has suffered illness or worse because of the exposure. I am hoping that with a mathematical assist, the number of people who are concerned increases beyond those who are clearly and immediately impacted.

In this and subsequent chapters I will be thinking of humans as biological beings, i.e., boxes, with various flows involving toxins, food, water, energy and so on. As we will see, to understand these flows at even the most elementary level requires a little mathematics. The total amount of synthetic substances in your body is called your *body burden*, especially chemicals known to be toxic and uninvited.

4.2 What's in the Synthetic Chemical Soup?

Are We All Polluted? In 1996 Theo Colburn, et al., [103, page 106], wrote: "Virtually anyone willing to put up the $2,000 for the tests will find at least 250 chemical contaminants in his or her body fat, regardless of whether he or she lives in Gary, Indiana, or on a remote island in the South Pacific. You cannot escape them."

Unfortunately Colburn is being repeatedly proven correct; we are soaking in a synthetic soup. From the Web site for the nonprofit Environmental Working Group (EWG), http://www.ewg.org/reports/bodyburden/, we once read:

> "In a study led by Mount Sinai School of Medicine in New York, in collaboration with the Environmental Working Group and Commonweal, researchers at two major laboratories found an average of 91 industrial compounds, pollutants, and other chemicals in the blood and urine of nine volunteers, with a total of 167 chemicals found in the group. Like most of us, the people tested do not work with chemicals on the job and do not live near an industrial facility. Scientists refer to this contamination as a person's *body burden*. Of the 167 chemicals found, 76 cause cancer in humans or animals, 94 are toxic to the brain and nervous system, and 79 cause birth defects or abnormal development. The dangers of exposure to these chemicals in combination has never been studied."

The EWG/Mt. Sinai/Commonweal study tested the blood and urine of the nine volunteers for 210 chemicals—the largest collection of industrial chemicals ever surveyed at that point in time. Of the 91 compounds found on average, most did not exist 75 years ago. As mentioned, in aggregate, 76 chemicals that cause cancer were found. They found a total of 48 PCBs, which were banned in the U.S. in 1976 (but are used in other countries) and persist in the environment. The list of participants, together with their test results, are at the EWG Web site mentioned above.

The Centers for Disease Control and Prevention (CDC) tested the blood and urine of a statistically representative sample of some 2,500 (noninstitutionalized) volunteers in the U.S. population for the years 1999 and 2000, cf., http://www.cdc.gov/exposurereport/. The CDC looked for 116 chemicals to which people in the U.S. are exposed via pollution or consumer products. The report (referred to as the 2nd report) found positive results for 89 chemicals, including PCBs (polychlorinatedbiphenyls), dioxins, phthalates, selected organophosphate pesticides, herbicides, pest repellents, disinfectants, and so on. The complete report is available in hardcopy or on the Web site noted above for the CDC. It is over 250 pages long. The CDC plans to issue a new report every two years, expanding the number of chemicals tested for.

The CDC released its 3rd National Report on Chemical Body Burden July 21, 2005. Look it up and compare it with the second report. As soon as the 4th report comes out, repeat, and so on. For those interested in this subject from the point of view of trying to minimize their body burden: http://www.beyondpesticides.org and the Web site of Pesticide Action Network North America, http://www.panna.org are helpful.

The tests done by the EWG and CDC mentioned above, imply that you likely have a large number of synthetic chemicals in your body known to be correlated with cancer and/or endocrine disruption and/or reduced intelligence and/or teratogenic effects and/or other maladies. Let me make this less abstract by pointing out one long-term development.

A survey of the scientific literature done in [103, 93] indicated that over the last half-century (in much of the world, including the U.S. and Europe) human sperm counts have been falling, breast and testicular cancer have been rising and human breast milk is contaminated.[3] At the time these books were published in the mid to late 90s (a survey of the media[4] at that time—which you can do—reveals that) a media campaign denounced as "junk science" the research upon which [103, 93] are based. There are likely many causes of breast cancer, but one wonders why the reported increases. A class of chemicals, organochlorines, is statistically implicated as among the possible causes (not proven, of course), see [307, p. 213], [103, 93]. Project Censored, in [533], rated as the number two censored story of 1999: "Chemical Corporations Profit Off Breast Cancer." This article points out, as do [93, 307], that some of the corporations that manufacture organochlorines also manufacture drugs to treat cancer, breast cancer, in particular. Also pointed out: some of these

[3]From [93, page 54], in October 1993 testimony before the House Committee on Energy and Commerce, we read: "...The breast milk of many American women has higher levels of DDT and its metabolites than allowed in cow's milk by the FDA; cow's milk contaminated at similar levels would be seized as adulterated and banned from interstate commerce." Despite such contamination doctors still recommend breast feeding.

[4]See also [533, p. 36].

same corporations created and continue to fund and promote "Breast Cancer Awareness Month" each October. Races for the cure, early detection and other worthy sentiments abound. But what about searching for and attacking ALL possible causes of breast cancer. One word is conspicuoulsy not in evidence: prevention.

One might think that in a country raised on Ben Franklin's adage: An ounce of prevention is worth a pound of cure; that people would be asking: When are we going to march/race for the *prevention* of breast cancer (as well as the cure)?[5]

Mark Schapiro points out in [602], among many things, that everyday products in the United States contain toxins that have been banned for some time in the European Union. Among those everyday products are cosmetics. Cosmetic industry critic, Stacy Malkan asks, in [439]: Why do companies market themselves as pink ribbon leaders in the fight against breast cancer, yet use hormone-disrupting and carcinogenic chemicals that may contribute to that very disease? (For those who use cosmetics and personal care products the following Web site of the Environmental Working Group should be of interest: www.cosmeticsdatabase.com.) Why do products marketed to men and women of childbearing age contain chemicals linked to birth defects and infertility?

Finally, I call your attention to [178]; the dark side of the politics of cancer involves some of the groups dedicated to fighting it.[6]

Exercise 4.1 Sperm Counts are Dropping. Breast and Testicular Cancer Rates are Rising One of several relevant references is [604].

(i) A report[7] found that sperm counts have fallen an average of 1.5% per year since the 1930s in the United States. Twice this rate of decline was found in Europe. What would be the total percentage drop from 1930 to 2000 in the U.S.? In Europe?[8]

---

[5]I want to dedicate this discussion of breast cancer to the late journalist, Molly Ivins, who died from breast cancer on January 31, 2007. She was one of very few journalists (perhaps the only one in the mainstream) to write about this subject as presented here, cf., her syndicated article which appeared in my local paper on October 25, 1997. Her work also taught me a great deal about how economics and politics really work. She is sorely missed.

[6]On August 26, 2000, an Associated Press article by Charley Gillespie pointed out that former American Cancer Society executive, Daniel Wiant, 35, pleaded guilty to embezzling nearly $8 million from the charity.

[7]On November 24, 1997, newspapers reported on the research findings of Dr. Shanna Swan, chief of the reproductive epidemiology section at the California Department of Health and the principle author of the sperm-count report in *Environmental Health Perspectives*, a monthly health journal of the National Institute of Environmental Health Sciences, a branch of the National Institutes of Health (NIH). The report found that sperm counts have fallen an average of 1.5% per year since the 1930s in the United States. Twice this rate of decline was found in Europe. Results were not conclusive in the rest of the world. Dr. Swan serves on the National Academy of Science's panel looking into hormonally active agents in the environment. She says that falling sperm counts in the last half-century are associated with higher rates of certain types of cancer of the male reproductive system and a growing number of unusual birth defects. Data on the decline in sperm counts can be found in *Vital Signs*, published by the WorldWatch Institute, 1999, pages 148–9.

[8]Hint: If you start with a quantity X in year one and it decreases 1.5% in one year, then in year two you have $X * (1 - .015) = X * .985$. In year three you have $X * .985 * .985$.

(ii) Saying that the average rate of decline of sperm counts is 1.5% per year is different from actually listing the rate of decline for each year. Take a five year period, for simplicity. Find a list of five rates of decline not all equal to 1.5% (one for each of five years) whose average is 1.5%. Compute the total decline over five years given your list of annual declines. Compute the total decline over five years if the decline in each of the five years was exactly 1.5%. Are these two totals different? Can they be? The point of this exercise is to find out how the cumulative decline in sperm counts depends on the actual list of yearly declines, as opposed to the average rate of decline.[9]

(iii) From [93, page 12] we read: "From 1940 to 1980, breast cancer rates increased by an average of only 1.2% each year. But more recently, rates have skyrocketed, according to the ACS[10]. Since 1980, the rate of diagnosis in women has increased about 2 percent a year, reaching a level of about 108 per 100,000 (*Cancer Facts and Figures, ACS, 1994*). From 1980 to 1987 alone, the number of breast cancer cases reported in the U.S. rose by 32 percent."[11]

In [103, page 182] we read: "Fifty years ago, a woman ran a one in twenty risk of getting breast cancer. One in eight women in the United States today will get breast cancer in her lifetime."

Are the two statements from [93, page 12] and [103, page 182] *consistent*?[12]

(iv) Perhaps more subtle than reproductive cancers is the topic of multiple chemical sensitivity (MCS), the existence of which is debated in the medical community. Some authorities say the MCS does not exist, other experts say that MCS sufferers are the "canary in the coal mine." What does the phrase "canary in the coal mine" mean, and do you think MCS really exists as a disability? Relevant references are [18], [567].

(v) Synergy is, mathematically speaking, a *nonlinear* phenomenon. Briefly, in a *linear* situation if you double an input you double the output; or if you halve the input you halve the output. In a nonlinear situation it can happen that a doubling, say, of a small input can have an enormous effect on the output. A list of a few known negative synergies to avoid among drugs and foods is in [251]. Can you find examples of synergy in your life or on the Web?

(vi) Does lipstick sold in the United States contain lead at the time you read this?

Scandal Revealed. I end this section with an update on Erin Brockovich, who fought a battle against chromium-6 water pollution in California similar to the battle of Lois Gibbs at Love Canal, [222, 223].[13] While it is to be expected that there are public relations efforts to carry out disinformation campaigns through the popular media, it is yet another level to tamper with academic and scientific literature. In [715, p. 4] I learned that an influential article downplaying the link between hexavalent chromium water pollution and stomach cancer was exposed as a fraud by the Environmental Working Group, EWG, resulting in the article's public retraction from a prestigious

[9] What is the biggest possible difference between these two answers? This is a pure math question for extra credit.

[10] American Cancer Society.

[11] Could any of this dramatic increase be attributed to better diagnosis, detection, and/or reporting? Can all of the increase be so attributed?

[12] Two statements are consistent if they can both be (simultaneously) true—without contradicting each other. Fuzzy (measured) logic allows a little more wiggle room here than sharp, Aristotelian logic. In this case, however, I am just asking if the two statements are saying roughly the same thing numerically in two different ways.

[13] If you are totally unfamiliar with the story of Erin Brockovich, there is documentary movie, *Erin Brockovich*, about the case available on DVD, (2000).

medical journal, *The Journal of Occupational and Environmental Medicine*, in July 2006. This now retracted article was used by the EPA and separately by California public health authorities to weaken pollution standards regarding chromium-6. By the way, do you think this scandal is unique?

4.3 Synthetic Flows and Assumptions

Now that it is established that our bodies are probably polluted, i.e., the box consisting of humans contains stocks of many different toxic, synthetic chemicals, what are the flows? Where did this stuff come from, and how did it get in my body? Who made the decisions that resulted in widespread human contamination? What criteria were (are) used in making these decisions?

Synthetic Flows. Pick any synthetic chemical: people created it; people marketed/advertised it; and people bought it—usually to do just one thing, most likely, thought to be beneficial. Kill some insects with DDT. Kill some dandelions in a lawn with 2,4-D. (Note 2,4-D is short for 2,4-dichlorophenoxyacetic acid.) Kill weeds on farms in much of the U.S. with glyphosate and/or atrazine (plus secret inert ingredients). Kill some fungus with a fungicide. Make a PVC pipe with vinyl chloride, or use vinyl chloride as a propellant in an aerosol can of hair spray (until this use was banned in 1974). Make a plastic water bottle with phthalates. Improve engine performance with tetraethyl lead, cf., Section 4.5. Make a safer more efficient refrigerator with Freon, cf., Section 4.5. Increase agricultural production with a chemical fertilizer like ammonium nitrate, cf., [81]. Cure an illness with .... This list could go on for thousands of pages. I could also introduce slightly involved mathematics (diffusion processes) that help(s) explain how pollutants ended up in isolated South Pacific islanders, Arctic Innuits, as well as you and me. (You will be able to understand some of the details of how this works after we have studied spreadsheets in V. But briefly for now, if something is put into the environment that lasts a long time, it is bound to end up everywhere.)

Recall the Principle of Unintended Consequences, page 48, which says that despite the intention to do just one thing, e.g., kill dandelions, one cannot. Since everything is connected to everything else, doing any one thing has a ripple effect throughout Nature. There are *unintended consequences* of any act.

Thus the introduction of a synthetic chemical into the environment to do one thing, will undoubtedly do other things. It would seem prudent to find out and understand what some of these other things are. For example, if a chemical is invented to kill one form of life, what is its effect on other forms of life—like you and me?!

According to [627], the children of baby boomers have higher rates of birth defects, asthma, cancer, autism and other serious illnesses than previous generations and [627] links these problems to various pollutants. A report in *onearth*, Volume 27, Number 4, Winter 2006, titled: "Hundreds of Man-Made Chemicals—In Our Air, Our Water, and Our Food—Could be Damaging the Most Basic Building Blocks of Human Development," states its thesis clearly in the title, cf., www.onearth.org. For example, humans' greatest advantage, intelligence, may be in the process of being lowered by our synthetic soup. The closest thing to a proof of this fact that I know, with respect to agricultural pesticides, is the remarkable study of anthropologist Elizabeth A. Gillette comparing two groups of preschool children, one exposed the other not. The group exposed to pesticides over time showed lower mental and motor skills and increased aggressiveness. See Gillette, Elizabeth A., Meza, Maria Mercedes, Aguilar, Maria Guadalupe, "An anthropological approach to the evaluation of preschool children exposed to pesticides in Mexico," *Environmental Health Perspectives*, v. 106, no. 6, June 1998, pp. 347–53.

Of course, the theses of the these and other similar studies are not universally shared, the contrary view being that there really is no problem, or any problem of significance. Actually, given the *variability* of living systems, humans in particular, it is quite possible that what is a truly incapacitating problem for one person is (at least in the short-term) ignorable for another. Although some pollution is unavoidable, for those who would like to minimze their exposure the following references might be of interest, [729, 18, 382, 316, 471, 385, 481, 604, 702, 284, 143, 119, 120, 567, 566, 565].

Hidden Assumptions. The next two exercises might challenge some assumptions commonly held.

Exercise 4.2 Less Poison Can Be More Deadly than More, Endocrine Disruptors in the Rain. It is a common assumption, justified by more than one rigorous, scientific observation, that biocides are more deadly the larger the dose. However, it appears that there are cases where this assumption does not hold. (See *Science News*, July 10, 2004, Vol. 166, No. 2, page 20, for a report of experiments done by Sara Storrs, University of Missouri, Columbia, and Joseph Kiesecker, formerly of Pennsylvania State University, State College. Their experiment is written up in the July, 2004, issue of *Environmental Health Perspectives.*)

The most commonly used herbicide in the U.S., as I write, is atrazine (glyphosate may become number one). In fact, some rain water has been shown to often contain atrazine; and atrazine occurs in some domestic drinking water. (In fact, the EPA has set limits on the maximum level of atrazine in drinking water, see part (i) below.) Somewhat surprisingly to many, an experiment showed that atrazine is more likely to kill developing amphibians (frogs and toads, for example) when it is highly diluted than when much more concentrated—at least in aquatic environments.

In the experiment more than 800 toad and frog embryos and tadpoles were left to grow for about a month in four aquatic environments: (1) no atrazine; (2) 3 ppb (parts per billion) atrazine; (3) 25 ppb atrazine; and (4) 65 ppb atrazine. In six of seven cases, premature death occurred more frequently among tadpoles exposed to 3 ppb atrazine than among those not exposed to the chemical. The death rates in the 25 and 65 ppb environments were less than the death rate in the 3 ppb environment, but greater than the death rate in pure water.

In other experiments this effect was also observed for herbicides mecoprop and dicamba. (See *Science News*, October 12, 2002 page 228.)

(i) In [385, p. 227] it is stated that the maximum contaminant level of atrazine allowed in drinking water in the U.S. is .003 mg/liter. Recall: A liter of water is 1000 cc (cubic centimeters). One cc of pure water has a mass of 1 g (one gram). The symbol mg stands for one milligram, or one one-thousandth of a gram. The .003 mg/liter standard is from the EPA (Environmental Protection Agency) Region 5, Water Division, 1-30-90, and is still in effect as I write. Is this the same concentration as the 3 ppb used in the experiment described above?

(ii) Can you give an explanation for this "less is more lethal" phenomenon? (For those interested in information on atrazine and its effects I highly recommend the research papers and presentations of Professor Tyrone Hayes at the University of California, Berkeley. In other work done by Hayes and reported in the October 21, 2006 issue of *Science News*, p.270, pesticide exposure, in the form of runoff from agriculture in Salinas Valley, Calif. in this particular study, compromised the immunity of frogs, making them highly susceptible to fungal infections, for example. His work and discoveries continue.)

(iii) Is atrazine an *endocrine* disruptor? (For a project expand this topic into a research paper, there is a great deal to be discovered.)

(iv) Research a list of endocrine disruptors, especially mimics of variants of the hormone estrogen, that you personally are exposed to routinely.

(v) While you are doing (iv) do not forget to check into whether the waste treatment plants that provide part (or all) of your drinking water are removing hormone mimics from the excretions of natural hormones and drugs that are flushed down toilets upstream from you. Has anyone checked the fish in your local water supply for gender modifications due to hormone exposures, for example? If you are drinking bottled water, are there endocrine disruptors in those bottles leaching into your water?

(vi) Find documentation that atrazine has been found in rain water. Can the atrazine in rain water originate from an atrazine application hundreds of miles away? A thousand?

In mathematics we are always on the lookout for the truth value of statements/assumptions, hidden or not.

Exercise 4.3 Popular Beliefs

In [385], the late Marc Lappé presents the following ten statements as popularly believed hypotheses. Note: An hypothesis is a special type of assumption. It is a statement *tentatively* offered for the purpose of investigation. It is in this context a guess why, an explanation of why, some phenomenon is thought to be observed. Before looking up Dr. Lappé's analysis of these hypotheses, assign a truth value between 0 and 1 to each of them.

(i) The body's defenses are adequate (regarding exposure to synthetic toxins).
(ii) Toxic effects not seen will not occur.
(iii) All effects of toxins disappear as doses diminish.
(iv) The fetus develops out of reach of toxic danger.
(v) "Nonreactive" chemicals, like silicone breast implants, lack adverse effects.
(vi) The body's own chemicals are safe.
(vii) Naturally occurring substances cause most cancer.
(viii) If it comes out of the water tap, it's safe to drink.
(ix) The environment is resilient.
(x) The problem of toxins is localized.

4.4 The Flow of Information about Synthetic Flows

When trying to understand anything in Nature, with or without mathematics, you need information/data about that which you are studying. The usefulness of any study is immensely impacted by the quality/accuracy/honesty of the information on which it is based. More often than not, given any particular set of data, it has been my experience that there is a constant struggle between those who wish to know the data and those who try to prevent "others" from knowing the data.

Thanks to the sustained efforts of environmental and labor activists we have to some extent "the right to know" about the toxins that have been and are being released in to our communities[14] and workplaces; the name Tony Mazzocchi merits mention in this regard, cf., [391]. Thus if you do an internet search for "Toxic Release Inventory," (TRI), you will find Web sites such as http://www.epa/gov/TRI/tridata, where "epa" refers to the Environmental Protection Agency (EPA), which hopefully still maintains data bases on the releases of toxic chemicals in the United States. I say hopefully, because every such regulatory agency of the government is subject to the political climate at any given moment in time. For example, the so-called USA PATRIOT ACT of 2001 made it more difficult for citizens to get toxics information; and in late 2006 the U.S. President started the process of closing EPA's network of research libraries across the nation. To follow the current status of issues such as these involving government employees and environmental issues the Public Employees for Environmental Responsibility (PEER), www.peer.org, is an organization worth consulting. It's motto is "Protecting Employees Who Protect Our Environment."

A user friendly Web site for accessing TRI information is www.scorecard.org; all you need is your zip code to get started. It is useful to compare current data with a data set from the past, say, 1991, such as is found in [232], where you will find a wealth of data—county by county across the U.S.—on causes of death, levels of exposure to various toxins in the air, water and other information such as geographically correlated diseases. Mathematics plays an important role here.

Exercise 4.4 Toxic Releases Near Your Home

(i) For a project you might want to investigate the data available from the Toxic Release Inventory concerning the toxic exposure of people in the zip code(s) where you spend most of your time.

(ii) Investigate whether any of the toxins you are exposed to are correlated with any particular diseases or health problems. For a brief, intuitive explanation of *correlation* see Section 5.11.

---

[14] For example, the Emergency Planning and Community Right to Know Act (EPCRA) was enacted by Congress on October 17, 1986, in response to public concerns over the protection of the public from chemical emergencies and dangers.

(iii) How many sources of data are there on toxic releases in the zip code(s) of concern to you?

In the following exercise we look at two points of view of the widely used herbicide 2,4-D. After this exercise we look at one of the hurdles that a critic of 2,4-D had to deal with regarding the flow of information about 2,4-D.

Exercise 4.5 Kill Your Dandelions (and What Else?) with 2,4-D.
Synthesized for military purposes in 1942, the two phenoxy herbicides, 2,4-D and 2,4,5-T, (Note: 2,4,5-T is short for 2,4,5-trichlorophenoxyacetic acid. 2,4,5-T has been outlawed, [673, Chapter 2]; but 2,4-D is still a popular herbicide.) were mixed together and used as Agent Orange[15] between 1962 and 1970 by the U.S. to defoliate rainforests and destroy crops in Vietnam.[16]

The herbicide 2,4-D has become one of the most popular weed killers in agricultural fields and forests, lawns, gardens, and golf courses. Regarding golf courses you may find the following story of interest.

"Two years ago, a new golf-related hazard made it into *The British Medical Journal*: 'golf ball liver.' In Ireland, a 65-year-old retired engineer who played golf every day experienced lethargy and abdominal discomfort, had dark urine and jaundice, and was finally diagnosed with acute hepatitis. The cause was a mystery, until it was discovered that he often licked his golf balls to clean them, a habit that exposed him to the weedkiller used on the greens (in this case 2,4-D). This is not an isolated event, given 'the propensity of golfers to lick their golf balls and the widespread use of weedkiller on golf courses,' in the words of researchers." ... "Also, if you take your shoes off while walking the green, it's probably a good idea to wash your feet and change your socks afterwards." This information is from the July, 1999 entry of the *Wellness Engagement Calendar* prepared by the Editors of the University of California, Berkeley Wellness Letter.

(i) It is not difficult to find respectable scientists, teachers and others minimizing the health threats of chemicals like 2,4-D. Can you find one?

Here is one that I found. In 1992 *The Restless Biosphere*, by Donald R. Eaton pages 145–6 says: "Herbicides are equally important economically, and have proved just as contentious in their usage as insecticides. The best known are the phenoxy herbicides, 2,4-D and 2,4,5-T, the structures of which are shown in .... This type of herbicide was the basis for the infamous Agent Orange used as a defoliant in the Vietnam war. These particular compounds are *not* particularly poisonous, they are about comparable to aspirin in this respect, but the earlier products contained dioxin as an impurity which is *very* poisonous and the reputation of these herbicides has never recovered. There are a number of alternatives on the market these days, a couple of which are shown in .... They are all organic molecules of intermediate complexity, the synthesis of which presents no difficulty to the modern chemical industry." (Emphasis in the original.)

Is 2,4-D comparable to aspirin with regard toxicity? What definition of toxicity are you using and how do you measure it? The EPA re-registered, i.e., approved, of 2,4-D in 2005. The herbicide 2,4-D is one of the most widely used in the world. Is its use associated/correlated with any health problems? Is dioxin still present as an impurity in 2,4-D? Is there a negative synergistic reaction of 2,4-D with any other commonly found substance? Are there any unintended consequences to the use of 2,4-D? Are there methods of agricultural production that do not rely on synthetic herbicides? Is the manufacture of 2,4-D dependent on access to fossil fuels?

---

[15]The barrels containing the mixture of 2,4-D and 2,4,5-T had orange paint identification to distinguish it from other chemicals, such as Agent White, Agent Blue, and so on.

[16]A long battle between Vietnam veterans and the U.S. government over "Agent Orange Syndrome," see [724], is being replayed with the "Gulf War Syndrome," see [302].

(ii) The U.S. Forest Service uses 2,4-D to kill "weeds" on public land. Do the people living near or in these forests have a right not to be sprayed? For a first hand account of one who was sprayed and felt she had a right not to be, see [673, 295].[17]

(iii) Steingraber in [667] briefly discusses health maladies associated with use of 2,4-D. She also recounts some peer reviewed literature. One of particular significance, [667, p. 145], was the occurrence of reproductive tumors in clams exposed to (2,4-D) herbicide runoff from blueberry bogs and herbicide drift from commercial forests in Cobscook Bay, Maine. This is a *control* experiment since otherwise this area is pristine. Note: A *control* experiment is one in which all the variables are under the control of, or at least known to, the investigator, and a particular variable of interest is varied. In this study, 2,4-D was the only synthetic toxin measured in that environment; 2,4-D was the variable being studied.

Is this "experiment" important for evaluating the effects of 2,4-D in the environment? Is such data taken into account in the EPA approval process of 2,4-D?

(iv) Do your own investigation of the toxicity of 2,4-D. In particular, find the *half-life* of 2,4-D.[18] Does the half-life depend on whether it is exposed to air or not? Whether it is in soil or water? Does the estimate of the half-life depend on the source of your information?

(v) Jessie De La Cruz, a farm worker all her life, first woman from the fields to become a United Farm Workers organizer, said (during an informal speech): *"What really got me going was the pesticides. I've seen those children born without limbs. I've seen them die of cancer, of leukemia, because of the pesticides. I've been there at the funerals. I've seen their parents cry. I've cried along with them, and it's something very sad to see."* (Here "pesticides" includes a wide range of chemicals.) Do you believe her? Explain your answer.

A Bad Review. Let's see how Steingraber's book, [667], was professionally greeted. A review of this book appeared in the *New England Journal of Medicine* (11/20/97)—a decidedly uncomplimentary review which the journal subsequently apologized for!

What readers of this review were not initially told is that the reviewer, Jerry Berke, was at the time he wrote the review (is he still?) the director of medicine and toxicology for W.R. Grace & Co., a major chemical company notorious for its (alleged) release of carcinogenic substances into the environment, cf., the next exercise. *The New England Journal of Medicine* identified Berke as having an M.D. and a Master of Public Health degree, but not as an employee of a company whose interests are directly threatened by the thesis of Steingraber's book—namely, that industrial pollution is a major cause of an increase in cancer.

Exercise 4.6 W.R. Grace & Company and the Environment

(i) Read the non-fiction, suspense-filled story of the court case *Anne Anderson, et al., v. W.R. Grace & Co. et al.* as told in Jonathan Harr, *A Civil Action*, Vintage Books, New York, 1995. This book, which was made into a movie of the same name, tells the story of what it was like for citizens in Woburn, Massachusetts to do battle (for their health and/or lives) in court with a large corporation like W.R. Grace & Co, (and Beatrice Foods).

[17]The late Rachel Carson's answer to this question is: "If the Bill of Rights contains no guarantee that a citizen shall be secure against lethal poisons distributed whether by private individuals or by public officials, it is surely only because our forefathers, despite their considerable wisdom and foresight, could conceive of no such problem."

[18]The *half-life* of a substance is the length of time it takes for half of a given quantity to change into something else.

(ii) Go to www.jimhightower.com and search for May 2000, and find the one-page article: "Dying to Help W. R. Grace & Company." For book length accounts of what this company meant for the lives of many people in Libby, Montana see [608], [523].

Deceit and Denial: The WACU Pattern of Behavior. The general public is likely not aware of the extent to which some organizations will go to discredit or suppress the flow of even extremely well-documented information that said organizations determine is not in their economic interest.

Volumes could be devoted to this subject; however, here I bring up one example, viz., the book *Deceit and Denial: The Deadly Politics of Industrial Pollution*, by Gerald Markowitz and David Rosner, [443]. This book specifically deals with the "lead industry" and the "vinyl chloride industry"; both lead and vinyl chloride (VC) are certainly documented to have quite negative health effects on humans, cf., [284]. Why is the involvement of *historians* Markowitz and Rosner pivotal? Historians are uniquely suited to answer the question: *Who* knew *What*, e.g., that "X" was toxic, and *When* did they know it?

Largely due to the determination of one woman, Elaine Ross, who filed a wrongful death suit against the vinyl chloride "industry" on behalf of her husband Dan, who worked in that industry, many tens of thousands of pages of industry documents[19] became available for inspection by attorneys for Ross (Billy Baggett, in particular) and historians. These documents established that the "industry" knew as early as 1959, for example, that 500 ppm of of VC in the workplace would cause "appreciable injury." Not to mention that VC was used in PVC plastic bottles (with the possibility of leaching) and as a propellant in many spray cans for drugs, pesticides, cosmetics (like hairspray), for example. Concentrations of 1000 ppm (VC) were documented in hair salons.[20] I cannot come close to effectively recounting this story. For that you can begin by reading [443] and see if you agree with the following review written by Keith Kloor of Audubon: "*Deceit and Denial* is so muscularly researched that it reads like a scholarly criminal indictment. The authors have marshaled an impressive body of evidence—from archival materials to legal documents—in depicting industry's disregard for worker safety and public health."

What is important for our look at the flow of information, however, is the fact that twenty of the biggest chemical companies in the United States: Dow, Monsanto, Goodrich, Goodyear, Union Carbide and others, hired lawyers who subpoenaed and deposed the five academics who reviewed and then recommended that the University of California Press publish [443]. Professor Philip Scranton of Rutgers University, enlisted by the chemical companies, wrote a 41 page critique of [443] and the ethics of its authors. (Professor Scranton also testified for the asbestos companies in their liability litigation.) Scranton

[19]Some of which are posted at www.chemicalindustryarchives.org/dirtysecrets/vinyl/1.asp

[20]By the way, the current Occupational Safety and Health Administration (OSHA), created in 1971, has currently set a limit of 1 ppm (VC) in the air in the workplace.

charges that Markowitz violated "basic principles of academic integrity, historical accuracy, and professional responsibility" and engaged in "sustained and repeated violations" of the official "Standards" of the American Historical Association. Scranton doesn't claim to be an expert on the postwar chemical industry, but he wrote as an expert on ethics, Markowitz's ethics in particular. But Markowitz is an expert on what is most relevant here: what the chemical companies knew, and when they knew it.[21]

Such legal attacks on the people involved with [443] costs these individuals time (that could be otherwise used) and the effort of finding and hiring legal counsel to defend them from attorneys from fifteen different chemical companies. Reviewers were asked, for example, if they had checked all 1,200 footnotes, many referring to multiple sources, a task for which the authors alone are traditionally held responsible. In fact, attorneys for PBS (Public Broadcasting System) and HBO (Home Box Office) had already checked the entire manuscript, footnotes included. Why? Bill Moyers ran a PBS documentary, "Trade Secrets," on March 26, 2001 (available 1-800-336-1917) based on the research in [443]; and HBO ran a documentary, "Blue Vinyl," in 2002 based on the same research—both documentaries quite worth anyone's time.

One of the most potent observations of Markowitz and Rosner, however, is to be found on p. 300, [443]:

"The history of the lead and vinyl industries gives us a window into why the relationship between industry and the public is so strained today. These industries responded to potent evidence of the danger of their products by hiding information, controlling research, continuing to market their products as safe when they were known to be dangerous, enlisting industrywide groups to participate in denying that there was a problem, and attempting to influence the political process in order to avoid regulation. There are those who find the actions of the lead and vinyl industries so egregious as to constitute a subversion of democracy. They believe that by promoting secrecy, interfering with scientific research and thereby inhibiting the free exchange of ideas, by buying the loyalty of elected officials with donations to political action committees and with soft money contributions, by threatening economic abandonment and unemployment if communities insist upon safety and health regulations, these industries posed a serious threat to political democracy in the United States."

"The question is this: How representative are lead and vinyl of general corporate behavior?"

Later on, see pages 300–301, [443]:

"As with asbestos and tobacco, the lead and vinyl industries knew of dangers from their products but chose to ignore or conceal them. In fact, they actively deceived the public about the safety of their products. While we may not yet know the actions of all industries with regard to industrial toxins, by now we do know that at least four or more major industries engaged in very similar activities to keep information from the public and to prevent regulation of products that they knew to be dangerous."

---

[21] For a fuller account see "Cancer, Chemicals and History," by Jon Wiener for *the Nation*, posted at; http://www.thenation.com/doc.mhtml?i=20050207&s=wiener.

I spent much of my life with advertisements extolling the virtues of smoking, e.g., "Now ... Scientific Evidence on Effects of Smoking! Much Milder Chesterfield is Best for You," (with a picture of Arthur Godfrey, a celebrity of the day, and his signature on a testimonial); "According to a recent Nationwide survey, More Doctors Smoke Camels Than Any Other Cigarette," (complete with a picture of man in a white coat and stethoscope who looks very much like a doctor); "Scientific tests prove Lucky Strike milder than any other principal brand!" (with a signed picture and testimonial of a celebrity of the day, Rex Harrison: "I smoke Luckies—they're mild and smooth." As the evidence of ill effects mounted over the years, so did the denials from the tobacco industry. Lucky for me, I did not believe the tobacco companies.

Exercise 4.7 A Corporate Pattern of Behavior: WACU

(i) I refer to the *pattern of corporate behavior* associated with tobacco, lead, vinyl chloride and asbestos, outlined in the above quotes from Markowitz and Rosner as the **WACU** Pattern of Behavior, i.e., "We Are the Center of The Universe" Pattern of Behavior. The following question takes up the challenge of Markowitz and Rosner: Is the WACU Pattern of Behavior representative of corporate behavior for the majority of corporations? For the majority of large corporations? This question is difficult to answer in either the affirmative or negative, since most corporations are not transparent; do not allow inspection.

(ii) Although the financial industry (in the U.S. and elsewhere) does not produce a chemical product, has it exhibited the WACU Pattern of Behavior?

(iii) Does the health insurance industry exhibit the WACU Pattern of Behavior? (See statements for the record of Wendell Potter, and his book, *Deadly Spin*, [549]. In particular, what was the health industry reponse to Michael Moore's documentary, *Sicko*?) Did the energy company ENRON exhibit the WACU Pattern of Behavior? (See Chapter 9.) How about WorldCom?

(iv) Can you find any large corporation that you can convincingly demonstrate does not exhibit the WACU Pattern of Behavior? How many such can you find?

(v) Look up advertisements from the tobacco industry and statements regarding the safety of cigarettes from the last hundred years. Can you estimate a time when you believe the tobacco industry started lying to the public?

(vi) Are there any characteristics of a corporation that might naturally lead it to exhibit the WACU Pattern of Behavior? Size? Lack of regulation, effective feedback from society? What social institutions, processes, activities, or structures might minimize the WACU Pattern of Behavior on the part of corporations?

(vii) Do you think that the experience of Markowitz and Rosner, and their associates, including the publisher, has any effect on future authors contemplating being critical of powerful industries?

(viii) What is the relationship of the WACU Pattern of Behavior to the Bio-Copernican Axiom?

Think as Though Your Life Depended on it. The following exercise involves being critical of what you read in the papers, doing a little thinking; and it ends with a choice.

Exercise 4.8 Love Canal and Other Things: How Safe?

(i) From two articles by Jane E. Brody in *The New York Times*, Science Times, July 13, 2004 pages D5 and D7 we read the following quote from Dr. Robert L. Brent, a distinguished professor at Thomas Jefferson Medical College in Philadelphia who has been studying environmental toxicology for nearly 50 years, specializing in the effects of environmental factors like radiation, drugs and chemicals on the developing embryo and child: "Love Canal[22] was an example of a terrible environmental problem that should be cleaned up, but there was no evidence of risk to the people who lived there. Many fears are irrational."

Lois Marie Gibbs, a then resident of Love Canal, would likely give the part of the above quote following the comma a truth value of 0, [222, 223]. What truth value do you give Dr. Brent's statement? Investigate as thoroughly as you can.

(ii) Bjørn Lomberg would more likely agree with Dr. Brent, [408].[23] Does this change or confirm your answer to (i). You might peruse some comments of the many scientists who have criticized [408]. See, for example, [610, 538]. Lomborg defends himself and returns criticism at his Web site www.lomborg.com.

(iii) Jane E. Brody, in one of the two articles mentioned in (i), says: "Many parents worry that their children may be harmed by exposure to environmental factors they cannot avoid or control, including pesticide residues on fruits and vegetables, approved food additives, chlorinated drinking water and hormones in milk."

"They fear electromagnetic fields as a cause of childhood leukemia, a mercury preservative in vaccines as a cause of autism,[24] alar,[25] a growth stimulant on apples, as a cause of cancer."

"None of these are actual hazards. But even if they were, they are hardly the main threats to the health and lives of fetuses, infants, children and adolescents, says Dr. Robert L. Brent..."

J. Brody then goes on to discuss what she refers to as main threats: Sudden Infant Death Syndrome, falls, vehicular accidents, burns, poisoning, drowning, choking, guns, electrocution, secondhand smoke, sunburn, sports injuries, power tools, and obesity.

Discuss the extent to which Brody is comparing longer term, abstract-mathematical death to short-term, real-time death.

Pick at least one of the topics Brody mentions in her quote above and make your own assessment of risk, based on some study. If you pick pesticides, you might consider [481, 382, 702].

(iv) You can assess your level of belief in the present safety (or risk) of living at Love Canal. Houses are again being sold in the neighborhood. Would you buy?

---

[22]Love Canal, in New York state, was used as a toxic dumping ground in the 1940s and 1950s by Hooker Chemical Co. In 1978 the residents (then living on this dump site) became aware of the situation, and a long battle ensued. Lois Marie Gibbs, a then resident of Love Canal, led citizen protests and authored [222, 223].

[23]Lomberg's book has many, many references and footnotes. Do such guarantee the truth?

[24]Eli Lilly, inventor of thimerosal (a mercury preservative in vaccines), was granted protection from lawsuits by parents of autistic children under a short-lived provision slipped into the Homeland Security Act in November 2002, [207]. Relevant to this discussion, an article by Robert R. Kennedy, Jr., details how the American government rushed to conceal data and prevent parents from suing drug companies, cf., "Deadly Immunity," *The Progressive Populist*, July 15, 2005, and Salon.com. Also relevant, the British Journal *The Lancet* recently retracted the controversial 1998 study that first set off widespread fears about a vaccine autism link because the editors found serious flaws and false claims. It was also discovered that the lead author had been paid by a lawyer suing the vaccine makers, see *U.C. Berkeley Wellness Letter*, June 2010. As was pointed out in a letter to the editor, *In These Times*, August 9, 2004, as thimerosal is phased out of vaccines it may become evident whether or not the autism-thimerosal correlation is with or without a cause. Of course, even if mercury is a cause of autism, there are other sources of mercury pollution, e.g., burning coal. And there are likely multiple causes of autism, cf., page 80.

[25]From www.nrdc.org, the Web site of the Natural Resources Defense Council, which led a campaign against alar: "Alar was a pesticide used on apples. Uniroyal withdrew it from the market following a *60 Minutes* story viewed by 40 million people on the health dangers it posed. The story was based on an NRDC study called *Intolerable Risk: Pesticides in Our Children's Food.* Alar was later banned by the EPA."

Regulation and Enforcement Are Feedback. There are a number of agencies of the Federal Government that were created to *regulate* some aspects of society, such as the Securities and Exchange Commission (SEC), Environmental Protection Agency (EPA), and the Food and Drug Administration (FDA). One important fact to remember is that any such government regulatory body was created, usually over stiff opposition, to address some compelling need of society. Thereafter, a "game" ensues (amenable in part to mathematical analysis) where those who are regulated do their best to "neutralize" or "capture" the agency in question, and various groups of citizens fight to prevent that from happening. Thus given any such agency, there is invariably a long and interesting story involving the history and contemporary politics of the struggle between these opposing forces. I can only hint at these struggles and suggest that the reader pursue their study as projects. I will, however, very briefly discuss the FDA, since it was the first citizen-protection agency of the federal government. The FDA's modern regulatory functions began with the passage of the 1906 Pure Food and Drug Act.

The FDA is relatively small for the tasks it faces, and it does not carry out tests of its own, relying primarily on studies carried out by the industries it regulates. One success story is its regulation of thalidomide, the dangers of which were missed by many other countries. For this story and general history of the FDA see [309].

On the other hand, the FDA has its critics. See www.citizen.org/hrg, for example, an organization inspired by Ralph Nader, which is one of the players struggling against the aforementioned "neutralization." A famous critique of the FDA is the 1970 book, *The Chemical Feast: The Ralph Nader Study Group Report on Food Protection and the Food and Drug Administration* by James S. Turner.

The following exercise is one among many possible exercises dealing with controversial decisions of the FDA. The FDA "changed its mind" about the artificial sweetener *aspartame* and in 1983 approved it for use in liquids, e.g., diet sodas. We cannot herein devote the space necessary to evaluate this decision; however, we can raise some issues the reader may or may not be aware of.

Exercise 4.9 Aspartame: Pro or Con

(i) An advertisement from Tufts University's Friedman School of Nutrition Science for its health and nutrition publication, the Tufts Letter, begins with "If you've been told ..." It then gives a list of 10 statements one of which is "aspartame sweetener is dangerous." The advertisement then asserts that "you might be interested to learn that ALL are FALSE" (referring to the 10 statements). Thus the advertisement asserts that aspartame is not dangerous. Nowhere in the ad does it discuss PKU, i.e., phenylketonuria. Of course, if you have this disease, you no doubt have read the following warning on any can of soda sold in the U.S. that contains aspartame: "Phenylketonurics: Contains Phenylalanine." Why the warning?

(ii) We learn interesting aspects of the history of the FDA approval of aspartame from Chapter 4 of [95], a book about Donald Rumsfeld, at one time the Secretary of Defense. Rumsfeld was the CEO of G.D. Searle & Company (owner of aspartame) from 1977 to 1985. It was during this period that Rumsfeld guided aspartame through the FDA approval process. What qualifications did Rumsfeld have for this task? In April 1981 Arthur Hull Hayes was selected to be FDA commissioner, replacing previous commissioner Jere Goyan, who had just been fired by an administration transition team (a rare occurrence). On July 18, 1981, Hayes approved aspartame for use as a sweetener in solid foods. Aspartame was cleared for use in liquids, e.g., diet sodas, in July 1983. Soon thereafter, Hayes resigned from the FDA and accepted a consultancy contract from Searle's PR firm, Burson-Marsteller. Searle was then sold to Monsanto for $2.7 billion. What would have been the value of Searle if aspartame had not been approved? Is there a "revolving door" involved here between industry and government? Note: Revolving door refers to the situation where persons from the industry being regulated by an agency later take positions in that agency and/or persons in a regulatory agency later take positions in an industry being regulated by said agency.

(iii) Some of the history in part (ii) is recounted as well in [235, p. 175]. Chapter 10 of this book discusses Miraculin, an extract of the naturally occurring "Miracle Fruit,"i.e., *Synseplaum dulcificum.* Though petitioned, the FDA would not approve the sweetener Miraculin. Why do you think this is the case? Miracle fruit, native to Africa, is grown in Florida, cf., www.miraclefruitman.com. What very unusual property does the Miracle Fruit have?

(iv) Research the pros and cons of aspartame as deeply as you have time for. Do you eat aspartame?

A Note on Biomonitoring: Some Signs of Hope. Biomonitoring consists of using living organisms to measure various aspects of the environment, such as the presence and quantitative level of toxins. In one such biomonitoring study called "Mussel Watch,"[26] begun in 1986 and ongoing, NOAA[27] has found that restrictions on some toxins have helped reduce their levels in the environment. Mussels and oysters are filter feeders and concentrate assorted toxins which are present in their aquatic environment. Among the banned or restricted toxins monitored are PCBs (Polychlorinated Biphenyls), chlorinated hydrocarbons and cadmium. A recent review of 17 chemicals at 246 different sites showed 108 increased concentrations, 830 decreased concentrations. Thus, regulations have had a measurable and positive effect.

4.5 You Cannot Do Just One Thing: Two Examples

Fluorine: Atomic Bombs and Teeth. There is a fuzzy boundary between "natural" chemicals and those new, human-made, i.e., synthetic ones not previously found in Nature. For example, fluorine, F, is one of the elements. See F on page 253 in the periodic table: one of Nature's basic building blocks. But fluorine reacts immediately and strongly with almost everything, and hence

[26] This is the longest continuous contaminant monitoring program in United States coastal waters. Over 250 sites are monitored including Atlantic, Pacific and Gulf coasts, the Great Lakes, Alaska, Hawaii and Puerto Rico. Recent results have been published in the journal *Marine Environmental Research.*

[27] The National Oceanic and Atmospheric Administration

is not normally found on Earth as a pure element. During World War II, however, large quantities of pure fluorine were produced as one step in the production of atomic weapons. Some chemicals, called fluorides, are added to toothpaste and some municipal drinking water, but not without controversy, however, see [62]. I will soon have more to say about F.

Putting the Lead in, then Taking It Out: The Story of Leaded Gasoline. Lead, Pb, is another element in the periodic table. For eons lead was largely held separate from most living things by rock formations in the Earth's crust. Humans have brought Pb in contact with living beings for thousands of years by mining. Lead is or has been used in pipes, radiation shielding, bullets, paint, and many other things including ethyl gasoline. We have also known lead is a poison for hundreds of, if not a couple thousand, years of human history.

On December 9, 1921 the brilliant inventor, Thomas Midgley, Jr., found what he had been seeking for a decade. At a ratio of 1:1,300 mixture of lead to gasoline, car engines showed increased compression, greater fuel economy, 25% increase in horsepower, and most of all the annoying "knock" was eliminated. In addition, lower grades of oil could be used for gasoline, since performance was enhanced by the almost magical gasoline additive known (to be precise) as tetraethyl lead, [703, page 119][28]

Charles Kettering,[29] Midgley's boss, went to his boss, Pierre Du Pont, the president and chairman of the board of General Motors, with the good news. Presumably since lead was (is) a known poison, the new additive was named "ethyl." General Motors and the Standard Oil Company of New Jersey formed the Ethyl Corporation with Kettering as president and Midgley as vice president and general manager. Du Pont, the chemical corporation, got the contract to provide the tetraethyl lead. The deaths of workers in the tetraethyl plant momentarily set back the public relations for ethyl; however, using their power with politicians, government, communications, and universities[30] ethyl gas was sold in the United States until a brief surge of politically effective environmental concern swept the United States in the 1960s and 1970s. At that time a number of environmental laws were passed, including the regulation of lead in gasoline.

The fascinating history of how lead got in gasoline is found in [66, 703, 443]. For some details of the science and politics that got the lead out of gas, see [563, 703, 443]. Pay particular attention to the role of citizen activists and

[28] The book [703] gives an in-depth discussion of occupational, pediatric, and environmental lead exposure/poisoning.

[29] The man after whom the Sloan-Kettering Research Institute is named.

[30] Notably Robert A. Kehoe, Director of the Kettering Laboratory of Applied Physiology at the University of Cincinnati, defended the safety of lead in gasoline for decades. Kehoe similarly defended fluoride on behalf of a group of corporations that included Du Pont, Alcoa, and U.S. Steel, all of which faced lawsuits for industrial fluoride pollution. Professor Yandell Henderson of Yale led academic opposition to lead in gasoline, but lost the battle.

scientist Herbert Needleman who risked his career when he published his research on the observed effects of lead pollution on children.[31] Science and citizen action were pivotal in getting the lead out of gas.

Exercise 4.10 The Lead Now in Your Body and the Environment. From [563, page 98] we read:

"Even today, however, the average North American carries between 100 and 500 times as much lead in his or her blood as our preindustrial ancestors. In cities where there has been a high density of automobile traffic, adults have blood levels of about 20 to 25 [micro]grams per deciliter—roughly half the level at which lead exposure leads to impairment of peripheral nerves. No other toxic chemical has accumulated in humans to average levels that are this close to the threshold for overt chemical poisoning. . . .

What we do know is that the lead industry continues to lobby, even today, against measures such as an excise tax on lead that would discourage its use and generate funds to help clean up its toxic legacy. Cleanup is needed because some three million tons of lead remain on the walls of homes that were built and painted prior to 1970. Another five million tons is found in the soil near busy roadways."

From [563, page 93] we read:

"In fact, even Henderson's warning turns out to be a gross underestimate. By the mid-1970s, 90 percent of the gasoline used for automobiles in the United States was formulated with ethyl. During the 60 years that leaded gasoline was used in the United States, some 30 million *tons* of lead was released from automobile exhausts. 'When many cars were getting just ten miles to a gallon in stop-and-go traffic, a busy intersection might have gotten as much as four or five tons of lead dumped on it in a year,' notes Howard Mielke, an environmental toxicologist and lead expert at the College of Pharmacy at Xavier University of Louisiana, in New Orleans. 'That's roughly equal to having a lead smelter at every major intersection in the United States. As a result, there is a very, very large reservoir of lead in soil.' "

(i) By doing your own estimates verify or debunk the statement that four or five tons of lead could be (have been) dumped at a busy intersection in a year.[32]

(ii) By doing your own estimates verify or debunk the statement that another five million tons (of lead) is found in the soil near busy roadways.[33]

(iii) What form does lead take when it comes out of the exhaust pipe of a car? How biologically active is it in this form? How would it enter a mammal's body?

(iv) Where is all that lead now?[34]

(v) Are there any commercially edible plants that take up lead? For example, see [728] and [731].

---

[31] Needleman found a correlation: for every 10 parts per million increase of lead in a child's (baby) tooth there was a two-point drop in IQ. With the help of others, including some honest scientists and professors, Needleman survived and eventually "won" a battle with other professors and scientists (supported by industry money) and with the public relations firm of Hill & Knowlton.

[32] The ratio 1:1,300 of lead to gas should be helpful. You now need to estimate the amount of gas.

[33] You will need to estimate how much gas is involved. You may have to look up some relevant data.

[34] For example, if there has been 30 million tons of lead released from auto exhaust and there are 5 million tons along the busy roadways and five tons per year at busy intersections, does that mean that *all* the lead is piled up along roads and at "several" busy intersections.? Has lead been found in the Arctic that came from auto exhaust? Is there lead in the bodies of animals (including humans)? See also [493].

(vi) If you know what a deciliter is (see Table Greek Prefixes, page 60), estimate the how much lead is in the blood of Americans. An adult male of "average" weight has about 5 liters of blood.

CFCs: The Story of Refrigeration, Spray Deodorant, and Bug Bombs. In June of 1918, General Motors entered the refrigeration business. They bought the Guardian Frigerator Company and renamed it Frigidaire. In those days the concept of pumping heat from inside an airtight enclosure via a compression and expansion cycle of a "suitable" gas was a novel idea, an alternative to the ice box.[35] Problems arose, however, with the "suitable" gas. Ammonia gas, though efficient and nonflammable, was toxic if breathed in large quantities. Its stinging odor caused consumer complaint whenever even a small leak occurred—fouling the air and spoiling the refrigerator's contents. This is, of course, an improvement over a toxic refrigerant gas that smelled bad and might explode. In the 1920s General Motors switched to methyl chloride, which was efficient and odorless. The main problem was its extreme toxicity. For example, on May 15, 1929, 125 patients and employees at a Cleveland hospital died in an accidental release of methyl chloride fumes.

General Motors' Kettering again called on their inventor, Midgley; and with a stroke of genius he invented a fluorine based compound, dichlorodifluoromethane, or CFC-12, which was ideal for refrigerators. This chlorofluorocarbon was christened Freon, and by 1935 it was the standard refrigerant gas in America's household refrigerators.

Freon was extensively tested and found to be inert. It does not burn or smell; it is nontoxic. It was a brilliant solution to an important problem, and it found applications beyond refrigerators. It was discovered that Freon was an ideal propellant and aerosolizer for insecticides like DDT.[36] Spray cans containing compressed Freon and insecticide, i.e., Bug Bombs, were quite popular and were used extensively by U.S. troops in the Pacific during World War II. Also, spray cans dispensing Freon-propelled deodorant eventually became nearly universal.

The Ozone Layer: Who Needs It? In 1970 I was finishing up my studies at the then new campus of the University of California, Irvine. UCI was small and beautiful with the math department sharing a building with the chemistry department. If you had mentioned the ozone layer to me at the first Earth Day, April 22, 1970, I could have easily recited the problems with the SSTs (SuperSonicTransport)—airplanes traveling faster than the speed of sound through the stratosphere, impacting negatively on on the Earth's protective ozone shield. It was a hotly debated topic.

However, I was completely unaware that the amiable F. Sherwood Rowland, i.e., "Sherry," the director of chemistry at UCI, together with his postdoctoral

[35]Literally an ice box was a box wherein a block of ice was periodically placed to keep the interior of the box cold.

[36]DDT is short for dichlorodiphenyltrichloroethane.

assistant, Mario Molina, would in June of 1974 publish a theoretical atmospheric chemistry paper of unparalleled importance. Their paper predicted that CFCs, though inert in the lower atmosphere, broke apart when exposed to ultraviolet light in the upper atmosphere—liberating chlorine atoms that efficiently[37] destroyed O_3, ozone. This ozone shield protects life on Earth from excessive ultraviolet radiation and its associated deleterious effects.[38]

This theory had its critics in industry for obvious financial reasons as well as among scientists. Most notable was critic James Lovelock, of Gaia hypothesis fame, who had published a paper in *Nature* in 1973 reporting that he had measured CFCs virtually everywhere in the atmosphere with an instrument that he had invented. Lovelock denounced Rowland and Molina's theory, and he also happened to have received research funding from the Manufacturing Chemists Association, cf., [66, p. 201].

Nevertheless, the U.S. banned CFCs in spray cans in 1978 over protests that it would put lots of people out of work. (It did not.) The production of CFCs was far from being banned, however. For example, U.S. Interior Secretary, Donald Hodel, was famously quoted as saying that damage to the ozone layer would be no problem if people would just wear broad-brimmed hats and sunglasses when they went outside.

Experimental evidence of CFC destruction of the ozone layer started pouring in, such as the discovery of the "ozone hole" over Antarctica; but politicians did not act. Susan Solomon, a NOAA[39] scientist in Boulder, Colorado, headed up the National Ozone Expedition to Antarctica in August 1986. Combining her knowledge of atmospheric physics and chemistry, she proposed a theory which explained the precise mechanism by which CFCs could lead to the dramatic thinning of Antarctic ozone. Still no CFC ban was forthcoming.

The United Nations Environment Program (UNEP), directed by Mustafa Tolba, hosted and prodded the international political process, and in October 1987 in Montreal, Canada, politicians from around the world were shown the "smoking gun graph," see Figure 4.1, of data collected September 16, 1987, on a NASA[40] plane which flew from Punta Arenas, Chile directly toward the South Pole and into the ozone hole. In this graph rises and drops in ozone are almost exactly mirrored by drops and rises in chlorine monoxide, ClO. Other

---

[37] By efficiently I mean that via a "chemical recycling process" just one chlorine atom, Cl, liberated from a CFC molecule by ultraviolet light, can destroy a multitude of ozone molecules; before other processes render that chlorine atom less harmful.

[38] In the July 13, 2004 issue of *The New York Times—Science Times* it is stated that we are in the midst of a reversal of the Earth's magnetic field, i.e, this magnetic field is in the process of collapsing and reappearing with north and south magnetic poles reversed. This could cause further damage to the ozone layer. Additionally, there are other chemicals (synthetic and not) I have not discussed that destroy ozone. You might want to look up what ultraviolet solar radiation at various levels does to living things.

[39] National Oceanic and Atmospheric Administration.

[40] National Aeronautics and Space Administration.

measurements showed that about two thirds of the chlorine in the stratosphere was coming from human-made chemicals.

Although Du Pont had put on hold its search for CFC substitutes when Ronald Reagan was elected in 1980, in 1989 it pledged to phase out CFC production completely.[41] Nations around the world signed the famous Montreal Protocol regulating CFCs. Led by UNEP, representatives from 92 countries met in London in 1990 and agreed to phase out all CFC production by 2000. The ozone layer continues to thin and the effects of CFCs will continue for a long, long[42] time; but it could have been much worse. For a great many more details see [66, 450, 579].

An Argument that Might Sound Good but Is Not Sound. I invite you to test your ability to analyze an argument critically in the following exercise based on [404] which indirectly says that most of the science, data and theory I have discussed above about CFCs and ozone (and much else) is rubbish. At the time I write this, Rush Limbaugh is a popular radio commentator with an apparently large following that believes much of what he says and votes accordingly.[43]

Exercise 4.11 Rush H. Limbaugh, III Says There is Nothing to Worry About. From [404] we read:

"... ozone is *created* by the sun, particularly ultraviolet sunlight. And yet these dunderhead alarmists and prophets of doom want us to believe that because there are occasional reduced levels of ozone over Antarctica (which, incidentally, always rebound to normal levels), our own activity, based purely on our natural behavior and technological advancement, is responsible for what they predict will be the destruction of the ozone layer. Poppycock. Balderdash.

Mount Pinatubo in the Philippines spewed forth more than a thousand times the amount of ozone-depleting chemicals in *one* eruption than all of the fluorocarbons manufactured by wicked, diabolical, and insensitive corporations in history. So much so that respected scientists now say that a 4 percent to 6 percent ozone loss–*could*, but may not–occur over the Northern Hemisphere in the next two or three years. Now, wait–before you think I have just destroyed my own argument, remember this: volcanoes have been doing this for 4 billion years. And guess what? We still have a healthy ozone layer! Isn't it wonderful? Aren't you thrilled? Hmmm. You still don't get it? Read it again, folks. One eruption in 4 billion years of eruptions–a thousand times as destructive as all man-made CFCs–and a temporary maximum loss of 6 percent of the ozone. Conclusion: mankind can't possibly equal the output of even one eruption from Pinatubo, much less 4 billion years' worth of them, so how can we destroy ozone? In other words, Mother Nature has been attacking her own stratospheric ozone for millions of years and yet the ozone is still there, and in sufficient quantities to protect Democrats and environmentalist wackos alike from skin cancer."

(i) There is enough information on the preceding pages to offer a tentative rebuttal to Limbaugh's argument above. Can you create such a rebuttal?[44]

---

[41] Apparently relevant patents were running out as well.

[42] If the atmosphere is a box and CFCs are the stock with flows, what is the residence time of CFCs in the atmosphere?

[43] For two references critical of Mr. Limbaugh see [343, 572].

[44] Just to complicate things I mention that organisms in the world's oceans create enormous quantities of chemicals that can also destroy ozone! Much of what Mr. Limbaugh says is true, so why worry about CFCs? Why then were (legal) CFCs then phased out?

(ii) If you are at all familiar with the pronouncements of Rush Limbaugh, written or verbal, can you find any substantive differences between the arguments/discussions in this (my) book and those of Limbaugh?

(iii) An environmentally concerned acquaintance of mine was concerned about chlorination of our municipal water. He thought that chlorine was not good for citizen health and that using chlorine in the water supply was helping deplete the ozone layer. There might be some positive and negative health effects of water chlorination that one can ponder. However, does chlorination of city water supplies measurably impact the ozone layer? Explain.

(iv) On October 3, 2006, Alex Morales reported the following for *Bloomberg News*. The European Space Agency, ESA, announced that 2006, in terms of the mass of ozone lost, 40 million metric tonnes, set a record for ozone loss over the South Pole. The previous record was 39 million metric tonnes ozone loss in the year 2000, cf., the Web site of the ESA, (data was collected from Envisat, the largest earth observation spacecraft built so far). The loss was in part a result of the lowest Antarctic temperatures recorded since 1979. Would you say humans have solved the ozone thinning problem?[45]

(v) In December 2009 Rush Limbaugh said: "Climate change is a lie and a hoax." Discuss.

Figure 4.1 is from J. G. Anderson, W. H. Brune, and M. J. Proffitt, "Ozone Destruction by Chlorine Radicals within the Antarctic Vortex: The Spatial and Temporal Evolution of ClO-O_3 Anticorrelation Based on in Situ ER-2 Data," *Journal of Geophysical Research* 94 (30 August 1989) 11,475. Copyright 1989 American Geophysical Union; Reproduced by permission of American Geophysical Union.

This graph is often referred to as the "smoking gun graph," since the graph of ozone, which is falling, is nearly the mirror image of the chlorine monoxide graph, which is rising. Thus it was strongly implied that chlorine byproducts resulting from the breakdown of CFCs were responsible for the thinning of the ozone layer over the Antarctic. The data shown above were collected on September 16, 1987 by instruments on NASA's ER-2 research airplane as it flew from Punta Arenas, Chile (53 deg S) to 72 deg S. As the plane flew into the ozone hole over the Antarctic the concentration of chlorine monoxide increased to about 500 times normal levels while ozone levels declined drastically.

[45] On November 4, 2006 an Associated Press article by Rita Beamish announced that the United States lobbied successfully against objections from European nations and the Montreal Protocol treaty's own technical committee to continue the use of methyl bromide, a potent destroyer of the ozone layer. The U.S. not only is continuing to use U.S. stockpiles of 11,000 tons of methyl bromide, it is manufacturing more that 5,000 new tons of the pesticide—despite proof that alternative methods and chemicals can replace methyl bromide.

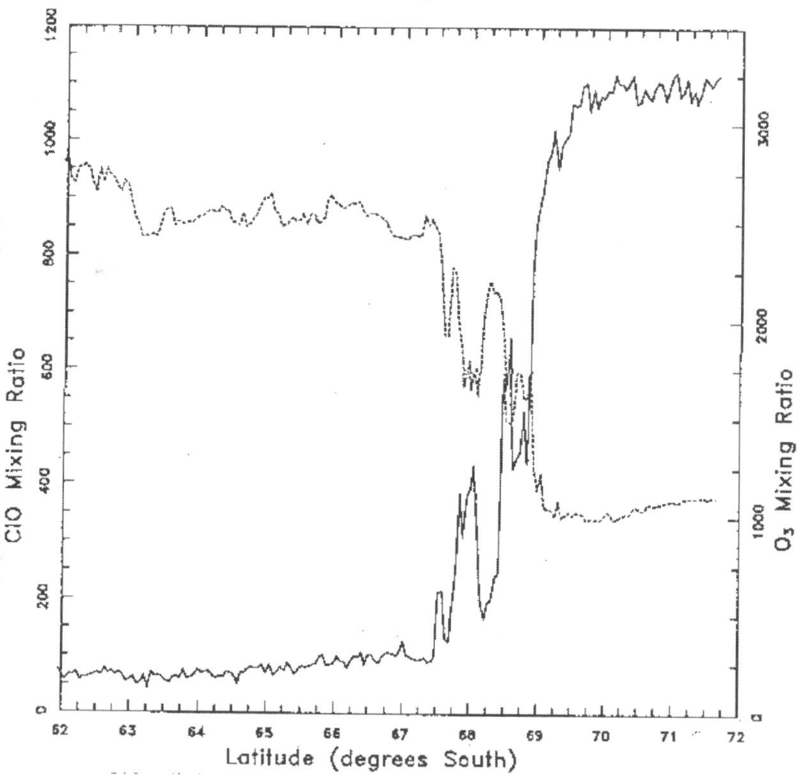

FIGURE 4.1: CFC-Ozone Smoking Gun Graph

Chapter 5

Mathematics: Food, Soil, Water, Air, Free Speech

As I write there are reports of yet another outbreak of *food borne disease* with some hospitalizations and deaths. This time it is *Escherichia coli* O157:H7, a virulent variant of the benign E. coli bacteria that resides in all of our digestive tracts. Extremely rare until 1982, 0157:H7 is not unusual today, [202]. According to a page at www.cdc.gov in 1999, more than 75 million Americans get sick each year from food, of which about 325,000 require hospitalization. At least 5,000 to as many as 9,000 die. What is the situation when you read this? I am not sure how this death rate compares to that of hunter-gatherers going after woolly mammoths, but it is not comforting. Let's do a U.S.A. warm-up exercise.

Exercise 5.1 Food-Borne Disease Roulette

(i) The population of the U.S. on July 1, 1999 was estimated to be 279,040,168. What are the chances of not getting sick from food that year? Hint: Consider $1 - \frac{75000000}{279040168}$. What does this number mean?

(ii) Assuming "the chances" do not change from year to year, what are the chances of going two years in a row without getting sick? Hint: Consider $(1 - \frac{75000000}{279040168})^2$. What are the chances of going a lifetime of 75 years in a row without getting sick from food?

(iii) What are the chances of going a lifetime of 75 years and not requiring hospitalization for food-borne disease?

(iv) What are the chances of going a lifetime of 75 years and not dying from a food-borne disease?

(v) Attempts to remove E. coli O157:H7 from meat involve treating with chemicals and/or irradiation, which likely will result in pathogens resistant to each. What is an emerging consensus as to the cause of the appearance of E. coli O157:H7 in the food supply, and what simple protocol would remove most of it? (Regarding the chemical treatment of ground beef with ammonia, see the front page article by Michael Moss, *The New York Times*, Dec. 31, 2009, pp. A1, A14. In particular I found the term "pink slime" describing processed ground beef interesting.)

(vi) A "typical" beef hamburger contains meat from how many different animals?

5.1 The "Hour Glass" Industrial Agriculture Machine

There are several serious food-borne diseases besides O157:H7, cf., [202]. Such food-borne illness can, for example, end in excruciating death or resign one to a lifetime of kidney dialysis or transplants. Less dramatic, but nevertheless impactful, say in the United States, are "Western diet" related maladies such as obesity, diabetes, cardiac disease and cancer. Whether or not you believe that diet has anything to do with health or disease, there is a definite *pattern* that has come to dominate the "food system" of industrialized entities such as Europe or the U.S.A. This is called the *"Hour Glass" Industrial Agriculture Machine*, cf., [521, p. 13]. If this machine provides your food, and it probably does, you should take note that it has an "Achilles heel;" it runs on fossil fuel. Fossil fuel supplies follow a Hubbert's Peak graph, cf., Section 1.7.

The "Hour Glass," for the U.S. for example, has at its top, as an order of magnitude, 10^6 farmers, those who grow the "inputs." At the base of the hour glass are an order of magnitude 10^8 consumers. In between there is a narrow neck in the machine consisting of, in order of magnitude, 10^2 to 10^4, *central decision makers.* In some subsets of this machine things are even more extreme. For example, in 2009 in the U.S. the four largest beef packers controlled about 84% of the market, and close to half of all supermarket food was sold by five corporations, Wal-Mart being the largest by far.[1]

You are most likely to encounter this machine in a supermarket; unless, for example, you happen to belong to the approximately 49 million Americans living with hunger or food insecurity, or the 1.02 billion people worldwide who

[1]In 2000 the top ten agrochemical companies (seven of which are U.S. corporations) controlled a majority of the market in global food production, see [71, page 187], [489, 291]. The seven U.S. corporations in 2002 were Philip Morris, ConAgra, Mars, IBP, Sara Lee, Heinz, and Tyson Foods. (Tyson has since bought IBP.) Ranked first was Nestlé, of Switzerland, ranked third Unilever, of the U.K./Netherlands and ranked sixth Danone, of France. In the U.S. in 1997 just three companies, Philip Morris, ConAgra and RJR-Nabisco accounted for nearly 20% of all food expenditures, [489]. Philip Morris (at one point was reorganized as a subsidiary of the holding company Altria Group, Inc.) owns hundreds of food brands such as Nabisco (which it bought in 2000 for $14.9 billion) and Kraft. Nabisco and Kraft products include Post cereals, Ritz, Triscuit, Waverly, SnackWell's, Honey Maid, Premium Saltines, Planters, Nutter Butter, Chips Ahoy!, Newtons, Oreo, Cool Whip, Jell-O, Kool-Aid, Capri Sun, Miracle Whip, Philadelphia cheeses, Velveeta, Cracker Barrel, Maxwell House coffee, Starbucks, Grey Poupon, A-1, Oscar Meyer, and Tombstone Pizza, [348]. (On July 9, 2004 the British Broadcasting Corporation (BBC) announced that Philip Morris paid the European Union (EU) $1.25 billion to settle legal action of the EU against Philip Morris accusing them of collaborating to smuggle cigarettes into the EU to avoid taxes and duties. See also Associated Press article by Paul Geitner, July 10, 2004. Other tobacco companies were similarly accused.) This information is getting more difficult to find as the proportion of a market controlled by a single firm in a very concentrated market is considered proprietary information. Mergers and acquisitions frequently change (and usually further concentrate) ownership. I entered "consolidation" + "agriculture" into Google on July 7, 2004 and the first URL was [291]. In 2002 we had the following: Only two companies, Cargill and Archer Daniels Midland, controlled about 75% of the grain and corn that's traded in the world (Heffernan says in [292]); the three largest beef processors sell about three-fourths of the beef in the United States; the four largest pork processors handle 60% of the country's pork; and four companies process half the nation's broiler chickens. And so on through virtually every agricultural sector.

are undernourished.[2] Many poor do not have easy access to "the supermarket" with its apparent cornucopia of options, cf., [1].

The relative few in control of most food have made a number of decisions of which we are not all aware. The U.S. 1938 Food, Drug and Cosmetic Act said that any product that was an imitation of a standard food, like bread, milk, cheese ... had to be labeled "imitation." The FDA (without a vote of Congress) in 1973 repealed the 1938 rule, and an imitation food did not have to be so labeled as long as it was not *nutritionally inferior* to the food it was imitating. You can guess how "nutrition" was (is) defined.[3]

In 1958 an amendment to the Food, Drug, and Cosmetic Act of 1938, The Delaney Clause, named after Congressman James Delaney of New York, said: "the Secretary of the Food and Drug Administration shall not approve for use in food any chemical additive found to induce cancer in man, or, after tests, found to induce cancer in animals." The Food Quality Protection Act of 1996 exempted pesticides from the Delaney Clause, as long as the pesticide amounts were "safe."

In the U.S., by means of the "revolving door," cf., Exercise 4.9, between industry and the FDA, genetically engineered entities were declared "substantially equivalent" to their unengineered counterparts.[4] Thus there never has been a public discussion in the U.S. or vote of Congress about this declaration of equivalence. Of course, there is a potential logical contradiction here, since all genetically engineered entities have patents which can only be granted if the thing patented is substantially "new." Are GMOs new enough to be tested for safety or anything else, say nutritional content? Are GMO foods/ingredients labeled?

The food industry does lobby the U.S. Congress, for example, and every 5 to 7 years a "farm bill" is passed which has profound consequences on our food supply—and international relations![5] One of these consequences was brought to the public's attention in [543], viz., a majority of the products in the typical supermarket are *some form of corn.* (See also www.kingcorn.net.)

---

[2]The 49 million figure was announced by the United States Department of Agriculture (USDA) on November 16, 2009, cf., *The New York Times*, Nov. 17, 2009, p. A14. There were officially 307,958,472 people in the U.S. on Nov. 18, 2009. The 1.02 billion figure was announced by the United Nations Food and Agriculture Organization (UNFAO) on Oct. 14, 2009.

[3]A solid argument can be made that "nutrition science" is still quite primitive and that it cannot now, and perhaps never will be able to, understand all the essential properties of food—that it cannot, for example, really make the determination that imitation bread is equal as a food to real bread.

[4]Thus GMOs, genetically modified organisms, are GRAS, generally recognized as safe, without special review. For an entertaining documentation of this fact see: the DVD "The World According to Monsanto," a Marie-Monique Robin film, and the DVD "Food, Inc.," a Robert Kenner Film. Also [488, Chapter 7].

[5]When subsidized U.S. corn, selling for below the cost of production, is imported into, say Mexico, thanks to NAFTA (the North American Free Trade Agreement), small farmers cannot compete and end up going to cities (or some other country) looking for work. This is one force contributing to the world-wide growth of slums, cf., [129].

When corn and soybean prices are kept below the costs of production by farm-bill policies—read subsidies—they are an attractive resource. Massive U.S. government subsidies have been paid to corn farmers, e.g., $9.4 billion in 2005, and $50 billion over a decade. And the majority of subsidies in general do not go to "small family farmers," e.g., in 2003–2005 two-thirds of subsidies went to one-tenth of U.S. farmers. While the number of "microfarms" have recently begun to increase, perhaps out of necessity; large farms still dominate food production. In 2002 144,000 farms produced 70 percent of the nation's food. In 2009, 125,000 farms did so—about 6 percent of the total number of farms. Just one of the implications for the cash strapped consumer is that, for example, between 1985 and 2000 the real price of fruits and vegetables increased by 40% while the cost of "liquid candy," i.e., soft drinks with high fructose corn additive, decreased by 26%. Small farmers in the U.S. and elsewhere cannot compete with such a system, unless they are very clever and lucky (more on this later), cf., www.usda.gov, www.ewg.org, [323].

Exercise 5.2 Supermarket Surprises

(i) If you live in the U.S. estimate the amount of money you have already paid in taxes that is embodied in the "real cost" of a 2 litre bottle of soft drink (with high fructose corn syrup).

(ii) From [544, p. 117], corn contributes 554 calories (Note: We will discuss calories in more detail in VII.) a day to America's per capita food supply, soy 257 calories, wheat 768 calories, and rice 91 calories. Compare your diet and estimate the fraction of your daily calorie consumption that comes from these 4 "seeds." An easier version of this problem is to check the ingredient lists of the foods you eat for the presence of these four, or substances derived from these four. For example, Ghirardelli Premium Baking Double Chocolate Bittersweet Chips contains "soy lecithin—an emulsifier," but no corn on the list.

(iii) Whether or not you are a vegetarian, estimate the amount you have already paid in taxes that have become subsidies for a pound of beef, a pound of pork, and a pound of chicken. Assume that all three are "factory farmed," cf., [352].

(iv) Count the total number of varieties of apples, oranges, and other fruits available to you at your local supermarket. Do the same for vegetables. Is the production of fruits and vegetables subsidized in the U.S.?

(v) Count the number of varieties of breakfast cereals available. Count the number of varieties of cookies and other baked goods. Count the number of varieties of candies. Compare their ingredient lists.

(vi) How many manufactured foods in your supermarket are "imitation food" as defined by the original 1938 act mentioned above? How many ingredients are synthetic chemicals? How many such synthetic ingredients did not exist, say, 100 years ago?

(vii) What fraction of the products in your supermarket have genetically engineered contents? (The answer may surprise you.) Why are they not labeled as such in the U.S.? Note that all of Europe, Great Britain, Australia, Japan, and Russia have strong labeling laws, cf., [497, p. 63]. What supermarket items are you fairly confident do not contain genetically engineered ingredients?

(viii) I claim that if a processed food in the U.S. contains corn or soy in some form, then the chances that you are eating genetically engineered food (at least partially) is quite high. Investigate this claim. Investigate what is known about potential effects of genetically engineered corn and soy, cf., [645].

(ix) Artificial flavors are key ingredients in manufactured food items. Have you ever heard of the company *Givaudan*? It apparently manufactured the vanilla flavor for Coca-Cola, cf., "The Taste Makers: The secret world of the flavor factory," by Raffi Khatchadourian, *The New Yorker*, Nov. 23, 2009, pp.84–99. This fact was apparently a carefully guarded secret for some time until it was accidentally revealed to a reporter. That is not the only closely guarded secret; try to find out the chemical composition of any of the flavorings used in manufactured food. What is the role of the FDA here? *Virginia Dare* is another "flavorist" corporation. How many flavorist corporations can you find? There must be several, since there exists the Society of Flavor Chemists. Is there a difference between "natural" and "artificial" flavors? According to a flavorist quoted in the article cited, it is not practical or economical to use "real" foods to add flavor. What proportion of your diet consists of "real" food?

(x) Does the length of the supply chain for particular foods affect the content of those foods?

5.2 Industrial Agriculture Logic vs. the Logic of Life

The reader might think perhaps that only mathematicians could be concerned with the subject of this section; however, I will show that Nature indicates that logic, in a form to be discussed, matters. Thus humans who eat should pay attention. We start with one difficulty; humans cannot define life, cannot create life (despite patents to the contrary) from scratch, and do not understand life in all of its complexity. In short, for the time being, *life* is bigger than mathematics; it is beyond human understanding. However, some facts are settled and generally accepted by scientists at this time.

CHNOPS: Atoms, Molecules, Cycles, Photosynthesis and Trophic Levels. Matter, i.e., "stuff," is composed of chemicals; the smallest unit of a chemical is called a *molecule*. Chemicals are composed of *elements*, which are listed in the Periodic Table, page 253. The smallest unit of an element is called an *atom* of that element. The typical human body has at least 60 of these elements; but as noted in [68, page 30], there are six elements that seem particularly important to life: C, carbon; H, hydrogen; N, nitrogen; O, oxygen; P, phosphorus; and S, sulfur; CHNOPS for short.[6] (The alert organic chemist will point out that since the time of Justus von Liebig in the mid 19th century, N, P *and* K, potassium, have been considered essential nutrients/elements for life.) Life creates proteins, fats and carbohydrates, as has been known since the early 19th century. There is a large nuclear fusion reactor, called the *sun*, which provides the energy for almost all life on earth, via a process called

---

[6]A widely cited paper, viz., "A Bacterium That Can Grow by Using Arsenic Instead of Phosphorus," by Filisa Wolfe-Simon (and 11 other authors), was published in *Science* on December 2, 2010. As the title indicates a bacterium strain GFAJ-1 of the *Halomonadaceae*, isolated from Mono Lake, California can substitute arsenic, As, for phosphorous, P, to sustain its growth. The organism can live on phosphorus, but it can use arsenic as a substitute! Perhaps someday the thus far science fiction substitution of silicon, Si, for carbon, C, may occur in some form of life somewhere. We are living in interesting times scientifically speaking, there is much left to learn about exactly what "life" is. In any event it appears that whenever possible: *Life will find a way!* Do follow developments based on this discovery!

photosynthesis; whereby green plants take inputs of carbon dioxide, CO_2, and water, H_2O and solar energy, and produce outputs of $C_6H_{12}O_6$, glucose, and molecular oxygen, O_2, viz.,

$$6CO_2 + 6H_2O + Solar\ Energy \longrightarrow C_6H_{12}O_6 + 6O_2.$$

I said "almost all life" in the sentence preceding because there are forms of life that do not depend on photosynthesis but on *chemosynthesis*. For example, *chemoautotrophic bacteria* create the outputs for their life, viz., sugar and sulfur compounds, with inputs of carbon dioxide and hydrogen sulfide in hot thermal vents in certain places at the bottom of the ocean. You can follow developments such as the recent discovery of 170 new chemosynthetic marine species at www.coml.org, the Census on Marine Life.[7] Although some believe that life on earth may have originated "chemosynthetically," today most life, including human life, ultimately depends on photosynthesis. Humans also use energy extracted from fission reactions and fossil fuels, but fossil fuel energy is actually fossil solar energy; and fission and fossil fuel energy are subject to Hubbert's Peak mathematics. One heretofore miminally exploited source of energy is geothermal, i.e., heat from earth.

Plants are participants in the *food web*, as are all living organisms. Any living organism is edible by (or otherwise can be assimilated by) other living organisms. So the food web has many *cycles*. The Axiom on Matter Cycles, page 56, thus can be refined into many subcycles: the carbon cycle, the water (hydrological) cycle, the nitrogen cycle, and so on. When one organism, call it the predator, eats another, call it prey, only a fraction of the prey's energy is useful for the predator. Thus a constant external supply of energy (from the sun) is required to keep the food web running.

In [503, Chapter 13] and [270, Chapter 1], and many other references, we see that food webs are organized according to *trophic levels*, based on who eats whom. In a simple model a green plant is assigned trophic level 1, a herbivore a trophic level 2, a carnivore that eats herbivores a trophic level 3 and so on up a food chain in the food web. Complications can easily arise, as in the case, for example, wherein a herring's diet may be half algae and half herbivorous crustaceans. Such a herring would be assigned a trophic level of $2\frac{1}{2}$. As a rough rule of thumb, which is often not accurate except to the nearest order of magnitude, it takes 10 units of energy, say 10 calories, at one trophic level to support 1 unit of energy, or 1 calorie, at the next higher trophic level. In many diverse ecosystems and in many predator-prey relations there are about

[7]It was reported in the media in November 2010 that a team from Oregon State University, with the Integrated Ocean Drilling Program, bore into the gabbroic layer by starting at the ocean floor in the mid-Atlantic. They found evidence of life 4,500 feet beneath the surface, where temperatures are higher than the boiling point of water. Unique types of bacteria which feed off hydrocarbons like methane and benzene were found there. These micororganisms are similar to those present in underground oil deposits.

90 kilograms of predator to every 10,000 kilograms of prey. When bacteria "eat" a lion or a wolf a cycle is completed. Are bacteria predators?

Interestingly there are other mathematical invariants associated with the food web. For example, as the CHNOPS elements flow from one organism to another, they remain in a fixed ratio, [68]. (You might investigate the closely related subject: Redfield Ratios.) For every C there are so many Hs, and so many Os and so on. This is an indication of a close connection among all living organisms.

Exercise 5.3 CHNOPS and the Elements of Life

(i) In the chemical equation (which chemists call a stochiometric equation) for photosynthesis, check to see that the number of atoms of C input (on the left side) equals the number of atoms of C in the output (on the right side). Do the same for O and H.

(ii) Find the relative ratios of the elements C, H, N, O, P and S in living organisms.

(iii) Do the same ratios apply to both photosynthetic and chemosynthetic life forms?

Soil is a Living Community. First, whatever life is, *soil* is a living community. As hunter-gatherers we could probably skip being concerned about the soils of the earth and just let them take care of themselves, cf., [537, Chapter 6]. However, about 10,000 years ago, humans invented *agriculture.* The process was not entirely smooth, from [236, p. 100]: *"After the invention of agriculture, most people were* ***worse off*** *than their ancestors. From skeletons we know that when agriculture arrived, average height declined about 4 inches, from 5′9″ to 5′5″ for men and from 5′5″ to 5′1″ for women. Compared to hunter-gatherers, farmers work harder, eat a less nutritious diet, and get sick more often."* The social structures and power relationships that agriculture enabled are a fascinating subject of study, and understanding them would probably shed much light on the current social structures associated with the Hour Glass Industrial Agriculture Machine—consider such a study an exercise.

Although non-soil based agriculture exists, for example, hydroponics and aquaculture (like fish and shrimp farming), they are resource intensive and face fundamental problems likely to prevent them from significantly replacing soil based agriculture. Consider it an exercise (for which I do not have a solution!) to figure out a way to "feed the world" without soil. (One candidate is to live off of farmed algae!) I will proceed under the assumption that we need topsoil for agriculture.

Humans cannot create life from scratch, neither can we create soil. Under tropical and temperate agricultural conditions, it takes 200 to 1000 years to form a 2.5 cm (about 1 inch) depth of topsoil. This one inch of topsoil is equivalent to about $\frac{340\ tons}{ha}$, where ha stands for one hectare, or $10^4\ m^2$, cf., [537, p. 152]. We can encourage topsoil formation, but we remain dependent on processes of Nature to create soil, just as we are dependent on processes of Nature to create life. There is a conflict between the logic of industrial agriculture and the logic of life/Nature: *Industrial agriculture mines topsoil. After each farming cycle, there is less topsoil than there was at the begin-*

ning. Unfortunately, the collection of folks destroying topsoil faster than it is forming includes many who are not totally part of industrial agriculture.

In most nations the average rate of topsoil loss is between 20 to 40 $\frac{tons}{ha-yr}$. (Note: $tons/ha/yr$ is the same as $\frac{tons}{ha-yr}$, see Exercise 5.4.) However, $\frac{3600\ tons}{ha-yr}$ has been observed. In the U.S. the topsoil on average is being eroded at a rate of $\frac{13\ tons}{ha-yr}$, while it is being regenerated at a rate of $\frac{1\ ton}{ha-yr}$. Over all, at least one-third of the topsoil in the U.S. has been lost during over a century of farming—it is estimated that Iowa, a state with some of the best soils anywhere, has lost half its topsoil during this time. Worldwide more than 10^6 ha of agricultural land are lost every year due to soil degradation; and during about the last 40 years of the last century 30% of total world arable land has been abandoned due to lack of productivity (10^8 ha of this in the U.S.). There are other processes ending in loss of agricultural land. For example, in the 30 year period 1945–75 an area of ag-land in the U.S. the size of Nebraska was covered with roads, homes, and factories. And this process continues year after year. See [537, 173].

This cannot continue indefinitely, consider the following exercises.

Exercise 5.4 Loss of Topsoil and Agricultural Land

(i) Topsoil regeneration in the United States is on average 1 $ton/ha/yr$. How much is this in inches or centimeters? Do you think you would notice this? Can you show that $\frac{1\ ton}{ha-yr} = 1\ ton/ha/yr = \frac{(\frac{1\ ton}{ha})}{yr}$? Are any of these the same as $\frac{1\ ton}{(\frac{ha}{yr})}$? See II.

(ii) If topsoil loss is $\frac{13\ tons}{ha-year}$, what is this in inches or centimeters? Do you think you would notice this?

(iii) Would you notice a rate of topsoil loss of 3600 tons/ha/year?

(iv) In the U.S. about 50 *liters* of oil equivalents are expended per ha to compensate for cropland degradation, cf., [537, p. 293]. If fossil fuels become unavailable, what do you propose will happen (or should have happened)? What are fossil fuels used for in industrial agriculture?

(v) If half the topsoil in Iowa has eroded away in the past century or so, what will happen in the next century? Will the agricultural pressures on the lands in Iowa increase or decrease in the next century?

(vi) Any process that removes organic matter from agricultural land probably contributes to soil loss. What does this imply for biofuels, i.e., making substitutes for oil using corn, wheat, soy, even grass? If food grains are used for biofuels, what effect will this have on food supplies, food prices, availability of food for the poor (or the more well off)?

Exercise 5.5 Some Global Numbers

In [537, p. 155], we read that in 1992, the year I started teaching this course, worldwide agricultural ecosystems covered about 50% of the world's (ice-free) land area. Forest ecosystems covered about 25% and human settlements covered about 20%. The total area of the earth is $5.10(10^{14})\ m^2$. The total area of ice-free land is $1.33(10^{14})\ m^2$.

(i) About how many hectares were used by humans for agriculture in 1992? What percentage of the total area of the earth is this? Do you think the land best suited for agriculture was cultivated first? What does this say about the future?

(ii) About 10^7 ha of agricultural land is abandoned each year due to serious soil degradation; is this significant? Where do you think people have gone to find replacements for this abandoned land?

(iii) What percentage of ice-free land is left unmanaged by humans, i.e., "wild"?

(iv) The human population of the earth in 1992 was about 5.5 billion, i.e., $5.5(10^9)$. In 2009 it was about $6.8(10^9)$. What do you think human use of the world's ice-free land area would look like if the world's human population were $11(10^9)$? Do you foresee any possible problems?

(v) Around the world, in North America, Europe, and parts of Asia, honey bees have been experiencing dramatic declines. The term CCD or colony collapse disorder has been used to describe this phenomenon. Research the health of the world's bee population at the time you read this. Have definitive causes for CCD been arrived at? Bees are important for the role they play in pollination, in fact, about a third of the modern diet relies on bee pollination. (For example, dairy cows are fed alfalfa which is pollinated by bees.) How are non-domesticated bees doing by the way? (Not well, last time I checked.)

The *Green Revolution* of the last century brought fossil fuel dependent (herbicide, insecticide, synthetic fertilizer) agriculture to the entire world. Global trade and institutions such as the World Trade Organization (WTO) together with the subsidies that industrial agriculture enjoys have made it difficult to grow food outside of the industrial model. This is because the central decision makers in the neck of the "Hour Glass" have written the rules. Small farms in the U.S. and around the world are under increasing pressure from industrial agriculture. For example, the industrial system is making it increasingly difficult for farmers to keep their own seed, a practice that is over 10,000 years old and once the source of literally boundless variety of locally adapted plants, and individual freedom for farmers. Amartya Sen, Nobel Laureate and economist, points out that 80% of all malnourished children in the developing world in the early 1990s lived in countries that boasted food surpluses—surpluses of luxury foods grown for export on land that could have grown subsistence crops to be eaten locally. Farmers increasingly find a limited number of buyers for their products. Farmers around the world go into debt, work hard, yet lose land that has been in their family for generations. Some become contract workers on land that was once theirs; thousands each year choose suicide. Many are displaced to urban areas, often slums. A few choose rage, cf., [159]. For the moment, industrial agriculture dominates the world food system. How many farmers or consumers are free to choose to be part of a food-system that is independent of industrial agriculture? Perhaps surprisingly there still remain millions of small farmers around the world (Latin America with 17 million farms, Africa with 33 million farms, Asia with well over 100 million farms, almost all less than 2 *hectares*) who survive using their intimate knowledge of the land, their locally adapted seeds and animals, and their cultural connections with domestic markets, cf., [36, pp. 16–7]. The total yield per hectare of these completely integrated small farms is substantially greater than that of any monoculture enterprise, cf., [36, p. 139]; but their strength—diversity—makes them practically invisible statistically in terms of output of any single crop. The logic of diverse, integrated agriculture more closely mimics the logic of Nature. The remaining sections of this chapter continue to address the logic of Hour Glass Industrial Agriculture machines.

5.3 Fast Foods, Few Foods, and Fossil Fuels

Food and the Air. It is fairly well known that industrial agriculture is one of the largest sources of global warming gases, cf., Exercise 1.10. For example, according to the IPCC, agricultural land use contributes 12% of global greenhouse gas emissions just from methane and nitrous oxide (each more potent in trapping heat than CO_2). In the United States, agricultural use of chemical fertilizers, herbicides and fossil fuels contributes nearly 20% of U.S. CO_2 emissions, cf., [247, p. 204]. These are important data, but what really makes an impression on me is seeing the local effects on the air of industrial agriculture. Take a prodigiously productive area like the San Joaquin Valley in California, home to some of the dirtiest air in America. From [113, pp. 155–6], (Fresno is a city in the Valley), *"The San Joaquin Valley—the most prolific farm belt in America—may be the most dangerous place in the United States to breathe."* (*The Fresno Bee*, December 2002) From the investigative series by *The Fresno Bee* we learn that: Farming *"creates more lung-searing air pollution than the Valley's eight highest-polluting large businesses combined."* Farms produce more than half of the region's particulate pollution, with heavy machinery plowing huge tracts of eroded monocrop land. Not even Los Angeles has more violations of daily or long-term air safety standards. Fresno County has California's highest rate of childhood asthma. *"The air, laced with some of the nation's highest concentrations of chemical particulates and dust, is a serious health hazard ... Medical experts, who have connected these particles to higher death rates, fear these specks are more dangerous than ozone or smog."*

Food and Water and Free Speech. A *dead zone* in the coastal waters of a continent (or in a lake) is an aquatic environment which lacks sufficient dissolved oxygen to support fish life. Although some occur naturally, the number of dead zones has been increasing, approximately doubling every decade since the 1960s, along with the increasing intensity of industrial agriculture and urbanization. (In fact, animal-factory farms *are equivalent* to large cities, some as large as Los Angeles, say, in terms of sewage generated. Farming in the United States is responsible for 70% of water pollution, cf., [91, p. 29].) Chemical fertilizer and sewage runoff contain soluble nutrients (such as nitrogen and phosphorus) which cause *eutrophication*; fertilized water gives rise to algal blooms which deplete the oxygen. As of 2008 there were 405 dead zones worldwide, cf., http://www.epa.gov/msbasin/pdf/diaz_data.pdf. Dead zones vary with time in size depending on environmental conditions. Perhaps the most famous, but unfortunately only one among many, dead zones in the U.S. is the one at the mouth of the Mississippi River. In the news media its size is often compared to the size of the state of New Jersey, 8,729 square miles. The scientific literature records a size of "more than 22,000 km^2." Are these roughly the same?

In a factory farm the animals, be they pigs (hogs), chickens, turkeys, or cows, by definition live their lives remaining approximately in place, eating food laced with antibiotics, urinating and defecating. The antibiotics are administered regardless, even when the animals are not sick, since animals gain weight slightly faster on a diet laced with these drugs. Aside from the fact that some consider this a form of pollution of the resulting product, such a practice has already resulted in bacteria which are immune to an increasing number of antibiotics. For example, some pathogenic bacteria in ground meat have been found to resist treatment by as many as 12 antibiotics, cf., [488, p. 177]. It is a near mathematical certainty that more and more bacteria will become immune and/or resistant to larger and larger classes of antibiotics, if we continue to expose them routinely and broadly to antibiotics—that's how Nature works![8] This process of bacteria becoming antibiotic resistant cannot be stopped entirely, but it can be slowed down considerably by greatly restricting the number of antibiotic exposures bacteria are given from which they can "learn" resistance. Review some history to see what life was like before antibiotics; it will likely motivate you.

Did you ever wonder what happens to animals that die before, or get too sick to make it to, slaughter? Most often they are *rendered*. Typically a fork lift or loader puts the carcass on a truck; and it is hauled to a rendering plant, where the remains are processed into a wide variety of products. So far, this is economical and rational. However, lifelong rancher Howard Lyman pointed out to Oprah Winfrey on her April 16, 1996 show on "Dangerous Foods" that sometimes sick cows are rendered and fed back to cows—in effect making cows cannibals! Since Europe had not long before had a "Mad Cow" experience, i.e., BSE, bovine spongiform encephalopathy, cf., [562, 588, 618], where people were infected with a *prion* disease, the study of which gained Stanley Prusiner the 1997 Nobel Prize in Medicine, beef prices dropped for a short time. This is because if a cow infected with BSE is rendered, and fed to other cows, the disease could spread exponentially. (This practice was banned by the U.S. government in the late 90s; however, rendered cows could still be fed to pigs and chickens. I am not sure how many inspectors there are monitoring/preventing illegal acts in this regard. Have there been any prosecutions?)

Now in 13 states, including Texas and Colorado, it is illegal to make disparaging comments about food. These laws are clearly unconstitutional, an abridgement of free speech and the press. Under these laws Rachel Carson, and her publisher, could have been sued for *Silent Spring*, same for Upton Sinclair and *The Jungle*, or *Consumer Reports* for telling us about pesticide residues on foods measured in their labs. Oprah was sued in Texas for millions of dollars, cf., *Texas Beef Group, et al. v. Oprah Winfrey*. Luckily Oprah Winfrey is a billionaire with principles who fought the lawsuit for years; it

[8]Overuse of antibiotics in humans and animals led to drug-resistant infections that killed more than 65,000 people in the U.S. in 2009—more than prostate and breast cancer combined, cf., Associated Press article by Margie Mason and Martha Mendoza, Dec. 29, 2009.

cost her millions of dollars, but she finally won. Unfortunately the legal process stopped short of the Supreme Court where such laws could be declared unconstitutional—hence they are still on the books! Since Colorado has one of the most repressive "Food Disparagement Laws" in the country—making "disparaging" comments about food is a crime with possible incarceration. Truth and documentation matter little if you are sued and you do not have the money to fight it—so much for the First Amendment to the Constitution of the United States of America for most of us. Estimate how many column inches in newspapers, or how many books or radio and television shows addressing food safety have not appeared because of these and other threats.

But let's get back to sewage. The amounts generated are enormous and concentrated. A 1997 estimate of total U.S. cattle, hog, chicken and turkey manure was 1.3 billion tons a year. That is 5 tons for every U.S. citizen—130 times the excrement American humans produced per year, cf., [113, p. 179–80]. You could make a project of documenting pollution of ground water, rivers, lakes and coastal oceans from this factory farm manure. For example, factory-farm-caused fish kills have become epidemic in 10 states, with more than 1,000 documented manure spills wiping out 13 million fish between 1995 and 1998. Look up *Pfiesteria piscicida*, an organism which can be found in manure, which the EPA estimated has killed more than 2 billion fish in the rivers, estuaries, and coastal areas in the Chesapeake Bay region of North Carolina, Maryland, and Virginia, cf., [113, p. 179]. As early as the 1970s one of the first dead zones was identified in the Chesapeake Bay region.

No amount of academic citation of statistics can match the visceral experience of visiting a factory farm—a "chicken house," a "hog farm," or a cattle feed lot—and taking a deep breath. It does not have to be this way. Farm animals can be a pleasure to be around under more "natural" conditions. (The word "natural" has little, if any, legal meaning.)

Monocultures are Risky. The logic of industrial agriculture embraces *simplicity* vs. the *complexity* of Nature. Nowhere is this more evident than in the ever larger areas planted in single crops, called *monocultures*, managed with fewer people, fewer steps, and bigger, faster machines. Example: Humans have selected as a principal source of food among the boundless varieties of food plants possible, a genetically engineered (more on this in the next section) variety of soybean which can be sprayed with a broad spectrum herbicide, most likely glyphosate or Roundup^{TM}. Such soybeans do not in general increase yield, cf., [265], but such a field of soybeans can be sprayed with glyphosate and all other plants, i.e., "weeds," are killed, leaving just the soybeans. In the span of a few years they now, as of 2009, are planted on 90% of the land in the U.S. devoted to soybeans, cf., [265]. The small army of laborers who just a few years ago were employed pulling weeds in soybean fields have been replaced by fast machines, often aircraft, spraying herbicide. A large area of land with only one species of plant growing is relatively speaking an ecological desert, where, for example, bees can no longer find as many "weeds" as they once "farmed" for pollen and nectar. Despite claims that herbicides do not significantly negatively affect the community of life that is soil and other organisms, there is evidence to the contrary.[9]

Studies have shown that glyphosate can kill fish in concentrations as low as 10 parts per million, that it reduces the growth of earthworms and increases their mortality and that it is toxic to many of the micorrhizal fungi which help plants take up nutrients from soils, cf., [9, pp. 24–5]. Even the long used synthetic nitrogen fertilizers have unintended consequences on the soil, in that they stimulate not only plant growth but the growth of carbon-hungry bacteria in the soil—decreasing soil carbon content, cf., [247, p. 212].

Also, 63% of the U.S. corn crop is genetically engineered, cf., [265]. Some of the details of a critique of this corn differ from the one above of Roundup^{TM} resistant soy; but the basic conflict with Nature's logic remains the same. *Humans are selecting food crops based on an extremely limited number of variables.* This locks the agricultural system into perpetually engineering new plants adaptable to changing environments, while simultaneously reducing the diversity of plants to work with. The system is also locked into chemically dependent cultivation regimens that require fossil fuels. It locks the industrial farmer into dependence on a decreasing number of corporations and diminishing fossil fuel reserves. It is highly unlikely that such a system can be sustained in the long run.

Most folks have heard of the Irish potato famine of 1845, which resulted because the Irish planted a monoculture of potatoes; they relied on one variety. When a serious, lethal disease arrives to attack vast fields of identical plants, they all die together. The usefulness of diversity is obvious. The potato famine was not unique in the annals of major monoculture crop failures; there was at least one such in each of the following years: 900 A.D., 1845, 1860, 1865, 1890, 1916, 1954, 1969, 1970 (twice), 1984, 1989, ... , cf., [386, p. 100].

A multitude of small independent farmers over many generations have developed literally a boundless variety of food crops specifically adapted to their local areas: corn(s) in Mexico, potato(es) in South America, rice(s) in India and so on; and luckily millions of small farmers remain around the world, cf., page 107. However, such diversity is under pressure. With each group of independent "peasant" farmers that loses their land, the varieties of food and the accompanying knowledge handed down to them over generations is lost as well. As previously undeveloped land is transferred to "development," wild species that might help diversify our food supply are often lost. The chill-

[9](http://www.esajournals.org/doi/abs/10.1890/04-1291) Roundup heavily impacts amphibians, THE LETHAL IMPACT OF ROUNDUP ON AQUATIC AND TERRESTRIAL AMPHIBIANS by Rick A. Relyea; (http://farmindustrynews.com/news/farming_multiplying_microbes_glyphosate/) Multiplying microbes—Glyphosate boosts Fusarium levels in Missouri study. Apr 24, 2001 12:00 PM, Gil Gullickson; (http://www.organicconsumers.org/ge/monsanto_fungus.cfm) Monsanto's Roundup Herbicide May Be Spreading Deadly Fungus August 23, 2003 by Jeremy Bigwood; and http://www.abcbirds.org/abcprograms/policy/pesticides/Profiles/glyphosate.html.

ing story is told in *The Last Harvest: The Genetic Gamble that Threatens to Destroy American Agriculture*, by Paul Raeburn, [560].

Just like the production lines at the meat packing plants that are processing animals at ever increasing rates, the Hour Glass Industrial Agriculture machine runs faster and faster. The term *Fast Food* gains a whole new meaning. The central decision makers manufacture food faster and faster at the expense of complexity, diversity, and, I claim, sustainability, because they can and it appears profitable—for now.

Exercise 5.6 Industrial Agriculture Converts Fossil Fuel to Food

(i) List all the ways that you can think of that fossil fuels are used to produce food in the industrial agricultural system. Then make your own estimate of the number of calories of fossil fuel energy needed to put one calorie of food energy on your plate. Does the order of magnitude of your answer agree with mine, viz., order of magnitude is one, hence the ratio is $10^1 = 10$? That is, *it takes 10 calories of fossil fuel energy to put one calorie of food on your plate.*

(ii) There once was a disagreement between two students in my class: one was from a ranch in Wyoming, another from the city of Denver. For a project one calculated the fossil fuel energy needed to put one calorie of deer meat on his plate (in his home at the ranch), where the deer was hunted on his ranch and stored in a freezer. The other (to make the math simple) calculated the fossil fuel energy used to put one calorie of potato on his plate (in Denver), where the potatoes were grown in Idaho within the industrial agricultural system. (Regarding storage, you can periodically ship potatoes, i.e., buy them at the supermarket, or process them.) Who do you think used more energy? If the potatoes were grown in a backyard garden in Denver, how would that change the calculations?

(iii) Which makes the greater contribution to global warming emissions: U.S. industrial agriculture or the U.S. transportation system (including cars, trucks, buses, and trains)?

(iv) Rich healthy soils are dark because of high carbon content. Green plants like grasses and trees with the help of the community of life in the soil (bacteria, fungi, and other organisms) sequester carbon from the atmosphere via photosynthesis. Suppose that it is true that the carbon in fossil fuels were mostly sequestered into the earth by green plants during the Carboniferous Period, from 360 million years ago to 290 million years ago. To the nearest order of magnitude this is 10^8 years. To the nearest order of magnitude we are burning these fossil fuels in 10^2 years. Thus, roughly, we are burning fossil fuels about 10^6 times faster than they were sequestered. If this is true, would restoring carbon to our soils globally by planting forests of trees and implementing restorative agricultural practices completely solve our "global warming problem?" Could the process of sequestering carbon in the soil be sped up with *biochar*, encouraging mycorrhizal fungi, rhizombium bacteria and subsidies to farmers that build soil? For example, see [389]. Are there other problems that planting forests and soil-restoring agriculture address?

(v) This exercise will help you get started on (i). Consider just one aspect of the *(synthetic) chemical dependency of industrial agriculture.* Globally over half of the nitrogen taken up by agricultural crops comes from *ammonium nitrate* fertilizer, created using natural gas energy using the *Haber-Bosch process*, cf., [81, p. xiv]. Natural processes such as bacterial action making nitrogen available to plants have been matched and exceeded, for the time being. Humans have thus created a global agricultural system that is not only fossil fuel dependent, it is (synthetically) *chemically dependent.* Many of the crops now grown were selected precisely because they respond well to synthetic fertilizers.

Plant fertilizers focus on three main ingredients: nitrogen, phosphorus, potassium. Estimate, or look up, global use of fertilizer for agriculture.[10] In particular, how much of this fertilizer is a synthetic chemical?

[10]For example, cf., *Vital Signs*, 2001, p. 33, WorldWatch.

(vi) This will also help with (i). Estimate the distance traveled for each of the items of food on your plate. Most estimates of an "average" for food on an American's plate are over 1,000 miles. See, for example, [440, p. 425], [490, pp. 28–32].

(vii) What was the principal chemical Timothy McVeigh used to blow up the federal building in Oklahoma City in 1995? See [159].

(viii) There is an agency of the federal government that earlier was called "Animal Damage Control," or the ADC, and is now called "Wildlife Services." At the behest of agricultural interests this agency kills bears, bobcats, coyotes, wolves, eagles, birds of prey, pet dogs and so on, some by accident. Methods used include shooting from aircraft and poisoning devices using compound 1080. How many thousands of wild animals are killed each year by this unit? (You might be surprised that there are that many left to kill.) U.S. taxpayers paid $120 million for this "service" in 2008.

Exercise 5.7 Find an Industrial Polluter Near You

(i) Find the nearest sources, in time and space, of industrial pollution near you. Using the EPA's toxic release inventory (www.scorecard.org) you should be able to find one.

(ii) Find the nearest, in time and space, industrial agricultural polluter near you. For example, "Cargill fined $200,000 for clean water violations at Fort Morgan packing plant," by Howard Pankratz, *Denver Post*, posted to the Web 11/13/2009; "Cargill plant pleads guilty to wastewater charge," Associated Press, *Denver Post*, Web-posted 09/29/09.

(iii) Does an answer to (iii) also count as an answer to (i)? Can you quantify the pollution in some way? For example, from the Associated Press article mentioned in (ii): "Court documents show that Cargill violated its permit because it discharged more than 2,875 pounds per day of total suspended solids and more than 400 coliform colonies per 100 milliliters." (Question: How much more than 2,875 pounds/day?)

(iv) If you are feeling adventurous, find an industrial polluter that has not been exposed in the media.

(v) Industrial agriculture *adds* to the world's food supply and *subtracts* from the world's food supply in so far as it contributes to dead zones, such as in the Chesapeake Bay area, the mouth of the Mississippi river, off the coast of Oregon-Washington, the coast of California, and elsewhere (in the U.S. and world). Pick a dead zone and estimate the amount of food, e.g., fish, shrimp, oysters, etc, not produced because of dead zone conditions. Are there forms of agriculture that would not contribute to dead zones in bodies of water?

5.4 Genetic Engineering: One Mathematical Perspective

I was recently looking at a large ad in *The New York Times* from the biotech industry promoting genetically engineered crops—although the term *genetic engineering*, GE, was not mentioned. In fact, this terminology, along with *genetically modified organism*, GMO, have all but disappeared from printed ads. The ad referred me to the Web site www.whybiotech.com where you can see a sequence of videos and comments from quite credentialed folks in support of *biotechnology*. From a marketing standpoint who can be critical of *bio*, meaning life, and *technology*, which is almost synonymous in the public mind with progress and the "good life." But as with all advertising it is necessary to look beneath the surface and study the claims that are being made or implied. (Note that in the following the word *transgenic* is synonymous with genetically engineered.)

The Question of Which Crops Have Higher Yields. A key attribute of any food crop is *yield*. By definition, yield is how much of the food in question is produced per unit area of cropland, measured in terms of kilograms or tonnes per hectare, or pounds or tons per acre. Yield is often measured in terms of volume, e.g., bushels per acre or cubic meters per hectare. The biotech industry clearly and consistently over the years has implied that its GE crops have better yields than non-GE crops with the suggestion spelled out for us that this will help the poor and hungry of an ever increasing global human population.[11] No one I know is against solving the problem of world hunger! Biotech crops have been around in commercial production since 1996, so we should be able to look at the record and see if genetic engineering has increased yield. Unfortunately, it appears not.

A careful study must distinguish between two types of yield: *intrinsic yield* and *operational yield*. Intrinsic yield refers to the ideal (or maximum possible) yield of a plant under ideal conditions. Operational yield is the yield that is actually obtained by farmers in real life. A fairly comprehensive 2009 study of peer-reviewed research on yield of transgenics and non-transgenics by the Union of Concerned Scientists, www.ucsusa.org, [265], concludes: *"So far the record of GE crops in contributing to increased yield is modest, despite considerable effort. There are no transgenic crops with increased intrinsic yield and only Bt corn exhibits somewhat higher operational yield. Herbicide-tolerant soybeans, the most widely utilized GE crop by far, do not increase either operational or intrinsic yield."* A couple other references not mentioned in the above study included the following of Lappé and Bailey, [386, pp. 83–4], who checked up on Monsanto's 1995 claim that there was no "yield penalty" whatsoever in its GE soy: *"In 30 out of the 38 comparisons, the conventional variety outperformed the Roundup Ready^{TM} variety. The likelihood of such an outcome occurring by chance is less than 1 in a hundred. (Chi^2 = 6.95, df 37). Overall yield was down an average of 4.34 bushels for Roundup Ready^{TM} varieties, a statistically significant loss of just under 10% compared to conventional types. In only 4/38 instances did the Roundup Ready^{TM} crop approximate the yields of the highest yielding conventional soybean varietal type grain in the region."*

Ed Oplinger, Professor of Agronomy at the University of Wisconsin, has carried out yield trials on soybeans for 25 years. GE soybean had 4% lower yield than conventional varieties on the basis of data he collected in 12 states which (at the time) grew 80% of the U.S. soy.[12] I invite the reader to do

[11]At the biotech Web site mentioned above we read (Nov. 28, 2010): "Scientists recognize need for GM to feed growing population"; and the newspaper ad mentioned above contains the line: "And, it's helping provide ways for developing countries to better feed a growing population." The question remains for me as to how poor people living on a couple dollars a day would find the money to buy genetically engineered seeds.

[12]Oplinger, E. S., Martinka, M.J., and K. A. Schmitz, "Performance of Transgenic Soybeans in the Northern U.S." (1999) Accessible in Adobe Acrobat format at http://www.biotech-

his/her own literature search on comparative yields of GE corn, soy and Bt cotton, as well.

In the August 2009 of *Scientific American*, p. 28, there appears an editorial that I found surprising. The title of the editorial is "A Seedy Practice: Scientists must ask seed companies for permission before publishing independent research on genetically modified crops. That restriction must end." We read:

> ... *"But agritech companies such as Monsanto, Pioneer and Syngenta go further. For a decade their user agreements have explicitly forbidden the use of the seeds for any independent research. Under the threat of litigation, scientists cannot test a seed to explore the different conditions under which it thrives or fails. They cannot compare seeds from one company against those from another company. And perhaps most important, they cannot examine whether the genetically modified crops lead to unintended environmental consequences."*
>
> *"Research on genetically modified seeds is still published, of course. But only studies that the seed companies have approved ever see the light of a peer-reviewed journal. In a number of cases, experiments that had the implicit go-ahead from the seed company were later blocked from publication because the results were not flattering."*

From the same article we read that a group of 24 corn insect scientists (who have remained anonymous) sent a statement to the EPA—which could set open inquiry as a condition for approval of new seeds—protesting that

> *"as a result of restricted access, no truly independent research can be legally conducted on many critical questions regarding the technology."*

A concise global overview of genetically engineered crops, with useful references, is [426, pp. 18–20]. Globally as of 2007 $114.3(10^6)$ *hectares* of crop land, half of that in the United States, was planted in GE crops, up 12% from 2006. Four cash crops account for virtually all GE production: soybean (51%), corn (31%), cotton (13%), and canola (5%). Globally 63% of GE crops are herbicide resistant, 18% insect resistant, with a combination of the two traits (called "stacked") accounting for the rest. In the United States GE crop production increased pesticide use 4% between 1996 and 2004. Monsanto GE crop traits are found in more than 85% of global GE crop hectares, and Monsanto controls 23% of the global proprietary seed market.

Note that reference [645] contains the results of studies on the health risks of genetically engineered food. Do you think this book would be thicker if the restrictions on research mentioned above were not in place?

Genetic Engineering: Some Basics. I am going to treat this subject rather abstractly so you do not need a degree in molecular biology to read it. I will introduce a minimal vocabulary needed to discuss the subject; but you

---

info.net/herbicide-tolerance.html#soy. Benbrook, C., "Evidence of the magnitude and consequences of the Roundup Ready soybean yield drag from university-based varietal trials in 1998," (1999) Benbrook Consulting Services, July 13, 1999. RR-Soya-Yield-Drag.htm. These two references along with many others were found via (http://www.facebook.com/note.php?note_id=11293182509) "Soil Association: GM crops do not yield more—sometimes less," which was posted by The Soil Association.

can go on to pursue the subject in deeper detail, indeed you can major in it! Somewhat surprisingly, abstract mathematics probably has something very important to say about GE (genetic engineering), and the implications might not only be grand intellectually and ecologically but economically as well.

You probably have heard of deoxyribonucleic acid, DNA, in a biology class or elsewhere. To a biologist a molecule of DNA is made by attaching building blocks of four types: Adenine, Thymine, Guanine, Cytosine—together in a "string." To a mathematician a molecule of DNA can initially be thought of as a (long) "word" spelled out with the letters A,T,G and C (with some geometry, including knots, included later). We will have some fun with these "words" by doing some DNA computing in IV.

Distinct regions of DNA contain distinct bits of information, and these specific regions of information are called *genes*. Living organisms are built up from cells (sometimes just one cell); and if said cells have a substructure, a membrane-bound bag, called a *nucleus* which in turn contains "packages" of DNA (plus the proteins that organize and compact the DNA) called *chromosomes*, then the organism is called a *eukaryote*. If the DNA in a cell is not packaged within a nucleus, the cell belongs to an organism called a *prokaryote*, of which there are two classes: Bacteria and Archea.[13] *Viruses* are "infectious agents" that can only replicate inside the cells of other organisms, and a virus has its own genes made from DNA or RNA (ribonucleic acid). *Prions*, which have no RNA or DNA are "infectious agents" that are essentially proteins that propagate by transmitting a mis-folded protein state. Thus geometry (and topology) arise in biology, not only in the double-helix of DNA, but in prions, and elsewhere! Prions are implicated in BSE, "mad cow disease."

The genetic information in the nucleus of a eukaryote is called its *genome*. Interestingly, most eukaryotes have one or more other little "membrane-bound bags" besides the nucleus, called *mitochondria*; and these have their own DNA, hence their own genome. The DNA in plant mitochondria can be 10 to 150 times larger than that of human mitochondrial DNA, [152, p. 276]. (Green plants also have *chloroplasts* with their own DNA.) Genes have a physical location, *gene locus*, and alternative DNA sequences in that locus are called *alleles* of that gene. You probably have heard of "dominant" and "recessive" genes, which are alleles. Later we will give a simple example of the mathematics involved and how this affects heredity. The *genotype* of an organism is the complete genetic information of the individual, including specification of alleles.

[13]The term prokaryote is a "failed term," as Norman Pace discusses in his paper, *Nature* 441, 289 (18 May 2006). Bacteria and Archea did not come "before," i.e., pro, eukaryotes, as gene sequence comparisons prove. Rather, Archea, Eucarya, and Bacteria are separate classes of life that radiate out from the "origin of life." They are not linearly ordered. Also defining something in terms of lacking a property, e.g., lacking a membrane containing genetic material, is not a sound practice. We have used the term prokaryote here with this warning and in the hope that it will be dropped soon.

The *phenotype* of an individual organism is the collection of its observable characteristics, sometimes called *traits*. The genotype of an individual is said to *express* itself to produce the individual's phenotype. The definition of "express" is not completely clear. This is a very important point, as we shall see.

Genetic Engineering: Lab Techniques. The idea is this. You have found an organism that has a trait you would like another organism to have. For example, a soil bacterium, *Bacillus thuringiensis*, *Bt* produces a crystalline protein which is toxic to certain organisms that attack corn. You would like corn to have this property. Using by now standard lab techniques, you can cut up the DNA of the *Bt* and isolate "the gene(s)" that "give" *Bt* the desired "trait." Copies of the gene(s) are cloned using established techniques, cf., [152]. Then a virus or bacterium is used to infect the corn plant and insert this gene into the DNA of corn. *Agrobacterium tumefaciens*, which produces galls in plants, is commonly used to do this step. Another method is to coat tiny gold or tungsten "microbullets" with the desired gene(s) and "shoot" them at corn cells with a "gun." Some of the bullets may make a lucky hit and get incorporated into the corn's DNA. In any event, a process that *randomly* inserts the desired gene into the DNA of the target can be expected to have a very low rate of success, i.e., most of the time the desired gene will not be taken up to produce a viable cell with the desired properties.

So there needs to be a method for finding any viable products of the above process. One common method is to attach a "marker gene" that confers antibiotic resistance if successfully inserted. In one case ampicillin resistance was used for transgenic corn, and for a transgenic tomato a kanamycin resistant marker was used. The cells are then grown in a medium containing the specified antibiotic, and only the ones with a "successful" implantation will survive.

There is one more thing: it has been found that "success" is assisted if a *promoter* is also inserted during the above process. The intention is that the promoter, very often coming from the DNA of *CaMV*, Cauliflower Mosaic Virus, will "turn on" the introduced gene and get it to express itself at high levels, [588, 9].

The whole process above of "engineering" is a lot less precise than the word engineering usually connotes to me. A lot of questions come to mind. For example, genes direct the production of proteins, and food allergies are reactions to proteins. The natural question to ask is, "Does any of this genetic engineering inadvertently create proteins that I am allergic to?" The politics, economics and science involved in this question are investigated in [497]; and I am not comforted when I read: *"Healthcare visits for food allergies in children nearly tripled between two time periods studied: 1992 through 1997 and 2003 through 2006. ... Asthma, eczema and hay fever are also going up."*[14] Before listing any more of the

[14] "4% of U.S. children have food allergies, analysis finds," by Shair Roan, *The Los Angeles*

questions regarding GE food, let us look at the "logic" involved.

The Logic of Genetic Engineering. First of all, what is a "trait?" This is a very flexible (or fuzzy) term. It appears that there are traits, i.e., observable characteristics, that are simple and some not so simple. For example, "tall," "blue eyes," "red flowers," "produces *Bt* toxin" might be candidates for simple traits. Candidates for not so simple traits might be "aggressive behavior" or "yield" or "quality of seed." Through experimentation you can find genes that affect these traits, but what exactly is the relationship between a particular gene and a particular trait?

Not only is it an assumption, it is claimed to be a fact by some that—for example—inserting the genetic material to create *Bt* corn, or glyphosate resistant soy does nothing else (negative) of significance. This is an example of, or at least borders on, simple *genetic determinism.* For each trait there is a gene, or group of genes, that express that trait in the phenotype—and it (they) does (do) not significantly affect other traits. (At least this is the hope if one does not mess with too many genes at once.) But we have seen earlier in this chapter evidence that the trait of yield is negatively affected in some cases, and intrinsic yield is not increased.

To my knowledge no one has isolated "the yield gene(s)" so it is not surprising that intrinsic yield has not uniformly gone up with GE crops. One advertised object was to get a plant to manufacture a pesticide, in the case of *Bt* corn, (*Bt* corn is registered as a pesticide); another object was to engineer a plant resistant to a herbicide. One can argue on general principles that the model of genetic determinism as described above is not complex enough to describe observed reality.[15]

Times, November 17, 2009. Increased levels of hygiene have been hypothesized by some as a cause; but rigorous long-term studies have not been done on this or the GE hypothesis.

[15]From http://www.genome.gov we learn that human DNA, as a "word" in A, G, C, T is $3(10^9)$ letters long with 20,500 genes. (Initially it was predicted that there would be more than 100,000 genes in the human genome, but as more research was done this number has gotten smaller and smaller, with 20,500 the one available to me in 2009.) We also learn that the genomes of any two people are more than 99% the same. Said differently, about 200 genes must account for the variability of *all* humans. This would be difficult to explain if one assumed the "linear," simple genetic determinism above. However, think of the genome *in context*, that is the genes are part of a cell where a true multitude of interactions of various combinations of genes among themselves and with other parts of the cell (don't forget the mitochondria, for example) are possible. Then, as we will see in IV, there are truly an astronomical number of possibilities; and it is no longer difficult to believe that all humans can share 99% of their genome, yet vary quite widely. (As an additional complication consider that it is known that "switches" exist that direct when and for how long given genes express themselves, with the result that a given gene can yield different phenotypic characteristics depending on when and for how long it is "turned on." Introducing *time as an additional variable*, with "on" and "off" states for genes allows for a great deal of additional complexity.) I call this the *genome in synergistic context hypothesis.* I also observe that if "playing" with 200 genes can result in observed human variability, playing with one or two genes in a plant might just have a lot more than one or two consequences. Maybe our connection axiom's corollary holds here with a vengeance: in a complex system (like a cell) you cannot do just one thing!

There are many things to consider, and it would take an entire book to deal with a fraction of them. So I will stick with the following elementary observation that has to be explained away by believers in (simple) genetic determinism as described above. This version of genetic determinism is *reductionist* in the sense that the study is concerned with isolating single genes, attempting to reduce complexity to something far less complex, and then studying the simpler system. While this is a valuable tool in science (from which we have learned much, even about the genome), it perhaps is not adequate for understanding the big picture when studying truly complex, non-linear systems with synergy. In these cases, the "whole" is not only more than the "sum of its parts," it is sometimes totally different than the "sum of its parts (in this case, genes) taken individually."

I thus replace the simple view of genetic determinism with the *genome in synergistic context hypothesis*: the genome expresses itself by means of a complex, non-linear synergism with all other parts of the cell, including the genes themselves.

I not only propose that the cellular context is important, I propose that to understand cells completely they must be studied in the context of the whole organism. And, that to understand the whole organism completely, it must be studied in the context of the environment of which it is a part! A grizzly bear in a zoo is a different system than a grizzly in the wild!

Thus the *context* in which genes express themselves ultimately includes the environment of which they are a part. It is entirely possible that this synergy is not a one-way street; I believe the environment can act on organisms, cells, and the contents of cells, including genes. (For example, consider environmentally induced mutations.)

Thus I have set before you two paradigms for understanding how the genome expresses itself: simple genetic determinism where it is at least approximately true that given a "trait" you can find a gene, or a group of genes, that cause said trait to be expressed; and the "synergistic context" paradigm. The reader is free to choose, or come up with a third. The science will have to catch up with the existing laboratory techniques somewhat before a rigorous determination can be made as to which view is closer to the truth. However, there are some interesting implications of the "synergistic context" view.

If the "synergistic context" hypothesis is true, then the whole basis of the patents held by the biotech industry can be questioned. For sure, the ingenious laboratory techniques are worthy of patent. But in my opinion the products of those techniques are not only not understood completely, they likely cannot be legally defined. (For example, try to define "life" itself, since what is being patented is a form of life.)

Consider just one complication. Since there is evidence that the phenotype (and hence presumably the genotype) of a GE organism resulting from the insertion of some gene can depend on *where* the gene ends up, and the lab techniques I am aware of do not control[16] this geometry, there is a different organism for each geometric configuration. Since the resulting organism is a geometrically dependent synergy between a long previously existing system and the inserted gene(s), none of which were invented but found in Nature, by the way; it is not clear to me that anyone can at this time even define rigorously what exactly the resulting GE organism is, let alone own the entire life form and patent it.[17] Of course, fortunes depend on these patents, thus any related lobbying of law makers or litigation in the courts would be (is?) intense!

A Short, Incomplete List of Potential Problems with GE.

Food Safety Concerns. If one wants to do research on GE food one needs to be knowledgeable and funded—and, as of the decade prior to 2009, get the approval of the GE industry to publish if one gets their genetically engineered products from the GE industry, the only legal source, cf., page 115. There is the famous case of the research of Dr. Arpad Pusztai, on GE potatoes, conducted at the Rowett Research Institute in Aberdeen, Scotland—ending in 1998. You can read the details in [488, 644, 645], and in [588], where author Andy Rowell is an investigative journalist who was closest to the "action" in the U.K. The one line summary is: eminently qualified researcher, Dr. Pusztai, with a long and distinguished career conducts research on GE potatoes, finds evidence that they may be dangerous to the public health, is so alarmed that he says so in a public forum, and is immediately terminated. (The DVD "The World According to Monsanto," has an interview with Dr. Pusztai.) I refer you to the three references just mentioned, to [9, 313, 386, 497] and others, that address, in part, the issues of potential health problems of GE food.

Genetic Pollution. Then there is the case of Dr. Ignacio Chapela, University of California, Berkeley, whose research in 2001, published in *Nature*, showed that GM corn had contaminated important sources of native corn in several rather remote areas of Mexico, viz., 15 of 22 areas in Oaxaca and

[16] According to a Dec. 18, 2009, news article, "As Patent Ends, a Seed's Use Will Survive," by Andrew Pollack, *The New York Times*, as the patent expires on Roundup Ready^{TM} soy, "Monsanto said it was confident that most farmers and seed companies would move to Roundup Ready 2, which uses the same bacterial gene but places it in a different location in the soybean DNA ...." Much of the information about this technology is a trade secret; however, I wonder just how carefully the location of an inserted gene can be controlled. Even if we assume that the geometric location can be specified exactly, it seems clear to me that *all* of the effects of this genetic modification are difficult (impossible?) to predict, and can only be learned after the fact by extensive research.

[17] After some investigation it appears unclear to me whether or not the courts would accept this argument. I have noted, however, that on March 29, 2010 *The New York Times* reported that United States District Court Judge Robert W. Sweet issued a decision invalidating seven patents related to genes whose mutations are associated with breast cancer, BFCA1 an BFCA2. The judge argued that the patents were "improperly granted" because they involved a "law of nature." How have appeals courts, businesses, and Congress reacted to this decision?

Ixtlán. Thus he established that genetic contamination, or pollution, was occurring and remains a problem for those, such as organic farmers, who wish to remain GE free, and anyone with an interest in maintaining the genetic seed banks of "original corn" intact and uncontaminated. Dr. Ignacio also opposed a large corporate tie-up between U.C. Berkeley and a large multinational biotech company, Novartis (later Syngenta). There followed a remarkably nasty public relations campaign against Dr. Ignacio and his research.[18] He survived, but the campaign managed to divert attention from the fact, now confirmed by other researchers, that the genetic contamination Dr. Ignacio discovered was real, cf., [488, p. 235] and [588].

Freedom to Farm GE Free. If you farm you may want to remain GE free so that you can save your seeds from year to year, a practice going back thousands of years. If you are an organic farmer genetic contamination is a real threat to your product. Even if you are not organic, note the following. Consider the case of Percy Schmeiser (http://www.percyschmeiser.com/, wikipedia, and www.monsanto.com), a farmer specializing in the breeding and growing of canola in Bruno, Saskatchewan, Canada for over 40 years. If you buy GE seed from Monsanto you need to sign a 'Technology Use Agreement' stating that you will not save seed, that you will allow Monsanto detectives to come on your property at any time for purposes of enforcing this agreement. Schmeiser did not buy seed from Monsanto and did not sign such an agreement. Nevertheless, Schmeiser has stated; Monsanto detectives trespassed on his land, gathered seeds that were found to be Roundup Ready^{TM} canola, and sued him, for example, for a $37 per hectare fee for the use of such seeds. Schmeiser claimed genetic contamination. Most farmers give up when confronted with a multimillion dollar legal team. He did not. Read about his ongoing efforts on the Web. Monsanto has a view that differs from Percy's, available at their Web site. The Center for Food Safety, CFS, has views different from Monsanto's. For example, they have done a preliminary report on the number of farmers that have been sued by Monsanto, the amount of money awarded Monsanto by the courts, and its effect on agriculture. How many farmers have been sued by Monsanto, anyway? See www.percyschmeiser.com/MonsantovsFarmerReport1.13.05.pdf for the CFS report, "Monsanto versus U.S. Farmers."

Pest Resistance. With near mathematical certainty long term, frequent use of a given pesticide can be expected to result in the evolution of pests that are resistant to that pesticide, be it glyphosate or *Bt* or another. Use of herbicides gives rise to so-called "super weeds." For example, glyphosate-resistant weeds have been on the rise since GE crops started gaining momentum; and these weeds now total 15 species—up from 2 in the 1990s—that cover hundreds of thousands of hectares in the United States alone [426, p. 19].

Resistance to *Bt* would pose major problems for organic farmers who now

[18] I personally cannot prove who or what entities organized and implemented this campaign.

can spray the bacteria and infect pests, who then die when the bacteria infects its host and produces its signature toxin. Every cell of genetically engineered *Bt* corn produces this same toxin at many times the level found in the organic application. This could hasten the appearance of resistant pests, and presents a plant residue every cell of which contains the *Bt* toxin. Since such biopesticides come from soil bacteria, and such bacteria likely have a natural function in pest control, there is a concern that some natural balances may be upset, permanently, cf., [313, p. 108]. You might want to check the Web, e.g., YouTube, for claims of some farmers that their pigs have suffered ill effects from eating *Bt* corn. (Note that Monsanto is apparently seeking to patent certain pig lines as I write.)

Monocultures and Stability and Loss of Diversity. There is concern that GE crops are not stable[19] from one generation to the next, and GE has certainly contributed to the rise of monocultures over enormous total areas. Independent farmers who have over millenia selected a diverse collection of food plants adapted to local conditions are being replaced by monocultures. Their knowledge and their plants have been largely lost as they disappear. Classical methods of plant (and animal) breeding and selection differ from GE in several ways. Classically, species barriers are usually preserved; and *whole organisms in context are selected.* This makes it possible to select for complex traits such as "yield," where it may not yet be possible to isolate the precise genetic mechanisms (and other possible mechanisms, such as actors in the environment) that affect yield.

Control. Ever larger portions of our food supply are coming under the control of corporations that do not appear open to dialog with folks who disagree with their designs. Unfortunately even independent scientific investigators are not fully free to investigate the situation and communicate their findings to the rest of us.

The "Computer Code" of Life. I have personally heard scientists, who participated in the discovery of the human genome, refer to that project as "reading the computer code of life." Anyone who has written a computer program of any complexity knows that if you make changes to a complicated program there will be bugs, i.e., unintended consequences. From experience then, we should anticipate unintended consequences if we start manipulating the genome of various organisms—and then introduce massive numbers of these organisms in the environment, while preventing free and independent scientific investigation of the products.

Exercise 5.8 Playing With the Building Blocks of Life. Several references in this book deal with potential problems with genetically engineered life forms, [386, 588, 644, 570]. Are they generally ignored?

---

[19] For example, consider http://www.ecoglobe.org/nz/ge-news/stab2214.htm and the statement of Dr. Mea Wan Ho therein; or see http://www.i-sis.org.uk/secretGMcrops.php. If GE plants are not stable from one generation to the next, for me this raises further questions as to what exactly is being patented when such plants/seeds are given patents.

(i) A common misconception is that humans have been doing genetic engineering for centuries via plant and animal breeding. How is genetic engineering decidedly different from the classic breeding of animals and plants, selecting for certain traits deemed desirable? (Note: Usually a rat cannot mate with a corn plant. Via genetic engineering rat genetic material can and has now been introduced into corn plants. Thus consider the role of species in ecosystems.)

(ii) It has been estimated that at least 60% of packaged food in the supermarket contains genetically engineered ingredients. Make your own estimate of this percentage at the time you read this.

(iii) How much choice do you have in whether or not you eat genetically engineered food?

(iv) Life is perhaps the most complex phenomenon in Nature. What does the Principle of Unintended Consequences say about genetic engineering of life forms? Does this Principle argue against humans doing anything?

(v) On the April 21, 2008 edition of www.democracynow.org we have: "In other food news, a new study by the University of Kansas has found that genetic modification reduces the productivity of crops. *The Independent* of London reports the study undermines repeated claims that a switch to the controversial technology is needed to solve the growing world food crisis. Researchers found that genetically modified soya produces about ten percent less food than its conventional equivalent." Did this story appear in, say, *The New York Times*? Examine major news media in the U.S., including National Public Radio, for any bias for GE.

Patterns of Inheritance and Mathematics. You no doubt have heard of Gregor Mendel who was an Augustinian monk who did experiments with heritable traits of peas in the 1860s. He discovered certain patterns were approximately followed as one went from one generation of peas to another. *A classical example.* There were some pea plants which when mated amongst themselves always produced tall offspring. These are referred to as "pure" or "true" breeding peas with respect to "tallness." Assign to such pea plants the label TT. (Why I do this will become apparent in a minute.) Likewise, there was a class of pea plants that when mated amongst themselves always produced short pea plants. Assign to these plants the symbol tt.

When Mendel bred "pure tall." i.e., TT, peas, with "pure short," i.e., tt, peas, he got all tall pea plants—no short plants and no "in between short and tall" plants. He thus called "tallness" a *dominant* trait, and "shortness" a *recessive* trait. The parent-plants in this experiment are referred to as the P (for parent) generation. The offspring of the above mating of two parents with a decided difference in the one trait, "tallness," is called the F_1 (first filial) generation. After many trials Mendel observed that in the F_2 generation, produced from mating among the F_1, that $\frac{1}{4}$ of the plants were "pure short," $\frac{1}{4}$ were "pure tall," and $\frac{1}{2}$ were tall, and did not breed true tall. But, again after many trials, on average if two plants from this last "half" were mated, $\frac{1}{4}$ were "pure short," $\frac{1}{4}$ were "pure tall," and $\frac{1}{2}$ behaved like the F_1 generation.

Mendel developed a simple mathematical "coin toss" model that was consistent with his results, i.e., provided an "explanation." Assume each pea plant has two genetic "particles," and there are three possibilities: TT, tt, and Tt. Assume that each parent contributes one "particle" to a given offspring. If a TT is mated with a TT, then each parent contributes a T, thus the offspring is again a TT. The same pattern holds if each parent is a tt. If one parent is a TT and the other is a tt, then each offspring (in the F_1 generation) must be a Tt, hence "tall" if T is the dominant "particle."

Now comes the interesting part. If two offspring in the F_1 generation are mated, what are the possibilities? Let's say we have two parents: "father," $T_f t_f$, and "mother," $T_m t_m$. What are the possibilities for the offspring now? Since father contributes "one particle," either T_f or t_f and mother contributes "one particle," either T_m or t_m, we have *four* possibilities: $T_f T_m$, $t_f T_m$, $T_f t_m$, $t_f t_m$. One-fourth will breed true tall, $\frac{1}{4}$ will breed true short, and $\frac{2}{4} = \frac{1}{2}$ will be tall but breed like the F_1 generation.

In the following exercise I will let you think about why this model gives the same result as "tossing a coin" twice, if T is heads and t is tails. Now it turns out that life is not always this simple. There are plants with traits where neither is dominant; matings give a blended result. A plant, the "four-o-clock," sometimes has white flowers, sometimes red. If a white and red are mated the offspring are pink, cf., [152, p. 266].

Independent, "coin toss," models apply if "particles," which we now call genes, are *independent*, viz., they are on different chromosomes and act as if other genes do not exist. We will look at such an example in part (viii) below. If two genes are on the same chromosome, they will tend to be inherited together and not act independently, cf., [152, p. 266].

Exercise 5.9 The Simplest Patterns of Inheritance

(i) Consider two coins with sides H (heads) and T (tails). If you toss one coin twice, what are the possible outcomes written as a pair of letters?

(ii) If you toss both coins at once, what are the possible outcomes written as a pair of letters?

(iii) Is your answer to (i) the same as your answer to (ii)?

(iv) In (i) is the first toss "independent" from the second toss? (Or are they linked in some way?)

(v) In (ii) is the outcome for each of the coins independent of the outcome of the other coin?

(vi) Compare your possible outcomes for either (i) or (ii) and compare them to the possible outcomes of mating two pea plants from the F_1 generation with each other. Are all of these three sets of outcomes basically "the same?"

(vii) Suppose p and q are two positive numbers that sum to 1, i.e., $p+q=1$. Suppose a gene has two alleles, A and a, with A dominant. Suppose that in a large population, where mating is random, that a fraction p of the population has allele A, and fraction q has allele a. See if you can that the fraction of the population that is AA is p^2, the fraction that is aa is q^2 and the fraction that is Aa is $2pq$. If $p = q = \frac{1}{2}$, you recover the F_2 generation discussed above with respect to "tallness." (Note the F_2 generation results from mating among the F_1 above.) The Web site http://www.k-state.edu/parasitology/biology198/answers1.html may be helpful. Note the binomial expansion: $(p+q)^2 = p^2 + 2pq + q^2$.

(viii) Suppose that c is the gene for cystic fibrosis. Assume everything is simple, i.e., you exhibit cystic fibrosis only if your genotype is cc. The "wild type," or normal allele, C, is dominant and if you are CC or Cc you do not exhibit the disease. Suppose that hair color is determined by a gene D for dark and dominant, d for fair and recessive. Assume that hair color genes and cystic fibrois genes are inherited independently. Suppose the genotype of "father" is: $C_f c_f D_f d_f$; and the genotype of "mother" is: $C_m c_m D_m d_m$. Thus both parents have dark hair and do not exhibit cystic fibrosis.

Since we are assuming independence of the disease and hair color there is equal chance that the father will contribute one of the four possibilities to his child: $C_f D_f$, $C_f d_f$, $c_f D_f$,

$c_f d_f$. Replacing the subscript $_f$ with $_m$ we get the same for the mother. What are the chances that their child will exhibit cystic fibrosis and be fair? What are the chances that their child will exhibit cystic fibrosis and have dark hair? What are the chances that their child will exhibit cystic fibrosis? Can you answer this last of the three parts by ignoring hair color and looking at the example on "tall vs. short peas" already worked out?

5.5 Toxic Sludge Is Good for You!

A Major Design Flaw in Our Civilization. The flush toilet is a candidate for the single greatest thing that differentiates "modern living" from the "primitive." Before the industrial revolution and crowded cities, human waste invariably ended up in the soil, via outhouse or emptied bed pan, not far from where it originated. With cities eventually came the invention of flush toilets that moved human waste, using water for transport, to the nearest lake, ocean, river or stream to prevent the obvious health problems of letting "it" pile up within city limits. This, of course, resulted in water pollution—but it was out of sight of the city. In time this problem could have been solved with sewage treatment plants that removed human waste from water and returned that human waste to the soil—where it would have gone in the first place. But developing industry, a driving force for the formation of cities, also had its waste—a usually toxic brew.

The *great mistake* was to do the easiest thing, let industry dump its toxic waste into the same sewer system that carried human waste. Along with this came the common practice of citizens dumping all manner of unwanted waste into this sewer system as well. A *second great mistake*, made in hundreds of U.S. cities, was to combine rainwater runoff with sewage. One of the functions of rainwater is to recharge the groundwater; a job that is thwarted by lack of absorbent urban landscaping and insufficient use of porous paving materials. Thus "When it Rains it Pollutes" since many sewer systems cannot handle the volume of rainwater runoff that occurs when it rains. *"In the last three years alone, more than 9,400 of the nation's 25,000 sewage systems—including those in major cities—have reported violating the law by dumping untreated or partly treated human waste, chemicals and other hazardous materials into rivers and lakes and elsewhere, ..." "More than a third of all sewer systems—including those in San Diego, Houston, Phoenix, San Antonio, Philadelphia, San Jose and San Francisco—have violated environmental laws since 2006 ..." "As cities have grown rapidly across the nation, many have neglected infrastructure projects and paved over green spaces that once absorbed rainwater. That has contributed to sewage backups into more than 400,000 basements and spills into thousands of streets, ... Sometimes, waste has overflowed just upstream from drinking water intake points or near public beaches."*[20] In all of the industrialized countries, populations

[20]Quotes are from: "Sewers at Capacity, Waste Poisons Waterways," by Charles Duhigg,

grew to the point where the sewage from the city upstream was no longer "out of sight" of the city downstream. In the United States the tipping point came in 1972 with the passage of the Clean Water Act. Sewage treatment plants were built to "clean up the water" before it was discharged to its next user. Since the purpose of the treatment plant is to remove as much pollution from the water as practicable, what remains is pollution—a mixture of human waste, household waste, and toxic industrial waste—toxic sludge.

The problem then becomes: what does a city do with its toxic sludge? To be honest, I had not thought much about this until I read *Toxic Sludge is Good for You!*, cf., [561]. This book is primarily about how the Public Relations Industry, (PR), manipulates our society; but Chapter 8 deals specifically with toxic sludge. This book provides a wealth of details on how the EPA, various corporations and municipalities participated in a public relations campaign to rename *toxic sludge* as *beneficial biosolids*. There are five methods for dealing with toxic sludge; one of which, ocean dumping, has been banned. The method of gasification (using sludge to make methanol or energy) is most environmentally sound, but the most expensive, according to EPA *whistle-blower*, Hugh Kaufman,[21] The EPA opted for the least expensive method, encouraging toxic sludge to be spread on farm fields.

The farms around the small Colorado town of Holly (population less than 2000) received 25 tons of New York City sludge per day for a couple of years until a citizen "revolt" stopped it. Over the objections of local residents, farms around the impoverished Texas town of Sierra Blanca (population about 500) on the Mexican border were chosen as the final resting place of New York City toxic sludge. Hugh Kaufman was sued for going on the record with the following comments, cf., [561, p. 118], (he then countersued): *"This hazardous material is not allowed to be disposed of or used for beneficial use in the state of New York, and it's not allowed to be disposed of or used for beneficial use in Texas either. What you have is an illegal 'haul and dump' operation masquerading as an environmentally beneficial project, and it's only a masquerade. . . . The people of Texas are being poisoned."*

More recently in 2002 we read:[22] *"Earlier this year, Synagro, the nation's largest*

---

p. A1, *The New York Times*, Nov. 23, 2009. Data originates with the EPA and state officials.

[21] A whistleblower is one who speaks truth *to* power about the powerful as opposed to a snitch who informs the powerful about the powerless. Hugh Kaufman blew the whistle on toxic contamination at Love Canal, and Times Beach (Missouri), and is known for his fearless honesty. He had to withstand enormous pressures to maintain his integrity. He blew the whistle on EPA administrator, Christie Todd Whitman, just after the attacks on the World Trade Center in 2001: *"There's enough evidence to demonstrate that Mrs. Whitman's statement to the brave rescue workers and the people who live there [in lower Manhattan] was false."* Kaufman is referring to Whitman's statement of Sept. 18, 2001: *"I am glad to reassure the people of New York ...that their air is safe to breathe and their water is safe to drink."* See [238, p. 4]. Some of those brave rescuers have already paid with their lives, many with their health.

[22] http://www.organicconsumers.org/Toxic/sewage_sludge.cfm. The Web site of the Organic Consumers Association, and the Environmental Working Group, www.ewg.org are

sludge producer, paid a Greenland, New Hampshire family an undisclosed amount of money to settle a wrongful death suit—the first payment to an alleged victim of sludge induced sickness."

Box-flow models can be applied here. Large boxes containing water are "the oceans," "the atmosphere," "ground water." Candidates for subboxes are the watersheds of any group, e.g., city, a collection of people; a collection of animals or people; a collection of industrial facilities, etc. Candidates for flows are flows of water and pollution. With such boxes and flows in mind a little thought leads to at least one conclusion: overall the intervals of time between the moment water leaves a pesticide plant, a cow or a toilet until the moment someone drinks it are growing shorter and shorter, cf., Exercises 20.1 and 5.10.

A similar analysis works if you substitute the word air for water. As we shall see in Chapter 7, the supplies of water, especially fresh water, and the envelope of air that sustains us minute by minute—these supplies are far smaller, thinner more vulnerable to human impact than at first you might be able to imagine. How urgently you might be concerned right now about the sentiments above depends on quantitative details: Just how long is that interval of time between toilet and water fountain, car exhaust to lungs? The sufficiently motivated can take the mathematics of the following exercise far beyond what is required for this book.

Exercise 5.10 We All Live Downstream

(i) Would a ban on placing nonbiodegradable waste in the sewer system help make toxic sludge less toxic? Can you think of a system for handling all nonbiodegradable waste other than putting it in sewers?

(ii) Research the *composting toilet*. Estimate the number of flush toilets in some area you are familiar with that could be converted to composting toilets—which require no water, and produce compost for soil. Estimate the amount of water saved per year if such conversions took place. What are the minimal energy requirements of a composting toilet?

(iii) What are some of the *pathogens* in toxic sludge? Why are they a problem in sewage sludge? Are they a problem in soil?

(iv) Consider your last drink of water. Estimate the interval of time that had elapsed since it was flushed down a toilet or sent out as a discharge from an industrial facility? Note that bottled water is quite often just processed tap water—even some "famous" brands. Pick a couple brands of bottled water and research until you find the sources of that water.

(v) Confirm, debunk or refine the conclusion that in today's world the more people there are the shorter is the time between the moment water is one person's effluent to the moment it is a source for someone else.

(vi) Is the following statement true? "Nearly half the people in the world don't have the kind of clean water and sanitation services that were available two thousand years ago to the citizens of ancient Rome."[23]

---

two sources for current information on toxic sludge/biosolids from the point of view of organic consumers/farmers, respectively, environmentalists.

[23]This quote is from "The Last Drop: Confronting the possibility of a global catastrophe," by Michael Specter, *The New Yorker*, October 23, 2006, pp. 60–1. The United Nations Development Program issued a report in November of 2006, cf., *The New York Times*, Nov. 10, 2006, titled "Beyond Poverty: Power, Poverty and the Global Water Crisis," Kevin Watkins, the lead author. About 2.6 billion people have no access to a "bathroom,"

(vii) Compare the intervals of time from (iv) with the half-lives of pollutants. For example, how much time lapses between the moment water runs off an agricultural field (or a lawn) sprayed with your favorite biocide (or water soluble fertilizer) and the moment someone else drinks it? Compare this with the estimated half-life of the biocide. Note that the half-life of a substance is the length of time for half of a given quantity of a substance to "decay" or change into something else (also possibly toxic). Half-life of a toxic chemical can depend on environmental conditions. Half-life, as it refers to radioactive decay of a substance, is a physical invariant of that substance.

(viii) Investigate the prevalence of *endocrine disruptors* in your drinking water, food supply and in sewage effluent, both upstream and downstream from you. For a research project investigate the prevalence (they are everywhere), effective dosages (sometimes in parts per trillion), and effects (there are very, very many) of endocrine mimics and/or disruptors. I predict that if you do this project you will probably be surprised at your exposure, the possible effects—to the point you might get upset.

(ix) Pick a river and estimate the interval of time from the moment a bit of water is flushed down a toilet, or discharged from an industry, in city A until the moment it is taken from the river downstream as drinking water by city B.

(x) Estimate the amount of energy per liter city A, in (ix), expends processing its sewage, and the amount of energy city B expends processing drinking water (so that the water meets EPA standards, both as a sewage discharge and as drinking water.) Where does this energy come from?

(xi) From the point of view of physics and mathematics, air is a fluid, just as is water—only less dense and viscous. Repeat (iv), (v), (vii) with air substituted for water. Note that many pollutants may start out as liquids or solids but that they become airborne as vapors or small particles.

(xii) Can you document particles of dust and/or pollution traveling in the atmosphere from Asia to the U.S.? From Africa? Can you document any type of pollution traveling in the atmosphere from the U.S. to the Arctic or northern Canada? From the U.S. to another continent?

(xiii) Which fertilizers sold for home gardening contain toxic sludge?

(xiv) Starting say with the article by Katherine Tweed, "Sewage's Cash Crop," *Scientific American*, November 2009, p. 28, investigate the process described there where slow-release phosphorus fertilizer pellets are processed from Portland, Oregon's sewage by Durham Advance Wastewater Treatment Facility. What is the toxic content of these pellets, if any? Could such a process be applied to animal waste, e.g., sewage lagoons on industrial pig farms? Do humans eventually have to figure out a way to recycle phosphorus, a key ingredient in plant growth? To what extent does removing phosphorus diminish "dead zones" and algal blooms? Note that a key word associated with the above process is *struvite*.

Exercise 5.11 Two Boxes of Water: Oceans and Atmosphere

(i) Estimate the residence time of water in the box, the earth's atmosphere. Note: From [283] (a 1988 book) there is $1.3(10^{13})\ m^3$ of water in the atmosphere, and the global precipitation rate is $5.18(10^{14})\ m^3/yr$. Do you think these figures have changed due to global warming? Note that [91, pp. 20–2] gives precipitation on salt water to be $4.58(10^{14})\ m^3/yr$ and precipitation on land to be $1.19(10^{14})\ m^3/yr$.

(ii) Estimate the residence time of water in the earth's oceans. Note: In [283, p. 26–7] the exercise is done for you. However, all you need is the amount of water in the oceans, which is $1350(10^{15})\ m^3$ (in 1988), and the global precipitation rate, cf., Section 3.3. Any global warming changes? Note that [91, p. 20] gives $1351(10^{15}\ m^3$ for global salt water volume.

---

and more than a billion people drink, wash in, and cook with water contaminated with human and animal feces.

(iii) A very important topic which we have no room to deal with here is the present and looming shortage of clean, fresh water for much of the world's population. Given the stresses of population growth and pollution will water sources that are now managed by public utilities be privatized as were the trolleys? See [632, 33, 296, 694, 640, 699, 98, 587, 337].

(iv) There is another topic, quite important to a complete model, or any study, of pollution. Even with the strictest of laws intended to protect the public health, which we have not yet achieved by the way, there is the matter of enforcing those laws. Investigate the role of *organized crime* in the disposal of toxic and/or hazardous waste, cf., the next few paragraphs. With regard to Sierra Blanca mentioned above, Hugh Kaufman further endeared himself to many with the following quote: *"We're talking about government basically taking a dive for organized crime during an open criminal investigation."* See [561, p. 119].

Hazardous Industrial Waste in Your Garden or Farm Fertilizer? Toxic waste from sewage sludge is one thing, but can hazardous industrial waste be deliberately added to farm and garden fertilizers (as a cheap method of disposal, for example)—legally? From [728, p. 165], *"If some of the chemicals in a toxic waste can be used as a plant food, the toxic chemicals go along for the ride."* Thus the solution to pollution is dilution. This shocking state of affairs came to my attention through a series of newspaper articles[24] first pointed out to me by Adrienne Anderson.[25] These articles were written by investigative reporter Duff Wilson who later wrote the book *Fateful Harvest*, [728], about the subject.

You might be as surprised as I was to learn that substances deemed unsafe in our air and water—lead, cadmium, other heavy metals, radioactive products, assorted poisons, and so on—can be legally considered safe to spread as fertilizer on the soil of our farms and home gardens. Might a possible source of some of the toxic substances found in our blood, cf., Chapter 4, be the "recycling" of hazardous waste as farm (home garden) fertilizer? This practice is thought to be expanding, yet I have not found anyone who claims to have the "big picture" telling us just how much hazardous waste is being applied to our farms and to what effect. Among the enablers of this situation

[24]Thursday, July 3, 1997, "Fear in the fields: Part 1: How hazardous wastes become fertilizer." Friday, July 4, 1997, "Fear in the fields: Part 2: How hazardous wastes become fertilizer." Friday, July 4, 1997, "Throughout the country, example after example of hazardous wastes being turned into fertilizer." Friday, July 4, 1997, "Tag-along toxics." Thursday, July 3, 1997, "From factories to fields." Friday, July 4, 1997, "Two approaches to toxins in fertilizer." (The Canadian and American ways.) Thursday, July 3, 1997, "Heavy Metals in Fertilizers." July 3, 1997, "Here's what's known, and not known, about toxics, plants and soil." July 4, 1997, "Experts: How to reduce risk." July 3, 1997, "Resources on the World Wide Web." These articles were available on *The Seattle Times* Web site, and other places on the Web, at the time of writing.

[25]An instructor in Environmental Studies at the University of Colorado who was "let go" and her classes cancelled in 2005, against her wishes and over student protests. After fully investigating the case, the university's Faculty Privilege and Tenure Committee unanimously ruled that the university's academic freedom rules and her rights had been violated and called for her reinstatement. The university's administration refused to do so. She and her classes investigated the polluting practices of some of the world's largest corporations. She once won a $450,000 judgment (regarding a pollution case) which was reversed by the 10th Circuit.

are large corporations, government agencies, unsuspecting farmers and others, cf., [728].

Duff Wilson's exposé would not have been written were it not for the courageous whistleblowing of Patty Martin, the mayor of a small Washington (state) town, Quincy. She was led to discover this method of toxic waste dumping by failed crops, sick animals and rare diseases in and around her town. For her efforts she was almost run out of town.

The author of [718], Eileen Welsome, wrote a series of articles[26] about the "plutonium in your pancakes project" of then Environmental Studies instructor, Adrienne Anderson, wherein plutonium was shown to be one of the likely pollutants in certain fertilizers.[27]

Exercise 5.12 Fertilize Your Food with Hazardous Waste and/or Toxic Sludge.
(i) Investigate the contribution of toxic fertilizer to the stock of toxins in your body. Relevant references include [731, 728].

(ii) As of 2004 there were about 9,700 approved agricultural toxins. If you tested each of these individually there would be at least 9,700 tests (maybe more if other variables than the number of toxins are taken into account). I once had a conversation with a pharmacist who told me that if a person interacts with a number of chemicals, the order in which one interacts with those chemicals can be important to the outcome. If order is taken into account, what is the least number of tests needed if pairs of (different) toxins were tested for interaction/synergistic effects?[28]

(iii) Do you think the tests discussed in part (ii) have been done? How long would it take?

(iv) One-half of one millionth of one gram, i.e., .5 μg, of plutonium, Pu, is considered an amount sufficient to cause cancer in a human. How big is this amount? Note that Pu has a density of $\frac{19.8\ g}{cm^3}$, cf., [284]

(v) Explain why pollutants which persist for long periods, i.e., they have long half-lives, tend to be found throughout the biosphere. List all of the mechanisms for the transport of pollutants that you can think of.

(vi) Research César Chávez and Dolores Huerta. What does their work have to do with toxins in the U.S. food supply—with civil and human rights in general?

---

[26] "Dirty Secrets," *Westword*, April 12, 2001, Vol. 24, No. 33; "A Matter of Trust," April 19, 2001; and "Board Games," April 26, 2001. *Westword* is a weekly newspaper published in Denver, Colorado.

[27] If allowing toxics into fertilizers strikes you as "an experiment on the people," actual medical experiments on unsuspecting populations are not as rare as they should be. Besides the famous experiments of Nazi doctors in World War II, see: the plutonium experiments in [718]; as well as syphilis experiments in [574] (and Susan Reverby's work on experiments in Guatamala); and [318], for starters.

[28] What are the number of permutations of 9,700 things taken 2 at a time? See IV. The same question for three (pairwise different) toxins at a time. The same question for four (pairwise different) toxins at a time, and so on. How many tests are there (at a minimum) if you add all these answers up, from 1, 2, 3, 4, 5, . . . , (all the way up to) 9,700 toxins at a time?

5.6 Media Concentration

Essential information, this is food for the mind. I will deal in more detail with the global flow of information in VIII. In 1983 ownership of the *media*, i.e., TV, radio, other electronic media, music, newspapers, magazines, movies, books, journals, was distributed over 50 corporations. By 2004 more than half of all media was in the hands of 6 or fewer corporations, cf., [20]. I call this collection of media corporations, with some interlocking agreements/boards, the *megamedia.* It is a common assumption that if something happens that we need to know about, the megamedia will tell us in a timely fashion. I wish to challenge this assumption throughout this book. For example, there are some important, easily understood things in this book that many of my students see/learn for the first time—which they should have learned in elementary or secondary school. This is evidence for my thesis that: *The range of debate in the megamedia is so narrow that much important information does not reach sufficiently many people in a timely*[29] *fashion—resulting in the unnecessary perpetuation of problems.*

Exercise 5.13 The Megamedia

(i) What are the 5 or 6 corporations making up the megamedia at the time you read this? What is(are) your source(s) for this information? See VIII.

(ii) Give evidence either for or against my thesis that the megamedia range of debate is too narrow.

(iii) How many people actually have ultimate control over the information that is carried by the megamedia? For example, is there any example of information that all of these people did not want to disseminate and it was not disseminated—at least in a timely manner?

(iv) Give evidence for or against the proposition that the megamedia is deferential to power. Make a list for and against, and assign weights to each example of their relative importance to society.

(v) Is there any important information that the megamedia did not convey to the general public in a timely fashion in the recent past? Did anyone gain financially because of it?

5.7 Oceans: Rising Acidity and Disappearing Life

One of the first lectures I attended as a new faculty member decades ago was on *"CO_2 Emissions and the Mean pH of the World's Oceans,"* by the late Dr. Roger Gallet of NOAA. The picture he painted was something akin to a science fiction horror film, except that it was science nonfiction. Briefly, the story goes like this. As humans burn fossil fuels, on average about a tonne of

[29] For example, if you get correct information after it is too late to do anything about it, you did not get that information in a timely fashion.

C, carbon, in the form of CO_2, carbon dioxide, is put into the air per person per year.[30] More CO_2 in the air means more CO_2 dissolves into the oceans. Everyone, know it or not, is familiar with the result: carbonated water—the way soda pop is made by forcing CO_2 gas under pressure into water. The more CO_2 in the air, the greater the (partial) air pressure of CO_2 on the world's ocean water, the greater the amount of CO_2 goes into the oceans. Chemists tell us that when CO_2 goes into water, carbonic acid, i.e., H_2CO_3, is formed, i.e., we get "fizzy water."[31]

Now the more H_2CO_3 the more acidic, i.e., the less alkaline, is the ocean. Dr. Gallet put up several scenarios on the board: 10% of the CO_2 emitted goes into the oceans, 25%, on up to 50%, the worst case scenario he considered. By the way, pH is a measure of how acid (or alkaline, the opposite of acid) something is.[32] A pH of 7 is neutral, neither acid or alkaline. The pH of pristine seawater is 8 to 8.3, slightly alkaline.[33]

So what does this mean for life in the world's oceans? There are at the base of the food chain very small organisms that happen to need calcium compounds in order to build their tiny bodies.[34] The bones in many whales and other marine mammals as well as the bones in fish contain calcium which can only get there if these larger organisms have these smaller organisms to eat directly or indirectly. These smaller organisms need to extract and hold onto calcium from the ocean, and this becomes difficult to impossible if the ocean is not sufficiently alkaline. The little bodies' calcium parts essentially dissolve if the pH is too low. Most of the familiar life forms in the oceans, including coral reefs, cease to exist. This was the picture Dr. Gallet laid out in his lecture long ago. Possible death of the oceans as we know it, if the pH gets too low.

So what do we know now? The best science available, as reported[35] in [146],

---

[30] This is an average, with some people putting in a lot more, some a lot less. These days the average is a bit more than a tonne per person.

[31] Well, not entirely. The pressure used to make soda water, i.e., fizzy water, is higher than the atmospheric pressure of CO_2.

[32] The pH is measured on a logarithmic scale, like the Richter scale for earthquakes, cf., Section V. If the pH goes down one unit, say from 6 to 5, acidity increases by a factor of 10. If pH increases by one unit, say from 8 to 9, then acidity decreases, i.e., alkalinity increases, by a factor of 10.

[33] The pH of common substances are: milk (6.5), acid rain (5 or less), coffee (5), beer (4.5), orange juice (3.5), cola (2.5), lemon juice (2.4), stomach acid (1.5–2.0), battery acid (.5), blood (7.34–7.35), hand soap (9–10), household ammonia (11.5), bleach (12.5), lye (13.5).

[34] For example, a type of phytoplankton called coccolithophorids, which are covered with small plates of calcium carbonate and which float near the ocean's surface doing photosynthesis. Another, planktonic organisms called foraminifera, related to amoeba. Another, small marine snails called pteropods.

[35] This *Scientfic American* article, [146], is one of very few popular articles (see also the article of Kolbert, footnote [36] on the next page) I have seen dealing with the acidification of the world's oceans. Read this article for a deeper, more detailed description of what is going on than I have room for here. This problem is not well known to the public! Coupling global acidification together with other pollution and overfishing, cf., footnote [7], Chapter

tells us that *"The ocean has absorbed fully half of all the fossil carbon released to the atmosphere since the beginning of the Industrial Revolution."* This was Dr. Gallet's worst case scenario. Currently about one-third of *current* CO_2 emissions are going into the oceans. (Incidentally, as a smaller percentage of our carbon emissions go into the oceans in the future, if we continue with business as usual, the CO_2 in the atmosphere will increase even faster than it has in the recent past, accelerating global warming.) Thus roughly 40% of CO_2 emissions stay in the atmosphere, 30% are taken up by vegetation on land and about 30% goes into the oceans. The resulting overall drop in the pH of surface waters is about 0.1, which may not seem like a lot; but ecologically it is a relatively large change. A change in pH of 1 means a change in acidity by a factor of $10 = 10^1$. We will see that a change in pH of .1 means a change in acidity by a factor of $10^{.1} \approx 1.2589$, or a change of about 26%. If we go with a fossil fuel century, the pH will drop an additional 0.3 by 2001, and $10^{.3} \approx 2$. Predictions of ocean pH several centuries from now show a pH lower than at any time in the past 300 million years.

As the pH of the ocean decreases entire oceanic systems shift. Coral reefs, half of which were dead or dying in 2009, could literally all dissolve if the pH gets too low. Thus as Gallet warned, everything with a calcium carbonate shell/skeleton is vulnerable: oysters, clams, snails, many sponges, sea urchins, and on and on, including many of the aforementioned planktonic organisms at the base of the the food chain. All the fish and sea mammals that have calcium in their bones will have to find another way to get that calcium if that calcium does not enter at the base of the food chain as it does now; and it is not clear that such a pathway exists. But maybe the "worst" will not happen. I really do not know how the base of the food chain in the oceans will shift as the pH goes lower, or how the planet will respond as we continue to emit CO_2 in vast quantities. As the planet warms maybe there will be a rather rapid release of methane, cf., the PETM exercise below. (Why would this be an important event?) Mathematically speaking one thing is clear: we are dealing with an extremely complicated non-linear system with a multitude of synergistic interactions—and we are giving this system a strong "kick." Non-linear systems often respond unpredictably when perturbed! Indeed, some experts fear dire consequences if we go with business as usual regarding the burning of fossil fuels.[36] It appears that a smart goal for humanity would be zero net carbon emissions as soon as possible, but I see little indication that humanity is at all close to even discussing this possibility.

One question that has occurred to me to ask is this: What effect will this acidification of the world's oceans have on oxygen production? As you know, quite a bit of the world's oxygen is produced by photosynthetic activity in

---

7, it appears that sea life is at extreme risk.

[36]The popular article, *The Darkening Sea: What Carbon Emissions are Doing to the Ocean*, by Elizabeth Kolbert, in the November 26, 2006 issue of *The New Yorker*, pp. 66–75. See also [359] and [161].

the ocean. For example, it was discovered in 1986(!) that the "blue-green" bacteria, *Prochlorococus*, the most numerous life form on the planet with 10^{27} individuals, contributes about 20% of the oxygen in the atmosphere, cf., [161, p. 54]. Thus the O_2 in every fifth breath you take is thanks to the *Prochlorococus*. Other ocean photosynthsizers contribute an additional 50% of atmospheric oxygen. From the "box-flow" point of view there are three "boxes" that contain most of the oxygen atoms/molecules: the earth's crust and mantle contains most, 99.5%, an estimated $2.9(10^{20})$ kgs of O_2; the biosphere, 0.01%, an estimated $1.6(10^{16})$ kgs O_2; and the atmosphere, .36%, an estimated $1.4(10^{18})$ kgs O_2, cf., http://en.wikipedia.org/wiki/Oxygen_cycle. The residence times of O_2 in each of these boxes is, respectively, 500,000,000 years (crust and mantle); 50 years (biosphere); 4,500 years (atmosphere). About $3(10^{14})$ kgs O_2 flow between the biosphere and the atmosphere each year. Only $6(10^{11})$ kgs per year flow in/out of the earth's crust/mantle. How will this system change as global warming proceeds and the oceans become more acidic? I have not found any definitive predictions as to what will happen. However, the article on global phytoplankton decline, [54], does not make me optimistic. For example, the oceans phytoplankton account for approximately half of the production of organic matter on Earth, as well as about half of Earth's atmospheric oxygen. This article estimates oceanic phytoplankton biomass as a function of time since 1899. The author's estimate a global rate of decline of about 1% of the global median per year, with an approximate 40 % decline in phytoplankton biomass since 1950. Critics of [54] believe that increasing CO_2 in the atmosphere, and increasing nutrient runoff from lands into seas will increase phytoplankton biomass.

There are three possibilities: the new situation will produce more, the same, or less oxygen. Of course, less oxygen would be troubling. (Just how are those phytoplankton going to react, anyway; will any other organisms fill in for them if they fall by the wayside?)

If there is more oxygen something very interesting might happen. About 300 million years ago, during the Paleozoic Era, insects were much larger. Dragonflies had wingspans of 2.5 feet! It so happens the time of large insects coincided with a time when oxygen content of the atmosphere was 35%; it is 21% today. It has been proposed that the only factor preventing insects of today reaching relatively gigantic proportions is the oxygen content of the air.[37]

In summary so far, we are initiating changes in the earth's biosphere, including the oceans, rather quickly; and "recovery" of the system more to

[37]This theory was proposed in research done by scientists from Midwestern University, Arizona State University and Argonne National Laboratory presented in October, 2006, at meeting of the American Physiological Society. They arrived at this conclusion by studying the way oxygen is transported into and around an insect's body, via spirales and tracheae. In today's situation, for example, the largest a beetle can grow and properly oxygenate is 15 *cm*.

humans' liking may take hundreds, thousands, or hundreds of thousands of years—if such "recovery" is possible at all.

Exercise 5.14 Rising Ocean Acidity, Acid Rain, pH, and Menhaden Fish

(i) If going from a pH of 8 to 7 means multiplying acidity by a factor of 10, then going from a pH of 8 to 7.9 means multiplying acidity by what factor?[38]

(ii) Consider this quote: "Scientists measure acidity by the 'pH' scale familiar to every high school chemistry student. Since 1800, ice core measures show the ocean's average pH level has dropped from 8.2 to 8.1, making it 30% more corrosive, Feely says. Expected emissions will likely drop it to a pH of 7.9 this century, a 150% increase in acidity since 1800, he says." (Quote from : http://www.usatoday.com/tech/science/2006-07-05-ocean-acidity_x.htm, which is a news article about a study, "Impacts of Ocean Acidification on Coral Reefs and Other Marine Calcifiers: A Guide for Future Research," co-authored by Richard Feely of the NOAA Pacific Marine Environmental Laboratory in Seattle.) Do you understand this statement, and is it correct?

(iii) The effects of CO_2 on the world's oceans will likely not be uniform. Investigate the variability of this phenomenon.

(iv) How much of the biosphere's oxygen flow is generated by photosynthesis in the oceans? How does this compare with the oxygen flow generated by the green plants on land?

(v) One of the causes of acid rain is sulfur dioxide, i.e., SO_2, emissions from coal-fired power plants. If you want to begin a study of how such emissions affect the pH of rain water see [283, pp. 120–3].

It has been reported that the largest source of mercury pollution in the United States is the burning of fossil fuels. You can make a project out of examining all known toxic emissions of coal-fired power plants and examples of effects in specific cases, e.g., molybdenumosis. There are many reasons besides global climate change for switching from fossil fuels. Is there a body of water near you that is contaminated by mercury from burning coal to make electricity?

(vi) Investigate a previous episode of global warming that occurred 55 million years ago, the Paleocene-Eocene Thermal Maximum (PETM). Was the biodiversity of the oceans affected? Did the system recover? How long did it take and under what circumstances? For example, see http://currents.ucsc.edu/04-05/06-13/ocean.asp, a newspaper article about a research paper that appeared the week of June 13, 2005 in *Science*, by James Zachos, et al.

(vii) What is the status of the Menhaden fisheries at the time you read this. Menhaden have been called the most important fish in the sea. Why? See [204].

Oceans: Two Common Patterns of Destruction. There is a common *pattern* of exploitation of a resource that will surprise no one. Humans take the most valuable, easily available part of a resource first. We saw this in Chapter 1 with regard to oil. (This phenomenon is related to the concept of *entropy*, cf., VII. Resources with the lowest entropy, i.e., most order or value, are taken first.) We have seen it with regard to trees, where the biggest are gone first. Ore deposits that are the richest (and accessible with the technology available) are gone first. For example, an ounce of virtually pure gold could be found 150 to 200 years ago in many a gold nugget in the creek beds of Colorado or California; today over 400 tons of ore are mined and processed to get that one ounce of gold.

Thus, it has been with the resources in the ocean. According to [161] 90% of the big fish/mammals in the oceans are gone today, i.e., the populations of

[38] If you cannot do this problem now, try it again after you have studied V.

large beings have been *decimated*, i.e., reduced to one-tenth. Among these are the several species of whales, many species hanging on after having been driven almost to extinction. Various varieties of dolphins, porpoises, seals, walruses, sea otters, manatees, dugongs, polar bears and on and on, all mercilessly exploited, most decimated or driven to extinction. "Modern humans" have been working on this project for hundreds of years, all chronicled in [547, pp. 151–4] or [161], or for the "new world" by Farley Mowat in the classic *Sea of Slaughter*. Cape Cod was aptly named because of a resource which today is gone from Cape Cod. The 500-year-old Canadian cod fishery collapsed in the 1990s. John Steinbeck wrote *Cannery Row*, about a sardine industry that pulled 600,000 tonnes of sardines a year off the California coast in the 1930s; today gone. This last example is interesting in that small fish can become a large resource when they travel in enormous schools. And we rarely stop short of, at least economic, extinction. Example: the magnificent Atlantic Bluefin tuna, close to biological extinction, still has price on its head—in the 1990s one 200 pounder could fetch $100,000, cf., [161, p. 65].

And we don't even eat all that is caught. I was not really familiar with the term "by-catch" until one of my students, who had worked for an extended time on a fishing boat in the Bering Sea, a "bottom dragger" or trawler, wrote a paper on it. Scraping the bed of the sea, destroying immeasurable amounts of life in the process, hauling up the catch and then throwing away an immense portion in a deceased state because it is not the desired, or legally allowed, form of life—is routine. The official global estimate of by-catch is 8%, and for shrimp trawlers the official figure is 62%, said to be an underestimate by some. Longliners with thousands of hooks on central fishing lines of up to 100 kilometers (60 miles) are estimated to kill some 4.4 million sharks, sea turtles, seabirds, billfish, and marine mammals in the Pacific each year.[39] Even more sadly, nets, many-mile long lines with hooks, and lobster traps are often lost at sea and continue "ghost fishing," killing untold amounts of wildlife for unknown lengths of time, cf., [161, p. 61].

After peaking in the year 2000 at 96 million tons, the wild fish catch fell to 90 million tons in 2003, then rose to 92 million tons in 2006, but see the following exercise. Though the total catch may seem stable, we are "fishing down the food chain," in that as more popular species become rare, substitute species are turned to. Even the remaining "large" fish are getting smaller as the older (sometimes 50, 100 and even 200 year old!) fish are fished out. For example, in the 1950s the average blue shark weighed 52 *kgs*, in the 1990s 22 *kgs*, cf., footnote [39]. Aquaculture is growing rapidly (from 1 million tons in 1950 to 40 million tons in 2003 and growing by more than 3 million tons annually), but it is not without complications. Increasingly, wild fish are rendered into fish meal for carnivorous farmed fish, and sometimes wild habitats are essentially destroyed in the process of farming fish. Note that

[39] http://www.earth-policy.org/index.php?/indicators/C55/

farmed salmon consistently have significantly higher levels of dioxin than their wild counterparts, cf., [426, p. 22].

I will say no more here except that if we continue what we are doing the science is clear. In [736] an in depth analysis was done of the probable consequences of our current relationship to life in the sea. The conclusion simply put is that species have been disappearing from ocean ecosystems at an accelerating rate. If continued, by 2048 *all* fish and seafood species will have been decimated—officially collapsed, cf. also Chapter 7. We will have extended decimation from the large all the way down the food chain. We can avoid this fate by using the $15 to $30 billion spent per year (in current national government subsidies for the fishing industry) to instead do two things: (1) create "new job training" for unsubsidized fishers; and (2) create (by expanding upon the existing global marine sanctuary system) a global network of marine reserves protecting up to 30 percent of the world's oceans which would cost around $13 billion, cf., footnote [39]. If we do nothing but continue business as usual we will be left with algae to cultivate as the last "sea food." Nature will indeed have shown forbearance if we successfully rescue this last algal remnant, because of our second deadly common pattern.

The *second common pattern* is our propensity to take the easiest way out in disposing of what we consider garbage. Nature knows not the concept of waste; every being's effluent is a resource for another form of life—except for synthetic products not previously known in Nature which are not biodegradable, e.g., plastics. If you have never heard of the immense floating islands of human-made debris in our oceans you should do a little research, cf., [161, 92–107]. One of the largest is the Great Pacific Garbage Patch, estimated to be twice the size of Texas, in the northeastern part of the Pacific gyre, with a counterpart in the western portion. Plastic kills wildlife. Plastic mistaken for food clogs digestive tracts and kills.

Add to this some additional toxic waste: nuclear waste, chemical waste, farm and urban effluent and on and on. The oceans are not as immense as we might like to think, cf., Chapter 7. Of course, the planet and the oceans will go on, but they may go on without us as a survival adaptation! There has been an overall trend for hundreds of years, from forests to fish. In anti-Bio-Copernican fashion "modern man" has taken for himself, to the point of exhaustion, all things within reach; and what we are finding out is that there is less and less left to support our lives.

Exercise 5.15 Living Resources in the Oceans

(i) Does the term "resource," when referring to wild, living beings, indicate an implicit assumption about the relationship of humans to these beings in the grand web of life?

(ii) How many "garbage patches" in the oceans of the world can you find? (For example, Kara Lavendar, lead investigator of a two-decade long study by Woods Hole, Massachusetts-based Sea Education Association, found considerable plastic garbage off the Atlantic coast of the U.S.A.) How big are each of them? How are they geographically distributed? What effect does garbage, lost nets, lost traps and other lost fishing gear, and toxic chemicals have on the wildlife in the oceans?

(iii) It has been asserted the official statements of global fish catches, such as stated above, are vast underestimates—in some ecosystems the real catch is double the official figure. This is due to the fact that recreational and "small" fishers are not reported, cf., http://ipsnews.net/news.asp?idnews=43118, and www.seaaroundus.org. How well do we know the actual fish catch each year? Are there any mathematical techniques that could be used to take into account, at least partially, the concerns expressed? Complicating even the official fishing industry stats is the following quote from, cf., footnote [39]: "Note: Taking into account probable overreporting by China, the world's largest fisher, as well as climate-related fluctuations in the large catch of Peruvian anchoveta, the global wild catch has actually been falling for longer than the official records reveal—dropping 660,000 tons per year since 1988. For more information see Reg Watson and Daniel Pauly, 'Systematic Distortions in World Fisheries Catch Trends,' Nature, vol. 414 (29 November 2001), pp. 534–36." Discuss.

(iv) For a project, which requires a wee bit of calculus, investigate the concepts of Maximum Sustainable Yield (MSY) and Optimal Sustainable Yield (OSY) and how they are (and should be) used. For example, it is one thing to take the MSY as a *limit* which should not be exceeded under any circumstances, and quite another to take it as a *goal* to be achieved annually. A good place to start is with [90] which deals thoroughly with the mathematics involved. To what extent does taking whole ecosystems into account, rather than looking at each species individually, affect the mathematics?

(v) What is the Sea Shepherd Conservation Society, www.seashepherd.org? What has been their role in the world's oceanic ecosystems?

(vi) How many kilograms of fish (official estimates anyway) were caught per person in the world in 2000? In 2003? In 2006?

5.8 Stocks, Flows and Distributions of Food

A reasonably complete accounting of the stocks and flows of all food on the planet is a daunting exercise worthy of a book of its own. However, globally we have come to rely principally on three grains, i.e., cereal grasses: corn, wheat, rice, (which account for 85% of global grain production by weight) and marginally on others such as barley, sorghum, millet, oats, quinoa, amaranth etc. (Soy, an important food source, is a legume, hence in a separate category. It actually competes with grains for crop area.) Studying data on these grains is a solid measure of the world's food supply, since humans have gotten roughly half of their calories from grains for the last 40 years. The task of accumulating into a reliable three page summary the essential information (for the interval 1960–2007) on grains, globally, has fortunately been done, cf., [426, pp. 12–14]. From the box-flow point of view, if the matter in "the box" is the *global stock of all grains*, then the "flow in" is *global grain production* and the "flow out" is *global grain consumption*. A useful number to know each year is the *global grain production per person*.

Let's first look at the "flow in." In 1961 world grain production was $285\ kilograms/person * 3.069 * (10^9) persons = 874.38 * (10^9)\ kilograms$. In 1986 a peak per person production of 376 *kilograms* was reached, thus total production was $376\ kilograms/person * 4.918 * (10^9)\ persons = 1849.2 *$

(10^9) kgs. For the last two decades the growth in grain production has matched population growth, with roughly $350 * (10^9)$ $kgs/person$. So total world production has gone from approximately $350 * 5.267 * (10^9)$ $kgs = 1843.45 * (10^9)$ kgs to $350 * 6.614 * (10^9)$ $kgs = 2315 * (10^9)$ kgs, (actually $2316 * (10^9)$ kgs). After several years of declining harvest, 2007 was an all time record harvest.

One might think then that total stock of grain at the end of 2007 would also be high. Unfortunately, demand was so high that grain stocks fell to their lowest level in 30 years, namely $318 * (10^6)$ $(metric)$ $tons$. This might seem like a lot, but it is only 14% of annual consumption, or about 51 days' worth. World grain stocks generally tended upward from 1960 reaching a peak in the mid 1980s of a little less than $600 * (10^6)$ $(metric)$ $tons$, dipping for a while, regaining about the same peak in the late 1990s, and since then there has been a more or less steady decline. Contributing to this is the growing share, up to 17% in 2007, used to make biofuels like ethanol, cf., [426].

Consider some of the following details. In 2002, the world grain harvest of 1,807 million tons[40] fell short of world grain consumption by 100 million tons, or 5 percent. This shortfall, the largest on record, marked the third consecutive year of grain deficits, dropping stocks to the lowest level in a generation, [58, page 7].[41] In 2003 grain production was less than consumption again. In 2004 global grain production was $2.015(10^9)$ tons, breaking the 2 billion ton mark for the first time in history, and exceeding consumption for the first time in 5 years. World cereal stocks, however, continued their long-term decline, standing at about 80 days' worth of consumption in 2004, cf. [662, pp. 22–3].

Thus given all the above, it seems that there just might be limits on the amount of food we can grow, especially given the fact that most of our spectacular gains thus far depend heavily on fossil fuels.

Exercise 5.16 Grain Stocks and Flows and Population Growth

(i) What was responsible for the large growth in the global, monoculture grain harvests from 1961 to 2007?

(ii) How many kilograms are in a metric ton? Express as a power of 10.

(iii) Is the box containing the "global stock of grain" in steady state?

(iv) Approximately what percentage of land used for primary crop production in 2007 was devoted to genetically engineered grains, cf., page 115?

---

[40] Note that reference [662, p. 23] gives $1.850(10^9)$ tons of grain production in 2002. Is this a significant difference?

[41] Note that on July 6, 2004 the British Broadcasting Corporation's (BBC) science program interviewed Ken Cassman, of the University of Nebraska, and the International Rice Research Institute, Phillipines. Reporting on research results released that week Cassman stated that a 1 degree (Celsius) rise in nighttime temperature resulted in a 10% decrease in rice yields. Lester Brown, [58], notes that eroding soils, deteriorating rangelands, collapsing fisheries, falling water tables, and rising temperatures are converging to make it more difficult to expand food production fast enough to keep up with demand.

Distribution of Food: What are Necessary and Sufficient Conditions for Eliminating World Hunger? At the beginning of the 21st century there is (was) clearly enough food to feed all humans on Earth, *if* it were approximately uniformly distributed. (Uniform distribution of food means each person would get an equal share of what food there is.) This can be considered proven if one accepts the data in [674], viz., that just half of the food currently being *thrown away* in the United States could feed the world's approximately 1 billion hungry people. As if to confirm this there is a growing movement of "dumpster divers," or *freegans*, who live off the discards of American supermarkets, restaurants and so on. In particular, freegan Daniel Suelo, who lives in a cave an hour's walk from Moab, Utah, has abstained from all forms of money, including barter, for 10 years, cf., "A Simple Life," by Jason Blevins, *The Denver Post*, November 22, 2009, p. 1A.

But even as food production and stocks rose to unprecedented levels a decade ago at the end of the 20th century, malnutrition spread more widely than at perhaps any time in history. The World Health Organization (WHO) estimated in 2000 that fully half of the human family, some 3 billion people, suffered from malnutrition of one form or another, [211, pp. 6–7].

Consistent with these facts is the argument put forward by, for example, Nobel Laureate and economist Amartya Sen, that poverty, rather than food shortages, is frequently the underlying cause of hunger. Walden Bello, in *Food Wars*, [36], analyzes the root causes for the spike in prices of basic food commodities from 2006 to 2008, putting the cost of essential foodstuffs out of reach for vast numbers of people worldwide—with accompanying social unrest.

The industrial agriculture model is not working for billions of people; and if we continue along the present path *it will not be safe and it will not be pleasant to live in a world where a majority of people have nothing left to lose.* Abstractly, to the central decision makers (my term), food for people, feed for animals, and food into biofuels are interchangeable investment opportunities. Satisfying people's need for food first is not a priority of the economic model. Thus speculation on food futures—as a commodity, shifting food to the biofuels sector, and so on, all helped make food unaffordable for billions. When your income is one or two dollars a day, there is not much room to maneuver. Policies of "structural readjustment" enforced by the World Bank and International Monetary Fund (IMF) plus trade rules of the WTO weakened various nations' ability to grow subsistence food for internal consumption. Thus people around the world are desperate for an alternative, for "*food sovereignty.*" Bello suggests that that alternative might be the Via Campesino movement, which is a marriage of numerous small independently managed farms with judicious input from science attending to local food needs first.

If the global financial system implodes fully, as it came near to doing in 2008, cf., Chapter 2, as the fossil fueled industrial ag model "runs out of gas," we may all be looking for "an alternative," which I have indicated exists in this chapter.

Thus even in the presence of plenty, distribution for many reasons leaves many people hungry and malnourished. Imagine for a moment that by magic we could implement a distribution system that gives everyone in the world a sufficient diet year after year. If that problem were solved could we sit back and rest—mission accomplished?

How Many People Can the World Support? In *How Many People Can the Earth Support?*, [102], we find that the answer "depends." It depends on the aggregate ecological impact of the world population; the lower the impact per person the more people possible. If the current fossil fuel based system were sustainable I would guess, based on agricultural considerations thus far in this book, and consideration of human consumption of the Net Primary Productivity of plants (NPP) in Chapter 6, 10 billion is the max—since everybody needs to eat! But [74] argues persuasively that we already have more people than we can support if we do not have access to abundant fossil fuels.

At this point values and the level of understanding of where humans fit in Nature come into play. For example, over the long term Nature leaves no known niches that can support life, unoccupied by life. There are no "bio-vacuums." The antithesis of the BioCopernican Axiom, cf., Section 3.2, is "I am the only life form that counts." A slightly weakened form: "Life forms do not count unless they are a lot like me." Thus, the western hemisphere, what was to become the U.S. in particular, was fully occupied with life in 1492; and we should know how that turned out, although it is not always part of the curriculum. Indigenous peoples and ecosystems all over the world historically have similar tales to tell upon encountering "modern man."

By many measures, humans already occupy over 90% of the "space" available for life on earth. We are in the midst of the 6th great extinction of living species, this time at the hand of humans, cf., [388]. If "we" truly valued the diversity of life that remains on the planet, "we" would figure out a way to make room for it by controlling our numbers and our planetary impacts so that this diversity could survive. The problems Nature has put in the face of humanity are not all that subtle, and "we" just do not "get it"; we strive for more economic growth as "good" and refuse to build a modern culture around an approximately steady state economics. Unfortunately this leaves Nature the one "solution of last resort": *Death will be (is?) Nature's way of telling us to slow down!*

To avoid collapse we might study the examples of collapsed civilizations that have gone before, [547, 137, 678]. We might study examples of human societies that maintained stable populations, cf., [285], and did not suffer the simplistic "nasty, brutish, and short" experience. In [435] Kerala, India and western Europe are offered as examples of numerically stable populations. In [409] it is argued that connecting with non-human-centered Nature is essential for a balanced life. To avoid collapse I think we must acknowledge that there are limits imposed by Nature, [450], and avoid bumping up against them.

I will throw out the following for debate as a minimal list of candidates for necessary and sufficient conditions for a stable population with a steady-state economic system.

(1) *Universal education* that includes a deep understanding of just how we as a group and as individuals are sustained by our environment. Every stable indigenous population is intimately aware of the aspects of Nature that give life and where they as a people "fit in" to Nature. "Modern man" by use of technology, urban living, and specialization has managed to create a people that has no fundamental understanding or conscious connection with Nature or the processes that their lives ultimately depend upon. Some are so isolated that they are not even aware that Nature exists; and if aware, they consider not only Nature as expendable, but a multitude of people as disposable as well. Just in case it is not clear, universal includes women and men, girls and boys.

(2) *Universal health care* which includes adequate nutrition, freedom from hunger, disease and abject poverty to the extent that a deep sense of community can provide. Parents are more willing to "stop at one or two" if they are confident their children will outlive them and go on to belong to a just society, see (3).

(3) *Equal civil rights and equal justice for all.* Just in case it is not clear, this includes women. The system of justice must be real, a fundamental part of culture, and perceived to be fair by all.

(4) Evolve a culture of dispute resolution within society which *minimizes violence*, hopefully to zero. Settle disputes with competing societies with a minimum of violence.

Exercise 5.17 What is Your Model for Achieving a Stable World Population with a Steady State Economy?

(i) Are current rates of population growth a problem in some parts of the world? Which parts? Find an example of a society, likely indigenous, that was (is) numerically stable. How did (do) they manage this?

(ii) In [137, 547] it is argued that living beyond the capacity of available natural resources leads to collapse. In [678] this reason is discarded as too simplistic. Instead a more complicated lack of ability to technologically respond to challenges is part of the key to understanding the collapse of complex societies. What do you think is happening to your society?

(iii) Richard Louv, in [409] discusses the consequences of "Nature deficit disorder." Does your society exhibit the symptoms? Do Louv's arguments support the Bio-Copernican Axiom?

(iv) Is the United States a country with a stable population? Does the U.S. have a steady-state economy? See V.

(v) Discuss, debate, embellish, improve, expand, delete, study, or change the four conditions for a stable population with steady-state economy. Create a Bigger Picture in which we can solve the problems of population growth and an exponentially expanding economy.

5.9 My Definition of Food

I am not completely happy with the following definition of food, because I can already see how a clever public relations person might contort it to include some substance in the definition which I do not consider food. More likely, I can see how such an ingenious individual could make the definition look completely surreal and silly. It is probably impossible for me to define food so precisely that I would agree with whatever a marketer might do with it in the future. Michael Pollan's motto is: "Eat food, not too much, mostly plants." In his books he makes a clear distinction between "food" and "edible, foodlike substances." Perhaps I should leave it at that, but I am a mathematician and I must try to define it. Thus, my definition of what I consider to be *real food.*

Definition of Food: Food is a living or part of a recently living being (minimally processed) that it is culturally permissible for me to eat.

For me "recently" means that the quality has not deteriorated, so freezing extends the notion of recently. Some nuts and grains store well for a while before going rancid, and so on. Minimally processed allows for cooking or freezing, possibly home canning, but that is about it. Of course, I live in the real world and sometimes find myself eating, as Pollan would say, "edible, foodlike substances" which are too highly processed, overpackaged and contain ingredients which I have never heard of and cannot pronounce. The reference to "culture" rules out cannibalism, at least for the time being.

Exercise 5.18 Your Definition of Food
(i) Come up with your own definition of food.
(ii) How much of what you eat satisfies my definition of food? Your definition of food?
(iii) Why did I wait until now to introduce my definition of food?

5.10 Choices: Central vs. Diverse Decision Making

Folks who do not care if their food is genetically engineered, irradiated, "fertilized" with toxic sludge, grown on poor soils, "factory-farmed," or doused with herbicides-insecticides-fungicides-rodenticides, et cetera, can skip this section. However, if you would rather avoid, or at least diminish your exposure to, one or more of the foregoing list, fortunately, there are options.[42] A couple

[42]More than one study correlating pesticide metabolites in children's urine with children's diet found that eating organic food reduced exposure to pesticides in food from an "uncertain" level to a "negligible" one, [206, pp. 2–3], [702].

ways are to grow your own food, or be a hunter-gatherer. These options are not available to all of us, the latter because people wildly outnumber the wild beings on the planet (and wild beings are here and there polluted even more than humans are). Surprisingly the former is not available even to some farmers—who are caught working more than full time in the industrial model in an attempt to keep ahead of financial debts. However, just about anybody can grow some of their own food, if only some herbs in a window box in an apartment, or vegetables on a roof top or in a vacant urban lot.

Co-operating at some level with others in growing at least some of your own food is another option; for some ideas see, for example, [495, 263]. One of these ideas is CSA, Community Supported Agriculture. In one model of CSA a local farm has subscribers who pay an initial fee and then get an agreed-upon number of boxes of food per week during the growing season. Usually the option of working on the farm in lieu of some or all fees is offered. Another option is to just get to know some local farmers (are there any left in your area?) at their farms or at a farmer's market, and buy food directly from them. All of the options discussed so far provide the eater with more information than is typically available about food being eaten. You can know what the growing conditions of the food you eat actually are.

Most of us will be buying food at a market, which could be a farmer's market. If it is not, then there is the option of buying organically certified food. Like the 40 hour work week, weekends off, abolition of child labor—the offical USDA, United States Department of Agriculture, definition of "organic" was obtained only after a hard fought battle. Now it is possible to define organic in such a way that almost no food qualifies; but at a minimum most "organic folks" do not want food that has been genetically engineered, irradiated, and fertilized with "sludge," for example, to be labeled "organic." The USDA's original proposed definition of "organic" showed the heavy influence of industrial agriculture; and, by February 1998, two months after the USDA's publication of this definition in the *Federal Register*, 4,000 comments had been submitted—an unusually high number for a USDA notice. The USDA postponed the comment deadline and scheduled public hearings, and within a month the number of comments was up to 15,000! For example, the Rocky Mountain Farmers Union, (not to be confused with the Farm Bureau) weighed in with a cogent analysis, since it speaks for many small farmers (organic or not). By the time of the deadline the USDA had 275,603 letters, mostly negative—the most it had ever received during any comment period in its history. In response, antibiotics, sludge, GE, and irradiation were eliminated from the definition of "organic."

Maintaining a rigorous definition of organic will likely be an ongoing battle; but if the meaning of the word is lost, reliable access of the public to truly organic food will be lost as well. Definitions are very important (plus inspection and enforcement).

A "quasi-industrial" model of agriculture can participate in the production of organic food, as it is defined; and to a limited extent it has. There are relatively large operations which adhere to organic standards. There are also co-operatives consisting of groups of farmers who provide the inputs for large buyers. As oil and natural gas become more expensive, it will be interesting to see how our system of agriculture evolves in response. A perspective of organic agriculture is found in [206], including origins, principles, and "business issues."

One thing more, would an agricultural system alternative to the industrial model be affordable? The Economic Research Service of the USDA [43] has statistics on the overall percentage of the retail price to the consumer of domestically produced food that goes to the farmer, from 1950 to 2006. In 1950 it was 41% $(\frac{18}{44})$, in 1951 it was 42% $(\frac{20.5}{49.2})$, in 1952 it was 40% $(\frac{20.4}{50.9})$. From this point on the farmer's share of the retail cost of food declines, almost monotonically. For example, in 1960 it is 33% $(\frac{22.3}{66.9})$, in 1970 it is 32% $(\frac{35.5}{110.6})$, in 1980 it is 31% $(\frac{81.7}{264.4})$, in 1990 it is 24% $(\frac{106.2}{449.8})$, from 2000 to 2006 it is 19% (except for 2004 when it was 20%), i.e., in 2000 we have 19% $(\frac{123.3}{661.1})$, and in 2006 we have 19% $(\frac{163.2}{880.7})$.

You can think of the numbers in parentheses as follows. In 1951, 42% $= .42 = (\frac{20.5}{49.2})$, and in 2006, 19% $= .19 = (\frac{163.2}{880.7})$, where the number in the denominator is price/expenditures in billions of dollars paid by consumer, and the number in the numerator (also on billions of dollars) is the amount paid to farmers. The difference is the amount taken by what I have been calling the *central decision makers*. This number in 1951 was $49.2 - 20.5 = 28.7$. In 2006 this number was $880.7 - 163.2 = 717.5$. Now the only parts of the agricultural equation that are absolutely essential are the growers and the eaters. In the middle are the buyers, sellers, supermarkets, restaurants and so on who all perform various tasks. But they have been taking a larger and larger share of the food dollar as time has passed. This difference between what farmers are paid and what consumers pay is called the *consumer-farmer price spread.* This number has gone from 28.7 billion dollars to 717.5 billion dollars, which represents a change of 58% to 81% of the consumer's food dollar going to the "middle men." (On rare occasion some central decision makers are caught and found guilty in court of price fixing, bribery, cover-up, and more, cf., [400, 170].) So when you ask: "Can I afford organic?" think of doing a reverse squeeze on the middle men, to help you pay for your groceries. Not all middle men are going away, nor should they. But the bottom line today, is that you the consumer could save about 81% of your food dollar by dealing directly with farmers! This may not happen overnight, and there will always be some costs involved in getting consumers and farmers to connect—there will still be middle men for most of us—but there is plenty of room for reverse squeezing if times get tough.

Consider the following comments of an activist dairy farmer from Wisconsin, Jim Goodman, in the *Progressive Populist*, Dec. 1, 2009, Vol. 15, No. 21,

[43] http://www.ers.usda.gov/Data/FarmToConsumer/Data/marketingbilltable1.htm

page 6.

"Even in the toughest economic times, the corporate buyers and sellers profit while the farmers loose (sic)." ... "Agribusiness spends multi-millions on lobbyists. Their lobbying efforts are aimed at increasing their profits, not farmer income or benefits to the consumer. They lobby for more cheap raw imports, less labeling, less restrictions on pesticide use and weaker environmental standards."

Earlier in his article we read:

"... a recent study by the Lieberman Research Group showed that organic food sales account for only 3.5% of all food product sales in the US." ..."Conventional farm milk prices have dropped by nearly 50% over the past year. Dean Foods controls 80% of the fluid milk market in some states and 40% of the market in the US; their net profits more than doubled in the last year."

"Conventional hog farmers have experienced losses for two straight years. Tyson, the second largest food company in the US, controls 40% of the US meat market. They reported a profitable third quarter for every segment of their business, including pork."

"When the farm price for beef cattle dropped $0.08 per pound, consumers were paying $0.17 more per pound at the supermarket. Average retail beef processing margins across all companies, increased 13% over 2008."

"And guess what, none of that was caused by organic farmers."

"Corporate agribusiness has a problem with organic farmers because they haven't yet figured out a way to totally bleed them like they have conventional farmers. But as surely as corporate agriculture is working its way into the organic market, we suffer from their growing control."

Exercise 5.19 Common Assumptions About Industrial Agriculture. Determine as best you can the truth value (between 1 and 0) of the following assumptions. I believe the majority of mainstream media sources state or imply that the truth value of these assumptions is closer to 1 than 0. See [348] for arguments that they are false. Briefly discuss how you arrived at your answers. When you are done you might want to think through what you would consider an ideal system (or systems) of providing food, for yourself, your family, your community, the world.

(i) Industrial agriculture will feed the world.
(ii) Industrial food is safe, healthy, and nutritious.
(iii) Industrial food is cheap.
(iv) Industrial agriculture is efficient.
(v) Industrial food offers more choices.
(vi) Industrial agriculture benefits the environment and wildlife.
(vii) Biotechnology (including genetic engineering) will solve the problems of industrial agriculture.

Exercise 5.20 We Are the Center of the Universe Pattern of Behavior: WACU

(i) Do you detect any corporate behavior in the Hour Glass Industrial Agricultural system that follows the WACU Pattern of behavior, cf., Exercise 4.7.

(ii) If you answered yes to (i), can you name any of the corporations exhibiting the WACU Pattern of Behavior? Feel free to research the current state of affairs when you read this.

(iii) The global food supply has largely been corporatized. Will water supplies around the world be corporatized as well? What would be the consequences? The book [33] raises some issues worth thinking about, such as: Who would make such decisions? What does the WTO have to do with it? and so on. We will deal more with the "public ownership" vs. "private corporate ownership" discussion in VII.

(iv) Discuss the implications of corporate ownership of your genome? How about the air?

(v) Compare and contrast the government centralized control of agriculture in the former Soviet Union to corporate centralized control of agriculture in the United States and elsewhere as it exists at the time you read this. Make a list of similarities. Make a list of differences.

5.11 Correlations

Correlations. When he was 84, I had the good fortune of having lunch at a Berkeley, Calif., café with Professor Jerzy Neyman, one of the greatest statisticians of all time. As he smoked one cigarette after another I ventured to ask: "Decades ago didn't you create some of the foundational statistics used to link lung cancer with smoking?"[44] He replied that he indeed had done such work. Then, taking advantage of the fact that Jerzy Neyman was quite kind and polite I pointed to his lit cigarette and asked: "Don't you believe in that work?" He replied: "Of course, but that is statistics; I am a special case!"

Statistics Helps Us Find the Cause—But Cannot Prove It Alone. Much might now be noted: the intense addictiveness of nicotine, the variability of human nature, the fact that Jerzy Neyman probably would have lived 10 more years, not 3, if he had not smoked. But Professor Neyman's point was mathematically correct: Pure mathematical statistics never proves anything, especially about individual cases. What the math did show was that the more you smoke the greater are your chances of getting lung cancer.[45] One suspects causality, but to come closer to proving it you have to do more science and understand more deeply.

For example, statistical studies surely would show that rooster crowing strongly correlates with the sun rising. But from a purely mathematical point of view you don't know whether rooster crowing causes the sun to rise or vice

[44] Richard Doll in the early 1950s is credited with being the first to establish this statistical link between smoking and lung cancer. Self-taught statistician, Lawrence Garfinkel, and his colleagues at the American Cancer Society also deserve mention for their pioneering work beginning in the 1950s. Garfinkel was an ardent anti-smoker.

[45] Lung cancer is just one of the diseases smokers have to look forward to. In the *Los Angeles Times*, October 23, 2006, by Susan Brink, there is a report of the Copenhagen City Heart Study which is ongoing, started in 1976. Therein it has been found that smokers have a 1 in four, i.e., 25% chance, of developing chronic obstructive pulmonary disease, such as bronchitis or emphysema. Smokers lose an average of 6 to 10 years of life span. The study was published in the online journal *Thorax*. Note that lung cancer can be caused by other environmental factors such as radon gas.

versa—or neither. You need more information to conclude what causes what. For example, just because one class of events A correlates with another class of events B, a causal relation is not guaranteed. There may be a third class of events C, yet unobserved, that causes both A and B. Statistics cannot tell us everything we want to know, but it certainly is a most powerful tool in showing us where to look.

Thus, not long ago cell biologists documented that a single component of cigarette smoke, benzo[a]pyrene, causes genetic mutations in lung cells that are identical to those seen in the tumors of smokers with lung cancer. These mutations occur not only in the exact same gene, called p53, but in the exact same location within this particular gene, cf., [667, page xv]. Do we now have the proof that we seek? Scientists would say, not absolutely—but with high probability we've got it. Those with conflicts of interest can use this rigorous scientific position to confuse folks who do not do science into thinking that smoking is safer than it actually is. Once a smoking habit is established, however, rational thought is not dominant, since nicotine in some people is more addictive than heroin. (I leave verification of this last fact to the reader as an exercise!)

Correlation Coefficient. Given two classes of "events" or "variables" such as soda consumption and body weight, what would the statement "these two random variables are correlated with a *correlation coefficient* of $+1$" mean? Well, we are now dealing with a subject in statistics called *linear regression analysis*,[46] and it turns out that correlation coefficients are always numbers between $+1$ and -1.[47] A $+1$ would mean there is a perfect (linear) correlation. That means if you graph one variable with respect to another on a pair of co-ordinate axes, as you may have done in some mathematics class before, the graph would come out a straight line with all of the points of your experiment (studying the relationship between soda consumption and body weight) lying on this line.

Positive and Negative Correlations. In the actual experiment with these two variables the correlation coefficient was not $+1$, it was a bit less. However, the study revealed that for each additional soda consumed (per day) the risk of obesity increased 1.6 times, cf., [489, p. 200].[48]

---

[46] Nonlinear regression analysis is also possible. One way to do this is to *transform* data that appears to not lie on any straight line into data that does, using a trick—like applying a logarithm function. We will do this in the chapter on population models. This technique is also useful in studying many natural phenomena, like earthquakes, cf. [614].

[47] If you want to read more, [237] is as gentle an introduction to the subject for the non-mathematician as I am aware of.

[48] The correlation was not a perfect $+1$, it rarely is. Thus the points of the experiment did not all lie perfectly on the line in question, but the points were very close to this line. How do you find this line? It turns out you can find the best line graph approximation to a collection of data points with little effort by using a computer. Given a bunch of points, just tell the computer to do a *linear regression analysis* to get the line which best fits the data points.

What does a negative correlation coefficient mean? In our example above, if body weight were negatively correlated with soda consumption,[49]then the risk of obesity would have *decreased* 1.6 times for each additional soda consumed per day. The real experiment did not come out this way, however.

Positive correlations have been measured in many experiments and studies involving food. The correlation just mentioned is of interest since educational institutions have been promoting soda drinking via exclusive campus contracts.

Hours of TV watched correlates positively with obesity, it also correlates positively with increased blood cholesterol levels. Soda consumption is positively correlated with tooth decay. Perhaps more surprising, adolescents who consume soft drinks display a risk of bone fractures three to fourfold higher than those who do not, cf., [489, p. 200]. Product purchases of TV watchers positively correlate with the number of ads for that product shown on TV.[50] Hours watched of commercial TV correlate positively with caloric intake.

There is a positive correlation between high female status and low fertility across 128 different countries, cf., [426, pp 84–5].

Exercise 5.21 Positive, Negative, and Zero Correlation

(i) If two variables have a correlation coefficient of 0, what do you think this means?

(ii) Can you give an example of a pair of variables, not mentioned in the text, that have a positive correlation coefficient? Can you give an example of a pair of variables that have a negative correlation coefficient?

(iii) Can you give any health conditions that positively correlate with the "western diet?" For example, high meat consumption vs. colon cancer or heart disease; consumption of high fructose corn syrup vs. obesity; exposure to certain endocrine disruptors vs. obesity, do these positively correlate?

(iv) Can you find a chemical with the following property. Corporation X manufactures chemical C. Corporation X funds research on possible negative effects of chemical C. Such research funded by X never finds any negative effects. Similar research funded by non-corporate sources nearly always finds negative effects of chemical C. Discuss the notion of correlation as it relates to this situation.

(v) If you have access to a computer with a spreadsheet program, cf., V, you can easily start computing correlations, i.e., the *correlation coefficient*, between two variables represented by two sets of data. What you need to do is compute a *regression line*, a best linear fit to the set of points in a plane obtained by plotting one variable vs. the other. Given such a data plot, called a scatter plot, the spreadsheet program will compute the best linear fit (the regression line) and the correlation coefficient for you. Intuitively the correlation coefficient measures how tightly the scatter plot fits the regression line. If all the points in the scatter plot sit on the regression line, then the coefficient is ± 1, i.e., 1 if both variables increase together, -1 if one decreases as the other increases. See http://illuminations.nctm.org/LessonDetail.aspx?ID=U135 by the National Council of Mathematics Teachers for lessons easy enough for grades 9 through 12 or above. Another source explaining correlations is [237].

[49]Thus the correlation coefficient would have been -1, say, or a number like $-.8$ of $-.901$.

[50]Contrary to what some believe, advertising/propaganda does work. Otherwise, no one would spend money on ads.

Chapter 6

Mathematics and Energy

With the exception of humans and some chemosynthetic ecosystems powered by geothermal energy, all other known ecosystems on earth are powered either directly or indirectly by the sun, mostly via photosynthesis, which usually converts about 1% to 2% of incident solar energy into "plant energy." If we include geothermal with solar energy, then humans stand alone in trying to use sources of energy other than these.

In this chapter I will, among other things, outline an argument for the following claim: *The most economical (as in "cheapest"), fastest, and most reliable way to provide non-carbon based energy for human societies is to make use of solar energy in its various forms.* About half the solar energy that reaches the ground drives the hydrologic, i.e., water, cycle. This is truly an immense amount of energy. Wind, solar thermal, solar photovoltaic, ocean thermal, ocean wave, ocean (moon caused) tides, and "cool" geothermal (available at shallow depths nearly everywhere) are all sources of energy. "Hot" geothermal, that issues from the earth due to radioactive processes ("nuclear energy") within the earth, is plentiful in select locations. The good news is that it is technologically possible to power our societies using these energy sources. It is also possible to make the transition to these sources in measured steps over the next twenty years. This incremental process would give us and the rest of the earth and its ecosystems time to adjust to whatever unintended or unforeseen consequences this process might entail. It is also possible that "subunits" of society can start this process without waiting for national leadership, although the most efficient transition could be effected by enlightened leaders. Biofuels may have a role to play, but there are some major difficulties we will discuss. Nuclear power, which is the favorite of some, is not competitive with or as reliable as a well designed, sustainable, solar-driven system; and we will examine this assertion as well.

6.1 How Much Solar Energy Is There?

We need to be able to measure energy in order to discuss it. I go into more detail in VII about various forms of energy: electric, mechanical, chemical, heat, and how the units of measurements for each interrelate. In this chapter I will measure energy in terms of the metric unit of *joules*, i.e., J. I explain in some detail what a joule of energy is in VII. For now the important thing to know is that *energy* is different from *power*. The unit of power that I will use in this section is the *watt*, i.e., 1 W. By definition 1 *watt* is 1 J per *second*, i.e., $1\ W = \frac{1\ J}{sec}$. Thus "energy per unit time is power." Note that "power multiplied by time is energy." A unit of energy you are no doubt familiar with is the kWh, i.e., the kilowatt-hour. For electric energy you pay about 10 cents for 1 kWh. You can burn a 100 W light bulb for 10 *hours* with 1 kWh. (Do you see this?)

A kWh is a unit of energy, so it should be equal to some number of joules. For practice let's calculate that number as follows: $1\ kWh = 1\ 10^3\ W * 1hr = \frac{10^3\ J}{sec} * 3600\ sec = 3600(10^3)\ J = 3.6(10^6)\ J$.

I am going to need to use the units of *joules* and *watts* to discuss solar energy flows. Recall that $1\ TW = 10^{12}$ *watts* (a terawatt). From [411], a book from 1981, I get the following information:

"The sun radiates energy at a rate of about $3.9 * 10^{26}$ *watts (W). Of this, some half a billionth, or 172,500 TW* $= 172,500 * 10^{12}$ *W, falls on the top of the earth's atmosphere, and about 81,100 TW reaches the ground. The world's hydrologic cycle is driven by an energy of about 41,400 TW, and the total energy of wind (1200 TW), waves, and ocean currents is several thousand TW. Solar energy is fixed in plants by photosynthesis at the net rate of about 133 TW. The total flux of geothermal heat to the earth's surface is about 32 TW. In contrast, the world's population 4.3 billion people directly used energy—not counting the indirect use of solar energy embodied in food and fiber—at a total rate of about 9 TW (with a probable error less than 10%). This is a great deal of energy in human terms, equivalent to the energy that would be consumed as food by an average of 15 full-time slaves (each eating 3000 Cal/day) for each person on earth; yet it is only a ten-thousandth of the rate at which solar energy reaches the earth's surface. Because of the way in which humankind converts energy, however, that 9 TW is rapidly becoming a significant force in the workings of global climate—the greenhouse effect and global warming are the result.*

"In particular, approximately 8 TW of the roughly 9 TW is derived from burning fossil fuels—solar energy stored millions of years ago when a tiny fraction of the total plant matter was trapped by freak conditions in an anaerobic swamp where it was protected from oxidation. Through unique conditions extending over geologic time, pockets of fossil fuels totaling over 70,000 TW − y were trapped beneath the earth's surface." (Note for a more detailed account of how sunlight became stored as fossil fuels see [130].)

So what has changed since 1981? I will leave it to you to determine at the time you read this if the sun is coming up each day with approximately the same intensity it had in the 1980s. Two things have definitely changed: world population, and global energy consumption. As 2009 turned into 2010 the world population was about $6.8(10^9)$, up from $4.3(10^9)$ in about 1978. Above it states that the rate of consumption of (non-food/fiber) energy of these 4.3 billion people was about 9 TW, give or take 10%. In about 2009 that number has increased to 12.5 TW, according to [326, p. 60], where this number is a maximum not an average. Wikipedia estimates 15 TW, and that as an average. For 2005 the World Resources Institute, www.wri.org, gave global energy consumption as $11,433,918$ $ktoe$, where $ktoe$ means *kilotonnes of oil equivalent*. I will leave it to you in Exercise 6.1 to figure out what this means, since this is a common unit of measurement. In other words, is the wri.org estimate closer to 12.5 TW or 15 TW?

Exercise 6.1 Energy Numbers Where applicable in the following do a little research and determine your answers as carefully and quantitatively as you can, for the year you are reading this.

(i) Express the amount of solar power that reaches the top of earth's atmosphere in *petawatts*, i.e., PW, and *exawatts*, i.e., EW. See Table Greek Prefixes, page 60.

(ii) From 1981 to 2010, or until the year you read this, has the energy output of the sun changed? Increased, decreased, stayed the same, oscillated? By how much measured in *watts*?

(iii) Vaclav Smil, [641, p. xv], states that the solar radiation reaching the earth per year is 5500000 EJ. Is this in approximate agreement with the $172,500$ TW figure given above? Hints: $EJ = 10^{18}$ J, and how many seconds are there in a year?

(iv) From 1981 to 2010, or until the year you read this, is the earth absorbing more solar energy or less, or the same? By how much measured in *watts*?

(v) If $4.3(10^9)$ people have 15 slaves each, each eating 3000 Cal/day, how many *joules per second*, i.e., W, of power is that? Note: $1\ Cal = 10^3\ cal$, and $1\ cal \approx 4.18\ J$. Is your answer consistent with the text above?

(vi) Is it true that fossil fuel energy represents solar energy that was stored over a period of about 10^8 *years*?

(vii) One (metric) *tonne* of oil has 41.868 *gigajoules* $= 11,628$ kWh of energy, according to www.wri.org. First of all, check to see if this last equation is (approximately) correct. Next, compute how many joules of energy $11,433,918$ $ktoe$ represents. Then if this was the amount of energy consumed globally in 2005, what was the average power consumption in *terawatts*, i.e., TW?

Putting aside for the moment the possibility that there is enough geothermal energy by itself to more than satisfy global human needs for energy, [247, Chapter 5], let's see how far we can get with just direct solar and wind energy resources. The starting point is roughly $80,000$ TW of solar power reaching the ground, about half of which drives the hydrologic cycle—from which every hydroelectric power plant draws its energy, by the way. So let's start with $40,000$ TW of solar power. If we take the 15 TW figure and more than double it to 40 TW, we are looking at a little less than $\frac{1}{1000}$ of the solar power not already driving the hydrologic cycle. This represents roughly 8 or 9 hours of sunlight to capture one year's worth of energy for all global human uses. I am being very conservative, because you can find correct quotes in the literature, from 2007 for example, which say that the sunlight falling on the earth about every 70 minutes equals the total annual energy use of human's worldwide. Now although there is plenty of direct solar power/energy, it is prudent to get by on as little as possible; thus minimizing the chances of unintended consequences. Sunlight is actually doing many things, and we probably do not understand all of the details. We do know that about 133 TW is driving photosynthesis, so keeping global human use to a small fraction of that number is likely wise.

Now [326, p. 60] estimates that 580 TW of direct solar is still available after subtracting areas of the earth that should be protected, areas that are inaccessible and/or otherwise less than desirable to develop. Included in this excluded area are the open oceans. Consistent with these restrictions is the following calculation of the National Renewable Energy Laboratory (NREL). NREL determined that in the United States urban areas and residences cover $140(10^6)$ *acres*, and that putting solar photovoltaic collectors on just 7% of this area would provide all of the U.S.A.'s current electricity requirements, cf., [413, p. 16]. This is a rather remarkable fact that a small portion of areas already heavily impacted by human activities such as roofs, parking lots, highway walls and so on collect enough sunshine to power America's electric grid.

What about wind? In [326, p. 60] global wind power is estimated to be $1,700$ TW, somewhat more than the $1,200$ TW given in the article above. Applying the same restrictions we applied to solar, [326] estimates that 40 to 85 TW of wind power remains available for human exploitation. Again, wind is a vital part of the global ecology, just as is sunshine; so it would be prudent to take as little as possible to avoid unintended consequences. But that said, I think it is reasonable to assume, especially if we proceed incrementally, assessing impacts at each stage, that there exists an immense amount of available wind power that can be exploited with minimal upset to global ecology.

Thus it appears that there are many times more solar and wind resources than would be required to power a global human civilization with more than twice the 2009 population of $6(10^9)$, viz., 14 or 15 billion people. And there are geothermal, and ocean (thermal, wave action, and tidal) power as well, each potentially huge energy sources. As I have indicated earlier in this book, however, it is unlikely that human population will exceed $10(10^9)$, since limits of land, water and other resources will constrain the growth of human numbers before the limits imposed by the availability of energy are reached.

6.2 Solar Energy Is There, Do We Know How to Get It?

In a word, the answer is YES! The proof, of course, is in the actual doing. I discovered this the hard way while building a house, operating a farm, or doing experiments in physics and chem labs. Often the biggest obstacle that must be worked around in order to implement a "clear vision" turns out to be the behavior of fellow humans with whom you must work in order to accomplish a goal.

The complete paper proof of my claim that we can "do it" with solar requires more details than I can put in one chapter. I will, however, outline some of the major points in a moment. For those who really want to get deeply into this subject, I claim that sufficiently many of the rest of the details can be found in the literature. A short summary is "A Path to Sustainable Energy by 2030," by Mark Z. Jacobson and Mark A. Delucchi, *Scientific American*, Nov. 2009, [326]. A 235 page book that "proves" my claim is by Arjun Makhijani, *Carbon-Free and Nuclear Free: A Roadmap for U.S. Energy Policy*, available at www.ieer.org, the Web site of the Institute for Energy and Environmental Research, cf., [438], (to read this you need to know that $3414\ Btu = 1\ kWh$, where Btu stands for *British Thermal Unit*, a unit of energy). Finally, and perhaps by geographic accident, I am most familiar with the work of the Rocky Mountain Institute, in Snowmass and Boulder Colorado, www.rmi.org. The co-founder, Chairman, and Chief Scientist, Amory B. Lovins has as deep a grasp of the math, physics, economics, and practical problems of generating useful energy as I have ever encountered. He also has the hands on experience of having consulted with over 100 utilities, including coal, gas and nuclear facilities. There is a wealth of information about energy on the rmi Web site. In any event, these three different sources demonstrate, each with a slightly different perspective, that we have the technical knowledge to make the transition to a solar/wind driven economy rather quickly. Do sufficiently many of us have the imagination and political will to do it? That remains to be seen. Not to overstate the case, but Nature is testing humanity. We must pass the test, for if we fail—well I recommend Cormac McCarthy's book, *The Road*, for a taste of a possible post-apocalypse, pre-extinction future!

Increased Efficiency Lowers Consumption of Energy Without Reducing Useful Work Done. The amount of energy we derive from sustainable sources is growing, and the lower our total energy demands the sooner these renewable sources will meet our needs. We can decrease our demands for energy now by being more efficient. Amazingly, just within the United States, if each state used electricity as efficiently as the top ten most efficient states (in 2005) we could shut down 62% of U.S. coal-fired electric generation with no lowering of productivity. Since one of these states is California, you might say, well, they have a mild climate. The foregoing statement has already been adjusted to take into account the economic mix and climate of each state, cf., [426, p. 3].

According to [412, p. ix],

"The U.S. today wrings twice as much work from each barrel of oil as it did in 1975; with the latest proven efficiency technologies, it can double oil efficiency all over again."

From www.rmi.org:

"Denmark just grew its economy 56% without using more energy." (statements made in 2009) "Japan wrings 2 – 3 times more work from its energy than the U.S. does, ..."

As long ago as 1991 I was in a math office in Tokyo, Japan, during the summer. The university was connected to a "smart grid" which allocated electricity according to an "optimality algorithm." My energy use was much less than were I at home (in the U.S.), but I did not notice much difference in the office—while outside it was quite hot and humid. Things have come a long way since then, but the point is that whole-system design integration can often make very large (sometimes even tenfold) energy savings! (www.rmi.org)

One could fill many pages with examples of increased efficiency. The main lesson, however, is simple. We can use our brains to design systems of cities with houses, businesses, industries, that use much, much less energy than we are used to at present in the U.S. with the same level of comfort and productivity—and save money at the same time. The lower our total energy consumption the easier it will be to satisfy those demands with sustainable energy sources—and the sooner it can happen.

One observation from [326, p. 60] is that if we ran on wind and solar electricity, we could achieve additional efficiencies over fossil fuel or biomass combustion. For example, only 17% to 20% of the energy in gasoline is used to move a gasoline-powered vehicle, the rest of the energy is waste heat. An electric motor can convert 75% to 86% of electric energy into motion. The technical reason for this is part of the Second Law of Thermodynamics, which I explain in VII. There are theoretical and practical limits on the efficiency of any "heat engine" like a gasoline-powered car or a coal-fired power plant. Electric motors are not subject to the same limitations. While we are still using coal generated electricity, however, Lovins observes that saving one unit of electric energy saves three units coal energy, again, in part, because of the Second Law. Thus the *first step* in our program is to increase efficiency!

Decentralized, Modular Systems vs. Centralized Systems. Now I want to present a mathematical principle with a large range of applicability. For example, in the next exercise the modular system could be a combination of solar and wind generators vs. a large coal-burning plant, nuclear plant, or even a large concentrating solar collector plant! Electric generating systems built in stages consisting of several smaller units (or modules) have an *intrinsic mathematical advantage* over a single, large plant with the same total size. The large plant would need to have some built-in "economies of scale" if it were to have a chance of besting the system of smaller units; and it turns out that these economies of scale are rarely realized and do not compensate for additional advantages of the "modular system." The following exercise was inspired by [663, Chapter 4].

Exercise 6.2 Building Power Sources One Module at a Time vs. Big Central Power Plants: The Modular Method vs. the Centralized Method In this exercise we create 100 MW of new electrical generating capacity, and we are going to do this in two ways. Let us suppose that it takes 10 units of time to build one big central plant that generates 100 MW of power after 10 units of time (from beginning to end of construction), but it generates nothing during construction. I will call this the *"centralized method."* On the other hand, suppose it takes 1 unit of time to build a smaller power source which delivers 10 MW after 1 unit of time has passed from start (to finish) of construction. If we repeat this process, then after 2 units of time we will have our original first 10 MW power plant, which will have been in operation for 1 unit of time, plus an additional 10 MW just coming on-line. We can then repeat, adding a third 10 MW plant during a third unit of time, and so on. I will call this the *"modular method."*

(i) Suppose that building 1MW of generating capacity costs the same, whether in the 10 MW units or the big centralized 100 MW unit. What is the total cost of building 100 MW of generating capacity by either method?

(ii) To make the math easy, suppose that all power plants are generating electricity at full capacity (once construction is completed) all the time. After 2 units of time how much electricity has the modular method produced (and hence sold)? Same question for the centralized method. (If "unit of time" is too abstract, assume the unit of time is some fixed number of hours. So your answer will then be in MWh, megawatt-hours.)

(iii) Answer the same question as part (ii) after 3 units of time, 4 units of time, ..., 10 units of time, 11 units of time.

(iv) What is the total amount of electricity produced by the modular method after 11 units of time? Same question for the centralized method?

(v) Same questions as part (iv) after 20 units of time. Will the centralized method ever "catch up" to the modular method?

(vi) Which system is more reliable? In other words, what are the chances of all 10 modular units failing all at once vs. the chances of the big centralized unit failing ("all at once")?

(vii) All systems need maintenance. Discuss how the modular units can be maintained according to a routine maintenance schedule while never falling below 90% generating capacity. Can you say that for the big centralized system? Does this mean that during planned maintenance the centralized system economically falls even further behind the modular system?

(viii) Which system is easier to finance? Which offers the quickest return on investment?

(ix) If you represent a poor developing country, which method do you think will be most attractive?

(x) Can the modules be geographically distributed so as to minimize transmission costs and electrical transmissions losses? Does this option increase the reliability of the modular system? Is the modular method more efficient?

It turns out that in the U.S. about 98% to 99% of power failures originate in the grid! So the modular method, with many electric generating units geographically dispersed, will be more reliable from the customer's point of view just because on average there is less grid between a given customer and some source of power, cf., [413, p. 7].

For billions of people in the world their energy crisis is not being able to readily gather enough fuel-wood to cook or heat. I can clearly recall the image of a woman carrying a bundle of (cooking/heating) wood on her head: an image repeated in China, India, Nepal, rural Africa, Mexico, Peru, Honduras and doubtlessly in many places I have never been. One of the books that deals with the fuelwood crisis from the point of view of those suffering it is [474]. For people in a remote village a modest bit of assistance will enable them to build a small scale, independent solar or wind generating facility, and in a short time. Such folks will likely never be "on the grid," and would have to wait forever before a large scale energy facility of any kind were made available to them. Trading deforestation for renewable sun and wind energy that they can control benefits these populations in immeasurably important ways.

6.3 Four Falsehoods

Among folks who accept the fact that we must phase out fossil fuels, there are those who propose solutions that include the "nuclear option" and those that do not. Fortunately for the purposes of this book there are two bonafide "environmentalists" who are on opposite sides of the question as to the efficacy of relying on nuclear power, at least in the near future. These contrasting views are pedagogically useful because we must confront a reality. Labels such as "green" and "environmentalist" will likely become increasingly meaningless without at least a cursory examination of what exactly it is that is being proposed as being "green." As with all popular labels, they become adopted by folks with intentions quite independent—often exactly opposite, consider *greenwashing*—of the original meaning of the label. It is also not unheard of for people to argue sincerely for positions which later on have consequences quite opposite of what was intended. This is called a "mistake." Thus for any presentation of an important topic it is important to get beneath the superficialities and symbols as soon as possible and do one's best to understand the meaning of what is being proposed and the likely consequences. Looking for mathematical structures is often (though perhaps not always) of help.

Thus in October of 2009 Stewart Brand's book, *Whole Earth Discipline*, appeared which argues for nuclear power as essentially the only viable option to burning coal, if we are to honestly confront global warming. Stewart Brand was a co-author of the original *Whole Earth Catalog*, a classic of the early environmental movement. Amory Lovins, mentioned above, has published many articles on all aspects of energy which are available at www.rmi.org. In particular, his article, "Four Nuclear Myths: A Commentary on Stewart Brand's *Whole Earth Discipline*" and on similar writings, [413], does just what the title says. I have taken the liberty of calling the four nuclear myths just "Four Falsehoods," since the misconceptions therein are held not only by proponents of nuclear power but by many others as well; and myths, as I have used the term, often contain enough truth to make them useful—but these four are not helpful. For a more detailed and extensive discussion than I can provide here, please read as much as you can of the "pros" and "cons" of nuclear, solar, and wind power. The four falsehoods that we will discuss now, however, are at the heart of the disagreement.

Variability is not the same as Reliability: Solar/Wind Can be the MOST Reliable. The *First Falsehood* goes like this. *Since the sun does not shine all the time, solar photovoltaic panels are not reliable. Since the wind does not blow all the time, wind generated power is not reliable. Thus solar and wind cannot be relied upon to power our grid, since they are not "on all the time."* (Being "on all the time" is often equated to the term "baseload electric power," but this is technically incorrect, so we will not use this terminology, cf., [413, p. 5].) Brand's book asserts that fossil fuels, hydro, and nuclear are the only sources that can be "on all the time." Thus a grid cannot rely on solar or wind, unless massive energy storage facilities are built.

The picture in the wind/solar skeptics mind is ONE set of wind generators at rest because the wind is not blowing, or ONE set of solar panels at night. This picture applies to one individual or small community whose system is off the grid, isolated from society. Such a simplistic system can be made to work, I know of examples; but that is NOT what we are dealing with, with respect to, say, the electric grid of the United States, or any industrialized country.

I need to define two crucial terms: *variability* and *reliability*. The term reliability can be applied to individual electric generation plants, and it can be applied to the *entire grid.* Reliability (or more precisely unreliability) is measured by the amount of downtime a facility experiences due to *technical failure.* Variability, as it applies to wind and solar generators, refers to variation in output determined by variations in wind and sunshine. (There could be variation as experienced by the customer in the long (or near) term for coal, nuclear or hydro plants due to shortages or price spikes in fuel or water, but we will ignore this.) Let's look first at reliability in the U.S. of various types of generators and the grid from 2003–2007. Coal plants were down for scheduled or unscheduled maintenance 12.3% of the time, 4.2% of the time without warning, cf., [413, p. 5], [326, p. 63]. Nuclear plants were down 10.6% of the time, 2.5% without warning. Gas-fired plants were down 11.8%, 2.8% without warning. The technical failure rates for solar photovoltaics and wind (on land) is less than 2%, and less than 5% for wind turbines at sea, cf., [413, p. 6], [326, p. 63]. Since existing plants go down from time to time already, back-up contingency plans exist already as well, to keep the grid functioning in case of plant failure. But as noted above, 98% to 99% of power failures experienced by customers result from failures in the grid! Thus solar and wind generators very reliably convert sunshine and wind into electricity whenever available!

So is it possible to design a solar/wind system that compensates for the variability at any fixed location? The answer is yes, with very modest storage, not "massive" storage. According to [326] 3.8 million large wind turbines, each rated at 5 MW, distributed strategically worldwide would generate 51% of global energy. Another 40% of global energy would come from photovoltaics and concentrated solar plants, with $\frac{3}{4}$, i.e., 30%, from solar panels on rooftops of homes and commercial buildings. Solar energy is more evenly distributed than is wind. The model in [326] includes 900 hydroelectric plants worldwide, 70% of which are already in place. The analysis at www.rmi.org and the one in reference [438], though not identical with [326] reach the same conclusion, i.e., an economy based on solar and wind is technologically possible. The interesting thing to note is that by having "smart grids" the parts of the grid that are "up" at any given time can perform backup for those parts that are "down" to such an extent that storage of energy can be quite modest. A smart grid can also manage demand, creating new efficiencies and lowering costs. New transmission lines would have to be built, but the solar/wind system would displace the need for 13,000 new central coal or nuclear plants that would have to be built over the next 20 years, requiring their own transmission line upgrades of greater extent than that required by a solar/wind system. As pointed out in [326], the world manufactures 73 million cars and light trucks every year, so producing the wind turbines and solar panels needed is well within current manufacturing capabilities. Since reliability is a *statistical function* of the entire grid and its embedded generators, by being "smart" we can design a solar/wind electric system that is just as reliable as the one now, likely more so, cf., www.rmi.org.

By the way, storage technologies exist right now; and improvements can surely be expected. Right now, pumped storage (pumping water uphill when excess power is available, letting it run downhill to generate electricity when needed) exists, some not far from where I live. Compressed air storage, molten salts are other alternatives. Electrolysis, using excess capacity to break down water into H_2 and O_2, stores electric energy as chemical energy. (Not that there would be many, but current gasoline engines can be converted rather cheaply to run on natural gas or H_2.) There is a known chemical process for making various alcohols with the following inputs: water, carbon dioxide, wind electric energy. This is an additional method of energy storage, cf., Exercise 6.4. Alcohols can also be used as transportation fuels, cf., ethanol. A predominantly electric motor, battery powered, vehicle fleet could provide an immense storage capacity, charging while parked or giving energy back to the grid as needed. A national electric rail system could utilize immense amounts of energy in real time without the need for storage.

Of the four falsehoods, the first is the one too many people see as "obviously true," and it has a subtle mathematical content. The other three I will mention briefly and leave a more complete analysis as an exercise, cf., [426]. Thus the *Second Falsehood* is that *wind and solar systems require enormous amounts of land, more than central power plants like nuclear generators, hence are environmentally unacceptable.* I have already mentioned that sufficient solar energy can be captured on roofs of buildings, parking lots, and so on, without more ecological impact than already exists. Concentrating solar facilities could be more impactful. For example, they should not use water, which is often scarce in places with the most intense sunshine, rather molten salts or oil or some substitute for water. The necessity of the simple process of cleaning dust from solar collectors should not be overlooked. But solar clearly need not take up "a lot of space." Regarding wind, giant wind turbines are quite compatible with agriculture. Animals happily graze and tractors operate right up to the base of supports and between turbines. Income from wind turbines can have the beneficial side effect of helping farmers stay on their land. There is sufficient wind resource so that sensitive areas such as wilderness or major bird migration routes can be avoided. Improvements in design and siting have reduced bird kills such as occurred at Altamont Pass, Calif. More improvements can and should be made; but for those concerned about birds (such as myself): regulation of house cats, pesticides and cars; reducing light pollution at night (turning off skyscraper lights at night for the benefit of the many species of night migrating birds, for example), putting decals on all commercial and residential plate glass windows; would have more effect, cf., [247, p. 84], www.audubon.org.

The *Third Falsehood* is that *all options, including nuclear, are needed to combat climate change.* Nuclear power is not necessary, since solar and wind (or more generally a diverse portfolio of sustainable energy sources) can do the job more cheaply and safely and be installed more quickly. We will say a bit more about this later. The *Fourth Falsehood* is that *nuclear power's economics matter little since governments must nevertheless use it to protect the climate.* See [426]. If this same logic were applied to solar and wind we would do it first and be done with it. Wind power in the 1980s was prohibitively expensive at 30 cents per kWh. Today wind joins geothermal and hydro in the "less than 7 cents/kWh" category, and can be as low as 3 cents/kWh wholesale. Solar is relatively expensive at the moment with costs trending downward. Solar would be competitive if all subsidies were dropped for all forms of power. For example, private insurance for a nuclear power plant is not available at any price; more on this momentarily. And let's not forget that wind and sunshine are free energy, nobody owns the sun (yet).

Exercise 6.3 A Solar Powered Future is Possible: Inevitable?

(i) Go as deeply as you have time for into the details of a "smart grid" and how it can compensate for the variability of sunshine and wind at fixed points. What does a "smart grid" have to do with mathematics? What is the role of the management of demand? How does our knowledge of wind and weather patterns, satellite monitoring of the earth and weather prediction capabilities relate to a smart grid?

(ii) Investigate the costs of wind and solar electric generation at the time you read this. Innovations are frequent. What innovations in efficiency have not yet been deployed as you read this?

(iii) What methods of storage of energy are available or commercial when you read this?

(iv) Investigate the second, third and fourth falsehoods as deeply as you have time for.

(v) What sustainable sources of power have been commercially exploited and where, when you read this. For example, cool and hot geothermal, algae biodiesel, ocean (thermal, wave and tide) energy.

(vi) I happen to believe that for the most part we humans only do what Nature forces us to do. If we build a fleet of nuclear plants, how long do you think it will be before Nature forces us to build a renewable/sustainable energy infrastructure? Will it be easier or more difficult to build such a sustainable system after a "nuclear age?"

6.4 Nuclear Power: Is It Too Cheap to Meter?

My initial response is that it is too expensive to matter, especially when sustainable alternatives do exist.

Costs are High: Both in Dollars and to Democracy. No private investors can (will?) put up the capital to build a nuclear power plant, cf., [426, p.

18]. From [426, p. 19]: "...nuclear power *requires* governments to mandate that *it* be built at public expense and without effective public participation—excluding by fiat, or crowding out by political allocation of huge capital sums, the competitors that otherwise flourish in a free market and a free society." If this quote seems a bit harsh, come up with a list of counterexamples. France is not one of them. In the U. S. I would venture to guess that nuclear power is as divisive politically as, say, the abortion issue. Consider the fact that the American public is forced to provide "insurance" for nuclear power plants, since none exists otherwise. This is done via the Price Anderson Act, which transfers the bulk of responsibility of any nuclear plant accident to the general taxpayer. Even so, if you do lose your property, health, or life to a nuclear plant accident, you (or your estate) are likely to get no more than pennies on the dollar. Anyone who complains about the possibility of a catastrophic nuclear plant accident runs the risk of being called "irrational." It is not irrational to ask the nuclear industry and participating utilities, including shareholders, to put their money (and their financial existence) on the line and lobby the U.S. Congress for repeal of the Price Anderson Act—and take full responsibility for any nuclear plant they own or operate. It is very telling that *if most liability for nuclear power plants were not transferred away from industry, by say the Price Anderson Act, there would be no chance of another nuclear power plant being built in the U.S., period.* In fact, the ones that presently exist would likely be decommissioned if their owners/operators were liable for an accident.

Thus in the U.S., taxpayers must subsidize any nuclear plant construction and then insure it, with about as much effective public participation as was had in the "bailout of Wall Street" in 2008–9, cf., page 41. Since the people will end up paying for whatever is done, one way or another, given a choice, folks would most likely want to buy the cheapest, fastest, safest, and most reliable option. Associated with each of these four "variables" or qualities is empirical data that leads to a clear choice.

Nuclear Power is Not Carbon Free. It is asserted that nuclear power is "carbon free." This is not actually true if you consider the entire life-cycle of a nuclear plant. Nuclear power results in up to 25 times more carbon emissions than wind energy, cf., [326, p. 59] and [247, p. 165]. The construction of a nuclear plant, which uses steel and concrete, results in carbon emissions. (There is a new, experimental, low-heat cement making process, *green cement* using *biomineralization* that might be used to reduce carbon emissions from concrete, a significant achievement if it works commercially.) The mining and enrichment of nuclear fuel uses fossil fuels, at least currently, to a considerable degree. Transport and storage of nuclear waste and decommissioning of the plant also result in carbon emissions. Consider the following, [426, p. 1]: "Expanding nuclear power ... *will reduce and retard climate protection.* That's because—the empirical cost and installation data show—new nuclear power is so costly and slow that, based on empirical U.S. market data, it will save about 2 – 20 times less carbon per dollar, and about 20 – 40 times less carbon per year, than investing instead in the market winners— ...." Briefly the "market winners" are: efficiency and "micropower" such as we have discussed, and cogeneration.

Nuclear Power and Proliferation of Nuclear Weapons. Nuclear power is inextricably linked with nuclear weapons. In about 2009 and beyond there was (is) an international "problem" with Iran and their nuclear power program. Why? Because the process of enriching nuclear fuel, if taken further with the same technology, creates nuclear weapons grade material. The Iranians do not want "foreigners" violating their national sovereignty and inspecting what they are up to. They probably would like to have nuclear weapons to counter the nuclear arsenal of Israel. Tensions mount. This situation is not unique: consider India and Pakistan, North Korea and South Korea (backed by the U.S.), Russia and the United States, China and Russia or India, and so on.

There is the sentiment that the U.S. government promoted "Atoms for Peace" after WWII, at least in part, to morally make up for the fact that we vaporized ("nuked") some of our opponents in that war, cf., [82]. There has been an enormous subsidy for nuclear power originating in this connection with the military, without which it is doubtful any nuclear plants would ever have been built.

Building thousands more nuclear power plants increases the probability that nuclear weapons will be found in more and more hands. This is called nuclear proliferation, and it increases the chances of a nuclear attack or exchange at some level. Having witnessed the reaction of the U.S. and the curtailment of civil rights via the Patriot Act after being attacked with box-cutters and civilian airplanes, what sort of reaction would you expect after detonation of a "suitcase" nuclear bomb in a major city? One has every right to at least suspect that a nuclear fueled world will eventually be not only much more dangerous, but much more politically repressive.

Nuclear Waste. Nuclear waste, from mining, military activities, and nuclear power plants needs to be isolated from the biosphere, else serious consequences can result. The world has been waiting for over 50 years and no such isolation or disposal mechanism has been found and/or implemented. There are many claims and promises, but no implemented solution. There are many technical problems and social ones as well. How does one design a containment strategy that will function longer than, say, the length of time human civilization has existed—let alone the length of time any nation state has managed to "keep it together?"

Exercise 6.4 Nuclear Power, Renewable Power, Embodied Energy, Subsidies, and Democracy

(i) Estimate the *embodied fossil fuel energy* in a nuclear power plant. In other words, fossil fuel energy is used to mine and process all of the material that is used to build and fuel a nuclear plant: uranium mining and enrichment, iron/steel, concrete and so on. Fossil fuels are used to transport all of these materials to the construction site, and fossil fuels are used to do the actual construction—which takes years (and decommissioning). How much fossil fuel energy is embodied in the full life-cycle of a nuclear plant? What is the shortest length of time it takes to build a nuclear plant? What is the longest time it has taken so far (at the time you read this)?

(ii) How much time does it take for a typical nuclear power plant to have generated an amount of electrical energy equal to the embodied fossil fuel energy of the plant?

(iii) Is a nuclear power plant CO_2 emissions free? Eventually a nuclear power plant will start displacing CO_2, but at what cost? How much more CO_2 is displaced per dollar invested if that money is invested in cogeneration, renewable energy and efficiency instead of in a nuclear plant?

(iv) Repeat the analysis done in parts (i), (ii) and (iii) for wind power generators and solar collectors. How quickly can a 5 MW wind turbine be installed on land? At sea? How quickly can an array of solar collectors, say 1 MW, be installed?

(v) Is France, which gets the bulk of its electricity from nuclear power, a counter-example to any or all of the arguments given against nuclear power in this section?

(vi) Is the fuel supply for nuclear (fission) power virtually unlimited? What is the status of "Fourth Generation" nuclear power plants which are claimed to consume over 99% of the energy in their fuel and to produce waste with a half-life of hundreds of years (as opposed to millennia)?

(vii) What is the status of nuclear fusion power (on earth) at the time you read this?

(viii) At the time you read this and for a couple decades before, look up and compare the status of U.S. and foreign government subsidies, in their entirety, for fossil fuels, nuclear, and renewable energy.

(ix) Vandana Shiva gave up a promising career as a nuclear physicist in 1972 saying that: *"...nuclear power as much as nuclear war are systems where you cannot have democracy, they're inconsistent with democracy, and I love democracy too much."* (See the video/audio archive www.democracynow.org December 13, 2006.) What does she mean? Do you agree or disagree? Comment.

(x) What is cogeneration?

(xi) Given a free choice, in what form of energy generation would you like to invest your own money?

(xii) What is the status of research on nuclear fission reactors when you read this? What is the status of research on nuclear fusion reactors when you read this? Are nuclear reactors subject to Hubbert's Peak mathematics? (Note: Claims that nuclear reactors can make more fuel than they use are based on breeder reactor technology and fuel reprocessing. Are there any additional problems with breeder reactors or fuel reprocessing?)

(xiii) Robert Zubrin, in [744], points out that H_2O+CO_2 + wind electricity can produce various alcohols, which reduces carbon dioxide in the atmosphere and produces fuel. How old is this technology? Any improvements by the time you read this? Is this a way of storing wind energy? (Note: Zubrin emphasizes biofuels, which is not what I am talking about in this exercise, for reasons that will soon be apparent.)

(xiv) Has a solution of the problem of what to do with nuclear waste been solved to your satisfaction at the time you read this?

(xv) Compare the gallons of water consumed per megawatt-hour of energy produced for nuclear, coal, gas, solar, and wind generation, cf., [247, p. 167].

(xvi) A commonly held/promoted view is that the March 28, 1979 accident at the Three-Mile Island nuclear plant did not result in significant health effects on people. There are contrary views, for example, see Harvey Wasserman's interview on March 27, 2009 (www.democracynow.org), and his interview with www.fair.org. Research at least two opposing views of the accident and decide which, if any, have rigorous documentation backing up assertions. Where there is lack of documentation, what does that tell you? There are 2400 families who have filed a class-action lawsuit against responsible parties involved in the 1979 accident; as of 2009 they have not had access to a federal court.

(xvii) Has the Nuclear Regulatory Commission (NRC) ever taken illegal gratuities from a nuclear plant operator while at the same time failing to find license violations by said operator which were later determined to exist? See the interview with Arnie Gundersen, February 24, 2010, www.democracynow.org. How was the legal system used in an attempt to silence him?

6.5 Net Primary Productivity and Ecological Footprints

I would like to focus on sentences from the long quote in Section 6.1, that may lead to some concern, viz., "Solar energy is fixed in plants by photosynthesis at the net rate of about 133 TW. ... people directly used energy—not counting the indirect use of solar energy embodied in food and fiber—at a total rate of about 9 TW." At this time I do *want to count* how much of that solar energy being fixed in plants by photosynthesis is being used either directly or indirectly by humans. The answer is important, educational, and alarming.

There is a concept that attempts to describe the job photosynthetic plants are doing on the planet, viz., *net primary productivity, NPP*.

There are at least two ways to measure net primary productivity of plants: (1) the rate at which solar energy is fixed by plants, recalling that solar energy per unit time is solar power; and (2) the rate at which *kilograms* of carbon are fixed by plants. (Carbon is used because of its key role in photosynthesis, cf., page 104; and the fact therefore that "life on earth as most of us know it" is carbon based.) An estimate of (2) can be found in several places, one of which is [283, p. 257]. The estimate of (1) given above is 133 TW. The estimate of (1) given by [283, p. 240] is 75 to 125 TW. Elsewhere in [411] they talk of net photosynthesis being around 100 TW. I could argue that deforestation since the 1980s has decreased NPP, others may argue that global warming has increased it. Let's just agree to look at the order of magnitude, 10^2 TW as our working estimate of NPP. I now ask the question: What fraction of global NPP do humans appropriate (use directly or indirectly) to support their activities?

In [695] it is stated that "Nearly 40% of potential terrestrial net primary productivity is used directly, co-opted, or foregone because of human activities." A closer look at the paper reveals the following estimates: 3% NPP directly "eaten" by humans or their animals, 19% NPP eaten or directly used, and up to 40% eaten, directly used or indirectly used. For example, clearing a forest, building a city or road displaces plants, and so on. Such were the estimates of appropriation of NPP in 1986 when the global human population was $4.94(10^9)$.

Many articles have followed the classic paper [695], too many for me to discuss here. For example, more detailed studies have compared "NPP for specific geographical areas" with "human appropriation of NPP in that area." In some areas people consume far more than the NPP in that area, in other areas people consume far less than the NPP of the area in which they live. If some people are consuming far more of the planet's plants' productivity than exists in the "neighborhood" where they live, they must be consuming NPP in areas where other organisms live. Since almost all living things depend ultimately on the NPP of the planet's vegetation, some organisms will have to stop eating, that is *die*, for this to happen. This is because, as we have noted before, Nature "abhors a vacuum." Thus if it is possible for organisms to live in some area they will. There are no "blank" places on the globe where NPP can be taken without taking the food from something already living on that NPP.

Thus, we are led to the concept of *ecological footprint*, cf., [696].

Exercise 6.5 Ecological Footprint and NPP

(i) Can you describe some geographical area that is clearly consuming more of the earth's NPP than exists in that area?

(ii) If about 5 billion people appropriated 40% of global NPP, i.e., 40% of NPP was not available for use by any organisms other than those 5 billion people, what might be the implications for a world population of 10 billion people? 15 billion people?

(iii) If NPP is about 100 TW, and about 5 billion humans appropriated 40% of this NPP, how much is this appropriation measured in TW? How does this compare with the 9 TW of direct energy use of the 4 to 5 billion people in about 1980? Discuss.

(iv) Compare Exercise 7.16 and see if you can calculate the global ecological footprint of the human race. Feel free to research the papers/books/Web sites that have been written on this subject, cf., for example, [248, 688]. What role do fossil fuels play in facilitating the implementation of this footprint?

6.6 NPP, Soil, Biofuels, and the Super Grid

Whenever I hear about any program to produce *biofuels*, my first thought is: What will it do to the soil? If it comes down to a choice between eating (which requires soil to a large extent) and manufacturing biofuels which negatively impact soil—I choose eating. Fortunately there are, at least theoretically, attractive alternatives which do not harm soil.

Biofuels, Soil, and Food. If a biofuel is made from a food stock like corn or wheat, the price of that food will increase due to the economic "law of supply and demand." There are already large populations of people who are hungry at any given moment in time, even in the United States; any increase in food prices can thus result in extreme hardship for some, and possibly social instabilities. Indeed there are billions of people living on $1 or $2 a day who have no room to maneuver when it comes to buying anything, especially a necessity like food.

There is another "bottleneck" for biofuels. While solar energy is plentiful, it is inefficient to get it via photosynthesis if all you want is energy, not the biomass associated with photosynthesis. All solar panels and wind turbines have higher efficiency than the typical 1% to 2% energy-efficiency of green plants. Biomass energy is, however, in chemical form; thus it is a form of energy storage.

From [641, Chapter 2] the theoretical peak efficiency of the process of photosynthesis is 11%, but no plant comes close to this. On ecosystemic scales tropical and temperate marshes and temperate forests are about 1.5% efficient. Arid grasslands are around 1.0% efficient. The best rates for highly productive natural formations, e.g., wetlands, and crop fields are 2 to 3%. The highest recorded field values of efficiency under optimum conditions for short periods of times are 4 to 5%. What this means is that if 1 kWh of solar energy is utilized by a plant with 1% efficiency, the result is .01 kWh of energy stored as biomass.

Thus solar energy tapped via biofuels first faces the low efficiency of photosynthesis. Then any process that converts biomass to ethanol, biodiesel or other biofuel further reduces net efficiency. As mentioned above, much higher efficiencies are already available with photovoltaic and solar thermal collectors and wind turbines, for example. Biofuels, such as biodiesel, however, have the advantage that they are portable liquid fuels that store energy in stable form and can be used in already existing engines with minor modifications; whereas electricity is, at the moment, more difficult to store and use in some applications—such as tractors, trucks, airplanes, individual automobiles. (One airline has already successfully flown a jet internationally, partially powered by liquid biofuel.) A relevant concept here is *energy density*. Biodiesel stores more joules per kilogram than say an electric battery. (Energy density is also measured in units of joules per unit volume.) I note in passing that restaurant grease is a potential source of raw material for biodiesel that is actually so used at the University of Colorado. It turns out, however, that more money can be made turning such grease into soap and cosmetics, for example.

Exercise 6.6 How Much Biofuel Can the World Possibly Produce? This exercise is a really rough estimate of the actual situation, but it is worth doing.

(i) If you make biofuel out of a plant whose photosynthesis is 5% efficient and you have a process that creates biofuels from this plant's biomass that is 20% efficient, what is the net efficiency of this two-step process?

(ii) If 81,100 TW reaches the earth's surface, and the area of the earth is $5.1(10^{14})\ m^2$, on average how much solar power reaches 1 ha? (Note: Clearly, the answer for a particular area on the earth depends on its position, i.e., equator or pole, time of year, time of day, and so on. Feel free to ignore these details if you wish. Also recall that 1 ha = 1 $hectare$ = $10^4\ m^2$ = 2.47 acres.)

(iii) To get 20 TW how many hectares would be required if we used the plants from part (i)? How does your answer compare to the area of the earth? What fraction of the earth's surface is required? What fraction of the earth's land area, not counting Antarctica or Greenland? (Note: Ice-free land area on earth is about $1.33(10^{14})\ m^2$.)

(iv) How sensitive are your answers above to the initial efficiencies assumed in part (i)?

(v) What difficulties might be encountered in actually implementing a biofuel program on this scale? What might be the consequences?

(vi) Look up the energy densities of diesel, gasoline, wood, electric batteries of various types and compare them.

Let's now take a detailed look at a biofuel like ethanol from wheat. I have already mentioned that this is a process fraught with problems: tendency to increase the cost of food, tendency to deplete soils. Is the energy return so great that it is worth it? The following discussion and exercise will answer that question.

Lester Brown, cf., [59, p. 29], makes the following observation. From 1950 to 1973 a *bushel* of wheat could be traded for a *barrel* of oil. In 1975 the ratio became $\frac{3\ bushels}{barrel}$, in 1990, $\frac{6\ bushels}{barrel}$, in 2000, $\frac{9\ bushels}{barrel}$ and in 2005, $\frac{13\ bushels}{barrel}$. From the Earth Policy Institute Web site we see: in 2006, $\frac{12\ bushels}{barrel}$, in 2007, $\frac{10\ bushels}{barrel}$, in 2008, $\frac{11\ bushels}{barrel}$.

Exercise 6.7 Wheat to Oil, Oil to Wheat

A *barrel* of oil contains 42 U.S. *gallons* and will yield 19.5 *gallons* of gasoline. According to [59, p. 34], under ideal conditions (in France) wheat yielded 277 *gallons* of ethanol per *acre*. (Ethanol has 67% the energy content of gasoline.) Suppose under ideal conditions you can get 60 to 70 *bushels* of wheat per *acre*. Suppose a typical farm operation uses 12 *gallons* of fuel per *acre* (assume it is gasoline, whereas it most probably is 8.75 *gallons* of diesel).

(i) How many *gallons* of gasoline is used to get 60 to 70 *bushels* of wheat? What is this in *gallons* per *bushel*?

(ii) Assuming these 60 to 70 *bushels* of wheat are converted to ethanol, what is the gasoline equivalent?

(iii) How many *gallons* of gasoline equivalent are we getting per *bushel* of wheat?

(iv) If you trade 13 *bushels* of wheat for a *barrel* of oil, then convert the *barrel* of oil to gasoline, how many *gallons* of gasoline are you getting per *bushel* of wheat?

(v) If the farmer wants a *gallon* of gasoline equivalent in fuel which is cheaper: trading wheat for oil or making wheat into ethanol?[1]

(vi) What implications might this exercise have for the world's food supply? How might things change, economically and otherwise?

(vii) Look up corresponding facts about corn, say grown in the midwest of the U.S. and do this exercise over to see the energy yield in terms of ethanol made from corn.

(viii) Investigate how much water it takes to make a gallon of ethanol from wheat, from corn. It takes water to grow the grain, and it takes water in the plant that converts the grain to ethanol.

(ix) What are the net carbon emissions of producing ethanol from wheat, or corn?

(x) One of the reasons I did the above calculations involving wheat, and indirectly petroleum, was due to the "intrinsic" value of a food commodity such as wheat. However, in this modern age of financial "innovation" the monetary value of an essential commodity such as wheat can be manipulated as a pawn in a much larger financial game. The same is true of oil, or any other commodity. See Exercise 2.7, page 38. Estimate the manipulation of the price of wheat for some ten year interval of your choosing. .

Exercise 6.8 The United States and Biofuels[2] Of the $2.3(10^9)$ *acres* in the United States, in 2004 about $450(10^6)$ *acres* were cropland, $580(10^6)$ *acres* were pasture/range land. These figures can change as water supplies and other environmental factors change. In 2004 for transportation the U.S. consumed about $60(10^9)$ *gallons* of diesel fuel, and $120(10^9)$ *gallons* of gasoline, both from petroleum. Taking into account that biodiesel energy density is slightly less than that of petroleum diesel and that gasoline engine systems are significantly less efficient that diesel engines, the equivalent in biodiesel consumption is

[1] I was in Brazil in 2006 and biofuel from sugar cane (650 gallons/acre) was sold at fueling stations at about half the cost of gasoline. It is more likely that corn, with 354 to 400 gallons of ethanol per *acre*, would be made into ethanol in the U.S.; that is yet another exercise which you can do, see part (vii) above.

[2] Information for this exercise and the next can be found in publications/Web sites for the United States Department of Agriculture, USDA, the Department of Energy, DOE, the National Renewable Energy Laboratory (Golden, Colorado), NREL, and university biodiesel programs such as the University of New Hampshire Biodiesel Group.

about $140(10^9)$ *gallons*. Rapeseed, a potential source of biodiesel fuel, can yield 100 to 145 *gallons* of rapeseed biodiesel oil per *acre*.

(i) How many *acres* of U.S. farmland are required to produce $140(10^9)$ *gallons* of biodiesel?

(ii) What fraction of U.S. cropland is your answer to (i)? Is this possible?

(iii) NREL, the National Renewable Energy Laboratory, did a study which indicated the possibility of producing $7.5(10^9)$ *gallons* of biodiesel from 500,000 *acres* of land under ideal conditions. These ideal conditions referred to farming algae ponds with high flows of nutrients and sunlight: the solar power found in the deserts of the American southwest, with nutrient levels such as are found in agricultural runoff, for example. How much land is required for $140(10^9)$ gallons of algae-biodiesel?

(iv) What promise and problems do you envision with algae-biodiesel?

(v) Switchgrass and miscanthus (elephant grass) are perennial grasses that grow on soils considered too poor for food crops. They also are credited with being able to improve soils, building carbon content, for example. Such plants can be used as sources of *cellulosic biofuels* (from 1000 to 1250 gallons/acre). What is the status of such a potential source of fuel at the time you read this?

(vi) I have not discussed the role biomass might play in the generation of electricity. Could biomass be used to provide quick on-off electricity as local backup for variable wind and solar? In the U.S. is methane gas from landfills a viable source of energy? Can cogeneration (of heat and electricity) increase the efficiency of a biomass energy plant? See [247, Chapter 6]. What might be the role of biochar, cf., [389], and page 112?

Exercise 6.9 The Geometry of Algae: The Power of Being Small

(i) Assume for simplicity that an alga is a sphere of radius r. What is the ratio of the area of the sphere to the volume of the sphere? Recall that the area of a sphere of radius r is $4\pi r^2$ and the corresponding volume is $\frac{4}{3}\pi r^3$.

(ii) The type of algae used to grow oil/biodiesel is microalgae, less than .4 mm in diameter. What is the ratio of surface area to volume for a sphere this small? Why does being small give a large surface to volume ratio?

(iii) Such algae cells can grow 20 to 30 times faster than typical food crops, can produce 15 to 300 times more oil per acre (depending on the species), and have a harvest cycle of 1 to 10 days (again, depending on species). It is the volume of algae that grows (and divides), nourished by absorbing nutrients through its surface. Using parts (i) and (ii) explain why being small allows algae to reproduce so fast, i.e., absorb more nutrients through its surface per unit volume in a given period of time and thus grow? (Note that some types of algae are nearly *half* oil!)

(iv) Algae will grow in polluted (recycled) water. For example, plastic tubes of sea water polluted by agricultural runoff or human waste sitting in the sun on pavement next to a sewage treatment plant. Algae does not compete with food, can provide oil and useful organic byproducts. Estimate (or look up) the amount of oil that could be produced from algae, in the U.S., in the world. See Exercise 6.8 (iii).

(v) What is the status of algae biodiesel at the time you read this? Is it commercial or still experimental (or forgotten)?

(vi) At the time you read this what is the status of the research project to find or create bacteria that create oil from raw ingredients like CO_2?

As Human Population and Its Impacts Go Up, We Will Likely Go Down the Food Chain. The previous exercise indicates that by going down the food chain to a lower trophic level, higher yields of biomass can be obtained.

A question I have asked/discussed before: How many people can the earth support? See [102], for example. There likely is no precise answer to this question. Such an answer depends on the technological capabilities of humans at the time, the availability of resources in relation to that technology and the variety of cultures present on the planet—and how much other species are valued/needed. According to Section 5.2, the answer also depends on what trophic level humans feed at. For example, eating whales, fish, cows and so forth have humans feeding on at least a trophic level of 3, higher if carnivores are part of the human diet. A diet consisting entirely of algae would at least lower our trophic level to 2 or less.[3]

Exercise 6.10 The NPP Tradeoff: Eating vs. Everything Else

(i) Assume the "rule of 10" going from one trophic level to another, cf., Section 5.2, assume there are "X" people on earth all of whom eat nothing but meat. How many more people could the earth support if all "X" people switch to an algae-only diet? Assume that all humans do not move around and do not impact global NPP except via direct ingestion/eating.

(ii) Assuming the figures from Section 6.5, viz., humans (and their animals) directly eat 3% of NPP, 40% of NPP is either eaten, directly or indirectly used, does this change your estimate in part (i)? By how much, roughly?

(iii) Do you think a significant number of humans will ever be subsisting at trophic level 1 or less?[4]

(iv) Pick a year, say, 2004, where world oil consumption is "Y" barrels in that year. Assume the world's human population is constant. What fraction of these "Y" barrels of oil will eventually be replaced with biofuels like biodiesel and ethanol? What fraction of these "Y" barrels of oil will eventually not be needed due to conservation measures and increased efficiency? Which of these two percentages is the largest? What impact do you think biofuel production will have on the world food supply? Now, in fact, the human population is not constant. It is growing with increasing energy demands and increasing food demands. Do you see a possible problem?

Regarding part (iii) of the last exercise, humans certainly have driven some animal species to extinction at least in part because said animals were tasty. One other reason for extinction is habitat destruction, which is part of the appropriation of NPP discussed before. For example, the passenger pigeon of America is extinct, cf., [546, pp. 168–170]. Whales are not abundant, due in part to predation by humans for meat and whale oil. As I previously mentioned the cod off Cape Cod and the sardines once canned on Cannery Row, California, are economically, perhaps biologically, extinct. Major fish populations around the world could collapse if present trends continue. American bison nearly went extinct, partly because of being eaten, partly due to war.

[3]I am definitely not giving dietary advice here, I am not recommending a diet of algae—even if delightfully, artificially flavored. This discussion is for critical thinking purposes only.

[4]A press release from the *National Environmental Trust*, October 19, 2006, cf., http://www.net.org/marine/antarctica_briefing.vtml tells of illegal industrial fishing in the waters off Antarctica wherein krill, tiny shrimp-like creatures, are being vacuumed up in vast numbers. Krill are near the bottom of the food chain in these rarely patrolled southern waters, providing food for seals, whales, penguins. The krill are largely fed to farmed salmon, which somewhat complicates the trophic number assigned to humans in this case. These fishers are also taking fish higher on the food chain, like Chilean sea bass (an PR invented name, by the way—what is this endangered fish's real name?). Parts of these ecosystems are under much stress.

On October 22, 2006, the BBC[5] and print media carried the story that hippopotamuses in eastern Congo could be wiped out of Viriunga National Park. This national park was home to one of Central Africa's greatest hippopotamus populations, 22,000, in 1988, according to the Web site of the Zoological Society of London. The population has been reduced to about 400 in 2006 because of a combination of war and predation by humans for food. Because the list of such stories is quite long, some scientists have seriously tried to compute the year when most humans will be feeding on a trophic level of 1 or less.

I want to end this chapter with a new view of humanity and its ecological niche. In the future each nation-state can be viewed as an organism of a special new type. These new organisms will tread as lightly on photosynthetic processes as possible (appropriating a minimum of NPP), and will mainly be powered by sustainable energy sources other than plants! Direct solar energy is, of course, a leading candidate for a source. Wind; ocean thermal, wave, and tide; and geothermal energy are all additional candidates. These new organisms will have built a new "brain" and "neural network" that more optimally provides energy to the "cells" that make up these new organisms. We will have to upgrade the existing electrical grid to what some folks call *The Super Grid.*

Exercise 6.11 Cost and Benefits of the SUPER GRID

(i) Investigate the costs and benefits of upgrading the electrical grid in your country to Super Grid status, i.e., a "smart" grid, capable of backing up wind and solar generators that are "off line" with others that are "on line" and producing power—to such an extent that minimal storage facilities need to be built. You can get started by looking at [247, Chapter 13], or www.rmi.org.

(ii) Does the technology exist already to create the Super Grid?

(iii)Estimate the costs and benefits of NOT creating a Super Grid.

(iv) A smart grid holds the potential for gathering enormous amounts of detailed data on the electrical use of individuals. This information, in real time, can be used to greatly increase the efficient use of energy. It can also be used to invade privacy. What regulations should there be on the use of such data? Who owns this data? Are there any technological developments that might be used to protect privacy while still allowing for efficiencies? See VI.

(v) For the last century municipally owned utilities (MUNIS) which are owned "by the people" of a given municipality have on average provided power more reliably and more cheaply than investor owned utilities. Not only are not many folks aware that MUNIS exist, they are unaware of this last mentioned fact, cf., [706]. Given that a Super Grid would be connecting a multitude of small, decentralized generators, would it make sense to have a nationally owned utility company, owned by everyone? Or possibly a federation of local MUNIS, much like the United States is a federation of states? What political obstacles exist to creating any Super Grid? To creating a National Energy Company, or a federation of MUNIS? How might these obstacles be overcome?

(vi) There is an interesting possible synergy between geothermal and lithium-based batteries. It is projected (in 2009) that one ton of lithium per month can be extracted from the wastewater produced by a geothermal plant built on the San Andreas Fault southeast of Palm Springs, California. What is the status of this technological development at the

[5]British Broadcasting Corporation

time you read this? Has a new battery technology replaced lithium-based batteries, eg., nano or ceramic technology? Or, has lithium production from geothermal plants become commercial, and has it become a key component of an emerging Super Grid?

(vii) In 2010 the Marin EnergyAuthority, MEA, (www.marinenergyauthority.org), began providing participating citizens of Marin County, California, with energy purchased directly by the citizens and not through PG&E, Pacific Gas and Electric, the former monopoly utility. According to Marin County Supervisor, Charles McGlasham, the effort took eight years due in large part to opposition from the monopoly energy utility. The arrangement of the MEA is midway between having a monopoly utility on one hand or a municipally owned utility, MUNI, on the other. The MEA went on the market and purchased energy, which in 2010 was 78% renewable, thanks to a 2002 California law referred to as "community choice aggregation." The MEA still delivers power via PG&E's grid and maintains the infrastructure with PG&E workers employed still by PG&E. According to McGlasham, PG&E wants to make it more difficult (practically impossible?) for other communities to follow in MEA's footsteps by promoting to the tune of 10s of millions of dollars, Prop. 16, of June 2010. (This would be a constitutional amendment requiring a two-thirds majority of a political unit to vote for adoption of a similar energy authority.) Did Prop. 16 pass? How is the MEA doing?

(viii) Two versions of the Super Grid can be summarized briefly in the words "centralized" and "distributed (or decentralized)." Discuss the political forces for and against each of these versions. Discuss the economic and engineering pros and cons of each of these versions. Discuss and compare the impacts to land and health of each of these versions.

(ix) Can "the" electrical grid be built to withstand impact from extreme solar storms (on the sun)? How often do such solar events take place, and what are their potential impacts? Is the current electrical grid that serves you built to withstand such a solar event?

Chapter 7

The Brower–Cousteau Model of the Earth

7.1 How Heavily Do We Weigh upon the Earth?

I once heard the late David Brower[1] end one of his talks with the following: *"Earth's air, water and topsoil are the three things that distinguish our planet from every other dead rock in the universe, and humans are doing their best to obliterate that difference."*

The sentiment of the above quote is clear. If you have read the preceding chapters you have seen plenty of evidence that humans collectively are having significant impacts on the biosphere. Humans supply the world's crops with more nitrogen than the traditional source, bacteria, [81, p. xiv], page 112. Humans have polluted nearly every niche of the biosphere on the planet, cf., Chapter 4. Humans are changing the climate of the entire planet and precipitating a great extinction more rapidly than any of the five great previous extinctions, cf., Chapter 1, [388]. Humans are significantly depleting the seas of life and even altering the chemistry of the world's oceans, cf., Section 5.7. This list, which can be lengthened, might overwhelm with complexity.

In the following I want to do simple, direct calculations, allowing little room for disagreement, which give a picture of the earth not as some vast, imperturbable ball, but rather as a *small world with a thin skin of water and air and a small dot of topsoil*—upon which all of our lives depend.

How Many People are on the Earth? When I first started working on this book there were approximately $5.7 * 10^9$, or 5.7 billion people on earth (December, 1994). That amounts to about 285 million (metric) tons of human biomass, i.e., $2.85 * 10^8$ MT, assuming that the average person has 50 kg of mass. Norman Myers, [475, p. ix], notes that no other species on earth has a cumulative biomass larger than we humans, except perhaps ocean krill—

[1]David Brower was a charismatic conservationist living in Berkeley, California. He was the long time executive director of the Sierra Club, leading the fight to save the Grand Canyon and what is now Dinosaur National Monument from hydroelectric dams. He founded Friends of the Earth, and the Earth Island Institute as well as the League of Conservation Voters. Although he died on November 5, 2000 at his home at the age of 88, his prolific, heroic activist spirit lives on in many who work for the preservation and restoration of the natural world. David Brower voted absentee for Ralph Nader for U.S. president as, presumably, his last political act.

and since we are going after krill with a vengeance that may soon change, cf. Chapter 6, footnote 4.

Exercise 7.1 Global Human Population

(i) As of 2005 the global human population was $6.45(10^9)$, cf., [662, p. 75]. Estimate the increase in the biomass of humans in the approximate decade between December 1994 and 2005.

(ii) Do you think that the impact per person on the biosphere in 2005 was more, less or unchanged from that per person impact in 1994?

(iii) Repeat this exercise for the year you are reading it.

7.2 Mining and Damming: Massive Rearrangements

Human mining activity moves more soil and rock on the earth—an estimated 28 billion tons[2] per year—than is carried to the seas by the world's rivers. This is unarguably a global impact.

In this next exercise I want to examine a tool for figuring out if some claim, like that just made, is remotely true or not. I will say that two statements are *consistent* if they can both be true at the same time. Given the existence of fuzzy/measured logic, the meaning of truth is a bit more complicated than in Aristotelian/sharp logic, and thus deeper more careful analysis is almost always called for. We briefly encountered this concept of consistency before, cf., Exercise 4.1. I wish to call attention to and somewhat formalize what I am talking about by stating an assumption that I will suppose in this book.

Consistency Axiom: *Reality (Nature) is Consistent.*

Exercise 7.2 Consistency and Data

(i) From [283, p. 10], globally 10^{10} m^3 of soil and rock are carried away by river erosion each year. Is this consistent with the figure above, i.e., that 28 billion tons of soil and rock is more than the amount of soil and rock carried by rivers to earth's seas?[3]

(ii) If the numbers in (i) reveal (or do not reveal) a gross inconsistency in the data from our two sources, what do you conclude?

(iii) How do you define the term *consistency*?

(iv) Do you have to take into account the time the statements in (i) were made when determining consistency?

(v) Do you have to take into account the error which always exists in real measurements when determining consistency?

(vi) Can you find sources that are not consistent with the two we have just dealt with in regard to the moving/mining/eroding of soil and rock? Can you find sources of information that you know are independent, e.g., they all didn't just copy a common source?

[2]Note that a ton is 2000 *pounds*. See John E. Young, Aaron Sachs, *The Next Efficiency Revolution: Creating a Sustainable Materials Economy*, WorldWatch Paper 121, p. 11. Do you think that mining activity has increased, decreased or stayed the same since this paper was written?

[3]Hint: How many grams of soil and/or rock on average are in a cubic centimeter of same?

Human imagination made this possible; we invented tools, machines—big ones. The world's largest frontloader is a 1,300-horsepower 994 Caterpillar. It takes 35 to 40 ton bites from the earth. *"The Cat operator sits in a cab twenty feet off the ground, moving his machine's segmented body, on eleven-foot-high tires, in tight radial turns. His machine's radiator block is the size of a garage door. In four rail-car-sized scoops, he fills up a 150-ton mud-splattered yellow haul truck and is ready for the next. This is as big and rudely basic as man's industrial processes get."* See [295, p. 190]. In modern mines 150,000 to 500,000 tons of earth moved per day is not uncommon.

Exercise 7.3 Digging the Earth

(i) How many pounds (or *kg*) per year would each person on earth today have to dig with a pick and shovel, say, to move 28 billion tons of earth?

(ii) What is the answer to (i) in pounds per day?

(iii) How much of your day would be spent digging? Remember that much of this "earth" is solid rock that is loosened up with explosives. Assume you have no explosives, fossil fuel or machines, except the pick and shovel.

(vi) The quote just before this exercise is from 1994. Consider the following quote from June 25, 2010 (by Clayton Thomas-Müller on www.democracynow.org) concerning Canadian tar sands, one of America's main sources of oil: *"...they're using the biggest trucks on the planet to move this stuff twenty-four hours a day, seven days a week, 300 tons per truck carrying capacity. The biggest earth movers on the planet, ten stories high, 300 tons per scoop, are operating twenty-four/seven in Canada's tar sands. They have a workforce of 77,000 workers to drive this massive, massive development. And so, you know, to provide a scale for the viewers, they move enough earth every single day in Canada's tar sands to fill up the Toronto SkyDome. ..."* Comparing these two quotes do you see a quantitative increase in the largest earth moving operations? What impacts does this extraction of oil from Canadian tar sands have on: the boreal forests, water resources, production of global warming gases, and the health of indigenous people living nearby?

People have Changed the Length of the Day on Planet Earth. Perhaps it is not easy to believe the following, but human activities have even measurably altered the earth's rate of spin.[4] Eighty-eight of the reservoirs built since the early 1950s have impounded more water than is contained in all of the moisture in the earth's atmosphere. So massive is the amount of water that has been moved that the earth's axis has been tilted slightly and the planet's rate of spin has changed by about .2 millionths of a second per day. According to Goddard Space Flight Center's geophysicist, Dr. Benjamin Fong Chao, who made the measurements and did the calculations, our oceans would have risen 1.2 inches more than they did had it not been for these reservoirs. Chao also found that the earth's gravitational field has been altered.[5]

---

[4]Malcolm W. Browne, "Earth affected by dams," *The New York Times*, as reported in the *Denver Post*, 3 March 1996.

[5]The IPCC estimates sea level rose on average 1.8 *mm* (1.3 *mm* to 2.3 *mm*) per year from 1961 to 2003. From 1993 to 2003 estimated average rise was 3.1 *mm* (2.4 *mm* to 3.8 *mm*) per year. Melting ice, thermal expansion, and anomalies in the earth's rotation all play a role in these numbers. The recent long term rise in sea level: sea levels rose 2 *cm* in the 18th century, 6 *cm* in the 19th century, 19 *cm* in the 20th century, and for the 21st century 30 *cm* is the projected, cf., UNFPA State of World Population 2009, p. 14.

Of course, we would not know these things were it not for the incredible technology I discussed on page 58. It is not comforting to note, however, that we did not anticipate and seemingly cannot control these global consequences of our human activities. On the other hand I derive some small comfort to read that Nature has created even more massive freshwater storage in the past. In the November 5, 2002 edition of *Science News* there is an article about Lake Agassiz, which was a fresh water lake about 10,000 years ago located north of the current Great Lakes of North America. This lake contained more fresh water than all of our current bodies of fresh water combined. When the lake finally burst through the barrier that separated it from the Atlantic, the massive influx of fresh water altered the climate by altering what we call today the Gulf Stream, which brings warm tropical water as far north as Norway. The melting of the Greenland icecap, i.e., a massive influx of fresh water into the Atlantic, raises similar concerns.

7.3 Fish, Forests, Deserts, and Soil: Revisited

Fish Feel our Machines. "After decades of buying bigger boats and more advanced hunting technologies, fishers have nearly fished the oceans to the limits. Of the world's 15 major marine fishing regions, the catch in all but 2 has fallen; in 4, the catch has shrunk by more than 30 percent." [6] Indications from scientists and fishers alike are that humans are having a global impact on life in the world's oceans. Overfishing, pollution, habitat destruction and modification are all adding up and the sum is a loss in the world fish catch.[7]

The work of [736] and [94] indicate that human pressures on global fish populations are increasing—with increasing effect. In [94] we read of a scientific paper which calculated a global "fish peak" in the year 1989. Seventy-five percent of the world's fishing stocks are (in 2006) fully exploited or overfished.[8] Climate change/global warming and fisheries overexploitation are perhaps the clearest and most urgent examples of the "Tragedy of the Commons," cf., VII.

[6] Peter Weber, "Net Loss: Fish, Jobs, and the Marine Environment," *Worldwatch Paper* 120, Worldwatch Institute, Washington D.C., 1994, pp.5–6. Have things improved or worsened by the time you read this? See Section 5.7

[7] See also the article Carl Safina, "The World's Imperiled Fish," *Scientific American*, November 1995, pp 46–53. Are the world's fish more or less imperiled now, cf. [662, pp. 26–7]? Recent (2006) work of Boris Worm, [736], professor of Marine Conservation Biology at Dalhousie University, Nova Scotia, Canada, predicts that if global fishing practices/rates do not change, global collapse of all species currently fished is possible as soon as 2048. Collapse is defined as decline to 10% of previous historic populations, i.e., decimation.

[8] Acidification of the world's oceans due to carbon emissions is emerging simultaneously as a threat to fish populations, cf., Section 5.7.

Exercise 7.4 50 Plus Ways to Save the Seas

(i) David Helvarg's Blue Frontier Campaign, cf., www.bluefront.org, suggests 50 ways to save the ocean. What are they? See also [295, 296, 297].

(ii) Captain Paul Watson suggests a couple other ways. What are they? What is seaacide? See [708, 709] and www.seashepherd.org.

Forests Fall. French writer, François-René de Chateaubriand, of the late 18th century, summarized the relationship between civilizations and forests thusly: *"Forests precede civilizations and deserts follow them."* Such books as [546, 137, 531, 145, 575, 328] among many document Chateaubriand's sentiment. (The book, *Deforesting the Earth: From Prehistory to Global Crisis*, [726], gives a history from which a clear pattern can be deduced, while [145] provides graphic pictures, at least for the U.S. and Canada, of what it means on the ground.) What is new is the global scale of forest destruction. The global appetite for forest products leaves not a single tree on the planet safe from saws. If the pattern oft observed locally of tree/forest collapse, followed by soil depletion followed by civilization collapse, is repeated today, the collapse will likely be more global in nature.[9] If anyone anywhere tries to stand in the way of civilization's consumption of a particular resource, such as a local forest, that person must prepare to meet an iron fist.[10]

"The current pace of tropical deforestation is frightening. The World Resources Institute estimates that almost half of the original 3.7 to 3.9 billion acres of tropical mature forests have been cleared to accommodate other uses. Satellite observations suggest that the rate of tropical deforestation worldwide is 40.5 to 50.4 million acres a year."[11]

"Although efforts to save the world's tropical rain forests have rightly received widespread attention, another type of rain forest is perhaps even more threatened. Now estimated to cover less than half their original area, coastal temperate rain forests are an exceptionally productive and biologically diverse ecosystem. They include some of the oldest and most massive tree species in the world, and constitute some of the largest remaining pristine landscapes in the temperate zone."[12]

If you ever get a chance to visit some of the world's remaining pristine forests do so (and, of course, tread lightly). I would like to devote a chapter

[9] From [546, 137, 678] and similar studies we see that the collapse of earlier civilizations followed a pattern: trees/forests collapse first, then the soil and/or food supply goes, followed by the collapse of the civilization itself. This pattern is quite familiar to archaeologists.

[10] Those who dispute this claim almost invariably have never tried to save any part of the biosphere in its natural state. A group I helped found in 1989 in Colorado, Ancient Forest Rescue, AFR, successfully prevented the logging of Bowen Gulch old growth forest, on the western boundary of Rocky Mountain National Park. Many of the group's members gave up conservation work, however, when they found themselves under surveillance by private police, forest police, local police, state police and the FBI. Not much is written about AFR; however, the book *Powder Burn: Arson, Money, and Mystery on Vail Mountain*, Pubic Affairs, New York (2001) will have to do until I write my version.

[11] See [339, p. 16]. What is the state of tropical forests when you read this?

[12] Derek Denniston, "Conserving the Other Rain Forest" from *Vital Signs: 1994*, edited by Lester R. Brown, et al., Worldwatch Institute, W.W. Norton & Company, Inc., Publishers, New York, NY.

or two telling you what an indescribably fantastic experience this is, but I cannot for two reasons. First, you're not supposed to do that in a math book; and second, words cannot come close to doing justice to the feeling—it really is indescribable!

In the United States (lower 48) we have cut over 96% of our original forests, some of them have not grown back, some of them have been turned into tree farms and some have grown back but do not have the diversity and complexity[13] they once had since that takes time—often longer than the current age of our nation. The point is clear, humans have had global, devastating impacts on the world's forests. Hosts of statistics corroborate this point.[14]

Exercise 7.5 What is the Current State of the World's Forests?

(i) Russel Monson and Jia Hu, researchers at the University of Colorado, Boulder, published a paper in *Global Change Biology*, Jan. 8, 2010, with the conclusion: as the climate warms and the growing seasons lengthen, subalpine forests are likely to soak up less carbon. Jia Hu said, "Our findings contradict studies of other ecosystems that conclude longer growing seasons actually increase carbon uptake."

Quiver trees (aloe dichotoma), which can live longer than 300 years, have begun dying in parts of South Africa and Namibia in their native habitat. A Namibian government report in 2008 stated that temperatures during the last century have risen about 2 degrees Fahrenheit in that country. It is believed that climate change is affecting the quiver trees.

In *Science News*, 2/14/09, p. 8, researchers reported that small background rates of customary tree death have doubled in old-growth, western U.S. forests since 1955, possibly because of climate change.

What are the explanations for the above findings? At the time you read this what is the net effect of climate change on forests globally and their role in sequestering carbon?

(ii) What is the Rainforest Action Network, cf., www.ran.org, doing about the world's forests?

Deserts Follow. *"According to the United Nations Environment Program, 11 billion acres—35 pecent of the earth's land surface—are threatened by desertification and, with them, fully one-fifth of humanity. Three-quarters of this area has already been at least moderately degraded and an astonishing one-third has lost more than 25 percent of its productive potential. ... The four principal causes of land degradation—overgrazing on*

---

[13] Complexity is a concept that is amenable to being defined and studied by mathematicians. It is a worthy project for the interested student.

[14] On July 27, 1995, President Clinton signed Public Law 104–19. One of the effects of this law was the lawless logging of much of the last 4% of America's roadless, ancient forests. I say lawless, because the law expressly ordered unsustainable levels of logging *"notwithstanding any other provision of law."* At the end of Clinton's second term, after nationwide public hearings in support, he promulgated the "Roadless Rule" in an attempt to protect the last roadless forests in the U.S. Constant efforts to overturn this rule have followed. In 2002 the "Healthy Forests Initiative" (note the spin) of the Bush II administration delivered treatment of our national forests that conservationists found heartbreaking, not to mention the species trying to live therein. Thus environmental legislation, extensively debated and carefully crafted over decades, does not apply in some of the most critical places for which it was designed.

rangelands,[15] *overcultivation of croplands, waterlogging and salting of irrigated lands, and deforestation—all stem from human pressures or poor management of the land."* [16]

In this next exercise I want you the check up on some of the numerical claims about the earth that were just made.

Exercise 7.6 The Area of the Earth, its Lands and Oceans

(i) The area of the earth is $5.10 * 10^{14}$ m^2. One *hectare* = 10^4 m^2. One hectare is abbreviated as 1 *ha*, and 1 *ha* = 2.47 *acres* (an English system of area measurement, where there are 640 acres in one square mile). What is the area of the earth in acres?

(ii) Using the information in the paragraphs immediately above, viz., 11 *billion acres* equals 35 percent of the earth's land surface, calculate the total area of land on the surface of the earth in acres. In square meters. Compare this answer with the information in part (i).

(iii) What is the area of the world's oceans?[17]

(iv) What fraction of the earth's surface is ocean? Land? Compare these estimates with answers you look up in another source, like the Web, or an encyclopedia. How close are they? Explain any discrepancies.

Soils at Risk. In a speech at the University of Colorado, Jeff DeBonis[18] said that in the United States, during the life of the nation, we have gone from an average topsoil depth of 28 inches to a current average topsoil depth of 8 inches. Of course, to know exactly what this means one must ask questions about the lithosphere, the solid part of the earth—are we dealing with the O horizon, A horizon or B horizon of the soil and such, cf., [339, pp. 324–348], Exercise 5.4? Without getting into a deep study of the state of the world's topsoil, the evidence is clear.[19] Humans are impacting the soil on a global scale.[20]

Exercise 7.7 Topsoil Formation/Loss Calculations: Are They Consistent?

(i) Topsoil is a complex association of living beings without which you and I do not eat. Thus it should be of supreme importance to know the state of topsoil, especially any topsoil that you depend on for food. One book estimates say that it takes 10,000 years for the formation of 30 centimeters (about 1 foot) of topsoil. Another reference, [173], says that it takes 100 to 400 years for 1 centimeter of soil to form. Still another reference, *Vital Signs, 1995*, says that it takes between 200 and 1,000 years to form 1 inch, or 2.5 centimeters

---

[15]See [325, 737], two remarkable texts that document situations, some tragic, on the public lands of the American west.

[16]Sandra Postel, "Restoring Degraded Land," in *The Worldwatch Reader*, Lester R. Brown, editor, W.W. Norton & Company, New York, NY, 1991, p.27. Are things improved when you read this?

[17]The polar ice cap at the north pole, is not land—besides it is disappearing. Antarctica is a continent covered with ice, do you think it is being considered as land by Postel?

[18]The founder of AFSEEE, the Association of Forest Service Employees for Environmental Ethics, (now FSEEE) and the founder of PEER, Public Employees for Environmental Responsibility, www.peer.org and www.fseee.org.

[19]If you feel like reading a little more about soil, see [173, 308, 37, 537, 226, 536].

[20]There is a two page update on soil erosion in *Vital Signs 1995*, Worldwatch Institute, W.W. Norton & Company, 1995, p. 118. *Vital Signs* is a yearly publication and a valuable reference along with *State of the World*, also by the Worldwatch Institute.

of topsoil. Are all of these estimates consistent? Do you think the rate of soil formation depends on many variable factors?

(ii) The U.S. does one of the most extensive estimates of topsoil loss (condition) of any country. If DeBonis is right and we have lost an national average of 20 inches of topsoil in approximately 225 years, out of a total initial estimated average of 28 inches, how much longer will the 8 remaining inches remain if we do not change the way we do agriculture? What agricultural practices might slow or reverse the loss of topsoil? Do you think this soil loss should be of concern equal to, greater than, or less than, say, the threat of attack from a foreign nation or terrorist group? When do you think we should start doing something to halt topsoil loss?

(iii) Ellis and Mellor, [173], assume a bulk density of topsoil of 1.33 grams per cubic centimeter. Another source says that 1 inch of topsoil weighs about 140 tons per acre (here a ton is 2,000 pounds). Are these figures consistent?[21]

(iv) Hillel, [308], says the in the 1960s the U.S. Soil Conservation Service's guidelines for soil loss tolerance set a maximum of 5 tons per acre (12.6 tons per hectare) per year. In actual practice, soil losses 10 times as great are common throughout the U.S. How many inches (or centimeters) of soil loss is equivalent to a 5 ton per acre loss? To a 50 ton per acre loss? Do you think such soil loss is noticeable from year to year?

(v) Ellis and Mellor, [173], state that background soil loss is less than 1 tonne per hectare per year. (Here a tonne is 1,000 kilograms.) Is this less than or greater than or equal to the rate of formation of topsoil given in part (i)? Recall that 1 hectare $= 10^4\ m^2 = 2.47\ acres$.

(vi) Ellis and Mellor, [173], state that soil erosion accelerated by humans commonly exceeds 10 tonnes per hectare per year and sometimes exceeds 100 tonnes per hectare per year. In India and Nepal erosion of over 200 tonnes per hectare per year has been measured. Are these figures consistent with the figures of Hillel, [308], in part (iv)?

7.4 The Brower–Cousteau Earth Model

It is hard for one person to grasp intuitively what it means for the human population to be of global scale and impact. There are various ways to tackle this difficulty. One way—one model, most beautiful in its simplicity, is what I call the Brower–Cousteau earth.[22] I first heard this model described in a speech by the late David Brower. Brower said he first heard of the model from Jacques-Yves Cousteau.[23]

The basic idea of Cousteau and Brower is to use *scale*: If we scale the earth down to something one human can intuitively grasp, such as an egg or an apple, what would the water, air and topsoil look like? Once vast, unperturbable oceans, limitless skies and soil are all brought down to the scale of one human.

[21]Hint: Recall that 1 pound = 453.6 grams, 1 acre = 43,560 square feet, and 1 inch = 2.54 centimeters.

[22]If the you get stuck on some of the math in this last section of Chapter 7 you might want to read II first, then come back and try this section again.

[23]Jacques-Yves Cousteau, who passed away 25 June 1997, founded the Cousteau Society (www.cousteau.org), spent most of his life in a heroic effort to prevent destruction of life on earth—especially life in the world's oceans.

I just ate the apple on my desk, but a similarly sized, smaller, more spherical orange which remains is 4 cm in diameter. I want to mentally scale the earth down to the size of this orange and see what happens to the water, air and topsoil. The radius of the earth varies from $6.38 * 10^6\ m$ at the equator to $6.36 * 10^6\ m$ at at the poles. Let's just say that the radius of the earth is $6.37 * 10^6\ m$.

Exercise 7.8 Mathematics of the Area of a Sphere, Which the Earth Almost Is

(i) The area of a sphere of radius R is $4\pi R^2$. Using the value of $R = 6.37 * 10^6\ m$, how close do you come to the value of $5.10 * 10^{14}\ m^2$, used in Exercise 7.6, for the area of the earth? Note the following interesting geometric fact. The area of a sphere is 4 times the area of a "great circle" on the sphere. In the case of the earth an example of a great circle is the equator. In so far as the earth is spherical, the area of the earth is 4 times the area of the disk abstractly visualized by chopping the earth through the equator into two hemispheres.

(ii) Let A_{orange} be the area of our orange with a diameter of $4\ cm$. Calculate A_{orange}.

(iii) Calculate the fraction $\frac{A_{orange}}{A_{earth}}$ where A_{earth} is the area of the earth. Now calculate $[\frac{R_{orange}}{R_{earth}}]^2$ and compare it with $\frac{A_{orange}}{A_{earth}}$, where R_{orange} is the radius of our orange, i.e., $.02\ m$, and R_{earth} is the radius of the earth, i.e., $\approx 6.37 * 10^6\ m$.

How Much Water—Salty and Fresh—is on the Earth? The waters of the earth, called the hydrosphere, comes in salty and fresh liquid and frozen forms. There are various sources of information on water, including: [230, 115, 339]. From [339, pp. 289–290], we read that 97% of the earth's water is saline (salty) and less than 3% is fresh. Of the fresh water $\frac{3}{4}$ was found in polar ice caps and glaciers (1999),[24] i.e., it was (is?) solid ice. Almost $\frac{1}{4}$ of the fresh water is found underground in water-bearing porous rock or sand or gravel formations. This water is called groundwater. Only a small fraction, $\frac{1}{2}$%, of all water in the world is found in lakes, rivers, streams and the atmosphere. This water is called surface water.

From [230], we see that volume of the water in the oceans is $1.338 * 10^9\ km^3$, and this is 96.5% of the total water on the planet, $1.386 * 10^9\ km^3$. Does this contradict the above figures from [339]? No. There are other sources of saline water besides the oceans, and [230] also tells us that there is an unknown uncertainty in all of these figures.

Exercise 7.9 Mass and Mass Density and Water

(i) Recall that the mass density of pure water (at 3.98 degrees Celsius and normal atmospheric pressure) is $\frac{1\ g}{cm^3}$. What is the volume of $1\ kg$ of pure water (at this temperature)? One thousand kilograms?

(ii) The appendix in [283] gives the mass of the water in the world's oceans as $1.4 * 10^{21}\ kg$. Assuming all our figures are correct, what is the average mass density of the oceans? Explain any differences with the mass density of pure water.

(iii) Check the numbers for water given here against the numbers you find in some of the references in your library or on the Web. Explain any differences. How confident are you in these numbers?

[24]How much of this has changed by the time you read this?

(iv) Consider the statement: "97% of all water on the earth is salty." Does it make a difference if the percentage is in terms of volume or weight? Do changes in temperature and pressure effect this percentage? Are we doing calculations in this book to such a degree of accuracy that any of these considerations make a difference to us?

Exercise 7.10 Some Politics and Water.

(i) Comment on the following quote, taken out of context, from [295, p. 220]:

"Nothing disappears. People talk about 1 percent of the world's water is fresh water. All the water in the world is potential freshwater. It's a natural system of recycling. There's no scientific basis for what they're doing, so the only rationale must be to take us to utter socialism."

(ii) An open-ended project is to investigate the political future of water. In the year 2002, for example, in the United States over 85% of domestic water supplies to homes is provided by municipal water companies, i.e., the citizens via their city governments own their own water supply and delivery systems. There is a global effort on the part of some corporations to privatize these water supplies, i.e., transfer ownership of water systems to certain private corporations. Would such a change be in your self-interest? For an argument that privatization of water is not in the public interest see: [33, 632], [337, pp. 135–6, 237, 334–6].[25] What are the arguments for privatization of city water supplies?[26]

Lest it be forgotten that species other than humans also need water, see Don Hinrichsen, "Appropriating the Water: What's left for wildlife?" in the January/February 2003 issue of *World Watch.*

Related to the public vs. private ownership paradigms see also VII in this book.

Rounding off the figures in Gleick, [230], we again see that 97% of all the water on the earth is salt water, and it cannot be used for drinking or growing crops. The bulk of the remaining 3% is fresh, amounting to about $35 * 10^6\, km^3$. Gleick says that most of this fresh water is locked up in ice caps and deep underground aquifers which are at the moment technically and/or economically beyond our ability to exploit. Gleick concludes that $10^5\, km^3$, or just .3% of the total fresh water reserves on earth constitutes most of the fresh water available for our use. From the references we have mentioned so far, we see that between .3% and .5% are available for our use—where the .5% includes atmospheric water.

Exercise 7.11 Some Interesting Calculations for Fresh Water on Earth and a Rope Around the Equator

(i) If the 3% of the earth's water that is fresh were all liquid (it is not), and if it were spread out evenly over the earth's surface, how deep would the layer be?[27]

(ii) Assuming Gleick's figures are correct, what percentage of the earth's total water supply is usable fresh water?

(iii) What percentage of the earth's water is in the atmosphere?

---

[25] See also two articles on water privatization by Jon R. Luoma, Jon Jeter, respectively, in the December 2002 issue of *Mother Jones*, a magazine.

[26] See, for example, http://www.esrresearch.com/Theprivatewaterindustry.htm, http://www.mackinac.org/article.aspx?ID=3157.

[27] Hint: If you take into account the fact that we have the radius of the earth to no more than three significant figures you will eventually be forced to use a fact like $x^3 - y^3 = (x-y)(x^2+xy+y^2)$. See the next exercise if you have forgotten the formula for the volume of a sphere.

(iv) This is a classic mathematics problem. Assume the earth is a sphere. Suppose you have a rope snugly wrapped around the earth at the equator. Suppose you wanted to lengthen the rope enough so that anyone could easily walk under it anywhere on earth. For the sake of completeness assume that you wanted your newly lengthened rope to be exactly 7 feet above the equator all around the earth. How much would you have to lengthen the rope? (Before solving the problem carefully using mathematics, guess the answer and write it down so you can compare it with the actual answer.)

(v) In part (iv) do you really need to know the radius of the earth to do the problem? Is your answer the same if you replaced the earth with the moon? Would your answer be the same if you replaced the earth by a sphere of any radius?

(vi) **A Global Warming Exercise:** In an article, Nov. 2009, for the Associated Press by Seth Borenstein: "Warming's impacts have sped up, worsened since Kyoto," we read: "Measurements show that since 2000, Greenland has lost more than 1.5 trillion tons of ice, while Antarctica has lost about 1 trillion tons since 2002, according to two scientific studies published this fall." "...in the dozen years leading up to next month's climate summit in Copenhagen: The world's oceans have risen about an inch and a half." How much of this inch and a half rise is accounted for by the 2.5 trillion tons of melted ice? Hint: The area of the world's oceans is roughly $3.61(10^{14})\ m^2$. If you were curious, the volume of the earth's oceans is roughly $1.35(10^{18})\ m^3$, cf., [283, p. 236].

Recall that from Gleick's book all the water on the earth has a volume of $1.386 * 10^9\ km^3$.

Exercise 7.12 Some Mathematics of Volume

(i) The volume of a sphere of radius R is $\frac{4}{3}\pi R^3$. What is the volume of a sphere with a radius of $6.37 * 10^6\ m$? Is this the volume of the earth, V_{earth}?

(ii) What is the volume of our orange, V_{orange}? How does $\frac{V_{orange}}{V_{earth}}$ compare with $[\frac{R_{orange}}{R_{earth}}]^3$?

(iii) What is the ratio of the volume of all the water on the earth to the volume of the earth?

(iv) What is the ratio of the volume of all usable freshwater on the earth to the volume of the earth?

We are now going to pretend that all of the water on the earth is in the form of a gigantic sphere of water. This sphere of water, as we see above, has a volume of $1.386 * 10^{18}\ m^3$. (You have to convert from km^3 to m^3, can you do it?) What is the radius of this sphere? Let this unknown (for now) radius be R_{water}. Then we have the following equation:

$$\frac{4}{3}\pi R^3 = 1.386 * 10^{18} m^3.$$

Solving this equation for R^3 we get:

$$R^3 = \frac{3}{4\pi} * 1.386 * 10^{18}\ m^3 = .331 * 10^{18}\ m^3.$$

We will systematically review how to solve equations, like the one above, for R in Chapter 2. For now just observe that to solve this equation for R, all you have to do is raise both sides of the equation (feel free to use your calculator) to the power $\frac{1}{3}$ and get:

$$\begin{aligned} R = [R^3]^{\frac{1}{3}} &= [.331 * 10^{18}\, m^3]^{\frac{1}{3}} \\ &= [.331]^{\frac{1}{3}} * [10^{18}]^{\frac{1}{3}}]\, m \\ &= .692 * 10^6\, m. \end{aligned}$$

Exercise 7.13 The Relative Volume of Water on Earth
(i) The ratio of the radius of our "water ball" to the radius of the earth is $\frac{.692*10^6\, m}{6.37*10^6\, m} = .109$. Compare the cube of this ratio to the ratio from Exercise 7.12 (iii).
(ii) Are all of the results of your calculations consistent so far?

We are now going to scale everything down to the size of our orange. In order to scale the radius of the earth, $6.37 * 10^6\, m$, down to the radius of our orange, $2\, cm$, we must divide by $6.37 * 10^6\, m$ and multiply by $2\, cm$. If we take $.692 * 10^6\, m$ and do the same we get:

$$[.692 * 10^6\, m] * [6.37 * 10^6\, m]^{-1} * 2\, cm = .217\, cm.$$

Thus if the earth were the size of an orange (with a radius of 2 cm), and if all the water on the earth were in the shape of a sphere, that sphere would have a radius of $.217\, cm$.

All the liquid fresh surface water on the earth is about $35 * 10^6\, km^3$. This is $35 * 10^{15}\, m^3$. *Thus on an earth shrunk to the size of our 2 cm radius orange, all the fresh liquid water on the earth would be a droplet with a radius of .06 cm.*

Exercise 7.14 The Relative Volume of Fresh Water (Usable Water) on Earth.
(i) Verify that the radius of our droplet of fresh liquid water on earth-orange would be $.06\, cm$. How big is that?
(ii) What would be the radius of the droplet of usable fresh water, scaled down to the size of our orange?

The point of this exercise, thus far, is not to give you the impression that there is no fresh water on the earth; in fact there is quite a bit. The point is to leave you with the impression that the human population is of global scale and is big enough to impact the planet's water. Also the next time you are thinking of dumping some toxic nuclear waste in the ocean where no one will notice, think of that drop of water $.217\, cm$, i.e., a little over $2\, mm$, in radius.

A quick way to see that the water calculation is believable is as follows. The mean depth of the oceans[28] is $3{,}730\, m$, or $3.73*10^3\, m$. The radius of the earth is $6.37 *10^6\, m$. Thus the "fractional thickness" of the ocean on our orange would be $\frac{3.73*10^3}{6.37*10^6} = .000586$. If our orange has a radius of $2\, cm$ then the ocean on our orange would be $.000586 * 2\, cm \approx .001\, cm = .01\, mm$ deep. That is a pretty shallow sea! The next time someone tells you that the oceans are so

[28] Don't take my word for this, look it up somewhere and see if you get the same number.

vast that humans could never hurt them, tell them about this earth-orange with an ocean only one hundredth of a millimeter deep.

How Much Air is in the Atmosphere? How Much Soil is on the Earth? Let's take another, cf., Section 4.5, look at the earth's atmosphere. The mass of the atmosphere, cf., [283, p. 235], is $5.14 * 10^{18}\ kg$. The bulk of the earth's atmosphere that is closest to earth is called the *troposphere*. Atmospheric scientists define the troposphere as that part of the atmosphere closest to the earth in which the temperature drops with increasing elevation. Over 90% of the air in the atmosphere is in the troposphere. The average elevation of the top of the troposphere is 12,000 m. The next layer of the atmosphere is called the *stratosphere*. This is where the ozone layer that we examined briefly in Section 4.5 is located. In the stratosphere the temperature increases with increasing elevation (up to about 50 km).

If we look at the ratio of the thickness of the troposphere to the radius of the earth we get $\frac{1.2*10^4}{6.37*10^6} = .0019 \approx .002$. Thus on our 2 cm radius orange, the troposphere (over 90% of the air) would be in a layer $.002 * 2\,cm = .004\,cm$ thick. That is four times thicker than our shallow sea, but still not very thick. And, of course, air is far less dense than water.

Now if we "liquefied" all the earth's atmosphere, i.e., assumed that the mass of the air was replaced by an equivalent mass of pure water, we would have $5.14 * 10^{15}\ m^3$ of liquid. If this were in a spherical droplet, it would have a radius of 107,059 m. Scaling this down to our orange, we get that all the air on the earth if liquefied would form a droplet .034 cm in radius.

The next time you hear an argument about the cost of pollution controls on smoke stacks or automobile exhaust, think about how thin the earth's blanket of air actually is.

David Brower includes in his recitation of numbers on the Brower–Cousteau earth the statement: "*...and if all the topsoil on earth were put in one pile on our (orange), it would be a speck barely visible to the naked eye.*"

Exercise 7.15 The Relative Amounts of Air and Topsoil on Earth

(i) Verify our calculation of the radius of the droplet of liquefied air on earth-orange.

(ii) If all the topsoil on earth were rolled into one spherical ball, estimate as best you can the radius of this ball on earth-orange. You can use the references already mentioned in this book, or others.

(iii) If the atmosphere were 78% nitrogen and 22% oxygen how many grams of mass would there be in 1 mole of air?[29]

(iv) The air is actually 78.08% nitrogen and 20.95% oxygen. The rest of the air is a mixture of argon, carbon dioxide, and many other gases, including water vapor, cf., [283]. Using the information in II, show that 1 mole of air has a mass of 28.96 grams.

(v) Verify our calculation alluded to above that if the mass of the air were replaced by an equal mass of water, that water would occupy $5.14 * 10^{15}\ m^3$.

So the comments of David Brower at the beginning of this section are quite apt. If the earth were reduced to the size of a sphere you could hold in your

[29] For the definition of a *mole* and examples of calculating with it see II.

hand, the seas and atmosphere would be very thin indeed; and the soil nearly invisible.

Some folks will say that the above exercise is quite meaningless and "not scientific." whatever that may mean in this context. One could point out that if we scale all the humans down we will have a hard time seeing them on our orange. My immediate reply is that the effects humans are having on the planet far transcend their physical size. I contend that given the present documented impacts of humans on the planet it is important to have the notion of limited air, water and soil in each of our minds to compete with our natural tendency to think of these three things as limitless.

Calculating Your Ecological Footprint. My enthusiasm for the following exercise is not shared by all. Professor Rees, coauthor of [696], once gave a lecture which I attended on the concept of *ecological footprint*[30] that he and Mathis Wackernagel introduced. An economics professor in the audience announced at the end of the talk that he would give Professor Rees an F for his analysis.

Exercise 7.16 Calculate Your Ecological Footprint

(i) Calculate your personal and/or family ecological footprint, or that of a "typical" American. Note that http://www.rprogress.org/ and http://wwwfootprintnetwork.org are Web sites (as I write) that will help you calculate your ecological footprint on earth, i.e., "how much of the earth is required to support your lifestyle." You should do an internet search for "calculate ecological footprint" and compare results from a variety of footprint calculators. Of course, you could do the calculation all by yourself as well. This requires that you put together an analysis of all impacts your life has on the planet and estimate the area of the planet you impact (directly or indirectly).

(ii) Repeat part (i) replacing the American by a person from a very poor country, then by a person who has an income in excess of $1,000/day, then by a "middle class" person who earns between $750 and $7,000 per year, measured in 1992 dollars just to be precise. See VII.

[30]Roughly, a person's ecological footprint is an estimate of the area of the earth needed to supply all of the consumption of that person.

Chapter 8

Fuzzy Logic, Sharp Logic, Frames, and Bigger Pictures

8.1 Sharp (Aristotelian) Logic: A Standard Syllogism

Many elementary mathematics books with an audience similar to that of this book have a section on logic, the logic of Aristotle.[1] Thus a triplet of statements called a *syllogism*, soon appears, such as:

$$\textit{(Assumption) Hypothesis} : \textit{All humans are mortal.} \tag{8.1}$$
$$\textit{(Assumption) Hypothesis} : \textit{All mathematicians are human.} \tag{8.2}$$
$$\textit{Conclusion} : \textit{All mathematicians are mortal.} \tag{8.3}$$

The syllogism is then illustrated with some so-called set theoretic diagram:

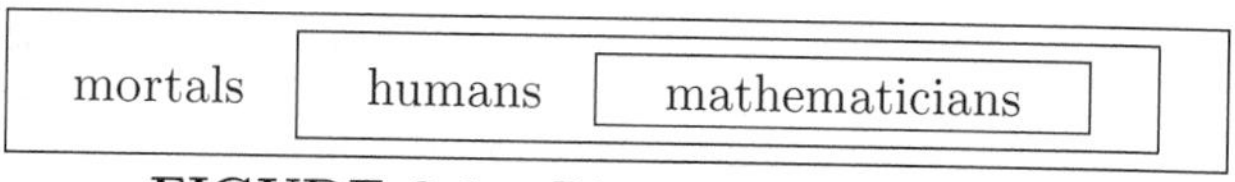

FIGURE 8.1: Diagram of a Syllogism

The set (box) of mathematicians is contained in the set (box) of humans is contained in the set (box) of mortals. Thus by the "transitivity of containment," an assumption in itself, we get that the set (box) of mathematicians is contained in the set (box) of mortals, cf. Figure 8.1.

Like most mathematics applied to an actual situation, Aristotle's logic is an abstract idealization, an oversimplification of reality at best—a mistake at worst.[2] I do not propose tossing out Aristotelian/sharp logic, in fact we will use sharp logic in creating the proofs of II. Such logic will remain a useful

[1] The Greek philosopher Aristotle (384 B.C.–322 B.C.) was a student of Plato and teacher of Alexander the Great.

[2] Which came first the chicken or the egg? This question becomes tractable if you abandon the implicit oversimplification.

tool for tackling problems that can be posed, at least approximately, in a sharp way. In fact, a common tool used by scientists is to look at idealized, oversimplified models to better understand whatever they are dealing with. Complications are added as it becomes possible to understand them. In the next section I will point out with a simple example, cf., Figure 8.2, some of the difficulties with applying sharp, "yes-no," "black-white" logic in the real world, and why fuzzy logic is necessary.

Before that, however, I want to review very briefly a couple concepts closely associated with syllogisms and "if ..., then ..." arguments/proofs. The following exercise will be useful in II and III. If you have not read Section 2.1 and done Exercise 2.2, do so now.

Exercise 8.1 The Concepts of Necessary and Sufficient

(i) Discuss the concept of *necessary* as it is used in precise logical reasoning. Also discuss the concept of *sufficient* as it is used in precise logical reasoning.

(ii) How does a statement of the form "If A is true, then B is true." relate to the concepts of necessary and sufficient?

(iii) Can you think of any statements P and Q such that in order for P to be true it is necessary that Q be true?

(iv) Can you think of any statements P and Q such that in order for P to be true it is sufficient that Q be true?

(v) Suppose for a statement A to be true it is necessary and sufficient for B to be true. What might you say about statements A and B?

(vi) Can you think of an example of two statements A and B such that for A to be true it is necessary and sufficient for B to be true?

(v) Discuss the truth of the following statement: In order to take effective action it is necessary to have accurate information.

(vi) History and Public Policy Professor Alexander Keyssar says in [347]: *"The current state of American politics makes clear that universal suffrage is a necessary but not sufficient condition for a fully democratic political order."* Discuss.

Explain what he might mean.

(vii) Charles A.S. Hall et al., in [270, p. 40], says: *"In other words, the availability of free energy[3] is a necessary, but not a sufficient, condition for the availability of labor, capital and technology."*

8.2 Measuring Truth Values: Fuzzy/Measured Logic

Aristotelian logic can be refined and expanded to better handle some of the "fuzzy" problems that come up in fields like history, sociology, biology, medicine, economics, engineering .... This expanded logic is called fuzzy

[3] Free energy here means that the energy is available and capable of doing useful work, at least for human purposes. Heat energy from a forest fire is, for example, not usually available to do useful work for humans. Fossil fuels, such as gasoline in the tank of a car, have some free energy.

logic, a name which is in some ways misleading. Mathematicians are not traditionally focused on public relations or marketing; the "fuzzy" in fuzzy logic does not mean muddleheaded. Fuzzy logic is more precisely called *measured logic.* What is being measured is the truth value of statements, and the measure or truth value of a statement can be any number between 0 for not true, i.e., (completely) false, and 1 for (completely) true. (Review Section 1.4, for example.) In Aristotle's logic statements are either true or false, and statements which are half-true and half-false, for example, are not allowed, even though such exist in Nature, cf., Exercise 8.2.

Why is Sharp Logic Often a Tool Too Blunt? Let's take a really close look at the following Figure 8.2.

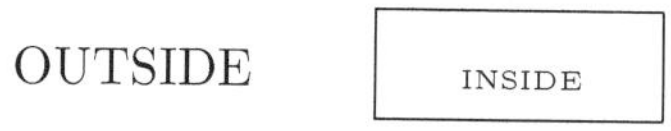

FIGURE 8.2: The Fuzzy Boundary between Inside and Outside

When I say close look I really mean it, use at least a magnifying glass, better yet a microscope—even an electron microscope. If you really do this the clear boundary between points inside the box and points outside the box melts away into a discontinuous bunch of different molecules making up the ink "lines" of the box. The perfect box boundary of zero thickness exists only as an abstract, idealized, sharp approximation of reality that we imagine in our minds.

So what does it mean for a point to be either inside or outside of the box? There are points clearly well inside the box, I can assign a truth value of 1 to the statement that such a point is inside. There are points that are clearly well outside the box, I can assign a truth value of 0 to the statement that such points are inside. If I were a bit smaller than the size of an atom and started walking from the inside of the box toward the outside it would be like walking through a "forest" where the "trees" were scattered molecules of ink. I could pass through the boundary without touching a single tree. I could make estimates and measurements and the truth value of "I am inside" would fall from a value of 1 to 0, attaining values strictly between 1 and 0 along the way.[4]

Are Some Definitions Fuzzy? What is a Tree? What is a Forest? A most prominent problem with Aristotle is not really his fault: What do the words in a syllogism mean? Most real life words and the concepts they stand for are fuzzy. I once was lucky enough to be sitting in the shade of a baobab tree

[4]Whether the set of values between 1 and 0 actually achieved is all of those numbers or a discrete "quantized" subset is an interesting question for quantum physicists. I will leave this for you to ponder.

in Zimbabwe looking for large mammals and reading a book on trees, [515]. I was struck by the following quote: *"The first, and perhaps the most contentious question that arises is, 'What is a tree?' This is a very difficult, indeed impossible question to answer for a 'tree' is a popular concept and not a scientific entity."*

There are plants that 100% of botanists agree are 100% tree. There are plants that 100% of botanists agree are 100% *not* tree. Then there are plants in between that some botanists consider to be trees and others do not, with percentages varying with the plant in question. I consider "tree" to be a concept which, although fuzzy, is more scientific than many other everyday notions. But perhaps you chalk up the fuzziness to the general level of imprecision in biology; surely things are more exact in, say, physics.

We Can Measure Something We Really Cannot Define. Guess what? Things are a bit fuzzy in physics as well. What is energy? It is a concept in physics which we will study in some detail later on. The physicist, Richard P. Feynman, said: *"It is important to realize that in physics today, we have no knowledge of what energy* **is**. *We do not have a picture that energy comes in little blobs of a definite amount."* See [187, p. 4–2 (second page of Chapter 4)].

Physics does tell us, however, that there are sharp mathematical formulas for the various types of energy and that energy can be accurately measured (with, of course, some fuzzy error). Abstract energy calculations can be done that are very precise, but only because energy is tantamount to being an undefined term of a mathematical system.

You might say that the inability to define what energy is is not the same as being fuzzy. You would be correct. It is a rather strange situation where scientists can measure physically but not philosophically. A clearer example of a fuzzy definition occurs in astrophysics with the attempt to define *planet.* In 2006 the popular press reported widely on the controversy regarding the planethood of Pluto.

Mix Biology and Politics and Things Can Get Very Fuzzy. Let's get back to biology—and political science. The United States Congress was debating the reauthorization and/or gutting of the Clean Water Act during 1995, and a part of this act deals with wetlands. In May of 1995 a committee of the National Academy of Sciences, chaired by a biologist from the University of Colorado, completed a $550,000 two-year study which made recommendations on how to define *wetlands.* The time and money involved indicate that the concept of wetlands is at least as fuzzy as the concept of tree. The people in Congress then took the definition and made it fuzzier, since they took into account not only science but the desire of some people to make money from wetlands—people some of whom also make sizeable campaign contributions.[5]

[5]What percentage of wetlands in the United States of 1776 no longer exist? In 1997 the following Web site of the EPA gave data on wetlands losses in the United States from 1780 to 1980: http://www.epa.gov/indicator/wetloss.html#chart1. A small project might consist of trying to find the information that was on this Web page since it was removed. Also, *follow the money.* Research the amount of: (1) money spent preserving wetlands in

Mixing money, power, science and politics indeed complicates the definition of the single word—wetlands—beyond belief. Such is in real life more common than Aristotelian sharpness. Mental models that claim to be meaningful need to take all of these complications into account.

Essential Habitat Definitions. Of potentially great significance is a debate that has been going on among conservation biologists for some years about how we should define an ESU, an Evolutionary Significant Unit.[6] This debate has been long and intense regarding endangered salmon species, for example, in the Pacific Northwest of the United States. An ability to look in detail at the DNA in the cells of organisms provides a great deal of extra information for the debate. A species without habitat, however, is not viable; and preservation of habitat for a given species (which is also desired by humans) is the most common obstacle in getting a political consensus to save a particular form of life.

Aldo Leopold made famous the following words in his classic essay "Round River" in *A Sand County Almanac*: *"The last word in ignorance is the man who says of an animal or plant: 'What good is it?' If the land mechanism as a whole is good, then every part is good, whether we understand it or not. If the biota, in the course of aeons, has built something we like but do not understand, then who but a fool would discard seemingly useless parts? To keep every cog and wheel is the first precaution of intelligent tinkering."*

Reductionist training that leads us, including me, to look at "the parts" that make up the whole sometimes makes us less capable of seeing the whole in our mind. As we have seen (and will see again), a complex system is one for which we cannot simply say that the whole is greater than the sum of its parts: a complex system is *different* than the sum of its parts, sometimes remarkably so. We may miss the boat entirely by not saving the whole of complex ecosystems. And seriously asking "What is it (a form of life) worth? is an indication of human-centered hubris and ignorance.

By the way, if you want a real challenge, try to define *life*. Define what it is, not just what it does, although the latter is sufficiently challenging. The language with which we do most of our logical reasoning is fuzzy at best and at times nearly meaningless when compared to the important fundamental of staying alive.

For further reading of a popular nature on fuzzy logic, cf., [365]. A serious mathematics book which clearly shows that fuzzy logic is a marriage of logic and measure theory, cf., [358]. See also [597].

---

the U.S.A. in some time period; (2) money made by turning wetlands into something else during the same time period; (3) money in the form political campaign contributions related to wetlands during the same time period. Wetland habitat is critical for what species? Are any threatened, endangered, extinct?

[6]Susan Milius, "What's Worth Saving? A fracas over a biological term could have huge consequences for conservation", *Science News*, Vol. 158, October 14, 2000, pp. 250–252.

Exercise 8.2 Fuzzy Logic in Real Life

(i) Someone says, "I have only one vote, it does not count." Criticize this statement from the perspective of fuzzy logic.

(ii) Someone says, "The city council should not concern itself with foreign policy, because the city can do nothing about it." Criticize this statement from the perspective of fuzzy logic.

(iii) Look at VII and analyze the government's definition of "the poverty line" from the perspective of fuzzy logic. Could fuzzy logic have a noticeable impact on people's lives were it actually used by the government to define poverty?

(iv) Someone says, "There are two sides to that argument." Critique this statement from the perspective of fuzzy logic.

(v) Try classifying all humans into male and female, or male and not male. Is there any fuzziness in this situation?

(vi) Try classifying all objects into living and not living. Is there any fuzziness in this classification?

(vii) Martin Luther King Jr. said: *"Injustice anywhere is a threat to justice everywhere."* Discuss this from the point of view of fuzzy logic.

(viii) The Telecommunications Act of 1996 replaced a communications act from 1934. It contained a provision, The Communications Decency Act which attempted to define indecent communications, i.e., pornography. How did the courts deal with this fuzzy definition? Note the fact that Congress in 1998 posted the (expensively produced) Starr Report (on the then President's sexual activities) on the internet. Large parts of this report were also reproduced in newspapers across the nation. Parts of the Starr Report were clearly indecent according to the 1996 definition of Congress. Discuss.

(ix) The Declaration of Independence states that "all men are created equal." Discuss the fuzziness in the definition of "man" and how the definition has changed incrementally from 1776 until today.

(x) Trace the "right to vote" in the United States. Is there any correlation with the definition of "man?" Reference [347] might be helpful.

(xi) Give an example of a situation where Aristotelian thinking "works" very well.

(xii) Give an example where thinking using Aristotelian logic does not "work" as well as thinking which uses fuzzy logic.

(xiii) Those who would like to see more commercial cutting of trees on America's public lands have justified this position, in part, by claiming that there are more trees on America's public lands now than there were in 1776. Assuming that this claim is true, how must the definition of a forest change if trees from one to several hundred years old are replaced by trees most of which are less than twenty to thirty years old? What are some fundamental ecological differences between these two types of forests? (For example, as early as 1990 more that 96 percent of the old growth forests in the lower 48 states had been cut.)

(xiv) Research as best you can the current uses of fuzzy logic in America, Japan and elsewhere. Why do you think eastern societies used fuzzy logic in business and engineering before it "caught on" in western societies?

(xv) The United Nations has a program, REDD, Reduced Emissions from Deforestation and Degradation, which became a topic of discussion during the Copenhagen Climate talks of December 2009, for example. The "saving of the forests" was called a breakthrough by some and greenwashing by others. What definition of "forest" was finally decided upon for the purposes of "offsetting" carbon emissions of industrial nations? For example, would clear-cutting of a diverse tropical rainforest and replacing it with the monoculture of a palm-oil plantation count as a "forest?"

(xvi) The definition of the term "murder" and its legal relatives are fairly well-defined in law. The concept does, however, involve fuzziness. For example, if operators of a coal mine consistently ignore, or dispute, hundreds of infractions called to their attention by mine inspectors (or workers brave enough to do so) declaring the mine unsafe and potentially a deadly work environment—might the mine operators be accused of some form of murder when the mine does indeed kill workers in precisely the manner predicted?

Making of Mental Models and Classical Concepts in Logic. None of us can replicate all of reality/Nature in our mental model of reality/Nature. We must simplify, discard some of the information available. This is what we do to discover patterns. The hallmark of genius is the ability to sift through the data available and concentrate on what is truly important to the subject of study—to simplify appropriately. Advice in this regard from [358] is: *"A good simplification should minimize the loss of information relevant to the problem of concern."*

Notice the word *problem.* When our models are not working, by definition we have a problem. We want or need to solve said problem. A reasonable first step is to gather information on the problem, via direct experience, indirect experience of others and literature/data searches, taking note of such details of where a particular source, if human, gains remuneration—always follow the money when money is involved.

Look for *extreme views* ardently held. There are reasons, not always explicit, for these views; there is at least a grain of truth or more than a grain to be found. *"Triangulate,"* that is, look at the problem from as many points of view, using as many sources of information as you can find and assimilate. This is sometimes referred to as doing your homework.

A Simplification Decision: Sharp or Fuzzy Logic. Among the first steps in simplification is to decide whether sharp logic is sufficient or is not sufficiently delicate to deal with the details. For example, stereotyping can be viewed as a form of Aristotelian oversimplification.[7] Thus all mathematicians are nerds. All fill in the blank are fill in the blank. In pure mathematics sharp logic is often very helpful. In social relations it rarely is.

8.3 Definitions, Assumptions and the Frame of Debate

Whenever a word is used it triggers a *frame* of mind which can be managed if you are conscious of this process. For example, in [380] we are told: "Don't think of an elephant." Of course, that is exactly what we are all thinking of when we hear that sentence! Masters of marketing, advertising, spin, propaganda, public relations and perception management surround us daily, to sell us ideas—to sell us products, wars, political candidates or the idea that professor so-and-so should be fired. None of this is new, it is just that technology has made these techniques ubiquitous and continuous, cf., Chapter 9 and VIII.

[7] Racial profiling, as in searching for criminals or terrorists, does not work. It is a form of stereotyping. For documentation of the type of police work that does apprehend law breakers see [281].

There are two useful countertechniques for dealing with the constant sales pitch. The idea of ferreting out the precise meaning of definitions, words, symbols—getting at the notions that notations represent—and continually challenging assumptions: these are two techniques developed to a fine art at least as early as the third century. Sextus Empiricus is famous as a 3rd century Roman, possibly Greek, skeptic who perfected these two techniques; and who, incidentally used them to attack mathematicians of his time. I was reminded of him while watching Harold Ickes quite successfully defend himself against a hostile congressional panel—using just these techniques. What do you mean by ...? Are you not assuming that ...?

Exercise 8.3 Definitions, Assumptions and Frames

(i) Titles of bills passed by Congress are selected to frame our consciousness. How do titles such as the Patriot Act, the Healthy Forests Initiative or the Clear Skies ...frame any discussion about them?

(ii) Take any bill that deals with terrorism, such as the Patriot Act or any such legislation passed since, and examine the definition of terrorism contained therein. Can just about any form of *peaceful* protest be considered terrorism? Are there pacifists who have been labeled as terrorists by the government? Why do you think this is?

(iii) Take any environmental piece of legislation, such as the Healthy Forest Initiative or one more recent, and discuss its actual—on the ground—effects.

(iv) Discuss the degree of fuzziness in one or more of the following sets: actions that are safe/not safe, plants that are trees/not trees, media that is pornographic/not pornographic, statements that are honest/not honest, actions that are environmentally sustainable/not sustainable, and actions that are legal/not legal.

(v) Discuss the following apparently self-contradictory statement from the point of view of fuzzy logic: "You only keep what you give away."

(vi) Discuss the following quote attributed to physicist Niels Bohr: *"A great truth is a truth whose opposite is also a great truth."* A classical book on quotations attributes this quote to Thomas Mann (1875–1955), *Essay on Freud.* (Note that to every proverb there is an equal and opposite proverb. For example, "Absence makes the heart grow fonder;" and, "Out of sight, out of mind.")

(vii) Discuss the following quote of Abraham Lincoln, (1809–1865): *"It's been my experience that folks who have no vices have very few virtues."*

(viii) H. L. Mencken was misquoted once as having said: *Every complex problem has an obvious, simple solution—which is invariably wrong.* The actual quote was: *There is always an easy solution to every human problem—neat, plausible and wrong.* Discuss what these quotes—or misquotes—have to do with fuzzy logic.

(ix) Ian McHarg, author of *Design with Nature* and [431] said: *"Brain is on trial."* Arthur C. Clarke, science fiction author, said *"It has yet to be proven that intelligence has any survival value."* Discuss what hidden assumption these quotes address.

(x) What hidden assumptions would you have to re-examine if electric power in your home and place of work (or your state) were not available for three hours? Three days? Three years? Answer the same question with petroleum substituted for electric power.

(xi) I once made the following comment to a person who did not know me very well: "I have to go now, I have a (romantic) date with a jet pilot tonight." I am a male. What possible hidden assumptions might be going through the person's mind to whom I was speaking?

(xii) Discuss how an implicit assumption that "the best science is reductionist" may promote or hold back scientific progress in some areas.

(xiii) Discuss the following comment: Whenever you communicate with someone else, you do so in a sea of intentions and hidden assumptions. Discuss how this "sea" affects the communication process.

8.4 Humans in Denial–Nature Cannot be Fooled–Gravity Exists

The definition of *denial*: an unconscious defense mechanism characterized by a refusal to acknowledge painful realities. When one's mental model fails, denial is common. Denial in the context of interpersonal relations is common, and I will not be dealing much with this. Denial of humans in their relationship to Nature is quite another thing, and it is this type of denial that concerns me here. Whether you "believe" in evolution or not, the subject does represent volumes of empirical evidence that one cannot successfully deny exists. Whether or not your senator "believes" in global warming, there is a body of evidence that is denied at our peril. The heart of "knowing" in science is careful observation of what is going on in Nature. After centuries of observations we have discovered some patterns in Nature, and thus it is quite disadvantageous to deny that laws of Nature operate. I like to say "GRAVITY EXISTS" to remind people that Nature's laws never go unenforced—whether any particular person "believes" in them or not.

Presumably any humans who were in denial about gravity tried to fly off cliffs and left no descendants. A given human may not fully appreciate Einstein's relativistic refinements of Newton's inverse square law for gravitation, but every human needs a basic understanding of gravity just to live each day. Gravity acts on us, for all practical purposes, instantly and continuously. There are less obvious laws of Nature that it is easier for humans to be in denial about—but the ultimate consequences of denial are the same.

An Example. Aerospace Professor Xinh's office was across the hall from mine. Ellison Onizuka was Dr. Xinh's graduate student. On January 28, 1986 I was talking to some students in the University Memorial Center when a friend ran up and exhaled in one breath: "The Challenger blew up! This morning on TV, they're all dead—DEAD!" It's one of those moments you never forget. Now, a short walk from my current office is a plaque with the following words:

In living memory of
ELLISON S. ONIZUKA
1946–1986
C U graduate, astronaut, and friend
who dedicated his life to the
enrichment of education and the sciences

Right after the space shuttle Challenger disaster there was a flurry of news, but not much information. Days passed. Then President Reagan appointed an investigatory commission chaired by William P. Rogers who immediately went on record saying: *"We are not going to conduct this investigation in a manner which would be unfairly critical of NASA, because we think—I certainly think—NASA has done an excellent job, and I think the American people do."* (NASA is the National Aeronautics and Space Administration.) The first of many statements in denial; "I guess I'll never know the truth," I cursed to myself.

Then, almost by accident, a last ray of pure, honest hope. Richard P. Feynman, Nobel Prize winning physics professor from Caltech, the precocious genius on the World War II Manhattan Project, was appointed to the President's commission charged with investigating the Challenger "accident." I had the good fortune of learning much physics out of the "Feynman notes" hot off a pre-Xerox copier. I had the even better luck of meeting him some years later. I *knew* Feynman was brilliant enough to see the truth, and I guessed that *nothing* would prevent him from telling the truth that he saw. He did not disappoint.

After a burst of intense investigation Feynman zeroed in on the physical cause of the disaster: O-ring failure, due in no small measure to the freezing temperatures surrounding the launch time. In fact, Feynman did a brilliant job of cutting through to the basic truth by doing an on-national-TV-camera experiment with a clamp, a bit of O-ring and a glass of ice water, showing that at 32 degrees Fahrenheit the O-rings did not perform properly. But Feynman did not stop there. He found out that engineers knew the O-rings would most likely fail and that they had issued warnings which were ignored. Feynman went on to analyze the whole way NASA did risk assessment and management. Management in relation to several subsystems was questionable; for example, before the O-rings failed the engine turbines could have failed due to cracks/metal fatigue. Feynman found the *fundamental reason* for the Challenger loss was that NASA had gotten complacent and sloppy—that NASA was *ignoring* hard scientific *reality*. For the full and interesting story, see [188]. While most were in denial, Feynman wrote an appendix to the official report. The last sentence of his report says:

"For a successful technology, reality must take precedence over public relations, for Nature cannot be fooled."

Had Feynman not been one of the twelve on the President's commission investigating the Challenger disaster, had Feynman not done a thorough, honest investigation on his own and succeeded in getting his report attached as an appendix to the "official report" of the commission, I am not sure when if ever Americans would have found out what happened.[8]

Astronaut, Dr. Kalpana Chawla, got her Ph.D. in aerospace engineering at the University of Colorado in 1988, in the same department Ellison Onizuka studied. On February 1, 2003 the space shuttle Columbia disassembled over Texas before its scheduled landing in Florida, killing all on board, including Dr. Chawla. NASA had requested satellite photos to check for damage to the Columbia while it was in orbit, before February. The U.S. government

[8]The findings of Feynman are corroborated in an NPR interview, April 29, 2001, with Chris Kraft, NASA's first flight director. Kraft was in the viewing room of Mission Control in Houston when the Challenger blew up. See also [371, 486]. For more on Feynman, cf., [229].

refused, due to its total preoccupation with Iraq.[9]

Fallacies. Classically a *fallacy* is an incorrect logical argument. There are many types of fallacy. They are categorized according to the type of logical falsehood, be it intentional or not. For example, Scott Nearing, cf., [483], lost his position in academia due in large part to a fallacy called *argumentum ad hominem*. An *ad hominem* argument attacks the person and not the argument that person is making. Nearing was labeled a Communist[10] and bounced out of academia as being "dangerous," thus diverting attention from the real reasons he was dismissed. Namely, Nearing argued unpopular views: he was anti-war and against child labor.

There are many classical books on fallacies, going back at least to Aristotle. For example, see [184, 271, 539, 76]. The ones below and others form the basis of some propaganda techniques.

Exercise 8.4 There are Many Kinds of Fallacies

(i) Can you give Aristotle's principal list of thirteen types of fallacy as he discussed them in his *Sophistical Refutations*?

(ii) Find examples of *argumentum ad temperantiam*, the moderate view is the correct one. Find examples of *argumentum ad verecundiem*, respect for authority, the advertising testimonial. Find examples of *post hoc ergo propter hoc*, after this, therefore on account of this, the fallacy of supposing that because one event follows another, then the second has been caused by the first. Find examples of *argumentum ad ignorantiam*, appeal to ignorance. Find examples of *argumentum ad antiquitam*, it is good or right because it is old. Find an example of *argumentum ad crumenam*, money is the measure of rightness; if you're so right, why ain't you rich? Find examples of *argumentum ad baculum*, arguing by intimidation. Find an example of *argumentum ad populum*, everyone believes it, so it must be true. Find an example of *argumentum ad misericordiam*, an appeal to irrelevant emotions. Find examples of *petitio principii*, circular reasoning. Find a red herring. Find an example of ex post facto statistics. Find examples of straw men, the gambler's fallacy, false precision, the undistributed middle, non sequitur, faulty generalizations, and so on—look them up. As C. L. Hamblin says in the book cited in [271]: *"We have no theory of fallacy at all, in the sense in which we have theories of correct reasoning or inference."*

(iii) In Chinese, accents, or more precisely tones, are part of the meaning of every word. In English they can be used slyly to mislead. I once saw a movie called *The Conversation*. The entire plot was balanced on the accent of one word. Gene Hackman was an electronic snoop, and he was recording conversations of the wife of a jealous husband. In one conversation the wife says to her lover: "He'd *kill* us if he had the chance." or was it "He'd kill *us* if he had the chance." How does emphasis change the meaning of the sentence? Find other examples of sentences wherein the entire meaning of the sentence changes if you change the word that is accented or emphasized. How can change of accent be used to mislead?

(iv) Create a new type of fallacy and see if you can fool a friend. (Then tell him/her the truth!)

(v) Discuss how racism can be exploited as a tool to prevent the formation of coalitions, e.g., divide and conquer. Coalitions might be able to solve certain problems that individual groups cannot solve separately. Can you think of any such problems? Can you think of anyone who would like to prevent the formation of certain coalitions among races, or classes?

[9] Listen to the audio archive of March 17, 2003, at www.democracynow.org.

[10] Labeling someone a Communist, end of discussion, was once the method of choice of removing persons from all walks of life, including academia, cf., [613].

8.5 The Bigger Picture Principle

Consider the following poem which artfully shows the human predicament. In this section I use the term "picture" interchangeably with "mental model."

The Blind Men and the Elephant

by John Godfrey Saxe

It was six men of Hindustan
To learning much inclined
Who went to see the elephant,
(Though all of them were blind);
That each by observation
Might satisfy his mind.

The first approached the elephant,
And happening to fall
Against his broad and sturdy side,
At once began to bawl,
"Bless me, it seems the elephant
Is very like a wall."

The second, feeling of his tusk,
Cried, "Ho! what have we here
So very round and smooth and sharp?
To me 'tis mighty clear
This wonder of an elephant
Is very like a spear."

The third approached the animal,
And happening to take
The squirming trunk within his hands,
Then boldly up and spake;
"I see," quoth he, "the elephant
Is very like a snake."

The fourth stretched out his eager hand
And felt about the knee,
"What most this mighty beast is like
Is mighty plain," quoth he;
"Tis clear enough the elephant
Is very like a tree."

The fifth who chanced to touch the ear
Said, "Even the blindest man
Can tell what this resembles most;
Deny the fact who can,
This marvel of an elephant
Is very like a fan."

The sixth no sooner had begun
About the beast to grope
Than, seizing on the swinging tail
That fell within his scope,
"I see," cried he, "the elephant
Is very like a rope."

And so these men of Hindustan
Disputed loud and long,
Each of his own opinion
Exceeding stiff and strong,
Though each was partly in the right,
And all were in the wrong!

The above poem is a metaphor for the state of humanity looking at Nature. We all have our own experiences, imaginations and hence our own mental models. We all would like to be right about our mental models, sometimes to the point of going to war.

In the poem we have six mental models and they are all consistent with and can be thought of as part of one larger model, which I call a Bigger Picture. In this case, the Bigger Picture is a model of an elephant as is commonly understood. Though difficult, if the six men got together and merged their models, subject only to consistency, a Bigger (more accurate) Picture results. This then is the essence of my Bigger Picture Principle. Recall the Axiom of Consistency, cf., Section 7.2, which asserts that Nature (Reality) is consistent.[11]

The Bigger Picture Principle. *Suppose there are given two or more mental models of some aspect of Nature. Merge these models into one model subject to the requirement that this new model is consistent.*[12] *Then this merged model,*

[11] The sharp logical concept of consistency means free from contradiction, A and not A cannot happen. Since I insist on allowing the option of fuzzy logic, statements that are say .1 A and .9 not A are allowed. What is not allowed is a statement that is 1.0 A and 1.0 not A. A statement that is .5 A and .5 not A is allowed. If you find all this confusing I suggest you read [365, 358, 597] or other books on fuzzy logic. I can cheat here and say that any model is consistent if it agrees with (observations of) Nature.

[12] I can take as a working definition of consistent: "agrees with all observations of and experiments in Nature." Thus the axiom of consistency is a bit of circular reasoning, or a

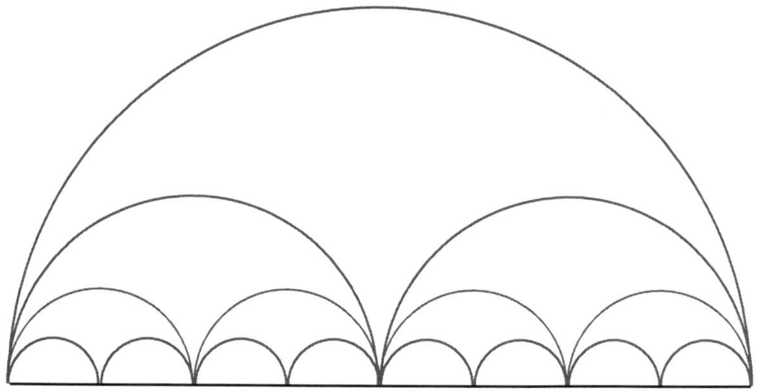

FIGURE 8.3: Two Curves Close in Distance but Not in Length

called a Bigger Picture, is more likely to be a more accurate model of the aspect of reality in question than each smaller model taken separately. (See also Section 1.4.)

The more you can explain with your model the better it is, by definition.

Examples of Bigger Pictures. What Does It Mean for Two Curves to Be Close? Consider Figure 8.3. The length of the straight line segment from 0 to 1 shall be taken to be 1 (unit). The length of the semicircle whose base is this same segment is $\frac{\pi}{2}$, since the circumference of a circle is given by the product of its diameter times π. (Recall that $\pi \approx 3.141592654$.) The length of the curve consisting of two semicircles is also $\frac{\pi}{2}$. Can you see this?[13] The length of the curve consisting of four semicircles is also $\frac{\pi}{2}$. Can you see this?[14] Perhaps you now see a pattern, and a possible problem with your intuition! There is a sequence of curves each of length $\frac{\pi}{2}$ converging uniformly to the line segment of length 1.

Your intuition might be telling you that if curves are "close" to each other, then their lengths should be close to the same number. This evidently is not true. Your picture, or mental model of the situation needs an overhaul. See if you can resolve this paradox by yourself by making your picture/mental model bigger and more at one with what is (before looking further).

tautology. Nature defines consistency. Thus any model, newer, bigger or otherwise, should undergo a reality check. Does it agree with Nature? A deeper discussion I leave to the reader to investigate as an exercise.

[13]You have two semicircles, each with a diameter of $\frac{1}{2}$.

[14]This is because you have four semicircles each with a diameter of $\frac{1}{4}$.

Exercise 8.5 A Bigger Picture Can Entail More Refined Definitions

Give yourself another day to contemplate the situation in Figure 8.3 before reading on. If you want to believe that curves that are "close" have lengths that are "close" you will have to be careful and make a precise definition of what it means for one curve to be close to another. If the points along one curve are uniformly close to the points of another, this is intuitively necessary—but not sufficient—for the two curves to be "close." Can you come up with a definition of "closeness" of curves which when applied to Figure 8.3 shows that as more and more semicircles are introduced the resulting curve moves farther and farther away from the straight line segment of length 1?[15]

The purpose of the above brain teaser is to give an explicit example of past mathematical research that creates a bigger more honest picture than existed before, and hence deepens understanding. There are many other examples, like Feynman integrals, which have defied a complete understanding for decades. Mathematical research is indeed very active today, making bigger more honest pictures.

Exercise 8.6 Counting Infinitely Many Points with Geometry

(i) Can you make an argument that a circle with one point removed has the same number of points as a straight line which is infinitely long?

(ii) Can you make an argument, similar to the argument in (i), that a sphere with one point removed has the same number of points as a flat plane infinite in all directions?

For an illustration of the Bigger Picture Principle involving the CIA see Section 23.4.

Exercise 8.7 A Bigger Picture of Climate Change

Critics of climate change science often make comments to the effect that last year wasn't so warm; or, the sun is getting warmer; or, climate scientists have an agenda in co-operation with proponents of world government; and so on. Can the Bigger Picture Principle be used in this context to incorporate and explain all the objections that the skeptics of global warming bring up?

(i) How many of the skeptics' objections are actually purposeful misrepresentations, cf., [317]?

(ii) Which skeptics' objections actually bring up true scientific phenomena, but the effects of these phenomena are not quantitatively significant?

---

[15]Hint: The astute observer will notice that the curves consisting of semicircles have more cusps (points at the end of a semicircle) the more semicircles there are. Thus such a curve is "far away" from the straight line segment from 0 to 1 in the sense that the tangent line to the curve (made up of semicircles) at each cusp is perpendicular to, i.e., orthogonal to, makes a ninety degree angle with, the corresponding tangent to the line segment from 0 to 1. (The line segment joining 0 to 1 is a flat curve, and the tangent to any point on this curve is the line through that point parallel to the line segment from 0 to 1. In other words, the line segment is essentially its own tangent at every point of itself.) Mathematicians can say that two curves are *close* if the points of one curve are uniformly close to the points of the other *and* the tangent at a point on one curve is close to the tangent at the corresponding point of the other curve. (Close tangents have slopes that are close, i.e., they are nearly parallel.) This definition can be made a great deal more precise by using calculus, and if you are interested there is a long discussion of the details in [379, pp. 325–334]. This book should be quite interesting to anyone who wants to learn or teach mathematics, or who is just interested in "where it comes from." Also of interest is a review of this book in the November 2001 issue of the Notices of the American Mathematical Society, pp. 1182–1188.

(iii) Has anyone proposed an effect that would quantitatively negate the effects of Arrhenius's CO_2 Greenhouse Gas Law, cf. Section 1.9?

(iv) Have any skeptics addressed the acidification of the world's oceans?

(v) How "Big" is the total picture of all climate change skeptics taken together?

(vi) Have the IPCC model's already addressed climate-change skeptic's objections? To what extent is climate-change skepticism just a form of denial?

Chapter 9

The Dunbar Number, Political Power, Public Relations, and Fear: Can Humans Act Sustainably?

9.1 The Sustainability Hypothesis: Is It True?

Most species eventually go extinct. For humans there are at least two possible types of extinction: collapse of complex civilization (a type of "super-organism"), and biological extinction of the species, *Homo sapiens* (or more precisely *Homo sapiens sapiens*). In this section I want to discuss from a mathematical perspective just a few fundamental weak links in the way society is now organized which could lead to its disintegration. Hopefully, suitable changes will be made so that social disintegration is made improbable.

Every organism, "super" or otherwise, needs to earn a living within the ecosystem it finds itself. Humans have changed their ambient ecosystems to an extent not practiced by other species, who largely adapt their behavior to the immediate requirements of Nature. But as Ian McHarg so aptly stated, cf., [431]: *"Brain is on trial."* Compared to bacteria, for example, humans have been around for a very short time on planet Earth. From the standpoint of survivability, bacteria present the most robust form of life. As the late Stephen Jay Gould said:

> *"On any possible, reasonable or fair criterion, bacteria are—and always have been—the dominant forms of life on Earth. Our failure to grasp this most evident of biological facts arises in part from the blindness of our arrogance but also, in large measure, as an effect of scale. We are so accustomed to viewing phenomena of our scale—sizes measured in feet and ages in decades—as typical of nature."*

Bacterial biomass on Earth is enormously larger than that of humans. "Brain" is a recent development, and humility is in order. We have a long way to go before we can claim superiority to bacteria as a resilient, sustainable form of life. Our complex civilizations are even more recent—and more fragile. For an argument (involving cognitive neuroscience and evolutionary psychology) that Nature's experiment, "the human brain," will be coming to an end relatively soon, see [551], by Marc Pratarelli. Given the history of human behavior, Pratarelli's arguments cannot be dismissed easily. In light of [551] *I will make an assumption* the truth of which is quite debatable: It is possible for some collection of humans to voluntarily create a culture which can live sustainably within the ecosystem it finds itself—surviving and preventing conquest by humans with unsustainable cultures. I will refer to this assumption as the *sustainability hypothesis*. The practices of war, subjugation, occupation, extraction (without recycling), and any other form of conquest for the purpose of acquiring resources, do not count as sustainable practices, since eventually one runs out of people or ecosystems to conquer.

We are thus led to a *tautology*, which is useful for focusing our attention. If human societies are to forestall extinction, sustainable cultures must be invented and prevail in the long term. We are left with defining "sustainable cultures." This definition is susceptible to analysis via mathematics and science. For example, a society that attempts to grow in numbers or consumption beyond all bounds is not sustainable, as has been pointed out by many folks. I will leave to you the exercise of stating the definition of a "sustainable culture" as precisely as you can. Hint: See Part VII.

Once having made a fundamental change in the ecosystem that supports our lives, such as inventing agriculture, we must make sure that the underlying logic of these changes is consistent with the logic of Nature and that we have social structures that can sustain these changes. Humans have the capacity to imagine. Imagination can be used to create temporary "realities" in contradiction to Nature's logic, or imagination can be used to create a logic for civilization that is consistent with Nature's logic and the technology to implement it. The "jury is out," it could go either way.

A fundamental problem is our focus on ourselves and our creations without ever thinking of how those creations fit into Nature. Is our society a comfortable, sustainable nest, compatibly nestled in Nature; or is it an irritant to Nature to be expelled—"spit out like a watermellon seed"—into the cold, darkness of eternal extinction? But I am getting ahead of myself. I want to first look at a number, a number I find endlessly fascinating since it likely is one of several human biological limitations, cf., [551, 360], that we must take into account in our quest for survival.

9.2 The Dunbar Number

Robin Dunbar, cf., [153], has developed a formula for computing what Malcolm Gladwell, cf., [225], might call the social capacity of various mammals. The social capacity, or maximum group size, for an animal is the largest number of intimates or reasonably close friends/acquaintances an individual can really relate to. Dunbar plugs in the size of the neocortex relative to the size of the brain and comes up with a number, which in the case of humans is 147.8, roughly 150. He looked at 21 hunter-gatherer societies for which reliable historical evidence exists—including the Warlpiri of Australia—and found that the average number of people in their villages was 148.4, cf., III.

There are various physical and biological constraints on humans, cf. [551, 360], but the *Dunbar Number* quite possibly may have far reaching implications. Consider for the moment a person like the President of the United States. He/she can have fewer than 200 close associates, but his/her decisions affect millions, perhaps billions, of people. How are those decisions reached? The President's close associates, one can safely assume, have a greater impact on the President than people the President does not know or has never met.

The same can be said for a senator or congressional representative. It might be important before voting for these folks to know who the "Dunbar associates" are of the person for whom you are voting, and what these associates think.

Exercise 9.1 Possible Implications of the Dunbar Number

(i) If the model proposed by Robin Dunbar is correct, then 150 is roughly an upper bound on the number of intimates a person can have. It has been claimed that it is common for a middle-aged men in the United States to have few to no close friends. How many close associates, i.e., intimates, do you think the typical human has? Does this number depend on the culture in which that person finds him/herself?

(ii) Is it possible to break the Dunbar number "barrier" using technology? For example, using media such as newspapers, books, the internet, television, radio, do you think it is possible to have more than 150 intimates? How do you define "intimate"? For example, if you send a personal email to someone and they return a curt (automated?) reply: "See my Weblog," do you consider this intimate?

(iii) Do you think the Dunbar number is a measure of our brain's ability to get on an intimate basis with a certain amount of data in a certain amount of time? By being intimate with data I mean familiarity to the extent one believes and/or trusts that data. Do you think that for data/information to lead to action that one has to be intimate with that information?

(iv) It is likely that a person is more deeply, personally, emotionally affected by things that happen to people in their Dunbar cohort than to people outside of this group. How might this influence a person's level of reaction to—and ability to sustain this reaction over time to—environmental tragedies or other events that appear to be localized in affect? (Consider at least two cases. One where the people affected are outside one's circle of friends and acquaintances, even outside of one's perceived social/economic class or regional/national identity. For example, when a tsunami or earthquake devastates a city or country, often a remarkable outpouring of assistance comes from a variety of folks not "Dunbar related" to the victims. But how long is this outpouring maintained? In the second case, assume the people affected are close family, friends, or associates. For example, if someone from group A uses a weapon, like a predator drone, to wipe out the family and friends of someone from group B, do you think the reaction of the remaining person in group B might be intense and long lasting?)

(v) Do you think the Dunbar number concept partially explains why associates of Goldman-Sachs quite often head the U.S. Treasury Department independent of which political party holds the presidency? Is a similar analysis relevant to the Secretary of Defense and the core leadership of the military?

(vi) If you think of each person surrounded by a "cloud" of Dunbar associates/intimates, does this change how you would model social interactions, communications?

(vii) Investigate the extent to which the Dunbar number is used in the organization of corporations? For example, are there advantages to limiting "subunits" of a corporation to fewer than 200 employees? Do any corporations do this?

(viii) Do Supreme Court justices in the United States traditionally come from a narrow cohort of the population?

Exercise 9.2 The Scale and Type of Representation Envisioned by the Founders

I have a poster-copy of the originally proposed 12 Bill of Rights hanging on my wall from the National Archives. There really were 12!

(i) When the original Bill of Rights was sent to the states in September of 1789, it contained 12 amendments not 10. The legislatures of the states voted on each amendment individually and ratified only 10 of the 12. The remaining two—the first two, as it turned out—failed to win approval in three-fourths of the legislatures, as required for ratification, and so did not become part of the Constitution.

As a result, Amendment Three of the original 12 became Amendment One, the "Free Speech" amendment, Amendment Four became Amendment Two[1] and so on.

Even though over 200 years have elapsed, the original first two amendments should still become part of the Constitution if three-fourths of the state legislatures ratify them—and guess what? The original Second Amendment was finally ratified on May 7, 1992 as the 27th Amendment. This 27th Amendment provides that no pay raise for senators or representatives could take effect until after the next election.

That leaves the original First Amendment. What did it say? It said: *There shall be at least one representative in the House of Representatives for each 50,000 people.* The current size of the House of Representatives is 435. What would be the size of the House of Representatives if this Amendment were ratified by three-fourths of the state legislatures? It is interesting to note that in my state one member of the Colorado State House of Representatives represents about 50, 000 people. What are the figures for your home state?

(ii) An interesting article appeared on February 23, 1997 in my local paper titled: "How to make Congress better: Make it Bigger." This article was a reprint of the original written for the Boston Globe by Jeff Jacoby. It is interesting to see that the scale of representation has appeared at least once in the majority media. Jacoby recounts that when the first Congress was seated in 1789, there were 65 members of the House of Representatives, with the U.S. population estimated at 2 million, i.e., there was one representative for roughly every 30,000 people. The 1790 census put the U.S. population at 3.9 million and the House of Representatives was increased to 105. In 1800 the U.S. population was 5.3 million, the House grew to 141. Jumping to 1910, the U.S. population was 92 million, and the next year President Taft signed a bill increasing the House of Representatives to 435 in 1911. The ratio of representatives to citizens was now roughly $\frac{1}{200,000}$. In 1929 the size of the House was officially frozen at 435, no matter how much the population grew. When I first considered this exercise in 1997 we had about 260 million Americans, with each of the 435 representatives representing on average almost 600,000 citizens. This is not a favorable ratio compared to many other countries. For example, (in 1997) in Japan each member of their Diet's lower house represented about 245,000 people; for the German Bundestag the ratio was $\frac{1}{100,000}$; for Canada's House of Commons, $\frac{1}{96,000}$; for Britain's, $\frac{1}{89,000}$. In 2010 the U.S. population soared past 300 million.

Discuss how passing the original first amendment would affect a representative's campaign expenses and current non-stop need to fund raise. Discuss other effects of passing the original first amendment. It is very interesting that the framers of the Constitution had a scale of representation in mind. That scale is determined by how many people one representative can adequately communicate with. Has this scale changed in the last 200 years?

[1]It is an interesting exercise to look up the history of the currently designated Second Amendment which is in part about guns. The original version of this amendment written by Jefferson proposed more of a Swiss model for an army which would be disbanded in a time of peace, but which would be an ever ready citizen-militia trained in the use of firearms and military matters. Imagine, no standing army, no military-industrial complex.

Estimate the number that would replace 50,000, if any, given modern communications technology and the current budget of each person in Congress allotted to communication with constituents?

(iii) There is an enormous army of lobbyists, about 33,000 to 38,000 "on Capitol Hill," prevailing upon the 435 representatives, 100 senators and various administrators. Do these lobbyists "fill up" the "Dunbar capacity" of all those who create and administer our laws? These lobbyists are equipped with an enormous weapon, not available to most groups of Americans. After the *Citizens United v. Federal Election Commission*, 130 S. Ct 876 (2010), decision of the U.S. Supreme Court on January 21, 2010, any (U.S., multinational, or foreign) corporation or union can spend an unlimited amount of money to elect or defeat a given candidate—a legislator, a President and so on. Does this affect, or intensify, the "Dunbar effect" of lobbyists?[2]

(iv) Thomas Jefferson was concerned that a common pattern would be repeated in the United States, i.e., that power would be consolidated into the hands of relatively few. He was also concerned that the judiciary would provide the mechanism for this consolidation to take place. Consider and comment on the following quotes from Jefferson, at the time you read this.

"The [federal] judiciary branch is the instrument which, working like gravity, without intermission, is to press us at last into one consolidated mass." Thomas Jefferson to Archibald Thweat, 1821, [407, 15:307].

"There is no danger I apprehend so much as the consolidation of our government by the noiseless and therefore unalarming instrumentality of the Supreme Court." Thomas Jefferson to William Johnson, 1823, [407, 15:421].

9.3 Public Relations, Political Power, and the Organization of Society

From one perspective civilization can be viewed as a big game, in the sense of game theory, involving acquisition or loss of political and economic power by the various players. One might expect that some players would be better at playing the game than others. The "game" depends on the "rules," either followed by mutual consent among (or imposed upon and by) the human players, or imposed by Nature.

A Possible Model of Social Organization. The building blocks of society are most naturally "Dunbar cohorts." Beyond this basic observation there are a number of ways these building blocks can be assembled into a society. Before outlining actual current structure, I would like to briefly describe an

[2]For example, just one corporation—Goldman Sachs—spends more annually to pay just its top employees than the combined assets of all the nation's major unions. University of Wisconsin communications professor Lew Friedland also points out that the nation's four largest banks would have to allocate a mere one-tenth of one percent of their assets – $6 billion – to counter a campaign in which the whole of the U.S. labor movement spent all of its assets. And some corporations, such as Exxon, have vastly more capital to work with than even Goldman Sachs, cf., http://www.thenation.com/blogs/thebeat/521020/unions_can_t_ompete_with_corporate_campaign_cash for these comparisons/assertions.

alternate model. Following a principle of *mutual respect* these Dunbar cohorts could *freely choose* to associate into larger units, e.g., neighborhoods, small to large cities, counties, states, nations, the global community. Vast networks of communication and trade, in ideas and materials could self-organize from "the bottom up." Toward guaranteeing the ability for subunits to leave larger associations when the larger system "does not work" for the subunit, the ability to provide basic necessities of food, water, energy, shelter, and social organization should be retained at the smallest, most local level possible. (For example, Los Angeles, via the citizen-owned Los Angeles Department of Water and Power, was able to provide its own energy during the 2000–2001 California energy crisis, and thus was not at the mercy of government deregulators and the likes of Enron, cf., pages 211, 492.)

Such a system would make maximal use of local knowledge, allow for innovation and experimentation, and be more robust than a centrally controlled "top down" model. For example, the so-called capitalist and socialist models of social organization could be but two among many models in practice at a given time, and "may the best model win." For example, some community could adopt nuclear power, another could adopt renewable solar, wind, pumped storage without either imposing their model on the other, or requiring one to subsidize the other against their will.[3] Hundreds of arrangements adapted to local geography and circumstances could be developed for solving the basic problems of life: eating, keeping warm or cool, staying healthy, seeking happiness and so on.

A key concept here is that the subunits freely and mutually agree to form larger associations—which any subunit can elect to leave if it decides the larger association is not in its best interest. There are (would have been) some interesting consequences of this concept. For example, the state of Florida would have had the right to ignore the top down decision of the Supreme Court in *Bush v. Gore* in 2000. Florida could have then followed its own laws and the fundamental principle: an election is determined by counting all the votes. The process of counting all the votes—a principle more fundamental than the Supreme Court itself—was arbitrarily interrupted by the Superme Court in this case. Of course, the "United" States would be a different sort of union, if such options were allowed.

Looking back over history, the Civil War could have been avoided if "the South" had been allowed to separate from the rest of the United States and continue with its abhorrent practice of enslaving humans. Over time slavery would likely have been abandoned in "the South" just as apartheid was abandoned in South Africa—without wars, since there exist much more popular and efficient methods of social organization. (Of course, slavery would not

[3]If the community that adopts nuclear power cannot manage the waste products produced or has an accident not confined to that community, then they are being subsidized by any outsider who is affected.

exist in the first place in a world that accepted the principle of associations freely entered upon at the individual level.)

The Bio-Copernican Axiom could play a role here. It basically has every individual realizing that "I am not the center of the (biological or social) universe." A Dunbar cohort that adopts this axiom would likely be quite successful and sustainable—if it is not eliminated or subjugated during a violent attack by others. And this points out one of the major weaknesses of *Homo sapiens*: the propensity to quickly devolve into violence as a means of asserting the will of one individual, group, or nation-state over the will of other folks. It would be preferable for such warring units to accept a universal principle of *mutual respect*—and just refuse to interact, to dissociate, perhaps forever. Peaceful "divorce" is preferable to violent war.

What is the Structure of Current Civilization? I will speak mainly about the U.S. situation here, although the model applies elsewhere. Briefly, some players of the economic and political game have temporarily "won," i.e., their wills dominate the society, and they maintain dominance by a quite sophisticated public relations apparatus, backed up by teams of lawyers, law suits, private or government "security" teams, and military assaults when and where necessary. (Especially for those who vigorously disagree with this last sentence please see Exercise 9.3, and cf., [477] and www.poclad.org, for a history of corporate power.) This, of course, is hardly a complete analysis; but if the reader will humor me I will soon outline some of the extreme negative consequences of a civilization run by a merger of corporations and the government while giving evidence that many (soon to be all?) of U.S. society's major decisions are made by relatively few folks in large corporations, either directly or by surrogates. There are alternatives.

From social psychologist Alex Carey's posthumous 1997 book: *Taking the Risk Out of Democracy: Corporate Propaganda versus Freedom and Liberty*, we read: *"The twentieth century has been characterized by three developments of great political importance: the growth of democracy, the growth of corporate power, and the growth of corporate propaganda as a means of protecting corporate power against democracy."*

Let me briefly review some of the immense negative consequences of past corporate behavior, which I abbreviate as the WACU (We Are the Center of the Universe) pattern of behavior, cf., Chapter 4, page 85. I accept as axiomatic that any pattern of behavior, no matter how positive or negative, will eventually be engaged in, absent effective feedback, i.e., regulation, from the rest of us.

Some Examples of Negative Consequences of WACU Behavior. For profit corporations are not easy to control, but they are easy to understand. They tend to follow the path to maximum profits in terms of cash while avoiding social or environmental costs, absent regulation, i.e., *effective* feedback from the rest of us. By means of well-funded public relations campaigns, i.e., *propaganda*, corporations can construct *alternative realities* which are professionally designed to get people to accept a narrow corporate vision of reality—which is often at odds with reality itself.

(•) For most of the last century the tobacco industry successfully created the false reality in people's minds that smoking is not harmful. Documentation now shows that the smoking habit in general has negative health effects. Thus, whether or not the tobacco industry actually believed its own advertising, its version of "reality" was false. (WACU) The campaign continues in various places around the world, many people believing that smoking does not cause health problems. As I write, even in the U.S., about 400,000 people a year die from smoking related illness, according to government statistics. If $\frac{400,000}{365} \approx 1,096$ people died per day from, say, airline crashes (a few passenger jets crashing *per day*) it would be newsworthy!

(•) Elsewhere, cf., page 85, corporations dealing with lead, vinyl chloride, and asbestos have been documented to exhibit the WACU pattern of behavior, creating virtual realities to maintain a narrow notion of profitability, costing some their lives. This documentation is extremely hard to get. It is even more difficult to prevail in court against such corporations, cf., [479]. Thus the question arises if there are many megacorporations that do NOT exhibit the WACU pattern of behavior.

(•) In the first chapter of this book we began by pointing out that "Big oil, auto, and tire" corporations conspired to eliminate a popular form of transportation in the United States, i.e., electric rail. The population at the time was kept in the dark as to the existence of this conspiracy; and since almost no one today is aware of this criminal conspiracy (they were convicted!), history has been (so far) successfully rewritten. (WACU) The reality is now upon us; there are halting attempts to reconstruct "fast tracks," "light rail" across the U.S. We would have continuously had a system of electric rail and roads if the natural processes of social and economic self-organization had not been purposefully derailed by a few corporate leaders for their own short term gain—at enormous costs to society in the long run.

(•) We now know, see page 25, that powerful corporations were (are) behind a disinformation campaign concerning global warming. (WACU) Their own paid scientists told them that global warming was real and that humans were the likely cause. These corporations chose to ignore the truth and disinform the public, and the propaganda continues. Society's response to global warming has been delayed for decades and continues to be "put on hold" due to this purposefully induced confusion in the public mind, greatly increasing the chances that the human response will be too little, too late. A tragedy is in the making, since alternatives providing more employment and less environmental impact have existed for some time.

(•) The financial industry, cf. Chapter 2, successfully managed Congress and the President into lifting government regulations on their activities. This industry knowingly gambled with other people's money and lost, nearly crashing the world economy, [669]. After trillions of dollars were transferred from the public to this industry—*thus privatizing profits while socializing costs*—they have shown no signs of remorse, of being thankful, or even educability. (WACU.) They immediately and successfully prevented government reinstatement of many essential previous regulations. This situation is the clearest and most undeniable demonstration that the U.S. does not have a capitalist system. As Senator Durbin said, the "banks own the place." It appears that in the eyes of these corporations America is managed first for their benefit, if economic benefits might remain for those they are chartered to serve or for the general population that would be nice but evidently not necessary. It should be noted that the special inspector general for the Troubled Asset Relief Program, Neil Barofsky, said: *"Even if TARP saved our financial system from driving off a cliff back in 2008, absent meaningful reform, we are still driving on the same winding mountain road, but this time in a faster car."* (See January 31, 2010 Associated Press article by Daniel Wagner and Alan Zibel.) Taxpayers have been forced to support failure. Wall Street is on public welfare, but such language is not SAFE, see page 591, in this context. See also Chapter 2.

Models for avoiding future tragedy in energy and finance exist. For example, the media did not cover the fact that while private investor-owned (and unregulated—see the next bullet) energy company Enron was wreaking havoc in 2000–2001, MUNIs, cf., page 492, municipally owned utility companies such as the Los Angeles Department of Water and Power and their public were doing quite well. Likewise, Canada did not experience the financial meltdown felt in the United States in 2008 and following. As Paul Krugman points out, cf., *The New York Times*, February 1, 2010, Canada's banks are too big to fail, and the "too low" interest rates in the U.S. were closely followed by the Canadians. The big difference was close regulation by Canada's independent Financial Consumer Agency which sharply restricted *predatory lending*, euphemistically and frame changingly referred to as "subprime lending," cf., Chapter 2. For the time being, American corporations and the U.S. government are apparently unable to learn from examples that work. Effective and honest feedback is necessary. This feedback might come positively from an informed public, or it might come negatively in the form of complete collapse.

(•) Research the behavior of the Enron Corporation and its role in the 2000–2001 energy crisis, particularly in California, cf., page 492. Tapes of disgusting and profane corporate Enron conversations concerning the apparent "joke" of causing people considerable pain have surfaced. (WACU) Here's an example.

A phone conversation somewhere between Los Angeles and Houston, July, 2000.

"Wassup down at the grid, bro?"

"Nothin'. Just f***ing with California, man. Wanna see something cool? Look outside your window."

"Hey, the lights just went out in Beverly Hills."

"Yeah. Let's see those rich b****es get liposuction in the dark. Watch this."

"The power just went down at Cedar Sinai Hospital."

"Cool, huh? We get bonuses for that one. Last night I was able to shut down a defibrillator at UCLA Cardiac Care from 1500 miles away."

"F'in A."

"Wait. The hospital lights just went on."

"F***ing generators."

"I know. I'll get someone on that right away.."

"F***ing sick people."

"I was just thinking. What if the Feds cap energy prices?"

(Pause, then hysterical laughter)

"Yeah. That'll happen...when Cheney tells Waxman who was in his energy meetings."

(More laughter)

"Wait a minute. What energy meetings?"

"Exactly."

"No, I'm serious. Cheney? Waxman? What energy meetings?

"Oops. Forget I said that. The Enron time/space continuum is still in experimental stage."

"You know, he steals money from California to the tune of about a million."

"Will you rephrase that?"

"OK, he, um, illegally abrigates, sucking every possible penny from any and everyone in California no matter how poor, no matter how elderly, no matter how sick, to the tune of a million bucks or two a day."

"Better. Who's he?"

"Well...not Lay or Skilling...wink-wink."

"They make Halliburton business practices look legal."

(Pause. Hysterical laughter)

"I'm only kidding."

"How can we get away with a f***ing mugging of a state's treasury without someone getting wise? I mean, we're stealing aren't we?"

Stealing? Nah. It's f***ing capitalism, man. First we manipulate the energy flow, they start rolling blackouts, prices go up and then we capitalize on it. That's why they call it capitalism. Stealing is when you get caught."

"You sure?"

"Look. It's all about denial. People say there might be a problem and we just deny it. Ooh, look. I just turned off all the street lights in Compton. BAM! Yeah! Y'see, people have faith in big business. We're like the rich uncle who you don't diss because you don't want to be written out of the will." (See http://www.americanpolitics.com/20040617Young.html, http://www.cbsnews.com/stories/2004/06/01/eveningnews/main620626.shtml?tag= contentMain;contentBody, http://en.wikipedia.org/wiki/California_electricity_crisis, there exists much more: search "Enron tapes.")

At the time the administration and media blamed environmentalists for the energy crisis, when in reality it was purposely caused by Enron and colleagues to increase profits. The megamedia did not cover this until after the inexorable march of mathematics and dishonest accounting brought Enron to bankruptcy. If it had not gone bankrupt, it would likely never have been stopped! Yet another example of megamedia's avoidance of information that is not SAFE.

(•) President and General Dwight D. Eisenhower warned us about "the military-industrial complex." It is now a military-congressional-industrial-educational-media complex with about a trillion dollars of economic impact per year reaching almost every part of the U.S. economy, [690, 286, 595]. I will let you decide who paid for the disinformation campaign, but the President and Vice President repeatedly gave us reasons to go to war with Iraq. The Secretary of State gave many reasons why the U.S. had to go to war with Iraq in a famous speech before the United Nations, February 5, 2003. All of those reasons were false. (WACU) Saddam Hussein was not involved with the 9–11 attacks in New York, there were no weapons of mass destruction, our GIs were not universally welcomed, and the rest of the world did not fall into line, and so on. In fact, worldwide protests were marginalized, if mentioned at all, by the megamedia. Again the megamedia deemphasized, or completely ignored, information that was (is) not SAFE. Information that is not SAFE might challenge the wisdom of the few who are making the big decisions that affect us all.

I could go on at book-length giving examples of decisions made by very few for their own narrow economic benefit, paid for by the many who are increasingly eliminated from the decision-making process.

All of the above examples were in a time *before* the Supreme Court decision in *Citizens United v. FEC* that gives corporations, whether domestic or foreign, the right to spend *unlimited* cash to influence elections. (WACU) Given the corporate track record above, I expect very clever campaigns that the general public will not associate with corporations at all, putting into office virtual corporate puppets. If a puppet acts without strings, he/she will immediately be disciplined by disinformation campaigns. Just the threat should be sufficient, and the occasional crucifixion will do the rest. Corporations will create a virtual reality for us with a logic quite at odds with the logic of Nature, and hence with disastrous consequences.

My hope is that a tipping point has been reached which will reverse the ridiculous legal fiction that corporations are people, cf., www.poclad.org, entitled to more freedoms than real people. A constitutional amendment to this effect coupled with public funding for elections (we all pay in the end anyway, cf., Exercise 24.3), hopefully will be the fruit of this Supreme Court decision.

Exercise 9.3 Government for Sale?

(i) Well before and up to the Supreme Court decision in *Citizens United v. FEC*, enormous amounts of money flowed into campaigns from corporate sources. (Much of this money then flowed on to other corporations, megamedia corporations, for political advertising.) Mathematically one can test the effect of this money on the legislative (judicial and executive) systems by seeing if there is any correlation between "money from X" and "decisions favorable to X." See, for example, [395, 396, 398, 636, 514, 668], www.votesmart.org, www.publicintegrity.org for data on financial contributions to the political process.

(ii) How do you think *Citizens United v. FEC* will affect the political process?

(iii) What actions can be taken to return the political process back closer to "one person, one vote" and away from "one dollar one vote?"

(iv) Research the use of force against citizens on behalf of corporations. For example, force has been used against labor on many occasions. How, and at what cost, did labor win the 8 hour day, 40 hour week, and 5 day work week, i.e., the concept of a "weekend"? Research the Ludlow massacre in Colorado, where the National Guard attacked mine workers on behalf of a corporation. Research Kent State and Jackson State, where force was used to kill students protesting the Vietnam War, for example. How many people went to jail for trying to get Congress to consider "single-payer" health care? (A concept supported by a majority of the population, yet there was not a single lobbyist for it.) Does a government that starts a war of choice when the majority of the population opposes the war actually represent that population or other interests? If a system ceases to work for a population, the only avenue available is (peaceful) protest, supposedly assured by the U.S. Constitution/Bill of Rights. Yet every protest of consequence is met with overwhelming force. Consider the first WTO protest in Seattle, cf., Exercise 2.10 part (iv), and every such protest since. Consider climate change meetings and protests, e.g., Copenhagen. Consider Democratic and Republican National conventions. All is not always simple. A peaceful protest can be made to look violent by means of *agents provocateurs.* What are they? I have had more personal experience than the typical citizen with regard to peaceful dissent with the goal of informing the public. The megamedia has almost never given coverage equally balanced between the dissenting citizen and corporate points of view; and significant, *effective* dissenters are "monitored," or "neutralized," (or worse) by police, the FBI, or some similar agency. Remember that environmentalists are often referred to as "ecoterrorists," even if they are dedicated to nonviolence, cf., Exercise 22.7. For those who doubt any or all of the above, try to peacefully change something in the world that seriously and negatively affects the flow of money to a powerful interest. Make sure you have many friends with you when you do. Especially for those who find this exercise unbelievable, see Chapter 22.

(v) Citizens in Arizona and Maine have managed to institute a system of public funding of elections. However, constant struggle is part of the process. Lawsuits against the system, in Arizona, have resulted in parts being declared unconstitutional by a lower court and the Supreme Court. What is the status of public funding of elections at the time you read this? (Of course, unlimited corporate funding of elections is constitutional!)

9.4 Political Uses of Fear

Hermann Göring was on trial at Nuremberg for Nazi war crimes during WWII. During a three-day Easter recess, on April 18, 1946, Göring made the following comments to Gustave Gilbert, a German-speaking intelligence officer and psychologist who was granted free access by the Allies to all the prisoners held in the Nuremberg jail. In Gilbert's book, *Nuremberg Diary*, Gilbert recorded Göring's observations that the common people can always be manipulated into supporting and fighting wars by their political leaders: (Göring's comments are in italics.)

"We got around to the subject of war again and I said that, contrary to his attitude, I did not think that the common people are very thankful for leaders who bring them war and destruction."

"Why, of course, the people don't want war," Goering shrugged. *"Why would some poor slob on a farm want to risk his life in a war when the best that he can get out of it is to come back to his farm in one piece. Naturally, the common people don't want war; neither in Russia nor in England nor in America, nor for that matter in Germany. That is understood. But, after all, it is the leaders of the country who determine the policy and it is always a simple matter to drag the people along, whether it is a democracy or a fascist dictatorship or a Parliament or a Communist dictatorship."*

"There is one difference," I pointed out. "In a democracy the people have some say in the matter through their elected representatives, and in the United States only Congress can declare wars."

"Oh, that is all well and good, but, voice or no voice, the people can always be brought to the bidding of the leaders. That is easy. All you have to do is tell them they are being attacked and denounce the pacifists for lack of patriotism and exposing the country to danger. It works the same way in any country."

Exercise 9.4 Uses of Fear in Politics and Assorted Topics

(i) Discuss the above quote in the sense of using fear to *frame* debate and sell a war, see George Lakoff, cf., [380] and Section 8.3. If everyone understood this exercise do you think wars would be less frequent?

(ii) Discuss the use of "fear of financial collapse" in politics, cf., Chapter 2. Such a fear was summoned while arguing for public bailout money to private banks. What measures should be taken to reorganize our political and economic system so that this tactic cannot be used? What is a *scapegoat*? When immigrants are made into scapegoats is this an example of a use of fear?

(iii) As Michael Moore pointed out in his movie, "Capitalism: A Love Story," many corporations take out life insurance policies on some (or all?) of their employees. They are referred to as *Dead Peasant* policies. Thus if, for lack of health care, for example, an employee dies, the corporation collects on the policy. Look up at least ten corporations that take out dead peasant policies on their employees. How good is the health insurance for employees of these corporations?

(iv) In the May/June 2009 issue of the *Multinational Monitor*, we read: "Drug Company Merck circulated a list of doctors that needed to be 'neutralized' because they had criticized the drug Vioxx, court documents revealed in April. Discovery for a class action lawsuit brought in Australia against the pharmaceutical giant alleging Vioxx caused heart attacks produced a series of e-mails containing the doctor 'hit list.' "

" '*We may need to seek them out and destroy them where they live,*' one e-mail excerpt read to the court stated. There were also indications that the company used intimidation tactics against researchers critical of the drug, including insinuating that funding to academic and research institutions would be rescinded."

Comment on this finding. How many more such "interesting" communications from such corporations can you find? See, for example, page 515.

(v) The U.S. Senate makes some effort to track money labeled as corrupt, e.g., money from foreign dictators, arms dealers, drug dealers, and so on, as it flows into American financial institutions. The Senate Banking Committee and Senate Homeland Security Committee, for example, issue reports from time to time on this topic, cf., Associated Press article by Jim Abrams, February 4, 2010. The practice of "laundering" corrupt money continues despite provisions in the law, such as the 2001 PATRIOT ACT, forbidding it, except for some "loopholes." (See, for example, www.pbs.org, the *Need to Know* program from 2010: "Getting dirty money clean: A Need to Know investigation reveals just how common it is for major U.S. banks to launder money from drug traffickers.") Given the fact that unlimited money can be spent on elections, even foreign money, due to the *Citizens United v. FEC* decision of the U.S. Supreme Court in January 2010, estimate the amount of such corrupt money being laundered in the American financial system at the time you read this. Of course, laundered money provides benefits to any financial system in the form of monetary profits, regardless, of its source. What activities are American financial institutions supporting when accepting this cash?

(vi) An understandable reaction to the following is immediate denial. However, the documentation presented cannot be easily dismissed. If you have the time and interest, consider the following challenge to conventional history and make up your own mind. In [670] Robert Stinnett lays out his proof, backed up by twenty years of research uncovering previously classified documents some unavailable for sixty years, that the attack of the Japanese on Pearl Harbor, Dec. 7, 1941, was not a surprise to the FDR administration. Stinnett demonstrates that not only did FDR have ample warning of the attack, but there was a plan to provoke the Japanese into war with the U.S. at the highest levels of the U.S. government. The motivation of the FDR administration was to get an isolationist America involved in World War II. Stinnett's purpose is not to question the wisdom of that decision, but to record/document what really happened. (I would like to thank the late Gilbert F. White, cf., [312], who was there, and the late Robert McFarland for bringing this subject to my attention.)

(vii) The movie "The Most Dangerous Man in America: Daniel Ellsberg and the Pentagon Papers, the Untold Story of a War, and the Story of the Man Who Told It" 2009, is a documentary film version of [176, 177] the thesis of which is that Americans were misinformed, i.e., lied, into the Vietnam War. Research and discuss.

(viii) Were Americans misinformed, i.e., lied, into the war in (and occupation of) Iraq, cf., [212]? Research and discuss.

(ix) This following exercise is very interesting because there is an abundance of material available putting forth widely divergent views of what happened on 9/11/2001 in the attacks on the World Trade Center towers in New York. First, there is the official view. Now investigations of the attack on Pearl Harbor, say, and the space shuttle disasters (Challenger and Columbia), were begun within a week. Following the attacks on 9/11/2001, it took 441 days to get the 9/11 Commission together and start an investigation. They produced an official report, [480]. Rebuttals to this official report appeared, [261, 2], pointing out such things as the remarkable omission of any explanation for why the *third* high-rise building, WTC-7, which was not hit by any airplanes, collapsed. Then rebuttals to these rebuttals appeared, [154]. These rebuttals were in turn responded to by further rebuttals from those who did not agree with the official report, [262, 624, 745]. There is a growing group of people who have done independent work, arrived at conclusions contradicting major parts of the official view, and who are generally quite as credentialed as those who support the official view. Among these are physicist, Stephen Jones, who has done some remarkable, original forensic work; Richard Gage, AIA, Architect; and Kevin Ryan, "the former Site Manager for Environmental Health Laboratories, a division of Underwriters Laboratories (UL). Mr. Ryan, a Chemist and laboratory manager, was fired by UL in 2004." Ryan publicly questioned the report being drafted by the National Institute of Standards and Technology (NIST) on their World Trade Center investigation. "In the intervening period, Ryan has completed additional research while his original questions, which have become increasingly important over time, remain unanswered by UL or NIST." (See http://www.ultruth.com/index.htm, http://www.ultruth.com/Kevin_Ryan.htm) For research articles by those unconvinced by the official view see http://911research.wtc7.net/index.html, and http://stj911.org/, and http://www.ae911truth.org/ and http://journalof911studies.com/. The megamedia and others often confuse the serious, rational dissenters with irrational folks who advocate violence and/or make assertions such as that there were no airplanes that hit the towers.

To dissent from the official view is regarded as ridiculous by most media. It is thus particularly challenging for anyone to dare question the official reports. It definitely is not a move one makes to enhance a career. But I have found that most people have not taken the time to investigate the serious questions that have been raised. (There are a lot of opinions and statements made by people who have not done their homework on this issue, but then this is not a new phenomenon.) Thus, as I asked in Section 1.4, how do you know what's true? If you have the time and want a challenge, some interesting research awaits you. I will leave you with some very elementary observations. The towers WTC 1 and 2 were 1,368 feet and 1362 feet, tall respectively. If one were to have thrown a bowling ball from one of the roofs it would have taken about $9\frac{1}{4}$ seconds for the ball to hit the ground, neglecting air resistance. Research as best you can the time it took these towers to fall once collapse began. Verify or debunk the following statement. Both towers fell in about 12 seconds, which is close enough to freefall to raise serious questions since that means that the structure of the buildings did not offer substantial resistance during collapse.

Also in the much delayed official explanation of the collapse of the third tower, WTC-7, do you find credible the NIST explanation of the full 2.25 consecutive seconds of freefall collapse in the 4 seconds of available video of the collapse of WTC-7? WTC-7 collapsed in about 6 and a half seconds. It was 47 stories, or 741 feet tall. It collapsed at freefall speed for at least part of the time, cf., search David Chandler and WTC7. (Check this data as best you can at the time you read this.) There are other more complicated exercises one can do, find them—or read the work of others who have done them. A short (and very partial) list of things to check about the destruction of the towers is: sudden onset of collapse; straight down, symmetrical fall; high velocity bursts of debris; sounds of explosions and flashes of light; relatively small rubble plies; total collapse; pulverization of concrete and other materials; huge dust clouds (resembling pyroclastic flows from volcanoes). Some of this research now is found in peer-reviewed journals.

9.5 Confronting Fear (and Apathy): Organizing Your Community for Self-Preservation and Sustainability

The only answer to organized money is organized people.

Bill Moyers[4] (March 26, 2010)

In Section 9.1 I divided humanity up into roughly two groups: sustainable and unsustainable. For a culture to be sustainable it must overcome inherent human biological limitations, as well as successfully resist all attacks or attempts at assimilation by unsustainable cultures. It is conceivable that islands of sustainable culture can exist within an ambient unsustainable one.

So what am I talking about in concrete terms? First of all, when I think of islands of sustainability I think of the Amish in America, who eschew technology like electricity and car ownership (though they will ride in one); they are apparently nearly self-sufficient. However, they have reportedly been averaging 6.8 children per couple (*Population Studies* (33): pp. 25576); a practice that is not sustainable.

Exercise 9.5 Are There Human Examples of Sustainability?

(i) Research globally for a dozen indigenous cultures that have existed historically. If they no longer exist, were they assimilated or annihilated by an unsustainable culture; or did they die out for ecological reasons? Were any of these cultures sustainable until encounter with an unsustainable culture? For those cultures that were sustainable, were they sustainable by choice or *de facto* for lack of technological alternatives?

(ii) How many indigenous, sustainable cultures on earth can you find at the time you read this?

(iii) Can you find any non-indigenous cultures that are sustainable? Are there islands of sustainable cultures in any unsustainable cultures?

(iv) Though perhaps not perfectly sustainable in every way, are there some contemporary examples of groups of humans who have achieved partial sustainability with regard to, say, energy, or food production, or population growth, etc.?

I suspect that sustainable human cultures have existed, at least *de facto*, and likely by conscious choice. For example, investigate the current status of the Atakapa-Ishak indigenous people of Grand Bayou, Louisiana and the Chugach indigenous people of Sleepy Bay, Alaska. Both of these peoples were (are?) subsistence cultures, which almost by definition implies sustainability, or at least the closest thing to it today. Why did I choose these two? The Chugach, decades after, still feel the effects of the Exxon-Valdez on their traditional food sources; and the Atakapa-Ishak are similarly severely impacted by the BP Gulf blowout of 2010. Is becoming collateral damage more like assimilation or an attack?

[4]This sentiment, possibly the quote, goes back at least as far as the work and writings of community organizer Saul Alinsky (1909–1972)

I have found it difficult to find a contemporary non-subsistence culture which is perfectly sustainable; nevertheless, some groups of humans are conscious of and try to minimize their ecological impact. Others, of course, could care less.

What is a Person to Do? It is an all too common occurrence that a person, or group, becomes acutely aware of the unsustainability of the society in which they are immersed when "the powers that be" come to take something precious away. For me that something was Bowen Gulch, a remnant of old growth forest just outside the boundary of Rocky Mountain National Park, Colorado. That experience, where I briefly had the heart to risk everything, taught me many things which inform the rest of this section.

For the people of Cochabamba, Bolivia, that precious something was local ownership of water. From [616, Chapter 1], we learn the following. In September, 1999, The national government of Bolivia, with involvement of the World Bank and the IMF (International Monetary Fund), signed a contract with Bechtel, one of the largest corporations in the world, and a Bechtel subsidiary, giving Bechtel and co-investors control of Chochabama's water for 40 years and guaranteeing a profit of 16% for each of those years. In return Bechtel said it would invest $50 million dollars in development. However, after $1 million invested and visible improvement not discernible, the local folks' water bills jumped upward and became unaffordable to many. This was a very efficient way to convey information to the masses. A protest was organized in 2000. The Bolivian military was called in. One died, Victor Hugo Daza, 17 years-old and unarmed; many were wounded; many were arrested. More protests. The heart of the Cochabambinos was strong enough to prevail. Bechtel left, but sued for $50 million. *Earthjustice*, formerly the Sierra Club Legal Defense Fund, defended the Bolivians and won.

The local population was then left with reviving its MUNI (Servicio Municipal) for waterworks. Cochabamba is not without problems, since there is a rapid influx of immigrants, swelling the population and the need for more water. The MUNI is not without ethical and management issues as well. However, Cochabambinos know where the MUNI lives, and can give active feedback! What is notable about the Cochabamba Water Revolt is that local, mostly poor people, were able to wrestle control of an essential for life, water, away from some of the most powerful political and economic forces on earth. This event has been an inspiration to many.

Suppose a large corporation sets down near you and starts pumping the ground water your community relies upon. For an example in Kerala, India, see http://www.corpwatch.org/article.php?id=7528. For an example in the United States see [590].

The book *Standing Up to the Madness: Ordinary Heroes in Extraordinary Times*, by Amy Goodman and David Goodman, cf., [242], details a number of inspiring success stories of seemingly ordinary citizens in the United States standing up for what they believe in and winning. I will not recount any of these here, but I will note that we have a lot to thank our nation's librarians for! Please keep in mind the fact that success is quite possible, as I bring to your attention some unpleasant realities below.

One of the first important things to realize is that whatever your concern, you are not alone. You must *communicate* with those of like mind. The goal is to *organize* as many people as you can around your issue. Technology has made possible a host of mechanisms for communicating and organizing not available a relatively short time ago. The same technology can be used to monitor and infiltrate your campaign, cf., Chapters 22 and 23. In times of repression governments and cooperating corporations can and have shut down internet and phone communications. Work with people whose egos are smaller than the issues you are working on. Do not be too picky, do not shun someone just because you do not agree on everything. Seek common ground with as many people as possible.

If your struggle is over the environment, seek out help from fellow environmentalists. There are, for example, *Sierra Club* chapters nearly everywhere, at least in the U.S. If there are none in your area, start one! One reason why I am a life member of the Sierra Club is its true grassroots organizational structure. If for some reason you do not think the Club is doing enough or doing the wrong thing, you can actually organize a grassroots campaign among the members and change things. This is not the case for many organizations. There are often local environmental organizations unknown elsewhere that can be very helpful. By the way, the *Audubon Society* is interested, of course, in birds, but often much more.

Different groups employ different methods, and all (nonviolent) groups have a productive niche. *Greenpeace* is a rather famous and effective environmental organization. But Captain Paul Watson did not find them aggressive enough in saving whales, for example; so he founded the *Sea Shepherd Conservation Society*. Captain Watson has been known to enforce laws of the sea to protect sea life, including ramming and sinking illegally operating ships. Different groups have different philosophies. Sea Shepherd's, for example, is discussed in [709, 708, 468]. Some groups practice non-violent civil disobedience, as did Mahatma Gandhi and Martin Luther King. Some practice monkeywrenching, cf., [197, 196].

I believe that if most folks know what is really going on in a particular battle of sustainable vs. unsustainable, the sustainable side will win over the popular opinion. (Education of the public is one of the most important missions of any environmental campaign.) That is one of the most important reasons non-violence is a most effective philosophy. The last thing you want is for a typical citizen to be afraid of your organization. That is also why corporations and governments often try to "neutralize" effective environmentalists by associating them with violence—using a host of tricks you must be prepared for. Also non-violence (including non-violent civil disobedience) is really the only viable option in the face of governments, corporations, and other sundry groups that are capable of overwhelming violence—and who,perhaps, will not hesitate to use it.

There are many methods and levels of action. For example, planning ahead, you can try to radically change the laws in your local community to prevent environmental destruction. Examples of this tactic can be found in *BE THE Change: How to Get What You Want in Your Community*, see [406]. While you are doing everything you can think of, contact your Senators and Representatives at the federal and state level and see if they will help you achieve your goals. Sometimes they actually are quite helpful. If they are antagonistic toward you, then you have one more project: organize and replace them with someone more accommodating.

Get a book, such as *The Democracy Owner's Manual: A Practical Guide to Changing the World*, see [615], which covers the basics of organizing. Don't reinvent any wheels. But you will have to be creative; there will likely be details in your campaign that are unique. If you are effective you will attract some attention. Do your best to control how you are represented in the media. When you are attacked in the media, respond as soon as possible. For everyone in your group: NEVER lie. Avoid all mistakes of fact, to the best of your ability.

I do not want to be unnecessarily alarmist, but look at Chapters 22 and 23 to get some idea of what you might be up against when it comes to your own culture, even in the United States. Arrest, torture, and assassination are possibilities just about anywhere in the world. For example, within a few months of my writing this line, two environmentalists were assassinated in Latin America, i.e., El Salvador. David Helvarg's *The War Against the Greens*, [295], gives a partial list of things enemies of the environment are willing to do to environmentalists. Environmental and labor activist Judi Bari was bombed in California, cf., www.judibari.org, cf., "Bari" in the index.

Hopefully you will never have to appear before a Grand Jury, but if you do be warned: *You will have no rights whatsoever!* If this should befall you, get a copy of *Representations of Witnesses Before Federal Grand Juries*, Volumes 1 *and* 2, by the National Lawyer's Guild. My library gladly displays my well-worn copy, since four friends of mine in Ancient Forest Rescue (AFR) were so summoned.

Actually, depending on the situation, you cannot count on having any rights. With laws like the USA PATRIOT ACT, just about anything you do to protect your environment can be interpreted as terrorism. I am *not* kidding. There are sympathetic police and judges, and there are unsympathetic ones. Reporters for Democracy Now! were roughed up and arrested by police for doing their job of reporting at the Republican National Convention in 2008. (A lawsuit is pending.) There are brutal police, and there are fair and just police; my dad was one of the latter. All law enforcement agencies from the FBI, state police, local police, private security, forest service police, and so on, even the National Guard, can and often do cooperate and work together–often for the unsustainable. Always be on the alert for *agents provocateurs* and other infiltration of your group by the police or other interests. This is one of the techniques used to get a group of non-violent protesters associated with illegal acts and/or violence. By adopting a campaign rule that there will be no drugs and no violence supported by the group, when an infiltrator (or not) proposes or introduces such they can be declared to be acting as if they wish to discredit the group. This way, a level of openness can be maintained while defending against infiltration. If you suspect infiltration, then a lack of trust has developed. Such suspects should not have access to any membership lists or sensitive information! By the way, you can be sued for speaking the truth by a mechanism called a SLAPP suit, Strategic Lawsuits Against Public Participation, cf., [554]. For example, Oprah Winfrey was sued for speaking ill of a food on her popular television show; see "Winfrey" in the index.

In the end it all comes down to courage, will power, determination, organization, and *heart*.

Think of your fight as being for sustainability, against unsustainability. If you are still reading this I assume you support living sustainably. There will be no human future unless you and people like you prevail. Remember that. If there are future generations they will thank you. Even though the famous economist, Kenneth Boulding, once said jokingly: "What's posterity every done for me, anyway?" Do it anyway. By definition, if unsustainability wins the struggles of today, soon there will be no tomorrows (for us).

Part II

Math and Nature: The Nature of Math

Chapter 10

One Pattern Viewed via Geometry and Numbers: Mathese

10.1 The Square Numbers of Pythagoras

Mathematics (as defined in this book) is the search for and study of patterns. I want to begin this chapter with a reasonably simple pattern that is interesting because it can be looked at in two different ways: geometrically and algebraically. Let's go back to the time of the Pythagoreans, a school of Greek mathematicians led by Pythagoras in southern Italy prior to 500 B.C.; among the numbers that held a special significance for them were the *square numbers,* illustrated in Figure 10.1, cf., [357, p. 13].

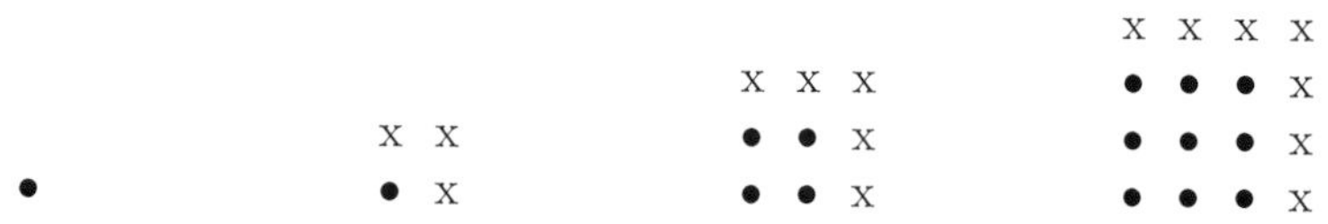

$1^2 = 1$ $\quad$ $2^2 = 4$ $\quad$ $3^2 = 9$ $\quad$ $4^2 = 16$

FIGURE 10.1: Square Numbers of the Ancient Greeks

Let's look at each square individually. There isn't much to say about the first one, just $1 = 1^2$. (Note given any number a, that a^2 equals aa or a multiplied times a. In English we say that a^2 is *"a squared."*) The second square has a total of 4 "marks" which we can *see* is the sum of 1 dot "mark" and 3 x "marks," thus we have the equation

$$1 + 3 = 2^2.$$

The third square has nine "marks." This third square can be obtained from the second (2 by 2) square by adding an "upside-down-backward L," consisting of 5 x's, on the upper and right sides of the 2 by 2 square. This "upside-down-backward L" is an example of what the ancient Greeks called a *gnomon* (knower). A gnomon is an object one can add to a figure to produce a larger figure of the same shape.[1] Now we can *see* that the nine "marks" in the third square can be written as the sum of 1 and 3 and 5. Thus we have the equation

$$1 + 3 + 5 = 3^2.$$

The fourth square has sixteen "marks," which we can *see* is the sum of 1, 3, 5 and 7. We thus see a pattern, viz., the sum of the first odd number (a whole number is *even* if it is a whole number multiple of 2, and a whole number is *odd* if it is not even) is one squared. The sum of the first two odd numbers is two squared. The sum of the first three odd numbers is three squared. The sum of the first four odd numbers is four squared.

Based on these examples we might *leap* to a conclusion and *guess* that *"the sum of the first n odd numbers is n^2"* is a true statement no matter what positive whole number we plug in for n, i.e., $n = 1, 2, 3, 4, 5, 6, \ldots$.[2]

As further evidence we note that in going from one square to the next, the "layer" of points, or gnomon, added (consisting of the top row and the right-most column) always has two more points than the immediately preceding "layer," or gnomon. Thus adding a "layer" of points (a gnomon) to a square array adds a number of points equal to the "next" odd number.

So it seems that no matter what the size of the square, say there are n dots on the bottom row, a rule applies. Namely, the square has n^2 dots total; and this total number of dots is the sum of n *layers* (one layer of dots for each dot in the square's bottom row) and that the first layer has 1 dot, the second layer (gnomon) has 3 dots, the third layer (gnomon) has 5 dots and so on—the n *layers* representing the first n consecutive odd numbers.

Succinctly stated, our rule is: *the sum of the first n odd numbers is n^2, where $n = 1, 2, 3, \ldots$* .

Exercise 10.1 A Variable and Another Pattern

(i) Think about the n that was just introduced until you are confident you understand the way it is used. The last italicized sentence above is an abbreviation for an infinite number

[1] See [112, p. 33] This is a great reference for getting acquainted with the magic of numbers. On page 283 there begins the story of surreal numbers and the values of combinatorial games—a recent development of the concept of number! Yes, even the ancient concept of number has just lately been expanded and is now better understood.

[2] At this point some students panic. What is n? For now, if you must, you can think of n as some particular number. What n represents is a number, (in this case a positive whole number) the specific value of which I do not want to tell you. What can you do with a number, n, if you do not know which particular number it is? Plenty, that is what this chapter is all about. If some symbols represent numbers, even if you do not know which numbers, these symbols must then satisfy some very definite patterns. In this chapter we will be studying what those patterns are. When a letter is used in the way that I am at this moment using n, it is called a *variable*.

of sentences, one for each value of n. Can you state ten of these sentences? Mathematicians say that n (as used above) is an example of a *variable.*

(ii) In the third (3 dots by 3 dots) square, draw all of the possible horizontal and vertical line segments that connect the nine dots, as in Figure 10.2. Do you see that the "big"

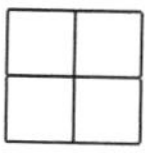

FIGURE 10.2: Pattern of Boxes (Not Dots)

square in Figure 10.2 is the "sum" of 4 smaller squares? Do the same exercise with the fourth (4 dot by 4 dot) square. Do you see a pattern in the number of small squares as you go from the 2 dot by 2 dot square to the 3 dot by 3 dot square and so on?

(iii) As you go from one square to the next (after drawing all the horizontal and vertical line segments) in Figure 10.1, what relationship, if any, do you see between the pattern in the number of small squares in each big square found in (ii) of this problem and the pattern in the number of dots in each big square array of dots?[3]

Exercise 10.2 Another Pattern from the Greeks and a Calculation of Gauss[4]

(i) See if you can find a formula for the sum of the first n whole numbers, i.e., $1+2+3+...+n$. Note that there is a way to do this problem without pictures. Consider

$$1+2+3+4+5+6+7+8+9+10=S_{10}$$
$$10+9+8+7+6+5+4+3+2+1=S_{10}.$$

If you add these two equations it is possible to see that $(10)(11)=2S_{10}$. Thus $S_{10}=\frac{(10)(11)}{2}$. See if you can mimic this calculation with 10 replaced by 100, and then with 10 replaced by n. All you are doing when you replace 10 by n is seeing if you can do the same calculation knowing only that n is a positive whole number (which satisfies all the laws of whole numbers) without assigning any *particular* value to n.

There is a way to do this problem with pictures, i.e., geometry, as well. Consider Figure 10.3 and the *triangular numbers* of the Pythagoreans,[5] cf., [357, p. 13] or [112, p. 33].

(ii) Given a square number n^2, can you find two triangular numbers which when added together give n^2?[6]

iii) Can a triangular number also be a square number?

---

[3]This exercise is relevant to our thinking about how many square centimeters are in a square meter, cf., Exercise 3.11.

[4]This exercise was done, as legend has it, by Carl Friedrich Gauss (1777–1855) when his age was a small single digit. At the age of three Gauss was checking and occasionally correcting the books of his father's business. One of Gauss's teachers, in an effort to keep his class busy, asked his students to add up the positive whole numbers from 1 to 100. Gauss astonished his teacher by coming up with the answer almost immediately. Gauss presumably used the above trick, but he was so fast that we cannot rule out the possibility that he just "added the numbers up in his head." Gauss went on to make some of the greatest contributions to mathematics/science ever made.

[5]Hint: Take two copies of the triangle with three points and make a 2 by 3 rectangular array.

[6]Hint: Doing this problem with pictures may be the easiest way. If you get stuck try a particular value of n^2, like 3^2, first; then consider the second and third triangular numbers.

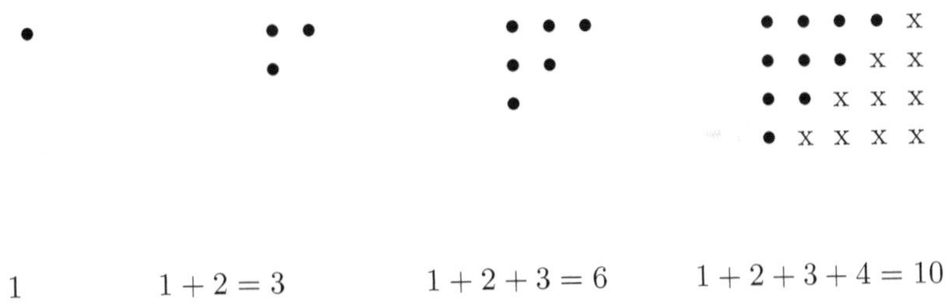

FIGURE 10.3: Triangular Numbers of the Ancient Greeks

10.2 The Language of Mathematics: Mathese

Let's focus for a moment on how a mathematician would write the statement: *"The sum of the first n odd numbers is n squared,"* in his/her own language—which I will call *mathese.* "Mathese" has symbols with specific meanings just like your native language, only the meanings of mathematical "words" are usually much more precise than the meanings of words in languages like English.

Hopefully you will study the few symbols in mathese that we introduce until you are fluent. This involves more than mere memorization. The words you use in your native language are not just memorized, they are tools you can use. The same goes for mathese. Get fluent and use mathese at least once in your life outside of class. Give the next two pages your best effort. Reread them as many times as you need to until you "get it." When you are done you will be speaking some real mathese, which will be very useful and save us all a lot of time later on.

10.3 A General Expression in Mathese: A Formula for Odd Numbers

The first thing we need is a "general expression" for odd numbers. What I mean by this is a formula with a letter in it that can stand for all the odd numbers. A formula that will do is $2m - 1$, where $m = 1, 2, 3, \ldots$. (In long-winded English this formula would read: "2 multiplied times the number m minus 1.")

Let's check it out. If $m = 1$, then $2m - 1 = (2)(1) - 1 = 1$, the first odd number. (Note that we write (2)(1) to mean two times one, which is 2. Sometimes mathematicians write the product of 2 and 1 as $2 \cdot 1$. In any event we must write something other than 21, which will certainly be taken to mean twenty-one, which is *not* what we want.)

If we let $m = 2$, then $2m - 1 = (2)(2) - 1 = 3$, which is the second odd number. (Note that we do the multiplication before subtracting the 1.) If we let $m = 3$, then $2m - 1 = (2)(3) - 1 = 5$, which is the third odd number, and so on. Thus our first word in mathese is $2m - 1$, and it stands for all the (positive, whole) odd numbers if we let $m = 1, 2, 3, 4, 5, \ldots$. Note that ... means "and so on," and the fourth . is the period at the end of the last sentence.

Exercise 10.3 Formulas Have Variables in Them

(i) Find a formula, or "general expression," for all even whole numbers. Note that even whole numbers are those which are multiples of 2.

(ii) Why did I use the letter m in the above formula $2m - 1$ for the odd numbers? Could I have used the letter n instead of m?

(iii) What does $2n$ mean? Write $2n$ in a different, yet equivalent way, using mathese but not using the number 2.

(iv) What does n^2 mean? Write n^2 in a different, yet equivalent way, using mathese but not using the number 2.

(v) If you are able to answer parts (iii) and (iv) of this question, state succinctly in words how it is that you know the answers.

(vi) Explain in a short sentence (in English) what the difference is between $2n$ and n^2?

10.4 An Important Word in Mathese: Σ

Our next word in mathese is Σ. Now Σ is the Greek letter capital sigma that corresponds to the English letter S, which is supposed to remind us of "SUM." Thus $1 + 3 + 5 = 3^2$ is written as

$$\sum_{m=1}^{3} (2m - 1) = 3^2.$$

Note that in English we would read this last equation as: "The sum, m goes from one to three, of two times the number m minus one is three squared." Do you see exactly what is going on here? The

$$\sum_{m=1}^{3} (2m - 1)$$

means first plug 1 in for m in $2m - 1$ and get 1. Next, plug 2 in for m in $2m - 1$ to get 3, then plug 3 in for m in $2m - 1$ and get 5. (*Start* with $m = 1$. *Stop* when $m = 3$.) Finally, add up the result and get $1+3+5$. In other words, the sum of the first three odd numbers is 3^2. Note also that the ( ) around the $2m - 1$ is there to emphasize that for each $2m$ we subtract 1. If we did not have the parentheses someone might add up the $2m$'s and just subtract 1 once at the end; see the next exercise.

Exercise 10.4 Some Sentences in Mathese

(i) Translate into mathese the English sentence: "The sum of the first five odd numbers is five squared."

(ii) Evaluate the following: $\sum_{m=1}^{4}(2m-1)$ and $\left(\sum_{m=1}^{4} 2m\right) - 1$. Are they the same or different?

(iii) Using a pattern we have studied, can you see instantly what $\sum_{m=1}^{4}(2m-1)$ is?

(iv) Using the calculation of Gauss (involving triangular numbers) can you see (almost) instantly what $\left(\sum_{m=1}^{4} 2m\right) - 1$ is?

10.5 Sentences in Mathese: Equations with Σ and a Dummy Variable

Now let me try something a little more ambitious. I will translate into mathese the sentence: "The sum of the first one hundred odd numbers is one hundred squared."

$$\sum_{k=1}^{100}(2k-1) = 100^2.$$

You probably noticed that I changed the letter m to a k. Does this really matter? No, not at all. You see the formula $2m-1, m = 1, 2, 3, \ldots$ has exactly the same *meaning* as the formula $2k-1, k = 1, 2, 3, \ldots$. Namely, both formulae stand for the set of all (positive, whole) odd numbers.

Now why did I change the m to a k? Just to be confusing? No, just the opposite. I don't want you to become emotionally attached to any one letter, so I changed the letter to break any emotional bonds that may have been forming with the letter m in the formula for the odd numbers. Instead, when you see the formula $2m-1$ or the formula $2k-1$, I want your mind to go right past the letter m or the letter k to "the set of odd numbers." Always focus your mind on the meaning of mathematical symbols and not on the superficial way they are presented.[7]

By replacing the number 100 with some other number, like 50 or 1001, you get a new equation, or sentence in mathese.

[7]Or as the mathematician Carl Friedrich Gauss put it, what matters in mathematics is "non notationes, sed notiones," i.e., "not notations, but notions."

Now I want to translate into mathese the sentence: "The sum of the first n odd numbers is n squared." This becomes

$$\sum_{k=1}^{n}(2k-1) = n^2. \qquad (*)$$

Can you see what confusion there might be if I replaced the k in formula $(*)$ with an n? We have *two* variables in our formula now. The variable k is called the *summation variable*. Some people call k, as it is used in $(*)$, a *dummy variable*; in part, because you can change the letter k to another letter, say l, and the meaning of the equation, i.e., the sentence in mathese, is not changed.

The variable n tells us how many odd numbers we want to add up. Note that on the left-hand side of $(*)$ the n above the capital sigma is the same n that is squared on the right-hand side of the equation. (It is incorrect grammar in mathese to use the same letter to stand for different things in the same equation—that would be very confusing!) We have used the (dummy) variable k to write down the n different odd numbers we are adding together on the left side of $(*)$.

Exercise 10.5 An Infinite Number of Formulas in One Mathese Sentence

(i) Formula $(*)$ above stands for an infinite number of formulas, one formula for $n = 1$, one formula for $n = 2$, and so on. Write down any five of the infinite number of formulas for which formula $(*)$ above (with two variables) stands. How many variables does each of the formulas you write down have?

(ii) Does $(*)$ say the same thing as: $\sum_{k=0}^{n-1}(2k+1) = n^2$?

10.6 Induction, Deduction, Mathematical Research, and Mathematical Proofs

The definition of mathematics is the search for and study of patterns. The search for patterns can require genius for the search to be successful, i.e., for new patterns to be discovered. This part of the process is called *inductive reasoning*. Briefly, induction is the process of seeing general patterns after looking at some particular examples.

Induction requires a leap of understanding, a *mathematical insight*, like the one we took after looking at Figure 10.1 and inducing that "the sum of the first consecutive n odd numbers is n^2."

Sometimes people think they are having a mathematical insight when in reality they are not. For example, one of my students once said that they thought that "the sum of the first n odd prime numbers is n^2." It turns out that this is wrong, and that is what this next exercise is all about.

Exercise 10.6 An Example of a False Insight/Leap

(i) A whole number is said to be *prime* if it cannot be evenly divided by any whole number except itself and 1. For example, 3 is a prime number, 4 is not prime (since it is evenly divisible by 2), 5 is prime, 6 is not prime (since it is evenly divisible by 3 and 2), and so on. For reasons I will not go into here, the number 1 is not regarded as prime by most mathematicians. For the purposes of this exercise only, however, I am throwing it into the collection of primes. Given this, can you list the first 5 prime numbers?

(ii) The number 2 is prime, but it is an even number. Is this statement correct?

(iii) Sometimes my students, when looking for a pattern in the "square numbers" do not first come up with the pattern "the sum of the first n odd numbers is n^2." Instead, they leap to conclude a different pattern: "the sum of the first n odd primes is n^2." Why is this false?[8]

Sometimes a student comes along and has what I consider to be a brilliant insight, something not obvious which turns out to be true. As an example I offer the next exercise.

Exercise 10.7 A Student Discovered this Pattern

The following pattern was discovered by a student after reading the previous pages. Group the odd numbers as shown:

$$(1)+(3+5)+(7+9+11)+(13+15+17+19)+(21+23+25+27+29)+(31+33+35+37+39+41)+\dots$$

Now note the following: The first odd number is 1^3. The group of the next 2 odd numbers is 2^3. The group of the next 3 odd numbers is 3^3. The group of the next 4 odd numbers is 4^3, viz., $(13 + 15 + 17 + 19) = 4^3$. The group of the next 5 odd numbers is 5^3. The group of the next 6 odd numbers is 6^3, viz., $(31 + 33 + 35 + 37 + 39 + 41) = 6^3$; and so on.

Can you prove that this pattern continues forever?[9]

10.7 What Is a Mathematical Proof?

Now you might find Exercise 10.7 a bit tough. One reason is that I have not told you what it means to create an acceptable proof. For example, if you could find a counterexample to the pattern of Exercise 10.7, i.e., an actual example where the pattern fails to be true, then you would have proven that the pattern does not hold forever, i.e., the pattern we thought we saw really does not always hold—it is false. On the other hand, unfortunately, just

[8]Find a *counterexample*. The statement "the sum of the first n odd primes is n^2" is true if $n = 1$, $n = 2$, $n = 3$, but is the statement true for all n? To find a counterexample to the statement, find a value of n for which the statement is *not* true.

[9]Hint: Look at the sum of consecutive odd numbers from 1 up to the nth triangular number of odd numbers. Look at the sum of consecutive odd numbers from 1 up to the $(n-1)^{th}$ triangular number of odd numbers. Take the difference of these two numbers and see if you get a cube. If you cannot do this problem now, try it again later after you have reviewed the properties of numbers, i.e., after you have studied all of this Part II.

looking at 1,000,000 cases and not finding a counterexample is *not* a proof that the pattern is true—maybe case number 1,000,001 will be a counterexample.

So just what is an acceptable mathematical proof? An answer that begs the question is this: A proof is a convincing argument. If you think you have a proof of the pattern in Exercise 10.7, then present it to someone who is a mathematician and see if he/she is convinced.[10] More precisely, if the mathematician is not convinced see if he/she will show you where he/she has found a mistake. Now that is a practical answer, but it is not a satisfactory one for obvious reasons.

What we are looking for here is the definition of a *formal proof.* This important definition relies on the concept of *deductive reasoning.* To give a formal proof that the pattern in Exercise 10.7 is true, you need to understand the *deductive system* of which numbers are a part, since Exercise 10.7 is about numbers. Thus to give a formal proof of any mathematical statement, you need to see that statement as part of a deductive system that contains it. Giving a formal proof of a mathematical statement/pattern (deductive reasoning) should not be confused with discovering a mathematical pattern (inductive reasoning). Coming up with new mathematical patterns that no one has ever seen before is part of mathematical research.

Mathematicians Pierre de Fermat (1601–1665) and Henri Poincaré (1854–1912) each discovered patterns known respectively as, Fermat's Last Theorem and The Poincaré Conjecture. No one could find a counterexample or a formal proof of either of these patterns for over three centuries in the case of the former and for one century in the case of the latter. Finally, in the past decade a formal proof of each has been discovered. Finding these formal (deductive) proofs is also part of mathematical research.

Fermat's Last Theorem is a pattern about numbers, just like Exercise 10.7 is. The Poincaré Conjecture belongs to another deductive system, geometry (actually a special kind of geometry that we will not discuss in this text). One reason I bring these two discoveries up is because many non-mathematicians have the impression that mathematics is static, dead, with an obituary written about 300 B.C. Nothing could be farther from the truth. There are more mathematicians alive today than ever before, due in part to the population explosion; and they are having mathematical insights at an astonishing pace. This last fact probably has something to do with the fact that both of the aforementioned long-standing problems have been recently solved.

[10]There is a theorem in mathematics that you cannot trisect an arbitrary angle with a straight edge and compass alone. Periodically someone shows up in a math department claiming to have accomplished the impossible; they claim to have trisected an angle. Invariably the straight edge has been used to measure lengths (not allowed!) or the straight edge is not straight, it is a French curve, or the angle is not arbitrary but a special one, and so on.

10.8 What Is a Deductive System?

The alert reader will have recognized that I have not yet completely defined a formal proof, since I have not told you precisely what a deductive system is. The most fundamental part, the core, the foundation of a deductive system is a *set of axioms.* The set of axioms for a deductive system is a set of fundamental patterns that are assumed to be true.[11]

A mathematical system then is developed by making precise definitions and proving that additional patterns, other than the axioms, are true by using only the axioms and fixed rules of logic (accepted patterns of thinking).

This may all seem a bit abstract, and it is. However, I am about to create in the following sections a deductive system containing numbers from which every statement you have ever learned about numbers can be deduced. In particular, Exercise 10.7 can be proved to be true in this deductive system. What I need to do is give you the 7 fundamental patterns of numbers, the 7 axioms for the deductive system that contains numbers. After that we will prove many things about numbers that may or may not be a mystery to you, for example, the rules for dealing with fractions and why you are not "allowed" to divide by zero.

Before I do this, however, I would like to leave you with a bit of philosophy on how to be a genius and come up with new mathematical patterns. Then I will finish this chapter with some exercises.

10.9 *Originalidad es volver al Origen*

A cornerstone of my philosophy for creativity is a quote from the famous architect from Barcelona, Spain, Antoni Gaudi: *"Originalidad es volver al Origen."* This has been translated by people who have studied Gaudi's thoughts and works as: "Originality (or creativity) is to return to Nature." People have powerful imaginations and can create intellectual edifices seemingly in a vacuum, but I contend that humans are at their most creative when they immerse themselves in and feel Nature. In Nature a dazzling array of patterns await a perceptive mind.

As Henry David Thoreau wrote in his *Journal* in the 19th century: *"It is the marriage of the soul with Nature that makes the intellect fruitful and gives birth to imagination."*

[11] The axioms can contain symbols or words that are *undefined.* In some real sense, the set of axioms "define" the undefined terms that the axioms contain—the axioms tell us how to deal with the undefined terms they contain.

Exercise 10.8 Some Practice with Sigmas

(i) What is $\sum_{n=1}^{10} n$? (Hint: See Exercise 10.2 part (i). Note that sometimes to save space we will write our sums like $\sum_{n=1}^{10} n$ instead of like $\displaystyle\sum_{n=1}^{10} n$.)

(ii) What is the expression $\sum_{n=1}^{10} 2n$ when written out with specific numbers?

(iii) Can you find a formula for $\sum_{n=1}^{N} 2n =$? Can you find this formula in two different ways? (Hint: Can you move the 2 to the left of the sigma without changing the value of the expression? Why? Also, try letting $N = 3$, or 5, or 10 or other concrete numbers, until you see what is going on.)

Exercise 10.9 A Detail about Sigmas

(i) What are $\sum_{k=1}^{5}(1 + 0k) =$?, $\sum_{k=1}^{4} 1 =$?, $\sum_{k=1}^{2} 3 =$? (Hint: Think of 3 as $3 + 0k$.)

(ii) What is $\sum_{k=1}^{5} a =$? (Hint: Think of a as $a + 0k$.)

(iii) What is $\sum_{k=1}^{m} n =$ (Hint: add m numbers each of which is equal to n) (Hint: Think of n as $n + 0k$, and note that multiplication is a shortcut to avoid repetitive addition.)

Exercise 10.10 Is this Mathese Sentence True?

Investigate $\sum_{j=1}^{n} j^2 = \frac{n(n+1)(2n+1)}{6}$. Do you think this formula is true for all $n = 1, 2, 3, \ldots$?

Exercise 10.11 Is this Mathese Sentence True?

(i) Investigate $\sum_{k=1}^{n} k^3 = \frac{n^2(n+1)^2}{2^2}$. Do you think this formula is true for all $n = 1, 2, 3, 4, \ldots$? What do you think a modern mathematician would accept as a proof?

(ii) Can you find any connections between this problem and Exercise 10.10 just above?

Exercise 10.12 Some Interesting Relationships

(i) What kind of number is the difference of two consecutive square numbers? That is, what kind of number is $n^2 - (n-1)^2$, where n is a positive whole number?

(ii) The following sum is called a *telescoping sum*: $(n^2 - (n-1)^2) + ((n-1)^2 - (n-2)^2) + ((n-2)^2 - (n-3)^2) + \cdots + 1 = \sum_{k=1}^{n}(k^2 - (k-1)^2)$. Do you recognize this sum? Have you seen it before (in a different form) in this text?

(iii) In Figure 10.1 we looked at square arrays of dots. If such a square array of dots has n dots on a side, then the whole square array has n^2 dots. We also noted how n^2 can be written as the sum of the first n odd numbers. Analogously for cubes, we can write n^3 as a (telescoping) sum with hex numbers. The nth hex number, h_n, is equal to $n^3 - (n-1)^3$ and can be visualized geometrically as a hexagonal array.[12] The sum of the first n hex numbers is n^3, i.e.,

$$h_1 + h_2 + h_3 + \ldots + h_n = n^3.$$

What are h_1, h_2, h_3 and h_4?

Again imagine a cubic array of dots with n dots on an edge, i.e., a cube with n^3 dots. This cube can be written in another way as the sum of the first n B&B numbers.[13] By this I mean

$$n^3 = \sum_{i=0}^{n-1}(2i(n+1) + 1).$$

(iv) If you are so inclined, investigate the nature of the following telescoping sum:

$$n^r = \sum_{j=1}^{n}(j^r - (j-1)^r),$$

[12] Roger B. Nelsen, *Proofs Without Words*, The Mathematical Association of America, 1993, p. 109.

[13] Robert Bronson, Christopher Brueningsen, see p. 111 of the reference in the footnote 12.

where r is $2, 3, 4, 5, \ldots$. Can n^r above be written in any other way as a sum? For example, in part (iii) above n^3 is written in two different ways as a sum.

Exercise 10.13 Thinking Big

Given the 9 dots in the following square array, Figure 10.4, draw four straight line segments that collectively pass through all 9 dots. Do so without lifting your pencil from the page while passing through each point exactly once.

FIGURE 10.4: Connect the Dots

If you liked this puzzle, see [250] for a whole book of them.

Exercise 10.14 The 400-Year-Old Voynich Book that Cannot Be Read?

In the Beinecke Rare Book and Manuscript Library at Yale University Library at Yale University there is a 7-inch by 10-inch 234-page parchment book known as the Voynich manuscript, after Wilfred M. Voynich, a book dealer who bought it in 1912 and tried for many years to read it. Some of the greatest brains in cryptography/code-breaking have not been able to translate this book. It is fairly certain the book was created by the early 1600s and was owned at one time by people in the Prague court of King Rudolph II of Bohemia (1552–1612), Holy Roman Emperor. For more information about this book see the following Web sites: www.geocities.com/Athens/Delphi/8389/voynich.htm, www.voynich.nu (a Web site run by Rene Zandbergen, who is trying to translate the book) and www.library.yale.edu/beinecke (this third Web site offers microfilm and print copies of the manuscript for sale). I found this information in an article by Michael Pollak in *The New York Times*, 20 September, 1999. Not mentioned in this article is the possibility that this book is written in a language that has been lost, a language not closely related to any known language. In any event, do you want to try your hand at translating it? If you solve the mystery, let me know. The answer is not in the back of this book!

Chapter 11

Axioms and Atoms

In order to discuss certain subjects of interest in Nature I have to briefly introduce a very few words and concepts from science. Namely, we need to be slightly familiar with one of the most important patterns in Nature ever discovered, the Periodic Table, cf., Figure 11.1, on page 253.[1] Don't worry, I am not going to do any serious chemistry. All we are going to do is learn to *count* atoms, and a couple of their parts. Along the way we will study fundamental patterns of *counting*!

11.1 Molecules and Atoms; the Atomic Number and the Atomic Mass Number of an Atom

Recall from Section 5.2, that matter is made up of *chemicals* and that the smallest unit of a given chemical is called a *molecule*. A chemical that cannot be decomposed into simpler chemical building blocks by chemical means is called an *element*, of which there are 88 that are "naturally occurring."[2] The

[1] Discovered by Dmitri I. Mendeleev and revised by Henry G. J. Moseley.

[2] According to [638], the idea that Earth, Air, Fire and Water are the fundamental elements with the two fundamental forces of love (harmony) and strife (discord) uniting and separating them has been traced back to Athens, Aristotle and eventually to Empedocles of Agrigentum in Sicily (c.500–c.480 B.C.). The idea that matter is ultimately made up of tiny indivisible particles (*atomos*) goes back to Leucippus. Before 1700 only 12 elements were known, only a couple dozen by 1800. By 1900 over 80 were known. Technetium, atomic number 43, Tc, was first synthesized/discovered in 1937. Neptunium, atomic number 93, Np, was the first element beyond Uranium, atomic number 92, U, synthesized/discovered and that happened in 1940. The elements with atomic numbers 43 (Tc), 61 (Pm, Promethium), 85 (At, Astatine), and 87 (Fr, Francium) were not discovered until they were synthesized in nuclear tests/experiments. They might occur "naturally," in nuclear reactions induced by cosmic rays, but they do not last long enough to accumulate in sufficient quantities to be "easily" observed—unless humans synthesize them. Two other elements that were not known until synthesized are Np, Neptunium, and Pu, Plutonium, atomic number 94. These two can arise from the absorption of neutrons during the spontaneous fission of U and Th, Thorium, atomic number 90 and (theoretically at least) exist as trace elements in uranium ore. Plutonium, one-half a microgram of which can be a lethal dose, cf., Exercise 5.12, is now present around the world due to nuclear weapons testing and aborted (exploded) space probes containing Pu. The biosphere never had Pu sprinkled

smallest unit of an element is called an *atom.* An atom consists of a *nucleus* surrounded by a cloud of *electrons.*[3] The nucleus is composed of *protons* and *neutrons.*[4] A proton has a mass of $1.672(10^{-24})$ *grams*, which is 1837 times the mass of an electron, so for the purposes of this text we will neglect the mass of the electrons in any atom. Also, for the purposes of this text a neutron has the same mass as an proton. An atom has an equal number of electrons and protons, and this number is the *atomic number* of that atom.[5]

The chemical properties of an atom are largely determined by the atomic number of the element, i.e., the number and structure of the electrons. For example, the simplest atom is hydrogen, H, atomic number 1, and the next simplest atom is that of helium, He, atomic number 2. Helium is an inert gas which means it does not react chemically, e.g., it does not burn. Hydrogen, on the other hand, is highly combustible.[6]

Isotopes. Two atoms of the same element must have the same number of protons in their respective nuclei, i.e., the same atomic number, but they may have differing numbers of neutrons. Atoms with the same atomic number but different numbers of neutrons are called *isotopes* of an element. For example, again consider the simplest atom, hydrogen, H, atomic number 1. This atom consists of one proton surrounded by one electron cloud. One of the most startling discoveries of the last century was that of *heavy hydrogen,* denoted D, also called *deuterium.* One atom of D consists of an electron cloud surrounding a nucleus of one proton and one neutron. Thus H and D have the same atomic number but different *atomic mass.* The *atomic mass number* of an atom of an element is the sum of the number of protons and neutrons in one atom of that element. Thus the atomic mass number of H is 1, the atomic mass number of D is 2. Hydrogen has a third isotope, T, tritium, with one proton and two neutrons in its nucleus. Thus T has an atomic mass

---

about before the invention of atomic weapons. For an interesting example see [679] for the story of 4 pounds of plutonium lost by the CIA near the headwaters of the Ganges near the mountain, Nanda Devi (it's still there). If isotopes are counted there are about 300 "naturally occurring" elements and more than 2500 synthetic (radioactive) isotopes.

[3]Think of a nucleus as a tiny ball made up of tinier balls, the protons and neutrons. The diameter of a nucleus is between 10^{-13} and 10^{-12} *cm.* How does this compare to the diameter of an atom? An electron is best *not* thought of as hard little sphere, but rather as a diffuse, fuzzy cloud that gets less dense the farther away from it center you get.

[4]The neutron was not discovered until 1932 by Sir James Chadwick. Physicists have elaborate explanations of the structure of atomic nuclei, but I will not discuss these models.

[5]Electrons and protons have equal but opposite charges, so an atom is electrically neutral. If an atom is missing one or more electrons, it carries a positive charge and is called a *positive ion.* Similarly, if an atom has one or more excess electrons, it is negatively charged and is called a *negative ion.*

[6]Remember the Hindenburg! Actually, hydrogen does not commonly exist in the form of a lone atom. Hydrogen gas is made up of molecules, each molecule consisting of a pair of hydrogen atoms bound together. Scientists write H_2 to denote one molecule of hydrogen gas.

number of 3, but an atomic number of 1.[7]

Consider now one of the chemicals that plants manufacture, a sugar called *glucose*. One molecule of glucose is composed of 6 atoms of carbon, C, atomic number 6; 12 atoms of hydrogen, H, atomic number 1; and 6 atoms of oxygen, atomic number 8. Scientists describe one molecule of glucose with the symbol $C_6H_{12}O_6$.

Exercise 11.1 Glucose and the Periodic Table

(i) Can you find C, H and O in the periodic table? (See page 253.) Find the atomic number of each of these three atoms in the periodic table. Also find the atomic mass number of each of these three atoms. Hint: For H the atomic mass number is "the other number" in the "H box," which is 1.008. The reason the atomic mass number is not 1 is because of the existence of the two isotopes of H, viz., D and T.

(ii) Explain why the atomic mass number of H is a number larger than 1, and a fraction to boot? The same for C.[8] (Note that a 2004-updated Periodic Table has the following numbers for: H, 1.00794; C, 12.011; O, 15.9994. The older Periodic Table on page 253 will be sufficient for our purposes. Check the internet for the latest developments.)

(iii) What is the sum of all the atomic mass numbers of all the atoms in one molecule of glucose? What is the total atomic mass in *grams* of one molecule of glucose? (Hint: Assume that one proton and one neutron have about the same mass, viz., $1.672(10^{-24})$ *grams*. Neglect the mass of electrons.) What is the easiest way to get the answer to the second question in this part (iii) if you know the answer to the first question in this part?

11.2 Scaling and Our First Two Axioms for Numbers

Atoms and molecules are usually too small for humans to deal with individually.[9] For most purposes you need an amount of a chemical or element that is perceptible on a human scale. This "scaling up" from one atom to a large enough number of them that you can measure with ordinary equipment is the subject with which I am now going to deal.

Let me pose a problem. How many atoms of H, i.e., how many Hs does it take to make one *gram* of H, i.e., 1 g of H? Neglecting the mass of the electron, this is the same question as: How many protons does it take to make 1 *gram* of protons?

Let N = the number of protons in 1 *gram* of protons. Then, since one proton has a mass of approximately $1.672(10^{-24})$ g, we have the following

[7] Isotopes of the same element usually have very similar chemical properties, since all the isotopes have the same atomic number. In the case of H, D and T, however, they have such large relative differences in mass that there is an appreciable difference in their chemical properties.

[8] The atomic mass number given is an average atomic mass number over all isotopes weighted according to their prevalence in Nature.

[9] Recently engineers/scientists have figured out how to manipulate single atoms!

equation:

$$(1.672(10^{-24})\ g) * \mathrm{N} = 1\ g. \qquad (\star)$$

In Equation ($\star$) for clarity we have used parentheses around the number multiplying the unknown number N. Here the $*$ stands for multiplication. Although this may (or may not) look like a formidable equation it is based on a very simple idea: The mass of one "marble" times the number of "marbles" is the total mass of the collection of "marbles." We assume the marbles are identical.[10]

11.3 Our First Axiom for Numbers

Now many readers will have an instinctive reflex and solve for N by dividing both sides of ($\star$) by $1.672(10^{-24})$ and be done with it. That is fine, and later on that is just how I would expect any reader to react to ($\star$). But remember, I promised to show you how numbers are part of a deductive system that contains every pattern/property of numbers that you have ever learned. There are 7 axioms at the heart of this deductive system, and none of them explicitly mentions division. There are reasons for this; see Exercise 11.5. So how are we to do this problem without explicitly resorting to an operation called division? It turns out that we can solve ($\star$) by using the first two of the 7 axioms. I will now introduce these two axioms and some needed definitions before returning to the solution of ($\star$) in Exercise 11.5. Thus our first axiom, which I refer[11] to as Axiom (A$*$), is:

ASSOCIATIVE LAW FOR MULTIPLICATION OF NUMBERS

The associative law of multiplication for numbers states that if A, B and C are (symbols representing) numbers then

$$\textbf{ASSOCIATIVE LAW FOR } * : \boxed{\mathbf{A} * (\mathbf{B} * \mathbf{C}) = (\mathbf{A} * \mathbf{B}) * \mathbf{C}} \qquad (\mathrm{A}*)$$

---

[10] A simpler equation to visualize, which is the same in spirit, is this. Suppose all doughnuts weigh one pound each. How many doughnuts does it take to have 10 pounds of doughnuts? Another: suppose all doughnuts weigh one-half pound each, how many doughnuts does it take to have 10 pounds of doughnuts? And another: suppose all doughnuts weigh 2 pounds each, how many doughnuts does it take to have 10 pounds of doughnuts? You can solve these problems in your head. If you think carefully and write down how you figured these out, you can easily see how to do the "proton equation," ($\star$). Another way to look at the proton equation is this. Think of N as having the units "protons" and think of the $1.672(10^{-24})$ as having the units $\frac{grams}{proton}$, i.e., grams per proton. Then if you multiply $\frac{grams}{proton}$ times *protons* you are left with the unit *grams*.

[11] The A$*$ in Axiom (A$*$) stands for Associativity of Multiplication.

where on the left-hand side of the equation we multiply B with C first, then multiply the result on the left by A, and on the right-hand side of the equation we multiply A with B first then multiply the result on the right by C. (Here $*$ stands for multiplication, and note that the order from left to right of symbols A, B and C is the same on both sides of the equation.)[12]

In the above axiom, as well as in all the other axioms I will state, look for the notions underneath the notations. Understand the *pattern* being presented without becoming fixated on the symbols used to represent the pattern. Also it is important to know what is NOT included in Axiom (A$*$), as it is easy to read into an axiom more than is actually there.[13] To give you a little practice in using the pattern of Axiom (A$*$) do the following exercise.

Exercise 11.2 Understanding Axiom (A$*$) as a Pattern

(i) Pick three numbers and verify that the associative law of multiplication works for them.

(ii) Given the three positive, whole numbers 2, 3 and 5, can you interpret the associative property: $2 * (3 * 5) = (2 * 3) * 5$ geometrically?

(iii) Can you visualize a geometric interpretation of $M * (N * P) = (M * N) * P$, where M, N and P are positive, whole numbers?

(iv) Can you deduce from the Associative Law for $*$ the law (or pattern): $(A*B)*(C*D) = A * (B * (C * D))$, for all numbers A, B, C, D?[14]

(v) Multiplication is an operation that you do two numbers at a time. Because of Axiom (A$*$) you can make sense of longer strings of multiplications. For example, deduce that $A * B * C * D$ makes sense because of Axiom (A$*$). By this I mean that you will get the same answer no matter how you *group* the four letters to do multiplication two numbers at a time.[15] Part (iv) just above shows two such groupings. Another such is $((A * B) * C) * D$. Are there any others?

11.4 Number 1: Its Definition, Properties, Uniqueness

We need to have a definition of that most important number, 1, and it turns out that there is more than one way to define it.[16] It is possible to

[12] Always do operations within parentheses or brackets first.

[13] For example, Axiom (A$*$) does say that $3 * (4 * 5) = (3 * 4) * 5$; but it does NOT say that $3 * (5 * 4) = (3 * 4) * 5$, even though we will see later on that this latter equation is true. Axiom(A$*$) says that results of multiplication are independent of how you *group* numbers. It does not say that you can change the *order* in which the numbers are written and get the same answer. That is a separate axiom which we will get to soon.

[14] Hint: Think of $(A * B) * (C * D)$ as $(A * B) * P$, where P happens to be equal to $C * D$. Then you can apply Axiom (A$*$) to $(A * B) * P$. If C and D represent numbers, then P will also represent a number so Axiom (A$*$) can be applied.

[15] Note that you are not yet allowed to change the *order* of the symbols being multiplied because Axiom (A$*$) does not address such a change in order.

[16] As an answer to the question "how many?" the number 1 can be defined to be the equivalence class of all sets in one-to-one correspondence with the set •. It so happens you

characterize the number 1 in terms of how it functions as a number. You certainly know from some previous math class that if x is any number then $1 * x = x * 1$, i.e., a number is unchanged if you multiply it by 1. Any number with this property I will define to be a *multiplicative identity*. Thus

Definition: A number U is called a *multiplicative identity* if for any number X the following are true: $X * U = X$ and $U * X = X$.

I am now going to prove, using only the definition of multiplicative identity that any two multiplicative identities, if they exist, must be equal. I describe this situation by saying that the multiplicative identity is *unique*.

Proposition. There is at most one number which can be a multiplicative identity.

Proof: Suppose there are two numbers that satisfy the definition of multiplicative identity, 1 and W. From the assumption that W is a multiplicative identity I can write $1 * W = 1$, why?[17] Also from the assumption that 1 is a multiplicative identity I can write $1 * W = W$, why?[18] Thus I now have $1 = 1 * W = W$. I conclude that $1 = W$.[19]

Exercise 11.3 The Definition of the Additive Identity, Zero

I will introduce the operation of addition of numbers in Chapter 12. In analogy with the above discussion of the multiplicative identity, can you invent and then state precisely the definition of the notion of *additive identity*? Characterize the additive identity in terms of how that number functions as a number (with respect to addition). Can you prove that if two numbers, say Z and 0, are additive identities then $Z = 0$, i.e., the additive identity is unique? (Hint: Mimic the proof of the uniqueness of the multiplicative identity above.) The common notation for the additive identity is 0, i.e., zero.

11.5 The Definition of Multiplicative Inverse

Our second axiom, which I state in the next section, makes use of two definitions, multiplicative identity, defined above, and multiplicative inverse, which we define now.

---

don't have to pick • as your set, you can pick any set with 1 object in it, but then ... . This possibly will make more sense after you have read IV.

[17]We are assuming that W is a multiplicative identity, thus from the definition, $X * W = X$ is true for any number X. Let X be the number 1 and we are done.

[18]We are assuming that 1 is a multiplicative identity and thus from the definition of multiplicative identity, for any number X, $1 * X = X$. Now just let $X = W$, and we are done.

[19]I am now cheating a little bit here, because I am using what I will call a *hidden axiom*, an axiom you have used so much yourself that I can sneak it in undetected. That axiom is the *transitivity* of $=$. Namely, if $A = B$ and $B = C$ then $A = C$. Can you see where I have used a second axiom, the *symmetry of equality*? This axiom states that if $A = B$, then $B = A$. Hint: I wrote $1 * W = 1$ and $1 = 1 * W$.

Definition: If X is a number, then a number Y is a *multiplicative inverse* of X if $X * Y = 1$ and $Y * X = 1$, where 1 is the multiplicative identity.[20] If number X has a multiplicative inverse, then that multiplicative inverse is denoted X^{-1}.

It turns out that a multiplicative inverse of a number, when it exists, is unique. By this I mean that if two numbers Y and W both satisfy the definition of being a multiplicative inverse of some number X, then $Y = W$, cf., Exercise 11.5, and thus $Y = X^{-1} = W$.

One way to think intuitively about the notion of the multiplicative inverse of some (nonzero) number X is this. Multiplication by X^{-1} undoes multiplication by X. Thus if you multiply some number V by X and get $X * V$, if you then multiply by X^{-1}, i.e., $X^{-1} * X * V = V$, you get back to where you started, namely, V. Note that we just used Axiom (A∗), since $X^{-1} * (X * V) = (X^{-1} * X) * V = 1 * V = V$.

You might be thinking that multiplication by X^{-1} is the same as "dividing by X." Eventually this thought of yours will be correct, but you may not yet think of division as an operation, since it does not occur in the axioms; and we have not properly introduced and defined the operation of division.[21]

Exercise 11.4 The Definition of An Additive Inverse
In analogy with the definition of multiplicative inverse, invent then state a definition of *additive inverse* of a number X. What symbol do you think is traditionally used to denote the additive inverse of X?

11.6 Our Second Axiom for Numbers

Our second axiom has two parts, the first denoted E1, for existence of 1; and the second denoted E∗I, for existence of multiplicative inverses (of nonzero numbers). I refer to this axiom as Axiom (E1,E∗I).

EXISTENCE OF A MULTIPLICATIVE IDENTITY, ONE, AND THE EXISTENCE OF MULTIPLICATIVE INVERSES

(i) *There exists for the multiplication operation for numbers a multiplicative identity, i.e., a number denoted by* 1. *This number* 1 *has the property that for*

[20] It turns out that one number does not have a multiplicative inverse, viz., 0, the additive identity. See Exercise 13.1 to see why this is the case.

[21] As you will see in Exercise 11.5, the operation of division does not satisfy any of the nice properties that multiplication does. Mathematicians get around this, eventually, by defining "division by X" to be the same as "multiplying by X^{-1}."

any number X *the following holds:*

$$\textbf{EXISTENCE OF 1:} \quad \boxed{\mathbf{X * 1 = X = 1 * X}} \; ; \qquad \text{(E1)}$$

(ii) *For any number* X *which is not equal to zero,*[22] i.e., $X \neq 0$, *there exists a number* X^{-1}, *called the multiplicative inverse of* X which satisfies the following equations:

$$\textbf{EXISTENCE OF * INVERSES:} \quad \boxed{\mathbf{X * X^{-1} = 1 = X^{-1} * X}} \; . \qquad \text{(E*I)}$$

11.7 If … , Then … . Our First Proofs

Before the next exercise, I want to tell you a little bit about how a mathematician uses the words *if* and *then*. In part (iv) of the next exercise, Exercise 11.5, you will encounter a sentence something like "if $x * y = 1$ and $y * x = 1$, then $y = x^{-1}$." The clause that begins with "if" is called the *hypothesis* of the statement and the (second) clause that begins with "then" is called the *conclusion*. (Note that you first encountered these concepts in Exercise 2.2, but it is not necessary for you to have done that exercise in order to understand this Chapter.)

What this means in operational terms is this. If you want to prove the statement "if $x * y = 1$ and $y * x = 1$, then $y = x^{-1}$," since x and y are numbers, you can assume whatever laws are known for numbers (which at this point are just the two axioms above, Axiom (A*), Axiom(E1,E*I)); and you can assume whatever is given in the hypothesis. After assuming all that, your goal then is to derive the conclusion. The way you do this is to start with a statement that is known to be true, for example, a statement assumed to be true in the hypothesis; then construct a sequence of statements, justifying each statement—or "move"—by an appeal to an axiom, to a definition, to the hypothesis or to some other statement previously proven to be true.

As you do any exercise, but particularly the ones coming up, focus on notions that underlie the notations. If you can do that, you really understand the material. Also, in the next few exercises do not use any formulas, patterns or axioms unless you are explicitly told you can use them.

Exercise 11.5 Having the Option of Replacing Division with Multiplication by Multiplicative Inverses was a Great Breakthrough!

In this exercise we give deductive proofs of properties of multiplicative inverses: (1) whenever x^{-1} exists it is unique; (2) if $(xy)^{-1}$ exists, then $(xy)^{-1} = y^{-1}x^{-1}$, where

---

[22]Zero, or 0, is the additive identity. See Exercise 11.3 and/or Chapter 12.

juxtaposition of x and y, for example, means multiplication;[23] (3) $(x^{-1})^{-1} = x$, if $x \neq 0$. Along the way we will find the number $(1.672(10^{-24}))^{-1}$, which is the multiplicative inverse of the mass of a lone proton (in grams); and we will define *division.*

(i) As the law above, Axiom (E1,E∗I), on the existence of multiplicative inverses states: every nonzero number, x, has a multiplicative inverse written x^{-1}. Now, either by hand or with your calculator, find the multiplicative inverse of the mass of a proton to one significant digit. (Note that you may have a "$\frac{1}{x}$" button on your calculator which will simplify your life at this point.) Does your answer when multiplied by $1.672(10^{-24})$ give 1?

(ii) Another way to view x^{-1} is to write the "fraction" $\frac{1}{x}$. (The symbol $\frac{1}{x}$ is just another notation for the notion "multiplicative inverse of x.") When one writes $\frac{1}{x}$ one is thinking "divide 1 by x." But in our axioms we have not explicitly introduced the concept of division. Thus we *define* (give meaning to the expression) $\frac{1}{x}$ by making the following definition, namely, $\frac{1}{x} = x^{-1}$. For practice, by hand or calculator write 10^{-1} in two other ways (as a fraction and as a decimal). Do the same for $(10^4)^{-1}$ and 2^{-1}. Double check your answers by seeing if $(x^1)(x^{-1}) = 1$ for the three values of x, viz., $10, 10^4$, and 2.

(iii) As we just mentioned above, our two laws for multiplication, Axiom (A∗), Axiom (E1,E∗I), do NOT mention division. None of the axioms for numbers mentions division, mainly because division does not satisfy nice laws like associativity and commutativity.[24] For example, $6 \div (3 \div 3) = 6$, while $(6 \div 3) \div 3 = \frac{2}{3}$. (Do you thus see that division is not associative?) Also $6 \div 3 = 2$, while $3 \div 6 = \frac{1}{2}$. (Do you thus see that division is not commutative?) We do not (explicitly) need the concept of division, however, if we have the multiplication operation and the existence of multiplicative inverses. Let's see how we can define division in terms given us by the axioms.

Definition of Division:[25] If $x \neq 0$ we write "y divided by x" $= \dfrac{y}{x} = \text{yx}^{-1}$.

Since we can write any given division operation using multiplication and multiplicative inverses we can always avoid division. Of course, sometimes you may not want to avoid division. To see what laws hold for division and fractions see the summary, near the end of this chapter on page 266. It may pleasantly surprise you to know that the laws for numbers are simpler if we *think of division by x as multiplication by x^{-1}*. With the above definitions, what is $\frac{y}{1}$? What is $\frac{y}{1}\frac{1}{x}$? What is $y\frac{1}{x}$? What is $\frac{y}{x}$? Express $\frac{y+3}{x^2+1}$ as a product of two expressions (assuming $x^2 + 1 \neq 0$).

SPECIAL INSTRUCTIONS FOR PARTS (iv), (v), (vi) and (vii): *While doing the rest of this exercise pretend that the only thing you know about numbers is what you have been told about them in the above two laws, Axiom (A∗), Axiom (E1,E∗I). Thus in doing the following exercises you may justify each manipulation, (or "step") that you perform by referring to one of the three equations which appear in the previous two laws for numbers. (You may also use the following "hidden" axiom: For numbers A, B and C, if $A = B$, then $A * C = B * C$ and $C * A = C * B$.)*

For example, if you had $z*(z^{-1}*w)$ and you wanted to say that this equals $(z*z^{-1})*w$ you would say that the associative law for multiplication, or equation (A∗), is the justification for this step or manipulation. If you wanted to follow this step with another step that says $(z * z^{-1}) * w = 1 * w$, you would have to appeal to equation (E∗I), to justify this step.

---

[23]Thus $x * y = xy$, both mean "x times y."

[24]Multiplication for numbers is associative, cf., Axiom (A∗). Multiplication is also *commutative*, see Axiom (C∗) later in Chapter 12. Briefly, the commutative law for multiplication states that for any two numbers A and B that $A * B = B * A$. Do NOT use this law in this exercise.

[25]This equality tells us that the definition of "y divided by x" is "multiply y by the multiplicative inverse of x." It will turn out that $yx^{-1} = x^{-1}y$, so there are two possible, equivalent definitions of "y divided by x." Do you know the name of the axiom, cf., Chapter 12, from which it follows that $yx^{-1} = x^{-1}y$?

Finally, if you wanted to say that $1 * w = w$, you would have to appeal to equation (E1), to justify this step. I am actually making your life less complicated for the moment. For the following four parts of this exercise you are not supposed to use all the mathematics that you know (or think you know). You only have to know—and for the moment you are only allowed to use—the three equations that appear in the two axioms above (plus our "hidden" axiom, definitions, hypotheses and statements previously proven, of which there are almost none at this point).

(iv) Equation (E∗I) tells us that if $x \neq 0$ then there is a number x^{-1}, called the multiplicative inverse of x, which satisfies $x * x^{-1} = 1$ as well as $x^{-1} * x = 1$. Given a number x, is its multiplicative inverse, x^{-1}, uniquely determined? Another equivalent way to ask this question is this: how many numbers y are there that satisfy the statement "y is a multiplicative inverse of x?" (By the definition of a multiplicative inverse of x, this means that y satisfies the two equations: $x * y = 1$ and $y * x = 1$.) The answer is "one;" and hence if y is a multiplicative inverse of x then it must be that $y = x^{-1}$.

Now how does a mathematician prove that the only multiplicative inverse of x is x^{-1}? Since you probably have not seen many mathematical proofs, I will guide you through two different proofs of this statement. I want you to do your best to answer each "Why?" question that I pose below in these proofs.

Every one of our mathematical proofs will begin with a statement that we know is true and end with the statement we are trying to prove is true. I will know my proof is done when I get to the statement I am trying to prove, which in this case will be $y = x^{-1}$.

My first proof will hopefully seem direct. The second proof might seem a bit more clever and mysterious. Each "Why?" means "Why do I know this step is true?" For each "Why?" give a reason why you know that particular step is allowed. The only acceptable answers refer to an axiom, a definition, a hypothesis, or a statement previously proven. (It is often, but not always, useful to ask yourself "What was done?" to a given line to get the next line in a proof.)

We will now prove the following statement:

If y is a multiplicative inverse of nonzero number x, then $y = x^{-1}$.

First Proof:

$$x * y = 1 \qquad \text{Why?}^{26}$$
$$(x^{-1}) * (x * y) = (x^{-1}) * 1 \qquad \text{Why?}$$
$$(x^{-1} * x) * y = (x^{-1}) * 1 \qquad \text{Why?}$$
$$(1) * y = (x^{-1}) * 1 \qquad \text{Why?}$$
$$y = x^{-1} \qquad \text{Why?}$$

Second Proof: The motivation for the first line of this proof is not obvious, but for any nonzero number x you should be able to tell me why it is true.

$$x^{-1} * (x * y) = (x^{-1} * x) * y \qquad \text{Why?}$$
$$x^{-1} * (1) = (x^{-1} * x) * y \qquad \text{Why?}$$
$$x^{-1} * (1) = (1) * y \qquad \text{Why?}$$
$$x^{-1} = y \qquad \text{Why?}$$

Warning: I often omit the $*$ when denoting multiplication. Thus we might have stated the first line of this second proof as $x^{-1}(xy) = (x^{-1}x)y$. (Note that the uniqueness of the

[26] Answer: This first line is a hypothesis. I am assuming y is a multiplicative inverse of x. This equation is part of the definition of "y is a multiplicative inverse of x."

multiplicative inverse can now be used in Exercise 11.6 to show that two possible definitions of x^{-n} are really equivalent.)

(v) Show that the multiplicative inverse of a product of two nonzero numbers is the product of their multiplicative inverses. For nonzero numbers x and y this is written $(xy)^{-1} = y^{-1}x^{-1}$. (Note that all we need to verify the validity of this formula for $(xy)^{-1}$ are the two axioms for numbers (A∗), (E1), (E∗I). Of course, it is also true that for numbers x and y that $(xy)^{-1} = x^{-1}y^{-1}$, but verifying this requires the commutative law of multiplication of numbers which we get to in Axiom (C∗).)

Again, since you are probably new to mathematical proofs I will guide you through this by making statements and having you fill in the reasons why each statement is simply true or follows from the previous statement. Thus we are now going to prove the following statement:

If $x \neq 0$ and $y \neq 0$, then $(xy)^{-1} = y^{-1}x^{-1}$.

First Proof:

$$(xy)^{-1} * (xy) = 1 \qquad \text{Why?}$$
$$[(xy)^{-1} * (xy)] * y^{-1} = 1 * y^{-1} \qquad \text{Why?}$$
$$(xy)^{-1} * [(xy) * y^{-1}] = 1 * y^{-1} \qquad \text{Why?}$$
$$(xy)^{-1} * [x * (y * y^{-1})] = 1 * y^{-1} \qquad \text{Why?}$$
$$(xy)^{-1} * [x * (1)] = 1 * y^{-1} \qquad \text{Why?}$$
$$(xy)^{-1} * [x] = y^{-1} \qquad \text{Why?}$$
$$[(xy)^{-1} * x] * x^{-1} = y^{-1} * x^{-1} \qquad \text{Why?}$$
$$(xy)^{-1} * [x * x^{-1}] = y^{-1} * x^{-1} \qquad \text{Why?}$$
$$(xy)^{-1} * [1] = y^{-1} * x^{-1} \qquad \text{Why?}$$
$$(xy)^{-1} = y^{-1} * x^{-1} \qquad \text{Why?}$$

Strictly speaking, in order to write the first line of the proof above I have to know that $xy \neq 0$. Why? (Read the Axiom on the Existence of Inverses carefully.) After doing Exercise 13.1 (vii) and (xii) it is easy to prove that if $x \neq 0$ and $y \neq 0$ then $xy \neq 0$.

I can, however, (without using Exercise 13.1) prove that for nonzero numbers x and y that $(xy)^{-1} = y^{-1}x^{-1}$ by using part (iv) above which says that multiplicative inverses are unique. Thus if I can find a number z that satisfies $z * (xy) = 1$ and $(xy) * z = 1$, then I must have that $z = (xy)^{-1}$. (Do you see why, using part (iv) above?) My candidate for z is $y^{-1}x^{-1}$. Thus $(y^{-1}x^{-1}) * (xy) = y^{-1} * [x^{-1} * (xy)] = y^{-1} * [(x^{-1}x) * y] = y^{-1} * [1 * y] = y^{-1} * y = 1$. Similarly, can you show that $(xy) * (y^{-1}x^{-1}) = 1$?

Second Proof: See if you can create a proof similar to the First Proof above whose first line is $(xy) * (xy)^{-1} = 1$.

(vi) Use the formula in (v) to compute $(1.672(10^{-24}))^{-1}$ as the product of $(1.672)^{-1}$ and $(10^{-24})^{-1}$. Does this answer agree with your answer to part (i) of this exercise?

(vii) Why is it that in our axiom on the existence of multiplicative inverses of a number X that we require that X $\neq 0$? See Exercise 13.1 (vii) and (xi).

(viii) Show that if $x \neq 0$, then $(x^{-1})^{-1} = x$. Again I will outline the proof for you, and you will provide the reasons for each step.

First Proof:

$$x * x^{-1} = 1 \qquad \text{Why?}$$
$$[x * x^{-1}] * (x^{-1})^{-1} = 1 * (x^{-1})^{-1} \qquad \text{Why?}$$
$$x * [x^{-1} * (x^{-1})^{-1}] = 1 * (x^{-1})^{-1} \qquad \text{Why?}$$
$$x * [1] = 1 * (x^{-1})^{-1} \qquad \text{Why?}$$
$$x = (x^{-1})^{-1} \qquad \text{Why?}$$

Strictly speaking there is a small problem with this First Proof, since in the second line I am tacitly assuming that $x^{-1} \neq 0$, which is true (see Exercise 13.1). The following Second Proof does not need this observation.

Second Proof: Do you see the role played in the following proof by the uniqueness of inverses proved in part (iv) above?

$$x^{-1} * x = 1 = x * x^{-1} \qquad \text{Why?}$$

$$x = (x^{-1})^{-1} \qquad \text{Why?}$$

Exercise 11.6 Deducing the Laws of Exponents in Our Deductive System

Special instructions: Read the special instructions that are given in the middle of Exercise 11.5. When doing parts (i), (ii), (iii), (iv) *and* (v) *of this problem you must justify each step of each answer you give by an appeal to one of the three equations, (A$*$), (E1), (E$*$I), in the above two axioms, to previous definitions, to hypotheses or by an appeal to previously proven statements such as parts (iv), (v) or (viii) of Exercise 11.5 or by an appeal to the definition of x^n, where x is any number and n is a positive whole number; or by an appeal to the definition (below) of x^{-n}, where x is any nonzero number and $-n$ is any negative whole number.*

Two laws for exponents and the definition of negative whole number exponents.

In this exercise we study why mathematicians choose the definition (for nonzero x) of x^{-n} and x^0 the way they do. Briefly, if we choose the traditional definitions, then the two laws of exponents below hold for all whole numbers n, m, k, whether they are positive, negative or zero. If we do not choose the traditional definitions our mathematical lives become an incoherent mess. Whenever possible we choose definitions that give us the most simplicity and order.

(i) For a positive whole number n and any number x, $x^n =$ (by definition) $x * \cdots * x$, i.e., n copies of the number x multiplied together. The number n is called the exponent of the number x. We say we have raised x to the power n, or we have exponentiated x by n. In a short statement explain how the axiom of associativity of multiplication plays a role in making sure that x^n is unambiguous. Hint: $x^2 = xx$ is unambiguous since the product of x with itself is uniquely defined for each number x. For $n = 3$, there are only two possible ways of computing x^3, $(xx)x$ or $x(xx)$. These are equal, why? Thus x^3 is uniquely determined by x. For $n = 4$, all the possible ways of computing x^4 yield the same answer. Why? .

(ii) Let's see if we can give a definition for x^{-n} where n is a positive whole number and hence $-n$ is a negative whole number. In other words, let's see what a negative exponent means. Suppose $x \neq 0$ and consider x^{-2}. If $x \neq 0$, then it can be shown that $x^2 \neq 0$ so that x^2 will have a multiplicative inverse by our axiom, (E$*$I), on multiplicative inverses.[27]

I will define x^{-2} to be $(x^2)^{-1}$. Now you might say, why not define x^{-2} to be $(x^{-1})^2$? It turns out that both definitions of x^{-2} are equivalent since I can (and we are about to) prove that $(x^2)^{-1} = (x^{-1})^2$. Answer each of the following instances of "Why?" as usual.[28]

We will now prove the statement:

If $x \neq 0$, *then* $(x^2)^{-1} = (x^{-1})^2$.

---

[27] I will prove that x^2 has a multiplicative inverse shortly without first showing that $x^2 \neq 0$. Do you see the claim that if $x \neq 0$, then $x^2 \neq 0$ is a special case of the claim made in Exercise 11.5 (v), and Exercise 13.1 (xii) that if $x \neq 0$ and $y \neq 0$, then $xy \neq 0$?

[28] The alert reader will note that this result we are about to prove, viz., $(x^2)^{-1} = (x^{-1})^2$, is a *special case* of Exercise 11.5 (v), viz., $(xy)^{-1} = y^{-1}x^{-1}$. Just let $x = y$. I am going to prove this special case anyway, a little more practice with proofs is worthwhile.

Proof:

$$\begin{aligned} x^2(x^{-1})^2 &= (xx)(x^{-1}x^{-1}) && \text{Why?}\\ (xx)(x^{-1}x^{-1}) &= x[x(x^{-1}x^{-1})] && \text{Why?}\\ x[x(x^{-1}x^{-1})] &= x[(xx^{-1})x^{-1}] && \text{Why?}\\ x[(xx^{-1})x^{-1}] &= x[(1)x^{-1}] && \text{Why?}\\ x[(1)x^{-1}] &= xx^{-1} && \text{Why?}\\ xx^{-1} &= 1 && \text{Why?} \end{aligned}$$

From the transitivity of "=" (footnote 19) it follows that $x^2(x^{-1})^2 = 1$. By a sequence of steps just like the one above you can also show that $(x^{-1})^2x^2 = 1$. It then follows that $(x^2)^{-1} = (x^{-1})^2$, why? Is $(x^2)^{-1} = (x^{-1})^2$ equivalent to $\frac{1}{x^2} = (\frac{1}{x})^2$? Said in English, for a nonzero number, is the multiplicative inverse of its square equal to the square of its multiplicative inverse?

I will now prove a similar result for similar equivalent definitions of x^{-3}.
Proof that if $x \neq 0$, then $(x^3)^{-1} = (x^{-1})^3$:

$$\begin{aligned} x^3(x^{-1})^3 &= [x^2x][x^{-1}(x^{-1})^2] && \text{Why?}\\ [x^2x][x^{-1}(x^{-1})^2] &= x^2[x(x^{-1}(x^{-1})^2)] && \text{Why?}\\ x^2[x(x^{-1}(x^{-1})^2)] &= x^2[(xx^{-1})(x^{-1})^2] && \text{Why?}\\ x^2[(xx^{-1})(x^{-1})^2] &= x^2[(1)(x^{-1})^2] && \text{Why?}\\ x^2[(1)(x^{-1})^2] &= x^2(x^{-1})^2 && \text{Why?}\\ x^2(x^{-1})^2 &= 1 && \text{Why?} \end{aligned}$$

From the transitivity of "=" (footnote 19) it follows that $x^3(x^{-1})^3 = 1$. By a sequence of steps just like the one above you can also show that $(x^{-1})^3x^3 = 1$. It then follows that $(x^3)^{-1} = (x^{-1})^3$, why?

Prove that $(x^4)^{-1} = (x^{-1})^4$ on your own. Do you see a pattern here? Do you see that you could eventually prove that $(x^n)^{-1} = (x^{-1})^n$ for any positive whole number n if you had enough time?[29]

(iii) I now give the following definition of x^{-n} where n is a positive whole number:

$$\boxed{x^{-n} = (x^n)^{-1}.} \qquad \textbf{(Definition of } x^{-n}\textbf{)}$$

Do you see that an equivalent definition would be $x^{-n} = (x^{-1})^n$? Is this equivalent to $x^{-n} = \frac{1}{x^n} = (\frac{1}{x})^n$?

(iv) You may have heard of some "laws of exponents." For example, if n and m are whole positive numbers then

$$\boxed{x^nx^m = x^{n+m}.} \qquad \textbf{(The First Law of Exponents)}$$

Explain why this First Law of Exponents, for whole positive n and m, is obvious.

(v) Let's now understand why the First Law of Exponents remains true, even if one or both n and m are allowed to be negative whole numbers. Keeping in mind the definition for (and above discussion of) x^{-n}, where n is a positive whole number, let's look at an example, say x^5x^{-2}. Let's be very clear about how our two axioms, Axiom (A∗), Axiom (E1,E∗I), let you manipulate this expression into x^3. We have $x^5x^{-2} = (xxxxx)[(x^{-1})(x^{-1})] = [(xxxxx)x^{-1}]x^{-1} = [(xxxx)(xx^{-1})]x^{-1} = [(xxxx)1]x^{-1} = (xxxx)x^{-1} = [(xxx)(x)]x^{-1} =$

[29] To prove this last statement rigorously requires an additional axiom called the axiom of mathematical induction. If you can "see" the pattern in the previous sequence of proofs, you have essentially done a proof by mathematical induction.

$(xxx)(xx^{-1}) = (xxx)1 = xxx = x^3$. Can you justify each of the above steps by an appeal to one of the above axioms or a definition? You might ask: Do we really need all these steps? The answer is yes and no. Yes, if you want to be very careful and precise and see exactly how $x^5x^{-2} = x^3$ follows from our two axioms above. The answer is no if you want to skip the details and just note that each x^{-1} "cancels" one x, leaving xxx. See if you can then come up with an argument for the general case, i.e., show $x^k x^p = x^{k+p}$, where the exponents can be any positive or negative whole numbers.

(vi) There is another law of exponents that is more or less obvious for positive whole numbers n and k, namely,

$$\boxed{(x^n)^k = x^{nk}.} \qquad \textbf{(The Second Law of Exponents)}$$

Why is this obvious if n and k are positive whole numbers?

(vii) Question: If we adopt the above definition for x^{-n} does it follow that the law $(x^n)^k = x^{nk}$ holds for all whole numbers n and k, positive or negative (we talk about the case when the exponent is 0 in part (xi) of this exercise)? Answer: Yes, but verify this for yourself, make sure you understand it. (At this point you may assume that a positive whole number times a negative whole number is negative, and that a negative times a negative is a positive. We will be discussing this "law of signs" pattern in more detail later.) The definition above for x^{-n} gives the simplest system of patterns I am aware of; in particular, the two laws of exponents above hold for all positive or negative exponents. *Simplicity* and *consistency* are two guiding principles in all of mathematics.

(viii) Is our definition of x^{-n} consistent with the definition of 10^{-n} we gave on page 59? That is, if $x = 10$ do you get the same results as we did on page 59?

(ix) Verify the following simple mechanical rules that you can apply without thinking, once you understand them: If $x \neq 0$ and n is any whole number, positive or negative, then $x^n = \frac{x^n}{1} = \frac{1}{x^{-n}}$. Is it also true that $x^{-n} = \frac{x^{-n}}{1} = \frac{1}{x^n}$?

(x) If n and m are positive whole numbers, we can write $x^n x^{-m} = \frac{x^n}{x^m}$. Does the "cancellation rule" that you learned in elementary school, i.e., wherein you cancel xs above and below the fraction bar until you can't cancel anymore, give you the same results as the First Law of Exponents?

(xi) If $x \neq 0$, I define $x^0 = 1$. Why? Hint: Consider $x^{-n}x^n$. What "increase in pattern" do I have with this definition as opposed to defining $x^0 = 2$? or defining $x^0 = x$? What "pattern" is lost if I pick some definition for x^0 other than $x^0 = 1$?

(xii) Can 0^0 be defined? One might think that one could define $0^0 = 1$. If that were done, what would be the consequences? If you defined 0^0 to be some other number, like 2381, what would be the consequences? The expression 0^0 is often called an *indeterminate* form. It arises, for example, in calculus as a limit of well-determined expressions; and its value can be anything!

11.8 Return to the Problem: How Many Protons in One Gram of Protons?

Now using the symbol $\approx$, which means approximately (equal to), we have

$$[1.672(10^{-24})]^{-1} = \frac{1}{1.672(10^{-24})} \approx 6(10^{23}).$$

Multiplying both sides of ($\star$), on page 240, by $6(10^{23})$ we get

$$(6(10^{23}))(1.672(10^{-24})\ g * \mathrm{N}) = (6(10^{23}))(1\ g). \qquad (\star\star)$$

Now using the associative law for multiplication of numbers on the left-hand side of ($\star\star$) we have

$$(6(10^{23})1.672(10^{-24})\ g)(\mathrm{N}) = (6(10^{23}))(1\ g).$$

We are led to

$$(1\ g)(\mathrm{N}) \approx (6(10^{23}))(1\ g).$$

Finally, we have

$$\mathrm{N} \approx 6(10^{23}).$$

11.9 What Is a Mole? Scaling Up from the Atomic to the Human Scale

The number, $6(10^{23})$ is called *Avagadro's number*, after Italian chemist Amedeo Avagadro (1776–1856). Now I must confess to you that I have not told the whole truth. Real chemists will tell you that Avagadro's number (to 3 significant digits) is $6.02(10^{23})$, and the multiplicative inverse of this number is $1.66(10^{-24})$. This last number is slightly less than the mass of a proton in grams and is called the *atomic mass unit*, or *amu*. One underlying reason for the discrepancy is the fact that when protons and neutrons form a nucleus the mass of the resulting union is a little less than the sum of the masses of the individual protons and neutrons involved. Some of the mass is converted into "binding " energy[30] that doesn't show up on the chemist's scales when mass is measured. This phenomenon complicates life a bit, and scientists before 1961 agreed to choose their atomic mass unit so that the most common isotope of oxygen came out with exactly 16 atomic mass units. After 1961 the atomic mass unit was chosen so that the most common isotope of carbon has exactly 12 atomic mass units. I am telling you this just in case you were curious. I also believe that when you use a tool you should know how sharp it is. You don't understand, say Avagadro's number, unless you have some idea of where it came from, how it is used and how precise it is.

In any event you will not need to know chemistry in much detail to read this book, since most of our calculations will only be accurate to one significant digit—or worse—to the nearest order of magnitude. Perhaps the most

[30] A very short range force, the *strong force* binds protons and neutrons together despite the repelling force protons exert on each other, due to the fact that like charges repel. Neutrons are necessary to hold any nucleus with more than one proton from flying apart.

important lesson to be learned from these complications is that there *always* appears to be more to learn when dealing with real Nature. Our scientific understanding of Nature seems to be growing and growing, but it looks like we will never know it "all." There always seem to be new patterns in Nature waiting to be discovered. That this has been well known for some time is indicated by the following quote from Shakespeare: *"There are more things in heaven and Earth, Horatio, than are dreamt of in your philosophy."*

That said, you certainly know what a *dozen doughnuts* means, i.e., 12 doughnuts. We now define a *mole* of doughnuts. A mole of doughnuts is "an Avagadro's number of doughnuts," viz., $6(10^{23})$ doughnuts. The concepts of a dozen and a mole are very much the same although the numbers are vastly different. You can think of a mole of some objects, that is an Avagadro's number of those objects, as a "chemist's dozen." Although we rarely if ever have occasion to think about a mole of doughnuts, we will frequently discuss a mole of H, which by our previous calculation has a mass equal to 1 gram of H. Thus the act of passing from one atom of an element, respectively, one molecule of a chemical, to a mole of atoms of that element, respectively, a mole of molecules of that chemical, is the act of "scaling up" from atomic to human scale.

Exercise 11.7 A Mole of Doughnuts Would Be Pretty Large

(i) How much mass would a doughnut have to have, to one significant digit, in order for a mole of doughnuts to have as much mass as the earth? Hint: The mass of the earth is approximately $6(10^{24})$ kg.

(ii) Why can numbers be used to represent numbers of atoms or numbers of oranges?

(iii) Discuss why the associative law for multiplication of numbers, as well as numbers themselves, can be used in the above example where we calculated the number of Hs in a gram of H.

(iv) Are the properties of numbers independent of the things that the numbers represent?

(v) Earth scientists tell us that our planet has a core which is 1,500 miles in diameter. (This core is at about 7,000 degrees Fahrenheit, yet it is solid due to the monumental pressure it is under. This inner core is surrounded by a molten core.) What fraction of the earth's volume is in this solid inner core?

Now in the *periodic table* in Figure 11.1 below we find the 92 "naturally occurring" elements,[31] the lightest of which is hydrogen with atomic number 1 and the heaviest of which is uranium. Uranium has atomic number 92; the atomic mass number of the most common isotope of uranium is 238. Uranium 235 is the isotope used in bombs. There are several (human-made) elements with atomic number higher than 92, the *transuranic* elements.

All molecules (of all chemicals) are composed of atoms of elements. We will use the following (simplified) rule:

The number of grams in one mole of a chemical is found by adding up the atomic mass numbers of all the atoms contained in one molecule of the chemical.[32] Remember this rule, it will be indispensable. It is also quite nifty

[31] See footnote 2.

[32] When doing precise chemistry one uses the above rule: except that the atomic mass

IA	IIA	IIIB	IVB	VB	VIB	VIIB	VIII			IB	IIB	IIIA	IVA	VA	VIA	VIIA	Zero
1 H 1.008																1 H 1.008	2 He 4.003
3 Li 6.940	4 Be 9.013											5 B 10.82	6 C 12.011	7 N 14.008	8 O 16.000	9 F 19.00	10 Ne 20.183
11 Na 22.991	12 Mg 24.32											13 Al 26.98	14 Si 28.09	15 P 30.975	16 S 32.066	17 Cl 35.457	18 Ar 39.944
19 K 39.100	20 Ca 40.08	21 Sc 44.96	22 Ti 47.90	23 V 50.95	24 Cr 52.01	25 Mn 54.94	26 Fe 55.85	27 Co 58.94	28 Ni 58.71	29 Cu 63.54	30 Zn 65.38	31 Ga 69.72	32 Ge 72.60	33 As 74.91	34 Se 78.96	35 Br 79.916	36 Kr 83.80
37 Rb 85.48	38 Sr 87.63	39 Y 88.92	40 Zr 91.22	41 Nb 92.91	42 Mo 95.95	43 Tc	44 Ru 101.1	45 Rh 102.91	46 Pd 106.4	47 Ag 107.88	48 Cd 112.41	49 In 114.82	50 Sn 118.70	51 Sb 121.76	52 Te 127.61	53 I 126.91	54 Xe 131.30
55 Cs 132.91	56 Ba 137.36	57 *La 138.92	72 Hf 178.50	73 Ta 180.95	74 W 183.86	75 Re 186.22	76 Os 190.2	77 Ir 192.2	78 Pt 195.09	79 Au 197.0	80 Hg 200.61	81 Tl 204.39	82 Pb 207.21	83 Bi 209.00	84 Po	85 At	86 Rn
87 Fr	88 Ra	89 #Ac															

*LANTHANIDE SERIES	58 Ce 140.13	59 Pr 140.92	60 Nd 144.27	61 Pm	62 Sm 150.35	63 Eu 152.0	64 Gd 157.26	65 Tb 158.93	66 Dy 162.51	67 Ho 164.94	68 Er 167.27	69 Tm 168.94	70 Yb 173.04	71 Lu 174.99
#ACTINIDE SERIES	90 Th 232.05	91 Pa	92 U 238.07	93 Np	94 Pu	95 Am	96 Cm	97 Bk	98 Cf	99 Es	100 Fm	101 Md	102 No	103 Lw

FIGURE 11.1: Periodic Table

for it allows us (at the human scale) to "look" at some things on the atomic scale with our mind's eye.

Let's do a practice problem: How much mass does 1 mole of the most common isotope of O_2, oxygen gas, have? Looking in the periodic table we see that the atomic number of O is 8, the atomic mass number, neglecting fractions, of O is 16.[33] Thus O_2 has a molecular mass number of $16+16=32$. Thus 1 mole of O_2 has a mass of $32\,g$.

Exercise 11.8 Moles and Mass

(i) Find the mass of one mole of (the most common isotope of) sulfur, i.e., S.

(ii) Calculate the mass of one mole of ammonia, i.e., NH_3.

(iii) Approximately how many molecules of glucose (a type of sugar) are in 1 gram of glucose, i.e., $C_6H_{12}O_6$? When you think about it, it is truly amazing that we can (fairly easily) calculate this number. We surely could not easily count these molecules one by one!

(iv) Find the mass of one mole of (the most common isotope of) carbon, i.e., C.

(v) How many grams of carbon are in 1 mole of glucose? What fraction (by weight or mass) of glucose is carbon?

(vi) How many neutrons are there in the nucleus of the isotope of uranium with atomic mass number 238? (The number of neutrons in excess of protons is of paramount importance to the stability and/or radioactivity of an element. Deuterium is stable and tritium is radioactive, for example.)

---

number is replaced by the mass in atomic mass units. This comes down to worrying about the difference between the mass of a proton, $1.672(10^{-24})$ g, and an atomic mass unit, $1.66(10^{-24})$ g. Also when doing chemistry it is important to know the relative abundance of the various isotopes of an element when converting between numbers of molecules (or atoms), i.e., moles, and mass. It will not be necessary for us to worry much about these details, for our simplifications will not qualitatively effect the conclusions that we arrive at in this book. As a last note, when dealing with atoms, chemists use the term "gram atom" instead of mole. We will not worry about this distinction either.

[33]The alert reader will note that there are at least two other isotopes of oxygen with atomic mass numbers 17 and 18.

Chapter 12

Five More Axioms for Numbers

Atoms are the building blocks of matter. Our axioms for numbers are the basic patterns from which the patterns of numbers, many of which you have learned earlier in your career, are built. It turns out that atoms can be torn down into even more elementary building blocks.[1] It also happens that our axioms of numbers can be broken down further into other patterns from which they can be constructed. Just as breaking atoms down leads us to advanced physics, further removed from everyday observations of patterns, breaking our 7 axioms for numbers down into patterns of set theory and logic leads us to more advanced mathematics and farther from the patterns of counting easily observed in everyday life. I am not going to explore set theory or logic in this text any more than I am going to study elementary particle physics. I just want you to know that there is more to learn, in case you get curious. There also exist other parts of mathematics with their own fundamental patterns that have little or nothing to do with numbers! I will introduce you to one such system when we study the Warlpiri Aborigines of Australia, cf., III.

We are studying numbers deeply enough so that we can use them with a solid understanding and without making too many technical errors. I won't be telling you everything that I know about numbers, which honestly, is less than is known—which in turn is a lot less than is knowable.[2]

I now proceed to state the last five axioms for numbers. Historically, addition is the most elementary operation for numbers. Multiplication, at least

[1] Physicists tell us that even protons and neutrons have internal structure and that there are many other particles they call *elementary particles* from which matter is built. For the curious, look up the word *quark* in a modern physics book. In Nature the natural building blocks of a system seem to depend on the scale of the system.

[2] In fact, with numbers and logic, just as in physics, biology and so forth, we never seem to know it "all." In mathematics our understanding of numbers, logic and the rest of mathematics grows every year. Consider this rather spectacular example from the last century. Mathematician, Kurt Gödel, proved in 1931 that there are statements about numbers that we cannot prove or disprove (starting from any clearly defined set of axioms that include numbers). Such statements are called *undecidable.* See Kurt Gödel, *On Formally Undecidable Propositions Of Principia Mathematica And Related Systems*, translated by B. Meltzer, Basic Books, Inc., New York, 1962. A more modern reference for Gödel's results is C. Smorynski, *The Incompleteness Theorems*, pp. 821–866, in the *Handbook of Mathematical Logic*, edited by J. Barwise, North Holland Publishing Company, Amsterdam, 1977. It is interesting to note that logicians have shown that the deductive system of "high school geometry," i.e., Euclidean geometry, is a decidable theory.

for whole numbers, can be viewed as repetitive addition. Were my intended audience learning about numbers for the first time, I might have introduced the following two axioms about addition of numbers before the above two axioms about multiplication. The next axiom I refer to as Axiom (A+), for associativity of addition.

12.1 Associativity, Identity, and Inverses for +

THE ASSOCIATIVE LAW OF ADDITION FOR NUMBERS

The associative law of addition for numbers states that if A, B and C *are numbers then*

$$\textbf{ASSOCIATIVE LAW FOR +:} \quad \boxed{\mathbf{A + (B + C) = (A + B) + C}} \quad \textbf{(A+)}$$

where we perform the operations in parentheses first.

Analogous to the corresponding axiom for multiplication we have the following axiom which I refer to as Axiom (E0,E+I), for existence of zero and existence of additive inverses:

THE EXISTENCE OF AN ADDITIVE IDENTITY, ZERO, AND THE EXISTENCE OF ADDITIVE INVERSES

(i) *There is a number,* 0, *called the additive identity, with the property that for any number* X:

$$\textbf{THE EXISTENCE OF 0 :} \quad \boxed{\mathbf{X} + 0 = \mathbf{X} = 0 + \mathbf{X}} \quad \textbf{(E0)}$$

(ii) *For any number* X *there exists an additive inverse*[3] *of* X, *denoted* −X, *which satisfies*:

$$\textbf{EXISTENCE OF + INVERSES :} \quad \boxed{\mathbf{X} + (-\mathbf{X}) = 0 = -\mathbf{X} + \mathbf{X}} \quad \textbf{(E+I)}$$

Exercise 12.1 Having the Option of Replacing Subtraction with Addition of Additive Inverses was a Great Breakthrough!

(i) We can define subtraction as follows. To *subtract* a number y from a number x, written $x - y$, add to x the additive inverse of y, $-y$, i.e., $x + (-y)$. Thus $x - y = x + (-y)$.

---

[3] In English the additive inverse of number X is defined to be the number which when added to X gives 0. This additive inverse of X is unique; see Exercise 13.1 (iii). Note that we have not yet demonstrated that −X = (−1)X, i.e., that the additive inverse of X is the additive inverse of 1 times X, but we will show this. See Exercise 13.1 (viii).

Note that our axioms do NOT mention subtraction explicitly, since "x subtract y" can always be realized by *adding* $-y$ to x. Believe it or not this simplifies life for the following reason: the operation of subtraction is not associative (and hence mathematics becomes simpler if we do not have to explicitly mention subtraction). We can see this from many examples. Try the following. Compute $6 - (3 - 2)$ and $(6 - 3) - 2$; do the operations inside parentheses first. Does this example establish the fact that *subtraction is not an associative operation?* Make up a few more examples on your own. Thus by creating the concept of an additive inverse, mathematicians were able to simplify the set of rules that govern numbers (the associative law for + holds for all numbers) without giving up the notion of "taking y from x."

(ii) In order to do subtractions with an operation, +, that always satisfies the associative law, we had to introduce additive inverses. Thus for the number 1 there must be a number -1, for 5 there must be -5 and so on. This required some imagination, for it is easy to imagine 2 oranges, but perhaps not so easy to imagine -2 oranges. Is there any concrete meaning that -2 oranges could have?

(iii) What is the additive inverse of -1? of -2?

(iv) Every number, y, has an additive inverse written $-y$, so that $y+(-y) = (-y)+y = 0$. Do you think that there is any number whose additive inverse equals its multiplicative inverse? If such a number exists, find one property that such a number must satisfy. Compare Exercise 13.4.

We now come to the commutative laws for the operations of addition and multiplication. Before we state these axioms let's note that not all operations in mathematics or in "real life" are commutative, i.e., the *order* in which the operations are performed makes a difference. See the next exercise, and see Exercise 15.2. about the non-commutativity of the Warlpiri kinship relations.

12.2 Commutativity of + and *

I refer to this next axiom as Axiom (C+), for commutativity of addition.

THE COMMUTATIVE LAW OF ADDITION

Given any numbers A *and* B *the following holds:*

COMMUTATIVE LAW FOR + : $\boxed{A + B = B + A}$ (C+)

I refer to this next axiom as Axiom (C*), for commutativity of multiplication.

THE COMMUTATIVE LAW FOR MULTIPLICATION

Given any numbers A *and* B *the following holds:*

COMMUTATIVE LAW FOR * : $\boxed{A * B = B * A}$ (C*)

When two or more expressions are multiplied together, for example, A∗B∗C, each of the expressions is called a *factor*. Thus A, B and C are factors in the product A ∗ B ∗ C. When two or more expressions are added together, for example, A + B + C, each of the expressions is called a *term*. Thus A, B and C are terms in the sum A + B + C.

Exercise 12.2 Commutativity of Addition and Multiplication of Numbers

(i) Let's illustrate the idea that, at least for whole numbers, multiplication is just repetitive addition. Draw a 3 by 2 rectangular array of dots, reminiscent of the arrays of dots in Chapter 10. You should now be looking at three rows each consisting of two dots. How many dots are there in the array? Compute this number by adding 2 dots in the first row, 2 dots in the second row and 2 dots in the third row. What is a quicker way of finding the answer? Yes, that is why you memorized your multiplication tables in grammar school. Now create at least three other rectangular arrays of dots, say a 4 by 7 array, and so forth. Can you imagine an "n by m" array of dots, where n and m are whole (positive) numbers? How many dots are in this rectangle?

(ii) Now imagine two arrays of dots: the first is the 3 by 2 array from part (i), and the other is a 2 by 3 array of dots. Since it is easy to see geometrically that these two arrays have the same number of dots, what can you say about $3 * 2$ and $2 * 3$? Now do the same with an "m by n" array and an "n by m" array. What can you say about mn and nm? (Note that mn means $m * n$, or m times n.)

(iii) The "axioms for numbers" that we have been studying hold if we take our numbers to be the rational numbers, the real numbers or the complex numbers. Rational numbers are easy to understand, they are just numbers of the form $\frac{m}{n}$, where m and n are whole numbers, with $n \neq 0$. Thus rational numbers are just "fractions." The concept of real number takes a bit longer to understand in detail. For example, $2^{\frac{1}{2}} = \sqrt{2}$, which is the "square root of 2," is a real number that is not a rational number. For a project, see if you can find and understand a proof of this fact. There are a lot of other real numbers that are not rational, like π, which is in a class of real numbers called "transcendental." For another project, see if you can figure out some properties that a transcendental number has. As for complex numbers, see Exercise 13.4.

(iv) Many operations are not commutative, i.e., order of execution makes a difference. For example, subtraction is not a commutative operation. Thus $6 - 3$ is not the same as $3 - 6$. In real life you can easily think of operations for which the order of execution affects the outcome. Suppose you have a gas stove without a pilot light. Think of lighting a match and holding it to a gas burner on the stove, which you then turn on for one hour. The result is a controlled flame which burns for one hour. Now reverse the order of execution: turn on the burner for one hour, then strike a match ... . Can you think of any other operations in mathematics or real life that "do not commute?"

12.3 Distributivity

We have so far seen axioms about addition of numbers and axioms about multiplication of numbers. The next axiom shows how these two operations interrelate; it is called the distributive law. Note that the right-hand side of the distributive law below could have been written as $(A * B) + (A * C)$. We instead have written $A * B + A * C$, since there is a convention in mathematics that multiplications are always done in a given expression before additions (and exponentiations are done before multiplications).

THE DISTRIBUTIVE LAW OF MULTIPLICATION OVER ADDITION

The distributive law of multiplication over addition says that if A, B *and* C *are any numbers then the following holds:*

DISTRIBUTIVE LAW : $$\boxed{\mathbf{A * (B + C) = A * B + A * C}} \qquad (\mathrm{D}{*}{+})$$

Exercise 12.3 Multiplication, $*$, Distributes Over Addition, $+$

(i) We can use the distributive law to see once again that, at least for whole numbers, multiplication is just repetitive addition. Thus if we add a number to itself, say $5 + 5$, we can use the distributive law to write: $5 + 5 = 5(1 + 1) = (5)(2)$. If we write $5 + 5 + 5$ how can we use the distributive law to understand it as 5 times 3?[4]

(ii) Can you see how to use the distributive law to understand that 5 added to itself n times, which we can write as (recall Exercise 10.9) $\sum_{k=1}^{n} 5$, is just $5n$?

(iii) In (ii), above, can you replace 5 by a number m and see that m added to itself n times is just mn?

(iv) How many terms are there in $\sum_{k=1}^{n} 5$?

(v) What law of numbers is implicitly being used when we write $5 + 6 + 7$? What other law is implicitly being used when we write $(5)(18)(3)$?

(vi) Can you find a geometric interpretation of the distributive law for numbers, involving rectangular arrays of dots? We found just such a geometric interpretation to illustrate the commutative law in Exercise 12.2.

[4]Hint: Use the distributive law on $5 + 5$, then use the distributive law again on $(5)(2) + 5$.

Chapter 13

What Patterns Can Be Deduced in Our Deductive System?

13.1 Playing the Mathematics Game

One of the goals of this chapter is to show that the rules for manipulating fractions are not random or arbitrarily imposed by authority figures. Such patterns are natural consequences of our 7 basic axioms for counting/numbers. I go through the details of deducing these patterns from the 7 axioms because a number of my students who had dutifully memorized correct patterns would sometimes for a variety of reasons imagine new patterns for dealing with fractions. How is one to tell if a new pattern that someone proposes is "OK" or it is wrong?[1]

There is another reason for showing that the familiar patterns for dealing with numbers learned in elementary and secondary school can be deduced from our 7 axioms. It demonstrates that arithmetic, and hence algebra, are manageable subjects. Knowing the 7 basic axioms, and how to deduce everything you learned in arithmetic and algebra as consequences of these 7 axioms, shows that everything learned in arithmetic and algebra[2] is part of a tightly organized logical system. Such is the product of thousands of years of humans looking at Nature and thinking. In a very real sense, the deductive system we are studying is an immense gift from the past—a gift from Nature and humans inspired by Nature. Established mathematical patterns represent cultural treasures, while the new patterns that await discovery/invention are gems often more valuable than diamonds.

[1] A pattern for numbers (or counting) is right or OK if it can be deduced from the 7 axioms. A proposed pattern is wrong if a counterexample can be found, showing that the pattern is not always true. You will not likely encounter an undecidable pattern—one for which neither a proof of nor a counterexample to can be found—but they exist.

[2] In secondary school algebra, in so far as the symbols used represent numbers, the subject belongs to the deductive system whose foundation is our 7 axioms of numbers. Thus all of the "rules" or patterns encountered in algebra are understandable in the same manner used to show that the patterns of fractions are deduced from our 7 axioms (which we do in Section 13.3).

13.2 Rules for Playing the Mathematics Game

One of the most frequent mistakes made in what I call playing the mathematics game, i.e., deducing patterns/facts from the 7 axioms, is justifying a step or "move" by appealing to something not allowed—a fact you feel (or know) is true but has not yet been proven, by us. As in the proofs we have already done earlier, every "legal" move or step in a proof *has to be justified by an appeal to one of the 7 axioms, a hidden axiom, a definition, a hypothesis or a pattern/fact previously proven using only legal steps.* If you cannot so justify some step you would like to make in some proof, then you have broken a link in the chain leading to our goal. This goal is to show that all the patterns of fractions and other patterns you have learned in arithmetic and algebra are contained in the deductive system whose axioms are our 7 axioms of numbers.

Keeping the above in mind, see if you can create proofs all of whose steps are "legally justified" as you do this next exercise.

Exercise 13.1 Try Creating Some Proofs on Your Own

This group of exercises shows some of the basic properties of numbers that are built-in to or, to be more precise, logically follow from, our 7 axioms. You probably already know that all of the following properties of numbers are true! The point of this exercise is to see if these properties that we know must be true can be derived using only our seven axioms for numbers. If they cannot be so derived, our axiom system is not complete, not big enough to tell us what we need to know about numbers. It turns out that our set of axioms is big enough. In fact, if you have an operation or pattern or fact dealing with numbers that you think is true, you had better be able to derive it from the 7 axioms.[3] If you can derive[4] it, it will be a "legal" operation or pattern or fact. Thus we find out that there are relatively few legal operations with numbers that you have to know. Life isn't as complicated as you might have feared.

(i) The number 0 is the additive identity, i.e., it satisfies equation (E0). Suppose that there is another number denoted by the symbol Z which is also an additive identity, i.e., it also satisfies equation (E0), viz., $X + Z = X = Z + X$ for all numbers X. Show that $0 = Z$. This shows that the additive identity, 0, is unique.[5]

(ii) Show that if 1 and U are multiplicative identities, then $1 = U$. This shows that the multiplicative identity, 1, is unique. See if you can do this without looking back at the proof in the text.

---

[3]Actually there are some "hidden" axioms of = that we use, like ones we have used before, without explicitly stating them. For example, if A, B and C are numbers and $A = B$, then $C + A = C + B$. A similar hidden property of = relative to multiplication was stated in the SPECIAL INSTRUCTIONS just after part (iii) in Exercise 11.5. These two properties of = plus the three encountered in part (iv) of this Exercise 13.1 are all the "hidden" assumptions that we (knowingly) use in this book.

[4]Recall, that every step of a legal derivation, i.e., proof, has to be justified by an appeal to one of the 7 axioms, a definition, a hypothesis, or a pattern or fact previously legally proven.

[5]Hint: Consider $0 + Z$.

(iii) Show that if x and x' are both additive inverses of the same number, say y, then $x = x'$.[6] Thus this exercise says that if x is an additive inverse of y, it must equal *the* additive inverse of y, i.e., $x = -y$.

(iv) Can you think of three properties of "=" that are so "obvious" that we often do not explicitly acknowledge the fact that we are using these axioms of equality? The three I am thinking of have nothing to do with numbers, + or ∗, per se.[7] One of the jobs of a mathematician is to keep us all "honest" in the sense that we make explicit all of our hidden assumptions. This is not always an easy job; and we often discover a hidden assumption, in physics or biology, for example, only when it leads us to a conclusion that Nature tells us is not so.[8] Pure mathematicians only have to worry about internal consistency, i.e., whether or not a set of assumptions leads to a contradiction.[9] This is one (but only one) of the secrets to the timelessness of mathematical truth. When we apply mathematics to the real world, however, we must worry about whether Nature actually obeys our axioms. Experience has shown me that Nature always has a trick up her sleeve; we never seem to completely understand the patterns in Nature.

(v) Show that if x'' and x' are both multiplicative inverses of the same non-zero number, say w, then $x'' = x'$. Thus the multiplicative inverse of a given non-zero number is unique. This is the same as Exercise 11.5 part (iv). Can you now do this exercise without looking back at Exercise 11.5? Thus this exercise says that if x is a multiplicative inverse of w, then x must equal *the* multiplicative inverse of w, i.e., $x = w^{-1}$.

(vi) What is the additive inverse of the additive inverse of a number x? In mathese, what is $-(-x)$? Can you show your answer follows from the axioms? Do the same problem for multiplicative inverses, cf., Exercise 11.5 part (viii).

(vii) Can you show from the axioms that $0x = 0$, i.e., 0 times x is 0, for any number x?[10]

(viii) If -1 is the additive inverse of 1, can you show from the axioms that the additive inverse of x, $-x$, is the same as $(-1)x$, i.e., "minus one times x" or more precisely, "the additive inverse of 1" times x?

(ix) If -1 is the additive inverse of 1, what is $(-1)(-1)$? Can you show that the answer is 1 using only the axioms and what we have proven so far?

(x) Show for any two numbers x and y that $(-x)(-y) = xy$, and $(-x)y = -(xy)$. This is as close as we will come to showing the *law of signs*, which states that the product of a negative number times a negative number is a positive number, a positive number times a positive number is a positive number and a negative number times a positive number is a negative number. The reason this exercise is not the same thing as the law of signs is

---

[6]Hint: You can construct a proof whose first line is $x + y = 0$, adding x' to both sides we have $x + y + x' = x'$. Note that we are implicitly using the "hidden" law about equality, i.e., if we add the same number to both sides of an equation we still have an equation. Keep making legal moves until you arrive at the last line of the proof which is $x = x'$. Second hint: We can construct a second proof of the uniqueness of the additive inverse in much the same way we proved the uniqueness of the multiplicative inverse in Exercise 11.5 part (iv). Namely, for the first line of the proof consider $(x' + y) + x = x' + (y + x)$.

[7]Hint: The three properties of = that I am thinking of are: (1) $a = a$ for any a (this is called *the reflexive property of* =), (2) if $a = b$, then $b = a$ (this is called *the symmetry property of* =), (3) if $a = b$ and $b = c$ then $a = c$ (this is called *the transitive property of* =). See also III.

[8] *"The entire history of science is a progression of exploded fallacies, not of achievements."* Ayn Rand (This is a bit harsh, since it is an—often tough—achievement to explode a fallacy!)

[9]A *contradiction* is a statement which asserts that a statement, S, is both true and false. A set of axioms is not consistent if a contradiction can be deduced from them.

[10]Hint: Consider constructing a proof whose first line contains $0 * x + 1 * x$ and whose last line is what you are trying to prove.

pointed out by the following question: if x is a number, is $-x$ a negative number? (Answer: Not necessarily!) Also, it may or may not come as a surprise to you that there are nonzero numbers which are not positive or negative. See Exercise 13.4 part (ii).

(xi) You no doubt recall the following statements from a previous mathematics class: "never divide by zero" or "you cannot divide by zero." I can now explain what this means. If I changed Axiom (E*I) on THE EXISTENCE OF MULTIPLICATIVE INVERSES to say that X^{-1} existed for all numbers X *including* 0, I would then be able to divide by 0 since then $\frac{1}{0} = 0^{-1}$. Then the following conclusions follow from part (vii) above. First $0 = 0^{-1} * 0 = 1$. Why? Then $X = 1 * X = 0 * X = 0$ for all numbers X. Why? Thus you have a choice. Assuming the usual patterns of numbers (as codified in the 7 axioms in this Part II), if you allow $\frac{1}{0} = 0^{-1}$ to exist as a number then all numbers are 0. There is only one number in the number system. Your other choice is to not allow $\frac{1}{0} = 0^{-1}$ to exist as a number, then there are lots of numbers besides 0, like 1, for example.

(xii) Prove that if $x \neq 0$ and $y \neq 0$ then $xy \neq 0$. [11].

13.3 The Usual Rules for Fractions Are Part of Our Deductive System

In this next exercise, supply proofs using only legal justifications. When you are done you will have shown that our 7 axioms deductively contain all the rules for fractions you have previously learned.

Exercise 13.2 The Rules for Fractions Follow from the Axioms

In this group of exercises we will see how the rules for manipulating fractions follow from our axioms for numbers. This exercise can serve as a review, and it also will give you some idea of what operations with fractions are "legal" and which are nonsense—and how to tell the difference!

(i) If x and y are two non-zero numbers, what is $\frac{1}{x} + \frac{1}{y}$ written as a single fraction (find a common denominator[12])?[13] Now the axiom on multiplicative inverses tells us that $xx^{-1} = 1$ and $yy^{-1} = 1$ and that any number multiplied by 1 remains unchanged. Thus

[11] Hint: If $xy = 0$, then $(xy)(y^{-1}x^{-1}) = 0$. But $(xy)(y^{-1}x^{-1}) = 1$

[12] Given a fraction like $\frac{m}{n}$, m is called the numerator and n is called the denominator.

[13] Hint #1: If you absolutely rebel at the thought of using letters for numbers, pick a couple of your favorite non-zero numbers in place of x and y and do this exercise. Repeat with other choices for x and y, keep doing examples until you see "patterns of symbol manipulation" emerge which do not depend on the specific choices you make for x and y. Hint #2: Recall that $\frac{1}{x} = x^{-1}$ and $\frac{1}{y} = y^{-1}$.

we have:

$$\begin{aligned}\frac{1}{x} + \frac{1}{y} = x^{-1} + y^{-1} = 1x^{-1} + 1y^{-1} &= yy^{-1}x^{-1} + xx^{-1}y^{-1} \\ &= yy^{-1}x^{-1} + xy^{-1}x^{-1} \\ &= (y+x)y^{-1}x^{-1} \\ &= (y+x)(xy)^{-1} \\ &= \frac{y+x}{xy}.\end{aligned}$$

What laws and definitions were used and where, precisely? For the next to last step do not forget Exercise 11.5 part (v). Recall the following definition: If z and w are two numbers and $w \neq 0$, then what we *mean* when we write $\frac{z}{w}$ is zw^{-1}, i.e., $\frac{z}{w} = zw^{-1}$.

(ii) Go through an argument similar to that of (i) to change $\frac{a}{b} + \frac{c}{d}$ to a single fraction with one denominator. Give a "legal"reason for each step.

(iii) From the axioms show: $-\frac{a}{b} = \frac{-a}{b} = \frac{a}{-b}$, where a is a number and b is a non-zero number. Note that the additive inverse of $\frac{a}{b}$, i.e., any of the three expressions above, is unique and that it is that unique number which when added to $\frac{a}{b}$ gives 0.

(iv) Show that if a number $b \neq 0$, then $(b^{-1})^{-1} = b$. This exercise is the analogue for the multiplication operation that Exercise 13.1 part (vi) is for the operation of addition. This is also the same as Exercise 11.5 part (viii).

(v) From the axioms show: $(\frac{a}{b})^{-1} = \frac{b}{a} = \frac{1}{\frac{a}{b}}$, where a and b are non-zero numbers. Note that the multiplicative inverse of $\frac{a}{b}$, i.e., any of the three expressions above, is unique and that it is that unique number which when multiplied by $\frac{a}{b}$ is 1.

(vi) Show that $\frac{a}{b} - \frac{c}{d} = \frac{ad-bc}{bd}$, where b and d are non-zero numbers, a and c are any numbers.

(vii) Show that $(\frac{a}{b})(\frac{c}{d}) = \frac{ac}{bd}$, where b and d are non-zero numbers, a and c any numbers; and $\frac{\frac{a}{b}}{\frac{c}{d}} = \frac{ad}{bc}$, where b, c, and d are non-zero numbers and a is any number.

(viii) What is a *percentage*, cf., Exercise 1.2? Since "cent" means 100, per cent means per one hundred. A percentage is just a fraction whose denominator is 100. Of course, this 100 is then taken out from the "bottom," twisted a bit until it looks like %, and put after the numerator of the fraction. Thus, for example, $\frac{1}{2} = \frac{1}{2} * \frac{50}{50} = \frac{1*50}{2*50} = \frac{50}{100} = 50\%$. Now just in case you missed the day on fractions in elementary school, another way to convert $\frac{1}{2}$ to a percentage is to get out your calculator (your calculator can still be your head!) and divide 2 into 1, getting .5, or .50 as your answer. Now .5 is just five tenths, but .50 is fifty one-hundredths! Which is 50%.

Convert $\frac{1}{4}, \frac{1}{3}, \frac{6}{7}, \frac{2}{1}$, 1 and .019 into percentages.

(ix) If there were once 2 million acres of ancient redwood forest in California, and if in 1994 less than 4% remained—at most how many acres of ancient redwood forest were there in 1994?

(x) According to RACHEL'S ENVIRONMENT & HEALTH WEEKLY #520, November 14, 1996, BRAIN CANCER UPDATE, roughly 17,500 Americans were diagnosed with brain cancer in 1994. This represents a 2.9% increase over the previous year. How many more Americans were diagnosed with brain cancer in 1994 than were diagnosed with brain cancer in 1993?

(xi) According to Shenk, David, *The Forgetting: Alzheimer's: Portrait of an Epidemic,* Doubleday, New York, 2001, the number of people in the United States afflicted with Alzheimer's in 1975 was 500,000. In 2001 the number was 5 million. Five million is what percentage of 500,000? Shenk states that from 2000 to 2050 an estimated 80 to 100 million people worldwide will succumb to Alzheimer's.

Summary: Counting/Number Patterns

Addition Axioms		Multiplication Axioms
$A + (B + C) = (A + B) + C$	Associativity	$A * (B * C) = (A * B) * C$
$0 + A = A = A + 0$	Identity	$1 * A = A = A * 1$
$A + (-A) = 0 = (-A) + A$	Inverses	$A * A^{-1} = 1 = A^{-1} * A$
$A + B = B + A$	Commutativity	$A * B = B * A$

Axiom on Distributivity of Multiplication over Addition

$$\boxed{A * (B + C) = A * B + A * C}$$

Note: A, B, C are arbitrary numbers in the above equations, except for the axiom on multiplicative inverses where $A \neq 0$.

The following facts/patterns can be derived from the axioms above.

Facts about Inverses and Identities

1. The additive identity, 0, is unique.
2. The multiplicative identity, 1, is unique.
3. Given a number A, its additive inverse, $-A$, is unique.
4. Given a number $A \neq 0$, its multiplicative inverse, A^{-1}, is unique.
5. Given a number A, $-(-A) = A$.
6. Given a number $A \neq 0$, $(A^{-1})^{-1} = A$.
7. Given numbers $A \neq 0, B \neq 0$, $(A * B)^{-1} = B^{-1} * A^{-1}$.
8. Given a number A, $-A = (-1) * A$.
9. $(-1) * (-1) = 1$.
10. Given numbers A, B, $(-A) * (-B) = A * B$, $(-A) * B = -(A * B)$.

Facts about Exponents

The following hold for any numbers A, B and whole numbers m, n (positive, negative or zero) with the restriction that at no time are we allowed to raise 0 to a negative or zero exponent.

1. $A^{-n} = (A^{-1})^n = (A^n)^{-1}$
2. $A^n * A^m = A^{(n+m)}$
3. $(A^n)^m = A^{(nm)}$

I did not prove the following in the text, but I state them here for your reference. If A and B are positive numbers, and p, q, r and s are whole numbers, with $q \neq 0$ and $s \neq 0$, then the following are true.[14]

1. $A^{\frac{p}{q}} = (A^p)^{\frac{1}{q}} = (A^{\frac{1}{q}})^p$
2. $(AB)^{\frac{p}{q}} = A^{\frac{p}{q}} B^{\frac{p}{q}}$
3. $A^{-\frac{p}{q}} = (A^{-1})^{\frac{p}{q}} = (A^{\frac{p}{q}})^{-1}$
4. $A^{\frac{p}{q}} * A^{\frac{r}{s}} = A^{(\frac{p}{q} + \frac{r}{s})}$
5. $(A^{\frac{p}{q}})^{\frac{r}{s}} = A^{(\frac{p}{q}\frac{r}{s})}$

Facts about Fractions

The following hold for any numbers A, B, C, D with the restriction that at no time is 0 to appear in any denominator.

1. $\dfrac{1}{A} = A^{-1}$.
2. $\dfrac{B}{A} = B * A^{-1}$.

[14] The laws of exponents are true when they make sense (This is the case, for example, when A and/or B are positive.) if the fractions $\frac{p}{q}$, $\frac{r}{s}$ or the numbers m and n are replaced by even more general numbers that cannot be written as fractions, for example the square root of 2 or π.

3. $\dfrac{A}{B} + \dfrac{C}{D} = \dfrac{A * D + B * C}{B * D}$

4. $-\dfrac{A}{B} = \dfrac{-A}{B} = \dfrac{A}{-B}$

5. $\left(\dfrac{A}{B}\right)^{-1} = \dfrac{1}{\left(\dfrac{A}{B}\right)} = \dfrac{B}{A}$

6. $\dfrac{A}{B} * \dfrac{C}{D} = \dfrac{A * C}{B * D}$

7. $\dfrac{\left(\dfrac{A}{B}\right)}{\left(\dfrac{C}{D}\right)} = \dfrac{A}{B} * \dfrac{D}{C} = \dfrac{A * D}{B * C}$

8. $\dfrac{A}{B} - \dfrac{C}{D} = \dfrac{A * D - B * C}{B * D}$

13.4 Can You Tell the Difference between True and False Patterns?

In this next exercise I want you to be able to either prove the stated pattern is true (meaning it is always true for any numbers you might plug in for the variables present) or is false (meaning that you can find an example of specific numbers which when plugged in for the variables show that the pattern does not work, i.e., it is false).

Exercise 13.3 Are Any of the Following Patterns True?
(i) Is the following statement true or false? For all (positive) numbers X and Y,

$$\sqrt{X+Y} = \sqrt{X} + \sqrt{Y}.$$

In English "the square root of a sum is the sum of the square roots." How do you propose to decide if this statement is true or false?
(ii) Is the following statement true or false? For all nonzero numbers X,

$$\frac{1}{X} + 3 = \frac{1}{X+3}$$

. How did you demonstrate your answer?

(iii) Is the following statement true or false? Assuming zero does not appear in any denominator,

$$\frac{a}{b} + \frac{c}{d} = \frac{a+c}{b+d},$$

for any numbers a, b, c, d.

(iv) When a pattern is not true, there are specific numbers that can plugged in for the variables that show the pattern is false. Given a false pattern, however, there may be specific values of the variables which when plugged in to the pattern make the pattern (in our three cases above our patterns are equations) true statements. This is often called "finding the solutions," in our case, to the three equations above. To the best of your ability describe the specific values of the variables involved that make the equation in (i) a true statement. Do the same for parts (ii) and (iii).

13.5 More Exercises

Exercise 13.4 The "Mysterious" Square Root of Minus One[15]

(i) You may have encountered the so-called "imaginary" number $i = (-1)^{\frac{1}{2}} = \sqrt{-1}$. There is a way to geometrically represent $\sqrt{-1}$ which is quite useful in many areas of mathematics (pure and applied!). Consider the figure below from Euclidean geometry:

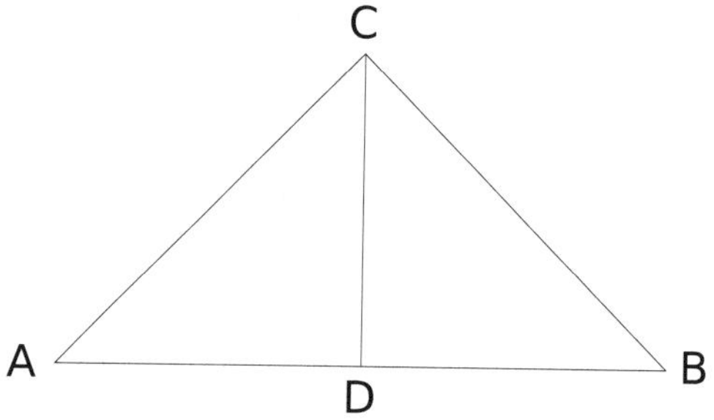

FIGURE 13.1: A "Picture" of $\sqrt{-1}$

The angle ACB is a right angle, as are the angles ADC and BDC. The line segments AC and BC are equal as are the line segments AD and DB. As you may or may not have learned in high school geometry, if you regard each of these lengths as positive numbers the following equation of proportionality holds (by looking at similar triangles ADC and BDC):

$$\frac{\textit{length of AD}}{\textit{length of CD}} = \frac{\textit{length of CD}}{\textit{length of DB}}.$$

[15]Compare Exercise 12.1 part (iv).

Thus

$$(length\ of\ AD)(length\ of\ DB) = (length\ of\ CD)^2.$$

If you think now of line segment AB as on the horizontal axis of a Cartesian co-ordinate system with point D at the origin, (0,0), and segment DC as on the vertical axis you can think of "directed lengths" as follows. The segments DC and DB have length 1 and the remaining segment DA has length -1. The last equation above then becomes

$$(-1)(1) = (length\ of\ CD)^2.$$

Thus CD becomes a geometrical realization of

$$i = (-1)^{\frac{1}{2}} = \sqrt{-1}.$$

Do your best to understand the above. Thus Euclidean geometry can be extended into complex-number-like Euclidean spaces called Hilbert spaces.

(ii) Is i a positive number, a negative number, or neither? Complex numbers are numbers of the form $a + i * b$ where a and b are real numbers. Recall that real numbers include the whole numbers and rational numbers (fractions), but that there are many more real numbers than these. Can you draw a picture of all the complex numbers? Where are the real numbers in your picture? Where are the positive numbers in your picture?

(iii) Is the number $i = (-1)^{\frac{1}{2}} = \sqrt{-1}$ any more imaginary than the number -1? Is i any more imaginary than the number 1?

(iv) For those of you who would like a real challenge, study Euler's equation:

$$e^{i\pi} = -1,$$

where e is discussed in V.

Note that an interesting discussion of this and other mathematical concepts from the point of view of a linguist and a psychologist can be found in [379].

Lest you might think all of this is too abstract to matter, I point out that engineers (especially electrical engineers) use $i = (-1)^{\frac{1}{2}} = \sqrt{-1}$ all the time. In particular, if you are using a light to read this by, or have ever used a computer, you have to thank the "square root of minus one" at least a little bit.

Exercise 13.5 Test Your Knowledge of Fractions

My Sunday paper[16] posed the following problem which can be done if you have a solid understanding of fractions and numbers.

There was a reunion of students at which fewer than 24 people attended. Two-thirds of the men were married to three-fifths of the women. Assuming all spouses were present, how many people were single?

Exercise 13.6 A Story About Chess, Exponential Growth and Human Nature

(i) Over the years I have heard various versions of the following account of the invention of the game of chess. (I have not researched the authenticity of this story.) A rich ruler of some state (presumably a Raja in India) was so happy with the game of chess that had been invented for him by one of his brilliant subjects that the ruler granted the inventor one wish. The inventor thought for a while and came up with the the following wish: "Please put one grain of wheat (or was it rice?) on the first square of the chess board. Put two grains on the second square. Double that and put four grains on the third square. Continue in this way putting on the kth square twice the number of grains that were placed on the previous square until all the squares have received their allotment of grain." The ruler agreed. The inventor was then, according to one version of the story, beheaded by the ruler when the

[16] *Parade Magazine*, 10/10/06.

ruler finally realized that he could not possibly fulfill his promise. Just how many grains would the ruler need to carry out the inventor's wish? How much wheat (or rice) is this?[17]

(ii) The game of chess is much simpler than the game of life. It is of great interest that even in this relatively simple setting the best human chess mind, Garry Kasparov, beat the world's best chess playing computer, IBM's Deep Blue, 4 to 2, in their encounter of February, 1996. If a game is significantly simpler than chess, like tic-tac-toe, and the computer can "see" all possibilities, then a human cannot win. The game of chess is complicated enough that the computer cannot yet come close "to seeing all possibilities." In a rematch, May 11, 1997, IBM's Deeper Blue beat Kasparov 3.5 to 2.5 (2 games won by Deeper Blue, 1 by Kasparov and 3 draws). Between games, however, humans upgraded Deeper Blue, based on performance in previous games. Can you estimate the number of possible chess games that can be played? You might want to return to this question after you have read this entire book.

Philosophically speaking the human brain, coupled with emotions, intuition and who knows what else, is fundamentally different from a digital computer. Although the IBM-6000 beat Kasparov at chess, as reported by Fred Kaplan in *The Boston Globe*, May 12, 1997, it is unlikely that digital computers will ever be better than humans at the game of living. Since it has taken eons of co-evolution within Nature to produce the human brain, it seems most likely that mimicking what Nature has produced will be the best human engineers will be able to do.

As time goes on and the problems of the human race mount, we might decide that the human mind is not as good at playing the game of life as we once thought. It might be of paramount importance to know what "makes our minds tick," and also to understand why we seem to be making some fundamental mistakes in how we humans comport ourselves on the planet. What specific improvements in *Homo sapiens*[18] behavior do you believe would improve humanity's overall relationship with Nature? See Chapter 9.

Exercise 13.7 Which Way Did the Bicycle Go?[19]

Assume that the graph in Figure 13.2 represents the track of a two-wheeled bicycle whose wheels only roll and never skid or slide. State clearly some assumptions (2 basic ones) about a bicycle, "Bicycle Axioms," if you will. From these axioms, prove that the bicycle traveled either from bottom left to top right or vice versa. Only one of the two possibilities is correct.

Exercise 13.8 Triangular Numbers Revisited

(i) Can you show that the triangular number of rank n, i.e., the sum $\sum_{k=1}^{n} k$, and the triangular number of rank $n-1$, i.e, the sum $\sum_{k=1}^{n-1} k$, when added together yield the square number of rank n, i.e., n^2? Can you show this using only the properties of numbers? Can you give a completely geometric (picture) argument? The Greeks discovered this fact by using geometric arrays of dots.

(ii) This problem is extremely difficult even for a professional mathematician. I thought you might like to know about it anyway. It is true that every positive whole number is the sum of three triangular numbers. It is true that every positive whole number is the sum of four squares. It is true that every positive whole number is the sum of five pentagonal numbers, and on and on. For the definition of pentagonal numbers, hexagonal numbers and so on see [112, p. 38]. For a discussion of the proofs of these facts you can begin with Emil

[17] Hint: Let $S = \sum_{k=0}^{63} 2^k$. Consider the quantity $2S - S$ written out. Most of the terms in the sums cancel out, and you are left with a short formula with only two terms for S.

[18] This is the name we have given to our own species, and it means "wise, intelligent" man. Do you see any relationship of this name assignment to the fact that humans see themselves at the center of the biological universe, as they once saw themselves at the center of the solar system?

[19] I first was introduced to this exercise by Professor John Horton Conway.

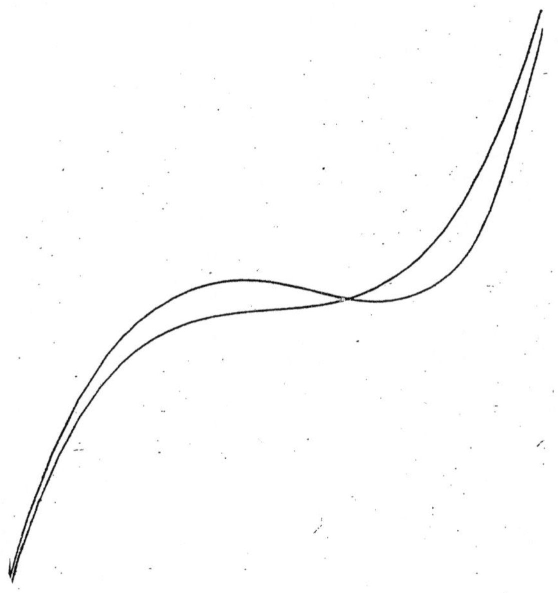

FIGURE 13.2: Which Way Did the Bicycle Go?

Grosswald, *Representations of Integers as Sums of Squares*, Springer-Verlag, New York, 1985, Chapter 3.

Exercise 13.9 One of the Biggest Fishing Disasters of All Time. A Metaphor for (some of) the Official Mathematical Models of our Time?[20]

In 1992 the cod population off the Grand Banks of Newfoundland, formerly one of the richest fisheries in the world, was officially pronounced depleted—economically extinct.[21] The Canadian government has imposed a total ban—indefinitely—on fishing off the Banks. The fishery may never recover. There are virtually no adult ("baby cod maker") cod left in this fishery. From the time of their discovery by John Cabot in 1497 (the Norse were earlier) until recently, Newfoundland cod were abundant. Cod of 100 to 200 pounds were once caught on the Grand Banks. As late as the 1960s children in Newfoundland could catch fish by scooping them out of the sea with a basket. Yet now the fishing jobs are gone, as well as all the jobs that depended on them. That this could happen in an advanced first world democracy which can avail itself of the best science, the best systems of communication and education, has served notice to the world. What went wrong?

The usual suspects turn up: weather (which has been around a long time), seals (ditto), foreign fishers, politicians who set the catch limits (too high), fishers who cheated on their quotas, and surprisingly *the scientific experts and their mathematical models which were pivotal in establishing the catch limits for the Grand Banks.* It is important to learn from this experience because the same problems, including the problems with the mathematical models,[22] are repeated around the world.

What went wrong with the model? First, the government sent out a research vessel to do random sampling of the cod population using reliable mathematical sampling procedures. The commercial fleets estimated total cod population using their catch as a guide. The two sets of numbers yielded similar estimates of age structure and total numbers of the cod population until 1989. In that year the commercial catch indicated that the total population was twice as large as the random sampling method. The scientists deferred and compromised, setting the estimated cod population at midway between their estimate and that of industry. *This compromise hastened the collapse* of the cod population. Instead of harvesting 16% of the cod, 60% of the cod were being caught. The fact that the catches in 1989, 1990 and so forth were increasing was due to the use of advanced technology (like sonar) to locate the fish—not due to increased numbers of cod! Discuss all aspects of this problem and how the collapse of the this cod fishery might have been prevented.

A recent history of cod fishing is: Mark Kurlansky, *Cod: A Biography of the Fish that Changed the World,* Walker and Company, New York, 1997.

Exercise 13.10 Compromises, n Steps

Suppose you compromise and save $\frac{1}{2}$ of a resource and give up $\frac{1}{2}$, for example, you save $\frac{1}{2}$ an old growth forest and $\frac{1}{2}$ is cut, or you dig up half a gold deposit and save half. Suppose after a while you save $\frac{1}{2}$ of the previously saved $\frac{1}{2}$, i.e., $\frac{1}{4}$, and give up for "development" another $\frac{1}{2}$ of a $\frac{1}{2}$, i.e., a $\frac{1}{4}$. Thus after two iterations you have saved $\frac{1}{4}$ and lost $\frac{1}{2}+\frac{1}{4}$.

[20] The global scope of the problem illustrated here is outlined in the article by Carl Safina, "The World's Imperiled Fish," *Scientific American*, November 1995, pp. 46–53.

[21] A reference for this problem is an article from a British magazine by Debora Mackenzie, "The cod that disappeared," *New Scientist,* September 16, 1995, pp. 24–29.

[22] A rather brisk (if you have not had calculus) introduction to the mathematics involved can be found in [90]. Please note that the author of this book, Colin Clark, is a pioneer contributor to mathematical models in resource management. It is not his mathematics which is to blame for the crash of the cod population but the abuse and misapplication of the mathematical modeling process by others.

(i) After n iterations of the above process of compromise, you will have saved $\frac{1}{2^n}$ of your "resource" or, for example, the old growth forest. What number does this fraction saved approach if n grows larger and larger?

(ii) After n iterations of the above process of compromise, you will have lost to development $\sum_{k=1}^{n} \frac{1}{2^k}$. What fraction of the original "resource" or old growth forest does this sum approach as n gets larger and larger?

(iii) Suppose you redo parts (i) and (ii) after making the following changes. Instead of saving $\frac{1}{2}$ of the resource and giving up $\frac{1}{2}$, save $\frac{9}{10}$ of the resource and give up $\frac{1}{10}$. Does this change your answers?

(iv) Discuss the difference between "renewable" and "non-renewable" resources and how these concepts affect your answers to the various parts of this problem. If a Bristle cone pine can live to be in excess of 4,000 years old with redwood trees not far behind, what does it mean that a redwood forest is "renewable?"

Exercise 13.11 How Big is an Oil Molecule?

(i) This exercise is from [283, p. 5–6]. According to history[23] Benjamin Franklin did the following experiment, which inadvertently measured the size of an oil molecule. Franklin put droplets of oil on the surface of a lake and noticed that there was always a maximum area the oil slick would cover—the oil slick could never be coaxed by any means to exceed this consistently achieved maximum. This maximum area was simply related to the volume of the original oil droplets: if the volume of oil was doubled then the area of the slick was doubled. Franklin did his experiments before science had discovered molecules; but with hindsight, Franklin's experiments are consistent with the (verifiable) hypothesis that such an oil slick is one molecule thick. Thus the thickness of such an oil slick is a measure of the size of the oil molecule involved. In Franklin's experiments he used units of teaspoons for volume of oil and acres for area of oil slick. Harte translated one of Franklin's observations into metric units and got: .1 cm^3 of oil spread to a maximum area of 40 m^2. Your problem is to compute the thickness of this oil slick. Hints: Assuming that the oil slick is square, then what is the volume of the oil slick in terms of its area and thickness? Will your formula for the volume of the oil slick (area times thickness) be valid if the oil slick has an irregular shape instead of a square one? (Think of the slick as approximately a bunch of small squares.) Based on the calculation you have just done, do you think that an oil molecule is *complex*? Do you think it is more complex than say a molecule of H_2?

(ii) In this problem we get an answer to the question: How far apart are two molecules of water? First, using your knowledge of moles, calculate the approximate number of molecules of H_2O in one gram of water. Imagine this gram of water as a cubic centimeter, i.e., a cube 1 cm on a side. Imagine that the water molecules are like little marbles all arranged neatly in a cubic, three dimensional version of the dots you encountered in Figure 10.1 with the Greek's square numbers. How many molecules of H_2O will be on one edge of this cube? The next paragraph is a hint.

An example of an exponent that is a fraction. Let $x =$ the number of water molecules on an edge of the cube. Then there will be x^3 molecules in our cubic centimeter of water. Why? Now from the first part of the problem we know what x^3 is. (In English we say that x^3 is "x cubed.") We now have the problem of determining what a number is if we know its cube. In some cases this is easy. If a number z when cubed is 8, then z must be 2, since $2^3 = 8$. (And there is no other "real" number whose cube is 8; can you see this?) Thus if $z^3 = 8$, then $z = 2$. Finding z in this case is called finding the cube root of 8. It turns out that the laws of exponents, page 267, hold if the n and m there are fractions as long as the number being exponentiated is positive. Thus we can solve our equation $z^3 = 8$

[23]Goodman, N., *The ingenious Dr. Franklin: selected scientific letters of Benjamin Franklin*, Philadelphia, University of Pennsylvania Press, 1956.

mechanically using a fractional exponent and the second law of exponents, thus:

$$\begin{aligned} z^3 &= 8 \\ (z^3)^{\frac{1}{3}} &= 8^{\frac{1}{3}} \\ z^{3\frac{1}{3}} &= (2^3)^{\frac{1}{3}} \\ z^1 &= 2^{3\frac{1}{3}} \\ z &= 2^1 = 2. \end{aligned}$$

To do this last calculation we needed to know a number which when cubed yields 8, namely 2. On your calculator there should be a way of finding the "cube root of a number;" namely, the cube root of a number N is that number which when cubed yields N. The cube root of N is denoted $N^{\frac{1}{3}}$. A note of caution: the cube (or any odd) root of a number is unique, but even roots, like square roots, of non-zero numbers fail to be unique. Thus $4^{\frac{1}{2}}$ could be 2 or −2. Unless otherwise stated, the square root of some positive number will refer to the positive square root of that number.

How far will it be from the center of one "marble," i.e, one H_2O molecule, to the center of the next? Now do you think that all of the H_2O molecules are sitting still? At what temperature do you think all the molecules will be sitting absolutely still? Do you think that all of the water will consist of molecules of H_2O, precisely two $H's$ "connected" to one O; or will some of the water be in "pieces" of H and OH?

Exercise 13.12 The Exxon-Valdez Oil Spill: How Big Was it? How are Things Today?

(i) In the March 24 1989 Exxon Valdez tanker spill it was reported that eleven million gallons of oil escaped. (This is Exxon's estimate, accepted by the press. Initial estimates ranged from 11 million to 38 million gallons. State-hired surveyors estimated the volume spilled to be closer to 30 to 35 million gallons, or three times what Exxon had reported, [511, pp. 193–4].) Find a theoretical maximum size for the area of an oil slick with a volume of eleven million gallons.[24] A cubic meter equals 1000 liters equals 264.2 U.S. gallons.

[24]Shortly after the Exxon Valdez tanker spill, a number of folks, including some scientists, commented that things were not so bad. However, On October 4, 1998, National Public Radio announced on "All Things Considered" that scientists had found that oil residues from the Exxon Valdez oil spill were still killing salmon. Exxon's scientists disputed this claim. On March 23, 2009, investigative reporter, Greg Palast, wrote an article, "Stick Your Damn Hand In It: 20th Birthday of the Exxon Valdez Lie," (see http://www.gregpalast.com/stick-your-damn-hand-in-it-20th-birthday-of-the-exxon-valdez-lie/) where he discovers from direct observation that if one sticks a hand below the gravel on some shores once prolific with food for subsistence of indigenous people, you come up with black goo—20 years after the spill. (I partially confirmed this for myself on a trip to Alaska in June, 2009). Palast also details why the common wisdom: that THE cause of the disaster was a drunk sea captain is a cover-up of the following facts. "Yes, the captain was 'three sheets to the wind"—but sleeping it off below-decks. The ship was in the hands of the third mate who was driving blind. That is, the Exxon Valdez' Raycas radar system was turned off; turned off because it was busted and had been busted since its maiden voyage. Exxon didn't want to spend the cash to fix it. So the man at the helm, electronically blindfolded, drove it up onto the reef." Palast also details further negligence on Exxon's part in not having the proper containment equipment "at the ready"—as required—so that the spill could have been contained and not become the disaster it did for the people who lived there. In a related development, the Supreme Court reduced the payment of Exxon to the aggrieved by 90%. For one first-hand account see [511, 510, 666]. In particular, in a 2009 twenty-year anniversary interview at www.fair.org, local Riki Ott stated that the herring (a species basic to the ecosystem) had not recovered in Prince William Sound. You might want to check on a lawsuit of some

(ii) In the November 1998 issue of *Scientific American*, pp. 56–61, there is an article titled "Natural Oil Spills." In this article we read: "Natural oil slicks—which range from less than .01 to one micron—may be only a few tens of molecules thick." Is this statement consistent with size of oil molecule you calculated in part (i)?

(iii) In June 1995, an industrial-agriculture-8-acre manure lagoon burst through a retaining dyke in North Carolina, releasing twice the volume of the Exxon Valdez oil spill—22 million gallons of manure—into a local river. Assuming the lagoon had a uniform depth, how deep was it?

Exercise 13.13 How Easy Is It To Take a Shower?

(i) Estimate your "access to a shower (or bath)." By this I mean the following. You probably take a shower or bath now and then. This takes a certain amount of time, say 5 minutes per shower (or whatever number applies to you). In addition, there are a certain number of bathing/showering facilities available to you daily: count them. Each of these is available to you for a certain length of time. Add up the total number of "available shower-hours," e.g., if the showers in the gym are available to you 12 hours a day, then the total number of shower-hours available to you at the gym is 12 times the number of showers in the gym. Now estimate the number of other people to whom the showers that are available to you are also available. Call this number the number of people in your shower pool. Suppose that (on average) everyone takes the same time for a shower that you do, say 5 minutes. Compute the shower-demand (in units of people-hours) by multiplying the number of people in your shower pool by $\frac{5}{60}$ hours. Now calculate the ratio $\frac{\text{number of available shower hours}}{\text{shower-demand}}$. How does this number change if the only showers available to you (no dorm or other showers exist) are those in the gym, and these showers are open 3 hours per day on Monday, Wednesday and Friday only? This latter situation exists at some universities in the world; and in some places in the world access to fresh, uncontaminated water is an absolute luxury.

(ii) At the University of Colorado's "environmentalist dormitory," Baker Hall, it has been proposed that either "shower timers" limiting the length of an individual's shower, or a buzzer indicating a shower has gone on too long (but not shutting off the water) will be installed. Discuss.

(iii) Pick some other "thing" such as personal transportation or public transportation, and repeat part (i) of this exercise.

Exercise 13.14 A Puzzle

Given three equilateral triangles formed from 9 matchsticks of equal length, and arranged as in Figure 13.3.

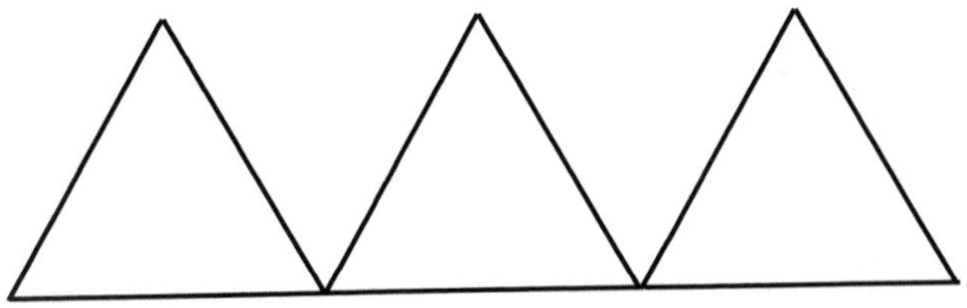

FIGURE 13.3: Three Equilateral Triangles

indigenous people of Ecuador against Texaco, which by some accounts spilled 30 times the Exxon-estimated volume of the Exxon Valdez spill in oil/waste in the jungles of Ecuador.

(i) Move exactly three matchsticks and place them without overlapping any other matchsticks (and without breaking them) so that you have (at least) four equilateral triangles.

(ii) Move exactly two matchsticks and place them without overlapping any other matchsticks (and without breaking them) so that you have (at least) four equilateral triangles.

Exercise 13.15 Money for Math

In a May 25, 2000 Associated Press article by Jocelyn Gecker, we learn that the Clay Institute, a U.S.-based mathematics foundation, is offering a total of seven million (U.S.) dollars for solutions to "the seven mathematical problems that stand out as great unresolved problems of the 20th century." (That is, a one million dollar prize for each of the seven problems.)

If you want to try your luck the problems can be found on the Web site of the Clay Mathematics Institute, www.claymath.org.

Exercise 13.16 A Possibly Counterintuitive Calculation Suppose a student bicycles from a city A to a city B, then returns along the same route from city B to city A. Suppose that on the trip from A to B the bicyclist is very energetic and travels at 20 miles per hour. On the trip back from B to A the bicyclist travels at 10 miles per hour. What is the average speed of the bicyclist for the entire trip?[25] Before you do the problem, guess the answer and compare your guess to the actual answer after you figure it out.[26]

[25] Hint: You do have enough information to work the problem. Just let the distance between city A and city B be d.

[26] Did you guess 15? Can you explain why the answer to this problem is not 15 miles per hour?

Part III

One of the Oldest Mathematical Patterns

Chapter 14

A Short Story and Some Numberless Mathematics

In the midst of a relatively dry, flat plain in the interior of Australia—suddenly and unexpectedly, there it is—Uluru, or Ayer's Rock as the Europeans call it. I saw it first from a bus, crammed with tourists taking pictures through the windows—lacking whatever it took to want to exit and touch the red earth! I couldn't stand it, so I grabbed my pack and contrary to National Park regulations I started a walk that would not get me back until well after "closing time." I communed with the huge, sometimes-orange red rock in glorious solitude for hours, walking completely around it with my only guide being memories of some scattered comments made to me by local Aborigines. The whole story, I will have to tell somewhere else. For now, let me say that this brief look into a small part of Aboriginal culture is a shortened summary of this entire book, both the mathematical and environmental parts.[1]

Here is a list of some of the major lessons to be learned from this Part III: (1) Essentially numberless mathematical structures[2] (besides geometry) exist; (2) Mathematical structures exist that satisfy an associative law but not a commutative law; (3) Creating and applying mathematical structures have had a survival value for humans for a very, very long time.[3]

---

[1]As for the mathematical part it was not until many years after my walk around the rock that I would more fully understand the significance of the fact that in a special place on the rock Aborigines had inscribed some of their kinship records. I actually saw these inscriptions, but I never dreamed at that time that I had been looking at a special case of a part of mathematics that was in my Ph.D. thesis—on locally compact groups. I was.

[2]This Part III, for example, is not part of the deductive system that contains numbers. It resides in the part of mathematics called *group theory*, cf., Exercise 15.11. Given numbers, mathematicians say that there are two groups associated with numbers, the additive group of numbers and the multiplicative group of numbers. This means that in some sense the theory of numbers is contained in the theory of groups, but not vice versa.

[3]This Part III answers a question that I posed to myself many years ago: Can I find human mathematics more complex than counting, which is older than that of the Chinese or Egyptians, Ancient Greeks or Romans, for example. Some mathematicians have vigorously told me that in their opinion the Warlpiri did not do mathematics. It all depends on how you define mathematics. My definition is the search for and study of patterns. This the Warlpiri certainly did, and one result was their kinship relations, which we are about to study. The Warlpiri did not write all this down in a format used by my contemporaries, but the patterns were definitely in their minds and an important part of their culture.

To explain Warlpiri mathematics properly, from the point of view of contemporary mathematical culture, I have to first introduce some fundamental notions. Note that this chapter is a bit dense with mathematical notation, so be patient and read it several times, if necessary.

14.1 Relations Defined as Collections of Ordered Pairs

In one approach to mathematics, the concept of *set* is taken as so fundamental that the term goes undefined. Thus we can talk about the *set of people on earth*, or the *set of tigers on earth*, or the *set of whole numbers* and so forth.[4] Intuitively, any collection of objects that you can think of is an example of a set.[5]

The particular set I want to study now is the set of Aboriginal people called the *Warlpiri*, who live in desert and coastal areas in Australia's Northern Territory. It is worth noting that some anthropologists believe that the Aborigines, of which the Warlpiri are but one group, have continuously lived in Australia for about 50,000 years![6]

Given any set, mathematicians often study *relations* between elements from that set. For example, if p_1 and p_2 are two elements in the set of Warlpiri people we can ask: Is p_2 the brother of p_1? We could ask: Is p_2 the mother of p_1? We could ask: Is p_2 the cousin of p_1? These are three examples of relations on the set of Warlpiri people. Now the *ordered* pair, written (p_1, p_2), is said "to be in the mother relation" if p_2 is the mother of p_1. The order makes a big difference here; because if (p_1, p_2) is in the mother relation then

[4] We denote sets with curly brackets. Thus {1,2,3} is the set consisting of the first three positive whole numbers. Note that the order in which the elements are listed is immaterial. Thus $\{1, 2, 3\} = \{2, 1, 3\} = \{3, 2, 1\}$ and so on.

[5] There are some difficulties with this statement (which will not concern us in this course) if you really examine the logical structure involved. For example, is the collection of all sets a set? Most of the time I will be leaving these types of problems to the logicians.

[6] In an article authored by Aisling Irwin, "Down Under discovery stands time on its head," that I obtained from the World Wide Web, more precisely a UK News story from 21 September 1996 put out by Electronic Telegraph, it is stated that there is strong evidence for the claim that Aborigines were present in Western Australia at least 76,000 years ago. Dr. Richard Fullagar, Dr. Lesley Head, David Price and Jenny Atchison (a doctoral student at the University of Wollongong) conducted the study, using such techniques as thermoluminescence and radio carbon dating. On October 22, 2002 a BBC (British Broadcasting Corporation) news story said that via a comprehensive global genetic mapping of humans, some scientists have concluded that the original Australians arrived on that continent approximately 50,000 years ago. See [169] for estimates that the Aborigines arrived in Australia between 40,000 and 60,000 year ago. Of related interest search for "Mathematical Approaches to Cultural Changes" by Colin Renfrew and Kenneth L. Cook, as well search Dennis Stanford, Chair of the Department of Anthropology, Smithsonian.

we know for sure that (p_2, p_1) is definitely *not* in the mother relation. Do you see why?

Definition: A *relation* R on a set S is a collection of ordered pair(s) of the form (a, b) where a and b are in set S.[7]

14.2 Symmetric Relations

There are mathematical properties that a relation can have. One of them is "symmetry," which I now define.

Definition: A relation, R, on a set, S, is said to be *symmetric* if whenever an ordered pair of elements from S, say (a, b), is in the relation R, then (b, a) is also in the relation R.

Said slightly differently, a collection R of ordered pairs is a symmetric relation if for every ordered pair in R the "transpose" or "flip" of that ordered pair is also in R. The transpose of an ordered pair is the ordered pair obtained by moving the element (or symbol) in the first slot to the second slot, and moving the element (or symbol) in the second slot to the first slot.

If you can find an ordered pair, say (p_0, q_0), that is in a relation R but such that (q_0, p_0) is NOT in R, then you have found a counterexample to symmetry, i.e., you have shown that relation R is NOT symmetric. A relation R either is symmetric or it is not symmetric.

An Example. Consider the relation $R = \{(1,2),(1,1)\}$, on the set $S = \{1,2\}$. If I transpose each ordered pair in R, I get the set of ordered pairs, i.e., the relation, $R' = \{(2,1),(1,1)\}$. Now $R \neq R'$ since $(1,2) \neq (2,1)$. Thus the relation R is NOT a symmetric relation on the set S.

Another Example. Consider the relation $R'' = \{(1,2),(2,1),(1,1)\}$. If I transpose each ordered pair in R'', I get $R''' = \{(2,1),(1,2),(1,1)\}$. Now $R'' = R'''$, since they are the same set of ordered pairs of elements from $S = \{1,2\}$. Thus R'' is a symmetric relation on set S.

Please note that the symmetry (or non-symmetry) of a relation is a property of the *whole* relation, i.e., a property of the entire collection of ordered pairs that make up the relation—NOT a property of *some* (subset which is not all) of the ordered pairs.

---

[7] Note that any collection of ordered pairs of this form is allowed to be a relation on set S, including one ordered pair by itself. In the ordered pair (a, b), a is called the first member (occupying the first "slot") of the ordered pair and b is the second member of the ordered pair (occupying the second "slot"). It is NOT necessary for each element of set S to appear among the slots of the ordered pairs of R, but the ordered pairs of R that do occur must have their first and second slots filled by elements of S in order that R be a relation *on* S.

Exercise 14.1 Symmetry is a Property Describing an Entire Relation—Not Just Parts of It

(i) Suppose a set $S_0 = \{1, 2, 3\}$. Is the following relation R_0 on set S_0 symmetric, where $R_0 = \{(1, 1), (1, 2), (3, 2), (2, 3), (2, 1)\}$?

(ii) In the definition of symmetric relation above (and in many other places in this and other books) you will see an "if" followed later in the sentence by a "then." This language triggers a response in a trained mathematician. In this case, when testing to see if a relation is symmetric, a mathematician looks at each ordered pair which is in the relation. *If* he/she finds a particular ordered pair, say (1,2) in the relation R_0 (as in part (i)), *then* he/she looks to see if its transpose, (2,1), is also in that relation R_0. The answer in this case is "yes." The mathematician does not stop there, but checks every ordered pair in R_0 to see if the corresponding transposed ordered pair is also in R_0. (The answer for each ordered pair is either "yes" the transposed ordered pair is present or "no" it is not.) In order for R_0 to be symmetric, the answer has to be "yes" for each ordered pair in R_0. If the transpose of one (or more) ordered pair(s) in R_0 fails to be in R_0, then R_0 is not symmetric. Did you do part (i) correctly?

(iii) Suppose a set $S_1 = \{A, a, B, 3, \#, w, \&\}$, then $R_1 = \{(A, a), (B, B), (\#, \&)\}$ is a relation on S_1. Is R_1 a symmetric relation?

(iv) With the same set S_1, $R_2 = \{(A, A), (A, a), (B, B), (B, \#), (a, A), (\#, B)\}$ is a relation on S_1. Is R_2 a symmetric relation?

Exercise 14.2 Testing Some Relations for Symmetry

Do not forget. Symmetry of a relation is a property of the entire collection of ordered pairs that make up the relation!

(i) Given the set of Warlpiri people, is the brother relation on this set symmetric? (For the sake of precision let me *define* person p_2 to be the brother of a person p_1 if the following three conditions are satisfied: (1) p_1 and p_2 are not the same person; (2) p_2 is male; and (3) p_1 and p_2 have the same parents. In this case I will say that (p_1, p_2) is in the brother relation.)

Define the sibling relation. (I usually define (q_1, q_2) to be in the sibling relation if $q_1 \neq q_2$ and q_1 and q_2 have the same parents.) Is the sibling relation symmetric on the set of Warlpiri people?

Is the "first cousin" relation on the set of Warlpiri symmetric? (You must first carefully define what it means for an ordered pair of persons to be in the "first cousin" relation. This can easily be done using our definition of the sibling relation—at the parent level.)

Is the father relation on the set of Warlpiri symmetric?

(ii) Given a set of numbers, call it N, we say that the ordered pair of numbers (s_1, s_2), where s_1 and s_2 are in N is in the "equal relation" if $s_1 = s_2$. Is the equal relation symmetric?

(iii) Given a set N' of real numbers (if you are not sure what real numbers are assume N' consists of whole numbers and fractions which are ratios of whole numbers) we say that the ordered pair from N', (s_2, s_1), is in the "less than relation" if s_1 is less than s_2. This is often written in mathese as $s_1 < s_2$. Is the $<$ relation on N' symmetric?

(iv) Why do you think mathematicians have defined the concept of "relation" and its properties using ordered pairs instead of ordinary English?

14.3 Transitive and Reflexive Relations

There are two other famous properties that relations can have: transitivity and reflexivity—which will be useful to know.

Definition: A relation, R, on a set, S, is said to be *transitive* if whenever two ordered pairs of elements from S, (a, b) and (b, c), are both in the relation R, then (a, c) is also in the relation R.

Said differently, a collection R of ordered pairs is transitive *if* whenever you find two ordered pairs in R such that the element (or symbol) in the second slot of one (call it the "first ordered pair") is the same as the element (or symbol) in the first slot of the other (call it the "second ordered pair"), *then* R contains the ordered pair whose first slot is filled with the element in the first slot of the "first ordered pair" and whose second slot is filled with the element in the second slot of the "second ordered pair." A word of caution is called for. If the ordered pairs (b, c) and (a, b) are in the relation R, then R needs to contain (a, c) in order to be transitive. Thus the order in which ordered pairs might be listed in R does not matter. All possible pairs of ordered pairs need to be investigated to determine transitivity.

Again, if you can find a pair of ordered pairs, say (p_0, q_0) and (q_0, s_0), that are in a relation R while ordered pair (p_0, s_0) is NOT in R, then relation R is NOT transitive. A relation is either transitive or it is not transitive.

Finally, the property of transitivity of a relation is a property of the entire relation!

Exercise 14.3 Testing Some Relations for Transitivity

(i) If set $S_2 = \{A, 5, \$, \&, >, b, B\}$, then $R_3 = \{(A, b), (b, b), (\$, \&), (\&, 5)\}$ is a relation on S_2. Is R_3 a transitive relation?

(ii) Given set S_2, $R_4 = \{(A, b), (b, b), (b, A), (\&, \$), (\$, \&), (5, \&), (\&, 5)\}$ is a relation on S_2. Is R_4 a transitive relation? Is R_4 a symmetric relation?

(iii) Let $R_5 = \{(A, b), (b, b), (\&, \&), (\$, \$), (5, 5), (b, A), (\&, \$), (\$, \&), (A, A), (5, \&), (\&, 5)\}$ be a relation on set S_2. Is relation R_5 transitive? If you answered no, what is the least number of ordered pairs which when "added to" R_5 make it transitive? What is (are) it (they)?

(iv) Can you give an example of relation on set S_2 which is transitive but not symmetric?

Definition: A relation, R, on a set, S, is said to be *reflexive* on S if for each a in S the ordered pair (a, a) is in the relation R.

For example, if the set $S = \{a, b, c\}$, then the relation $R = \{(a, a), (b, b)\}$ is NOT reflexive, but the relation $\{(a, a), (b, b), (c, c)\}$ is reflexive.

Exercise 14.4 Reflexivity of Relations

(i) Give an example of a relation on set S_2, above, which is reflexive.

(ii) Give an example of a relation on set S_2, above, which is not reflexive.

Exercise 14.5 Testing Some Relations for Transitivity and Reflexivity

(i) Repeat Exercise 14.2 replacing the word symmetric with the word transitive throughout. In particular, is there a definition of brother relation under which the brother relation is transitive? Is there a definition of brother relation under which the brother relation is not transitive?

(ii) Repeat Exercise 14.2 replacing the word symmetric with the word reflexive throughout.

(iii) Is the "=" relation on a set of numbers symmetric, transitive and reflexive? See Exercise 14.2 part (ii) above.

14.4 Equivalence Relations

Some of the most important relations in mathematics are very similar to the intuitive notion of "=" that we all share. We make this precise with the following.

Definition: A relation on a set S that is transitive, symmetric and reflexive is said to be an *equivalence relation* on S.

Transitivity, symmetry and reflexivity are the three fundamental properties of our intuitive notion of equality. Thus mathematicians use these three properties to abstractly define what they mean by "equality." I will use equivalence relations to help us do some counting exercises in IV.

Exercise 14.6 Some Relations Can Be Drawn in a Euclidean Plane

(i) Let $R_1 = \{(0,0),(1,1),(2,2),(9,9)\}$. Let S be the set of whole numbers {0, 1, 2, 3, 4, 5, 6, 7, 8, 9}. Is R_1 a relation on S?

(ii) Which if any of the three adjectives reflexive, symmetric, transitive apply to R_1?

(iii) Let $R_2 = \{(1,2),(3,2),(2,1),(3,3),(2,3),(1,1),(2,2)\}$ be a relation on the set {1, 2, 3}. Is this relation transitive, symmetric or reflexive?

(iv) We can draw a picture of, or graph, R_1 as in Figure 14.1 below; and we can graph R_2 as in Figure 14.1 below.

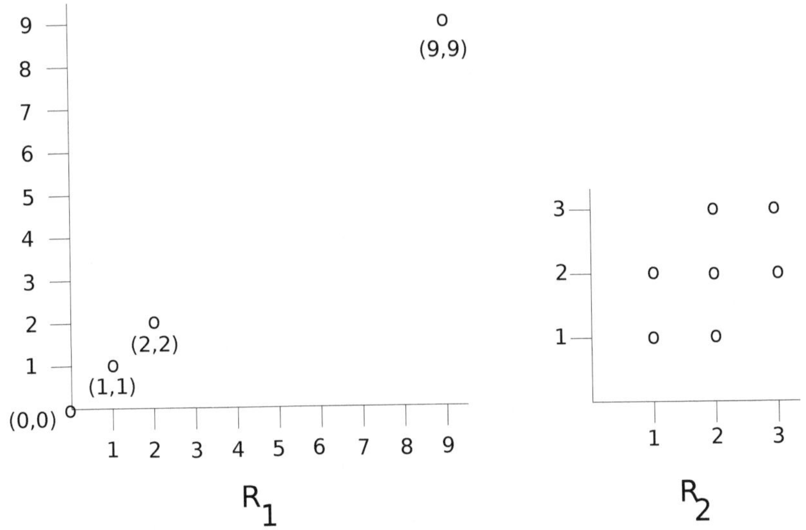

FIGURE 14.1: Graphs of Some Relations in the Plane

Suppose R is a relation on S, and suppose we plot, draw or graph the points of R in the plane as we did for R_1 and R_2 above. Can you give a geometric property that every reflexive relation must satisfy?

(v) As in (iv) can you give a geometric property that every symmetric relation must satisfy?

14.5 Relations That Are Functions

Thinking of Relations as Operations. Given a relation R on a set S, we can think of the relation R in another way. Namely, we can think of R as an *operation.* Thus, define the operation of "applying R" to an element p in S in the following way: R acting on p gives us the collection of elements q in S such that (p, q) is in the relation R. Let's look at what this means in some special cases.

Given a person p in the set of Warlpiri people we can apply the "mother relation" to p and obtain "the mother of p."

Given a person p in the set of Warlpiri people we can apply the "brother relation" to p and obtain "the brothers of p."

There are two differences between the "mother relation" and the "brother relation" that are of supreme importance to mathematicians. First of all, everyone has a mother, so if you apply the mother relation to a person p you will get a non-empty result. There will be a person who exists (or, at least, once existed) who is the mother of p. This is not the case for the brother relation. If you apply the brother relation to me, for example, you get an empty result since I do not have (nor can I ever have) a brother.

Second, and most importantly, the result of applying the mother relation to a person p is unique. Everyone has exactly one mother, "the mother of p" is a unique person, for every p. This situation does not hold for the brother relation, since a person can have any number of brothers from zero, one, two to .... Thus "the brother of p" is not a unique person for every person p (even though *some* people have exactly one brother).

Mathematicians say that the "mother relation" is a *function* defined on the set of all people, because of the two properties immediately above. Namely, the "mother relation" can be applied to any person and the (non-empty) result is unique. The "brother relation" is not a function on the set of all people.

We define a function precisely as follows. (For additional discussion of the concept of function see Exercise 15.10.)

Definition: A *function* on a set S is a relation F on set S which satisfies the following two conditions:

(1) For each element x in set S, there is an element, which we call $F[x]$ in S, so that $(x, F[x])$ is in the relation F.

(2) Given x in S, there is exactly one y in S such that (x, y) is in F, and $y = F[x]$.

Intuitively part (1) of the definition of a function on a set S tells us that function F "knows how to operate on" or "is allowed to operate on each x in S." The set of elements upon which F acts is called the *Domain* of F. In this case $Domain(F) = S$. Part (2) of the definition says that when a function F acts upon a particular x in its domain, it produces exactly one result, namely, $F[x]$.

Consider the following example: the "square root relation." Given a number like 4, we can apply the square root relation to it. This means find all numbers q such that the ordered pair $(4, q)$ is in the square root relation, i.e., q is a square root of 4, or said another way, $q^2 = 4$. The answer to this problem is either $q = +2$ or $q = -2$. Thus, the ordered pairs $(4, 2)$ and $(4, -2)$ are precisely the ordered pairs with the number 4 in "first position" which are in the square root relation; and thus if we apply the square root relation (thinking of it as an operation) to 4, we get the set of numbers 2 and -2. In mathese this set is written $\{2, -2\}$.

Recall that when we want to describe a set,[8] order is not to be considered. We write $\{\dots\}$, i.e., we put curly brackets around the list of elements in question. Thus $(2, -2)$ is an ordered pair which is not equal to, i.e., is not the same ordered pair as, $(-2, 2)$. Whereas $\{2, -2\} = \{-2, 2\}$ is a set, which is an unordered collection with two elements in it.

Exercise 14.7 Testing Some Relations to See If They are Functions

(i) If you apply the cousin relation to a person p, might you obtain a set of people with more than one person in it? With no people in it? Is the "cousin relation" a function on the set of all people?

(ii) Repeat part one but replace "cousin" with "father." How is the father relation, as an operation, different from the cousin relation? (Here and elsewhere in this Part III the terms "father" and "mother" are used/defined in the classical, biological sense. For the sake of simplicity I will ignore complications introduced by transgender, cloning, surrogacy and so on—all worthy topics for independent study.)

(iii) Is the "sibling of" relation a function on the set of people who have siblings? Is the "sibling of" relation a function on the set of all people?

(iv) Thinking of the cube root relation as an operation to be applied to (real) numbers, apply it to the numbers 8, 27 and -125. What (real number) answers do you get? How is the cube root relation (applied to real numbers) different, as an operation, from the square root relation?[9]

---

[8] See footnote 4.

[9] A cube root of a number X is a number Y which when cubed, i.e., $Y * Y * Y = Y^3$, yields X. We write $X^{\frac{1}{3}} = Y$. Thus Y is a cube root of X if $Y^3 = X$. In studying cube roots of numbers we must be careful. If, for example, you know about *complex numbers*, i.e., numbers of the form $A + iB$, where $i = \sqrt{-1}$, and A and B are so-called *real numbers*, then you probably know that there are three complex-number-cube roots of any complex number. Thus there are three complex number cube roots of real number 8: since 8 can be thought of as the complex number, $8+i0$. These three cube roots are 2, $-1+i\sqrt{3}$, $-1-i\sqrt{3}$. Only one of these three cube roots is real, i.e., has no "i" in it. When doing this exercise work only with real numbers. Thus there will be only one real cube root of each of the three numbers 8, 27 and -125.

Chapter 15

A Set of Social Rules for the Warlpiri People

I am now going to describe the actual set of social rules that the Warlpiri people enforce upon themselves, rules which among other things determine the kinship relationships among these people. [1]

15.1 The Section Rule

The first rule is that every Warlpiri person belongs to one and only one of eight, so-called, Sections, or as some Aborigines say in English, "Skins." The original labels in Warlpiri are without doubt more colorful and informative,[2] for each individual. Persons are held immediately responsible for the effects of their actions on other individuals and on the environment in general. It is a major weakness of my culture that individuals can escape responsibility for their actions, be it incorporating toxic waste into fertilizer for food plants, toxic dumping, all manners of abuse of other people and/or other species or of the earth itself. The assignment of a section or skin has significance beyond what we have room to discuss here. One reference for further information is W. H. Edwards, *An Introduction to Aboriginal Societies*, Social Science Press, P.O. Box 89, Wentworth Falls, New South Wales 2782, Australia, 1996. Of course, another way to learn about Aborigines is to experience them directly,

[1]This information can be found in [17, Chapter 3]. This book has information about mathematics in "other" cultures. I also learned this subject first-hand in Australia.

[2]It is important to note that we are just scratching the surface here of a deep and complex world view which evolved over thousands of years of intimate interaction with the environment. The actual names of the eight "skins" are: Napaljarri (Japaljarri), Napangardi (Japangardi), Nakamarra (Jakamarra), Nungarrayi (Jungarrayi), Napurrurla (Japurrurla), Nangala (Jangala), Napanangka (Japanangka), Nampijinpa (Jampijinpa), where "N" denotes a female and "J" denotes a male. One of the consequences of being assigned a "skin" at birth is that a person assumes responsibilities for some aspects of the environment, i.e., the person must learn all of the "stories," history, function, status and so on of that part of the environment—and they are responsible for maintaining the "health" of that part of the environment. In the language of Exercise 15.9, the traditional Aborigines have a cultural way of closing the *responsibility gap*

they are very much still here. but I shall apply a labeling method common in my culture and call these sections: $S_1, S_2, S_3, S_4, S_5, S_6, S_7, S_8$.

15.2 The Mother Relation Rules

These next rules determine how the "mother relation" (or mother function), M, relates to the eight sections. The mother of a person in S_1 is in S_2. Said another way, the Warlpiri assign all the children of a female in S_2 to be in S_1. This mother relation, acting as a function on S_1, can be denoted in (Western culture influenced) mathese[3] as: $M[S_1] = S_2$. A mathematician might read $M[S_1] = S_2$ in English as "M of Section 1 is Section 2."

The complete list of the "mother function rules" are:

$$M[S_1] = S_2, \quad M[S_2] = S_3, \quad M[S_3] = S_4, \quad M[S_4] = S_1.$$

$$M[S_5] = S_8, \quad M[S_8] = S_7, \quad M[S_7] = S_6, \quad M[S_6] = S_5.$$

Note that we have two cycles of length four. Anthropologists call each of these cycles a *matricycle*. See Figure 15.1, where the arrow from, say, S_2 to S_1 can be interpreted as "S_2 gives birth to S_1." The arrows going from S_8 to S_5 can be interpreted as "S_8 gives birth to S_5," and so on.

15.3 The Marriage Rules

A person in S_1 may only marry a person in S_5, and vice versa. A person in S_2 may only marry a person in S_6, and vice versa. A person in S_3 may only marry a person in S_7, and vice versa. A person in S_4 may only marry a person in S_8, and vice versa. Only married couples are allowed to have children.

We summarize the mother relation rules and the marriage rules in Figure 15.1 which is hopefully self-explanatory.

[3] The alert reader will notice here that if we take all the mothers of all the people in Section 1, the result most likely will not be all the people in Section 2. For example, there will possibly be males in Section 2. We could correct this obvious problem by using $\subseteq$ or "contained in" instead of $=$, but let us simplify our lives for the moment and just declare that the "mother of Section 1" is Section 2.

$$
\begin{array}{cccccc}
\leftarrow & S_1 & \longleftrightarrow & S_5 & \leftarrow & \\
\downarrow & \uparrow & & \downarrow & \uparrow & \\
\downarrow & S_2 & \longleftrightarrow & S_6 & \uparrow & \\
\downarrow & \uparrow & & \downarrow & \uparrow & \\
\downarrow & S_3 & \longleftrightarrow & S_7 & \uparrow & \\
\downarrow & \uparrow & & \downarrow & \uparrow & \\
\rightarrow & S_4 & \longleftrightarrow & S_8 & \rightarrow &
\end{array}
$$

FIGURE 15.1: Warlpiri Kinship Relation Rules

15.4 The Father Relation Rules

It follows logically from the rules above that the father relation, F, must satisfy certain rules. Let's figure out what these rules are. Suppose that you wanted to find out which section is the "father of Section 7." Well, we know from the mother relations that the mother of a person in Section 7 is in Section 6. From the marriage rules, the husband of the mother in Section 6 must come from Section 2. Why? Thus, the father of a person in Section 7 is in Section 2. We abbreviate, and/or simplify this, by stating in mathese that $F[S_7] = S_2$, i.e., that the "father of Section 7 is Section 2."

Exercise 15.1 The Father Relation Rules

(i) Deduce the father relation rule: $F[S_2] = S_7$. Said in perhaps a less intimidating way: is it true that $F[S_2] = S_7$? And if it is true, how do you know?

(ii) Deduce that there are 4 "father cycles," or *patricycles*, of length 2, namely: $F[S_1] = S_6, F[S_6] = S_1$; $F[S_2] = S_7, F[S_7] = S_2$; $F[S_3] = S_8, F[S_8] = S_3$; and $F[S_4] = S_5, F[S_5] = S_4$.

Exercise 15.2 Commutativity vs. Noncommutativity

(i) Pick a section, say S_1, and compute the "mother of the father of Section 1." In mathese we would write $M[F[S_1]]$.

(ii) Now compute the "father of the mother of Section 1." In mathese we would write $F[M[S_1]]$.

(iii) Compare your answers to parts (i) and (ii). Are they different? Does the order that you perform the mother and father operations make a difference? If the order in which two operations are performed makes a difference, then we say that the two operations do not *commute*. If the order in which two operations are performed does not make a difference then we say that the operations *commute*. Do the operations M, F commute or not?

(iv) Think of some operations that do commute.[4]

Exercise 15.3 Understanding the Warlpiri Multiplication Table

(i) Pick a section, say S_3, and compute the "father of the mother of the father of Section 3." Precisely stated in mathese, compute: $F[M[F[S_3]]]$.

(ii) Compute $M[M[M[S_3]]]$. In a shorthand mathese this last calculation is often written: $M^3[S_3]$.

[4]Hint: There are some operations you can do with numbers that commute.

(iii) Do you get the same result for (i) and (ii)?

(iv) Repeat the above three parts, but substitute in turn each of the other seven sections for S_3. In other words, compute $F[M[F[S_k]]]$, where $k = 1, 2, 3, 4, 5, 6, 7, 8$. Also compute $M[M[M[S_k]]]$, where $k = 1, 2, 3, 4, 5, 6, 7, 8$. Then see if $F[M[F[S_k]]] = M[M[M[S_k]]]$ for each value of $k = 1, 2, 3, 4, 5, 6, 7, 8$. If your answer to (iii) is "yes" for all eight sections, then we say that $FMF = M^3$. Is this true?[5]

There are 64 exercises that you could do like the last one. I summarize the results of these exercises in the following table.

TABLE 15.1: The Warlpiri Kinship Multiplication Table

	I	M	M^2	M^3	F	MF	M^2F	M^3F
I	I	M	M^2	M^3	F	MF	M^2F	M^3F
M^3	M^3	I	M	M^2	M^3F	F	MF	M^2F
M^2	M^2	M^3	I	M	M^2F	M^3F	F	MF
M	M	M^2	M^3	I	MF	M^2F	M^3F	F
F	F	M^3F	M^2F	MF	I	M^3	M^2	M
MF	MF	F	M^3F	M^2F	M	I	M^3	M^2
M^2F	M^2F	MF	F	M^3F	M^2	M	I	M^3
M^3F	M^3F	M^2F	MF	F	M^3	M^2	M	I

We have created the table above in such a way that the intersection of the ith row with the jth column contains the operation obtained by first performing the operation at the top of the jth column, followed by the operation at the left of the ith row. This table can be considered a type of "multiplication table."

Let's use the above table to multiply, say, M, with M^3, in that order, i.e., find MM^3. First find the row whose first entry (the entry furthest to the left) is M. Then find the column whose top (or first) entry is M^3. The entry in the box where "the row that begins with M" and the "column that begins with M^3" intersect is I. Thus $MM^3 = I$. Now I is the word that does not move any section, i.e., $I[S_k] = S_k$, for k=1,2,3,4,5,6,7,8. We say that I is the "identity operation." Indeed, we learned that $MM^3[S_k] = M[M[M[M[S_k]]]] = M^4[S_k] = S_k = I[S_k]$ for all sections S_k. (See the Mother Relation Rules in Section 15.2.)

For our next example let's calculate FM. Find the "row that begins with F" and the "column that begins with M," and we see they intersect in the box containing M^3F. Thus $FM = M^3F$. It turns out that this is a very

[5]It turns out that if two "words" in M and F, like FMF and M^3 agree on one section, for example, $F[M[F[S_3]]] = M^3[S_3]$, then they will agree on all eight sections automatically. This is not necessarily easy to see at this stage, but it is true. Can you see why?

important fact, which can be used to simplify any word with a finite number of Ms and Fs into one of the eight words: I, M, M^2, M^3, F, MF, M^2F, M^3F, cf., Exercise 15.5.

Consider the intersection of the row that begins with F and the column that begins with MF, which is a box containing M^3. Thus $F[MF] = M^3$, which is what we discovered in Exercise 15.3 above.

There is a property that is "hidden" in the above "Warlpiri multiplication table" that mathematicians call *associativity.* Associativity of the multiplication in this table implies, for example, that $F(MF) = (FM)F$. (See the next exercise for the complete definition of associativity. Formally, purely in terms of symbols, it looks like the associative law for the multiplication of numbers—although the symbols being "multiplied" no longer stand for numbers.) If our Warlpiri multiplication is associative then it must be that if $F(MF) = M^3$, then $(FM)F = M^3$ as well. This is, indeed, the case as we can see from the following calculation. First of all, $FM = M^3F$ as we saw above. Then $(FM)F = (M^3F)F = M^3$ as can be seen from the table (see the intersection of the bottom row which has first entry M^3F, with the column headed by F).

Exercise 15.4 Associativity of Warlpiri Multiplication

(i) Think of the table above, in Table 15.1, as an abstract multiplication table. Let X be any one of the eight symbols: $I, M, M^2, M^3, F, MF, M^2F, M^3F$. Similarly let Y be any one of the same eight symbols. Also, let Z be any one of the same eight symbols. Compute $X(YZ)$, and compute $(XY)Z$, noting that you perform the multiplications inside parentheses first. Do this for as many choices of X, Y and Z as you can stand to do. What do you observe?

(ii) If you can demonstrate that $X(YZ) = (XY)Z$ for all possible substitutions of the eight symbols in for X, Y and Z, then we say that the multiplication represented by our table is *associative.* Can you so demonstrate?

(iii) Can you calculate how many possible $X(YZ) = (XY)Z$ equations of the form encountered in (ii) there are? We will learn how to systematically answer this question in IV. One way to prove that the multiplication table of Warlpiri kin relations is associative is to check all of these possibilities.

(iv) Can you think of an abstract way to demonstrate that our "Warlpiri multiplication" is associative, without laboriously checking all possibilities individually? Besides being "slick," the answer to this exercise shows that a mathematical insight can save a great deal of computation and provide a deeper understanding of a given situation. Thus the so-called "abstract math" of today can be very useful today, and it becomes the "concrete (everyday) math" of tomorrow.[6]

Thus, Warlpiri multiplication is, indeed, associative. This means that $(XY)Z = X(YZ)$ no matter what "Warlpiri expressions" we plug in for X, Y and Z. You can get a better idea what this means by doing the following exercise.

[6]Hint: Consider $X(YZ)[S_k] = X[Y[Z[S_k]]] = (XY)Z[S_k]$, for k=1,2,3,4,5,6,7,8. Note that in Exercise 15.3 I calculated FMF with no parentheses, so the answer should be independent of how the symbols are grouped.

Exercise 15.5 Associativity Can Help Simplify Things

(i) Simplify $(MF)F$ to an expression with one letter.

(ii) Simplify $((MM)(MF))(FM)$ as much as you can.

(iii) In part (ii), explain why knowing that the associative law holds for "Warlpiri expressions" allows us to make unambiguous sense out of $MMMFFM$, i.e., the same expression as in part (ii) without all the parentheses.

(iv) Simplify $MMFMFMFMMFMFMMMFFFFMMFFMFM$ as much as you can. Hint: Associativity implies you can group the symbols (without changing their order!!) any way you wish. Also you do not need to refer to Table 15.1 if you remember to use $FM = M^3F$ repeatedly.

(v) It is very interesting to note that you have just been dealing with an equivalence relation[7] in the following sense. Any string of letters made up of a finite number of Ms and Fs can be simplified by using Table 15.1 to, i.e., is equal to, one of the eight strings: $I, M, M^2, M^3, F, MF, M^2F, M^3F$. Is this notion of "equal" transitive, symmetric and reflexive?

Exercise 15.6 The Spouse Relation

(i) There is a relation that we can call the *Spouse* relation. From the marriage rules we know that the "Spouse of Section 1" is Section 5 and vice versa. Similarly, the "Spouse of Section 2" is Section 6 and vice versa; the "Spouse of Section 3" is Section 7 and vice versa; the "Spouse of Section 4" is Section 8 and vice versa. Find a formula in terms of M and F for the operation "Spouse of." Hint: Find a sequence of M and F operations that take S_1 to S_5 and then check to see if it works in all other cases.

(ii) Is the Spouse relation, when thought of as acting on the set of eight sections, a function, assuming the Warlpiri multiplication rules?

15.5 Cultural Contexts in Which Mathematics Is Done

There are observations to be made, questions to be asked. There are things to be learned from the kin structure of the Warlpiri people. First of all, if you found the foregoing discussion the least bit complicated maybe you have found a new respect for the possible level of complexity of so-called primitive cultures.[8] If the anthropologists who claim that the Australian Aborigines have been around for nearly 50,000 years are even remotely correct, then it is highly likely that the Warlpiri ancestors were doing this relatively complicated "kinship mathematics" many thousands of years before the Greeks, the Egyptians, the Babylonians or the Chinese existed, as we currently define them. In fact, the last Neanderthals likely died out in Europe only 35,000

[7] This equivalence relation can be defined as follows. Any two "words," i.e., strings, W_1 and W_2, in the letters M and/or F, are equivalent if they can be transformed one into the other using the rules of the multiplication table. It turns out that you only need three of the rules of the multiplication table, viz., $FM = M^3F$, $F^2 = I$ and $M^4 = I$. Any "word" in M and/or F is equivalent to one of the eight words in the top row of the Warlpiri Multiplication Table.

[8] The Warlpiri kin structure based on 8 sections is not entirely alone. Certain people in Polynesia developed kin structures based on 6 sections, for example.

years ago. During the last ice age the ancient Australians were witness to a sea-level that was as much as 160 meters lower than it is today! Between 15,000 and 8,000 years ago the last ice age was retreating from Europe, cf., [192, p. 182].

I find the ancient Australian culture astounding if, for no other reason, than that it has lasted so long. Any culture that can maintain itself for many thousands of years "knows" something that the culture in which I find myself probably does not. Are there any radical changes in store for us in the near future?

Why Did the Warlpiri Develop Their Kinship Relation Rules? An obvious question is: Why did the Warlpiri develop such a complicated kinship system? I have presented only a tiny piece of the Warlpiri world filtered through a brain which undoubtedly does not fully understand the Warlpiri world view. Anthropologists tell us that the kinship system of the Warlpiri determined obligations and behaviors that linked these people to each other, to the past, to the future, and to the land and all its animate and inanimate inhabitants. I think that perhaps the most important lesson to be learned from this time-tested culture is that it is smart to think of humans as part of all life, not dominant over all life and certainly not the center of the biological universe, cf., the Bio-Copernican Axiom.

There is, however, one very simply stated and biologically important problem that the Warlpiri kinship system solves: inbreeding. The Warlpiri lived in groups of ten to sixty people, periodically coalescing temporarily into larger groups. If mates had been selected randomly there would have been a good chance one would marry someone too closely related for genetic comfort—as science now tells us.[9] Such a culture would not long have survived. It is tautological but not pointless to say that current Aborigines are the descendants of those who figured out how to survive, and a bit of mathematical structure was an essential part of the picture. The following exercise illustrates how the Warlpiri prevented inbreeding.

Exercise 15.7 Who Can a Warlpiri Not Marry?

In the following exercises do more than answer "yes" or "no." Give an argument that demonstrates why your answer is correct, based on the Warlpiri Kinship Relations (except for the last exercise which is independent of such relations).

(i) Is it possible for a Warlpiri male to marry his sister if all of the social rules are followed?

[9] There is some mathematics here. Each person inherits half his/her genetics from each parent. Let's look at one such possibly inherited pair of genes, call them "A" and "a" for dominant and recessive, respectively, with regard to some trait. Unless one inherits an "a" from both parents, the recessive gene is not expressed. When this recessive gene is expressed the individual suffers some mental or physical disability. Thus if father carries (A,a) and mother carries (a,A) there are four possibilities for the child: (A,A), (A,a),(a,A), and (a,a). Thus one out of four children on average will have a recessive gene related dysfunction. Genetic problems worsen with each generation. There is, as in every section of this book, a lot more to study, discover and understand. I will leave that for you.

(ii) Is it possible for a Warlpiri to marry a parent if all of the social rules are followed?

(iii) Is it possible for a Warlpiri to marry an aunt or an uncle if all of the social rules are followed?

(iv) Is it possible for a Warlpiri to marry a first cousin if all of the social rules are followed?

(v) Can you find a formula, or formulas, in terms of the letters M and F for the relation *cousin of*? By this I mean that if you apply one of these formulas to a section, say S_i, it will produce a section that contains some of the first cousins of the people in S_i.[10]

(vi) Is the relation *cousin of* a function when thought of as acting on the set of eight sections?

(vii) Is it possible for a Warlpiri to marry a grandparent? More precisely, can a male marry his grandmother? Can a female marry her grandfather? If one is allowed to marry a grandparent, is this likely to occur in real life?

(viii) Define the Earth Mother operation which assigns to each human the Earth as his/her Earth Mother. Is this operation a function on the set of humans?

The Warlpiri solved the problem of inbreeding by following a complex set of rules that at first sight may seem foolish to us. The Australian Aborigines apparently avoided other potentially lethal problems—some problems likely of their own making—by adjusting their culture to fit their environment.

Patterns in Fire Use in Australia and America. There is persuasive evidence that many tens of thousands of years ago the Aborigines hunted (possibly using fires as a hunting technique) giant, herbivorous marsupials, like the diprotodon, to extinction. Massive accumulation of vegetative material not fueling diprotodons exploded into eco-altering fires instead, cf., [192].

If the theory of my Colorado colleague in geology, Professor Gifford S. Miller, is correct, the flow of water, indeed, the very climate, over most of the continent was changed, at least in part, as a result of these massive fires. The Aborigines responded by developing "fire-stick farming," or what we might call "controlled burns."[11] This technique made the best of a diprotodonless landscape by creating a floral mosaic which preserved and promoted the remaining biodiversity that the Aborigines needed to survive—while preventing massive, destructive fires, cf., [387].

The influx of Europeans inexperienced with the poor Australian soils and the weather visited upon Australia by ENSO[12] did not see the wisdom of fire-stick farming. We periodically read about the results, as massive bush fires burn in Australia. Not well known is that Americans are only now slowly rediscovering our own Native American version of "fire-stick" forest management, after over a century of ignoring it in favor of fire suppression.

Thus a couple patterns are apparent. First, indigenous peoples on opposite sides of the earth reached similar conclusions, that periodic small fires tend

[10]Hint: I is one such formula, there is one other formula. It helps if you have already worked Exercise 15.6.

[11]Professor Miller believes that uncontrolled (pre-"fire stick") Aboriginal caused fires used for hunting and other purposes caused the extinction of many animals.

[12]El Niño Southern Oscillation is a complex Pacific ocean-atmosphere phenomenon which periodically alters weather worldwide, but particularly in Australia.

to avoid catastrophic fires. Next, this knowledge was displaced in favor of an attempt to control, what is ultimately, an insuppressible force of Nature. After over a century of avoiding the inevitable, controlled burning is slowly coming back into the picture. Are there any other contemporary conflicts between human culture and forces of Nature?

Exercise 15.8 Fire Suppression Versus the "Indian Way"

In a May 2001 interview with Ed Marston, then editor of High Country News (www.hcn.org), historian Stephen J. Pyne, cf., [558, Chapter 2], recounts the impact of European style forestry on America which is reminiscent of the situation in Australia mentioned above. See also [738] which addresses 10 assumptions about forest management.

(i) Look up the controversy, going back to the 19th century, between the following two philosophies: (1) the view of the U.S. Forest Service-Gifford Pinchot-Yale University-European school of thought which held that the way to manage/protect forests is to control/exclude/suppress all fire—all fire is bad; (2) the "Indian way" of forest management which held that proper care entailed lightly burning the landscape very regularly.

Interestingly, about 100 years ago many timber owners, novelists, poets, the state engineer of California, Southern Pacific Railroad and so on supported the "Indian way." The U.S. Forest Service won the argument with academic backing.

(ii) What has been the result of fire suppression for the past century?

(iii) Can fires be suppressed indefinitely? Why or Why not?

(iv) Can humans "manage" fires and forests? How would you change current practices?

(v) What is a pulaski? After whom is it named?

(vi) What effects are there from logging and/or application of herbicides in forests?

(vii) To get some idea of the role of fire in Ancient Forest Ecosystems read [445], and comment.

(viii) Look in the literature for mathematical models of forest fires. What kind of mathematics is used?[13]

(ix) Dr. Pyne was asked a question at the end of a lecture in 2009 as to why Greece had just experienced some dangerous fires that threatened urban areas—when such fires had not occurred in recent memory. Briefly, Pyne's answer was that the areas that had burned had been under intense human agricultural management for a very long time. Recently there had been an exodus from these areas to cities, leaving these areas unmanaged and susceptible to fire. Thus once humans decide to "manage" an ecosystem according to certain rules, humans are thus committed to continue such management; failure to do so could have unintended consequences. Can you think of any other current situations to which this logic applies?

Might there be some adjustments in our culture that should be made so we better fit our environment? What do you suppose these cultural changes might entail?

Everyone lives and thinks in some cultural context. Every culture is replete with assumptions, many hidden, that affect how life is lived. In order to live from day to day the problems of life have to be solved. Every culture has its own solutions—its own values and priorities. This was humorously brought home to me once as I ended a hike in an Australian national park, guided by two Pitjantjatjara Aborigines. Our hike ended near a spot with a good view of a sacred rock, Ulru; but about 100 yards from the bus parking lot where everyone on the hike, except the Aborigines, had to eventually end up.

[13] Hint: Words like percolation theory, stochastic process and others come up.

I asked why the Aboriginal trail that we were on did not go to the parking lot. The very logical answer was this: The trail has been in place performing its proper functions since the Dreamtime long before there were buses and parking lots. Soon the buses will be gone. The trail must continue to perform its functions relative to the natural landscape now and long after the buses and parking lots are gone, so there is no need to move it.

In some sense, since mathematics is a human creation, there is no such thing as "pure" mathematics, mathematics done in some Platonic isolation chamber devoid of cultural context.[14] There is abstract mathematics, fun mathematics and important mathematics; but one must always be aware of how the culture in which the mathematics is being done is affecting the mathematics. At the moment, in the culture in which I exist, the most appreciated mathematics, in fact the most appreciated thought, is that which can attract funding, i.e., money. This effect on modern mathematics, and modern thought, is nontrivial.

Exercise 15.9 Actions, Responsibility, Nature's Feedback and the Gap Between

(i) I was once told that a certain Aboriginal tribe believed in a "yam god." One of the practical consequences of this belief was this. When one found an edible yam, one was not supposed to eat more than, say, one-third of the yam. To eat more than one-third risked the wrath of the yam god, and the yam god might take away the yam, i.e., the yam plant might die. Discuss how the belief in a yam god is a cultural tool of the Aborigines that has helped the Aborigines survive.

(ii) One can think of the death of a yam plant which has been "over-harvested" as an example of Nature giving *feedback* to the Aborigines. How much time do you estimate might elapse between the act of over-harvesting a yam and the death of the yam?

(iii) An Aborigine who obeys the "one-third rule" of the yam god can be said to be *responsible* relative to the yam. Your answer to part (ii) above is the estimated time for Nature's feedback to be felt after irresponsible behavior relative to a yam. By (my) definition, a *responsibility gap* exists whenever a person does not very soon experience the consequences of an action taken by that person. We can attempt to measure the responsibility gap between an action and its consequences by using concepts of distance, time, money, social or personal suffering or benefit and so on.

Discuss how modern agricultural technology can lengthen the time for Nature's feedback to be felt, e.g., increase the size of the responsibility gap between our actions in growing food and the consequences of those actions.

(iv) Discuss how agricultural technology can disperse the effects of Nature's feedback among a large group of people.

(v) Are there any people committing irresponsible acts today while remaining insulated from the consequences of those acts?

(vi) Do you think that clever technology or social structures can prevent humans from feeling Nature's feedback for irresponsible acts indefinitely?

[14]For example, because of cultural differences "fuzzy" or "measured" mathematics, cf., Chapter 8, was accepted and applied sooner in Asian cultures than in western cultures. In a real sense the abstract mathematics itself is the same everywhere (up to isomorphism, cf., Exercise 15.12), but its role in a society depends on the culture of that society. Also, different cultures can emphasize different parts of mathematics. For example, in some cultures *set* is taken as a fundamental undefined term whereas in others *function* (or functor) is taken as a fundamental undefined term. This reflects a difference in cultural emphasis on "states" vs. "processes."

Exercise 15.10 Some Exercises with Functions

In this exercise I want to generalize the notion of relation slightly and then look at a special case of that generalization: the notion of a function. The function notion or idea has proved to be one of the most important in modern mathematics. In one approach to modern mathematics the term *function* is taken to be undefined. In this approach a function, F, is intuitively understood to be an "operation" or "formula" or "transformation." The key point is that if an object s is "fed" to F, then F operates on s, or acts on s or transforms s into a new object which is called $F[s]$; and this $F[s]$ is *uniquely* determined by s.[15]

To clarify what is going on here let's look at two examples. First, denote by H the "squaring" operation applied to numbers. Thus, if s is a number, like 5, then $H[5] = 5^2 = 25$. More generally, $H[s] = s^2$.

Second, let G be the operation of taking the square root of a number. Thus if s is a number, like 25, then $G[25] = 5$ or -5.

(i) Using the intuitive notion of function as an operation that produces a unique result each time it is "fed" something it knows how to operate on, tell me: Is H a function? Is G a function?

(ii) Is the relation F, "father of," a function defined on the set of all humans?

(iii) Is the relation "cousin of" a function defined on the set of all humans?

We do not have to accept the term function as undefined; we now show some of the details of how we can define the notion of function ultimately in terms of sets (and members of sets). Recall that we have taken the terms *set, member of a set* to be undefined in this course. If you do not want to know these details, you can still read the rest of this book if you at least understand the intuitive notion of function that I have just given you. (In fact, some students have successfully read most of the rest of these pages with a less than perfect understanding of even the intuitive notion of function.)

Definition: A *relation* R from a set S to a set T is a collection of ordered pairs of the form (s, t) where s is in S and t is in T. (Any collection of such ordered pairs is allowed. And, in particular, the set S can be the same as the set T.)

As we did in the Section 14.5, we can still think of such a relation R as acting on a member s_0 in S and producing the collection of all members t in T that occur as the second component of an ordered pair (s_0, t) which is in the R relation.

Definition: A relation F from a set S to a set T is a *function* from set S to set T, if the following two conditions hold:

(1) For each s_0 in set S, there is an ordered pair in the relation F of the form (s_0, t) for at least one t in set T; (Intuitively, F "acts on" each element of S. In particular, when F acts on s_0, it produces element t. We say $t = F[s_0]$. The set, S, upon which F acts is called the *domain* of F.)

(2) For each s_0 in set S, there is no more than one t in set T such that the ordered pair (s_0, t) is in the relation F. (Intuitively, when F acts on an element, s_0 in set S, it produces a *unique* result, t, that is, $t = F[s_0]$.)

If relation F is a function from set S to set T, then F "acts" or "operates" on each s in set S and each ordered pair in F can be written in the form $(s, F[s])$, where $F[s]$ is in set T and is uniquely determined by s.

When mathematicians want to emphasize the "action" or "operation" aspect of a function F from a set S to a set T, they write $F : S \longrightarrow T$.

(iv) Create an example of a relation that is NOT a function. Create a second such example.

(v) Create an example of a function. Create a second such example.

(vi) Is there any cultural bias involved in taking the terms "set" and "member of a set" as the fundamental undefined terms in mathematics? Is there a different cultural bias in

[15] In English, the symbols $F[s]$ are read "F of s."

taking the term "function" to be a fundamentally undefined term in terms of which the concept of set is defined?[16]

A function $F : S \longrightarrow T$ is by definition *one-to-one* if is satisfies the property: If s_1 and s_2 are in S, then $F[s_1] = F[s_2]$ implies that $s_1 = s_2$. If you find this definition of one-to-one confusing, it can be said in another way which is logically equivalent. Namely, if s_1 and s_2 are two distinct (that is, unequal or different) elements of S, then $F[s_1]$ and $F[s_2]$ are also distinct (that is, unequal or different). More succinctly, if $s_1 \neq s_2$, then $F[s_1] \neq F[s_2]$.

(vii) Explain why the defining property just given for a function F to be one-to-one, i.e., $F[s_1] = F[s_2]$ implies that $s_1 = s_2$, is or is not the same as part (2) in the definition of function given above.

(viii) Is the "mother of" function one-to-one?

(ix) Is the "squaring function," i.e., the function H that satisfies $H(x) = x^2$ for all numbers x, one-to-one?

(x) Is the function $K[n] = n^3$, where n is a whole number, a one-to-one function? (What are you taking as the domain of K?

A function $F : S \longrightarrow T$ is by definition *onto* set T if for each t in T, there is an s in S which satisfies $F[s] = t$.

(xi) Give a set which the "mother of" function is onto.

(xii) Give a set which the "squaring" function in (ix) is onto.

(xiii) Give a set which the function in (x) is onto.

Exercise 15.11 The Mathematical Concept of a Group

In this exercise we define and give an example of the mathematical concept of a group. It is not as hard as you might expect, since you already know at least one, probably more than one, example of a group—without, perhaps, ever having used the term.

Definition: A *group* is a set G together with an operation, $\circ$, (which we will call "multiplication," but in many examples this operation is called "addition") which satisfies the following properties:

(1) Given any two members g_1 and g_2 of G, $g_1 \circ g_2$ makes sense, i.e., is defined and is a member of G. This property is called the *closure* of operation $\circ$.

(2) There is an element, called I, in G which satisfies the following: $I \circ g = g = g \circ I$ for all g in G. This property is called the *existence of an identity* for operation $\circ$.

(3) For each member g of G, there is member g^{-1} of G that satisfies: $g \circ g^{-1} = I = g^{-1} \circ g$. This property is called the *existence of the inverse* with respect to $\circ$.

(4) For any three members g_1, g_2, g_3 of G (not necessarily distinct) the following is satisfied: $g_1 \circ (g_2 \circ g_3) = (g_1 \circ g_2) \circ g_3$. Note that operations within parentheses are performed first. This property is called the *associativity* of $\circ$.

(i) Let $G = \{I, M, M^2, M^3, F, MF, M^2F, M^3F\}$ where $\circ$ is the "Warlpiri multiplication" defined in Table 15.1. Show that G with $\circ$ is a group. The set G with $\circ$ is called the Warlpiri group.

(ii) Suppose $G = \{0, 1, -1, 2, -2, 3, -3, \ldots\}$, that is, G is the set of integers or otherwise said the set of both positive and negative whole numbers and 0. Also suppose $\circ$ is the usual operation of addition of such numbers, and that the inverse of g in G with respect to $+$ is $-g$, i.e., "minus g." Is this G a group?

Exercise 15.12 An Example of an Isomorphism

In this exercise we discover another example of a group. Consider a square as in Figure 15.2.

[16]Compare endnote [14].

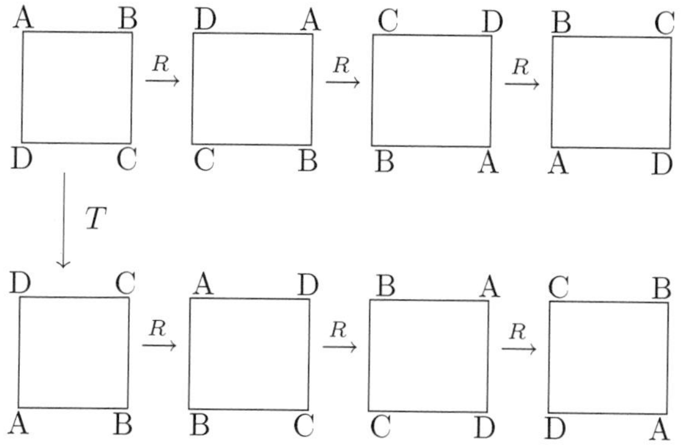

FIGURE 15.2: Isometries of a Square

Let R denote the operation of rotating this square ninety degrees clockwise about its center. Let T denote the operation of reflecting the square through a horizontal line which is drawn through the center of the square.

(i) Let us write $\circ$ between two operations to indicate that these operations have been performed one after the other. Thus $R \circ R$ (often written simply as R^2) is the operation of rotating the square clockwise by 180 degrees, $R \circ R \circ R$ (often written simply as R^3) is the operation of rotating the square clockwise by 270 degrees, and so forth. What geometric operation on the square is represented by $R \circ T$? By $R^2 \circ T$ (often written simply as R^2T)?

(ii) Let $G = \{I, R, R^2, R^3, T, RT, R^2T, R^3T\}$, where I is the identity operation on the square that leaves it unchanged, R^2 is the operation of rotating the square 180 degrees clockwise, R^3 is the operation of rotating the square 270 degrees clockwise, and T is the reflection mentioned above. What geometric (symmetry) operation on the square results from first performing R, then doing T, i.e., TR? What is the geometric interpretation of TR^2 and TR^3?

(iii) Show that this set of 8 geometric operations on the square form a group with $\circ$ defined as in (i).

(iv) Can you write out the multiplication table for this group? Hint: Make the first row of this multiplication table $I, R, R^2, R^3, T, RT, R^2T, R^3T$.

(v) Is this the "same" multiplication table as our Warlpiri multiplication table, Table 15.1?

(vi) If your answer to (iv) is yes, try to make your statement precise by identifying the rotation, R, with the appropriate letter in the Warlpiri group. Also, identify the reflection, T, with the appropriate letter in the Warlpiri group.

Hopefully, after doing this exercise you have made a remarkable observation. Namely, the set of 8 symmetry operations of the square is the same group from an abstract mathematical point of view as the Warlpiri group, with its kin relations of the Warlpiri sections. There is a word for this: *isomorphism*. It comes from the Greek: "iso" meaning "same" and "morph" meaning "shape." This group is called the *dihedral group* of order 8.

One of the most interesting things about modern mathematics is that the same, i.e., isomorphic, patterns can be seen to occur in wildly different contexts. This phenomenon is one of the reasons why modern mathematics is such a powerful tool for understanding some things.

Part IV

Counting

Chapter 16

Counting Exactly

16.1 Numeracy

Cultural anthropologist, Thomas Crump, in [116], has said:

Once again numbers succeed because they need no culture-specific support: no culture has an inbuilt defence against numeracy. Numeracy is a sort of Trojan horse, for, once it is admitted, the institutions it supports tend to become dominant in every domain, whether it be the local economy, leisure and play, religion or whatever.

From communications, cryptography and computers to phone numbers and addresses, from electronic mail and numbers of bits of information to wave lengths and magnetic resonance imaging, from clocks and calendars to a cook's recipes, from your salary to the stock market average, from latitude to longitude, from the number of degrees of heat, the velocity of wind, the Richter-scale number of an earthquake to the number of decibels of a sound, from miles per hour to miles per gallon, from how much you weigh to how old you are, from the number of points scored to sports statistics—on and on it goes—numbers truly have invaded our culture.

As I have shown, mathematics is much more than numbers. But numbers are a large part of mathematics. Having a reasonable command of a language in both a spoken and written form is called *literacy*. Having a reasonable command of numbers (and a little mathematics beyond numbers) is called *numeracy*, cf., [522]. If you cannot read, write and speak a language you are obviously missing out on a lot in life. I claim the same is true if you cannot read, write and speak, as well as think with, a little mathese.

16.2 Counting Social Security Numbers among Other Things

Let's start with a counting problem that is not very pressing, but instructive. Our government assigns each of us a social security number, SS#, of the form 123-45-6789. If all ten digits, 0,1,2,3,4,5,6,7,8 and 9, can be used in each of the nine positions of a SS#, what is the maximum population that the United States can have before we run out of SS numbers?

If our SS numbers were one digit long, obviously, only ten people could have different numbers.

Suppose our SS numbers were two digits long, how many would there be? If we put a 0 in the first position, then there would be ten choices for the second digit. So there are ten two digit numbers that have 0 in the first position. Similarly, there are ten two digit numbers that begin with 1. There are ten two digit numbers that begin with 2. There are ten two digit numbers that begin with 3, and so on. Finally, there are ten two digit numbers that begin with 9.

So how many two digit numbers are possible? There are ten groups of two digit numbers, each group containing 10 numbers. Thus the number of two digit numbers equals ten 10s added together, which is[1] ten times ten, i.e., $10 * 10 = 100 = 10^2$. Note that 10^2 in English is "ten squared".[2] Recall from page 59 that the 2 in 10^2 is called the exponent of ten, or sometimes people say 10 is raised to the power 2 to get 10^2.

This is all pretty obvious so far since we can list these 100 two digit numbers in almost the same way as we count. Thus we have 00, 01, 02, ..., 10, 11, ..., 20, 21, ..., ..., 98, 99.

We can draw a "branching" diagram of the two digit numbers as follows.

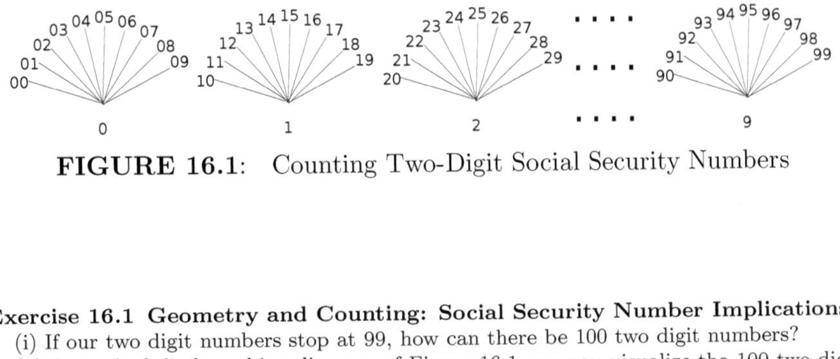

FIGURE 16.1: Counting Two-Digit Social Security Numbers

Exercise 16.1 Geometry and Counting: Social Security Number Implications

(i) If our two digit numbers stop at 99, how can there be 100 two digit numbers?

(ii) Instead of the branching diagram of Figure 16.1 can you visualize the 100 two digit numbers as a square array? Please note that the *branching diagram* in Figure 16.1 and the 10 by 10 *square array* of this problem are two geometric ways of visualizing the product of 10 times 10.

(iii) List all of the social implications that you can think of of everyone having a SS#. What are the advantages? Disadvantages?

[1] Can you prove this using the distributive law?

[2] Recall this from II, Section 10.1.

Exercise 16.2 Multiplication (of whole numbers) is a Shortcut for Doing Repetitive Addition[3]

(i) Suppose that m and n are positive, whole numbers. Can you geometrically represent, i.e., draw a picture, of $m * n$ two ways? Try to represent $m * n$ as both a rectangular array and as a branching diagram, analogous to Figure 16.1.

(ii) Geometrically represent the commutative law of multiplication.

(iii) Geometrically represent the distributive law.

(iv) Geometrically represent the associative law of multiplication.

If SS numbers were three digits long, how many would there be? This is pretty easy if you think about it. There are 100 two digit numbers, and for each one of them there are ten possible choices for the third digit. So for *each* of the 100 two digit numbers there is a group of ten three digit numbers. So we have a total of 100 groups of ten three digit numbers. We add up 10 a total of 100 times. Imagine $10 + 10 + 10 + ... + 10$, one hundred 10s added up. This, of course, is[4] just $10 * 100 = 1000 = 10^3$. Note that 10^3 is called "ten cubed." We say that we have raised 10 to the power 3. The number 3 is the exponent of 10 in 10^3. In English 10^3 is one thousand.[5]

Exercise 16.3 Geometry and Counting Again: More Math Vocabulary

(i) Can you visualize the 1,000 three digit numbers two ways? One way as a branching diagram, another way as a cubic array?[6]

(ii) Consider the sum $10 + 10 + 10$. What law(s) or axiom(s) do we need to justify the following:

$$10 + 10 + 10 = 10 * (1 + 1 + 1) = 10 * 3 = (10)(3) = 30?$$

(iii) Recall from page 258 that if you add two expressions as in $A + B$ then expressions A and B are called (fill in the blank).

(iv) Recall also from page 258 that if you multiply two expressions as in $A * B$ then expressions A and B are called (fill in the blank).

(v) Each 10 in the expression $10 + 10 + 10$ is called a (fill in the blank).

(vi) We have three (fill in the blank) in $10 + 10 + 10$.

(vii) Add up ten 10s (you have ten (fill in the blank)). Now use the distributive law to write this repetitive addition as a multiplication. What do you get?

(viii) Imagine a sum of one hundred 10s (you now have 100 (fill in the blank)). Use the distributive law to write this sum as a multiplication. What do you get? I hope you didn't actually write down one hundred 10s. This is a good time to use your imagination!

Do you see a pattern forming here? The number of one digit numbers is 10^1. The number of two digit numbers is 10^2. The number of three digit numbers is 10^3. I'm about ready to have a mathematical insight: *"The number of n digit numbers is 10^n"* is a true statement when $n = 1, 2, 3, 4, 5, 6, 7, 8, 9, ...$.

[3]Compare Exercise 10.9, Exercise 12.3, Exercise 12.2, Exercise 11.2.

[4]Can you prove this from the axioms for numbers?

[5]Starting with 10^9 the American and British use different words. For example, 10^9 in American English is one billion and in British English it is one milliard. See a good dictionary for more information on this topic.

[6]Hint: If you absolutely are stuck on this problem make it simpler. Assume there are only three digits, say, 0, 1 and 2. There are then 3 single digit numbers possible, 9 two digit numbers possible and 27 three digit numbers possible. Draw in complete detail the branching diagram, and do your best to draw out the 27 point cubic array.

For the record, 10^n is n 10s multiplied together. Thus, for example, $10^6 = 10 * 10 * 10 * 10 * 10 * 10 = 1,000,000$ (a one with 6 zeros after it), viz., one million.

If you multiply two things like A and B together to get $A * B$, then A is a factor of $A * B$ and so is B. In 10^6 there are 6 factors of 10. In 10^9 there are 9 factors of 10, and $10^9 = 1,000,000,000$ (a one with 9 zeros after it), viz., one billion.

Thus, since there are nine digits in our SS#, there are 10^9 or one billion different numbers. Since the population of the U.S. is roughly 300 *million*[7] = .3 *billion* = $.3 * 10^9$, the population of the U.S. can double about one and a half more times before we run out of SS numbers.

Exercise 16.4 Exponentiation

(i) Write $1,000,000,000,000$ using 10 raised to some power. What is the name of this number in (American) English? What is the exponent in this case? How many factors of ten are there in this number?

(ii) The number 10^n where $n = 1, 2, 3, ...$ when written out is one followed by how many zeros? How many factors of ten are in this number?

(iii) Recall from page 59 that the operation of raising 10 to the nth power to get 10^n is called exponentiation of n to (or by) base 10. We just made an exponent out of n. *Exponentiation (using whole numbers) is a shorthand for repetitive multiplication.*[8] If you exponentiate 3 to base 2 what is the result?

Think about the following: You can add 5 to n and get $5 + n$. You can multiply n by 5 and get $5 * n$. You can exponentiate n to or by base 5 and get 5^n. *These are three (different) fundamental operations in "arithmetic," please do NOT confuse them with one another!*[9] Now note that you can "undo" each of the above operations as follows. You can add -5 to $n + 5$ to get n. This is sometimes called subtraction. You can multiply $5 * n$ by $\frac{1}{5}$ and get n. This is sometimes called division. Do you know what operation you can do to 5^n to get back to n?[10]

(iv) Is the operation of exponentiation associative? If exponentiation is associative then the following equation holds (at least if a, b and c are positive whole numbers):

$$a^{(b^c)} = (a^b)^c.$$

If you think that the associative law for exponentiation holds, then you must either accept it as a new axiom or prove it follows from the axioms that we already have. If you think that the associative law for exponentiation does not hold, then all you need to do is find an example where it fails to be true.

(v) Can you think of anything we might run out of in the U.S. before we run out of nine digit SS numbers?

Some of you may be wondering why I went to all of this work counting SS numbers the hard way when we could have just *counted* them the easy way. Namely, the first SS# is 000-00-0000, the next one is 000-00-0001, the next

[7]Actually the U.S. population was over 304 *million* in 2008. By the time you read this there will probably be more, because when I wrote my first draft of this chapter there were 266 million Americans. What year was that? See V on population.

[8]Compare Exercise 16.2.

[9]All three operations, addition, multiplication, exponentiation either are or can be built out of addition, at least when dealing with whole numbers.

[10]Hint: See V Section 19.3.

one is 000-00-0002, and so on until we get to 999-99-9999. This last number is one billion minus one, i.e., 999,999,999. But we didn't start with 1, i.e., 000,000,001, we started with all 0's. So we really do have exactly 10^9 SS numbers.

Well, we did all of that hard work to get an easy answer in order to practice using a *counting principle* that we will use again. Since we have an easy way to count SS numbers, we have a way to double check to see if we were using our more complicated counting technique correctly. We also got to have some fun with the distributive law.

So what is this counting principle? It is:

THE FUNDAMENTAL PRINCIPLE OF COUNTING

If you have two positions and you can fill the first position n_1 ways, the second position n_2 ways, then you can (jointly or simultaneously) fill both positions a total of $n_1 * n_2$ ways.

The proof of this counting principle is immediate if you can visualize the corresponding branching diagram, as in Figure 16.1. You can iterate the counting principle and get a counting principle for any finite number of slots, as we explore in Exercises 16.5 and 16.6.

Exercise 16.5 The Counting Principle Extended

Let's extend our counting principle.

(i) Suppose we have three positions. We can fill the first position n_1 ways, the second position n_2 ways, the third position n_3 ways. Is it true that there is a total of $n_1 * n_2 * n_3$ ways of filling the three positions? (Answer yes or no.)[11]

(ii) Suppose we have k *positions*, where $k = 1, 2, 3, 4, \ldots$. Suppose we can fill the first position n_1 ways, the second position n_2 ways, the third position n_3 ways, ... and the kth position n_k ways. Then is it true[12] that we can fill the k *positions*

$$n_1 * n_2 * n_3 * ... * n_k \text{ ways?}$$

(iii) Draw a diagram analogous to Figure 16.1 which illustrates the counting principle. (Go back and read the counting principle if you need to.) To get started you might want to pick $n_1 = 3$ and $n_2 = 4$. Can you draw a branching diagram for the extension (to three slots) of the counting principle discussed in part (i) of this exercise?

You will recall that if we wanted to add k numbers, say, $n_1+n_2+n_3+...n_k$ we could write $n_1+n_2+n_3+...+n_k = \sum_{i=1}^{k} n_i$. See II, Section 10.4. Analogously, there is a Greek symbol used to write products, namely, capital pi, i.e., Π. Thus,

$$\prod_{i=1}^{k} n_i = n_1 * n_2 * n_3 * ... * n_k.$$

[11] Hint: There are $n_1 * n_2$ ways of filling the first two positions. Now apply the counting principle thinking of the first two positions as a "single" position and the third position as the "second" position. It also helps to visualize or draw the corresponding branching diagram.

[12] Yes. This is our *Extended Counting Principle.* But do not take this footnote's word for it. Understand it.

Exercise 16.6 Counting Using the Fundamental Principle

(i) Suppose we did not allow the first digit of a SS# to be 0. Then how many SS#'s would there be?

(ii) How many social security numbers would there be if you could use any letter from the English alphabet in the first slot and any digit from 0 to 9 in the remaining 8 slots?

(iii) Suppose you live in a state where the car license plates are of the following form: 123 ABC. You can have any digit in the first three positions, and any of the 26 letters in the last three positions. How many cars can have distinct license plates? How could you double the number of plates? How could you multiply the number of possible plates by a factor of 10? (Put another way: How could you get ten times the number of license plates?) How could you multiply the number of plates by a factor of 26?

(iv) How many cars are registered in your home state? What form(s) do the license plates take?

Exercise 16.7 Find the Pattern

(i) If you have 10 friends and 2 gifts, how many ways can you give out the gifts—assuming giving both gifts to the same person is allowed?

(ii) Given 10 Libertarians and 10 Greens how many 2-person committees with one Libertarian and one Green are possible?

(iii) How many ways can you touch one finger to one toe, assuming you have 10 each?

(iv) If a group of people consists of 10 males and 10 females, how many heterosexual couples are possible?

(v) If there are 10 politicians and 10 lobbyists, how many lunches with one politician and one lobbyist are possible?

(vi) Do you see a pattern in the above five parts to this problem?

16.3 Permutations: Order Matters

Opening Your Gym Locker: Order Matters, No Repetitions. Let's use our counting principle to do a slightly different problem. Suppose you get to your gym locker and you have forgotten your locker combination—but not completely. You are sure of the three numbers 11, 20 and 9; but you don't recall what order they are in? What are all possible rearrangements of these three numbers?

You can think of the locker combination as using three positions. In the first position you can try one of the three numbers, 11, 20 or 9. Once you put a number in the first position, it is not available for the second position. Thus, no matter which of the three numbers you put in the first position, there are only two choices remaining for the second position.

Once you have put a number in the first position, and another number in the second position, there is only one number left; and it goes into the third position. See the Figure 16.2.

Thus our counting principle applies in this new situation. Namely, you can fill the first position 3 ways, the second position 2 ways and the third position

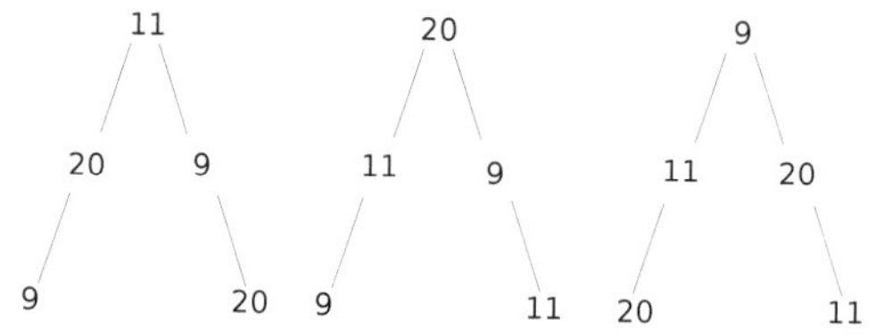

FIGURE 16.2: Opening Your Gym Locker

1 way. Thus there are $3 * 2 * 1 = 6$ possible rearrangements of the three numbers.

Now we used the same counting principle in the previous section on SS #'s as we used in this section. Do you see the big difference between the two examples? When we were counting SS #'s we allowed *repetition* and in the locker combination example *no repetitions* were allowed.

Let's go one more step and count how many ways you can rearrange 4 distinct objects in 4 positions. To be precise suppose we have the letters A, B, C and D and 4 positions. You can think of each rearrangement of the four letters as a "word" four letters long, and you can only use each letter once in each word. Well, you can fill the first position 4 ways, the second position 3 ways, the third position 2 ways and the last position 1 way. Thus there is a total of $4 * 3 * 2 * 1 = 24$ "words."

Suppose now that you have n pairwise[13] distinct objects, and you want to know how many ways you can rearrange them in n positions.

Definition of Permutation: Each rearrangement of the n (pairwise) distinct objects in n positions is called a *permutation* of the n (pairwise) distinct objects. The set of all such rearrangements is called the set of all permutations of the n objects. Note that the positions (or order) of the n objects is very important. Any change in positions (or order) of objects changes the permutation.

Exercise 16.8 Permutations are Functions

From III, Section 14.5 and Exercise 15.10, recall the definitions of *function, one-to-one function,* and *onto.*

(i) Suppose we define a permutation P of a set S to be a function $P : S \longrightarrow S$ which is one-to-one and onto S. Does this definition of permutation coincide with the definition of permutation given above if the set S is a set of n pairwise distinct objects?

(ii) Consider the set consisting of the three numbers $\{1, 2, 3\}$. How many distinct functions from this set to itself are one-to-one and onto?

(iii) Can you list all of the functions in part (ii)?

(iv) Repeat part (ii) but replace the set $\{1, 2, 3\}$ with the set $\{1, 2, 3, 4\}$.

(v) Are the mother and father functions M and F, respectively, from Chapter 15, permutations of the set of eight sections, S_k, $k = 1, 2, 3, 4, 5, 6, 7, 8.$?

[13] We might be tempted to just say n distinct objects, but what does this really mean? It means that given any two of the n objects, one in your left hand the other in your right, you can tell them apart. Thus mathematicians often write pairwise distinct, instead of just distinct. More succinctly, you have n distinct objects if no two of them are identical.

16.4 There are $n!$ Permutations of n Distinct Objects

How many permutations of a collection of n (pairwise) distinct objects are there? In the following discussion I want to show that:

The number of permutations of n distinct objects is n!.

First of all, what is $n!$? In English one does not say n exclamation, rather one says "n factorial" when reading the symbol $n!$. The definition of $n!$ is (in English) the product of all the positive whole numbers from 1 up to and including n, more precisely:

$$n! = n * (n-1) * (n-2) * (n-3) * \cdots * 4 * 3 * 2 * 1 = \prod_{i=0}^{n-1} (n-i).$$

For example,[14] $1! = 1$, $2! = 2*1 = 2$, $3! = 3*2*1 = 6$, $4! = 4*3*2*1 = 24$, $5! = 5*4*3*2*1 = 120, \ldots$.

Now suppose we have n objects and n slots, or positions. How many ways can I put the n objects into the n slots *without repetition*? Using our counting principle we see that the first position can be filled n ways, the second position can be filled $n-1$ ways, the third position (if there is one) can be filled $n-2$ ways, the ith position can be filled in $n-(i-1) = n-i+1$ ways[15], ... , the "third from the last position" or the $n-2^{nd}$ position can be filled $n-[(n-2)-1] = n-(n-2)+1 = 3$ ways, the "next to last position" or the $n-1^{st}$ position can be filled $n-[(n-1)-1] = n-(n-1)+1 = 2$ ways, and the last, or n^{th} position can be filled 1 way.

Exercise 16.9 Permutations: Order Matters

If all the letters in the immediately preceding paragraph (like n slots and i^{th} position) give you mental indigestion, just let $n = 3$ in the above paragraph. Do you see that when $n = 3$ the paragraph is just saying what we already figured out in the example of the locker combination? Re-read the same paragraph with $n = 4$. This is just a recap of our $ABCD$ example. If you can do that, try it again with $n = 5$. Keep going with bigger and bigger values for n until you feel it is all pretty obvious.

(i) If you have a jar with one red, one green, one blue, one black, one white and one yellow marble in it, how many ways can you pick all of the marbles out of the jar if you do so one at a time and the order in which you pick them makes a difference?

(ii) Suppose a lock has a dial with 40 numbers to choose from and that the lock "combination" consists of 3 numbers. Suppose that it takes you 3 seconds to try any given combination of 3 numbers. How long would it take you to try all possible combinations?

(iii) What is the maximum number of Greek organizations (fraternities and sororities) that can exist if each must have a name consisting of 3 pairwise distinct Greek letters, i.e., you cannot repeat a letter in a name but different names might have letters in common?[16] If you relax the condition that the 3 letters be pairwise distinct, i.e., you can repeat a letter in a name, how many names can there be?

---

[14] Your calculator may have a button that calculates factorials for you.

[15] Do you see this last step? Explain it.

[16] Hint: There are 24 Greek letters.

(iv) If you roll 2 ordinary dice there are the following possible outcomes for the sum of the two numbers that show up on the top of the dice: 2, 3, 4, 5, 6, 7, 8, 9, 10, 11 and 12. How many ways can each of these eleven sums, i.e., outcomes, be achieved? Be careful!

(v) Again consider two ordinary dice. When you roll these dice the first die has six different possible outcomes, as does the second, for a total of $6^2 = 36$ possible outcomes when two dice are rolled simultaneously. Add up the total number of possible outcomes from your answers to part (iv). What do you get?

(vi) The alphabet of the indigenous Hawaiian language has 12 letters. Redo part (iii) above replacing the Greek alphabet with the Hawaiian one.

Thus the number of permutations of n distinct objects is:

$$n * (n-1) * (n-2) * (n-3) * \ldots * 4 * 3 * 2 * 1 = \prod_{i=0}^{n-1}(n-i) = n!,$$

i.e., the product of all the positive whole numbers from 1 to n. Recall from the comment following Exercise 16.5 that the large pi stands for "product" just as the large sigma stands for "sum."

The above product appears often in mathematical formulas; that is why it is given a special name, viz., $n!$.

Exercise 16.10 Factorials!

(i) Compute $n!$ for $n = 1, 2, 3, 4, 5, 6, 7, 8, 9, 10$. What do you notice about how fast the value of $n!$ grows as n grows? For example, how do the values of $n!$ compare to the values of n^2?, n^3? If you have a calculator it may have a built in $n!$ button.

(ii) For the record $0! = 1$. This is a definition. You may need to know this when you read various formulas in some books. Why can we define $0! = 1$? What would happen if we defined $0! = 0$? Would mathematics become more complicated? Would we lose some patterns? See, for example, Exercise 17.5, and page 323.

(iii) How many permutations are there of a set with n points?

16.5 Counting Connections: Order Does Not Matter

Counting Small Numbers of Connections. Now I want to study a different kind of counting problem which, when you think about it, tells us something very important (and scary) about real life. Just to be concrete, suppose you are a highway engineer in some county which has a growing population of humans. Part of your job is to count roads and to make sure they are safe. You not only have to know the roads that exist today, you have to plan ahead a couple of years or more. For the moment you are concerned with big, multi-lane roads between big population centers. The current situation is illustrated in Figure 16.3.

Right now there are two big cities, *City*1 and *City*2, with equal populations. There is a big multi-lane road connecting them. But part of the roughly $3*10^6$ people added to the U.S. population every year are moving, not to *City*1 or

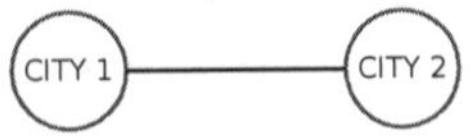

FIGURE 16.3: Two Cities

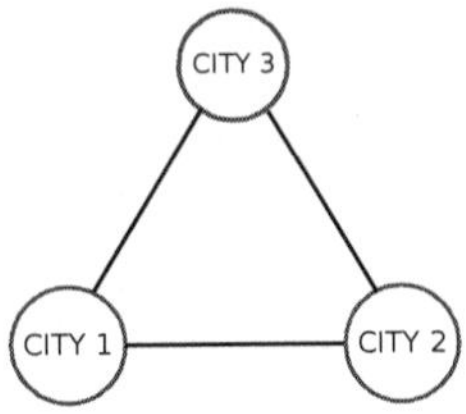

FIGURE 16.4: Three Cities

*City*2, but to a new boom town called *City*3, because housing costs are less in *City*3 than in *City*1 or *City*2.

Exercise 16.11 What is the Optimal Size for a City?

This problem requires some research and is somewhat open-ended.

(i) Find out as best you can the taxes paid per person in various cities. For each city note the total population of the city (and surrounding metropolitan area). Note what kind of services the taxes paid are buying.

(ii) Do you see any optimal size for a city in terms of taxes, i.e., a size for which the taxes paid per person is the least? Make a graph and plot the taxes paid per person vs. the size, i.e., total population, of the city, where city size varies from very small (less than 1000) to very large (millions). What shape is this graph?

(iii) Is there some minimum size a city must attain before it can support fire protection, police, schools, water, sewer and so forth? (Is there a minimum size of human settlement that requires such services?)

(iv) Is there some intermediate size for a city where a person gets the most service per tax dollar expended?

(v) Is your local chamber of commerce aware of your answers to this question? Do they agree with you? Why? Why not?

*City*3 is growing so fast that its population will be the same as that of *City*1 (or *City*2) in a few years. A major multi-lane highway will have to be built between *City*3 and *City*1 and between *City*3 and *City*2. See Figure 16.4.

Exercise 16.12 Roads and Population

In going from the two city model in Figure 16.3 to the three city model in Figure 16.4 answer the following questions:

(i) Quantitatively describe the increase in total population.

(ii) Quantitatively describe the increase in the number of new roads, in the numbers of miles of new roads. On what does your answer depend?

(iii) In our example, which is growing relatively faster: population or roads?

(iv) Quantitatively describe the ratio of "roads per person" and how it changes between the two city and three city scenarios.

Exercise 16.13 Optimal Length Roads in Mathland

(i) Consider the diagram in Figure 16.5 of an alternative road network connecting our three cities:

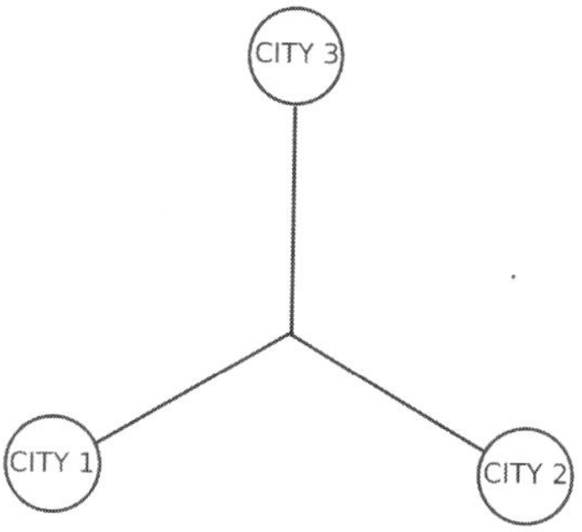

FIGURE 16.5: Soap Bubble Roads in Two Dimensions

Is the total length of this road network shorter than the road network in Figure 16.4? In both Figures 16.4 and 16.5 assume the (triangluar) cities are reduced to points. Can you estimate how much shorter this road network is?

(ii) In Figure 16.5 where would you place a fourth city with minimum extra road building?

(iii) Suppose you have four cities which are the vertices ("corners") of a square. How can you connect these four cities with roads so that the total length of the road network is as short as possible? The answer is actually quite surprising and is related to a host of mathematical problems called Plateau problems (after a mathematician named Plateau) and minimal surface—or "soap bubble"—problems.

I want to give you my answer to Exercise 16.12. If there are 10^5 people in *City*1 then there are 10^5 people in *City*2 as well, since they have equal populations. Then *City*3 will also end up with 10^5. We can express the growth in the region as a fraction:

$$\frac{\textit{growth in population}}{\textit{total initial population}} = \frac{10^5}{10^5 + 10^5} = \frac{10^5}{2 * 10^5} = \frac{1}{2}.$$

If we assume for the moment that the distance between all three cities is the same, then the growth in the number of roads is the same as the growth in the number of miles of roads and we have:

$$\frac{\textit{number of new roads}}{\textit{initial number of roads}} = \frac{2}{1} = 2.$$

In this case the number of roads is growing 4 times faster than the population! You often hear people ask: What is the percentage increase in whatever? I remind you that a *percentage* is just a fraction whose denominator is 100.[17]

[17]See Exercise 13.2 part (viii) for a review of percent.

In our case we can convert $\frac{1}{2}$ into a percentage in a couple of different ways.

$$\frac{1}{2} = \frac{1}{2} * \frac{50}{50} = \frac{1 * 50}{2 * 50} = \frac{50}{100} = 50\%.$$

Or divide 2 into 1, getting .5 or .50 as your answer, which is fifty one-hundredths, which is 50%.

With our other fraction, viz., 2, or $\frac{2}{1}$, proceed in the same way:

$$2 = \frac{2}{1} = \frac{2}{1} * \frac{100}{100} = \frac{2 * 100}{1 * 100} = \frac{200}{100} = 200\%.$$

Thus, we can say that the population grew by 50% while the roads to service the increase grew by 200%. Another way to look at this situation is to calculate "roads per person" both before and after. Thus before we have $\frac{1\ road}{2*10^5\ persons}$. After, we have $\frac{3\ roads}{3*10^5\ persons} = \frac{1\ road}{1*10^5\ persons}$. It is clear that the road burden per person has doubled, and somehow you might expect that this will eventually be reflected in the taxes paid per person.

You might say that if *City*3 were built between *City*1 and *City*2 that the roads wouldn't grow at all. This sounds good, but in today's world extra lanes are added to the existing roads to handle the increased traffic (usually from automobiles). In fact Figure 16.4 illustrates what is happening as I write, right here in the county in which I am writing.[18] Of course, it doesn't have to be this way. There are such concepts as trains, buses, bicycles and feet, living close to where you work and shop—and stable populations.

The above exercise sheds some light on the following statistic that holds in the county where I live and other places as well. For each vehicle we had in my county twenty years ago, there are (slightly more than) 2 now. For each vehicle trip taken twenty years ago in my county there are now 3. This may change someday if the cost of petroleum fuels gets high enough.

Exercise 16.14 Roads and Cities. Who Pays? Who Profits? Who's Responsible?

(i) Suppose that only the people in *City*1, *City*2 and *City*3 pay for the roads as shown in Figure 16.4. (Actually state and federal gas taxes often fund main arteries.) Suppose that only the people in *City*1 and the people in *City*2 pay for the road as shown in Figure 16.3. Suppose that *City*1 and *City*2 are both the same size and do not change size as *City*3 grows to become the same size as *City*1 and *City*2. Compare the cost *per person* of the road system in Figure 16.4 to that of Figure 16.3.

(ii) What do you think is the effect of having the roads paid for by state and federal funding? What are the advantages, disadvantages? What methods of funding road building promote growth? What methods of funding come closest to making growth "pay for itself"?

(iii) I discussed the idea of a *responsibility gap* in Exercise 15.9. Discuss the relative size of the responsibility gap for building two extra roads if the three cities pay for the roads out of city taxes vs. the situation if the roads are paid for by state and federal taxes.

[18] The county in question is Boulder County, Colorado. *City*1 is Boulder, City 2 is Longmont and City 3 is the Lafayette-Louisville complex. This triple is immersed, as you might expect, in a larger network of roads.

(iv) Under what circumstances could you reduce the responsibility gap to be as small as possible for everyone concerned?

(v) If you want to apply our three-cities discussion to a real situation you are familiar with, is road growth relative to population growth the most important thing to consider? Is it among the important things that should be considered?

(vi) Who paid for the roads in your area? Who profits from the roads in your area? Discuss the sets of people who paid for the roads and how much they paid per person. Discuss the sets of people who profit from the roads and how much profit per person the roads enable them to make.

(vii) Can you name some species (other than *Homo sapiens*) that have paid[19] for (or profited from) the construction of the roads in your area?

Counting Large Numbers of Connections. I would like to be a pure mathematician for a moment and jump into an abstract mathematics problem that contains our three cities discussion as a special case. The abstract problem is still a counting problem, but of a slightly different kind than we have encountered before.

The problem is this: Suppose you have n dots on a plane. If you rebel against abstraction think of the dots as cities, or individual people. Suppose you want to connect each pair of points with a line segment. Again, if you are thinking of each point as a city you can think of the connecting line segments as roads, if the points are people (or telephones) you can think of the line segments as telephone connections (with wires or wireless). The question is: How many line segments are there if every pair of dots has a line segment connecting them?

This problem has the same answer as the following problem. Given n points, how many pairs of points can be chosen from the n points (where the pairs are not ordered)? To clarify what I mean, let $n = 3$, i.e., we have three points (as in the example with three cities). Let's label the points P_1, P_2 and P_3. I can select the pair $\{P_1, P_2\}$, which is the same for our purposes here as the pair $\{P_2, P_1\}$ since *now order does not count!*

It is not hard to see that there are just three such pairs: First choose P_1; then you have two choices for completing the pair, getting $\{P_1, P_2\}$ and $\{P_1, P_3\}$. (Thinking of the points as three cities, you can put a road between P_1 and P_2 and a road between P_1 and P_3.) Next choose P_2. Then you have only one way to complete the pair to get a new pair; namely, you can choose P_3 getting $\{P_2, P_3\}$. (If you had chosen P_1 you would have obtained the pair, $\{P_2, P_1\}$, and you already have a "road" between P_1 and P_2.) If you now choose P_3 you are led to no new (unordered) pairs (since all possible "roads" have now been built).

Thus interpreting three points as three cities I am led to three roads as before, one road for each (unordered) pair. Returning to our original problem, if we have n points, how many (unordered) pairs of points are there? Well, this problem is easy if we do it in two steps.

---

[19] A species without money can pay with its life or diminished numbers, for example.

Step 1. How many pairs do I have if I count pairs that are ordered differently to be distinct? In other words, let's first solve an easier problem: how many *ordered* pairs of n points are there? I can use the counting principle to do this. Think of two positions. If I have n points, I can fill the first position n different ways; and I can fill the second position $n-1$ different ways. Thus, if order makes a difference I can fill the two positions $n*(n-1)$ ways, according to the counting principle.

Step 2. If P and Q are two of these n points, in Step 1, I will have counted both $\{P, Q\}$ and $\{Q, P\}$. (Can you see this?) Thus each pair of points was counted exactly twice. But I do not want to count each pair twice. (For instance, if the points are cities, then there is only one road for each pair of cities.) So the number $n*(n-1)$ tells me how many pairs of points I have where each pair is counted twice. Thus, given n points

$$\textit{the number of (unordered) pairs of points} = \frac{n*(n-1)}{2}.$$

Let's use our formula in a place where we already know the answer. Suppose $n = 3$ cities, i.e., "points." Then the number of roads, if we join each pair of cities with one road, is:

$$\frac{3*(3-1)}{2} = 3.$$

Let's use this formula in a different situation. Suppose you own an airline and there are 10 cities to which you want to offer service. If you offer direct, non-stop service between every pair of cities then you will have:

$$\frac{10*(10-1)}{2} = 5*9 = 45 \textit{ routes}.$$

That's quite a few routes. Now suppose, since you offer such good service, that ten more cities want to be in your network of cities. If doubling the number of cities roughly doubled the number of routes, then you would expect to have to offer about 90 routes. Is this correct? Let's see. With 20 cities the number of direct, non-stop routes would be:

$$\frac{20*(20-1)}{2} = 10*19 = 190 \textit{ routes}.$$

Multiplying the number of cities by **2** *multiplied the number of routes by more than* **4**. Now you know one of the reasons why you do NOT always get direct, non-stop flights. The number of possible routes between cities grows a lot faster than the number of cities.

Now let's imagine that our n points are people, and the lines between people are "connections." Suppose you live in a city with 50,000 telephones. Then, since every phone is "connected" to every other phone there will be:

$$\frac{50,000*(50,000-1)}{2} = 25,000*49,999$$
$$= 1,249,975,000 = 1.249975*10^9 \textit{ connections}.$$

Suppose the size of your city doubles and there are now 100,000 phones. We then have:

$$\frac{100,000 * (100,000 - 1)}{2} = 50,000 * 99,999$$
$$= 4,999,950,000 = 4.99995 * 10^9 \ \textit{connections}.$$

Thus, again after doubling the number of phones, the number of connections is multiplied by a number slightly greater than 4 (actually 4.000040001). Also note that the number of phones is relatively small, $5 * 10^4$. The number of connections is 5 orders of magnitude greater, or about 10^9.

The phone company is clever. Even before wireless phones it did not run a separate wire directly between every pair of phones. However, the phone company does have to create a connection of some sort so that every phone can "talk" to every other phone. And the number of possible connections increases a lot faster than the number of phones. We say that the number of connections increases, roughly, as the *square* of the number of "points," in this case, phones. All this has something to do with your phone bill.

Now note that in any population of people any (unordered) pair constitutes a possible "communicating couple." If you double the population the number of possible communicating couples is multiplied by a number a little bigger than 4; thus the number of couples behaves numerically more like the square of the number of individuals, than just the number of individuals. The potential demand for communication connections in a society grows much faster than the population itself grows.

With the increase in communication of information (or data) from one "phone" (or computer) to another via "phone lines,"[20] I was curious as to how the corporations in charge of data communications controlled costs and dealt with the problem outlined above. I called a friend of mine, Don Glen, who is a consulting data communications engineer. He told me that when confronted with a massive network, such as the phone and computer network that we now have, you use *time* as an extra dimension. Two kinds of mathematics get involved: queuing theory, as in "you are on hold," and blocking theory, as in "busy signal." Now if you are sending data via computer, as is the case if you send an e-mail message, then the data can be sent in "packets." Time delays with e-mail are not as crucial as in, say, voice communications over the phone. Thus you can line up, i.e., or queue, the packets which are the various e-mail messages and send them when the line is free. You don't have to create a dedicated connection between every two computers, that is, a connection that is there all of the time. In fact, if I want to send a message to your computer from mine I can do it as follows. I can send my message to a third computer which we can call a "hub." My message can wait at the hub until a

[20]Connections are often made via fiber-optic cables that carry light instead of electricity, wireless satellite connections or wireless cell-tower connections

line opens up between the hub and your computer. We might never have to actually connect my computer to yours at the same time.

Great savings can thus be achieved with electronic mail. Similar, but less impressive savings can be achieved with voice communications. The phone company has a good idea of how many connections, or links, it has to create at a given time; and it saves money by not building the capability to create more links than are actually expected to be needed. Of course, there will then be times when more connections are requested than can be supplied, as on Mother's Day or when there is an earthquake in Los Angeles or a hurricane in New Orleans.[21]

Exercise 16.15 Why Are We Too Busy to Say: Hi!?

(i) Discuss why life could tend to be more hectic in a large city vs. a small town. How many possible "interacting pairs" of people are there in a city with a population of n persons?

(ii) If you double the population of a city, what is the effect on the number of possible interacting pairs of people?

(iii) How would you take into account interacting triples? quadruples? quintuples? and so forth. (See the next chapter.)

[21] Joel, Jr., Amos E., editor, *Electronic Switching: Digital Central Office Systems of the World*, The Institute of Electrical and Electronics Engineers, Inc., 345 East 47th Street, New York, NY 10017, 1982. Also, for an introduction to the subject: Joel, Jr., Amos E., "What is Telecommunications Circuit Switching?," Proceedings of the IEEE, Vol. 65, No. 9, September 1977, pp. 1237–1253.

Chapter 17

Equivalence Relations and Counting

Let's recall the definition of equivalence relation which is supposed to capture the mathematical essence of the intuitive notion of "equality." We will use the concept of equivalence relation as a tool to help us solve some interesting counting problems.[1]

Definition: A relation R on a set T is an *equivalence relation* on T if it is reflexive, symmetric and transitive.

17.1 Using Equivalence Relations to Count

Let us revisit the problem of calculating the number of unordered pairs of points that can be chosen from a given set S, which we assume contains n points. The first step, as on page 317 and ff., is to consider the collection of all possible ordered pairs of points that can be selected from S.[2] As before, we will use our counting principle. We see that the first member of an ordered pair of points from S can be chosen in n ways. Once the first position is filled, the second position can be filled in $n-1$ different ways. Thus by the counting principle there are $n*(n-1)$ possible distinct ordered pairs that can be chosen from S. This is the same answer we got before on page 318.

We are now going to do something a little bit different than we did before. I want to define a relation on the set of ordered pairs from S, in particular, I want to define an equivalence relation, $\approx$, on this set of ordered pairs as follows: Two ordered pairs of points from S, say (P_1, P_2) and (Q_1, Q_2) are $\approx$, i.e., $(P_1, P_2) \approx (Q_1, Q_2)$, if (Q_1, Q_2) is a permutation (or rearrangement) of (P_1, P_2).[3]

[1] In III we studied relations on sets. We noted there and in Exercise 13.1 part (iv) that the notion of "equality" is a relation which is reflexive, symmetric and transitive.

[2] Recall that repetition or picking the same element twice is not allowed.

[3] If you really learned Chapter 14 well and remember it, you might be a little confused. The set upon which the relation $\approx$ is defined is not S but the set of ordered pairs of elements from S. Also, in the formal language of Chapter 14 instead of saying, for example, that $(P_1, P_2) \approx (Q_1, Q_2)$, we would say that the ordered pair of ordered pairs $((P_1, P_2), (Q_1, Q_2))$ is in the relation $\approx$. We will drop the latter terminology and use the simpler terminology,

Exercise 17.1 The Definition of $\approx$ for Ordered Pairs

Suppose that P_1, P_2, Q_1 and Q_2 are points in some set S.

Show that $(P_1, P_2) \approx (Q_1, Q_2)$ if and only if ($P_1 = Q_1$ and $P_2 = Q_2$) or ($P_1 = Q_2$ and $P_2 = Q_1$).

Let's first look in detail at a concrete example where $n = 4$. Suppose our set $S = \{A, B, C, D\}$. The problem, which we know how to do from the last section, is to count the number of unordered pairs that we can make from the set S. Toward that end let us list, in the shape of a rectangular array, all of the possible ordered pairs that can be made from our set S.

$$\begin{array}{c} (A, B), (B, A) \\ (A, C), (C, A) \\ (A, D), (D, A) \\ (B, C), (C, B) \\ (B, D), (D, B) \\ (C, D), (D, C) \end{array}$$

Did we list all possible ordered pairs that can be made from the set S of four letters? We have listed 12. Well, thinking of ordered pairs as resulting from the filling of two slots, with 4 letters to choose from (repetition not allowed), I see that the first slot can be filled 4 ways and the second slot can then be filled 3 ways. Thus by the fundamental principle of counting the total number of ways of filling the two slots, i.e., the total possible number of ordered pairs, is $4 * 3 = 12$.

Now given an ordered pair, say (A, B), what is the collection of ordered pairs that are equivalent to it according to the definition of equivalence I am using here, i.e., our definition of $\approx$? Well, do you see that $(A, B) \approx (A, B)$ and $(A, B) \approx (B, A)$, and that is all? That is, there are no other ordered pairs equivalent to (A, B)?

Given an equivalence relation on a set, call the set T, (in this case T is the set of ordered pairs from set S) the subset of all the elements in the set T equivalent to an element x in T is called the *equivalence class* of x.

Do you see that the each row of the two by six rectangular array is an equivalence class? Thus we have

$$\begin{array}{c} (A, B) \approx (B, A) \\ (A, C) \approx (C, A) \\ (A, D) \approx (D, A) \\ (B, C) \approx (C, B) \\ (B, D) \approx (D, B) \\ (C, D) \approx (D, C) \end{array}$$

You can think of the first row, for example, as capturing the essence of—or representing—the unordered pair $\{A, B\}$. Thus each row of the above rectangular array represents an equivalence class, or in other words, an unordered

i.e., $(P_1, P_2) \approx (Q_1, Q_2)$, in all that follows.

pair. Since we started with all possible ordered pairs, I must have found all possible unordered pairs.

Now how many unordered pairs are there? That is the same question as asking how many rows there are in the above rectangular array. The "area" of the rectangle, i.e., the number of ordered pairs, in our first rectangle is 12. The "width," i.e., the number of columns, of our rectangular array is 2. Thus the "length," i.e., the number of rows, of the rectangular array is $\frac{12}{2} = 6$. In some sense, the above rectangular array illustrates the division operation, dividing 12 by 2.

Thus the answer to our question has been found. The number of unordered pairs that can be created from the set $S = \{A, B, C, D\}$ is 6.

The following exercise is easy but subtle. There is some substance to this exercise because of what our definition of $(P_1, P_2) \approx (Q_1, Q_2)$ means. Recall that the relation $\approx$ does not mean "identical to."

Exercise 17.2 The Number of Unordered Pairs that Can be Chosen from a Set S with n Points

Suppose that A, B, C, D, E, F are all from some set S. Review Exercise 17.1.

(i) Show that any ordered pair of points from S is $\approx$ to itself. This says that $\approx$ is reflexive.

(ii) Show that if $(A, B) \approx (C, D)$, then $(C, D) \approx (A, B)$. This says that $\approx$ is symmetric.

(iii) Show that if $(A, B) \approx (C, D)$ and $(C, D) \approx (E, F)$, then $(A, B) \approx (E, F)$. This says that $\approx$ is transitive.

(iv) Imagine that you have picked one ordered pair of points from S, call it (A, B). Now imagine the collection of all ordered pairs of points from S that are $\approx$ to (A, B). This collection is called an *equivalence class* of ordered pairs. How many distinct ordered pairs will be in any equivalence class?

(v) Show that if any two equivalence classes of ordered pairs have an ordered pair in common, then the two equivalence classes are, in fact, the same equivalence class of ordered pairs.

(vi) If you are having trouble visualizing what is going on, pick a concrete set with, say, five points in it. Write down all of the possible ordered pairs, and then write out all of the equivalence classes of ordered pairs. Follow the example above this exercise and write out the appropriate rectangular array.

Hopefully what we learned from Exercise 17.2 is that no matter what the set S is, each equivalence class of ordered pairs of elements from S has two ordered pairs in it, viz., the two possible rearrangements (permutations). Do you see that we can think of such an equivalence class as an unordered pair? For any set S can you imagine or visualize the rectangular array of all ordered pairs of elements from S as we did above, where each row is an equivalence class? Thus there will be half as many unordered pairs as there are ordered pairs. We thus get the same answer as on page 318, that the number of unordered pairs of points that can be chosen from a set with n points is $\frac{n*(n-1)}{2}$.

Unordered Triples. Triangles and Three-Phone Conference Calls. It turns out that the argument of the last section on pairs can be generalized to triples. Thus, given a set S with n points, we want to find out how many unordered triples of points can be chosen from S, repetition not allowed. If S is a bunch of points on a sheet of paper, do you see that the number of unordered triples is the same as the number of triangles that can be drawn with vertices at points in S?

Before we tackle the next exercise let's warm up on a set S with 4 elements: $S = \{A, B, C, D\}$. The question is this: How many ways can I choose 3 letters from S if the order in which I choose them does not matter? I will do this problem in two steps.

Step 1. An easier question to answer is: How many ways can I choose 3 letters from S if order of choice does matter? Phrased another way: How many ordered triples of letters can I create using the 4 letters in S? By our counting principle there are 3 slots and there are 4 ways to fill the first slot, 3 ways to fill the second slot and 2 ways to fill the third slot. Thus the answer to this question in Step 1 is $4 * 3 * 2 = 24$ ways.

Exercise 17.3 Permutations Again

Let us momentarily review something we need to remember.

(i) Given three (pairwise distinct) letters, how many permutations (rearrangements) of these three letters are there?

(ii) Given k (pairwise distinct) letters, how many permutations of these k letters are there?

Now let us write down explicitly what these 24 ways of choosing *ordered* triples from the set $S = \{A, B, C, D\}$ are. The first row of 6 below are the 6 ways, if order matters, of choosing 3 letters from S involving the letters A, B, C. The second row of 6 are the 6 ways, if order matters, of choosing 3 letters from S involving the letters A, B, D. The third row of 6 are the 6 ways, if order matters, of choosing 3 letters from S involving the letters A, C, D. The fourth and final row of 6 are the 6 ways, if order matters, of choosing 3 letters from S involving the letters B, C, D.

$$(A, B, C), (A, C, B), (B, A, C), (B, C, A), (C, A, B), (C, B, A)$$
$$(A, B, D), (A, D, B), (B, A, D), (B, D, A), (D, A, B), (D, B, A)$$
$$(A, C, D), (A, D, C), (C, A, D), (C, D, A), (D, A, C), (D, C, A)$$
$$(B, C, D), (B, D, C), (C, B, D), (C, D, B), (D, B, C), (D, C, B)$$

We now define $\approx$ for ordered triples of letters coming from S. Namely, $(X, Y, Z) \approx (R, S, T)$ if (R, S, T) is a permutation, or reordering of (X, Y, Z). In plainer English, two ordered triples are $\approx$ if one is a reordering of the other. This relation on the set of ordered triples of letters from S is an equivalence relation. Can you see this? (Hint: Check to see that this definition of equivalence of ordered triples is reflexive, symmetric and transitive.)

Now looking at our collection of 24 possible ordered triples of letters from S we see that all the ordered triples in the first row are pairwise equivalent, i.e., the first row of 6 ordered triples are "all equivalent to one another." The same is true of the second row, the third row and the fourth row. More explicitly we have:

$$(A, B, C) \approx (A, C, B) \approx (B, A, C) \approx (B, C, A) \approx (C, A, B) \approx (C, B, A)$$
$$(A, B, D) \approx (A, D, B) \approx (B, A, D) \approx (B, D, A) \approx (D, A, B) \approx (D, B, A)$$

$$(A, C, D) \approx (A, D, C) \approx (C, A, D) \approx (C, D, A) \approx (D, A, C) \approx (D, C, A)$$
$$(B, C, D) \approx (B, D, C) \approx (C, B, D) \approx (C, D, B) \approx (D, B, C) \approx (D, C, B)$$

Now note that the first row above is an *equivalence class* of ordered triples, i.e., it is the set of all ordered triples of letters from S that are $\approx$ to, say (A, B, C). Note that $(A, B, C) \approx (A, B, C)$, since the permutation or rearrangement of (A, B, C) that leaves everything alone, the "identity permutation," is allowed.

Similarly, the second row is an equivalence class of ordered triples, as is the third row and the fourth and final row.

Step 2. How many unordered triples can I form from the letters in S? Said another way, how many ways can I choose 3 letters from S if order does not matter? The answer to this question should now be[4] clear: $\frac{24}{6} = 4$. The first row, the set of 6 ordered triples equivalent to the ordered triple (A, B, C), is meant to capture the notion of the *unordered* triple, or set, $\{A, B, C\}$.

The second row, the set of 6 ordered triples equivalent to the ordered triple (A, B, D), is meant to capture the notion of the unordered triple, or set, $\{A, B, D\}$. The third (equivalence class) row is meant to capture the notion of the unordered triple, or set, $\{A, C, D\}$. Finally, the fourth (equivalence class) row is meant to capture the notion of the unordered triple, or set, $\{B, C, D\}$.

Before leaving this example let's check our answer by looking at the same problem in a different way. The act of selecting (without regard to order and without repetition) 3 letters from S, a set with 4 letters, is the same thing as NOT selecting 1 of the 4 letters in S. So how many ways can you leave one letter in S behind, i.e., not select one letter from S? The answer is clearly 4.

Exercise 17.4 The Number of Unordered Triples that Can be Chosen from a Set S with n points

(i) How many ordered triples of points can be chosen from a set with n points?

(ii) Define two ordered triples to be $\approx$ if one is a permutation (or rearrangement) of the other. Show that this relation, $\approx$, is reflexive, transitive and symmetric.

(iii) Imagine that you have picked one ordered triple of points from S, call it (A, B, C). Now imagine the collection of all ordered triples of points from S that are $\approx$ to (A, B, C). This collection is called an *equivalence class* of ordered triples. How many distinct ordered triples will be in any equivalence class?

(iv) Show that if any two equivalence classes of ordered triples have an ordered triple in common, then the two equivalence classes are, in fact, the same equivalence class of ordered triples.

(v) If you are having trouble visualizing what is going on, pick a concrete set with, say, four points in it, like we just did above. Without looking at the example above, try by yourself to write down all of the possible ordered triples, and then write out all of the equivalence classes of ordered triples.

(vi) How many unordered triples can you choose from a set with n points?

Hopefully you now see that each equivalence class of ordered triples has $3! = 3 * 2 * 1 = 6$ ordered triples in it. For example, the equivalence class of

[4]The "area" of the rectangular array is 24 and the "width" of the rectangular array is 6, so the "length" of the rectangular array is 4, since $4 * 6 = 24$.

the ordered triple (A, B, C) will have 6 ordered triples in it, viz., the 6 possible rearrangements (permutations) of the ordered triple (A, B, C). Can you see that we can think of such an equivalence class as an unordered triple? Thus there will be six times as many ordered triples as there are unordered triples.

We are led to conclude then that the number of unordered triples that can be chosen from a set with n points is $\frac{n*(n-1)*(n-2)}{3!}$.

The number of unordered pairs of points that can be chosen from n points is called "n choose 2," and is written $\binom{n}{2}$. Thus, as we saw before, $\binom{n}{2} = \frac{n(n-1)}{2}$. The number of unordered triples of points that can be chosen from n points is called "n choose 3," and it is written $\binom{n}{3}$.

Exercise 17.5 Writing "n choose 2" and "n choose 3" in Different Ways

(i) Can you show that $\frac{n*(n-1)}{1} = \frac{n!}{(n-2)!}$?

(ii) Can you show that $\frac{n*(n-1)*(n-2)}{1} = \frac{n!}{(n-3)!}$?

(iii) Can you show that if k is a positive whole number between 0 and n that $\frac{n*(n-1)...(n-(k-1))}{1} = \frac{n!}{(n-k)!}$?

(iv) Can you show that $\binom{n}{2} = \frac{n!}{2!(n-2)!}$?

(v) Can you show that $\binom{n}{3} = \frac{n!}{3!(n-3)!}$?

17.2 Combinations: Order Does Not Matter

The number of connections/unordered tuples that we have been calculating in the previous subsections often goes by another name. We have the following definitions.

Definition: The number of unordered pairs, without repetition, that can be chosen from a set of n objects is called the *combinations of n things taken 2 at a time.*

Similarly we have

Definition: The number of unordered triples, without repetition, that can be chosen from a set of n objects is called the *combinations of n things taken 3 at a time.*

In the next exercise we generalize this from pairs, i.e., $2-tuples$, and triples, i.e., $3 - tuples$, to $k - tuples$.

Exercise 17.6 Combinations: Order Does Not Matter

(i) What are two formulas for $\binom{n}{3}$?

(ii) Try to figure out what "n choose k" is where k is a whole number between 0 and n. That is, try to find a formula for the number of ways of selecting k things from n things where the order of selection does not matter, i.e., "n choose k". This number[5] is written $\binom{n}{k}$.

(iii) The number of unordered pairs that can be chosen from n objects is called the number of combinations of the n objects taken two at a time. The number of unordered triples that can be chosen from n objects is called the number of combinations of the n objects taken 3 at a time. Looking back at Section 16.3, and thinking like a mathematician, would you say that a "locker combination" is a combination or a permutation?

(iv) Suppose you have a set of n "blue points" and a set of m "red" points. How many different line segments are there with different colored endpoints? In other words, what is the maximum number of line segments connecting points of different colors?

(v) In a city with 10,000 telephones, how many three-phone conference calls are possible?

(vi) In a city with 100,000 telephones, how many three-phone conference calls are possible?

(vii) How do the answers to parts (v) and (vi) compare?

It turns out that knowing about $\binom{n}{k}$ is a key to an intuitive understanding of the Second Law of Thermodynamics, and hence a key to understanding ecological economics, cf., VII.

17.3 Additional Counting Problems

Suppose you want to count how many distinct ways there are of permuting (or rearranging, or reordering) the five letters $TREES$. If all five letters were distinct then the answer would be easy: the number of permutations of 5 distinct objects is $5! = 120$. But two of the letters are not distinct. Since the number of permutations of the two identical E's is $2! = 2$, the number 120 counts twice as many distinct permutations as there really are. Thus the answer to our question is $\frac{120}{2} = 60$.

Said another way, if our letters were TRE_1E_2S, where we can tell the difference between the E_1 and the E_2 because of the subscripts, then five distinct objects have $5! = 120$ rearrangements or permutations. But then if we wish to consider the E's identical, we can define an equivalence relation $\approx$ which declares two permutations equivalent if the only difference between them is that the subscripts of the E's are permuted. For example, $TE_1RSE_2 \approx TE_2RSE_1$. Thus the 120 permutations are partitioned into or "divided up into" 60 equivalence classes each containing 2 equivalent permutations of the five letters.

---

[5]Hint: Think of k slots; you can fill the first slot n ways, the second slot $n-1$ ways, ..., down to the kth slot which you can fill $n-k+1$ ways. Now multiplying these numbers together gives you the number of ways you can fill k slots with n objects to choose from, where order counts (and repetitions are not allowed). Visualize the set of all ordered k-tuples of n elements as a rectangular array. What is the "area" of this rectangle? What is the "width" of this rectangle, i.e., how many of these ordered k-tuples "collapse" to, or are $\approx$ to, a given unordered k-tuple?

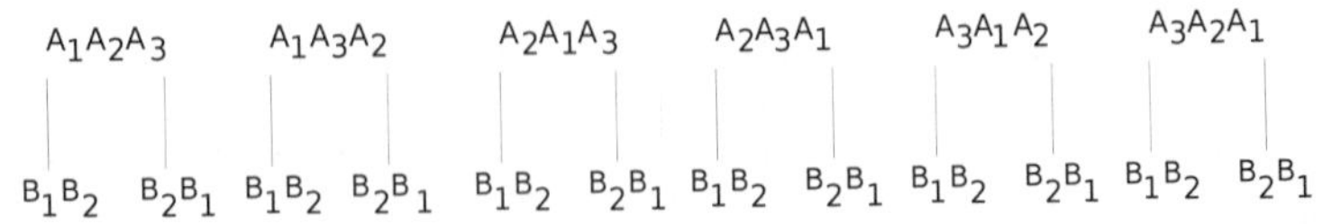

FIGURE 17.1: There are 12 Permutations in One Equivalence Class

Consider now the problem: How many distinct permutations of the five letters $AAABB$ are there? If we think of the five letters as distinct, say $A_1A_2A_3B_1B_2$, then there are $5! = 120$ distinct permutations of the five (now distinct) objects. Now I am going to define two of these 120 permutations to be $\approx$ if the only difference between them is a permutation of the subscripts of the A's or a permutation of the subscripts of the B's. For example, $A_1B_1A_3A_2B_2 \approx A_2B_2A_3A_1B_1$. This equivalence relation partitions or "divides up" the set of 120 permutations into equivalence classes of permutations that are equivalent according to my definition of $\approx$. How many permutations are going to be in each equivalence class? Well, given a permutation, say $A_1B_1A_3A_2B_2$, there are $3! = 6$ permutations of the subscripts on the A's. But for *each* of these 6 permutations of the three subscripts of the A's there are two permutations of the two subscripts of the B's, so there must be $3! * 2! = 12$ permutations in each equivalence class. You can see this, for example, by looking at the branching diagram in Figure 17.1: Thus there are $\frac{120}{12} = 10$ (pairwise) distinct permutations of $AAABB$.

Exercise 17.7 Interesting Permutations

(i) How many distinct permutations are there of the letters $FREEZE$?

(ii) How many distinct permutations are there of the letters $AAABBB$?

(iii) How many distinct permutations of the letters $AABBBCCCC$ are there?

(iv) How many distinct permutations are there of the letters $MISSISSIPPI$?

(v) Go back and do Exercise 1.6 again, deriving the necessary formula for doing the problem by yourself.

Here is a problem that involves combinations. How many committees consisting of 3 men and 4 women can be chosen from a class of 12 men and 15 women? The order in which people are chosen does not matter, just the eventual set of men and women. First of all, the number of ways of selecting the men is just $\binom{12}{3}$, or 12 choose 3. Now for each choice of men there will be "15 choose 4," or $\binom{15}{4}$ choices of the women. Thus we multiply $\binom{12}{3} * \binom{15}{4}$ to get the answer to our problem, which is $\frac{12*11*10}{3*2*1} * \frac{15*14*13*12}{4*3*2*1}$, which is $300,300$. Can you visualize a branching diagram with $\binom{12}{3}$ choices of the men, and for each of these there are $\binom{15}{4}$ branches, representing the number of choices of the women that can be made for each choice of the men?

Suppose we still have a class with 12 men and 15 women. How many committees of 7 can we choose on which there is at least one woman? This problem can be done at least two ways. One way is longer than the other. I will do the shorter way first. There are two types of 7 person committees:

those with at least one woman and those which are all men. Since there are $12 + 15 = 27$ people in the class, the total number of possible committees of 7 is $\binom{27}{7} = 888,030$. The number of possible committees which are all men is $\binom{12}{7} = 792$. Thus the number of committees which are not all men, i.e., which have at least one woman, is $888,030 - 792 = 887,238$.

The longer way to do this problem would be to figure out how many committees there are with exactly 1 woman, exactly 2 women, ..., exactly 7 women and add them up. Do this and see if you get the same answer, viz., 887238.

Exercise 17.8 Interesting Combinations

(i) In the class with 12 men and 15 women, how many committees of 5 are possible which have at least one man?

(ii) In the same class with 12 men and 15 women, how many committees are possible with at least two men?

17.4 DNA Computing

It's a hundred times times faster than the best serial supercomputer. It's a billion times more energy efficient. It's a trillion times denser than the best storage media. It's a teaspoonful of DNA that's a computer! And Leonard Adleman invented it.

Thomas A. Bass, *Wired Magazine*, 1995

You probably have heard of deoxyribonucleic acid, DNA, in a biology class or elsewhere. To a biologist a molecule of DNA is made by attaching building blocks of four types: Adenine, Thymine, Guanine, Cytosine—together in a "string." To a mathematician a molecule of DNA can be thought of as a "word" spelled out with the letters A,T,G and C. We will be using lower case a,t,g and c interchangeably with the capital letters at times for clarity.

Biologists can create very long (medium or short) "DNA words" with any spelling. Biologists have also discovered the following interesting property of DNA: every "DNA word" has a *complementary* DNA word formed by replacing A with T, T with A, G with C and C with G. If a DNA "word" gets close to its complementary DNA "word" they can stick together or link up, but A can only link with T and G can only link with C.

Exercise 17.9 Math and DNA

(i) Write down the DNA "word" that is complementary to ATGCTACCGTA.

(ii) Suppose you have the following two DNA "words": CTAGTACCTGA and ATGGA. If these two molecules got close together what possible way(s) can they link up?

The first problem that Leonard Adleman tackled was the so-called *traveling salesman* problem. Let's imagine that our salesperson is young and is selling ideas and needs to visit 100 cities in the United States. The boss of our idea-salesperson directs him/her to visit each city exactly once, under the mistaken impression that such a path—if found—will necessarily minimize costs to "the firm" and/or the environment. Being a loyal employee, our salesperson sets out to plan an itinerary on an ordinary computer, using a standard algorithm—and finds that the computer cannot finish the problem until well after he/she retires!

In late 1993 Professor Adleman, an interdisciplinary mathematician, invented a way to tackle this traveling salesman problem using the above mathematical properties of (and the technology to manipulate) DNA "words." Since we are just trying to grasp the basics here let us suppose that we are dealing with just four cities: Los Angeles (LA), San Francisco (SF), Denver (DEN) and New York (NY). Let's suppose that due to bad weather or bad labor-management relations that on the days our salesperson wishes to travel he/she can only get the following flights: LA $\longleftrightarrow$ SF; SF $\longleftrightarrow$ NY; SF $\longleftrightarrow$ DEN; DEN $\longleftrightarrow$ NY. The problem now is to find a route from, say, LA to NY which passes through each of the other two cities exactly once.

When Adleman did his initial experiment he synthesized 30 trillion molecules of DNA for each of his cities[6] and another 30 trillion DNA molecules for each route. Let us arbitrarily assign six letter DNA names (Adleman used 20 letter names) to each city and route as follows (lowercase is just used for clarity and has no special significance):

LA: AACagt
SF: TGActc
DEN: CCGgta
NY: GTAaag
LA $\longrightarrow$ SF: agtTGA
SF $\longrightarrow$ NY: ctcGTA
SF $\longrightarrow$ DEN: ctcCCG
DEN $\longrightarrow$ NY: gtaGTA.

Exercise 17.10 Itineraries and DNA

(i) Assuming the preceding two paragraphs (which tell us which cities are connected by direct flights at the moment), how could you fly from LA to NY and pass through each of our four cities exactly once? (This is easy to see in this case.)

(ii) Write down the complementary DNA word for each of the four cities.

Adleman mixed the (complementary) DNA names for the cities and the DNA names of the flights in a glass container, and some "magic" transpired.

[6]Adleman actually synthesized the complementary DNA molecule for the name of each city for reasons we will soon see.

To illustrate a small piece of this biomathematical magic, let's examine what happens when the DNA molecules for the flights from LA to SF and SF to DEN get "close" to the DNA molecule which is the complement of the name of SF.

First of all, picture two DNA molecules, one representing the flight from LA to SF and the other representing the flight from SF to DEN:

$$\begin{array}{c} \text{LA} \longrightarrow \text{SF} \longrightarrow \text{DEN} \\ \text{agtTGA} \quad \text{ctcCCG} \end{array}$$

Now visualize the *complementary* molecule for SF, i.e., the molecule formed from the SF molecule by replacing each T (or t) with an A (or a), replacing each G (or g) with a C (or c), replacing each A (or a) with a T (or t), replacing each C (or c) with a G (or g). Do you see that this molecule is ACTgag?

The molecule representing the flight from LA to SF is made up of the last part of the molecule for LA and the first part of the molecule for SF. The molecule representing the flight from SF to DEN is similarly made by taking the last part of the molecule for SF and following it with the first part of the molecule for DEN. These two molecules representing flights into and out of SF can bond to the molecule complementary to the molecule for SF as follows:

$$\begin{array}{c} \text{agtTGActcCCG} \\ \text{ACTgag} \end{array}$$

where the A links with T, C with G, T with A, g with c, a with t and g with c as shown.

Since there are so many agtTCA, ctcCCG and ACTgag molecules in the solution in the glass container we can be almost certain that some of these triples will link up as shown above.

Hopefully you can see now how when you mix all of the (complementary) DNA names of the cities with the DNA names for the (directed) routes between the cities that the trillions of molecules of DNA most likely "try out" or "compute" all of the possible routes just because of the overwhelming numbers of molecular encounters involved. The "answer" DNA molecules, which in this case are those DNA splices that begin in LA and end in NY and pass though each intervening city precisely once—should be floating in the soup. One then must use laboratory techniques from molecular biology to find this "answer molecule."

Exercise 17.11 Solving a Problem with DNA

(i) Can you write down a "solution molecule" for our problem of finding a route from LA to NY which passes through LA, NY, SF and DEN exactly once?

(ii) Can you write down the molecule (consisting of a string of molecules complementary to the molecules of the cities involved) to which the molecule in part (i) is "paired," i.e., bonded via A-T and C-G pairs?

(iii) For want of a better place to mention it, research the status of *quantum computing* at the time you read this.

We note in passing that Adleman is the "A" in RSA public key encryption systems, which was invented while Adleman was a co-researcher with Ron Rivest and Adi Shamir at MIT in the late '70s, cf.,VI.

17.5 More Exercises.

Exercise 17.12 Combinations and Permutations

(i) Suppose you have a class of 36 students, half male, half female. How many study groups of 6 students can be selected from this class of 36 such that each study group has *at least* one female?

(ii) If you have n objects and k slots (where $k \leq n$) how many ways can you fill the k slots with the n objects if repetition is not allowed and order matters?

(iii) If you have n objects, how many ways can you select k objects from the n objects where $k \leq n$ and the order of selection of the k objects does not matter?

(iv) How many distinct rearrangements of the letters in TREE are possible? In FOLLOW? In AAABBBBCDDDDDD?

(v) If you have n letters with k letters identical to one another, and j letters identical to one another (but different from the k letters) how many distinct "words" can you make (assuming $j + k \leq n$ and there are thus $n - j - k + 2$ pairwise distinct letters)?

(vi) Suppose a class has 40 females and 12 males. How many three person study groups can be formed with exactly two females?

(vii) Suppose a class has 40 females and 12 males. How many four person study groups can be formed with exactly two males?

(viii) Suppose a class has 36 students with half male, half female. How many study groups of 4 can be formed with *at most* one male?

Exercise 17.13 A Dice Game with Distributions

Consider the following dice game that can be (and sometimes is) introduced to first year high school students who enroll in an interactive mathematics program. This game is then the subject of analysis in each of the following years of high school mathematics.

Here is the game. You have 11 positions, one for each number from 2, 3, up to 12. So do any other number of persons who wish to play the game with you simultaneously. Everyone has the same number of identical chips. You then distribute your chips among your 11 positions any way you like; everyone else does likewise. You then roll a pair of dice and add the outcomes on the two dice. If you have a chip on the number so obtained you remove one chip, otherwise you just pass the dice to the next player without touching your chips. The first person to remove all of their chips wins. Note that mathematically it is smart to distribute your chips according to the most likely distribution of outcomes of many rolls of the dice. What is this most likely distribution?

Exercise 17.14 An Introduction to Chance and $\binom{n}{k}$. A Random Walk

Consider the following game, which mathematicians call a random walk. In Figure 17.2 below, imagine starting the game by standing at point $(0, 0)$.

You now flip a fair coin. If the coin comes up "heads" you move one step to the right (after the first move you would then be at point $(1, 0)$). If the coin comes up "tails" you would move one step down (which after the first move would land you at the point $(0, 1)$). Flip the coin again and repeat, i.e., if heads take one step to the right, if tails take one step down.

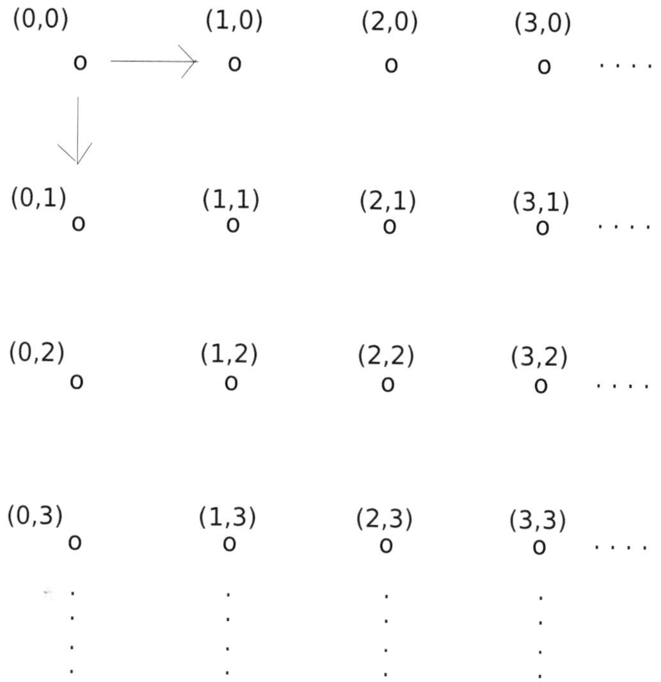

FIGURE 17.2: Discrete Random Walk

(i) Prove that the number of ways you can arrive at a given point is the sum of two numbers, viz., it is the sum of the number of ways you can arrive at the point immediately above and the number of ways you can arrive at the point immediately to the left.

(ii) There is 1 way of arriving at $(0,0)$, $(1,0)$ and $(0,1)$. Using this information and (i), assign to a few other points the number of ways of arriving at these points.

(iii) Do you see any relationship between the array of numbers that you have generated in (ii) and $\binom{n}{k}$ for various values of n and k?

Exercise 17.15 Pascal's Triangle

Do you recall Pascal's Triangle from a previous mathematics course? It is generated in the following way, see Figure 17.3 below. On the first line put a 1. On the second line put a 1 and a 1. On the third line put a 1, then a 2 then a 1. Note that the 2 is the sum of the two 1's closest to and above it. Proceed to the third line, with a 1, 3, 3 and 1. Note that each 3 is the sum of the two numbers, a 1 and a 2, closest to and above it. Continue in this fashion. Do you see any connection with Pascal's Triangle and the Random Walk of the previous problem?

Exercise 17.16 The Fibonacci Sequence

(i) The Fibonacci sequence is obtained in the following way: Start the sequence with the number 1, followed by the number 1. Now use the rule that the next number is the sum of the previous two, viz., $2 = 1 + 1$. Continue in this way obtaining $3 = 2 + 1$, $5 = 3 + 2$,

```
              1
            1   1
          1   2   1
        1   3   3   1
      1   4   6   4   1
    1   5  10  10   5   1
  .   .   .   .   .   .   .
```

FIGURE 17.3: Pascal's Triangle

8 = 5+3, and so on. Thus the first few numbers in the Fibonacci sequence are 1,1,2,3,5,8,13, ... Can you find the Fibonacci sequence in Pascal's Triangle?

(ii) The Fibonacci sequence is connected to numerous other subjects, so many in fact that there is a journal devoted to the subject. Look up as many subjects as you can that are related to the Fibonacci sequence.

Exercise 17.17 The Binomial Theorem

(i) Consider the following sequence of calculations: $1x+1y$; $(1x+1y)^2 = 1x^2+2xy+1y^2$; $(1x+1y)^3 = 1x^3+3x^2y+3xy^2+y^3$; $(1x+1y)^4 = 1x^4+4x^3y+6x^2y^2+3xy^3+y^4$; and so on. Do you see any connection between these calculations and Pascal's triangle?

(ii) The numbers $\binom{n}{k}$ are called *binomial coefficients* because they appear in the following formula called the binomial formula. Thus:

$$(a+b)^n = \sum_{k=0}^{n} \binom{n}{k} * a^{n-k} * b^k.$$

Write this formula out for $n = 1, 2, 3, 4, 5, \ldots$. Do you see a pattern in the binomial coefficients? If you get stuck on this problem, look at Pascal's Triangle and the Random Walk above.

(iii) Can you show using a sequence of "legal moves" that

$$\binom{m+1}{k} = \binom{m}{k} + \binom{m}{k-1}?$$

Do you see any connection between this last equation, Pascal's triangle, cf., Exercise 17.15, or the Random Walk, cf., Exercise 17.14?

(iv) *An Example of a Cellular Automaton.* A one-dimensional cellular automaton consists of a row of cells, each cell containing an initial number, and a set of rules specifying how these numbers are to be changed at each step. Consider such an automaton with an initial state of:

...0100000000000....

Suppose that there is just one rule which specifies at each step how to go from one state to the next, and this rule is "replace the number in each cell by the sum of itself and its left neighbor." Thus following the above initial state we have the next state:

...0110000000000....

The next states in succession are:

...0121000000000...

...0133100000000...

...0146410000000...

and so on. Do you see any connection between this cellular automaton and Pascal's Triangle? The Binomial Theorem?

Exercise 17.18 Counting Triangles with Mathematics

This is a favorite problem of Tony Starfield, cf., [660, 661]. In the two figures below, how many triangles are there? You can use whatever mathematical tools you have, starting with "brute force" counting. It turns out that there is a slick way to do the problem with mathematical objects called matrices.

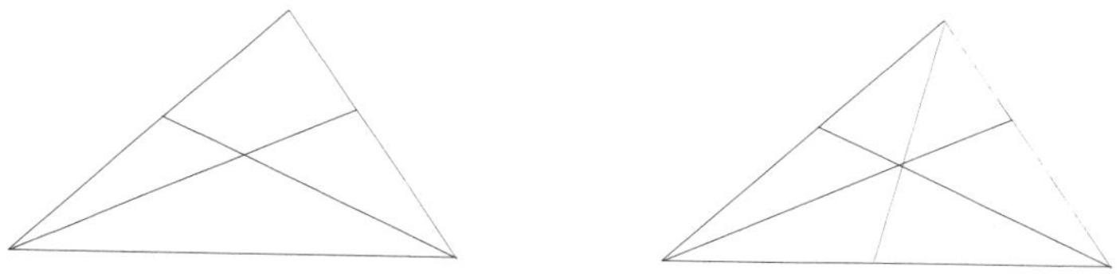

Exercise 17.19 Synergistic Effects on You of Chemicals in Combination

A friend of mine read[7] that as of 2001 there were 75,000 synthetic chemicals registered in the U.S. for use, and about one-tenth of these make up the bulk of chemicals produced and marketed at large levels. (This number needs to be substantially revised upward each year.) Many have been tested individually for carcinogenicity but not in combinations and not for such things as ability to disrupt endocrine systems.

Unfortunately we have discovered that life is more complex than some of us first imagined. For example, if two or three or more chemicals are present simultaneously things can happen that are not observed when the same chemicals are observed one at a time. This phenomenon is an example of *synergy*.

(i) Your first question is this. How many experiments would have to be done to test all possible combinations of 75,000 chemicals taken two at a time?

(ii) How many experiments would have to be done to test all possible combinations of 60,000 chemicals taken three at a time?

(iii) How many people-days do you estimate it would take to do all of the experiments in parts (i) and (ii)?

(iv) If we did all of the experiments in parts (i) and (ii) would we have complete information about all possible synergistic effects of the 75,000 chemicals? What fraction of information about possible synergistic effects do you estimate would remain unknown?

(v) How many possible experiments are there if you did toxicity experiments (in which order of the chemicals looked at does not matter) [8] with each of the 75,000 chemicals 1 at a time, 2 at a time, 3 at a time, . . . , up to and including 60,000 chemicals at a time? Hint: See Exercise 17.17 part (ii).

(vi) Redo parts (i) through (v) replacing 75,000 by 7,500. Is the problem of testing all combinations manageable? Is the problem of testing all permutations manageable?

(vii) On the bright side, I have heard it said that there are 12 main pesticides that are used on the apple crop in the United States. How many experiments would have to be done to test all possible combinations of 12 chemicals taken two at a time?

---

[7]See http://www.rachel.org for lots of information on pesticides and other toxic substances and for all back issues of *Rachel's Environment & Health Weekly*. If you do not have access to the Web you can write or fax Environmental Research Foundation, P.O. Box 5036, Annapolis, MD 21403 USA; phone 1-888-2RACHEL; fax: (410) 263-8944.

[8]A pharmacist told me that in many instances the order of exposure to a set of chemicals does matter!!

(viii) How many experiments would have to be done to test all possible combinations of 12 chemicals taken three at a time?

(ix) How many experiments would have to be done to test all possible combinations of 12 chemicals taken, 1, 2, 3, ... , 10, 11 or 12 at a time?

(x) Just one of the 75,000 chemicals mentioned above is formaldehyde. Try to estimate what your daily exposure to formaldehyde is. You can start by finding out how many things you come into contact with on a regular basis have formaldehyde in them. A friend of mine who is chemically sensitive (to formaldehyde, for instance) has a difficult time finding products that do not contain this chemical. Just to start the list, formaldehyde is contained in building materials, cosmetics, home furnishings, textiles and so on. "There is no identifiable safe level" of formaldehyde, says Peggy Jenkins, who works for the California Environmental Protection Agency's Air Resources Board, which funded a report on formaldehyde published in the January 1, 1999 issue of *Environmental Science & Technology*, a journal published by the American Chemical Society. This study is the first of its kind in more than a decade. Since repeated exposure to even low levels of formaldehyde may increase a person's risk of developing cancer, why so few studies?

(xi) Find at least one other chemical of the 75,000 to which you might be routinely exposed and which may cause ill health effects. How many new chemicals (largely untested?) are introduced commercially every year?

An example of synergy[9] of an over-the-counter medication and a pesticide that had horrific effects on one man's life, see Section 4.1.

Exercise 17.20 Inequalities and Binomial Coefficients

This exercise requires that you know a little about how to manipulate inequalities. First of all, we write things like "5 is greater than 3" in mathese as $5 > 3$. A statement in mathese, such as $5 > 3$ or $m > 3$ where m is a number, is called an *inequality*.

If an inequality is multiplied on both sides by the same positive number, or if the same number is added to both sides of an inequality, then the sense (the direction of the $>$) does not change.

(i) Using what we have just stated can you show that in a population the number of pairs increases by more than a factor of 4 if the population doubles? In mathese can you show that

$$\binom{2n}{2} > 4\binom{n}{2}.$$

(ii) Can you show that

$$\binom{3n}{2} > 9\binom{n}{2}.$$

(iii) Can you show that for any positive, whole number $k > 1$ that

$$\binom{kn}{2} > k^2\binom{n}{2}.$$

Exercise 17.21 A Theorem of Erdös

Paul Erdös, was a very prolific and unusual mathematician. He wandered the world visiting one group of mathematicians after another, proving theorems, befriending students and professors and writing joint papers—more papers than any other mathematician ever. A mathematical result, of the type for which he was famous, is the following.

Suppose you have 6 or more dots and that any pair of dots are joined with a line segment which may be either blue or red. Then there is either a triangle consisting of red line segments or a triangle consisting of blue line segments.

[9] See, for example, the newsletter of the Center for Public Integrity, Vol. 4, No. 3, August 1998, Web site www.publicintegrity.org, for the complete story.

Consider the case of 3 dots. It is easy to draw an example for which the above conclusion fails. All we need do is draw a triangle with two sides of one color (say red) and the third side of a different color (say blue). Find a similar example for the case of 4 dots. Find an example involving 5 dots for which the conclusion above fails.

Amazingly, for 6 dots you cannot find an example where the conclusion fails. Can you find a proof of the above statement in italics for the case of 6 dots or more? You can rephrase the above statement as:

If six or more people enter a room, then there will be three people who already know each other or there will be three people who do not know each other.

For more information on this subject see Graham, Ronald L.; Rothschild, Bruce L.; Spencer, Joel H., *Ramsey Theory*, Wiley, New York, 1990.

It turns out that this problem is closely related to extremely difficult problems. For example, what is the smallest number of dots necessary (all pairs joined by either a blue or red line segment) such that one is assured of having a quadrilateral with all 4 sides the same color? One can repeat this question with quadrilateral replaced by pentagonal (5-sided) figure and so on.

Exercise 17.22 A Horse Race

Suppose you have a horse race with four horses, A, B, C and D. How many finishes are possible if all possible ties in any position(s) are allowed?

Exercise 17.23 The Game Show Problem

Suppose there are three doors. Behind one door there is one-hundred dollars, behind each of the other two doors there is only one dollar. To play the game you pick one of the doors. Then the master of ceremonies opens one of the other two doors that you did not pick and shows that behind that door there is only one dollar. Your goal is to pick the door with the one-hundred dollars behind it. After your initial selection of doors, and after the master of ceremonies discloses one of the "one-dollar doors," you are allowed two options: to keep the door you initially chose or to change your selection to the other remaining door.

(i) What are your chances of picking the one-hundred dollar door if you stay with your initial selection?

(ii) What are your chances of picking the one-hundred dollar door if you switch?

If you do not believe the answers to this problem you can simulate this game—either by playing it many times by hand or many times on a computer—and observing the probabilities empirically.

Exercise 17.24 Silver and Gold Coins in Chests of Drawers

This problem is similar to the previous one. Suppose you have three chests each with two drawers. In one chest there is a silver coin in each drawer, in one a gold coin in each drawer and in the third chest there is a silver coin in one drawer and a gold coin in the other drawer. Suppose that you open one drawer in one of the chests and you see a silver coin. What are the chances that the second drawer in that same chest has a gold coin?

Exercise 17.25 Can the Number of Combinations Taken Two at a Time Equal the Number of Combinations Taken Three at a Time?

Suppose you have n points in the plane. Suppose you join each point with all others, i.e., you draw "all the possible roads between cities." Is there a value of n for which the number of line segments, i.e., "roads," equals the number of triangles formed by the "roads"?

Part V

Box Models: Population, Money, Recycling

Chapter 18

Some Population Numbers

18.1 Counting People in the World

How many people were on the planet in any given past year, how many are there on earth today? These questions are not easy to answer and any numbers should properly be intervals of numbers. For example, pick a time like July 1994; the CIA *World Fact Book 1994* stated that at that time the world human population was 5,643,289,771. The implied accuracy is certainly misleading. Why?[1]

There are various sources that give estimates of the world population, such as the CIA and the Worldwatch Institute, cf., [662, p. 75]. The fact that these and others agree can be attributed to the fact that almost everyone relies on some of the same key sources for data. There are many possible reasons for errors, not the least of which are occasionally unreliable counting methods and political motivation to alter the data one way or the other for a host of reasons. I have watched the estimates change over the years. For example, when did the world's human population reach its first billion? Those estimates have varied from 1850 to 1800 over the past decade or two. At the moment the year 1802 appears to be the accepted estimate. Thus every population number we give must be assumed to have an error associated with it, whether stated or not.

The annual increase in world population has varied from about $38(10^6)$ in 1950 to $75(10^6)$ in 2006, peaking at $88(10^6)$ in 1989. The world population hit $6.5(10^9)$ in late 2006, having added the last 1.5 billion since 1987. At the end of 2009 the world human population was about $6.8(10^9)$. Are these numbers large, small or just right? A person's individual view depends on a host of personal values, such as depth of concern over threatened and/or endangered species and loss of habitat. However, there are less arbitrary ways to think about these population numbers. For example, in 2006 the time it took the world population to double was about 50 years up from a doubling time of about 46 years in the first half of the last century. If the standard of living

[1]To start with, the population would not be constant for one month, and the data upon which estimates like this are based are not accurate down to the individual person. How many figures do you think are really significant in such a CIA estimate?

of everyone is not to decrease then everything from water supplies, housing and medical care to food production has to double in 50 years. Can this be done? This is doubtful, given the considerations of Part I. The answer to any question such as: How many people can the earth support? depends on the standard of living/quality of life used in the model, cf., [102]. For example, if the main ingredient in the world diet is algae, more people can be supported than if people have access to grains like rice, corn and wheat.

Human Population Numbers in History. In Table 18.1 I give some working estimates of world human population at various times in history, culled from [168, pp. 2–23], [102] and the Web.[2] I have simplified things a bit by rounding off some of the numbers.

TABLE 18.1: Estimated World Human Population in History

Year	8000 BC	1 AD	1600	1802	1928
Population	$5(10^6)$	$2.5(10^8)$	$5(10^8)$	10^9	$2(10^9)$
Year	1961	1974	1987	1999	2006
Population	$3(10^9)$	$4(10^9)$	$5(10^9)$	$6(10^9)$	$6.5(10^9)$

Exercise 18.1 Doubling Times of the World's Human Population

In this exercise I want you to estimate the time in years that it was taking the world human population to double in 1600, 1800, 1930, 1975 and 2000.

(i) The time it takes some quantity to double is called the *doubling time* of that quantity. The Ehrlichs estimate, cf., [168], that the doubling time of the human population just prior to the agricultural revolution (about 8000 B.C.) was about 34,000 to 35,000 years. They estimate that the total human population in 8000 B.C. was about 5 million. In 1 A.D. the population was around 200 to 300 million. To estimate the doubling time of the the human population between 8000 B.C. and 1 A.D. first figure out how many doublings it takes to change 5 million into 200 or 300 million, say 250 million. Then divide 8000 years by this number of doublings to get the "average" number of years per doubling.

(ii) The human population increased to $\frac{1}{2}$ billion in about 1600, about the first year there was any historical data on which to base a population estimate. Estimate the doubling time of the total human population between 1 A.D. and 1600 A.D.

(iii) Estimate the doubling time of the total human population between 1600 A.D. and 1800 A.D. Hint: Look at Table 18.1.

(iv) Estimate the doubling time in 1930, in 1975 and in 2000.

(v) What is the doubling time of the world human population at the time you are reading this? How does that compare to the doubling time throughout human history? Can you think of anything that humans depend on that will not double during the next doubling time of the world human population?

(vi) The Ehrlichs, cf. [168], say: *"The story of human population growth is not primarily a story of changes in the birth rate, but of changes in the death rate."* What implications does this statement hold for us today?

[2]See, for example, the sources cited in http://en.wikipedia.org/wiki/World_population.

18.2 A Fundamental Axiom of Population Ecology

I was once reading a famous little ecology book, cf., [106], wherein I came across what might possibly be a very important idea. In this book is the statement that

"the reproductive effort (of a population of a species) makes no difference to the eventual size of the population (of that species)." In other words, if a particular population of some species breeds very fast—or very slowly—it has no effect on the eventual size of the population.[3]

So if rate of reproduction does not determine eventual population size, what does? The key concept is the term *niche* (of some living being), defined as the "job" or functional place of the being in question in the environment. The very important idea is this: the eventual size of a population of some species is determined by the number of niches in the environment available for occupation by that species. I find this idea so important that I shall enshrine it as an axiom.[4] The total number of niches of a species available in an environment is the *carrying capacity* of that environment for that species.

Ecology Axiom on the Size of a Population: *The eventual size of the population, i.e., the number of individuals, of a species is determined by the number of niches available for occupation by that species, the carrying capacity, not by the rate of reproduction.*

Think of a glass of water. It can be filled by a dripping faucet, very slowly. It can be filled with the faucet fully on, very quickly. The rate at which the glass is filled does not determine the final volume that the glass holds. The

[3] Of course, we rule out the case where a population does not breed at all; since in that case the population would eventually have zero individuals in it.

[4] My enthusiasm is not shared by all biologists, cf., [655], for example. Some biologists see the following axiom as a mere tautology, nothing more than circular reasoning. I contend that some useful estimates can be made. For example, given a population of grizzly bears living in some part of the world (some ecosystem), each bear needs to eat an approximate minimal amount in a season. Given a list of the things known to be considered food by grizzly bears, the part of the world, i.e., ecosystem, in question can produce some estimated amount of food: salmon, berries, other bears (cannibalism), and so on. Known estimates of how productive a given ecosystem can be of food sources for the bears can lead to estimates on the maximum number of bears that particular ecosystem can support. That number can be adjusted due to fluctuations in climate, competition from other species and so on. The number of niches available for a given species depends upon (is a function of) *everything* in the environment: available resources, competition/support from other species, etc. In fact, the number of niches for the bears is likely not computable (not estimable) without simultaneously estimating the number of niches for each of a number of related species. Admittedly actually estimating simultaneously the number of niches for each of a collection of species could be a difficult problem. I take the position that the abstract concept of niche, whether calculable/useful in the field or not, is very useful conceptually.

only question for biologists is: "Can the eventual or final volume of 'the glass' be defined?" That is, can the number of niches available to a given species be determined.

Whether or not the carrying capacity of earth for humans can be precisely determined, it can be argued that said carrying capacity—whatever it is—will be *decreasing* in the near future due to the impending shortages of various resources and the lack of planning for changes such as climate change that science has warned us of.

Exercise 18.2 How Many Niches for Humans Are There, Were There, Will There Be?

(i) Do you believe the following statements? Discuss.

(a) Table 18.1 tells us the approximate number of humans on earth at various times.

(b) Table 18.1 tells us the number of niches in the global environment filled by humans.

(c) Table 18.1 measures the growing ability over time of humans to create niches for humans in the world's environment, using ideas, inventions, technology, resources, taking niches from other species, and so on.

(ii) Discuss the following statement. Humans have continually created niches for themselves that did not previously exist for humans. These new niches are all destined to be filled according to the above ecology axiom on populations.

(iii) Discuss the statement. While creating more niches for humans, humans have decreased the number of niches available for many other species.

(iv) Discuss. Many species that compete with humans for niche space, such as lions, tigers, elephants, grizzly bears, wolves, prairie dogs, certain plants and so on, will only exist in the future in niches specifically set aside for them by humans. For example, in Royal Chitwan Park, Nepal, poachers are prevented from poaching by an armed force.[5]

At the entrance to Matopos Park in Zimbabwe a sign declares that poachers will be shot on sight. (The sign then goes on to tell visitors what they must do to not be taken for a poacher. This sign is usually read quite carefully!) On December 18, 2000 the BBC News carried a story about the rhinos in Tsavo National Park, Kenya. At one time there were thousands of rhinos in this park, now there are a few dozen each of which have a constant military escort (at a discrete distance so that the rhinos have some privacy). This has been necessary to keep the poachers from killing the rhinos.

(v) Animals in a park preservation area that are poached are clearly put to a use (necessary or unnecessary?) in some human's niche. Are animals that are allowed to live in a park preservation system also being put to some human use?

(vi) Discuss the following statement. Since many humans have not developed social mechanisms for keeping their numbers at a comfortable level below the maximum human carrying capacity, all technological innovations ultimately lead to increased numbers of humans living in misery.

(vii) Are Western Europe and Kerala, India examples where humans have stabilized their population within carrying capacity?

(vii) William R. Catton, Jr., [74], argues that humans have used fossil fuels—coal, oil, natural gas—to create far more niches for humans than were possible prior to the widespread use of said fuels. He argues that the human population will crash as the fossil fuels run out and the abuse of the environment that has accompanied their use takes its toll. Discuss.

(viii) "The greater the human population, the fewer de facto freedoms remain to individuals." Discuss this comment.

---

[5]Social unrest in Nepal may have affected this capacity of the society to protect the park's habitat for animals and plants.

It appears quite likely that the recent growth in human population is not sustainable; and is, in fact, beyond the long term carrying capacity. Barring some unforeseen innovations, Nature will be eliminating human niches and the humans that occupy them—even in the United States, as the next section indicates.

18.3 Counting People in the United States

The Fastest Growing Industrialized Country in the World. Before we get into the mathematical principles of population growth, I want to do a little counting of noses at home. I have often heard people say that population growth in the United States is not a problem. I would like to banish that thought from your consciousness once and for all. The U.S. human population in 1993 was about 260 million, and growing at about 3 million people per year. With mathematical precision the U.S. population hit 300 million in October 2006.[6] The United States is the fastest growing industrialized nation on earth.[7]

The Numerical History of the United States Population.

TABLE 18.2: U.S. Population According to the U.S. Census Bureau

Year	1900	1915	1950	1967	2006
Population	76,094,000	100,546,000	152,271,417	$2(10^8)$	$3(10^8)$

From Table 18.2 we see that the U.S. population hit 100 million in 1915. Then 52 years later, in the last half of 1967 the second 100 million was added. Then 39 years later the U.S. population hit 300 million.

The much mentioned "baby boom" that followed World War II created a peak of 69.9 million Americans under 18 years old, and this peak was hit in 1966. Well guess what? The Census Bureau tells us that near the end of 1998 the number of Americans under 18 reached 70.2 million, and unlike the

[6]The official government estimate, at any given time, of the population of the United States can be found at http://www.census.gov.

[7]For example, in 1992 there were 4.1 million babies born, with 1.9 million *more* births than deaths. In that same year there was an estimated net immigration of 1.1 million people. In 1992 the U.S. Census Bureau projected that by the year 2000, the U.S. population would reach 275 million, more than double the 1940 population. In fact, the 2000 year census counted 281,421,906 Americans. At the end of 2009 the census estimated more than 308 million Americans.

previous "boom" this one shows no sign of ending. In fact, the U.S. Census estimated the under 18 population to exceed 72 million in 2000, to exceed 73 million in 2003 and 2004.

Now even as the number of people under 18 rise, their percentage of the population has declined from 36% in 1960 to 26% in 1996. This means that there are a lot of middle-aged and old folks as well! This does not sound like a stable population to me. In fact, given the material demands to which Americans are habituated, this is an exploding population from the Earth's point of view.

Consumption. The reason U.S. population growth is so important is this: U.S. citizens on average hammer the planet hard as they satisfy their demands for energy and other resources. The good news is that there is much room for increased efficiencies. The bad news is that there is little movement toward those increased efficiencies in 2006, and many in the developing world are starting to follow the U.S. model of consumption.

Exercise 18.3 Average U.S. Cititzen Demand for World's Resources

(i) At the time you read this what percentage of the world's population lives in the United States? For example, in 2006 we have $\frac{3}{65} \approx .046$, i.e., 4.6%.

(ii) At the time you read this what percentage of world energy consumption is taken by Americans? For example, look at oil, coal, natural gas, electricity and renewable energy.[8]

(iii) At the time you read this look at U.S. consumption of key natural resources such as agricultural products, forest products, iron/steel, aluminum, copper and other metals and water.

(iv) Averages can be misleading. According to the U.S. Census Bureau: in 2004 11.9 percent of all U.S. households were "food insecure" because of a lack of resources. What this means is that these people are going hungry because they cannot afford to buy food sufficient to prevent malnutrition over time. About 31 million in the U.S. were hungry in 1999, compared to 33.6 million in 2001, compared to 38.2 million in 2004. Is the number of hungry in the United States growing faster than, slower than or the same as the total population is growing? There are a number of Web sites that deal with "Hunger in America." What is the status of hunger in the U.S. at the time you read this? Are these people consuming resources at a rate typical of the "average" American?

(v) As the developed world continues its consumption pattern while developing countries such as China and India increase their consumption per person, in some respects using the U.S. as a model to emulate, do you foresee any problems?

At the end of 1990 I read that: *"Each year, Japan uses 30 million 'disposable' single-roll cameras, and Americans toss away 18 billion diapers and enough aluminum to make about 6000 DC-10 jet airplanes."*[9] To put this in perspective I read in 1992: *"The amount of aluminum the United States throws away in beverage cans is greater than the total use of the metal by all but seven nations."*[10] In 2009[11]: *"... 1 year's worth of*

---

[8]In late 1993 the U.S. with about 5% of the world's population was consuming about 24% of the world's energy. This was equivalent respectively, to about 3 Japanese, 12 Chinese, 33 Indians, 147 Bangaladeshis, 281 Tanzanians or 422 Ethiopians.

[9]Alan Durning, "How Much Is Enough?," *WorldWatch*, Nov/Dec 1990, pp 12–19.

[10]John E. Young, "Aluminum's Real Tab," *WorldWatch*, Vol.5, No. 2, March-April 1992, pp. 26–33.

[11]www.thegoodhuman.com

America's trashed cans would provide enough aluminum to make more than 8,000 747s." Also note the quotes from 2009: *"Yes, that's every second of every minute of every day—Americans throw 1,500 aluminum cans in the trash every second. Not recycle—throw away. Back in the 1990s, we actually recycled more than 60% of our aluminum cans, but now we are only recycling 51.1%! ...And according to E Magazine, we threw away 11 billion cans in the 1970s, 29 billion a year in the 1980s, 35 billion a year in the 1990s, and 46 billion a year since 2000."* These numbers tell us that at least as far as aluminum cans in the U.S. go, the recycling effort is not at all keeping up with population growth. As a project you might want to check how the recycling effort on other fronts is going, in the U.S. and elsewhere.

Exercise 18.4 What is the World's Definition of Middle Class?

An interesting book,[156], breaks the world's human population (in the early 1990s) into three broad classes: the consumers, the middle income and the poor—defined by average annual income and life-style. "The world's poor—some 1.1 billion people—includes all households that earn less than $700 a year per family member.[12] They are mostly rural Africans, Indians, and other South Asians. They eat almost exclusively grains, root crops, beans, and other legumes, and they drink mostly unclean water. They live in huts and shanties, they travel by foot, and most of their possessions are constructed of stone, wood, and other substances available from the local environment. This poorest fifth of the world's people earns just 2 percent of the world income.

"The 3.3 billion people in the world's middle-income class earn between $700 and $7,500 per family member and live mostly in Latin America, the Middle East, China, and East Asia. This class also includes the low-income families of the former Soviet bloc and the western industrial nations. With notable exceptions, they eat a diet based on grains and water, and lodge in moderate buildings with electricity for lights, radios, and, increasingly, refrigerators and clothes washers. (In Chinese cities, for example, two thirds of households now have washing machines and one fifth have refrigerators.) They travel by bus, railway, and bicycle, and maintain a modest stock of durable goods. Collectively, they claim 33 percent of the world's income.

"The consumer class—the 1.1 billion members of the global consumer society—includes all households whose income per family member is above $7,500. Though that threshold puts the lowest ranks of the consumer class scarcely above the U.S. poverty line, they—rather, we—still enjoy a life-style unknown in earlier ages. We dine on meat and processed, packaged foods, and imbibe soft drinks and other beverages from disposable containers. We spend most of our time in climate-controlled buildings equipped with refrigerators, clothes washers and dryers, abundant hot water, dishwashers, microwave ovens, and a plethora of other electric-powered gadgets. We travel in private automobiles and airplanes, and surround ourselves with a profusion of short-lived, throwaway goods. The consumer class takes home 64 percent of world income—32 times as much as the poor.

"The consumer class counts among its members most North Americans, West Europeans, Japanese, Australians, and the citizens of Hong Kong, Singapore, and the oil sheikdoms of the Middle East. Perhaps half the people of Eastern Europe and the Commonwealth of Independent States are in the consumer class, as are about one fifth of the people in Latin America, South Africa, and the newly industrializing countries of Asia, such as South Korea. ...Just as the world's top fifth—the consumer class—makes the remainder appear impoverished, the top fifth of the consumer class—the rich—makes the lowly consumers seem deprived. In the United States, for example, the highest paid fifth of income-earners

[12] I asked Alan Durning in person if the $700 included informal economic inputs like backyard gardens. He said that it did not.

takes home more than the remaining four-fifths combined, and top corporate executives earn 93 times as much as the factory workers they employ."[13]

Along the lines of the last paragraph the U.S. Congressional Budget Office found for after-tax incomes that in 1980 the richest fifth of the U.S. had 8 times the income of the poorest fifth. By 1989, the ratio was *more than 20 to one.*[14]

From 1992 to 2009 the "wealth gaps" and "wealth/poverty ratios" have all gotten much larger. For example, in 2001 the wealthiest 1% of households in the U.S. owned 32.7% of the nation's household wealth. The 95th through 98th percentiles (the next 4%) owned 25%. The bottom 50% owned 2.8% of total net worth, [109, p. 6]! As a research project update Alan Durning's analysis for the year you read this. Mathematically it is interesting to note that the global distribution of wealth pattern is duplicated in the United States, but on a different scale. Does this "concentration of wealth" pattern hold in most countries in the world?

Mathematics of Counting Can Be Controversial. Before leaving this section I want to note that the simple act of counting the number of Americans is not as simple today as it was in the 18th century when there were about 3.9 million Americans. For example, the people who do the U.S. census estimate that about 4 million Americans were missed in the 1990 census. To correct for these errors the government's people-counters wanted to introduce some sophisticated mathematical techniques known generically as statistical sampling. Why did Thomas Petri, representative to Congress from Wisconsin, in the summer of 1996 introduce a bill that would ban the use of sampling? Because he believed it would negatively influence his state. How? The number of representatives to Congress that each state gets is determined by census data. A hundred billion dollars of federal money is apportioned among the 50 states based on census data. You see, there is a lot riding on the mere counting of heads.[15]

Immigration Can Be Controversial But Not Well Understood. A Web site that follows the numbers regarding population growth in the U.S. is www.numbersusa.com. They have modeled the future based on current and alternative trends, and they predict that if the U.S. had an immigration rate of about a quarter million a year along with the birth rate which has been

[13]The number 93 has been replaced by a much higher number in 2006. What is this ratio at the time you read this?

[14]I found this information in *EXTRA! The Magazine of Fair*, July/August 1994 Vol.7, No. 3, p. 12. *EXTRA!* points out the following quote from Rush Limbaugh, *The Way Things Ought To Be*, New York: Pocket Books, 1992, p. 70: "Don't let the liberals deceive you into believing that a decade of sustained growth without inflation in America [in the '80s] resulted in a bigger gap between the haves and the have-nots. Figures compiled by the Congressional Budget Office dispel that myth." I advise you not to take my word or that of Mr. Limbaugh. Go to the library and research the topic thoroughly; you may be amazed at what you find. Relevant to this topic is an article by Paul Krugman in the Sunday *New York Times Magazine*, October 20, 2002.

[15]As reported on the July 9, 1996 morning edition of National Public Radio. Actually some mathematicians thought that any errors in the sampling techniques contemplated by the U.S. Census folks would have been to Congressman Petri's liking. It is interesting to note that advanced mathematics gets involved in congressional debates

maintained for the last few years, the U.S. population would eventually stabilize. Congress, however, has allowed immigration rates of closer to 1 million a year. If this rate of immigration is maintained, the population of the United States is projected in their model to be greater than 390 million in 2050, and not stabilize but grow indefinitely. Immigration is both a positive and negative, and a politically charged phenomenon.

Thus, one need not be against all, or even substantial, immigration in order to achieve a stable U.S. population. It is a quantitative issue. However, if we continue current policies[16] the U.S. will continue to grow, equaling the current population of China, possibly within 100 years. Population growth has many causes; stabilizing populations requires a big picture.

There are those who have focused on our neighbor to the south, considering immigration from Mexico a problem. Individuals have proposed building a wall on the border between Mexico and the United States, and as of 2009 much of it has been constructed. However, I have not heard one such person mention the fact that the North American Free Trade Agreement, NAFTA, passed in 1994 destroyed, among other things, one of the major accomplishments of the Mexican Revolution, the Ejidos program. This program guaranteed land-rights to indigenous Mexicans. Destruction of the Ejidos program drove over 1.5 million Mexican families from their land and left them in search of work, cf., [337, p. 87]. This is, of course, just one of the pieces missing from the usual picture; for example, U.S. agricultural policies which subsidize U.S. corn to the extent that it can be exported and sold at less that the cost of production have made it difficult to survive as a corn farmer in Mexico, cf., Section 5.1. I leave it up to you to finish creating your own big picture.

Population Dynamics. According to the United Nations Population fund any increase in human population will exacerbate climate change challenges, cf., UNFPA State of World Population 2009. With business as usual, more people mean: more fossil fuels combusted, more forests cut, more cows. As sea levels rise and local climate patterns change, significant human migrations will occur. In this section we address the simplest of questions about how populations can behave mathematically. The more detailed study of how impacts are distributed geographically and socially I leave for you to pursue.

[16]The United States does not have an official population policy.

Chapter 19

Basic Mathematical Patterns in Population Growth

19.1 Schwartz Charts Are Box-Flow Models

I first saw the mathematical model[1] for population growth we are about to study, i.e., *the Schwartz chart*, in an article by Richard H. Schwartz, cf., [619, pp. 38–41].[2] Using Schwartz charts we can get a good understanding of how populations grow (or shrink)—amazingly good, considering that we employ only a few simple rules that require only addition and multiplication to fill out the chart. I have yet to meet a student who cannot get started and make real progress using this model.

Our goal is to find some fundamental patterns of counting that "explain" to some degree why populations grow the way they do. That there are any such patterns at all is one of the wonders of Nature. To begin with let's focus our attention on a specific population, and ask specific questions—and see where it leads.

Question #1: How long does it take a population to double, i.e., what is the *doubling time* of the population? I think it obvious that this is a crucial thing to know about any growing population. If a population doubles in say 20 years, food production (and/or food imports), teacher training (or imports), number of doctors and many more things all have to double in 20 years if the average standard of living is not to decline.

Question #2: What determines the doubling time? What variables[3] are most important in calculating the doubling time?

Just the numerical behavior alone of the population of any county, any country or the world, can be quite complicated. The trick is to think of (guess at) the most important variables, make a mathematical model of the population (in our case a Schwartz chart) that incorporates these variables and then see if the model fits the data we know.

[1] I will use Tony Starfield's definition of a model as a *purposeful representation* of some part of the real world.

[2] R. H. Schwartz gives some prior credit to L. Schaefer.

[3] Think of a variable as any quantity or entity that can be measured with numbers.

Picking Important Variables: Omitting Not So Important Variables. When making any mathematical model, decisions must be made. We must make an appropriate simplification of the real-life situation. Which factors are most important? Which are merely distracting details as far as our immediate goals are concerned?[4] What variables lie between the indispensable and the irrelevant? It is at this initial decision making stage that the values and hidden assumptions of the modeler can play a pivotal role. It is at this stage that ambient politics, physical, fiscal and social forces, being human, wanting the model to agree with what the modeler wants to be right about—all possibly come into play.

For a typical growing human population the variables that I will guess are important are: (1) *birth cycle*, which is defined to be the number of years from a female's birth to the beginning of her childbearing years; (2) *generation cycle*, which is defined to be the number of years from a female's birth until the end of her childbearing years; (3) *life expectancy*, which is the number of years any individual can expect to live; and (4) *family size*, which is defined to be the number of children per couple.

Exercise 19.1 Picking Variables for a Model that are Important at a Given Time

(i) Did I really pick the most important variables? Can you think of a population, real or imaginary, human or not, where at least one other variable would be extremely important?

(ii) The orbit of the "planet" Pluto can be considered a variable. Is this variable relevant to a mathematical model of the world's human population? If a planetoid about half the size of the earth were about to collide with the earth in 7 months, would the orbit of this planetoid be relevant to a mathematical model of the world's population?[5]

(iii) Can you think of a situation where disease would be an important variable in a population model?[6]

(iv) This part can become an open-ended project. Analyze various mathematical models or even models from other disciplines that use little mathematics. Look for any hidden assumptions, values or prejudgements that are built into the models. Look at old (new) studies on smoking and health. Is nicotine addictive? Does smoking cause cancer? Look at models of toxic waste dumps or other sources of pollution, do any modelers predict negative health consequences? Are any of these models discussed in the media? Also,

---

[4] Our purpose here is to see if we can find a model that captures the gross numerical behavior of the world's human population, or the gross numerical behavior of a particular country's human population.

[5] *The New York Times*, on 25 April 1996, reported on a paper published in the journal *Nature* by Italian and French mathematicians at the University of Pisa and the Observatory of the Côte d'Azur, respectively. The mathematicians calculated that within a million years or so there is a 50% chance that the planetoid, 433 Eros, will be nudged into an Earth-crossing orbit. This particular planetoid, or asteroid, is 14 miles in diameter and were it to hit the earth it would have about 10 times the destructive effect of the asteroid that is believed to have created the Chicxulub crater some 65 million years ago in the Yucatan Peninsula, cf., [8]. This asteroid collision is studied in [283, pp. 7–9]; this collision is among one of the possible causes for the mass extinctions (of dinosaurs, for example) at the end of the Cretaceous period. Such collisions are thought to have played a role in each of the five previous major global extinctions.

[6] We will look at a model of an epidemic, cf., Section 20.4.

look at models concerning the safety of the nuclear power industry, the fossil fuel industry, renewable energy.

A topic addressed in mathematical models is *risk assessment*. How safe is that pesticide? How dangerous is that pesticide? How risky is it flying, driving, eating artificial food additives, eating genetically engineered food? How risky is exposure to dioxin? An important situation to watch for is what I call the "zero-infinity" scenario. For example, you may be told that the risk in some situation (having an intercontinental nuclear missile in a silo next door to your home) is very small ("zero"), but the adverse consequences can be huge ("infinity"). Mathematics/science alone in such cases is not sufficient; building into the model the values and beliefs of the people involved is necessary.

Can fuzzy logic be useful (essential?) in making some models?

(v) Discuss the relevance for making mathematical models of the following quote from William James: *"The art of being wise is the art of knowing what to overlook."*

19.2 Our First Population Model: Simple Boxes and Flows

This first model will sacrifice some realism to achieve mathematical simplicity. I will introduce more realistic models later and in the exercises, but this simplest model contains a very important mathematical pattern, as it turns out.

The Schwartz chart is really a box-flow model written out on a sheet of paper.[7] Thus everyone in our population is put into one of four boxes according to age. The first box consists of Children, people 0 up to and including 15 years of age. The next box I will call the Reproducers, people over 15 up to and including 30 years of age. The next box holds Parents, people over 30 up to and including 45 years of age. The last box holds Grandparents, people over 45 up to and including 60 years of age. I now make the (unrealistic) assumption that everyone dies while a Grandparent, not sooner, not later. I also make the (unrealistic) assumptions that in each box half the occupants are male, half are female and that everyone marries while in the Reproducer box and has 4 children, viz., 2 males and 2 females. No children are produced by Children, Parents or Grandparents.

Exercise 19.2 Assumptions of Our First Model

(i) What are the numerical values (determined by the assumptions I have just made) for the four variables: birth cycle, generation cycle, life expectancy and family size?

(ii) State at least three changes you would make to our assumptions to make the model more realistic.[8]

---

[7] I will also discuss how write out a Schwartz chart on a computer using a tool called a *Spreadsheet*, such as EXCEL or OpenOffice.orgCalc.

[8] All manner of complications can be built into our model In the extreme, everyone can be in his/her individual box.

For the moment I will neglect many of the complications of real life: immigration, war, disease, access to food, water and so on. I now have to decide how many people to put in each box to begin with. I will start with a socially strange but mathematically simple setup, sort of a *Lord of the Flies* society with both genders equally represented. Assume there are 256 Children, 0 Reproducers, 0 Parents and 0 Grandparents for the first 15 years (see the first numerical row of Table 19.1). After 15 years, the 256 Children move to the Reproducer box and stay there from the 15th to the 30th year. In Table 19.1 this is represented by putting the number 256 in the second numerical row, Reproducer column. The 0 Reproducers move to the Parents box, thus we put a 0 in the second numerical row, Parents column. The 0 Parents move to the Grandparents box, thus we put a 0 in the Grandparents column, second numerical row. The 0 Grandparents leave the picture entirely. This and more is summarized in Table 19.1.

TABLE 19.1: Four-Child Family Model

Years	Children	Reproducers	Parents	Grandparents
0 to 15	256	0	0	0
15 to 30	512	256	0	0
30 to 45	1024	512	256	0
45 to 60	2048	1024	512	256
60 to 75	4096	2048	1024	512
75 to 90	8192	4096	2048	1024
90 to 105	16384	8192	4096	2048
105 to 120	32768	16384	8192	4096
120 to 135	65536	32768	16384	8192

The first row of Table 19.1 represents the number of people in each box for the first 15 years of our society. The second row of the table represents the number of people in each box from year 15 to year 30 of our society. The first column tells us which 15 year interval our society happens to be in. In the second row of numbers how do we know how many Children go in the Children box? Well, there are 256 reproducers in the second numerical row, thus there are $\frac{256}{2}$ couples. Each couple has four children, so the total number of children in the second numerical row is $\frac{256}{2} * 4 = 256 * 2 = 512$.

Exercise 19.3 Different Four-Child Family Model with the Same Basic Pattern

(i) Continue the process I have just started describing and check that each box in Table 19.1 is filled in correctly.

(ii) Instead of 256 for the Children and 0 elsewhere in the first numerical row, put any numbers you choose in this row and then fill out the rest of the chart using the same rules we just used. Save this calculation. At this time do you see anything that is the same about your new Table and Table 19.1?

Now I want to add one more column to Table 19.1, the Total Population (abbreviated to Total Pop) of a given row.

TABLE 19.2: Four-Child Family Model

Years	Children	Reproducers	Parents	Grandparents	Total Pop
0 to 15	256	0	0	0	256
15 to 30	512	256	0	0	768
30 to 45	1024	512	256	0	1792
45 to 60	2048	1024	512	256	3840
60 to 75	4096	2048	1024	512	7680
75 to 90	8192	4096	2048	1024	15360
90 to 105	16384	8192	4096	2048	30720
105 to 120	32768	16384	8192	4096	61440
120 to 135	65536	32768	16384	8192	122880

The first numerical row is easy: $256+0+0+0=256$. The next row gives: $512+256=768$. The next: $1024+512+256=1792$. I will let you finish the remaining six rows. Now let us use the following symbol, $P[15]$, to denote "the Total Population" at time 15 years. This is the Total Population of the first numerical row. Thus $P[15]=256$. Do you see then that $P[30]=768$? In summary we have:

$$P[15]=256; P[30]=768; P[45]=1792; P[60]=3840; P[75]=7680;$$

$$P[90]=15360; P[105]=30720; P[120]=61440; P[135]=122880.$$

Exercise 19.4 Calculating the Doubling Time in the Four-Child Family Model

(i) After 45 years all the 0s are gone from the rows in Table 19.2. Thus after 45 years the only number from the first row that survives into the fourth numerical row is the number of Children in the first row. Note that $3840*2=7680$, i.e., $P[60]*2=P[75]$. Is it true that $P[75]*2=P[90]$? Is it true that $P[90]*2=P[105]$? Do you see a pattern here? Describe it in your own words.

(ii) What is the doubling time of the total population in Table 19.2?

(iii) If $P[15]$ is "total population at time 15 years," ... $P[120]$ is "total population at time 120 years," what is $P[t]$?

(iv) In Exercise 19.3 part (ii) what is the doubling time after 45 years in the Table you created?

(v) Do you think the numbers used to fill in the first row of Table 19.2 have any effect on the doubling time of the total population after 45 years?

(vi) For what values of t in Table 19.2 is it true that $P[t+15]=2P[t]$?

(vii) Go back and redo the above model in Table 19.2, changing only one thing; namely, change the family size from 4 to 3. Can you estimate the doubling time? Finding the exact doubling time is not so easy as when the family size was 4. We will deal with this problem shortly. In any event, is your estimate of the doubling time greater or less than 15 years? Intuitively should it be greater or less than 15 years? Why?

(viii) If you replaced the 15 year birth cycle by 20 years, the 30 year generation cycle by 40 years and the 60 year life expectancy by 80 years, what do you get for a doubling time if the family size is 4? If the family size is 3? Can you think of a country that somewhat satisfies any of these new assumptions? Does this model account for immigration?

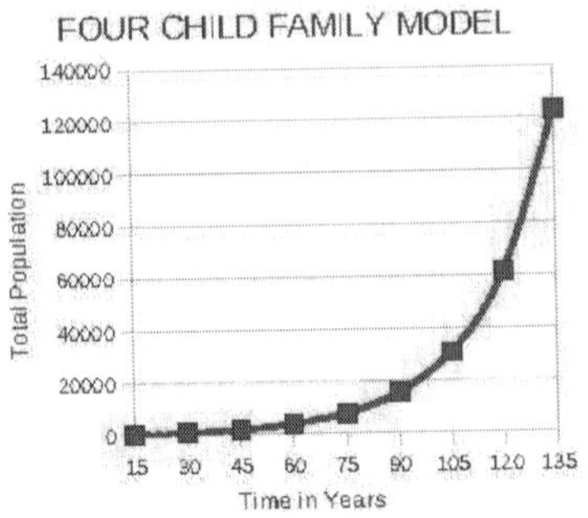

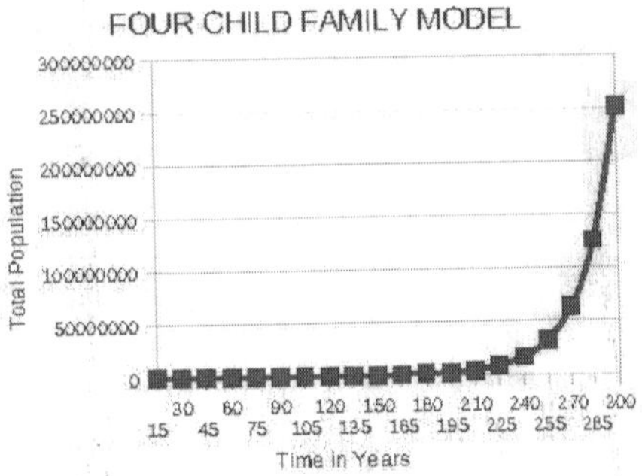

FIGURE 19.1: Total Population vs. Time

Now let's make a graph of the total population (of our 4 child, 15 year birth cycle society in Table 19.2) vs. time, cf., Figure 19.1.

Exercise 19.5 Making Some Graphs Yourself and One Way to Lie with Graphs
(i) Make a graph of world total population vs. time from Table 18.1.
(ii) Make a graph of United States Population vs. time from Table 18.2.
(iii) Compare the shape of the graphs from parts (i) and (ii) with the shape of the graph in Figure 19.1. Comment.
(iv) If one wanted to make the growth in population in the model of Table 19.2 look small, all that needs to be done is to extend Table 19.2 several rows to say row "285 to 300 years." Do this and draw the corresponding graph. What happens to the part of the graph between 0 years and 135 years?

More Population Models. In this next exercise we will explore some Schwartz chart (box-flow) models that are a little more complicated than what we have seen so far.

Exercise 19.6 Slightly More Complicated Models, Age Distributions and Momentum
Now that you know how to do Schwartz charts by hand (with a sheet of paper, pencil and/or calculator), I recommend that you find a computer that will do Schwartz charts with much less effort (after you have learned to work them). In my classes we have used computers with a program called EXCEL, which is called a "spreadsheet." (There are other similar spreadsheet programs which are called *open source* software, like OpenOffice.orgCalc.) You can still do the the following exercises without a computer, but it is a lot of fun and much faster to use a spreadsheet.
(i) Go back to our original Schwartz chart with birth cycle 15 years and family size 4. Let's change the generation cycle to 45. This means that not only will children be born to the reproducers, but to the parents as well. Suppose that all couples have 3 children while

in the reproducers column and 1 child in the parents column. Does this change the doubling time? Suppose instead that all couples have 1 child while reproducers and 3 children while parents. Does this change the doubling time?

(ii) You may find the answer to this problem surprising. Suppose we have two Schwartz charts with most of the same assumptions: 15 year birth cycle, everyone dies at 60. In one Schwartz chart every reproducer couple has 3 children (this is Exercise 19.4 part (vii)), and the parent and grandparent columns produce no children. In another Schwartz chart with the same overall assumptions let each reproducer couple have 1 child and each parent couple have 3 children. Compare the the doubling times of these two populations as best you can now. Come back and find the doubling times more precisely (to 3 significant digits say) after you have read the Section 19.5.

(iii) Let us suppose that not everyone dies at 60, but at 75. Suppose for simplicity that only reproducers have children (4 of them per couple). Add another column of great grandparents who are between 60 and 75 years old. Now assume that no children die, 10% of the reproducers die (before reproducing), 20% of the parents die, 30% of the grandparents die, and nobody lives past 75. Does this effect the doubling time?

(iv) Look at various combinations of the assumptions from (i) and (ii) and see how the doubling time is affected. Pick a set of assumptions and answer the following questions. What happens if all of the children in one generation die? What happens if all of the reproducers in one generation die? All of the grandparents? Change your assumptions and ask the same questions.

(v) *Can we stabilize the population, at least mathematically?* Play the following game. By making various assumptions, see if you can arrange a stable population, that is, one that neither doubles endlessly nor goes extinct. (A population is extinct if it has 0 members at some time.) Can you arrange it so that the population remains below some predetermined upper limit but never goes to zero? Do you think that any real population of humans will ever satisfy any of your sets of assumptions? What implications does this exercise have for real life?

(vi) *Age Distributions.* An important aspect of any population is age distribution. For example, if half the population of the United States were over the age of 65, then the implications for Social Security would be different than if a quarter of the population were over 65. If half the population were under 12, then the implications for schools, for example, would be different than if one quarter of the population were under 12. Pick a Schwartz chart and plot on a bar graph the number of people in each of the categories: children, reproducers, parents and grandparents. What is your graph like? What change in assumptions can you make that would change this age distribution bar graph?

(vii) In the immediately previous problem dealing with age distribution, which value of K, where K is the number of children per couple, gives an age distribution which gives the grandparents the most *stable, reliable* societal support, say via a Social Security System? This is different from constructing a society which has an ever growing population. For example, authors[9] like Gregg Easterbrook have argued for increasing populations: ... *"More traffic but plenty of employees to support retired people seems like a better deal than a stable population with a stagnant economy swamped by pension costs."* There are two logical flaws in this statement. Stable does not necessarily equate to stagnant. The statement proposes a Ponzi scheme. What is a Ponzi scheme? (Look in the index of this book, or a dictionary if you do not know.)

(viii) Again in the problem dealing with age distribution, which value for K, where K is the number of children per couple, gives the children the most societal support?

[9]The following quote is from Gregg Easterbrook, a fellow at the Brookings Institution and author of *The Progress Paradox: How Life Gets Better While People Feel Worse*, who wrote an article for the *Los Angeles Times* which was carried in my local paper on October 15, 2006 which (besides the quote in the text) clearly says not to worry about population growth in the U.S.

(ix) *Momentum.* Pick any Schwartz chart with a 4 or 3 child family size. Pick a row, say the tenth row, and from that row onward reduce the number of children per couple to one (leaving everything else the same). How long does it take for the total population to decrease to half of the total population of the tenth row?

(x) *Averages.* If in any Schwartz Chart above we replaced the requirement that each couple in some category, say the reproducers, have K children with the requirement that *the average* number of children per couple in that category be K children, would the outcomes in terms of population growth be changed?

19.3 Three Basic Operations: Addition, Multiplication and Exponentiation

The Inverse Operation of Exponentiation. If I asked you to extend the graphs in Figure 19.1 some thirty years into the future (assuming the assumptions of the model remain unchanged), you would probably come reasonably close to the correct extension—or extrapolation, as mathematicians call it. But any two students would probably make slightly different extrapolations. Now if we could transform our curved graph into a straight line, we would all extend this line into the future (with a ruler, no doubt) in the same way. We are now going to discover an operation that turns each of the curved graphs from Figure 19.1 into a line.

Using Logarithms. We know how to "undo" the operation of adding, say 10, to a number x, viz., add the additive inverse, -10 to $x + 10$ and you get back to x. We know how to "undo" multiplication of a number x by, say 10, viz., multiply $10x$ by the multiplicative inverse, 10^{-1}, and you get back to x. We will now find out how to "undo," or invert, exponentiating x to base 10, i.e, how to undo, or invert, 10^x and get back[10] to x. In this next exercise it will be most helpful for you to have a calculator with a button labeled "log" somewhere.[11]

Exercise 19.7 Find the Logarithm Function on Your Calculator or Computer

(i) Add one more column to Table 19.2, and label it "log of the total population," see Table 19.3 below. (Note that in Table 19.3 the "Grandparents column," "Total Pop column," and "Log Total Pop column" are right below the "Children column," "Reproducer column," and "Parents column," respectively, instead of extending to the right. This is because they would not fit on the page between the margins in a font size you could easily read! You, of course, do not have this problem if you are doing this on a sheet of paper—or with a spreadsheet on a computer.) Now your calculator will have one or more "log" buttons. One will have the label "log" or "log_{10}," and another will have a label "*ln*," "ln" or "log_e." If you do not now understand these buttons, don't worry about it for the

[10]In this section we will answer the last question posed in Exercise 16.4 (iii).

[11]It is possible, in fact recommended, to do many easy logarithm exercises in your head. It is possible to do all logarithm exercises without a calculator. However, the problems that cannot be done quickly in one's head are now done by machine by just about everyone.

moment. Just pick one of the "log" buttons and stick with it. Do not use one for part of this exercise and another for another part of this exercise.[12] In your column with "log of the total population" at its head I want you to do the following. Apply the log operation to the number 256 (the 256 in the total population column) and enter the result in the same row, one column over. On my calculator if I calculate log[256] I get 2.408239965. If I calculate ln[256] or *ln*[256], I get 5.545177444. Fill out the rest of the column, remembering to stick with just one of the logarithm buttons.

(ii) Now graph the results of (i). On the horizontal axis put t, the time, 15, 30, , ..., 135; and on the vertical axis put the log[$P[t]$]. I am pairing a total population number with the largest time in that row, 256 goes with 15 years, 768 goes with 30 years, 3840 goes with 60 years and so forth. Thus the first point you will graph will be the ordered pair (15, log[P[15]]) = (15, log[256]) = (15, 2.40823997). (If you used the other ln button you would get (15, 5.545177444).) Continue plotting at least 5 more points as accurately as you can. What kind of graph do you have?

TABLE 19.3: Four-Child Family Model

Years	Children	Reproducers	Parents
0 to 15	256	0	0
15 to 30	512	256	0
30 to 45	1024	512	256
45 to 60	2048	1024	512
60 to 75	4096	2048	1024
75 to 90	8192	4096	2048
90 to 105	16384	8192	4096
105 to 120	32768	16384	8192
120 to 135	65536	32768	16384
Years	Grandparents	Total Pop	Log Total Pop
0 to 15	0	256	2.40823997
15 to 30	0	768	2.88536122
30 to 45	0	1792	3.25333801
45 to 60	256	3840	3.58433122
60 to 75	512	7680	3.88536122
75 to 90	1024	15360	4.18639122
90 to 105	2048	30720	4.48742121
105 to 120	4096	61440	4.78845121
120 to 135	16384	122880	5.08948120

If you look closely at the graph in Figure 19.2 you will see that it appears to be a straight line, at least for time t greater than 45 years. There is a way to check numerically to see if we really do have a straight line for t greater than 45, and we do this in the next exercise.

[12]As for me, I am mainly going to use the *log* button for now and discuss the *ln* button later.

Exercise 19.8 Do We Really Have a Straight Line? Is the Slope of the Graph Constant?

(i) Look at Figure 19.2. If we really have a straight line for t greater than or equal to 60, then a number called the *slope* of the line will be constant. The slope of the line in Figure 19.2 (if it is a line) will be denoted by the letter r. Now remember that the slope of the line through the two points (t_1, p_1) and (t_2, p_2) is "rise over run," i.e., $r = \frac{p_2 - p_1}{t_2 - t_1}$. If $t_1 = 75$, then $p_1 = log[P[75]] = 3.88536122$. If $t_2 = 120$, then $p_2 = log[P[120]] = 4.78845121$. We then get $r = \frac{4.78845121 - 3.88536122}{120 - 75} = \frac{.90308999}{45} = .0200686664$. Check this computation. Do you agree that it was done correctly?

(ii) Pick another pair of points with t greater than or equal to 60 and see if you get the same number r for the slope of the line through that pair of points.

(iii) Pick still another pair of points and see if the slope r remains unchanged.

(iv) Do you think that we are looking at the graph of a line with slope $r = .0200686664$, (at least for t greater than or equal to 60)?

(v) What is the slope of the line through the first two points in the graph of Figure 19.2, i.e, $t_1 = 15$ and $t_2 = 30$?

(vi) What is the slope of the line through the next pair of points, $t_1 = 30$ and $t_2 = 45$? How many significant digits agree in the various answers to your slope calculations?

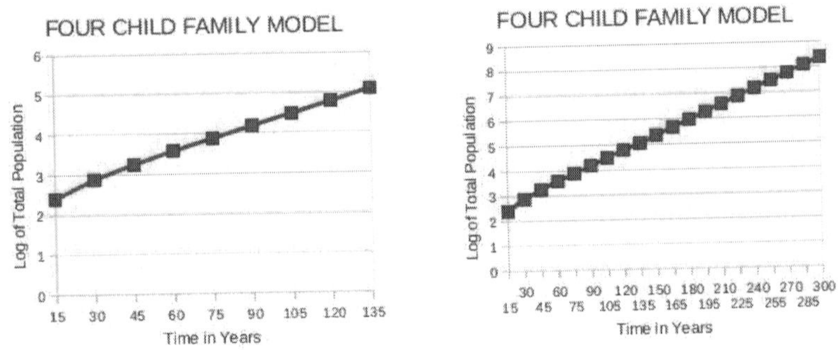

FIGURE 19.2: Log to Base 10 of Total Population vs. Time

So just what is this *log* button anyway? We have discovered one of its important properties. Whatever P does to t to make the graph of $P[t]$ vs. t a steeply rising curve (as t increases), log "undoes," or inverts, P to make the graph of $log[P[t]]$ vs. t a straight line.

Finding a Pattern in a Schwart Chart with Symbols. The first thing I want to do is understand a little better what exactly it is that P is doing to t to get $P[t]$. Then we will know what it is that log undoes, i.e., inverts. Let's go back to the Schwartz chart in Table 19.2. Instead of putting in 256 for the number of children in row one, let us put in the number P_0. See Table 19.4 below.[13] Thus P_0 represents the number of children we start with, and we

[13]The "Simple-15" in the caption for Table 19.4 is shorthand for the fact that the columns

are not telling ourselves what the particular value of P_0 might be. Thus, even though we do not know its exact value, we do know that P_0 is a number and that it has lots of known properties—it satisfies all the axioms for numbers, for example.

Now let us assume that in the population of our Schwartz chart, each Reproducer couple has K children. The number K might be 4, or 3 or 5, 1 or whatever. If this is the case, then the entries in the second row of the Schwartz chart from left to right are: $\frac{K}{2} * P_0$, P_0, 0, 0, and $P[30] = P_0 + P_0 * \frac{K}{2} = P_0 * (1 + \frac{K}{2})$. What axiom(s) for numbers did I use to obtain the very last equality?

TABLE 19.4: Schwartz Chart for a (Simple-15) K-Child Family

Time	Children	Reproducers	Parents
15	P_0	0	0
30	$(\frac{K}{2})P_0$	P_0	0
45	$(\frac{K}{2})^2 P_0$	$(\frac{K}{2})P_0$	P_0
60	$(\frac{K}{2})^3 P_0$	$(\frac{K}{2})^2 P_0$	$(\frac{K}{2})P_0$
75	$(\frac{K}{2})^4 P_0$	$(\frac{K}{2})^3 P_0$	$(\frac{K}{2})^2 P_0$
90	$(\frac{K}{2})^5 P_0$	$(\frac{K}{2})^4 P_0$	$(\frac{K}{2})^3 P_0$
105	$(\frac{K}{2})^6 P_0$	$(\frac{K}{2})^5 P_0$	$(\frac{K}{2})^4 P_0$
120	$(\frac{K}{2})^7 P_0$	$(\frac{K}{2})^6 P_0$	$(\frac{K}{2})^5 P_0$
135	$(\frac{K}{2})^8 P_0$	$(\frac{K}{2})^7 P_0$	$(\frac{K}{2})^6 P_0$

Time	Grandparents	Total Population
15	0	P_0
30	0	$(1 + \frac{K}{2})P_0$
45	0	$(1 + \frac{K}{2} + (\frac{K}{2})^2)P_0$
60	P_0	$(1 + \frac{K}{2} + (\frac{K}{2})^2 + (\frac{K}{2})^3)P_0$
75	$(\frac{K}{2})P_0$	$(\frac{K}{2} + (\frac{K}{2})^2 + (\frac{K}{2})^3 + (\frac{K}{2})^4)P_0$
90	$(\frac{K}{2})^2 P_0$	$((\frac{K}{2})^2 + (\frac{K}{2})^3 + (\frac{K}{2})^4 + (\frac{K}{2})^5)P_0$
105	$(\frac{K}{2})^3 P_0$	$((\frac{K}{2})^3 + (\frac{K}{2})^4 + (\frac{K}{2})^5 + (\frac{K}{2})^6)P_0$
120	$(\frac{K}{2})^4 P_0$	$((\frac{K}{2})^4 + (\frac{K}{2})^5 + (\frac{K}{2})^6 + (\frac{K}{2})^7)P_0$
135	$(\frac{K}{2})^5 P_0$	$((\frac{K}{2})^5 + (\frac{K}{2})^6 + (\frac{K}{2})^7 + (\frac{K}{2})^8)P_0$

and rows are built on 15 year intervals and that only Reproducers are having children, viz., K per couple. For the same reason as in the immediately previous table, the "Grandparents column" and the "Total Population column" appear below the columns for Children, Reproducers, and Parents instead of to the right them—for lack of room.

Exercise 19.9 A Schwartz Chart with Symbols that Represent Numbers: Calculations Still Possible!

(i) Using axioms for numbers (which ones?) show that $\frac{K*P_0}{2} = (\frac{K}{2})*P_0$. In other words, multiplying P_0 by K and dividing the result by 2 is the same as multiplying P_0 by $\frac{K}{2}$.

(ii) Using (which?) axioms for numbers show that $\frac{K*((\frac{K}{2})*P_0)}{2} = (\frac{K}{2})^2 * P_0$. In other words, doing the two part operation of multiplying by K and dividing by 2 twice in succession has the result of multiplying by $(\frac{K}{2})^2$ once.

(iii) What do you guess the next step of this exercise would be? And the next? And the next?

The entries from left to right in the third ("numerical") row of our Schwartz chart are: $(\frac{K}{2})^2 * P_0$, $\frac{K}{2} * P_0$, P_0, 0, $P[45] = P_0 + \frac{K}{2} * P_0 + (\frac{K}{2})^2 * P_0 = P_0 * (1 + \frac{K}{2} + (\frac{K}{2})^2)$.

The entries from left to right in the fourth row of our Schwartz chart are: $(\frac{K}{2})^3 * P_0$, $(\frac{K}{2})^2 * P_0$, $\frac{K}{2} * P_0$, P_0, and the last entry, the total population at time $t = 60$ years, $P[60] = P_0 * (1 + \frac{K}{2} + (\frac{K}{2})^2 + (\frac{K}{2})^3)$.

Exercise 19.10 Finding a Simple Pattern in the Schwartz Chart: Table 19.4

(i) Verify that in the fifth row of the Schwartz chart (assuming a family size of K children per family) we have for the total population:

$$P[75] = P_0 * (\frac{K}{2} + (\frac{K}{2})^2 + (\frac{K}{2})^3 + (\frac{K}{2})^4).$$

(ii) Verify that in the sixth row of the Schwartz chart we have for the total population:

$$P[90] = P_0 * ((\frac{K}{2})^2 + (\frac{K}{2})^3 + (\frac{K}{2})^4 + (\frac{K}{2})^5).$$

(iii) Verify that in the seventh row of the Schwartz chart we have for the total population:

$$P[105] = P_0 * ((\frac{K}{2})^3 + (\frac{K}{2})^4 + (\frac{K}{2})^5 + (\frac{K}{2})^6).$$

Do you see a pattern forming here?

(iv) Verify that in the nth row, n greater than or equal to 4, of the Schwartz Chart we have for the total population:

$$P[15 * n] = P_0 * ((\frac{K}{2})^{n-4} + (\frac{K}{2})^{n-3} + (\frac{K}{2})^{n-2} + (\frac{K}{2})^{n-1}).$$

(v) Verify that $P[15 * n]$ can be rewritten, in the case $K \neq 2$, as:

$$P[15 * n] = P_0 * (\frac{(\frac{K}{2})^{n-4} - (\frac{K}{2})^n}{1 - \frac{K}{2}}) = (\frac{K}{2})^n * C_K,$$

where C_K is a constant that depends on K and P_0. It is not crucial to know this exact formula, but

$$C_K = \frac{(\frac{K}{2})^{-4} - 1}{1 - \frac{K}{2}} * P_0.$$

In the case where $K = 2$, we have

$$P[15 * n] = 4 * P_0.$$

Thus, in case of a two child family the population is constant in time, as we would expect.

(vi) Verify that the above formula for C_K, for $K \neq 2$, gives a positive number. Can you show that $C_K = \frac{P[60]}{(\frac{K}{2})^4}$?

(vii) Verify that the following pattern holds in this Schwartz Chart: The total population of any row (at least for the fifth row and beyond) is $\frac{K}{2}$ times the total population in the immediately preceding row.

Using the Exponential Pattern of the Simple-15 K-Child Family Schwartz Chart. Thus the big conclusion we reached in the previous exercise is that there is a pattern in the Schwartz chart, Table 19.4. Namely, if our families are having K children per Reproducer couple and the birth cycle is 15 years, then the total population after $15 * n$ years[14] is

$$P[15 * n] = [\tfrac{K}{2}]^n * C_K,$$

that is, the nth power of $\frac{K}{2}$ times a constant, C_K, that depends on K and the number of children you start with, P_0.

The total population at time $15*n$ years, $P[15*n]$, is said to be *an exponential function of* n. The n is a variable, and it appears as an exponent; thus we have the name exponential function. Loosely speaking, mathematicians use words like "P is a *function of* n" whenever there is some formula, operation or rule into which you plug n and out pops a *unique* answer called $P[n]$, or in our case, $P[15 * n]$. We have seen this very same concept of function before; see Exercise 15.10 for details.

Before seeing what applying the log operation to P does, I want to perform a trick called *changing the variable.* This is not a fancy trick, it is mostly just a change of name or appearance. I want to get rid of the 15 inside $P[15 * n]$, so I will rename $15 * n$. I will call it t. Thus let us define t by the following equation:

$$t = 15 * n.$$

If $t = 15 * n$, then $n = \frac{t}{15}$. Thus we have:

$$P[t] = (\tfrac{K}{2})^{\frac{t}{15}} * C_K. \qquad (\Diamond)$$

Let's take a moment to look at the above equation and make sure we really understand what it says. The left-hand side, $P[t]$, in English says "the population P at time t" or "P of t." This is a special use in mathese for the parentheses ( ), or in this case the parenthetical brackets [ ], which is quite different from the use of parentheses in, say, multiplication. Mathematicians call P a *function* of t and write $P[t]$; P depends on t. Just how P depends on t is given by the right-hand side of the equation above. Thus, in this case, $P[t]$ means literally to "take the number t, divide t by 15, then raise $\frac{K}{2}$ to the power $\frac{t}{15}$ and multiply the result by the number C_K." We can think of P as an operation.

[14] At least for n larger than 3.

19.4 Defining Logarithm Functions

It turns out that you can think of *log* as an operation, i.e., a function, as well. If you take the *log* of a number x, you get another number, called $log\,[x]$, often written more simply as just $log\,x$. The fact that the graph of $log[P[t]]$ vs. t is a straight line means that the log function essentially undoes or inverts whatever the function P does. The exact inverse function of P is something we will soon be able to calculate, and this exact inverse function when applied to $P[t]$ yields t.[15]

Thus the graph of $P[t]$ vs. t is a curve that rises steeply as t increases. Applying the log operation to both sides of ($\diamondsuit$) yields a straight line graph as we have seen, viz., the graph of $log[P[t]]$ vs. t. How does this work? The secret lies in the definition of log that was built into your calculator. It is the same definition that you may have seen in high school or in another math class. Let's recall that definition in English.

Definition: The *logarithm*, or *log*, *to base 10* of a number[16] x is *the* power[17] to which 10 must be raised to get x. The log to base 10 of x is written $log_{10}x$ or $log_{10}[x]$ or just $log\,x$ or $log[x]$ if the base 10 is understood.[18]

Exercise 19.11 Do You Understand the Definition of Logarithm?

(i) Verify that $log_{10}10 = 1$.

(ii) What are $log_{10}100$, $log_{10}1000$, $log_{10}.1$, $log_{10}.01$, $log_{10}.0001$?

(iii) What is $log_{10}1$?

(iv) What is $10^{log_{10}x}$, where x is some positive number? (If you cannot do this problem, i.e., part (iv), go back and study the definition of log_{10}; and repeat this definition to yourself 10 times before breakfast every day until you can do this problem quickly—with understanding.)

(v) What is $log_{10}10^x$, where x is any positive or negative number or 0? If you cannot do this problem see the parenthetical comment in part (iv).

(vi) If you have a button on your calculator labeled *log* and you compute $log\,10$ and get the answer 1, what base is the calculator using?

Exercise 19.12 An Exercise With Logarithms

(i) On your calculator find your *log* button and calculate $log\,2$. You should get (to 10 significant digits) 0.3010299957; did you?

(ii) What is $10^{0.3010299957}$?

---

[15]The resulting graph would be (not just any straight line but) the straight "45 degree" line through the origin, i.e., the point (0,0).

[16]I will only apply the *log* operation to positive numbers.

[17]This power is a number which may be positive, negative or zero.

[18]There is a detail here which I am glossing over, cf., the end of footnote [23]. I am really claiming here that *log* is a function, hence given a positive number x, there must be a *unique* number, called $log\,x$ with the property that $10^{log\,x} = x$. This is true; and your calculator confirms this fact every time you plug in some number, like 2, and push the *log* button. Because when you do, you get a unique answer, in this case, $log\,2 \approx .301$.

(iii) Let us examine the meaning of these exponents just a little further. Just what, for example, does $10^{0.3}$ mean? Well, $.3 = \frac{3}{10}$, so $10^{0.3} = 10^{\frac{3}{10}}$. Thus, do you see that $10^{0.3}$ can be thought of as the tenth root of 10^3?

(iv) Can you give an interpretation of $10^{0.3010}$ similar to the interpretation in part (iii)?

(v) Is the number $10^{0.3010299957} - 2$ exactly zero? How big (or small) is this number?[19]

Exercise 19.11 points out two of the most important properties of the *log* function, and those properties are immediate consequences of the definition of *log* (if you understand that definition).[20]

THE FUNDAMENTAL EQUATIONS FOR log_{10}

$log_{10}[10^x] = x$, where x is any positive or negative number or 0,

$10^{log_{10}[x]} = x$, where x is any positive number.

What these two equations say in English is that "exponentiating x to base 10" and "taking the *log* of x to base 10" are functions that undo, or invert, one another. If you exponentiate x to base 10, i.e., form 10^x, then take the *log* (to base 10), i.e., $log[10^x]$, you get back to x, i.e., $log[10^x] = x$. In reverse order, if you take the *log* of x (to base 10), i.e., $log\, x$, and then you exponentiate that number to base 10, i.e., form $10^{log\, x}$, i.e., again you get back to x, i.e., $10^{log\, x} = x$.[21]

From Equation (◇) we have that the formula for the total population at time t (for t greater than or equal to 60 years and assuming a birth cycle of 15 years and K children per reproducer couple) is:

$$P[t] = [\tfrac{K}{2}]^{\frac{t}{15}} * C_K,$$

where C_K is a positive constant that depends on K and P_0, the initial population.[22]

Now if we apply the *log* operation to this formula we get:

$$\begin{aligned} log\, P[t] &= log\, [[\tfrac{K}{2}]^{\frac{t}{15}} * C_K] \\ &= log\, [\tfrac{K}{2}]^{\frac{t}{15}} + log\, C_K \qquad \text{(Why? Read on.)} \\ &= \frac{t}{15} * log\, \frac{K}{2} + log\, C_K \qquad \text{(Why? Read on.)} \end{aligned}$$

[19] Your answer to this question is dependent on how accurate your calculator is. The number is very close to zero, but not exactly zero.

[20] This is one of those cases where calculations are useless. There is nothing fancy to be done. Either one has read and absorbed the definition of *log*, in which case the two properties are easy, or one has not, in which case the two properties are impossible to grasp.

[21] Mathematicians say that the two functions $F[x] = 10^x$ and $G[x] = log_{10} x$ are inverse functions.

[22] Do you recall what Equation (◇) becomes if $k = 2$?

Thus we have:

$$log\,P[t] = \frac{t}{15} * log\,\frac{K}{2} + log\,C_K \qquad (\diamond\diamond)$$

Two Laws for Logarithm Functions. There are three things we have to understand about the above list of equations: Why does the last line mean that the graph of $log\,P[t]$ vs. t is a straight line? What are the justifications for the two steps leading up to the last line?

First, what is the justification for going from the first line to the second? This follows from a property of the *log* operation which in English is: The *log* of a product of factors is the sum of the *logs* of the factors (as long as all the *logs* are defined—for example $log\,0$ does not exist). In mathese we can write:

$$\boxed{log\,[A * B] = log\,A + log\,B,} \qquad \textbf{(First Law for } \textit{log}\textbf{)}$$

where A and B are expressions that represent (positive) numbers. These expressions can be simple or complicated. In the first of the three lines above, $A = [\frac{K}{2}]^{\frac{t}{15}}$ and $B = C_K$. Logarithms convert multiplications into additions. Read footnote [23] to see why this law of logs is just the "inverse" of the First Law of Exponents.[23]

If the First Law of Exponents led us to the First Law for *log*, you have every right to expect that the Second Law of Exponents would lead to the Second Law for *log*. And so it is:

$$\boxed{log\,[A^r] = r * log\,A,} \qquad \textbf{(Second Law for } \textit{log}\textbf{)}$$

where A is an expression that represents a (positive) number and r represents a positive or negative number or zero. Logarithms convert exponentiation (by, for example, r above) into multiplication (for example, by r, above).

The argument for this is similar to the argument for the First Law for *log*, read footnote [24] for the details.[24]

---

[23] Let's see why this first law of the *log* operation holds. First, what does $log\,[A * B]$ mean? It is the power to which 10 must be raised to get $A * B$. So, if I raise 10 to the power $log\,[A * B]$ (remember $log\,[A * B]$ is just a number, even though it has a long name with 8 symbols in it) I should get $A * B$. Thus: $10^{log\,[A*B]} = A * B$. Now what about $log\,A + log\,B$? If I raise 10 to the $log\,A + log\,B$ power what do I get? Well, $10^{log\,A + log\,B} = 10^{log\,A} * 10^{log\,B}$. Why? Because of the First Law of Exponents that we studied in Exercise 11.6. But now, $10^{log\,A} = A$ and $10^{log\,B} = B$. Why? Thus $10^{log\,A + log\,B} = A * B$. But $log\,[A * B]$ is the power to which 10 must be raised to get $A * B$, so it must be $log\,A + log\,B$. This is a nice argument and its conclusion is correct. However, this argument for the validity of the first law for *log* is not totally complete. The missing step is this: I need to show that if $10^x = 10^y$, where x and y are real numbers, then $x = y$. I will leave this for you to ponder in a future math class.

[24] Consider the following equalities: $10^{log\,A^r} = A^r = [10^{log\,A}]^r = 10^{[log\,A]*r} = 10^{r*log\,A}$. The next to last equality follows from the Second Law of Exponents, and we see that $r * log\,A = [log\,A] * r$ is the power to which 10 must be raised to get A^r, thus $r * log\,A$ must equal $log\,A^r$. There remains still one step to create an argument acceptable to professional mathematicians, see the end of footnote [23].

Thus applying the First Law and Second Law for *log* in succession we are led to: $log\, P[t] = \frac{t}{15} * log\, \frac{K}{2} + log\, C_K$, i.e. Equation ($\diamond\diamond$). We now have left to explain why the graph of $log\, P[t]$ vs. t is a line. I am hoping here that in a previous math class you learned that the graph of any equation of the form $y = m * x + b$ is a line, where x is the "horizontal co-ordinate" and y is the "vertical co-ordinate."[25] Suffice it to say for now that in Equation ($\diamond\diamond$), t is measured on the horizontal axis and corresponds to x in $y = m * x + b$. Also $log\, P[t]$ is measured on the vertical axis and corresponds to y in $y = m * x + b$. The slope of the line $y = mx + b$ is m, the number which multiplies x. Thus the slope of the linear[26] graph of the equation $log\, P[t]$ vs. t is $\frac{1}{15} * log\, [\frac{K}{2}]$, the number that multiplies t.[27]

Comparing what we just did with Exercise 19.8, where we calculated the slope of the line in Figure 19.2 to be $r = .0200686664$, we can ask if this new formula for the slope of our line gives the same result. Is it true that

$$r = \frac{1}{15} * log\, [\frac{K}{2}]?$$

What value of K shall we use? In Table 19.2 we were using a 4 child size, i.e., $K = 4$. If we plug in 4 for K into $\frac{1}{15} * log\, [\frac{K}{2}]$? we get:

$$\frac{1}{15} * log\, [\frac{4}{2}] = \frac{log\, 2}{15} = .0200686663.$$

Our formula gives almost the same answer as we got before! The two answers, each about $2(10)^{-2}$, differ by only 10^{-10}.

Exercise 19.13 Using Logarithms to Calculate a Slope

(i) As we did in Exercise 19.4 part (vii) go back to the Schwartz chart, Table 19.2, and change only one thing. Change the family size from 4 to 3 children per family. Keep the birth cycle at 15; everybody dies at age 60. Now calculate the slope of the line $log\, P[t]$ vs. t two waysr: (1) Find the slope r graphically as we did in Exercise 19.8; (2) Calculate the value of r from the formula above, namely, $r = \frac{1}{15} * log\, [\frac{K}{2}]$. Are these numbers about the same?

(ii) Assuming that A and B represent positive numbers, can you write $log\, \frac{A}{B}$ in another way? In English you are being asked to complete the sentence: The *log* of a quotient is ....

Exercise 19.14 An Intuitive Guess of How Many "Rows" it Takes the Total Population to Double

Before rigorously calculating the doubling time for the population in the Schwartz Chart given by Table 19.4, I want to look at that chart "intuitively" and make some educated guesses.

(i) Suppose $K = 3$. We have the rule that the total population in row n (for n greater than 3) is $\frac{3}{2}$ times the total population in row $n - 1$. Therefore (as we observed before) the

[25] If you did not learn this fact before, all is not lost, just assume it for the moment.

[26] If a graph is a line we call it a linear graph. If the graph of an equation is a line we say we have a linear equation.

[27] We say that m is the *coefficient* of x in $m * x + b$, and $\frac{1}{15} * log\, [\frac{K}{2}]$ is the coefficient of t in $\frac{t}{15} * log\, \frac{K}{2} + log\, C_K$.

time to double is more that 15 years and less than 30. Do you see how this is related to the following inequalities: $\frac{3}{2} < 2 < (\frac{3}{2})^2$?[28]

(ii) I now ask the following question, what power of $\frac{3}{2}$ will give me 2? If we call this power x, we have $\frac{3}{2} < (\frac{3}{2})^x < (\frac{3}{2})^2$.[29]

(iii) Repeat this exercise for $K = 4$.

(iv) If you are feeling very ambitious, repeat this exercise for $K = K$. That is, do not assume any specific numerical value for K.

19.5 Computing Formulas for Doubling Times

We are now ready to calculate the doubling time for any exponentially growing population, like the populations in our Schwartz charts. The doubling time is important. If a population is doubling in, say 20 years, then you will have to double available food and housing and water and everything else every 20 years; or else everyone (on average) has to try to survive on less.

Let us start with the familiar formula $P[t] = [\frac{K}{2}]^{\frac{t}{15}} * C_K$, i.e., Equation ($\diamondsuit$). Now $P[t]$ tells us what the total population is at any time t. Suppose some number of years after a given time t the population is double what it was at time t. How can we write this in mathese? The time it takes the population to double is what I want to find. Let's give this amount of time a name, t_d = the time it takes the population to double. (It will turn out, as we will see in a moment, that the time t when you start counting does not affect the doubling time.)

We have taken a very important step! We have labeled the thing we are looking for, the doubling time. We have labeled it in mathese, i.e., t_d. Now can we come up with a sentence or equation in mathese that involves t_d? In a word, yes.

The population at time t is $P[t]$. The population at time $t+t_d$ is $P[t+t_d] = 2P[t]$. Why? (Look at the definition of t_d.) It follows immediately from the formula for P, Equation ($\diamondsuit$) above, that:

$$[\tfrac{K}{2}]^{\frac{t+t_d}{15}} * C_K = P[t+t_d] = 2P[t] = 2 * [\tfrac{K}{2}]^{\frac{t}{15}} * C_K.$$

Which gives

$$[\tfrac{K}{2}]^{\frac{t+t_d}{15}} * C_K = 2 * [\tfrac{K}{2}]^{\frac{t}{15}} * C_K.$$

Multiply both sides on the right by C_K^{-1} and we get

---

[28]Hint: $\frac{3}{2}$ corresponds to waiting 15 years and $(\frac{3}{2})^2$ corresponds to waiting 30 years.

[29]Hint: Solve $(\frac{3}{2})^x = 2$ for x and see if you get the same answer for the doubling time that you estimated in Exercise 19.4 part (vii) or that we will calculate in the next subsection. Don't forget that $(\frac{3}{2})^1$ corresponds to 15 years and $(\frac{3}{2})^2$ corresponds to 30 years. So you should expect x to be a number between 1 and 2.

$$[\tfrac{K}{2}]^{\frac{t+t_d}{15}} = 2 * [\tfrac{K}{2}]^{\frac{t}{15}}.$$

Our goal is to do legal operations on this equation and end up with an equation that has the symbol t_d alone on one side with expressions on the other side involving only things that we already know—not explicitly involving t_d. Said succinctly, we want to solve for t_d, or get a formula for t_d.

Can you see ahead that our formula for t_d, whatever it may be, will not depend on C_K and hence not on P_0, the initial population, either?

Using the Distributive Law in the exponent on the left-hand side of the equation, and then using the Second Law of Exponents, we get:

$$[\tfrac{K}{2}]^{\frac{t}{15}} * [\tfrac{K}{2}]^{\frac{t_d}{15}} = 2 * [\tfrac{K}{2}]^{\frac{t}{15}}.$$

Now multiply both sides of the equation by $[\frac{K}{2}]^{-\frac{t}{15}}$ and we get:

$$[\tfrac{K}{2}]^{\frac{t_d}{15}} = 2.$$

What fact did I just use?[30]

Now operate on both sides of this equation with *log*. Then use the Second Law for *log* to get:

$$\frac{t_d}{15} * log\,(\tfrac{K}{2}) = log\,2.$$

Multiplying both sides by 15 and also by $[log\,(\frac{K}{2})]^{-1}$ we isolate t_d on the left-hand side of the equal sign, i.e, we have solved for t_d and have the following formula:

$$\boxed{t_d = \frac{15 * log\,2}{log\,(\frac{K}{2})}.} \qquad \text{(Doubling Time Formula)}$$

The doubling time thus depends on the birth cycle, 15 years, and the number of children per (reproducer) couple, K, but not on the initial population size or the time t we started counting. By the way, I have suppressed units of measurement in the above calculation of t_d; what are the units that t_d is given in?

Exercise 19.15 Formulas are Based on Assumptions

(i) The formula that we have just derived for the doubling time comes from a mathematical model. This model incorporates certain assumptions. State one of these assumptions such that if this assumption is changed then the doubling time is changed.

(ii) Change the assumption you have stated in part (i) and find the resulting change in the formula for the doubling time.

(iii) Can you find a second assumption on which the formula for the doubling time depends?

(iv) If $K = 1$, t_d is negative. What does this mean? Hint: t_d is a "half-life."

(v) Do you think that most formulas in mathematics depend on assumptions (explicitly stated or not)?

[30] Here we have used the fact that $A^{-x} = [A^x]^{-1}$ for a positive number A and any number x. Can you see that our formula for t_d will not depend on the time t when we start counting?

Exercise 19.16 Before You Use a Formula Make Sure that the Assumptions the Formula is Based on are Satisfied

(i) What is the doubling time for a population given by a Schwartz chart wherein the birth cycle is 15 years, everyone dies at 60, and the family size is 3 children? Compare your answer to the estimate you got in Exercise 19.4 part (vii), if you did it. (Can you use any formula recently derived?)

(ii) If you change the birth cycle to 20 years, everyone dies at 80 and the family size is 3 children, what is the doubling time (neglecting immigration)? This situation is closer to the situation in the United States. How do you think immigration effects the doubling time?

(iii) If you double the birth cycle, how does that effect the doubling time? If you halve the birth cycle, how does that effect the doubling time?

(iv) If you take the square root of the number of children per couple, how does that effect the doubling time? If you replace K with K^2, how does that affect the doubling time? If you double the number of children per (reproducer) couple does that change the doubling time? How?

(v) What do you think mathematicians mean when they say that (in our Schwartz charts) the doubling time of the population is directly proportional to the birth cycle?

(vi) Is the equation $t_d = \frac{15 * log\, 2}{log\,(\frac{K}{2})}$ equivalent to the equation $t_d = \frac{15}{\frac{log\, K}{log\, 2} - 1}$?

(vii) Is it true that $t_d = \frac{log\, 2}{r}$, where r is the slope of the linear (that is, straight line) graph of the log of the total population vs. time?

(viii) Go back and do Exercise 19.6 parts (i), (ii) as precisely as you can, i.e., calculate the doubling times asked for in these questions to at least 3 significant digits. Hint: Does the formula for the doubling time stated just before Exercise 19.15 apply? Does the formula for the doubling time in part (vii) just above apply?

(ix) Suppose $P[t]$ is given by Equation ($\diamondsuit$). Can you find a function F which is the inverse function for P? By this I mean F satisfies: $F[P[t]] = t$ and $P[F[t]] = t$.

(x) Suppose you create a population model which is more complicated than any we have thus far studied. And suppose that you graph the total population $P[t]$ vs. time. Now suppose that your graph of $log\, P[t]$ vs. t comes out to be a straight line. What can you say about $P[t]$? Can you calculate the doubling time?[31]

19.6 Natural Logarithms

We are now ready to understand what that other logarithm button on your calculator is all about. I am talking about the *ln* button. We have the following:

Definition: The number $ln\, x$ is the power to which the number e must be raised to get x, where $e \approx 2.718281828\ldots$. We say that $ln\, x$ is the *logarithm of* x *to base* e, or $log_e x$ or $ln\,[x]$; $ln\, x$ is also called the *natural logarithm of* x.

Now where did this number e come from? Why is it natural? One of several ways of understanding e starts with putting money in a bank (or credit union).

[31]Hint: See part (vii) of this problem.

Suppose you put \$100 in the bank at 5% annual *simple interest.* What this means is that at the end of one year the bank will add 5% of \$100 to your \$100. Now 5% of \$100 is $\frac{5}{100} * \$100 = \$5$. Thus at the end of one year you will have \$105. It turns out to be very useful to think about the calculation we just did "all at once" in terms of the following equation: $\$105 = (1 + .05) * \$100 = (1.05) * \$100$.

Now a second bank, in an effort to get you to deposit your money with them, will offer you a better deal. They will give you 5% annual interest *compounded* twice a year, i.e., every six months. What does this mean? Well, it means that you put your \$100 in the bank and after six months the bank will pay you $\frac{5}{2}\%$ interest on your \$100, and you can then earn $\frac{5}{2}\%$ interest on that interest (as well as on the \$100) for the next six months. Let's see how much money you get at the end of the first six months: $\frac{5}{2}\%$ of \$100 is $\frac{2.5}{100} * \$100 = \$2.50$. Thus after six months you have in your bank account a total of $\$102.50 = (1 + \frac{.05}{2}) * \$100$. But now we earn $\frac{5}{2}\%$ interest on \$102.50 (during the next six months). So we can repeat the calculation we just did. Thus, substituting \$102.50 for \$100 in the calculation we just did for the first six month period, we get for the second six month period $(1 + \frac{.05}{2}) * \$102.50 = (1 + \frac{.05}{2}) * (1 + \frac{.05}{2}) * \$100 = (1 + \frac{.05}{2})^2 * \$100 = \$105.0625 \approx \$105.06$.

Exercise 19.17 Money in the Bank Interest Compounded

(i) Think very carefully about why simple 5% annual interest leads to multiplying the money deposited, called the *principal*, by $(1 + .05)$. How much money would you have in the bank at the end of one year if your initial deposit was \$1,000,000 and you earned 8% annual simple interest?

(ii) Think very carefully about why 5% annual interest, compounded twice (once each six months), leads in one year to multiplying the principal by $(1 + \frac{.05}{2})^2$. How much money would you have at the end of one year on a principal of \$1,000,000 if you received 8% annual interest compounded each six months?

(iii) How much more money did your million dollars earn in part (ii) compared to part (i) above?

Not to be out done, a third bank advertises an offer of 5% annual interest compounded three times a year, or every four months. Thus, at the end of four months our \$100 becomes $(1 + \frac{.05}{3})\$100 \approx \$101.666666$. At the end of eight months our \$101.666666 becomes $(1 + \frac{.05}{3})\$101.666666 = (1 + \frac{.05}{3}) * (1 + \frac{.05}{3})\$100 = (1 + \frac{.05}{3})^2\$100 \approx \$103.3611$. At the end of the year our initial \$100 has become, after compounding for the third time, $(1 + \frac{.05}{3})\$103.3611 = (1 + \frac{.05}{3})^3\$100 \approx \$105.0838$. Thus far the more times the bank compounds a given interest rate the more money you earn.

Now if you start with \$100 the increased earnings are not dramatic. However, there are people with billions of dollars to invest.

Exercise 19.18 Multiple Compound Interest on One Billion Dollars

(i) Suppose you have a billion dollars to invest. Suppose the annual interest rate is 5%. After one year at simple interest how much money do you have?

(ii) Suppose the bank compounds the annual 5% interest twice, once each six months. At the end of one year how much money do you have?

(iii) Suppose the bank compounds the annual 5% interest three times, once each four months. At the end of the year how much money do you have?

(iv) Continue the previous three parts, compounding four times, five times , six times, ..., until you can see a pattern or a formula for computing the result if the bank compounds the annual 5% interest n times, once every $\frac{12}{n}$ months.

(v) How large does n have to be in part (iv) before the change in the money earned after n compoundings is less than \$100 more than the money earned with $n-1$ compoundings? (We are still working with an initial principal investment of one billion dollars.)

If you see the same pattern forming here that I do, then you will agree with the following statement. If you start with \$100 and a bank pays 5% annual interest and compounds that interest n times, once every $\frac{12}{n}$ months, then at the end of the year you will have $(1+\frac{.05}{n})^n\$100$.

This leads us to the following. If you start with P_0 dollars, earn annual interest at rate r, expressed as a decimal (for example, 5% corresponds to $r = .05$) and the bank compounds your interest n times (at equal intervals of time), at the end of one year you will have the number of dollars given by the formula:

$$\boxed{(1+\frac{r}{n})^n * P_0.}$$

A very interesting question now suggests itself, what happens if the bank compounds the interest continuously? It is not easy to explain rigorously what this last sentence means, but it is easy to understand it intuitively as follows. What happens to the expression $(1+\frac{r}{n})^n$ as n gets larger and larger without bounds?[32]

Mathematicians say the following: Does the limit as n goes to positive infinity of the expression $(1+\frac{r}{n})^n$ exist? You might think that the expression $(1+\frac{r}{n})^n$ just gets bigger and bigger and goes to infinity itself; but that does not happen. It turns out that we have the following miraculous fact,

$$\boxed{e^r = \lim_{n\to\infty}(1+\frac{r}{n})^n,} \qquad \textbf{(Definition of } e^r\textbf{)}$$

where e is the same number we introduced in the definition of natural logarithms.

In fact we can define this illusive number e as follows,

$$\boxed{e = \lim_{n\to\infty}(1+\frac{1}{n})^n,} \qquad \textbf{(Definition of } e\textbf{)}$$

Exercise 19.19 Some Calculations that Help Us Understand e

(i) Get out your calculator and see if you have a button labeled e^x. Use this button to evaluate $e^1 = e$.

(ii) Evaluate the expression $(1+\frac{1}{n})^n$ for $n = 1,2,3,4,5,6,7,8,9$ and 10. How close have you gotten to the value for e you found in part (i)? Evaluate $(1+\frac{1}{n})^n$ for $n = 100$. How

[32] Mathematicians say "n goes to infinity" instead of "n gets larger and larger without, or beyond all, bounds."

many decimal places of agreement with the value for e in part (i) do you have now? Repeat for $n = 1000$. How close are you now? Again for $n = 10000$. How close are you now? How large do you think you have to choose n in order for your value of $(1 + \frac{1}{n})^n$ to agree with the value of e in part (i)? Note that there is no value of n so that $(1 + \frac{1}{n})^n$ is exactly equal to e. However, given any degree of accuracy, say to 10 significant digits, you can find an n so that $(1 + \frac{1}{n})^n$ agrees with e to 10 significant digits.

(iii) If a bank pays a simple annual interest rate of r, and this interest is compounded m times per year, then how much money do you have at the end of 1 year if you put $\$1,000,000$ in the bank at the beginning of the year?

(iv) With the same assumptions as part (iii) how much money do you have at the end of 2 years?

(v) With the same assumptions as part (iii) how much money do you have at the end of 3 years?

(vi) Now with the same assumptions as part (iii) except this time let the money stay in the bank for t years. How much money do you have now? Can you write down a formula for the amount of money in the bank as a function of t? Does your formula have a factor $(1 + \frac{r}{m})^{mt}$ in it?

(vii) By letting $n = mt$, can you rewrite $(1 + \frac{r}{m})^{mt}$ so that it becomes $(1 + \frac{rt}{n})^n$? (Note that $(1 + \frac{r}{m})^{mt}$ is interest rate r compounded m times a year for t years. And $(1 + \frac{rt}{n})^n$ is interest rate rt compounded n times for 1 year.)

(viii) It follows from part (vii) that $\lim_{n\to\infty} (1 + \frac{rt}{n})^n = \lim_{m\to\infty} (1 + \frac{r}{m})^{mt}$, where t is any positive, negative or zero number. Is it true that $e^{rt} = \lim_{n\to\infty} (1 + \frac{rt}{n})^n = \lim_{m\to\infty} (1 + \frac{r}{m})^{mt}$? Why? (Hint: Look at the definition of e^r, then what is the definition of e^{rt}?) Is the following a correct statement: If you continuously compound an interest rate of rt for 1 year, you get the same result as if you continuously compound an interest rate of r for t years? Finally, is the following true: $(e^r)^t = e^{(rt)}$?

(ix) Consider the statement: "If r is a very small positive number, then $e^r \approx 1 + r$." Is 1 a small number? Is .000001? Is .0001? Is .01? Plug each of these 4 numbers in for r in the statement and see how accurate it is.

To understand this statement more fully, compare it to the statement: If r is a very small positive number, then $10^r \approx 1 + [ln\,10]r \approx 1 + (2.302585093)r$. You might also consider the statement: If r is a very small positive number, then $2^r \approx 1 + [ln\,2]r \approx 1 + (.6931471806)r$. It turns out[33] that "for very small r, that $a^r \approx 1 + [ln\,a]r$." Thus e is a very special number. It is the only number such that $ln\,e = 1$. Thus it is the only number such that for very small r, $e^r \approx 1 + r$.

Exercise 19.20 Replacing 10 by e

(i) Without looking back in the text, state the definition of "the logarithm of positive number x to base e." You can use the definition of $log_{10}\, x$ as your guide.

(ii) Recall the two fundamental equations for log_{10} on page 365. Restate these if the base of logarithms is changed from 10 to e.

(iii) Are the statements of the First and Second Laws of logarithms changed if we change the base from 10 to e?

[33] I am assuming in this sentence that a is a positive number.

19.7 Logarithms to any Base

What other numbers can be the base for a logarithm operation? Suppose you picked the number 2. You could ask what power of 2 yields 10, i.e., $2^x = 10$, for what x? Then this x would be the logarithm of 10 to base 2, i.e, $log_2 10$. How would you calculate this on your calculator? You most likely do not have a log_2 button, only log_{10} and ln buttons. We can get out of this dilemma by using the following formula:

$$\boxed{log_a b * log_b x = log_a x,} \qquad \textbf{(Change of Base of Logarithms Law)}$$

where a and b are positive numbers, neither of which are equal to 1.

See footnote [34] to see why this equation is true.[34]

There is a rather silly way to remember the equation above. Think of "rolling" the $log_a b$ over the $log_b x$. Then the b above "cancels (not in any rigorous sense)" the b below and you are left with $log_a x$.

An equivalent way to write our change of base of logarithms law is:

$$\boxed{log_b x = \frac{log_a x}{log_a b}.}$$

Thus, if you wanted to find $log_2 9$ with your calculator it is easy with this formula if you let $a = 10$ (or $a = e$). Thus

$$log_2 9 = \frac{log_{10} 9}{log_{10} 2} \approx \frac{.9542425094}{.3010299957} \approx 3.169925001 (\approx \frac{log_e 9}{log_e 2}).$$

Thus if we know how to calculate logarithms to a base a, then we can calculate logarithms to any other base b.

There is a corollary of our change of base law which we can get from that law by choosing $x = a$. We then get:

$$log_a b * log_b a = log_a a = 1.$$

This becomes

$$\boxed{log_b a = \frac{1}{log_a b.}}$$

Thus in case we want to know $log_2 10$, we have (in terms of numbers you can evaluate on your calculator):

$$log_2 10 = \frac{1}{log_{10} 2} \approx \frac{1}{.3010299957} \approx 3.321928094.$$

[34]The argument that establishes the above equation is this: $a^{(log_a b)*(log_b x)} = (a^{log_a b})^{log_b x} = b^{log_b x} = x$. Thus $log_a x$, the power to which a must be raised to get x, is also equal to $log_a b * log_b x$.

Exercise 19.21 The Base of Logarithms Does Not Have to Be 10 or e

(i) Without a calculator estimate $log_3 10$. Is it bigger than 2? 3? 4? Is it less than 4? 3? 2?

(ii) Calculate $log_2 e$.

(iii) Calculate $log_5 3$.

(iv) Why do we never consider $log_a x$ for $a = 1$?

(v) In the Doubling Time Formula on page 369, if you change the base of the logarithms used from 10 to e, how does the formula change?

(vi) Suppose a population starts with 100 people. How many times does the population have to double to get a population of a quarter billion people?[35]

(vii) Suppose you have one unprotected sexual encounter with someone who had sex with three other people in the previous year. Suppose each of those three had sex with three others the year before that and so on going back for twelve years. Assuming that each person (partner) appears no more than once in the resulting branching diagram, what is the minimum number of people you have had indirect sexual exposure to? If one percent have HIV, how many HIV positive people have you been indirectly exposed to?

(viii) Suppose that you have 3 sexual partners per year and that everyone you have either direct or indirect sex with also has 3 partners per year. What is the maximum number of sexual partners (both direct and indirect) you can have in 10 years? 11 years? 12 years? What is the latency period for AIDS?[36]

(ix) You have 2^1 parents, 2^2 grandparents, 2^3 great grandparents, and so on to the nth previous generation with 2^n individuals. Is there a value for n so that 2^n is greater than the earth's human population during that n^{th} previous generation's time on earth? Can you conclude that any two people have a common ancestor?

Exercise 19.22 The Difference Between Simple Annual Growth Rates and Growth Rates Compounded Continuously

(i) Suppose a population grows 50% per year. We say that the population has a simple annual growth rate of 50%. Thus if $P[t]$ is the population at time t, where t is measured in years, then one year later the population is $P[t+1] = (1+.5)P[t]$. Find a formula for $P[n]$, where n is a positive whole number of years, in terms of $P[0]$, the initial population.

(ii) If a population $N[t]$, t measured in years, experiences a simple annual growth rate r, then the population when $t = 1$ year is $N[1] = (1+r) * N[0]$, where $N[0]$ is the initial population. If we start with the same initial population, $N[0]$, growing at an annual rate s, *compounded continuously*, then after one year we have $N[1] = e^s * N[0]$. If you are given a simple annual growth rate, r, can you find a formula for the continuously compounded annual growth rate s that yields the same net result after one year as r, i.e., so that $e^s = 1 + r$? Apply your formula to part (i) above where $r = .5$. What is s in this case?

(iii) If you take a simple annual growth rate of r and compound it continuously for t years (see Exercise 19.19), try to understand why you end up with a population at time t years, $N[t] = e^{rt} * N[0]$, where $N[0]$ is the population you start with.

(iv) Pick a year, like 1985, and start your clock ticking. Thus $t = 0$ corresponds to the year 1985. Suppose that the population of the world at time t, measured in years, is $P[t]$. The annual growth rate of the world population in the mid 1980s was $r = .019$, compounded continuously. This gives $P[t] = e^{.019t} * P[0]$. What value of $P[0]$ should you use?[37] What does this formula say the world population in 1995 should have been? Compare this number with the actual population in 1995 and draw a conclusion.

[35] Hint: Consider the equation $100 * 2^N = .25 * 10^9 = \frac{1}{2^2} * 10^9$, and take the log_2 of both sides.

[36] Hint: This problem is a lot more involved than it might at first appear. For example, each of your partners in *each* past year gives rise to a branching diagram of past partners. Also, each past partner of each past partner gives rise to branching diagrams as well! Try solving the problem going back 2 or 3 years first.

[37] The world population in 1985 was $4.85(10^9)$. Which of our references has this data?

(v) Measuring time in years, the annual growth rate, compounded continuously, of the world population in the early to mid 1990s is estimated to have been $r = .016$, i.e., 1.6%. If the world population in 1995 was 5.69 billion, how long would it have taken for the world's human population to grow by one billion to 6.69 billion? What is the estimated doubling time (see part (viii) of this exercise and Exercise 19.23) of the world population if this 1.6% is correct?[38]

(vi) What would be the world population in the year 2285 A.D. if the continuous growth rate of part (v), i.e., 1.6%, remained unchanged? How many years would it take the world population to reach 1 trillion? Redo this exercise with the growth rate given in part (iv), i.e., 1.9%.

(vii) Under the assumptions of part (iv), i.e., annual growth rate 1.9% compounded continuously, how many times will the world population have doubled between 1980 and 2285? Redo this exercise assuming a growth rate of 1.4%

(viii) Under the assumptions of part (iv) what is the doubling time, t_d, of the world population?[39]

The last part of Exercise 19.22 leads us to one of the most important formulas in applied mathematics. Namely, if a population is growing at a rate r compounded continuously, then the doubling time t_d is given by

$$\boxed{t_d = \frac{ln\,2}{r}.} \quad \textbf{(Doubling Time if Continuously Compounded Growth Rate is r)}$$

For example, we have a growth control ordinance in Boulder, Colorado that limits residential growth to about 2% per year (compounded continuously). How long would it take Boulder's population to double? We have $\frac{ln\,2}{r} \approx \frac{.69}{.02} \approx \frac{70}{2} = 35$ years.

Think about the implications that doubling times of various sizes have for a given population.

Exercise 19.23 Calculating Doubling Times for Continuously Compounded Growth

(i) If a population is growing annually at 1% compounded continuously, i.e., $r = .01$, how long will it take the population to double?

(ii) Same as part (i), but $r = .025$.

(iii) Same as part (i), but $r = .03$.

(iv) Same as part (i), but $r = .10$.

(v) Based on what you computed for doubling times in parts (i) through (iv), do you think that it is possible or reasonable to have a 10% annual growth rate in electric power generating capacity for, say 100 years, in the United States?[40]

(vi) Is a 2% growth rate in world population high? low? medium? Would your answer have been different before you understood what this means in terms of doubling time?

(vii) If it takes 20 to 30 years (from birth) to train a doctor or mathematician, what can you say about the doctors and mathematicians in a population which is doubling every 10 to 20 years?

---

[38] The global human population growth rate for 1996 is given as 1.4% in the 1997 edition of *Vital Signs*, published by the Worldwatch Institute.

[39] Hint: $P[t + t_d] = 2P[t]$. Mimic the computation in Section 19.5 using the formula for $P[t]$, with continuous compounding, i.e., $P[t] = P[0]e^{rt}$.

[40] Magazine ads from the '70s said U.S. would collapse if it did not expand electrical generating capacity at an annual 10% continuously compounded growth rate.

(viii) What can you say about forests with a natural cycle of 1000 years which are being used by a population that doubles every 70 years?

(ix) If our population $P[t]$ satisfies the formula $P[t] = e^{rt} * P[0]$, the doubling time for this population is given by the formula $t_d = \frac{ln\,2}{r}$. Find a line with slope r which is related to our formula for $P[t]$.

Exercise 19.24 Problems with Interest on Money: Find the Interest Rate

(i) Suppose you borrow $1,000 from a bank for one year, and at the end of that one year you owe $100 in interest. Thus for the one year you paid 10 percent interest. What rate of interest r would the bank have to charge you, compounded monthly, i.e., 12 times a year, so that at the end of one year you pay 10 percent interest? What rate of interest r would the bank have to charge you, compounded continuously, so that at the end of one year you pay 10 percent interest?

(ii) Suppose you borrow $1,000 from a bank for two years, and at the end of those two years you owe $200 in interest. What rate of interest r would the bank be charging you, compounded monthly in this situation? Is your answer larger or smaller than your answer to the corresponding part of part (i)? What rate of interest r would the bank be charging you, compounded continuously, in this situation? Is your answer larger or smaller than your answer to the corresponding part of part (i)?

19.8 Further Study: More Complicated Models and Chaos Theory

Iterated Models: Simplest Case. If the population in birth cyle n is $P[n]$ then the simplest Schwartz Chart box-flow models we studied led us to equations of the form $P[n] = R^n * P[0]$, where we could take R, for simplicity, to be $\frac{K}{2}$. A little thought shows that this equation is equivalent to:

Iteration Equation for Simple Growth without Competition

$$\boxed{P[n] = R * P[n-1]}$$

where $P[n]$, the total population after n birth cycles, is expressed as a multiple R of $P[n-1]$, the total population after $n-1$ birth cycles, and $n = 1, 2, 3, \ldots$.

We call this the iterated form of the equation since it gives a formula for the population in one birth cycle in terms of the one just previous. In more complicated models, like the ones in the rest of this chapter, iteration equations are the best and simplest ways to formulate what is going on. Unlike our simplest case Schwartz Chart, there is no simple formula that gives $P[n]$ in terms of $P[0]$.

Iterated Model with Competition. In Chapter 17 we learned that if there are P points then the number of "connections" that can be made between these points, two at a time, is $\binom{P}{2} = \frac{P*(P-1)}{2}$. I am now going to make an assumption which is, like all applied mathematics, an oversimplification of reality. I am going to assume that the diminution of numbers in our population due to competition among members of the population depends in a simple way on the number of possible connections in the population, taking two individuals at a time. More precisely, I am led to the

Iteration Equation for Simple Growth with Competition

$$\boxed{P[n] = R * P[n-1] - S * \binom{P[n-1]}{2}}$$

where S is a constant and the term $S * \binom{P[n-1]}{2}$ is the number of individuals subtracted due to competition as we go from birth cycle $n-1$ to birth cycle n.

I will leave it to you to write down more general equations that take into account competition terms due to interactions 3 at a time, 4 at a time and so on.

I will now turn my attention to rewriting the above equation in an equivalent but simpler form. In this next exercise we convert the above iteration equation for simple growth with competition into a simpler form.

Exercise 19.25 The Logistic Iteration Equation

(i) Why can we rewrite our iteration equation for simple growth with competition as

$$P[n] = R * P[n-1] - S * \frac{P[n-1](P[n-1]-1)}{2},$$

$n = 1, 2, 3, \ldots$?

(ii) Show that the equation in (i) is equivalent to

$$P[n] = (R + \frac{S}{2})P[n-1] - \frac{S}{2}(P[n-1])^2,$$

$n = 1, 2, 3, \ldots$.

(iii) Define $R_p = (R + \frac{S}{2})$ and define $S_p = \frac{S}{2}$, then our equation becomes

$$P[n] = R_p P[n-1] - S_p (P[n-1])^2,$$

$n = 1, 2, 3, \ldots$.

(iv) From (i), (ii) and (iii) above do you see that we have the

Logistic Iteration Equation for Simple Growth with Competition

$$\boxed{P[n] = R_p P[n-1] - S_p (P[n-1])^2 = R_p P[n-1](1 - \frac{S_p}{R_p} P[n-1]),}$$

where R_p and S_p are positive constants, $n = 1, 2, 3, \ldots$?

Thus introducing competition among the members of the population P into our model leads us to the equation above, which is commonly referred to by mathematicians as the (discrete) *logistic equation.*

If we let $N[k] = \frac{S_p}{R_p} P[k]$ for $k = 0, 1, 2, \ldots$, then our equation becomes $\frac{R_p}{S_p} N[n] = R_p \frac{R_p}{S_p} N[n-1](1 - N[n-1])$, and after multiplying both sides by $\frac{S_p}{R_p}$ and changing the letter N back to P (I hope that is not too confusing) we get:

Normalized Discrete Logistic Iteration Equation

$$\boxed{P[n] = R_p P[n-1](1 - P[n-1])},$$

$n = 1, 2, 3, \ldots$, where we need to be careful to remember that $P[n]$ is no longer a whole number since we replaced P with $\frac{S_p}{R_p}$ times P. In fact, we will assume that $P[n]$ is between 0 and 1 for all n, including 0 and 1. Note that if $P[n_0] = 1$ or 0 for some n_0, then $P[n] = 0$ for all n greater than n_0. Can you verify this?

I am going to stop now, just when things are about to get very interesting. A mathematical biologist, Robert M. May, in 1976 discovered that populations that satisfied the "Normalized Discrete Logistic Iteration Equation" could exhibit *very* complicated behavior, cf., [448, pp. 455–461]. This was Professor May's introduction to chaos theory and it could be yours as well. See [162, 135, 228] and related works.

19.9 The World's Human Population: One Box

A simple model of the world's human population consists of one box, with the flow in, F_{in}, the birthrate, b; and the flow out, F_{out}, the death rate d. The possibility of humans departing the earth (alive) to go live elsewhere in the universe will not be considered here. Do your own estimates of the energy and economic costs of such an adventure and see if you agree with [277] that there will not be a significant flow of living humans off the planet any time soon.

At the moment b is larger than d. The only way to stabilize population is to get to the place where $b = d$. To do this now there are two choices, decrease b or increase d.

Births ⟶ [World's Human Population] ⟶ Deaths

FIGURE 19.3: One Box Model of World's Human Population

Controls on Births and Deaths. Consider the following table wherein we list under "Promotes Population Growth" some of the things that more or less increase b and/or decrease d. Under "Controls Population" are listed some of the things that more or less decrease b and/or increase d. Note that some things can act in both columns to some degree—there is some fuzziness here. Feel free to add your own entries in each column.

Controls Population Growth	**Promotes Population Growth**
Hunger, famine	Agriculture, food
Disease	Education
War	Sex
Economic decline	Health care
Political chaos	Political stability
Abortion	Economic growth
Birth control	Shelter
Health care for all	
Education for all	

Which list do you support? More to the point, every politician I have ever heard of supports *growth*. Have you ever heard someone running for office that supported a stable population? When you look at the list of controls few are vote getters. Some politicians support war, but elsewhere, not in the streets near your house. Abortion can easily bring the roof down, birth control is rarely discussed in political campaigns, except maybe to deny access to some. Affordable, universal health care is tough to bring up in the United States, with 40 to 50 million citizens lacking health insurance in 2006. (In 2009–10 an attempt at "nearly" universal coverage, with private insurance companies firmly in charge, may become law.) Even serious programs for universal education though lauded are unfunded.

If we do not stabilize the global human population Nature will do it for us. Which tools from above or your extended list do you think Nature will use?

Exercise 19.26 Demographic Transition[41]

In 1945 Princeton demographer Frank Notestein formulated the concept of *demographic transition*. There are three stages, each with widely different growth characteristics. In the first of the three stages the birth and death rates are both high, resulting in slow population growth. In the second stage, population growth increases rapidly—typically to about 3% per year—since in this second stage death rates are reduced (due to medicine, sanitation, and so on) while birth rates remain high. In the third stage, the society modernizes and birth rates come down, almost in balance with the death rates. The growth rates of such societies are between 0 and .4% per year.

(i) Analyze each stage of demographic transition as a box model.

(ii) Analyze the transition from stage 1 to stage 2 and the transition from stage 2 to stage 3 using your box models. Find examples of countries which have gone from stage 1 to stage 2 to stage 3.

(iii) Discuss the following scenario: going from stage 1 to stage 2 and then back to stage 1. Why would this happen? Give examples of countries which have followed this scenario.

(iv) Can you design a box-flow population model in which population is stable and nonzero? Would a poverty floor, universal health care, education, human rights and a functioning system of justice be sufficient to stabilize population?

[41] This and other interesting concepts and facts are discussed in [57].

Chapter 20

Box Models: Money, Recycling, Epidemics

20.1 Some Obvious Laws Humans Continue to Ignore

You might think it silly that I would find it necessary to bring to anyone's attention the self-evident implications of Figure 20.1, Figure 20.2 and Figure 20.3. Unfortunately, given the reality of the largest garbage dumps in history and the toxic contamination of organic human waste which needs to find its way back into food producing soil, it is necessary. The following box-flow model, Figure 20.1, has an obvious flaw, what is it? In 2001 Professor

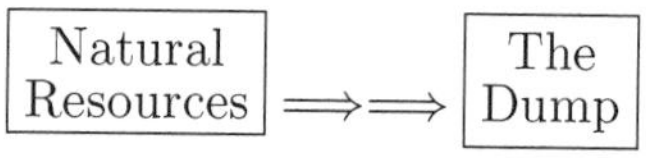

FIGURE 20.1: Impossible (in the long term): It Violates the Law!

Joel E. Cohen once asked his audience: "What is the largest structure ever created by humans?" The answer he gave was the Fresh Kills Landfill on Staten Island, New York. Whether or not this is still true, humans have been transferring Nature to the rubbish heap at an alarming rate for a long time, cf., [589, 568, 581, 249]. When the world's human population was much smaller, Nature could recycle our waste back into the resources box without our even taking much notice. The pace has picked up and we need a human assisted arrow from the dump back to the resource box; we need to recycle, as in Figure 20.2. I will discuss the mathematics of recycling in Section 20.3.

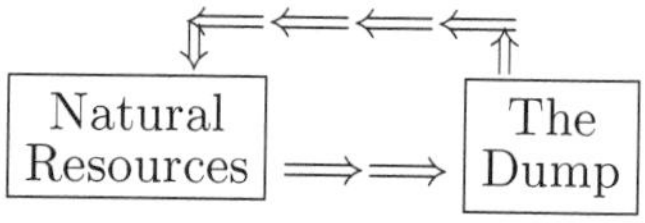

FIGURE 20.2: Inevitable: It's the Law!

There is a sub-box-flow model of the model above, cf., Figure 20.3 of some interest. Organic matter, such as animal wastes, for most of the history of such life on earth, has been distributed about, somewhat but not completely randomly, as nutrients for other forms of life. Such waste was thus recycled in the normal course of events.

Today, enormous amounts of food are shipped to cities, wherein is produced a corresponding amount of human waste. The natural cycle would be to move this waste back to where it could be used as fertilizer to grow more food. There are at least two problems with this cycle that need to be dealt with. First, it takes energy, now in the form of fossil fuels, to move all this biomass around—food to cities, sludge to farms. The second problem is that various classes of pollutants get mixed in with human waste to, more often than not, make toxic sludge, cf., Section 5.5.

Mixing toxic waste with human waste may prove to be one of the greatest design flaws of civilization.

A third problem, not usually even considered, is that we use considerable amounts of drinking water to move human wastes around within cities.[1] Thus not only is sludge contaminated with toxins that are best not found in fertilizer, but water is polluted as well. It is obvious to most folks that industrial wastes often contain heavy concentrations of toxins. What is just becoming recognized is that households are flushing a host of potentially powerful chemicals in the form of hormones and hormone mimics, so-called *endocrine disruptors.* All of the medicines, birth control pills, psychoactive drugs, and detergents, plasticizers, phthalates, antimicrobials, pesticides, caffeine, . . . create a potentially powerful brew which thus far escapes the processing abilities of most contemporary sewage treatment plants.

For example, local researchers have found that treated water when introduced back into our local waterway causes sex changes in fish. *"What we see in the fish downstream is as if they are taking birth control pills."*[2] Ongoing research on fish, other aquatic life, and with worms in sludge shows bioconcentration of a variety of the aforementioned chemicals (and others) with some known and other perhaps unknown effects. All of this is difficult to reconcile with organic agriculture, for example. The problem goes away if the term organic is changed to allow toxic sludge as a fertilizer, for example; and there are continuous efforts to allow toxic sludge, genetically engineered organisms, pesticides . . . in the definition of organic agriculture. A first step is to reframe the language. Thus "beneficial biosolids" is more often than not a mere semantic reframing of toxic sludge, cf., *Toxic Sludge is Good for You!*, [665]. If successful, more than language gets polluted.

---

[1]Composting toilets avoid this problem, but the resulting compost must eventually be moved and utilized. This is most easily done if gardens/farms are in close proximity.

[2]Quote from David Norris, Professor of Integrative Physiology at the University of Colorado.

Exercise 20.1 Toxic Sludge is Good for You

(i) What is the status of the U.S. government's definition of organic food at the time you read this?[3]

(ii) Estimate the amount of energy used in moving biomass (food) to cities and biomass (sludge) within and out of cities. If fossil fuels were not used, how would you accomplish the cycle in Figure 20.3, which is the law (eventually).

(iii) What is the status of research on water pollutants such as endocrine disruptors when you read this?

It would be smart to keep this loop non-toxic!

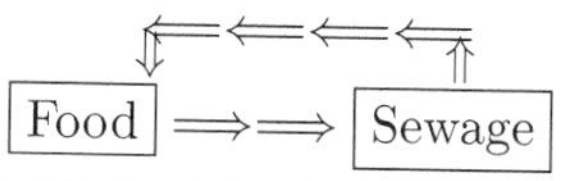

FIGURE 20.3: Food–Sewage–Food Cycle

20.2 A Linear Multiplier Effect: Some Mathematics of Money

Big Box Retail vs. Local Business. Imagine that you are living in a city and that you have earned a thousand dollars, 10% of which you will save and 90% of which you will spend for living expenses. You might think that that is pretty simple mathematics and that that is all there is to say about your budget—but there is a bigger picture. How you spend your money can have a profound effect on your community.

Follow the Money: Scenario 1. Let us imagine two scenarios. In the first you spend all your money at department stores, grocery stores and other businesses that you believe have the lowest prices and which have their corporate offices in some far away place; and the things you buy were manufactured or grown in far, far away places—by people that will work for far less than the minimum wage in the community in which you live. Imagine that these businesses in the first scenario hire as few people as possible from your local community, paying them the lowest possible wages. (Do you think this scenario reflects the reality in some cities, or parts of some cities?)

Let us trace some of the effects of your $1,000. You put 10%, or $100, in savings—either in a can in the backyard or in a bank. The remaining $900 is taken by those with whom you do business. Almost all of that money, save

[3]The USDA was deluged with hundreds of thousands of comments supporting a strict definition of organic food when the issue first came up, cf., page 144.

a local sales tax[4], leaves your local community. It is quite possible that you spent that money in a few hours, say 10 hours, and that you were served by people working at say, less than \$10 per hour, leaving less than $\frac{\$10}{hour} * 10\ hours = \$100$ in your local community and perhaps even less than that going to the production and transportation (locally) of your purchases. Also, depending on the policy of the place(s) where you put your savings, your savings may or may not be available for loans to people in your community. Also, the reason you think you are getting a good deal may be due to less than truthful advertising created in offices far from your city and brought to you by media corporations headquartered far from your city.[5] Anyway, the net result in this scenario is that quite likely less than 10% of your money remains in your local community. Feel free to play with the numbers in this first scenario so that they might more closely reflect what is going on in some city with which you are familiar.

In the area where I live, a local hardware store with competitive prices collected and paid a million dollars in city sales taxes which went, by a decision of city council, to buy and develop land for a big-box retail hardware store in direct competition with the local store.[6] Cities often offer subsidies in forms such as tax breaks to mega-retailers, subsidies not available to local merchants.[7] Also some mega-retailers generate a lot more police work, often creating a need for an increase in the number of police in a community, and an increase therefore in taxes.[8] Workers at mega-retailers often end up needing assistance in the form of community subsidized health care due to lack of benefits; food stamps for such workers are not unheard of. Locally owned businesses tend to increase the sense of community, businesses owned from afar tend to decrease sense of community [461, 138, 634, 635].

[4] Sometimes local subsidies/tax breaks divert even some of these sales taxes away from the local community

[5] Always be on the alert for "bait and switch" tactics. The advertised cheap item is not available, or in short supply, or is placed in the store amidst other "bargains" (which aren't) which you are encouraged to buy.

[6] Local merchants have traditionally been more closely connected and involved in the communities in which they live than executives thousands of miles away. This becomes apparent if you ever have to ask a community favor of a merchant.

[7] For example, as of 2004 www.goodjobsfirst.org documented over 1 billion dollars in taxpayer subsidies to Wal-Mart. Due to the secrecy and undisclosed private records, the actual subsidy figure is much higher but difficult to document, cf., [447]. In an Associated Press article by Don Babwin, Jan. 3, 2010, we read that a few cities and counties are building "clawback" provisions into the contracts which subsidize corporations. Thus if the corporation does not provide the number of jobs promised, for example, then they must pay increased taxes, i.e., their subsidy is reduced. DeKalb, Ill. & Target Corp., Monore County Indiana & Printpak, St. Louis County & Pfizer, and Detroit & General Motors, all have "clawback" provisions, cf., www.goodjobsfirst.org.

[8] Whereas a local retailer might unceremoniously kick a petty thief out the door, mega-retailers prosecute all offenses because of their business model. The anonymity of mega-retail often attracts bad checks/credit and shoplifting, cf., [461, 138].

One bottom line is this: as has been documented, once global corporations eliminate local competition they are free to raise prices and they do; and at that point consumer choice is greatly reduced. Antitrust legislation was created long ago to deal with such monopolies, but is little enforced.[9] For those local communities that have chosen to participate in power before it is too late, success has happened.[10] Of course, there is no law of Nature that prevents a global corporation from acting like a union of local merchants with the health of local communities its highest priority.

Keep Following the Money: Scenario 2. Now let us look at a second scenario. Suppose you put your \$100 savings in a local credit union, which deals almost exclusively with the economic lives of people in your community. Also, suppose you spend all of your \$900 on local goods and services. Suppose also that everyone else in your city behaves like you, saving and spending all of their money locally. (I am not saying this is possible—or even always desirable, but I am assuming this for simplicity in this second scenario.)

Let us see what happens to your \$900. You give this to people in your city in return for goods and services, and these people save 10%, or \$90, at local credit unions and spend 90% of that \$900, i.e., $.9 * \$900 = \$810$, locally. Suppose this process goes on *ad infinitum.* What is the cumulative effect on your community of the \$900 that you earned and spent locally? We are led to the following sum which tells us how much money was spent directly on goods and services locally as a result of your initial \$900:

$$.9 * 1000 + .9 * .9 * 1000 + .9 * .9 * .9 * 1000 + \cdots = (.9 + .9^2 + .9^3 + \dots) * 1000$$

$$= \sum_{k=1}^{\infty} (.9)^k * 1000 = \left(\sum_{k=1}^{\infty} (.9)^k\right) * 1000.$$

Exercise 20.2 The Geometric Series Trick

(i) Show[11] that $\sum_{k=1}^{N} (.9)^k = \frac{.9-(.9)^{N+1}}{1-.9}$.

(ii) If N is very, very large, how big is $(.9)^N$?

(iii) Show that $\sum_{k=1}^{\infty} (.9)^k = \frac{.9}{1-.9} = 9$.

Thus if you spend \$900 locally, and everyone else spends (at the same 90% rate) locally as well, there arises from your \$900 a net amount of $\$1000 * 9 =$ \$9000 spent locally!

What about the money saved locally at credit unions? Well, you saved \$100. At the next stage, 10% of \$900, or \$90, was saved. At the next stage, 10% of $\$810 = .9^2 * \$1000$, or \$81, is saved. Continuing on in this way, *ad*

[9] As with all laws, these would be enforced if the public demanded it to the extent that the public voted into office lawmakers and executives that took these laws seriously.

[10] See [461], www.bigboxtoolkit.com and www.newrules.org.

[11] Hint: Consider first $\sum_{k=1}^{3} (.9)^k = (.9)^1 + (.9)^2 + (.9)^3$. Obtain a second equation by multiplying this one by .9, then subtract this second equation from the first.

infinitum, we get

$$.1 * 1000 + .1 * .9 * 1000 + .1 * (.9)^2 * 1000 + \cdots = .1 * 1000 * \sum_{k=0}^{\infty} (.9)^k.$$

Exercise 20.3 Geometric Series with One More Term
(i) What is the difference between $\sum_{k=1}^{\infty}(.9)^k$ and $\sum_{k=0}^{\infty}(.9)^k$?
(ii) Show that $\sum_{k=0}^{\infty}(.9)^k = 10$.

We are thus led to an amazing result. If you earn \$1000, save 10% (or \$100) and spend 90% (or \$900) locally, and if everyone in your community does likewise, i.e., locally saves 10% and spends 90% of the money that passes through their hands that originated with you, your community ends up saving (in the local credit union)

$$.1 * \$1000 * \sum_{k=0}^{\infty} (.9)^k = .1 * \$1000 * 10 = \$1000 \; \textit{locally}!$$

Comparing the two scenarios above, we see that the effect of earning \$1000 and spending it mostly non-locally results in a local impact of about \$100. The local economic effect of your earnings is divided by a factor of 10.

If you and your neighbors earn and spend locally, your \$1000 has the local economic effect of \$10,000. The local economic effect of your earnings is multiplied by a factor of 10!

Thus the local economic impact of doing business locally is 100 times greater than the local impact of doing business non-locally—at least in our two scenarios.

Now think about this. Suppose some issue of great concern to you and your community arises, and you need the support of the businesses operating in your local community to get the result your community desires. If the businesses are locally owned and operated you might guess you would have more influence over them than if they are owned and operated from afar. This is obvious. What is not so obvious is that your local economic clout is "divided by 10" if businesses are not local, but your local economic clout is "multiplied by 10" if businesses are local. In order to effect non-local businesses you must co-ordinate with sufficiently many others in other places who also do business with the businesses in question.

From the standpoint of abstract mathematics, the term "local" people above can be replaced with any subgroup of people you wish to benefit from the multiplier effect. If a subgroup of people give the benefits of their multiplier effect to people "outside" their subgroup, there should be definite benefits or good reasons for doing so. For example, perhaps someone from outside can provide a service or product unavailable within the subgroup. To give away the benefits of the multiplier effect without gaining something in return does not appear to be smart economically.

Exercise 20.4 Some Extra Details in Following the Money

(i) Actually I glossed over some things in the first scenario. In that scenario 10% of your $1000 passed into the hands of other local people. Not all of that stays in the local economy. Why?

(ii) Suppose you pass 10% of your money to local people and 90% to "outsiders," and everyone else in your community does the same. How much "local" economic activity is generated by your $1000? (By local economic activity I mean money which passes through local hands on the way to other locals or non-locals.)[12]

(iii) How much of your initial $1000 stays in the local economy, under the assumptions in part (ii)?

(iv) In a biological system one can think of energy (say solar) as analogous to money. Where does this analogy lead you? (Note that the analogy is certainly not complete.)

(v) The media declared Wal-Mart the world's largest corporation, for example, the largest corporation in the Fortune 500, on or about April 1, 2002. Does the mathematics of this section help explain how Wal-Mart became the largest retailer in the world?

Exercise 20.5 The Two Scenarios: More Comparisons

(i) Compare the two scenarios above if the savings rate in each is 20% instead of 10%.

(ii) Redo the two scenarios above and compare the results if the savings rate is r.

(iii) Show that the total economic activity (savings plus spending) in Scenario 2 with savings rate r is $\$1000 * \frac{1}{r}$. Thus if $r = .1$, $\$1000 * \frac{1}{.1} = \$10,000$, as we found before.

(iv) What are the economic consequences of raising wages 5 cents per hour to: (1) a person in charge of thousands of sweatshops, (2) one who purchases products manufactured in said sweatshops?

(v) What are the consequences to the environment of local and non-local economic activity in the city with which you are most familiar.

20.3 Multiplier Effects Arising from Cycles: The Mathematics of Recycling

Imagine we live in a community that gets all of its paper from trees.[13] Imagine that 1000 tonnes of trees are cut per day which gives rise to about 1000 tonnes of raw pulp per day, manufactured at a pulping plant. Now imagine that each day 75% of this pulp is made into paper products that are politically/physically easy to recycle (like newspapers, junk mail and so on), and that each day 25% of our pulp is made into paper products that are not easy to recycle (like paper colored with luminescent dyes or like paper towels used to clean up toxic waste). If we in fact do no recycling then the situation is summarized below in Figure 20.4. We end up with a net production of 1000 tonnes of paper products each day, ending up rather quickly in the dump.

Now let us suppose that we actually recycle $\frac{2}{3}$ of the paper that is easy to recycle and none of the paper that is not easy to recycle. After one day we have

[12]Hint: Consider $\$1000 * \sum_{k=1}^{\infty}(.1)^k$.

[13]This is really not smart since there are many, many other sources of raw materials that can be made into paper, many of which are less destructive to the environment than the use of trees. Can you name some?

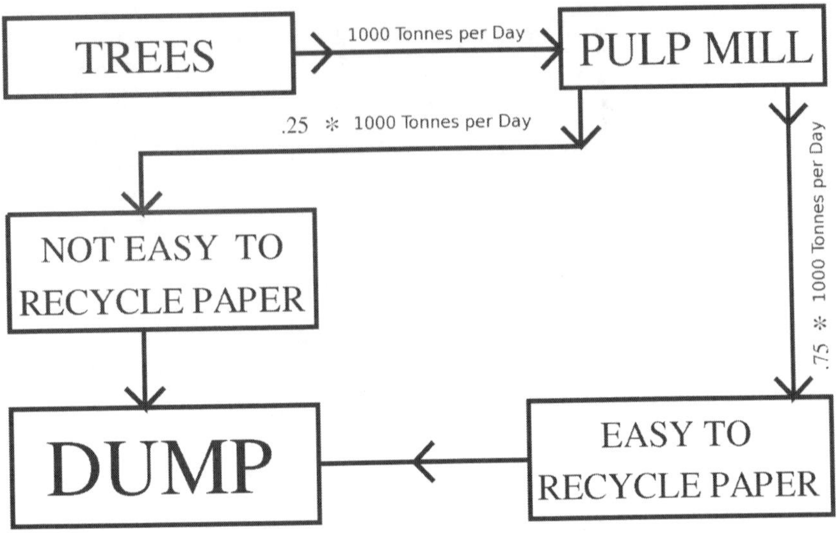

FIGURE 20.4: No Recycling

1000 tonnes of paper products as above, but at the beginning of the second day something changes. We have not only the 1000 tonnes of virgin paper pulp from the forest, but we have $\frac{2}{3} * .75 * 1000\ tonnes = \frac{2}{3} * \frac{3}{4} * 1000\ tonnes = 500\ tonnes$ of pulp from recycled paper to add to our inputs. (I assume that there are no losses in the recycling process for simplicity.) Thus the second day begins with 1000 *tonnes* of virgin pulp input plus 500 *tonnes* of recycled pulp input. This means that we end the second day with $.75 * 1500\ tonnes = 1125\ tonnes$ of easy to recycle paper and $.25 * 1500\ tonnes = 375\ tonnes$ of not easy to recycle paper. At the end of the second day we have 1500 tonnes of paper products, 500 tonnes more than if we did not recycle.

At the beginning of the third day we have $\frac{2}{3} * 1125\ tonnes + 1000\ tonnes = 1750\ tonnes$ of paper pulp as input. Seventy-five percent of this, or 1312.5 tonnes, becomes easy to recycle paper, and one fourth, or 437.5 tonnes, becomes not easy to recycle paper. Nevertheless, we end the third day with 1750 *tonnes* of paper products.

At the beginning of the fourth day we have $\frac{2}{3} * 1312.5\ tonnes + 1000\ tonnes = 1875\ tonnes$ of pulp input. This yields $.75 * 1875\ tonnes = 1406.25\ tonnes$ of easy to recycle paper and $.25 * 1875\ tonnes = 468.25\ tonnes$ of not easy to recycle paper. Thus we end the fourth day with 1875 *tonnes* of paper products.

Do you notice a pattern here? Each day the total amount of paper product in *tonnes* is greater than that of the day before—due to recycling and the constant cutting of 1000 *tonnes* of trees for pulp each day.

The situation with recycling is summarized in Figure 20.5 below. Do you

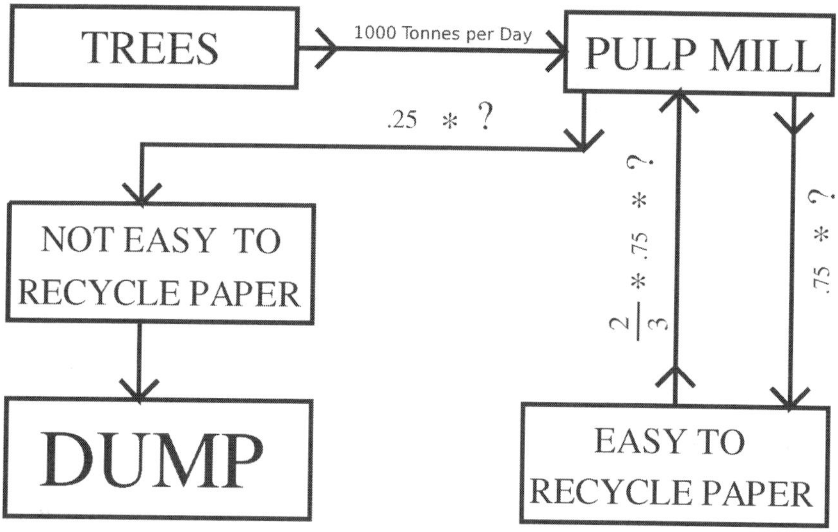

FIGURE 20.5: Recycling Increases Production

suppose the total amount of paper products produced in this scenario grows arbitrarily large, or do you think total paper production will approach some "limiting" value?

Exercise 20.6 Recycling: Our Concrete Case

(i) Calculate the total number of *tonnes* of paper products at the end of the fifth day.

(ii) Calculate the total number of *tonnes* of paper products at the end of the sixth day.

(iii) Keep calculating until you see a pattern and are willing to conjecture (or prove) that this recycling system will stabilize with a certain total tonnage of paper products produced each day.

Exercise 20.7 The Mathematical Patterns of Our Recycling Machine

(i) Suppose in the previous problem that all the assumptions are the same except that you start with X *tonnes* of virgin tree pulp per day, instead of 1000 *tonnes* of virgin trees per day, input into the pulp mill. The output at the end of the first day is then X *tonnes* of paper products. Verify that the input into the pulp mill at the beginning of the second day is X *tonnes* of virgin pulp plus $\frac{1}{2}X$ *tonnes* of recycled pulp. Is it true that the input on the second day is $X * (1 + \frac{1}{2})$ *tonnes*? Is the total output equal to the total input? Is the amount recycled equal to one-half of the total output?

(ii) Is it true that the input on a given day is X *tonnes* plus one-half the output of the previous day?

(iii) Is it true that the input on the third day is X *tonnes* plus $\frac{1}{2} * X * (1 + \frac{1}{2})$? Can this input on the third day be rewritten as $X * (1 + \frac{1}{2} + \frac{1}{2^2})$?

(iv) Is it true that the input on the nth day is equal to $X * (1 + \frac{1}{2} + \frac{1}{2^2} + \dots \frac{1}{2^{n-1}})$? Can you rewrite this using the Σ notation?

(v) Can the input (and hence the output) on the nth day be written[14] as $X * \frac{1-\frac{1}{2^n}}{1-\frac{1}{2}}$?

[14]Don't forget what you learned about the geometric series trick in Exercise 20.2.

(vi) If n is very, very large, what is the value of $\frac{1}{2^n}$?[15]

(vii) The symbol $\lim_{n\to\infty}$ means let n grow large beyond all bounds and "see what happens" to the value of the expression that follows the limit symbol. The expression $\lim_{n\to\infty}\ldots$ in mathese becomes "the limit as n goes to infinity of ...." Does the following make sense to you? $\lim_{n\to\infty}\frac{1}{2^n}=0$?

(viii) Does the following make sense to you? $\lim_{n\to\infty} X * \frac{1-\frac{1}{2^n}}{1-\frac{1}{2}} = 2 * X$?

(ix) Do you agree with the following statement: The "recycling machine" above produces about $2 * X$ *tonnes* of output (eventually) for every X *tonnes* of virgin input?

Exercise 20.8 Recycling Effectively Multiplies Inputs

(i) Can you calculate how much virgin pulp would be required in the above system to maintain a daily production of 1500 tonnes of paper products?

(ii) How does your answer to (i) change if we can increase the recycling rate of easy to recycle paper from $\frac{2}{3}$ to $\frac{3}{4}$?

(iii) Suppose *all* of the easy to recycle paper products are recycled, then for every X *tonnes* of virgin input, the "recycling machine" (eventually) produces about how many *tonnes* of output?

With the above example as a model, you should go on and create models with more than one cycle. Perhaps the most famous example from ecology is the tropical rain forest with multitudes of cycles reinforcing each other—vastly increasing the productivity of the forest over what it would be without cycles.[16] Compare this with the methods used in industrial countries to raise food crops. Are there any cycles in this process?

20.4 A Simple Model of an Influenza Epidemic

Infections and Invasions. From the point of view of patterns, infections are a special case of a more general phenomenon: *invasions*. If a plant, animal or more generally, a living creature is introduced, or introduces itself, into another living being or ecosystem, what happens after that is often of great mathematical (and social and economic and biological) interest.

For example, the Colorado potato beetle invaded England. The European spruce sawfly, *Gilpinia hercyniae* invaded Quebec and attacked the white spruce trees, *Picea glauca.* The Japanese beetle, *Popillia japonica* has invaded the United States, cf. [174]. The Zebra mussel, *Dreissena polymorpha,* is native to freshwater lakes of southeast Russia; it has invaded many fresh

[15] Hint: Exercise 13.6 part (i) involves calculating the meaning of 2^{64} grains of wheat (or rice). If you do this exercise you will have a better intuitive grasp of just how big 2^{64}, say, really is.

[16] From *Science News*, November 18, 2006, p. 333, "The African source of the Amazon's fertilizer," we see that if we take into account all of the cycles for the Brazilian Amazon, we need to include the estimated 40 million tons of Saharan dust from Africa that blows across the Atlantic every year, providing many nutrients including iron and phosphorous.

water lakes and rivers in the United States creating a sizable economic impact. The kudzu plant, *Pueraria lobata*, native to Japan and China, is referred to as "the plant that ate the South" (southern United States). Considerable sums of money are expended every year in clearing kudzu from roads, bridges, power lines and local vegetation. African "killer bees" were introduced to and then invaded the Western Hemisphere. Water hyacinth, *Eichhornia crassipes*, native to tropical South America, was introduced to and then invaded North America, Asia, Australia and Africa. Without controls it can double its biomass every two weeks. Asian carp were introduced from China into Mississippi in the 1970s, now they have invaded the entire Mississippi river ecosystem, starving out native species. They are within miles of the Great Lakes. If they get to Canada they can colonize much of Canada's freshwater habitat. Finally, a small, long-lived, jelly-fish like hydrozoan, *Turritopsis nutricula*, native to the Caribbean has been spread around the world via the ballasts of ships. It is growing in such numbers that it might disrupt various maritime environments.

Humans are probably the most successful exotic species of all, having invaded and colonized the habitat of just about every other plant and animal on earth.

Volumes could easily be filled with more examples/stories of exotic plants and animals intentionally introduced to and/or invading ecosystems or whole continents. Many of these invasions are on a scale and intensity with effects of longer duration than similar large scale invasions, like the Indonesian invasion of East Timor or the U.S. invasion (and occupation) of Iraq, to name just two of many, many such invasions in history. Human wars often involve invasions and there is a lot of interesting mathematics involved therein as well.

On a much smaller scale, the scale of the living cell, genetic engineering is just another form of invasion, [570, 386, 313, 9]. Obviously the genetic structure of one organism is being invaded by genes from another entity during genetic engineering, GE. What is not commonly known, however, is that the process is very much like being infected or shot with a gun (on the molecular level). One of the two principal methods of GE creates a "ferry" made with a piece of genetic material taken from a virus or bacterium. This "ferry" is then used to infect the, plant, say; and in so doing it smuggles the foreign gene into the plant's own DNA. Another principal method of GE is to coat large numbers of tiny pellets made of gold or tungsten with genes. These pellets are fired with a special gun into, say, a plant's cells. It is literally hit or miss, but after shooting enough pellets into enough cells genetic transfer usually takes place, cf., Section 5.4.

The mathematical study of invasions is potentially enormous. I will only deal with a very simple model of a very small part of the subject. Infectious diseases are the most deadly of all "natural" disasters; and history records this fact [97]. As documented in works such as [214, 215], tens of millions of people are dying[17] from untreatable forms of tuberculosis, malaria, strep, staph and other diseases.[18] The AIDS pandemic grows larger each year, cf., [662]. We have all heard of "bird flu," and "swine flu," and we wait patiently for the next pandemic [128, 151].

A Six-Rule Model of an Influenza Epidemic. There is quite a bit of literature on infectious diseases involving interesting mathematics.[19] The model we will present here is instructive, fairly realistic, simple, well known and can be found in [299, pp. 320–8]. You can study the following model of an epidemic with a pencil and paper, or paper and a calculator, or with a spreadsheet.

As in our study of population growth with Schwartz Charts we will assign each person in our model to a "box" or compartment. Instead of using age to determine a person's assigned box at a particular time, however, we will use the relationship of the person to the influenza disease. See Rule 1 below.

Imagine a typical influenza outbreak, such as occurs almost every winter. One fact that allows us to simplify our model is that the time for the epidemic to "play itself out" (usually three or four months) is short compared to the length of the generational cycles in the population (a couple decades). We thus will assume the following six rules for our epidemic model.

RULE 1. We assume that at any particular time every individual is in exactly one of the following "boxes:"

SUSCEPTIBLES: Defined to be those well persons who have not had the disease this season and have no immunity;

INFECTED: Defined to be persons who have influenza in some stage;

CURED: Defined to be those persons who are now well, but who have had the disease, who have recovered and who hence have short term immunity;

DEAD: Defined to be those persons who died from the influenza.

We can now sketch the following diagram, Figure 20.6, of our epidemic.

RULE 2. We assume that the total size of the population affected is constant during the epidemic. We will measure the time t in days (in our case this is an integer which will also denote the row in our spreadsheet, see Table 20.1). Let $S(t)$ denote number of people in the Susceptibles box at time t, $I(t)$ denote the number of people in the Infected box at time t, $C(t)$ the number of people in the Cured box at time t, and $D(t)$ the number of people in the Dead box

[17] For example, infectious diseases killed 16.5 million people worldwide in 1993, a number comparable to the combined worldwide number of deaths due to cancer (6.1 million), heart disease (5 million), cerebrovascular diseases such as stroke (4 million), and respiratory diseases such as bronchitis (3 million). See Chapter 7, "Confronting Infectious Diseases," by Anne E. Platt in *State of the World 1996*, W.H. Norton & Company, New York and London, 1996, by Lester Brown et. al.

[18] Note that as mathematics would predict, sometimes untreatable diseases become treatable if medication protocols are changed. For example, stop treating malaria with drug X and you stop selecting for malaria parasites that are resistant to drug X and such resistance can and has waned. At the moment half of all antibiotics in the U.S. are used in animal feed—a tragic, misguided use of incredible, life-saving medical tools for marginal economic gain.

[19] For an encyclopedic treatise see [11].

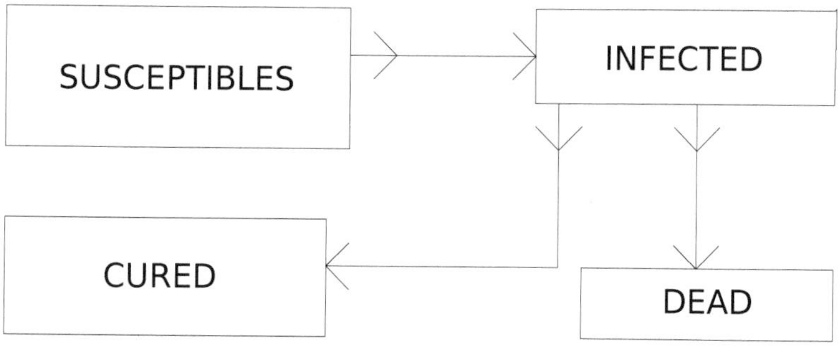

FIGURE 20.6: Box-Flow Diagram of an Influenza Epidemic

at time t. Our RULE 2 can now be written:

$$S(t) + I(t) + C(t) + D(t) = N,$$

where N is a constant and the equation holds for all times t.

RULE 3. We will assume that there is a number (which does not depend on time), R_i, called the *infection rate* which satisfies:[20]

$$S(t+1) - S(t) = -R_i S(t) I(t).$$

RULE 4. We will assume that there is a number (which does not depend on time), R_c, the *cure rate* which satisfies:[21]

$$C(t+1) - C(t) = R_c I(t).$$

RULE 5. We will assume that there is a number (which does not depend on time), R_d, the *death rate* which satisfies:

$$D(t+1) - D(t) = R_d I(t).$$

RULE 6. The rates of infection, cure and death also satisfy:

$$I(t+1) - I(t) = R_i S(t) I(t) - R_c I(t) - R_d I(t).$$

Exercise 20.9 Transmission of Disease

(i) Suppose you have m red dots and n blue dots. How many line segments can you draw that begin at a red dot and end at a blue dot? Answer: mn, or m times n. Why?

(ii) If m represents the number of Infected and n represents the number of Susceptibles, then mn represents roughly the number of "contacts" that can occur between these two groups. A certain fraction of these contacts will result in transmission of disease from the Infected to the Susceptibles. If $R_i = .00001$, then we are assuming that that fraction is "one

[20]See Exercise 20.9 for some justification for this assumption.

[21]See Exercise 20.10 for a short discussion of this assumption.

in one hundred thousand." (The number $R_i = .00001$, is an educated guess on my part. After to you have read and understood the model we are constructing feel free to experiment and see how the model responds to changes in the infection rate. The consequences are not dire since this is just a mathematical model.) Thus the infection rate tells us how virulent a particular flu is. Do you think this is a reasonable value for the infection rate?

(iii) We can rewrite Rule 3 as $S(t+1) = S(t) - R_i S(t)I(t)$. Do you see that this equation says that the number of Susceptibles in one row is equal to the number of Susceptibles in the previous row less the number of Susceptibles that have moved over to the Infected box?

Exercise 20.10 Equations for the Epidemic Model

(i) The equation $C(t+1) - C(t) = R_c I(t)$ says that the change in the number of people in the Cured box, in a given row (that is in a given day), is a certain fraction of the persons in the Infected box. If $R_c = .7$, what fraction of Infected become Cured each day? Is this a reasonable fraction based on your experience with the flu?

(ii) The equation in part (i) can be rewritten as $C(t+1) = C(t) + R_c I(t)$. Do you see what this says about the flow into the Cured box from the compartments (or boxes) from the previous row?

(iii) The equation $I(t+1) - I(t) = R_i S(t)I(t) - R_c I(t) - R_d I(t)$ tells us that the change in the number of people in the Infected box is given as a sum. Namely, the number of people who move from the Susceptibles box to the Infected box, plus (minus) the number of people who move from the Infected box to the Cured box, plus (minus) the number of people who move from the Infected box to the Death box. Explain each of the minus signs in this equation and the minus sign in the equation in Rule 3.

(iv) We can rewrite the equation in (iii) as $I(t+1) = I(t) + R_i S(t)I(t) - R_c I(t) - R_d I(t)$. Can you explain in words what this equation says?

(v) Does it follow from Rules 3, 4, 5 and 6 that

$$S(t) + I(t) + C(t) + D(t) = S(t+1) + I(t+1) + C(t+1) + D(t+1)?$$

(vi) My educated guesses for the values of the cure rate and the death rate are: $R_c = .7$ and $R_d = .001$, respectively. Do you think that these numbers are "reasonable"? You should feel free to experiment with these values as you were previously encouraged to experiment with the infection rate.

(vii) If the cure rate and the death rate were to add up to 1, what would this say about our epidemic model?

In Table 20.1 we have the complete output from a model (done on a spreadsheet) obeying the six rules above with the values I have indicated for the infection, cure and death rates. I assumed that the total population, N, was 100,000. The computer rounded off all the numerical entries to be integers to make the output look better (and so it would fit in the space allotted). Note that Suscept and Infect are abbreviations for Susceptible and Infected, respectively; again so the table will fit on the page.

In Figure 20.7 is a graph showing the rise and fall of the infected population and the death toll. This model was not created to fit the data from any particular real epidemic I have studied.

Exercise 20.11 Assumptions of the Model and Beyond

(i) What hidden assumptions are there in this model about births and deaths?

(ii) How could you take into account the existence and level of effectiveness of flu shots?

(iii) In a population which has not been inoculated against smallpox and for which the population is largely susceptible (as in the United States in 2001 or so) about 30% of those who make a single contact with an infected person become infected themselves. Also, of those who become infected approximately 30% die. Using this data create a model for a smallpox epidemic.[22]

(iv) If you want a challenge (beyond our current scope) try to build a model of an epidemic that takes spatial relationships like geography into account. See, for example, the *Notices of the American Mathematical Society,* Volume 48, Number 11, December 2001, and the article therein "Mathematical Challenges in Spatial Ecology," by Claudia Neuhauser.

(v) For another fun challenge look at epidemics from the point of view of networks. Networks allow you, the modeler, to incorporate in more detail the structure of society into the epidemic model. See [711, pp. 162–194], [710].

[22]Although smallpox has been eliminated from the human population it could be reintroduced via bio-warfare. See [507].

TABLE 20.1: Daily Population in Each Box during Epidemic

Day	Suscept	Infect	Cured	Dead	Day	Suscept	Infect	Cured	Dead
1	100000	1	0	0	41	60565	4980	34406	49
2	99999	1	1	0	42	57549	4505	37893	54
3	99998	2	1	0	43	54956	3940	41046	59
4	99996	2	3	0	44	52791	3343	43804	63
5	99994	3	4	0	45	51026	2765	46144	66
6	99991	4	6	0	46	49616	2237	48080	69
7	99987	5	9	0	47	48506	1779	49646	71
8	99982	6	12	0	48	47643	1395	50891	73
9	99976	8	17	0	49	46978	1082	51867	74
10	99968	11	22	0	50	46470	831	52624	75
11	99958	14	30	0	51	46084	635	53206	76
12	99944	18	39	0	52	45791	483	53651	77
13	99926	23	52	0	53	45570	365	53989	77
14	99903	30	68	0	54	45404	276	54244	77
15	99873	39	89	0	55	45278	208	54437	78
16	99835	50	116	0	56	45185	156	54582	78
17	99784	65	151	0	57	45114	117	54692	78
18	99719	85	197	0	58	45061	89	54774	78
19	99634	110	256	0	59	45022	66	54835	78
20	99525	142	333	0	60	44992	49	54881	78
21	99383	184	433	1	61	44970	37	54916	78
22	99200	238	562	1	62	44953	28	54942	78
23	98964	308	729	1	63	44941	21	54961	79
24	98659	396	944	1	64	44931	16	54976	79
25	98268	510	1221	2	65	44924	12	54987	79
26	97768	653	1578	2	66	44919	9	54995	79
27	97129	834	2035	3	67	44915	7	55001	79
28	96319	1059	2619	4	68	44912	5	55005	79
29	95299	1337	3360	5	69	44910	4	55009	79
30	94025	1674	4296	6	70	44908	3	55011	79
31	92452	2074	5468	8	71	44907	2	55013	79
32	90534	2538	6919	10	72	44906	2	55015	79
33	88237	3056	8696	12	73	44906	1	55016	79
34	85540	3610	10835	15	74	44905	1	55016	79
35	82452	4168	13362	19	75	44905	1	55017	79
36	79015	4683	16280	23	76	44904	0	55017	79
37	75315	5100	19558	28	77	44904	0	55018	79
38	71474	5366	23128	33	78	44904	0	55018	79
39	67838	5440	26884	38	79	44904	0	55018	79
40	63959	5306	30692	49	80	44904	0	55018	79

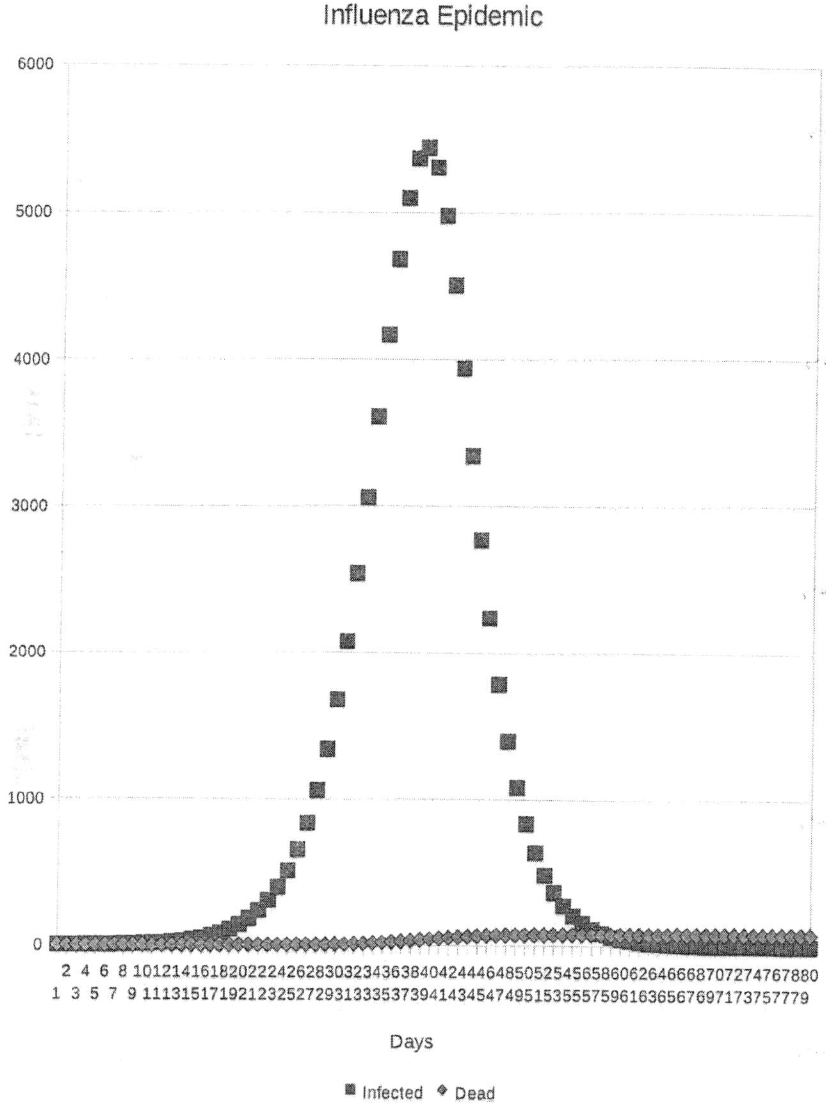

FIGURE 20.7: Daily Population of the Infected and Dead during Epidemic

Part VI

Chance: Health, Surveillance, Spies, and Voting

Chapter 21

Chance: Health and News

21.1 If You Test HIV Positive, Are You Infected?

The following question was once asked of 60 staff and students at the Harvard Medical School: "If a test to detect a disease whose prevalence is $\frac{1}{1000}$ has a false positive rate of 5%, what is the chance that a person found to have a positive result actually has the disease, assuming you know nothing about the person's symptoms or signs?"[1]

Nearly half of the people answered 95%. What is your answer? Write it down for future reference—before you look at the following footnote.[2] Only 11 people got the correct answer. For us to find the correct answer we need to do some "fancy" counting. It obviously isn't obvious, but these days it is worth figuring out!

Calculating Your Chances. Maybe you are a bit confused at the moment so let's slow down and consider the following about AIDS.[3] Currently about $\frac{1}{2}$% of the U.S. population is HIV infected.[4] Let's assume that the test for HIV is 99% accurate.[5] This means that if you were to test a "large" group of people known to be truly HIV infected it would give the correct, in this case a positive, result 99% of the time. We call the truly infected who test positive the *true positives*. Now the test is wrong 1% of the time. So 1% of the truly infected when tested would get an *incorrect*, in this case, negative test result. The truly infected who test negative are called the *false negatives*.

Now if you were to give the test to a "large" group of people known to be truly HIV free the test would give the correct, in this case negative, result 99%

[1]Bruce Bower, *Roots of Reason*, Science News Vol. 145, Jan. 29, 1994. The Harvard Medical School question was from a 1978 study.

[2]The correct answer is (a little less than) 2%.

[3]For global data on AIDS see [662].

[4]You might ask where this estimate comes from. To track the official U.S. statistics (at the time you read this) I refer you to the Web site for the Centers for Disease Control. Since the U.S. population is growing, this percentage will change depending on whether the rate of increase in HIV infection and/or AIDS is greater, less than or the same as the rate of population growth.

[5]New and faster tests continue to be developed. Each type of test will have its own accuracy rate which will affect the calculations in this section.

of the time. The infection free who test negative are called the *true negatives.* Now when testing the people free of infection you would get an *incorrect,* in this case, positive result in 1% of the tests. The people free of infection who test positive are called the *false positives.*

So the big question is this: *If you test HIV positive, what are the "chances" that you actually are HIV infected—assuming the one test is your only source of information?* Write your answer down for future reference.

This is not an idle question for a great many people these days; but if AIDS is "cured" by the time you read this, there are other diseases for which the same mathematics applies. This question is also of political importance. There has been at least one person of note who has called for mandatory HIV testing for everyone, with public disclosure of the results; and for individuals testing HIV positive, "isolation" in the form of concentration camps or execution was proposed. Some folks take this "final solution" seriously, hence we must take it seriously. Civil and human rights aside, I want you to ponder the question: Would such a program of testing and isolation work? But back to our original problem.

We have not defined what we mean by "chance." I want to do this now for this problem. If one tests HIV positive, the *chance* that the person actually is HIV infected is defined to be the following fraction:

$$\frac{number\ of\ people\ who\ test\ HIV\ positive\ who\ actually\ are\ HIV\ infected}{number\ of\ people\ who\ test\ HIV\ positive}.$$

So we now have to figure out what this fraction is. With a little thought you can see that the numerator of our fraction is the number of true positives. The denominator of our fraction, *all* the people who test positive, is the sum of the true positives and the false positives. Thus the above fraction we are looking for equals:

$$\frac{the\ number\ of\ true\ positives}{the\ number\ of\ true\ positives\ + the\ number\ of\ false\ positives}.$$

Let's suppose we are dealing with a population of a million people.[6] So suppose 10^6 people take the HIV test. The first thing we have to find is *the number of people who test HIV positive, which equals true positives + false positives.* This is a little tricky, but not very tricky. There are *two* kinds of people who test HIV positive: those who are (truly) HIV infected, the true positives, and those who are not, the false positives. How many are in each group? Well, we are told that $\frac{1}{2}\%$ of our population, in this case $\frac{1}{2}\%\ of\ 10^6$, are HIV infected. Thus there are $.005*10^6$ people who are indeed HIV infected. When they take the test which is 99% accurate we find that

$$.99*.005*10^6\ people\ = 4950\ people\ who\ are\ HIV\ infected\ test\ HIV\ positive.$$

[6]We could use the current population of the United States, but 10^6 is a little more convenient. It turns out that the calculation we are about to do produces the same answer no matter what initial nonzero population size we assume.

This is the number of *true positives.*

The rest of the population, i.e., $.995*10^6$ people who are HIV free, takes the test which is wrong (in this case, positive) 1% of the time. Thus the number of *false positives* is

$$.01*.995*10^6 people = 9950\, people\ who\ are\ HIV\ free\ who\ test\ HIV\ positive.$$

The total number of people who test positive is the sum of the true positives and the false positives. Thus

$$the\ number\ of\ people\ who\ test\ HIV\ positive\ = 4950 + 9950 = 14900.$$

We now can get our answer, since we have found the numerator of our fraction (the number of true positives) along the way; namely, *the number of people who test HIV positive who actually are HIV infected* (the true positives) = 4950. Thus we have:

$$Chance\ of\ being\ HIV\ infected\ given\ that\ you\ tested\ positive\ = \frac{4950}{14900} \approx .3322 \approx 33\%.$$

In summary, the chance of being a true positive given that you tested positive is $\frac{true\ positives}{true\ positives\ +\ false\ positives}$. This number might be a lot smaller than you might intuitively guess if the number of false positives is large.

Imperfect Tests, Imperfect Information, Public Policy. The above mathematical argument is relevant whenever you administer an imperfect test. Do you know of any test, or decision making process that is perfect? *The mathematical answer is quite surprising (to many) when the group the test is looking for is a relatively small fraction of the total population.* See Exercises 21.1 and 21.2. In Figure 21.1, we "draw a picture" of the above situation. Note the relative sizes of "infected who test positive," i.e., the true positives, vs. "the infection free who test positive," i.e., the false positives. Note also that the "infected who test negative," i.e., the false negatives, exist.

In Figure 21.1 we see that the entire population is divided by a vertical line into a thin rectangle of HIV infected people on the left and a fat rectangle of HIV free people on the right. A horizontal line divides the population into a skinny rectangle on the bottom of those with false test results and a fat rectangle on top of those with true test results. The net result is four categories of people: the true positives, false positives, true negatives and false negatives.

You actually read in the papers from time to time about some person who tested HIV positive who wasn't and such-and-such happened. With our assumptions, which are not very far from the truth for some tests, we see the chances are about 1 in 3 of actually being HIV infected if you test positive.[7]

[7]This is assuming that the test result is the only information known about you. If it is also known that you engage in risky behavior or get frequent blood transfusions from random sources, then we would have to recalculate using the additional information.

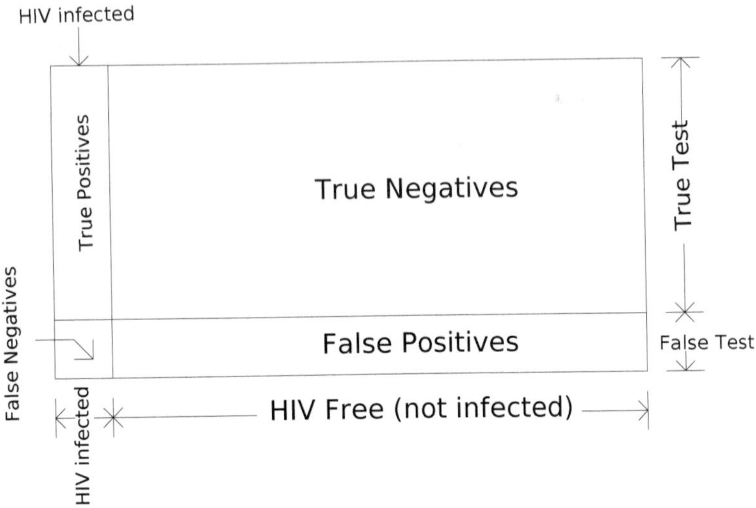

FIGURE 21.1: Boxes for Organizing Calculations

So what would be the results of mandatory public testing and "isolation"? To get the answer, we need only ask ourselves how we would react. Knowing that false positive tests happen, even if we were certain that we were not HIV positive we would not want to take the test, since doing so could lead to our "isolation." No one would have an incentive to take the test, many folks would resist. A lot of HIV free people, about 9950 per million, would end up "isolated." Though unconstitutional this sort of "isolation" of certain groups has already happened in the United States (during World War II), and some may be tempted to say: "Well, it may be unconstitutional but our civil rights are a small price to pay to stop the AIDS epidemic." But would it? Don't forget that 1% of the truly HIV infected persons will test negative and not be "isolated."

Someone may come back with: "Well, we can make the test much more accurate (see the next exercise)." Unfortunately, the test will never be perfect, and to make the test 99.9% accurate costs more than to make it 99% accurate. Costs escalate far more quickly than you might expect with each tiny increase in accuracy. To combat AIDS we must have a smarter strategy than that of "force-test and isolate." In any event, if you take a medical test of any kind you should ask your doctor how accurate the test is.

Before leaving this discussion and going to the exercises one important fact must be mentioned. The foregoing discussion should rationally lead you to vote against a public policy of "test and dispose" of those allegedly infected, but it should not lead you to be complacent if you do happen to test positive. Why? If you go get a test, you probably were "worried" about something. The 33% figure above applies to a blind test where no additional information is known about the persons being tested. The 33% figure does not apply to you if, for example, in addition to testing positive you know that you are an intravenous drug user who shares needles with friends and you have had unprotected sex with 100 different partners in the recent past.

Consider the following easy example. Suppose an individual tests HIV positive in the fall of 1995 and the only other thing you know about the person is that he is a young American male, say 30 years old. What happens to that 33% figure in this case? Well, one of the official statistics released by the U.S. government in November of 1995 was that approximately one of every 92 American males between the ages of 27 and 39 is HIV infected.[8] Suppose that this statistic is true, how does that change the 33% figure? The additional bit of information changes the $\frac{1}{2}\%$ (recall that was the percentage of the general population which is estimated to be HIV infected) to $\frac{1}{92} \approx 1.09\%$. If you go back and redo the calculation, we get (assuming the test is still 99% accurate) the following fraction: $\frac{\frac{1}{92}*.99}{\frac{1}{92}*.99+\frac{91}{92}*.01}$ which is about 52%.

Thus, it works out both intuitively and mathematically—what you know changes your chances. Regarding the subject of chance more generally, there is a vast literature on probability and statistics and related subjects, of which [133, 144] might be of interest.

Exercise 21.1 Tests with Variable Accuracy

(i) Redo the above calculation assuming the test is 99.9% accurate, calculate a "formula" for the answer with the accuracy rate as a *variable* so you don't have to do the whole calculation over again if the accuracy of the test changes.

(ii) Assuming 99% accuracy redo the above calculation with a whole range of estimates for the percentage of truly HIV infected persons, e.g., $\frac{1}{2}\%$, 1%, 5%, 10%, 20%, 50%, 100%. Can you calculate a "formula" for the answer with the infection rate as a **variable** so you don't have to do the whole calculation over again every time you change the infection rate?

(iii) In Figure 21.1 imagine a vertical line somewhere to the right of the vertical line already there. Imagine that only people who are infected and some percentage (say 50%) of those who are not infected get tested. (You can imagine that only people who are infected and some subset who think they might be infected get the test.) How does this affect the answer to the question: What are the chances of being infected given that you test positive?

(iv) Answer the question asked of the 60 staff and students at the Harvard Medical School, cf., the first paragraph of this chapter. You can assume that the test is in error 5% of the time.

(v) What strategy for combating the AIDS epidemic would you recommend if you were advising health officials and the government?

(vi) Other diseases have heavily impacted human populations; tuberculosis and the black plague come to mind. Pick one or more such major diseases of the past, study them, and compare their impact with the impact AIDS has had so far, cf., [11].

(vii) Using what you know about counting calculate the number of people to whom one unprotected sexual encounter exposes a man or woman, if the partner had sex with three

[8] Neergaard, Lauran, *AIDS cuts through ranks of U.S. men,* The Associated Press, Denver Post, November 24, 1995. On November 28, 2006 an Associate Press article by Maria Cheng noted that in 25 years AIDS is set to join heart disease and stroke as the top three causes of death worldwide. She was reporting on a paper by Dr. Colin Mathers and Dejan Loncar of the World Health Organization, who published their findings in the Public Library of Science Medicine journal.

other people in the previous year, and each of those three had sex with three others the year before that, etc., when the chain of exposures is traced back over 12 years—the time HIV can take to surface.[9]

Exercise 21.2 Examples from the Real World[10]

(i) This is another real situation where the mathematics of this section applies. Suppose that you are on a jury and you have to vote for conviction or acquittal of the defendant based on the following information. The county in which you live has a population which is 10% from Tribe 1 and 90% from Tribe 2. The defendant is from Tribe 1 and is accused of mugging Mr. Niceguy. Now it turns out that the entire case of the prosecution hinges on one assertion: Mr. Niceguy stated under oath (and truly believes) that his mugger is from Tribe 1. The defense, however, showed that under the circumstances of the mugging (at night, mugging from behind with gloves and mask etc.) that the victim can correctly identify the Tribe of the assailant 75% of the time. Given that the mugger was identified as being of Tribe 1, what are the chances of the mugger actually being from Tribe 1?[11]

(ii) If you were an innocent defendant, wouldn't you be hoping that the jury could "do math"? Just to illustrate the reality of this exercise I quote from an article by Robyn Blumner in the *St. Petersburg Times*, Dec. 2000.

"The sad truth is that the very evidence a jury is likely to find most convincing—a victim pointing across the courtroom at the defendant and saying 'He did it, I'd never forget that face'—is frequently unreliable and prone to error.

Recent exonerations due to DNA evidence drive that point home. According to the Innocence Project at Cardozo Law School, of the 77 men whose convictions were overturned after DNA testing proved their innocence, 65 were found guilty because of faulty eyewitness statements."

In the 25 years prior to 2001, 98 people were released from death row after DNA evidence established their innocence.

On January 31, 2000, the Governor of Illinois, George H. Ryan, suspended further executions under the Illinois death penalty until provided with a good explanation of why, of the last 25 people on Illinois' death row at that time, 12 were executed and 13 were exonerated.

In 2002, after the accused had served their sentences, the famous New York Central Park jogger case was resolved. The original 5 defendants were convicted without direct evidence linking them to the crime, but several of them "confessed to the crime." Finally, about a decade after the crime, the actual criminal confessed and was linked directly to the crime by DNA evidence. Research this case, or a similar one, as best you can and comment. This exercise can easily be made into an open ended project; use your imagination.

(iii) Pick your favorite tribe of humans, say the tribe you belong to, and try to define your tribe with a list of "characteristics." Now imagine a "police lineup" with one member of your tribe plus members of various other tribes with characteristics similar to yours in the lineup. Discuss the chances that someone using your list of characteristics as a definition of your tribe will correctly pick out the member of your tribe from the lineup. Did you define your tribe (only) in terms of "averages" or "stereotypes"; or did you define it incorporating the concept of variation?

An example of students making a big difference in people's lives, and perhaps eventually the American system of justice, is the Wisconsin Innocence Project, operated by the students at the University of Wisconsin School of Law. In an article by Terrence Stutz for the *Dallas Morning News*[12] we read

[9] *World·Watch*, Vol. 7. No. 6, November/December 1994, p 39.

[10] This exercise was inspired by an example in [522].

[11] Hint: Draw a picture analogous to Figure 21.1 and label the parts.

[12] Reproduced in *The Daily Camera*, Boulder, Colorado on January 17, 2001.

about a Texas man, Christopher Ochoa, freed after spending 12 years in prison for a murder he did not commit. The Innocence Project was persuaded by Ochoa to take on his case in 1998 and initiated proceedings that led to examination of DNA evidence from a 1988 murder at a Pizza Hut. It is important to note that Ochoa had confessed to the crime to avoid a death penalty conviction. Because Ochoa "confessed," his one-time roommate, Richard Danziger, was also wrongfully convicted—and has suffered permanent head injuries sustained in a prison beating.

Because of Ochoa's confession and his testimony implicating his roommate his case was largely forgotten until 1996, when another man in prison, Achim Josef Marion, wrote a letter confessing to the Pizza Hut crime to the Austin chief of police and to the *Austin American-Statesman*. In February 1998, after getting no response, Marion wrote another letter of confession to then-Gov. George W. Bush, also informing Travis County District Attorney Ronnie Earle. After the Wisconsin Innocence Project entered the case, Earle agreed to the DNA testing that eventually proved Ochoa and Danziger could not have committed the murder. Ochoa now feels sorry that he did not stand up to the police, who held the death penalty over his head to extract a confession. Of course, it is possible that Ochoa would have been executed if he had stood his ground.

This is just one of the cases the Wisconsin Innocence Project has successfully resolved, showing that concerned students can make a big difference in the "real world."

Exercise 21.3 Testing More Than Once in Succession

(i) This exercise is a bit ambitious, but you can handle it! Assume the original data, i.e., that $\frac{1}{2}\%$ of our population is HIV infected and that the test is 99% accurate. Suppose someone in this population takes the test twice and comes out positive both times. What are the chances that this person is HIV infected?[13]

(ii) If the 99.9% accurate test of Exercise 21.1 part (i) costs ten times as much as the 99% accurate test, which gives the most accurate results for the least amount of money: doing the 99.9% test once or doing the 99% test twice?

(iii) Suppose someone takes the test three times and comes out positive each time. What are the chances now that the person is HIV infected?

(iv) Suppose someone takes the test four times and comes out positive each time. What are the chances now that the person is HIV infected?

(v) Do you notice any patterns here? Are your answers to parts (i), (ii) and (iii) approaching a limit? What do you think the chances of being HIV infected are if you test positive many tests in a row?

Exercise 21.4 Chance Calculations with a Variable Population P

(i) Redo the first calculations of page 403 leading to 33.22%, but instead of assuming that the population is 10^6 assume that the population is some number P. First, write down

[13]Hint: Suppose you look at the sub-population of people that test positive after taking the first test. What fraction of these people are truly infected, what fraction are not? The test is still 99% accurate, so draw a new rectangle that represents this situation and calculate the appropriate ratio.

an expression for the number of true positives and the number of true positives plus the number of false positives.

(ii) Write down the ratio: $\frac{\textit{the number of true positives}}{\textit{the number of true positives + the number of false positives}}$. Is this number independent of the value of P? Can you use the axioms of numbers, i.e., legal operations to make the P disappear?

Exercise 21.5 Real Life Complications

(i) In the *New England Journal of Medicine* in April of 1998 were published the results from a survey on mammograms and breast exams. A total of 9,762 mammograms and 10,965 clinical breast examinations were performed on 2,400 women over 10 years.

Of those: 23.8 percent had at least one false positive mammogram.

In addition, 13.4 percent had at least one false positive breast examination.

Finally, 31.7 percent had at least one false positive result from either.

From these figures above, researchers estimated a general risk of false positive test results for all women of 49.1 percent. Do you agree with this estimate?

(ii) Using what you have learned in this chapter, do your best to make a mathematical model of the above breast exam/mammogram test, as we did for the AIDS test.

Exercise 21.6 Yearly Drug Tests

Suppose your employer requires you to take a drug test every year which is 99% accurate. Suppose you never actually take the drugs being tested for. After 10 years what are the chances (after taking the test 10 times) that you will never have had a false positive test?[14]

Exercise 21.7 Are All Tests Useful?

In the June 2002 issue of *Scientific American* I read about a test for ovarian cancer that always detects ovarian cancer when it is present. However, 5% of the women who are healthy also test positive for ovarian cancer. The incidence of ovarian cancer in the United States is 1 in 2,500 women who are over 35 years of age. If a woman over 35 tests positive for ovarian cancer what are the chances she actually has ovarian cancer? What do you conclude about the usefulness of this test?

Exercise 21.8 The Chances of a Flood

A friend of mine in Iowa was told, after his university was flooded, that the chances of a flood were just 1 in a 100, i.e. 1%. Thus over a long period of time, there would roughly be one flood every 100 years. If this is true, what are the chances of going 10 years in a row without a flood?[15] Hence, what are the chances of a flood in any decade?

Exercise 21.9 Shortcomings of Statistics

It is important to know how to apply mathematics toward finding solutions to problems, analyzing information and so on. It is equally important to know when mathematics is being misapplied. Quite often you can find news articles reporting on a "recent scientific finding" which involves the application (or misapplication) of statistics. The efficacy of a drug, the expected effects of a pollutant, the chances of this or that—such news invariably rests on an understanding or misunderstanding of statistics. Find and analyze, to the best of your ability, a few such articles. Is statistics being applied correctly or incorrectly in the articles you have chosen? One article you can use to start your discussion might be: "Odds Are, It's Wrong: Science fails to face the shortcomings of statistics," by Tom Siegfried, *Science News*, March 27, 2010, pp. 26–29.

---

[14]Hint: Assume that the 10 tests are "independent." What this means mathematically is this. If a finite number of events are independent, then the probability that they all happen is equal to the product of the probabilities of each happening.

[15]Assume flood events in different years are independent. The chances of not having a flood in a given year is $\frac{99}{100}$.

21.2 Chance and the "News"

From time to time, seemingly more frequently as of late, mathematically ridiculous presentations are put forward in the megamedia. As an example of this consider the following statement of Brit Hume, (8/26/03) made on his Fox News show "Statistically Speaking" with a picture of Iraq compared geographically to California in the background.

"277 U.S. soldiers have now died in Iraq, which means that statistically speaking U.S. soldiers have less of a chance of dying from all causes in Iraq than citizens have of being murdered in California, which is roughly the same geographic size.

"The most recent statistics indicate that California has more than 2,300 homicides each year, which means about 6.6 murders each day. Meanwhile our U.S. troops have been in Iraq for 160 days which means they are incurring about 1.7 deaths, including illness and accidents, each day."

Exercise 21.10 Find the Flaw in the "Statistical" Argument

(i) At the time Hume made the above presentation there were about 140,000 U.S. troops in Iraq and $35(10^6)$ people living in California. (In 2010 the U.S. Census Bureau estimated the population of California to be just under 37 million.) If you calculate deaths in units of "deaths per square mile per day" Hume was correct for the time period in question. However, are there other units you can calculate death rates in for which the death rate in Iraq for U.S. soldiers is considerably greater than the death rate of citizens living in California? How many times greater? At the time would you rather have taken your chances of living/dying in Iraq or California?

(ii) Are there any numerical errors in Hume's presentation of a more trivial nature than that pointed out in part (i)?

(iii) Continuing on this theme, seemingly similarly inspired, a local Boulder resident wrote the following letter to my local paper, *The Daily Camera*, published on November 15, 2003.

"Hardly a day passes that the Daily Camera does not carry news of the number of U.S. soldiers killed in Iraq since the war "ended" last May 1. In the Nov. 12 Daily Camera, an Associated Press story shocks us with the statistic that soldiers are dying at the average rate of one every 36 hours. In the Nov. 8 Daily Camera, columnist Christopher Brauchli adds the wind-chill index to this statistic, informing us that it's much worse than 144 lost lives. The addition of injured soldiers brings the casualty tally to the "magic number" of 1,737.

"While all these numbers may be helping Iraq war opponents win over some public opinion, the numbers that Mr. Brauchli and his compatriots "don't want you to see" are the U.S. national homicide statistics for major U.S. cities. Statistics compiled by Safe Streets DC (www.safestreetsdc.com) reveal that the top U.S. cities, ranked by homicides per capita, accounted for 3,101 deaths in 2002. The complete list of 32 cities, (U.S. cities with populations greater than 500,000) accounted for 5,172 deaths.

"Let's see—that's one homicide every 36 minutes.

"It would seem any one of us might be safer dressed in combat fatigues strolling downtown Baghdad armed with an assault rifle, than in Los Angeles (658 homicides), Chicago (647 homicides), New York (584 homicides), Detroit (402 homicides), or Washington D.C. (262 homicides)."

Do you agree? How's the math in this letter?[16]

[16]By the way, Chris Brauchli's column was terminated against his wishes.

(iv) According to the Johns Hopkins Bloomberg School of Public Health:[17] As of July 2006: *"As many as 654,965 more Iraqis may have died since hostilities began in Iraq in March 2003 than would have been expected under pre-war conditions,"* These studies used time honored methods of biostatistics, but were not well received, in fact they were ridiculed, by those in power. The "error bars" for this number were approximately plus or minus 250,000. Given this study and the approximately 3,000 U.S. soldiers killed as of the end of 2006, comment and put the Hume/Stewart comments above in context.[18] What would the total number of deaths have been in the United States, March 2003–July 2006, if the death rate had been the same as in Iraq? What have been the death and casualty tolls on all sides in the year you read this? Why was (is?) this war fought in the first place?

(iv) Can you find any other examples in the megamedia or elsewhere of mathematics being used to lead one to dubious conclusions?

[17]See http://www.jhsph.edu/publichealthnews/press_releases/2006/burnham_iraq_2006.html.

[18]Also from the Johns Hopkins study we read: *"According to the researchers the overall rate of mortality in Iraq since March 2003 is 13.3 deaths per 1,000 persons per year compared to 5.5 deaths per 1,000 persons per year prior to March 2003. This amounts to 2.5 percent of the Iraqi population having died as a consequence of the war."*

Chapter 22

Surveillance, Spies, Snitches, Loss of Privacy, and Life

It often happens that some mathematics is invented long before it gets "applied." For example, not long after the time of Newton the basic mathematical physics needed to send a rocket to the moon already existed. It was not until much later, of course, that a rocket did, in fact, land on the moon. What was needed (besides rockets!) was fast and vast computational power—provided by electronic computers. If you have ever had an MRI (Magnetic Resonance Imaging), basic mathematics (the Radon Transform and Inverse Radon Transform) was employed to create the image from the data collected; and you need a computer to do the calculations necessary to create the image. In both these cases, the basic equations and functions used had been known for a long time. However, solving these equations or evaluating these functions involve computations that cannot practically be done by hand.

And so it is in the business of surveillance—and avoiding surveillance, for example, via sending messages using codes or encryption. The "good news" is that perhaps the mathematics most important for you and me to understand is not "rocket science" and is simpler than that used for an MRI. We will look at these simplest ideas, the first of which you have already seen in the "mathematics of the AIDS test," cf., Chapter 21. What differentiates modern surveillance from that of old is the use of computers to assemble enormous databases which are used to create *mathematical models* of nearly everyone. I will need to spend some time documenting a few facts, which are either unknown, or rejected by most folks, in order to show that the material in this chapter implies profound consequences for our safety, for free inquiry, for the form of our government and the structure of our society.

22.1 Is Someone Watching You? Why?

The answer is almost certainly yes; and there are at least two basic reasons why: political and commercial. Whoever is observing you is basically performing a series of tests: Are you a terrorist? Are you involved with drugs? Are you likely to purchase a certain product? Are you an environmental activist? Are you a human rights activist? Are you an *effective* citizen? As with any test there will be both true and false positives and true and false negatives—because no test is perfect. Those who are viewing us via data sets and computers can only establish probabilities that, i.e., the "chances" that, we belong to this or that group. This is, as we have seen, a very simple mathematical observation; except that now the consequences range from the not so serious to the serious. They range from determining the type of junk mail/e-mail and other advertisements you get, to whether you get insurance, to whether you are allowed to fly, to whether a swat team smashes into your house in the middle of the night, to whether you live or die. People from the government and corporations have assembled and are assembling databases which include much information about you, mathematical representations of you, with or without your knowledge, that can profoundly affect your life. There are deep implications for the very structure of society. It is wise to be aware of this.

Let me begin with a fact that took me 15 years of patient observation—watching sufficient documentation unfold—so that I can comfortably present it now. *There are estimated to be between 15 and 20 (probably more) secret wiretapping rooms across the United States, a co-operative effort of the U.S. government, e.g., the NSA (National Security Agency), and telecommunication companies such as AT&T, paid for by taxpayers, monitoring and storing all electronic communications, including but not limited to phone, e-mail and financial, e.g., credit card, transmissions.* These operations are unconstitutional, but the U.S. Congress has (as of 2009) given telecommunications corporations retroactive immunity from prosecution for these activities (why go to the trouble if they aren't doing it?), likely nullifying a number of lawsuits. Although the U.S. Constitution "guarantees each American" protection against unreasonable search and seizure, of say our private information, it would be most naive to assume that any of us have any privacy at all in actual fact, regardless of the country in which one resides. Also, given the deference the current U.S. Supreme Court shows to corporate power, it is quite likely that this aspect of the Constitution will not be restored to the people in the U.S. Probably the Court's reason will involve "matters of national security." Unfortunately, there is some mathematical justification for Benjamin Franklin's oft quoted words:

Those who would give up Essential Liberty to purchase a little Temporary Safety, deserve neither Liberty nor Safety. You can give up liberty; but deserved or not, safety will remain illusive. There seem always to exist mechanisms for "getting around" our constitutional rights. For much of the last century you had no rights if you were successfully labeled a communist. Now the drug laws and anti-terrorist laws, e.g., The PATRIOT ACT, suffice. (From a mathematical perspective, terrorism has a fuzzy definition, being somewhat greater than a "common crime," but far less than outright war with another nation-state. Upon analysis the phrase "war on terror" is at best a war on crime, which includes at the top of the domestic list, cf., Exercise 22.7 (i), "environmental extremists," and at worst a war on an illusive noun.) We can give up all of our rights as citizens, and we will eventually not be any safer. Since abstractly—but, of course, not emotionally, especially for those directly involved—we are "playing a game." For each action we take, the "other side" creates a countermove; it is the nature of life. The U.S., for example, would likely become more secure if it reconsidered its military presence which operates in most of the countries in the world, cf., [335, pp. 155–160]; by actually solving the Israel-Palestine conflict; by actually resolving the India-Pakistan conflict in the Kashmir; by neutralizing the "hot buttons" used to inspire suicide bombers.

Outline of Documentation of Total Surveillance. In the mid 1990s while at a conference of math department chairs I met some folks from the NSA during a speech on employment opportunities for mathematicians held at the National Academy of Sciences. The NSA is one of the world's largest employers of mathematicians, cf., [23, 3]; and their work on "code cracking" and related subjects I found fascinating. About that time I ran across an article in *CovertAction Quarterly*, no. 59, Winter 1996–7, pp. 17, by Nicky Hager, "The New Age of Surveillance: Exposing the Global Surveillance System," http://mediafilter.org/caq. I then read [268], a book length treatment of the global surveillance system: ECHELON. This system is run by the United States, New Zealand, Australia, the United Kingdom, and Canada and has been around for many decades. It is capable of monitoring and storing all electronic communications in the world, with some supposed legal limitations; for example, communications taking place entirely within the U.S. were legally off limits without a warrant—but later developments make this point moot. Further confirmation of the existence and function of ECHELON arrived on May 26, 2001 when the *Guardian* of London published (and BBC reconfirmed on May 29, 2001) an article by Stuart Millar, Richard Norton-Taylor, and Ian Black titled, "Worldwide Spying Network is Revealed: MEPs confirm eavesdropping by Echelon electronic network." This article (available on the Web) describes the results of an in-depth study of the European Parliament. Among many things it states that ECHELON was set up by secret treaty in 1947, and that it was originated to monitor military and diplomatic communications—but that it had expanded to include industrial and commercial targets, and private individuals. The French, in particular, had suspicions that acts of industrial espionage had been carried out against them. The article mentions that the British had brought in (in 2000) the Regulation of Investigatory Powers Act, which allowed authorities to monitor email and internet traffic through "black boxes" placed inside service providers' systems. It gave police the authority to order companies or individuals, using encryption to protect their communications, to hand over the encryption keys. Failure to do so was punishable by a sentence of up to two years!

About this time (or before?) the United States was building some "black boxes" of its own to monitor all electronic communications of Americans. With the advent of fiber-optic cable communications, ECHELON could no longer monitor all signals "in the air." Devices had to be built within the telecommunications system in order to intercept, monitor and store transmitted information. Since this is unconstitutional, the project was kept ultra secret. There were a number of "insiders" who were apparently greatly upset by these illegal developments. As we later found out, one of these *whistleblowers* was Thomas Tamm, then an attorney with the Department of Justice. Reporters, James Risen, [577], and Eric Lichtblau, [399], got the story; and it was published on December 16, 2005, on the front page of *The New York Times*. The title of the article was: "Bush Lets U.S. Spy on Callers Without Courts." Although the story was known about a year before it was published, the Bush administration applied pressure to delay publication, cf., www.democracynow.org, Feb. 12, 2007 and April 1, 2008. (Coincidentally (?) there was a presidential election in the interim.)

On April 16, 2008 we learned that federal investigators had obtained the phone records of James Risen with the object of uncovering the identity of the whistleblower. (It turns out that Tamm was very careful, using pay phones in subway stations; but he was likely discovered via analysis of the net of emails of government employees.) In the December 22, 2008 edition of *Newsweek* Michael Isikoff's article: "The Fed Who Blew the Whistle," about Thomas Tamm appeared, cf., also Isikoff's interview on www.democracynow.org, Dec. 16, 2008. Thomas Tamm is an award winning prosecuting attorney with an impeccable family history of FBI "insiders." Nevertheless, Tamm's efforts to get his government to comply with the law made him the object of a relentless criminal investigation by the FBI, including a raid of his home in August 2007 by 18 FBI agents in flak jackets with guns. The FBI took his computers, his children's laptops, many of his personal possessions, his books, even his Christmas card list. The ordeal left Tamm deeply upset and in debt. I will let you read the rest. But Isikoff's article exposed the name of the illegal wiretap operation as "Stellar Wind." From what I have been able to read in other sources, the program was continued under Obama with the code name "Einstein."

I need to go back and explain one word, "Courts," in the title of the Risen-Lichtblau article of Dec. 16, 2005. This refers to the FISA court, the Foreign Intelligence Surveillance Act Court. The FISA courts were an outgrowth of the *Church Committee*, a U.S. Senate committee chaired by Idaho's then Senator Frank Church during the mid 1970s. It is worth your while researching the findings of this committee, which looked into abuses of the FBI and CIA among other things. The FISA courts were set up to allow wiretaps of U.S. citizens—with a warrant, i.e., some judicial review! From their creation in 1978 to 2004 the FISA court turned down 5 applications for warrants out of 18,748. If the government was in a hurry they could do a wiretap and go and ask for a retroactive warrant within a specified period of time. All of this has been effectively thrown out with the current, apparently unregulated, surveillance system. It looks like we are back to the "pre-Church" committee days, and I want to leave you with the following quote from that time to indicate what we might have to look forward to.

Unfortunately the following words of U.S. Senator Philip A. Hart, Senate Select Committee on Intelligence, 1975, are still relevant, cf., Section 22.3:

"Over the years we have been warned about the danger of subversive organizations that would threaten our liberties, subvert our system, would encourage its members to take further illegal action to advance their views, organizations that would incite and promote violence, pitting one American group against another... . There is an organization that does fit those those descriptions, and it is the organization, the leadership of which has been most constant in its warning to us to be on guard against such harm. The FBI did all of those things."

The final bit of documentation of the government's program of total surveillance of the population was provided by another whistleblower, Mark Klein, cf., [356]. Klein is now a retired AT&T communications technician, who read the December 2005 article by Lichtblau and Risen. He knew that he had documentation of precisely how the system worked because he helped build the infrastructure for one of the "wiretap rooms," Room 641A (SG-3), Folsom Street, San Francisco, to be precise. No one wanted to hear what he had to say or look at his documents: not Congress, not the *Los Angeles Times*, until he met up with a lawyer for the Electronic Frontier Foundation (EFF). His evidence was of importance in the EFF lawsuit against the telecommunications companies that violated the law in order to cooperate with the NSA program of total surveillance of electronic communications. Klein finally got to break the story in a series of interviews with: Amy Goodman, Democracy Now!; Robert Siegel, NPR All Things Considered; Keith Olbermann, MSNBC Countdown; Hedrick Smith, PBS Frontline; and Brian Ross, ABC Nightline. The details in Klein's book are interesting; the EFF continues on with its law suit, trying to get a court to restore some hope for privacy.

What the government is doing with all the information it collects is not known to me in detail. However, one thing that is customarily done with such information is *data mining*. Doing searches for key words and phrases helps the government build a model of you, a profile. Looking at to whom you talk, what you say, viewing your financial transactions, what amounts, where, what is purchased, all this helps build a data model of a given person. At the base of it all, however, is a series of "tests" which are far from perfect. There will be false and true positives, false and true negatives. Information is power, and as we shall see; the government is capable of using this information—not only to protect you and me from evil—but to promote favored political and economic interests, and to create a social harm of its own. The following is an observation from a mathematician's perspective: *In just about any system that is not adequately regulated by feedback mechanisms, any behavior that is possible is likely to eventually occur.* Just as the financial sector is not adequately regulated, cf., Chapter 2; just as many corporations are not adequately regulated, cf., Chapter 4; our executive branch, including the FBI, CIA, NSA, and assorted police units, are, I contend, not adequately regulated (by courts, Congress or us)—thus diluting the essential effort to catch criminals. We will look at some examples shortly, cf., Section 22.3. It is rather unpleasant to bring up examples of attempted assassinations, successful assassinations, and other forms of suppression of dissent by our government—without due process of law, without any process of law. Because such evidence—I will let you judge if it is sufficient "to convict"—rarely sees the light of day, I will give an outline. It is, unfortunately, necessary to not skip these examples. It is important to understand reality, especially for those who would peacefully change it.

Exercise 22.1 Yet Another Attack on Secure Communications

On September 27, 2010 an article in *The New York Times*, by Charlie Savage, "U.S. Tries to Make It Easier to Wiretap the Internet," appeared. Entities such as the FBI and the NSA propose that all communications be made available to them in a form which they can intercept and read, listen to, or watch. This is not the first time such a request has been made. Steven Levy, in his 2001 book *Crypto: How the Code Rebels Beat the Government—Saving Privacy in the Digital Age*, [394], tells the story of how this attempt to "wiretap" digital communications in the 1990s was beaten back. Briefly, in this "new" proposal all encryption software would be required to have a "back door," a built-in weakness, that the FBI or the NSA, for example, could use to intercept and decode any encrypted communication. This would include all e-mail transmitters like BlackBerry, all social networking Web sites like Facebook, all "peer-to-peer" communications like Skype—all communication systems! The very architecture of the internet would likely be changed, channeling communications through "central hubs" to facilitate surveillance.

(i) Given the capabilities of government surveillance outlined in this section, why do you think this proposal was made?

(ii) If it becomes law that all encryption software in the United States has a built in "back door," estimate the increased vulnerability of our communications systems to criminals.

(iii) The internet was designed to be decentralized and virtually impervious to disruption, even a nuclear attack. How does the proposed change in architecture to facilitate government surveillance greatly weaken the overall security and stability of the internet?

(iv) If encryption software without a "back door" is made illegal, what will be (are?) the penalties for creating/using real encryption? The story of Phil Zimmerman, cf., Chapter 23, is very instructive. (In other words, we have already "been there" and "done that.")

(v) What is the status of this attempt to "monitor" the internet, yet again, at the time you read this?

22.2 Living with a Police Escort?

Everything you do electronically, e.g., on a computer or phone, or someone else does electronically for (to?) you can be and hence is being or likely will be monitored by government agencies, police or corporations, marketers, your employer, private detectives or just plain snoopy folks. For example, it is obvious that search engines like Google, Yahoo and Amazon collect massive amounts of information on us and that this information is privately owned by said corporations. What is done with this information?[1] Have they, do they, will they share this information with other corporations, with governments, with whomever, for a price?[2]

To what extent do search engines censor? The possibility exists, of course; and details of its practice would be not only of interest but of supreme importance to anyone investigating anything. As a test I once did Google and Yahoo searches while in China on topics politically sensitive in China, but not in the United States—like Mao, for example. I got markedly different search results in China than I did in the United States. As an exercise you should try to find out which Web sites are being censored in the United States, or whatever country you happen to be in, at the time you read this. International co-operation is useful in this exercise.

You, eBay and the Police. In a very modest form of technological *jujitsu* I entered, at noon on December 6, 2006, "eBay"+ "Haaretz" on Google and the top entry was an article in the newspaper *Haaretz* by Yuval Dror. Dror writes about a law enforcement conference, "Cyber Crime 2003," held in Connecticut, closed to reporters, wherein Joseph Sullivan, director of the "law enforcement and compliance" department at eBay[3] gave a lecture—which Dror managed to get an audio recording of.

> *...Sullivan tells the audience that eBay is willing to hand over everything it knows about visitors to its Web site that might be of interest to an investigator. All they have to do is ask. "There's no need for a court order," Sullivan said, and related how the company has half a dozen investigators under contract, who scrutinize "suspicious users" and "suspicious behavior."*
>
> *...Sullivan says eBay has recorded and documented every iota of data that has come through the Web site since it first went online in 1995. Every time someone makes a bid, sells an item, writes about someone else, even when the company cancels a sale for whatever reason—it documents all of the pertinent information.*
>
> *"... We want law enforcement people to spend time on our site," he adds. He says he receives about 200 such requests* (for information on customers) *a month, most of them unofficial requests in the form of an email or fax.*

eBay will even impersonate fictitious persons with made-up histories, and solicit information, going so far as entrapment, to assist law enforcement,

---

[1]Yahoo assisted the Chinese government in sending four dissidents: Li Zhi, Jiang Lijun, Shi Tao, and Wang Xiaoning, to prison for as much as 10 years! Google's relationship with Chinese authorities, while allowing censorship for years, has been different, cf., page 438. In fact, the *Los Angeles Times*, Jessica Guynn and David Pierson, reported that Google shut down its search engine, google.cn, on March 22, 2010, over disputes regarding censorship and (Chinese sourced) hacking into Google. Google directed Chinese users to their uncensored search engine in Hong Kong, but the Chinese government has blocked access to Google Hong Kong with its censorship machinery known as the Great Firewall.

[2]Not to pick only on these three, these same queries need to be put to any database. Large libraries these days provide access to many databases, but not secret ones.

[3]At least at the time of writing, eBay did not have operations in Israel, where this newspaper resides.

according to the article. eBay bought PayPal, Inc. in July 2002 for $1.45 billion. In 2000 eBay bought Half.com, which specializes in CDs and books. From the Dror article:

Sullivan explained that these acquisitions help eBay to provide lawmen with a full picture. "Every book or CD comes with a bar code. So we know who bought what. The acquisition of PayPal helps us to locate people more precisely. In the old days, we had to trace IP addresses (unique address given to computers linked to the Internet), to locate the buyer, but now Paypal supplies us with the money trail."

Also of interest from this article:

Attorney Nimrod Kozlovski, author of "The Computer and the Legal Process" (in Hebrew), heard the lecture, and could not believe his ears. ...

Kozlovski is part of the Information Society Project group at Yale Law School, in which he and his colleagues consider the effects of the new media on the structure of society. ...

Kozlovski feels that eBay's practice should be seen as part of a worrisome trend in the West to curtail protection of individual rights. In communist regimes, he says, the state would assign watchers to follow every citizen, who would pass incriminating information on to the authorities. Now the state doesn't have to do a thing. People come to it of their own free will. This is also the case for eBay, which exploits its stature in the market to have users accept contracts that strip them of their privacy. Perhaps the regime is different, but the outcome is most assuredly the same.

Just One More Revelation about Phone Companies. The Electronic Frontier Foundation carried the following article on its Web site[4] December 1st, 2009: "Surveillance Shocker: Sprint Received 8 MILLION Law Enforcement Requests for GPS Location Data in the Past Year."

From the article we read:

"Sprint received over 8 million requests for its customers' information in the past 13 months. That doesn't count requests for basic identification and billing information, or wiretapping requests, or requests to monitor who is calling who, or even requests for less-precise location data based on which cell phone towers a cell phone was in contact with. That's just GPS. And, that's not including legal requests from civil litigants, or from foreign intelligence investigators. That's just law enforcement. And, that's not counting the few other major cell phone carriers like AT&T, Verizon and T-Mobile. That's just Sprint."

The rest of the article is very interesting, especially the method used to get this information! I think we are beginning to get a Bigger Picture of the world we are all living in. Sprint replied to EFF asserting that the 8 million figure does not represent 8 million distinct individuals. Sprint's reply to EFF, available at the Web site cited, raises more questions than it answers. Some of those questions are also available at the Web site for EFF.

Hold that Smile: You are on Camera. There is increasing video surveillance everywhere. In England surveillance cameras are ubiquitous, the U.S. is catching up. For example:[5]

---

[4]http://www.eff.org/deeplinks/2009/12/surveillance-shocker-sprint-received-8-million-law

[5]February 1, 2001 *Washington Post* article by Peter Slevin.

Superbowl fans never knew it, but police video cameras focused on their faces, one by one, as they streamed through the turnstiles Sunday in Tampa, Fla. Cables instantly carried the images to computers, which spent less than a second comparing them with thousands of digital portraits of known criminals and suspected terrorists.

You might think you are safe from video surveillance (on the ground if not via satellite) when you are camping in an American national forest—think again![6]

"Last month, Herman Jacob took his daughter and her friend camping in the Francis Marion National Forest. While poking around for some firewood, Jacob noticed a wire. He pulled on it and followed it to a video camera and antenna. The camera didn't have any markings identifying its owner, so Jacob took it home and called law enforcement agencies to find out if it was theirs, all the while wondering why someone would station a video camera in an isolated clearing in the woods. He eventually received a call from Mark Heitzman of the U.S. Forest Service. In a stiff voice, Heitzman ordered Jacob to turn it back over to his agency, explaining that it had been set up to monitor "illicit activities." Jacob returned the camera but felt uneasy. Why, he wondered, would the Forest Service have secret cameras in a relatively remote camping area? What do they do with photos of bystanders? How many hidden cameras are they using, and for what purposes? Is this surveillance in the forest an effective law enforcement tool? And what are our expectations of privacy when we camp on public land? Officials with the Forest Service were hardly forthcoming with answers to these and other questions about their surveillance cameras. When contacted about the incident, Heitzman said "no comment," and referred other questions to Forest Service's public affairs, who he said, "won't know anything about it." Heather Frebe, public affairs officer with the Forest Service in Atlanta, said the camera was part of a law enforcement investigation, but she declined to provide details. Asked how cameras are used in general, how many are routinely deployed throughout the forest and about the agency's policies, Frebe also declined to discuss specifics. She said that surveillance cameras have been used for "numerous years" to "provide for public safety and to protect the natural resources of the forest." Without elaborating, she said images of people who are not targets of an investigation are "not kept." In addition, when asked whether surveillance cameras had led to any arrests, she did not provide an example, saying in an e-mail statement: "Our officers use a variety of techniques to apprehend individuals who break laws on the national forest." Video surveillance is nothing new, and the courts have addressed the issue numerous times in recent decades. The Fourth Amendment guards against unreasonable searches and seizures, and over time the courts have created a body of law that defines what's reasonable, though this has become more challenging as surveillance cameras became smaller and more advanced. In general, the courts have held that people typically have no reasonable level of privacy in public places, such as banks, streets, open fields in plain view and on public lands, such as National Parks and National Forests. In various cases, judges ruled that a video camera is effectively an extension of a law enforcement officer's eyes and ears. In other words, if an officer can eyeball a campground

[6] "Charleston man surprised when he found one while camping with daughter," by Tony Bartelme, postandcourier.com, published Monday, March 15, 2010.

in person, it's OK to station a video camera in his or her place. Jacob said he understands that law enforcement officials have a job to do but questioned whether stationing hidden cameras outweighed his and his children's privacy rights. He said the campsite they went to—off a section of the Palmetto Trail on U.S. 52 north of Moncks Corner—was primitive and marked only by a metal rod and a small wooden stand for brochures. He didn't recall seeing any signs saying that the area was under surveillance. After he found the camera, he plugged the model number, PV-700, into his Blackberry, and his first hit on Google was a Web site offering a "law enforcement grade" motion-activated video camera for about $500. He called law enforcement agencies in the area, looking for its owner, and later got a call from Heitzman, an agent with the National Forest Service."

National Security Letters. Just in case the FBI wants to know even more information about anyone they can issue (an administrative subpoena called) a National Security Letter (NSL) to: your university, your employer, your library, your doctor, your scuba association, ..., anyone at all. They can ask just about anything (they are not supposed to use an NSL to get the content of communications, maybe the NSA program can take care of that?), and the recipient is *gagged*, i.e., under penalty of imprisonment and fines they can *never* tell anyone that they have received such a letter, let alone what was in the letter. Without a subpoena from a court, presented with an NSL most businesses and organizations just hand the information over: scuba diving associations handed over disks with information on 2 million members, airlines cooperate, at least 175 universities have handed over information. It was a rare event when librarians revolted, sued, and won a case based on the First and Fourth Amendments to the Constitution! See [714, 242]. In 2000, a year before the PATRIOT ACT, 8,500 NSL requests were sent out. In 2004 there were 56,000, [242, p. 55].

Black Bag Jobs: Secret or Not. If the FBI wants to get more personal, under Section 213 of the PATRIOT ACT, government agents can enter your home secretly—search and seize your documents, computers etc, where a court "finds reasonable cause to believe that providing immediate notification of the execution of the warrant may have an adverse effect." There is more legalese, but there are cases where black bag jobs were discovered only after successfully suing the FBI under the Freedom of Information Act, cf., [629, p. 263], [714, p. 118]. Secret search and seizure is probably preferable to having police smash down your door, guns drawn in your face, and then search and seize, cf., [629, p. 162].

There is some instructive video[7] of the FBI and local police (assault rifles, guns drawn) making a pre-emptive raid, (and detainment, with objective to confiscate computers, video equipment, etc—August 30, 2008) on Eye Witness Video folks in Minneapolis during the 2008 Republican National Convention (RNC). Eye Witness Video provided a great deal of evidence for successful legal actions against police behavior at the previous (2004) RNC in New York

[7] http://www.youtube.com/watch?v=Vi1eluuDGss&feature=player_embedded

City. Amy Goodman and Democracy Now! journalists (with full press credentials hanging from their necks—until ripped off by police) were "roughed up" and arrested at the 2008 RNC as well. Attitudes were similar at the Democratic National Convention (DNC) held near to me in Denver, Colorado. Asa Eslocker, a producer for ABC News, was investigating the role of corporate lobbyists and high donors at the DNC. He was arrested on a public sidewalk outside a hotel in Denver that was hosting some Democratic party functions and charged with trespassing (on a public sidewalk!), interference, and failure to follow a lawful order.[8] Furthermore, the Denver Police Protective Association created and distributed at least 2000 commemorative black T-shirts featuring a baseball-bat wielding cop and the slogan "WE GET UP EARLY to BEAT the Crowds 2008 DNC."[9] There were mass arrests at both the DNC and the RNC.

Pre-emptive suppression of dissent *and* the press is not uncommon and international in scope. The Danish police made pre-emptive raids and arrests of peaceful climate activists during the Copenhagen climate conference (COP15), Dec. 7–18, 2009; these peaceful activists were distinct from the few who did property damage.[10] Rackspace is a Texas-based internet company that has hosted the Web sites of Independent Media Centers. Two of their computer servers were seized in October 2004 by the FBI from their office in *England*, reportedly the result of joint United Kingdom-United States-Switzerland-Italy cooperation, cf., [714, p. 180]. I have observed a common pattern, not only of the suppression of dissent at any major national or international meeting, but of suppression of any *independent* media that tries to report on such dissent.

Of course, when drugs are alleged to be involved, whether or not there really are any drugs involved, you can lose your life, cf., Section 22.3.

Exercise 22.2 Biometric Identification

(i) Estimate the number of times your image was taken and stored by cameras in the last week. How many business establishments, city, county, state, or federal agencies took your picture? Traffic cameras? How long are these images stored and with what databases are they shared?

(ii) Via biometric technology: your picture, fingerprints, retinal scan, description of your height, weight, and other physical parameters, are scheduled to be embedded on a chip and placed in your passport, if you get one. How accurate is this going to be? Let's look at "facial recognition technology" as referred to in the above Superbowl example. There are false negatives, i.e., the technology fails to recognize you. There are false positives, i.e., the technology matches you to someone else, incorrectly. According to [714, pp. 99–100], U.S. government tests have shown that when a person is compared to their own biometric data, using a recent photo, there is a 5% false negative rate and a 1% false positive rate.

---

[8]Defending Dissent Foundation Letter, September 2008, p. 1.

[9]See picture at http://coloradoindependent.com/9276/denver-police-beat-the-crowds-t-shirt-no-laughing-matter-protesters-charge

[10]http://www.climate-justice-now.org/civil-society-groups-at-cop15-denounce-preemptive-arrests-of-climate-justice-protestors-in-copenhagen/

What does this mean? Will the reliability rates increase or decrease if the technology is comparing you to a random individual?

(iii) The reliability rates degrade to 15% if a picture just 3 years old is used, cf., [714]. What cost implications does this imply for keeping biometric identification updated?

(iv) With these reliability/accuracy rates, suppose the world's governments try to monitor one billion travelers. Roughly how many people will be identified incorrectly assuming the data from part (ii), and then part (iii)?

(v) No biometric technology is yet 99.99% accurate. But suppose it were. How many false identifications would there be, assuming (2009 figures of) 1.17 billion Indians, 1.3 billion Chinese, .83 billion Europeans, .52 billion North Americans, .4 billion South Americans, 1 billion Africans, etc.? Will it be possible for some organized crime operation to counterfeit documents, even sophisticated ones?

(vi) Does the concept of reliability rate, which includes rates of false positives and false negatives, apply to the mathematical models of each of us created by the "total surveillance system" described in this chapter? What are (might be) the consequences of not having any form of "judicial review" of this system? (See Section 22.3 for some ideas.)

Exercise 22.3 The REAL ID ACT of May 2005

(i) When Congress created the Department of Homeland Security the legislation made it perfectly clear that no National Identity system was to be created. In the fall of 2004 a bill creating a national identity card (de facto) was voted down by Congress. However, in May 2005, without discussion or debate on the floor or in any committee, the REAL ID Act was attached to a bill authorizing money for the war in (and occupation of) Iraq. This act mandated a universal ID driver's license with embedded data which, for example, was to be linked to immigration data. Thus each state's Department of Motor Vehicles becomes a de facto immigration authority. It appears that the government has plans to demand that states insert RFID (Radio Frequency Identification Chips) into drivers' licenses, since the government specifications for biometric RFID chips embedded in passports include details for embedding the same into drivers' licenses, cf., [714, p. 97], Exercise 22.4. What do you think the error rate of this system might be?

(ii) What if it becomes a crime to not have your National ID card with you at all times? Is a police officer stopping you to "ask for your National ID" if you have been doing nothing wrong, OK with you? (It is OK with the Supreme Court, cf., [714, p. 97].)

(iii) How many states have passed laws saying they will not comply with the REAL ID ACT? Why did they do this? (The deadline for compliance was postponed from Dec. 31, 2009 to May 2011, since 46 of 56 states and territories were out of compliance on Jan. 1, 2010.) What is the status of REAL ID when you read this?

I don't have space to go into details; besides, the technology is changing. However, [213] indicates that it is possible to monitor every aspect of everyone's life. In [465] we see that we can all be monitored from satellites above, while [4] lays out the plan to monitor us more closely using RFID, Radio Frequency Identification Chips, implanted in everything we buy, wear, read, etc. In fact, there is a program of the USDA, U.S. Department of Agriculture, called NAIS, the National Animal Identification System, already in place, which will ultimately require every livestock animal owner in the U.S. to register each of his/her animals, have an RFID chip implanted, and report any "event" in the animal's life, such as if your chicken "comingles" off of your property. If you fail to report "events" you will be subject to a fine. Of course, if you have more than 30,000 chickens you only have to buy one tag/registration for the whole flock.[11]

Note that willingly many have already had devices installed in their cars, such as EZpass, which allows tolls to be paid electronically instead of stopping to drop money at a toll booth. Such devices also allow complete tracking of a car's whereabouts.

How many of us are so fearful of terrorists, crime, enemies, or whatever, [656, 517], that we are willing to have every aspect of our lives monitored by the likes of convicted felon, John M. Poindexter,[12] cf., Section 23.4?

Exercise 22.4 Just One Example: RFID Chips in Our Animals

(i) What is the status of the USDA plan to track all animals in the United States using RFID technology at the time you read this? Each tag is to carry a 15-digit number. How many tags can be issued?

(ii) Which entities have lobbied for this NAIS for a long time? Who will profit monetarily from the system, how much? Who will pay, how much?

(iii) If you have just one pigeon (or horse, cow, pig, chicken, etc) you will have to register your premises with the government. How much of the human owner's information, like GPS co-ordinates, must be provided to register? As of December, 2006 about one fourth of an identified 1.4 million premises have been registered.

(iv) If you agree that such registration will help track disease, why is it that folks with fewer than 30,000 chickens must tag each one, but for flocks of more than 30,000 one tag will suffice? Who has more than 30,000 chickens? Where is the greatest danger of disease to the public coming from?

(v) The plan is to make this mandatory by 2009 (but no system of penalties are in place for non-compliance as of 2009). What are the relative costs—in money and time—involved for hobby farmers, pet owners, local organic (or not) farmers, small farmers vs. industrial agriculturists dealing with massive numbers of animals?

(vi) How many people have gone the extra step and had RFID chips implanted in themselves or their children by the time you read this? If people accept NAIS, do you think they will also accept a requirement to have RFID chips implanted in all people to keep track of children, criminals, demented folks, regular folks? Why? Why not? Is this a foolproof identification system?

(vii) If the system is kept voluntary are there market advantages for farmers who use the RFID technology?

Slavery. Christian Parenti begins his book *The Soft Cage: Surveillance in America*, [517], with the history of slavery in America and the methods used to

[11]For all the reasons you should sign up right away go to www.usda.gov/nais and for the downsides go to stopanimalid.org or jimhightower.com.

[12]In January 2002 the U.S. Military unveiled a program called Total Information Awareness, renamed Terrorism Information Awareness (TIA) in May of 2003. Convicted felon, John M. Poindexter (Admiral Poindexter was convicted in 1990—several felony counts—for his role in the Iran-contra affair, cf., Section 23.4. His conviction was reversed in 1991 by a 2 to 1 vote of a federal appeals court because Poindexter had been granted immunity for his testimony before Congress about the case. By the way, the 2 votes were from Nixon appointed judges, the 1 vigorously dissenting judge was a Democrat.), was chosen to head the TIA. The stated goal was to review and analyze all information on everyone, including but not limited to bank accounts, credit card transactions, computer and phone activity, medical records, and on and on—and not limiting this just to Americans. This alarmed the public sufficiently that Congress officially requested that TIA be discontinued. It looks like "Stellar Wind" took its place. Maybe there are plans to track us via our DNA as well, cf., [353].

control slaves. It is instructive to review this history, since one fundamental way for one group to exert power over another is to require "identification on demand." I will begin by defining *slavery*, i.e., "not free," by giving the definition of *freedom* as stated by Marcus Tullius Cicero (106–43 B.C.)., cf., [381].

Definition: *Freedom* is participation in power.

Exercise 22.5 Are You a Slave?

(i) On a scale of 0 = slave to 1 = free, rate yourself.

(ii) For a slave to go anywhere in the "Old South," i.e., pre-civil war slave states in the U.S., a pass written by the slave's owner was required. If a slave learned to read and write, they often were able to write their own passes. Thus it was common practice to forbid slaves from learning to read and write. Compare this to modern efforts forbidding, or making it illegal, to encrypt electronic documents, without giving the encryption key to the government, cf., Chapter 23. Is encryption just another level of "literacy?"

(iii) If the government requires you to carry a National Identity Card which must be shown to any official on demand, how will your situation be different from or similar to that of a slave in the "Old South?"

(iv) "Cisco is required by law to include technology in its networking products that allows investigators to tap the hardware for information." (Quote from "Hackers Who Leave No Trace," by John Markoff and Ashlee Vance, *The New York Times*, January 20, 1010, pp. B1, B4.) Discuss.

22.3 I'm Not Worried, I've Done Nothing Wrong

" 'IF YOU HAVEN'T DONE ANYTHING WRONG, you don't have anything to be afraid of.'

"That line has been used for countless years by law enforcement officials to trick people into submitting to warrantless searches and interrogations without their lawyers being present. ... ... "Innocent people are routinely arrested, tried, convicted and sent to jail. Hundreds of convicts have been proven innocent and set free in the last few decades, including close to 90 who were sitting on Death Row waiting to be executed for murders they did not commit."[13]

A Few of the Known False Positive Death Sentences. On September 29, 1999 Ismael Mena, a father with a family to support, was relaxing at home after work when the police—unannounced—burst into his home under a "no-knock" entry warrant. Unfortunately, Mena had an aluminum beverage can in his hand that the police say they mistook for a gun—and the police shot Mena eight times—he died. The warrant was obtained on the word of a drug informant. No illegal drugs were ever found in Mena's home or body. All evidence indicates that Mena was an honest, hard-working, law-abiding, family man. But, of course, he never got to present any evidence—drug enforcement

[13]From [569, p. 14].

pre-empts constitutional rights. (In the 1980s the Colorado and U.S. Supreme Courts eliminated any requirement that police corroborate accusations from a drunk, or one of your enemies, or "an informant, with undefined credentials," before getting a warrant.) Later the police thought they may have raided the wrong address.

(Innocent) California millionaire, Donald Scott, was killed trying to defend his family during a midnight police raid of his mansion, a raid—based on false information. Scott Bryant was killed during a "no-knock" raid of his trailer in Wisconsin.

The late 70 year-old reverend and drug counselor, Acelyne Williams, died of a heart attack after being chased through his home by masked intruders who forced him to the floor, screaming at him, while putting guns to his head—another "no-knock" police raid; this time by the Boston police drug squad, based on a warrant based on the word of a drunk, a confidential informant.

The DEA, Drug Enforcement Agency, impounded Dr. James Metzger's Lexus before charging Metzger with anything, let alone convicting him. The DEA thought Metzger may have illegally written some drug prescriptions. This is just part of the drug law's forfeiture provision, a convenient way to avoid one's constitutional rights. All these examples come from one article written for *The Denver Post* by Ari Armstrong and Dave Kopel, Dec. 30, 1999. There are many more examples. Find an example where the police deliberately planted drugs in someone's car and then confiscated it under the above forfeiture provision. What is more important here: the statistics of how many such false positives there are, or the logic of the system in which they occur?

One False Positive Can Ruin Your Decade. You do not have to be killed or sent to jail in order to suffer from a "false positive" test result. Very briefly, for details see [213, pp. 25–8], Steve and Nancy Ross through no fault of their own ran afoul of the IRS, Internal Revenue Service, which placed a lien on their house. With some work they got the IRS to acknowledge the IRS's error, they got an official letter from the IRS to that effect. The long and complicated story, the details of which I have skipped, were also skipped by TRW and Equifax, corporations that maintain credit information[14] on us all—of that you should educate yourself immediately if you have not already. When the Ross' MasterCard came up for renewal the bank canceled it, based on the bad credit history in the TRW and Equifax files.

The Ross couple complained, and TRW and Equifax did respond by putting a copy of the IRS letter in their file, after confirming authenticity with the IRS, along with the copy of the IRS lien on their house. (Mistakes in your credit record are not expunged, but your explanation is, hopefully, added to the record.) But all was not well, since TRW and Equifax had already sold the credit data to at least 187 independent bureaus. Then, behaving math-

[14]TransUnion, Equifax and Experian are the three credit bureaus in 2006.

ematically like a disease, information from the independent credit bureaus kept reinfecting TRW's computer with the incorrect information about the IRS lien.

With some effort, the Ross family managed to convince their original bank to reissue them a MasterCard. However, for the next seven years, they could not obtain a new credit card from any other financial institution; they were rejected for bank loans, could not refinance their house, could not move, could not get a mortgage on a new home. Had Steve not been able to convince his original banker to issue him a MasterCard he would not have been able to rent a car or hotel room or, in fact, been able to make a living at his job which required lots of travel.

The Ross family would probably still be in credit-record-hell, save for a provision in an act of Congress, the Fair Credit Reporting Act (FCRA), which required that the record of the lien be removed from credit reporting data banks after 7 years.[15] The credit card industry has thus far prevented federal legislation that would allow consumers to more fully protect themselves from fraud by freezing access to their credit records, if they so chose. This, of course, would prevent a store from issuing a store-branded credit card without the customer unfreezing their information. It would also restrict credit bureaus from selling consumer credit files!

Effective Peaceful Citizens on the "No Fly List." The "No Fly List"(and the longer "Terrorist Watch List") amounts to a test administered to airline passengers with many false positives. A room full of men with the same name as a person on this list appeared on the CBS TV program "60 Minutes" one Sunday to detail their predicament. They are not allowed to give feedback to the administrators of this list to correct the errors in the list. They are not allowed to know why they are continually encountering trouble every time they try to fly. The late Senator Edward Kennedy was once on a no fly list, but he did have the power to get the record corrected, cf., [714, p. 179].

Dave Lindorff writes[16] that Jesuit priest, John Dear; Nancy Oden and Doug Stuber of the Green Party; Virgine Lawinger, a 74-year-old nun; Barbara Olshansky, author of [505]; Nancy Chang, senior litigation attorney at the Center for Constitutional Rights and author of [80], all have great difficulty flying or have been banned outright. They all have one thing in common, they are outspoken about pacifism, the U.S. Constitution or current government policies. The *Wall Street Journal* reported that the FBI has made its lists available to a wide range of private companies, from banks and casinos to rental car

[15]The Fair Credit Reporting Act of 1970, the same year the Clean Air Act was passed unanimously by the U.S. Senate, is still in effect, http://www.ftc.gov/os/statutes/031224fcra.pdf, as amended. This act gives people whose credit reputation has been damaged by a credit-reporting company only two years from the date of the mistake to file a lawsuit against the credit agency. Thus everyone should check their credit report at least once a year.

[16]See "No Fly List: Is a federal agency systematically harassing travelers for their political beliefs?" in the November 22, 2002 issue of *In These Times*. Also see [405].

companies. Thus getting on such a list can have far reaching consequences, for employment, say.

While Doug Stuber was being detained at Raleigh-Durham airport, TSA (Transportation Security Administration) security guards called in Secret Service agents to do a retinal scan and interrogation. He claims to have been able to get at a look a page in a loose-leaf-binder that was open during the questioning. The page had a long list of progressive political organizations on it, including Amnesty International, Green Party, Green Peace, Earth First!. . . .[17]

Exercise 22.6 What is Going on with the No Fly List and Terrorist Watch List?

(i) *False Positives?* Estimate the chances that a significant number of people with no history of violence, but with a common characteristic of being critics of current government policy, would be selected by accident/randomly to be on a "No Fly List" of suspected terrorists? This research project has to be approached like a "black box" experiment where you are not allowed to look inside the box (unless you have great connections), but you must rely on the externally visible manifestations of what is going on inside the box. Decide if the selection of peaceful activists, environmental, human rights, etc., is an accidental false positive (since they intend no physical harm to anyone and do not carry weapons) or whether they actually are intended to be selected as true positives. In the latter case what is the operational definition of *terrorist* being used by the government? This then would say much about the internal logic of the administration.

(ii) Consider the following case and what it implies about the logical structure of the "no-fly list," or more accurately, the longer "terrorist watch list." Michael Hicks, also known as Mikey Hicks, was first singled out as a potential terrorist for additional screening and patted down when he attempted to board a plane at Newark Liberty International Airport. His age at the time? TWO! He, of course, was then with his parents. This first time might be reasonable; for determined terrorists might pack a two year old with a remotely controlled bomb without his parents' knowledge—or the parents could be involved. (It turns out that Mikey's parents were not and are not on the "list;" since, for example, Michael Hicks's mother once had Secret Service clearance to fly on Air Force II, the vice-president's plane, as a photojournalist.) But Mikey's trouble in traveling has continued unabated until the time I write this in January 2010, and Mikey Hicks is now 8. The "terrorist watch list" is thus a potentially "dumb" list with no capacity to absorb information in the form of "feedback." One also wonders what sort of data is in this list, besides *names*. If there is, indeed, a potential terrorist among the 1,600 Michael Hickses out there, adding age, height, weight—you know, the information on a standard driver's license—would get this particular Mikey off the hook. At the time you read this, how "smart" are the terrorist watch list and the no-fly list? One person was able to get off these lists simply by changing his name! (See *The New York Times*, January 14, 2010, front page article by Lizette Alvarez, from which this information comes.)

[17]It was revealed in late 2006 that as early as 2002 the Department of Homeland Security expanded its Automated Targeting System (originally applied to air cargo) to all air passengers who have flown into or out of the U.S. (and maybe to more folks than that). Again we see a pattern: no one can find out if they are on the list; no one can challenge the list; no one can challenge the terrorist risk level assigned to them; the list is shared with foreign governments, private employers, but not with the individual "charged." A general pattern of authoritarian regimes: they want to know everything about you, but everything about them is secret—for reasons of national security. This is nothing new; it is a time honored pattern of a system uncleansed by appropriate feedback mechanisms, see the chapter on voting, Chapter 24.

(iii) *A False Negative.* The following was reportedly known[18] about Nigerian Umar Farouk Abdulmutallab *before* he boarded a plane on Christmas Day 2009 headed for Detroit with a powerful explosive and a syringe full of acid in his underpants which he tried to detonate and thus destroy a fully occupied passenger plane: (1) On November 19, 2009, his own father had reported to two CIA officers at the U.S. Embassy in Abuja, Nigeria that he was worried about his son's recent radical activities; Umar spoke of "sacrificing himself." Embassy officials thus sent a warning to the National Counterterrorism Center, and Umar's name was added in November 2009 to the U.S.'s 550,000-name Terrorist Identities Datamart Environment. Umar was not put in a smaller database of people tagged for extra scrutiny at airports, nor was he put on a no-fly list; (2) Umar paid for his ticket in cash (supposedly a "red flag" to the CIA and FBI); (3) Umar did not check any bags (for an international flight, another "flag"); (4) Umar's visa renewal had been denied by the British who put Umar on a watch list to prevent him from re-entering Britain; (5) Umar was in a database of people with suspected terrorist ties, on a counterterrorism watch list; he had studied Arabic in al-Qaida sanctuary Yemen; (6) There were about a dozen other bits of intelligence, some involving NSA phone intercepts, mentioned in the articles referred to in footnote 18 which linked Umar to potential terrorism.

Explain how Umar was not even singled out for extra screening at airports, while pacifist nuns, Green Party officers, legal activists in *support* of the U.S. Constitution, some environmentalists and other activists were so singled out—some prevented from flying at all?

Another False Negative. On December 22, 2001, Richard Reid tried to blow up a commercial flight to the U.S. with explosives in his shoes. Review the details if you do not know them. This act led to the requirement that would-be airline passengers must remove their shoes for preboarding inspection. Would it not be logical that the Umar incident lead to a rule regarding the inspection of underwear? What might be the response of security systems if suicide bombers start using body cavities, or surgical implantation, to hide explosives?

No system/test is perfect, false negatives will always be with us; and the system in place is certainly a significant obstacle for would-be airline terrorists. However, given parts (i) (ii), it is reasonable to ask if there have ever been true positives? If so how many?

(iv) The FBI has offered no example of a terrorism case in which a National Security Letter has helped them, [714, p. 116]. Discuss.

(v) A curious Associated Press article: "Explosive Was on Dublin Flight In Bungled Slovak Security Test," *The New York Times*, Jan. 6, 2010, p. A11, said that Slovak authorities placed real bomb components in nine passengers' checked bags as part of a security test. Eight were detected, but a ninth bag containing 3 ounces of plastic explosive, made it through inspection, onto the plane. The man carrying the explosive was unaware (and had never been informed) of the situation until Irish police, acting on a Slovak tip, raided his apartment thinking the man was a terrorist—until Slovak authorities admitted that they had planted the explosive. The man was released without charge after several hours of detention. (A longer AP story by Veronika Oleksyn and Karel Janicek appeared Jan. 7. 2010.) Discuss.

(vi) Does the apparent march toward increasing surveillance occur independent of what political party is in control of the executive, legislative (and judicial?) branches of the U.S. government? For example, see [288]. How effective is this surveillance in meeting its stated objectives?

(vii) On February 18, 2010 Joseph Stack flew his small plane into a building in Austin, Texas that housed the IRS (Internal Revenue Service). (He was upset.) He mimicked the 9-11-01 attacks on the World Trade Center, and he did so for a political purpose with the hope of inspiring others to take up violence to continue his struggle. This, of course, precisely satisfies the definition of terrorism as defined by the U.S. government, e.g., the USA PATRIOT ACT. My brief survey of the media coverage of this event, however, found

[18]See news articles from Jan. 2010, including one by Eric Lipton, Eric Schmitt, and Mark Mazzetti, p. A1, A14, *The New York Times*, Jan. 18, 2010.

the word *terrorist* rarely used, if at all, to describe Stack and his act. On the contrary, a peaceful environmentalist who commits civil disobedience or not is easily called a terrorist, cf., Judi Bari. Do your own media analysis and compare the use of the word terrorist in describing strident environmentalists vs. say the frequency of the use of the word terrorist to describe Stack. Why do you think this is?

The "no fly" list and "terrorist watch list" are just the tip of an iceberg. There are a number of immigration and terrorist watch lists from: the State Department, ICE (Immigration and Customs Enforcement), DEA (Drug Enforcement Administration), the FBI—which have a reported total of 13 million names, cf., [714, p. 183]. Reminiscent of the troubles of Steve and Nancy Ross mentioned above, some of the names on these "watch lists" have been circulated among some private corporations, with many false positives. It is easy to see how the situation of computers talking to one another about these lists can take on a life of its own, generating headaches or serious trouble for innocent folks, cf., [714, p. 184]. As international lists are generated, the practice of repressive governments to designate as "terrorist" or "terrorists sympathizer" anyone who has a dissenting view, with no avenues of recourse, greatly dilutes the supposed objective of civilian safety—becoming in part an authoritarian terror of its own—see below.

Disturbing Facts. From [557]:

In March 1967, shortly after I was assigned to teach law at the Army Intelligence School, I had the occasion to take a book down from my office shelf. Inside the cover was the imprint of a rubber stamp. It said:

THIS PUBLICATION IS INCLUDED IN THE COUNTER INTELLIGENCE CORPS SCHOOL LIBRARY FOR RESEARCH PURPOSES ONLY. ITS PRESENCE ON THE LIBRARY SHELF DOES NOT INDICATE THAT THE VIEWS EXPRESSED IN THE PUBLICATION REPRESENT THE POLICIES OR OPINIONS OF THE COUNTER INTELLIGENCE CORPS OR THE MILITARY ESTABLISHMENT.

The book was the Constitution of the United States.

As the son of a former (local) policeman, I respect the essential, often difficult, sometimes dangerous work that police must do. The immense power of the police, FBI, CIA, nationally and internationally, however, requires effective feedback from the civilian populations they serve. Effective feedback depends on information. Without this feedback, innovative individuals, conscientious groups, society and our environment in general suffer—possibly to the breaking point. The following abbreviated, far from complete, list of executive excesses will be unbelievable to some, but all too softly stated for those who have had direct experience.

Fred Hampton. I will let the title of [287], *The Assassination of Fred Hampton: How the FBI and the Chicago Police Murdered a Black Panther*, speak for itself. This assassination and another, Mark Clark, are corroborated by retired FBI agent, M. Wesley Swearingen, [677]. See also [88, 89].

Elmer Gerard "Geronimo" Pratt. After 27 years in jail (8 years in solitary confinement), Mr. Pratt's murder conviction was reversed by Orange County California Superior Court Judge Everett Dickey in May of 1997. Why? The judge found that the chief witness against Pratt was a felon and a police and FBI informant, Julius Butler, (a fact not revealed in the 1972 trial) who lied under oath. The FBI also had an agent in Geronimo Pratt's lawyer's office posing as a secretary; and (according to an interview on www.democracynow.org) the FBI also placed one of their agents, Dennis Romo, on the jury—and he actively lobbied (as jury foreman) for a guilty verdict despite the fact that such activities were illegal. Also, the FBI had Pratt under surveillance 300 miles away from the scene of the crime he was accused of at the time he was supposed to have committed it.

So the FBI knew that Pratt was innocent of the crime of which they accused him. The FBI "lost" this evidence. (See the audio archives of www.democracynow.org for March 4, 1999, June 11, 1997 and October 5, 2000.) I noted with much interest a comment made by former Black Panther, Bobby Seale, who mentioned that some key evidence in Pratt's case was "discovered" in the files of the Department of Agriculture!

After all of the FBI's illegal activities were exposed in court—including the outright framing of an innocent man, incarcerating him for most of his 20s, all of his 30s, all of his 40s, for a total of almost 27 years—the FBI did not give up and sought to put Mr. Pratt back in prison by appealing Judge Dickey's decision. The Court of Appeal denied that appeal. If for no other reason than to clear his name, Geronimo Pratt countersued. In April 2000, Pratt's lawsuit for violation of his civil rights and *false imprisonment for over a quarter century* was settled out of court. The City of Los Angeles agreed to pay $2,750,000, the FBI agreed to pay $1,750,000, cf., [504]. Note that FBI officer Richard Held was in charge of putting Mr. Pratt in prison in 1972. (About 1998 Held took the job of being in charge of security for Visa, the credit card company.)

Judi Bari and Darryl Cherney: www.judibari.org. Judi Bari (1949–1997) is dead, but her spirit lives on in many environmentalists' hearts. On May 24, 1990 Judi and Darryl, on an organizing tour to save the redwood trees, were victims of a car bomb in Oakland, California. Judi was an effective organizer of both environmentalists and timber workers. Although Ms. Bari was long known as a practitioner of non-violence (to the point of being against tree spiking), she was arrested (while not conscious) in the intensive care ward of a hospital immediately after the explosion. She was charged by the FBI with carrying an explosive device. *The New York Times* reported in May 1990 misinformation provided to it by such sources as the FBI, e.g., that Judi Bari bombed herself. Despite much evidence now to the contrary (see court decisions mentioned below) *The New York Times* has not to my knowledge corrected its errors. Charges were never filed against Bari by the District Attorney for lack of a case. No other suspects were or have been investigated even though Ms. Bari had received numerous death threats prior to the bombing from prologging and anti-abortion parties.

The FBI had Judi Bari under almost continual surveillance and, in fact, was on the scene of the bombing minutes after it happened. It is thus believed by many that the FBI saw who put the bomb in Bari's car. The FBI also handled evidence, such as the bombed car, in such a way that it became nearly useless as evidence—contrary to established procedures. Judi Bari died on March 2, 1997. A counter-suit against the FBI was pursued by the estate of Ms. Bari, and Mr. Cherney.

Ms. Bari and Mr. Cherney contended that the FBI is either protecting the bomber or was itself involved directly in the bombing. Mainstream credibility of the Bari/Cherney case was garnered indirectly via the testimony of FBI lab whistleblower, Frederic Whitehurst (beginning with Whitehurst's August 14, 1995 testimony at the World Trade Center bombing trial, and an article in the *Washington Post*, September 14, 1995, and continuing with occasional articles thereafter). Whitehurst claimed among other things that the FBI altered reports to come to desired conclusions.

After over ten years of police and FBI delaying tactics, the 9th Circuit Court of Appeals upheld a lower court's decision that the FBI and the Oakland Police had to stand trial for conspiracy to deny Bari and Cherney their civil rights. The trial took place.

On June 11, 2002 a jury awarded Cherney and the estate of Bari $4.4 million in their civil suit against four FBI agents and three Oakland police officers (80% of the $4.4 million total damage award was for violation of their First Amendment rights to speak out and organize politically in defense of the forests), despite the fact that the Bari-Cherney legal team was not allowed to bring up previous COINTELPRO activities of the FBI. Please see www.judibari.org for enormous amounts of information I cannot reproduce here. (For example, research the role of Richard Held.) The jury award has been upheld. It is also instructive to do a search of the media, say *The New York Times* for its coverage of the Bari case. See, for example, the article on page A14 of the June 12, 2002 edition of the *The New York Times*. Also see the post bombing coverage after May 24, 1990. Judge for yourself if there is any "spin" involved. Do not miss the long April 7, 2002 (Sunday) *The New York Times Magazine* article titled "The Color of Domestic Terrorism Is Green" by Bruce Barcott. This article ran (coincidentally?) the day before jury selection in the Bari v. FBI case, and does not portray her in a positive light.

Frank Wilkinson. Why did the FBI create a file on Frank Wilkinson with 132,000 pages (that we know about)? Frank Wilkinson is best known for being a life-long defender of the First Amendment of the Constitution. On two of those pages (which I once saw) there is "discussion" of an assassination attempt on Wilkinson, cf., [629, pp. 15–6]. Wilkinson learned of the assassination attempt while reading his file decades after the fact. Wilkinson founded NCARL, Nation Center Against Repressive Legislation, which has been renamed the Defending Dissent Foundation.

Leonard Peltier. Richard Held was the FBI officer in charge of putting Leonard Peltier in prison, where he still remains. The case stems from a shoot-out between FBI agents and American Indians on a remote part of the Pine Ridge Indian Reservation, near Wounded Knee, South Dakota. Peter

Matthiessen wrote a book about the Peltier case, [446], *In the Spirit of Crazy Horse*, which was kept off book shelves for 8 years because of one of the most protracted and bitterly fought legal cases in publishing history. The case, in part, involved a libel suit by an FBI agent against the publisher and author, which the FBI eventually lost on appeal. How many books are not written because authors and publishers do not or can not sustain such a legal effort?

The Case of Steve Kurtz: www.democracynow.org, June 16, 2008. Steve Kurtz is an acclaimed art professor at SUNY Buffalo who was working with Robert Ferrell of the University of Pittsburgh Graduate School of Public Health on a joint art-science project dealing with the health effects of germ warfare programs. The work is considered cutting-edge art, and Kurtz had scientific equipment in his home in order to perform *harmless* and *legal* experiments. On May 11, 2004 Kurtz's wife died in her sleep of natural causes. Kurtz called 911. The responders were suspicious of the death and the science experiments; so they called the police, who called in the FBI, leading to a raid of his home by the Joint Terrorism Task Force and Homeland Security. His belongings, his cat, and even the body of his wife were seized.

While Kurtz was making arrangements for his wife's funeral four FBI cars came up and did a "soft-rendition," (we will discuss extraordinary rendition shortly, cf., section 23.2). They detained him in a hotel and interrogated him, until Kurtz managed to make a call to a lawyer on his cell phone. The lawyer managed his release.

Apparently unable to understand the Kurtz-Ferrell work, the government pursued charges of bioterrorism against them. To make a long story short, the government could not get a grand jury to indict. They could not make a case that held up in any court, but the government kept up the attack on Kurtz and Ferrell for four years! Finally the government used their "go-to" law, which is written very broadly, and they charged Kurtz and Ferrell with mail fraud. The government tried to show a conspiracy, but the suppliers of the harmless bacteria culture and the University of Pittsburgh opted to not participate.

Under the PATRIOT ACT Kurtz and Ferrell were looking at 20 years in prison if convicted. The end result is that Ferrell, who in the interim had developed cancer and suffered a number of strokes, plea bargained for a lesser charge. Kurtz fought to the end and won when the judge essentially threw the case out. We all owe Kurtz a debt, for had he lost the FBI would have greatly broadened the scope of applicability of the mail and wire fraud law in question. Instead, the Kurtz win narrowed the range of applicability of that law. Related works on academic freedom and the "police" are [613, 722, 553, 147, 363]. Related works on the influence of business on academia are [721, 705, 45, 651].

Exercise 22.7 How Big Is Your FBI File?

(i) Via application of the Freedom of Information Act folks have uncovered the FBI files of a great number of famous folks (http://foia.fbi.gov/famous.htm), like Albert Einstein, [332], John Lennon, [723], Martin Luther King (whose file contains 16,659 pages that we know about), John and Robert Kennedy, Elvis Presley, Studs Terkel,[19] Groucho Marx, and on and on, all known for reasons other than being criminals. From the Web page for the organization founded by Frank Wilkinson, http://www.defendingdissent.org we read:

"In 2005, the FBI testified to Congress that animal rights and environmental 'extremists' are the number one domestic terrorism threat in the U.S. Recently-leaked documents indicate that FBI and other police informants have unjustifiably infiltrated mosques as well as anarchist, peace, anti-death penalty and other activist groups. Leaked documents indicate that police agencies at the federal, state, local and tribal level are tracking activists of all stripes based on their legitimate First Amendment-protected activities."

Given the state of surveillance that now exists in the U.S., and the likely fact that databases of various agencies are shared, try to find someone who has voiced an opinion about anything, via a letter to the editor, an email, belonging to a legal, law-abiding organization, and so on, who does NOT have an FBI file. Hint: the Web site of the Defending Dissent Foundation has information on getting one's FBI file.

From [584, p. viii], we read: *"White-collar criminals cause more pain and death than all 'common criminals' combined."*

Estimate the amount of (tax-payer) money spent on catching and prosecuting white-collar criminals—who have actually committed crimes—vs. the amount of money spent on tracking, infiltrating, monitoring, creating disinformation, and so on, about folks who have committed no crimes and who, incidentally, are against violence.

Hints: You can get started by seeing how many of the folks who helped collapse the economy in about 2008–9, cf., Chapter 2, were pursued by the FBI to conviction and sentencing. Compare, for example, the whistleblower case of Bradley Birkenfeld who blew the whistle on about 14,000 of the richest and most powerful Americans—who committed felonies, many multiple times over decades. What happened to the whistleblower? What happened to the 14,000? See www.democracynow.org January 7, 2010. Again, what is more important here: the statistics or the logic of the system?

(ii) Study the book, *An Act of State: The Execution of Martin Luther King*, by William F. Pepper, cf., [526, 527, 528]. Pepper is an attorney who has devoted much of his life to finding out how Dr. Martin Luther King was killed. He was the lead attorney in a civil action suit on behalf of the King family against those responsible. From the book we have (about this trial which lasted almost 4 weeks): *"...Seventy witnesses set out the details of a conspiracy in a plot to murder King that involved J. Edgar Hoover and the FBI, Richard Helms and the CIA, the military, the local Memphis police, and organized crime figures from New Orleans and Memphis. The evidence was unimpeachable. The jury took an hour to find for the King family. But the silence following these shocking revelations was deafening. Like the pattern during all the investigations of the assassination throughout the years, no major media outlet would cover the story. It was effectively buried."*

Please study this and come to your own conclusions. What you determine is true could be very important in your real life.

(iii) Why did Martin Luther King say: *"The greatest purveyor of violence on earth is my own government."*? Also, why did he say: *"A nation that continues year after year to spend more money on military defense than on programs of social uplift is approaching spiritual death."*? When did he make these statements and in what context? Hint: The magazine *Time* called the speech from which these quotes come: "demagogic slander that sounded like a script for Radio Hanoi." *The Washington Post* said King, "diminished his usefulness to his cause, his country, his people." (I wrote this on Martin Luther King Day; there were few references to King's stances on war, class, and poverty. For a perspective of one who was there see [278, 279].)

(iv) What was Operation Mockingbird, see, for example, [241, Chapter 5]?

[19]See New York City News Service Nov. 15, 2009 by Valerie Lapinski, http://nycitynewsservice.com/2009/11/15/fbi-tracked-working-man-studs-terkel/. Studs Terkel was tracked by the FBI from 1945 to 1990, and under a FOIA request in 2009 a year after Studs died the FBI only released 147 of 269 pages (the we know about) of his file.

Chapter 23

Identity Theft, Encryption, Torture, Planespotting

23.1 Encryption Mathematics and Identity Protection

Encryption Mathematics for Everyone! While I was writing this book—the early years—a war for our privacy was being waged between fortunately brilliant "nobodies"—computer jockeys, mathematicians, academics—and some of the most powerful people in the world—presidents, generals, ... There is a branch of mathematics/computer science called encryption. In particular, a fellow, Whitfield Diffie, discovered public key encryption which allows two people to communicate in private via computer e-mail, for example. It is analogous to putting your message in a sealed envelope as opposed to a postcard that anyone along the way can read. One of the heroes of this battle is Phil Zimmerman, who lived in Boulder, Colorado at the time. What Zimmerman did was create encryption software called PGP, or Pretty Good Privacy. Using this software anyone can "put their e-mail in an envelope" and send it to someone else who can then easily "open the envelope with the appropriate key."

Now certain folks, for example the NSA (National Security Agency, cf., [23, 24]) and the FBI, have programs like Carnivore and Stellar Wind (and Einstein), that monitor all electronic information, e.g., e-mail and telephone messages. They want to have easy access to it all. Encryption makes it more difficult for third parties, be they police or thieves, to read your stuff. Every encryption program can be broken into, i.e., decrypted; however, it takes more computer time to read an encrypted message. The folks at the NSA are probably the best in the world at this, and if they want to read your stuff, they will (unless you are really, really talented). However, if everyone encrypted their messages the NSA would have to decide where to put their resources: decrypt the work of a pacifist nun or some fellow with the habit of blowing things up. Also there are an increasing number of just plain old criminals who would like to empty your bank account. By encrypting all your communications you decrease the chances that such folks will be able to steal enough information about you to cause you financial pain.

Amazingly, Phil Zimmerman had to wage a lengthy legal battle to keep out of prison. His crime? He gave his encryption program away for free on the internet, cf., Exercise 22.5. The U.S. government charged Zimmerman with trafficking "munitions," and vigorously tried to convict him. The government equated his "giving PGP away for free" to "shipping guns and ammo to foreign enemies"—like Denmark. The whole farce became moot after several improved versions of PGP were created in several countries around the world. The cat was out of the bag. For more on this epic battle see [394]. As an exercise find an encryption program (open source programs are free), making as sure as you can that it does not have a "backdoor," and use it to send an encrypted message. A lot of people worked very hard, some risked everything they had, to preserve this right for you!

Exercise 23.1 A Simple Encryption Exercise and a Real One

(i) Three people are to fly on a plane—the total weight of all three must be known to the pilot. The weight scale will hold only one person at a time and each of the three jealously guards his/her privacy and does not want anyone else to know his/her weight. How can this total weight be found while protecting the privacy of each of the three persons?[1]

(ii) Search for "public key encryption." Try to understand it in terms of an analogy with mailboxes with physical keys. On a deeper level, if you have the time and inclination, investigate what the factoring of large numbers with two prime factors has to do with it. Many years ago some number theorists developed a method, using several small computers in parallel, of factoring very large numbers very quickly; and they published their results in the *Fibonacci Quarterly*. They got a visit from the government demanding that their results be classified secret (apparently they accidentally got ahead of the government spies for a while).

(iii) Search for "encryption software" (open source) for your computer. Use it to send an encrypted email to a friend.

(iv) The National Security Agency, NSA, have employees who are masters of encryption. They also are a top secret spy agency of the federal government, referred to as "no such agency" for decades. By the time I discover the existence of a spy agency of the federal government I assume there are others already created that I do not know about. A few years ago I ran across, quite by accident, a U.S. spy agency, the NRO. In a conversation with a colleague I asked him if he had ever heard of the NRO. He told me that some of his research was funded by the NRO, but that he was not allowed to acknowledge their existence anywhere in his research. What is the NRO? What does it do? What comes next after the NRO? (I do not know.)

Identity Theft, Identity Protection. Identity theft is one of the fastest growing crimes anywhere. Your identity has been stolen when someone else acquires sufficiently many critical bits of information about you, e.g., your Social Security Number, your mother's maiden name, your credit card or bank account number(s), etc., to enable them to take any number of actions detrimental to your finances, your job prospects, your FBI, file and so on. It is a fact of life these days that you must treat your personal information as you would cash, or a pile of gold coins, or something you find very valuable—like your reputation or ability to get a job. (The story of the Ross family, page 429, of what havoc one mistake in your files can create gives us some idea of what a stolen identity can mean.)

[1]I got this exercise from the Will Shortz's NPR Puzzle, February 29, 2004.

To steal your identity a criminal must have *access* to your information. So we should all plan to make this as difficult as possible.[2] So do not put your personal information in places where criminals can easily get it: on the Web, in unencrypted email, in the trash. Never put recoverable personal information in the trash, for example. For the sake of the environment, shred sparingly, since shredding reduces or destroys the ability to recycle paper. It leaves the fibers too short. But do shred when necessary.

Assuming you have the habit of being very stingy with your personal information, there remains much that is not in our control. There are a great many databases in cyberspace that contain our identities. I have already mentioned the fact that we are all susceptible to total surveillance. After the Glass-Steagall Act was repealed in 1999, cf., Chapter 2, your bank account information became accessible by whatever entities your bank merged with or bought. Any number of government and commercial enterprises have an interest in maintaining a database on you and me, for purposes from targeted advertising to assessing of criminal intentions. In my opinion it is not safe to assume that any database is confidential; not that anyone should worry excessively. For example, the Internal Revenue Service (IRS) is supposed to keep our tax returns and the information it stores about us confidential. We assume it is reasonable to make exceptions in the effort to contain organized crime, but few would agree that it is OK to use IRS records to harass people for their political beliefs. But from [584, p. 321–2]:

> *"During the 1960s, the IRS served as the Kennedy administration's 'hired gun' against chosen targets, mainly right-wing extremist groups. In 1963, President Kennedy used the IRS to develop an aggressive plan, called the Ideological Organizations Project, to examine the tax-exempt status of 10,000 political organizations."*
>
> ... *"Between 1966 and 1974, the IRS and the FBI engaged in a persecutive alliance, which even a high-ranking IRS official termed 'probably illegal.' Tax records were used to 'disrupt and neutralize the political threat of the New Left.' "*
>
> ... *"In 1969, the IRS created a Special Service Staff (SSS) to examine what it termed 'ideological organizations.' "*

Targets of the SSS were primarily provided by the FBI. By 1973 the SSS had surveillance files on over 2800 organizations, such as the Americans for Democratic Action, and on individuals such as Linus Pauling (*Nobel laureat*), Mayor John Lindsay of New York City, and so on. Richard Nixon used the

[2]Hints on how to do this are available at http://www.ftc.gov/index.html (click on "identity theft"), the Web site of the Federal Trade Commission (FTC), and at a number of other places such as www.elevationscu.com (click on "reality check"). By the way, if your identity is stolen there is a standard procedure we are advised to follow; see the FTC Web site; and call 1-877-IDTHEFT and your local police as soon as possible, since time is not on your side! Caution: it is not always obvious when your identity has been stolen. For example, many credit card accounts can be opened in your name with account information sent to some address you are unaware of. Huge amounts of debt can be racked up in your name before you even find out about the existence of these accounts. Thus it is a wise idea to check annualcreditreport.com or call 1-877-322-8228 at least once a year.

IRS to pursue folks on his "enemies list." There is a lot more, but the point has been made about databases.

A perpetual mathematical game is being played between cyber thieves and cyber security; for every defense there is an offense that can breach it, which is then countered with a more sophisticated defense and so on. For example, in January 2010, there were reports of the famous "cyberwar" between the Chinese government and Google. Google discovered that, for example, Gmail accounts of Tibetan activists at Stanford University had been hacked. A counter-attack by Google revealed that the attack on Google originated in China, and that information was being gathered on people regarded by the Chinese government as dissidents, as well as apparent economic espionage against about 30 other corporations. This, by the way, is a good argument against electronic voting machines, cf., Chapter 24. (Google may end its relationship with China over this cyberwar and Chinese demands of Web censorship.) It turns out that often the weakest link in cyber security is usually a human being.[3]

In an Associated Press story by Marcy Gordon, September 14, 2000, we learn that about 500,000 (to one million) people a year in the United States have their identity stolen. (At www.ftc.gov we read that in 2010 an estimated 9 million Americans were having their identity stolen each year.) Identity theft is a federal crime, but Gordon's article indicates that this law is not being vigorously enforced. In fact, Congressman Jim Leach of Iowa, Banking Committee Chairman (in 2000) did an experiment. His staff phoned several information brokers and private investigators around the country to see how many would sell them bank account information and under what circumstances. In less than three hours, the first 10 companies they reached were willing to sell detailed account data likely only to be obtained through deceptive means. None turned them down.

On Dec. 4, 2002 I met Kevin Mitnick, once labeled by the FBI as "the most wanted computer criminal in U.S. history." He was out of prison and giving a talk on his then new book [462] which is about methods of breaching cyber security by exploiting human nature. I will let you read about the host of ingenious tricks Mitnick details. Mitnick reminded me of some of the very bright young students I have had. He got into trouble, not because he wanted to steal money or do harm; but very much like mountain climbers and wilderness adventurers, he wanted to prove himself by doing "impossible" things, going places where no one had gone before—very much like scientific research. The forward of his book was written by Steve Wozniak, another genius, who was expelled from my university early in his career for "computer abuse." Steve

[3]For example, *pretexting* (using a pretext such as masquerading as ...), was famously used in Fall 2006 by investigators to obtain the call records of the board of directors of Hewlett-Packard and the call records of various reporters. In another famous case the governor of an eastern state had his social security number sold for $30 and for $125 all of his credit card numbers and much other personal information were purchased.

went on, as you may know, to co-found Apple Computer Company; and the rest is history.

Exercise 23.2 How Safe Is Your Identity?

(i) Go to the FTC Web site and make a plan which you can quickly implement if you discover that your identity has been stolen.

(ii) Estimate the number of databases that contain your "identity." Estimate the level of cyber security of each.

(iii) How many people in your home country have their identity stolen, say per year, in the year you read this?

(iv) Which of the databases in (ii) have been used in the past to harass groups or persons for their religious, political, or environmental beliefs? How difficult is it to get "up to the minute" information to this question?

(v) What are some of the methods which exploit human nature to get around some of the best "firewalls" of cyber security? See [462]. It is wise to be aware of these, if for no other reason than to protect yourself to the extent possible.

23.2 Extraordinary Rendition = Kidnapping and Torture

"Maher Arar was the first person to be 'rendered' in the 'war on terrorism' and come back alive to speak to the public fully about his experience," [714, p. 40]. And what an experience it was. Arar's flight from Tunis arrived at 2 p.m. on September 26, 2002 in John F. Kennedy International Airport. He had a few hours to wait for his connecting flight to Montreal, Canada, where Arar was (is) a citizen. He was taken by U.S. authorities, interrogated until midnight, denied a lawyer, then kept in solitary confinement for two weeks. October 8, 2002 Arar was awakened at 3 a.m. and was soon deported to Syria. His wife and family were not informed, they did not know why or where he had been taken—Arar just disappeared. Arar's "accommodation" in Syria for ten months and ten days was a grave-like cell with no light: 3 feet wide, 6 feet deep, and 7 feet high. Arar saw the sun three times during his imprisonment. He was *repeatedly* tortured. Due in no small part to the political movement his wife, Monia, facilitated, Arar was returned home October, 2003, physically and psychologically broken. There was a change in Canada's prime minister; there was an official inquiry (which it took a "political movement" to obtain), Arar was found innocent of any wrongdoing or links to terrorism. Canada later settled out of court for 10.5 million dollars Canadian. Arar has been given a number of awards in the interim, one by *Time* magazine. The U.S. would not allow Arar entry into the United States to accept the award; the U.S. keeps Arar on its terrorist watch list and denies culpability. Arar sued the U.S. government, but all courts, including the court of appeals, threw his case out "based on issues of national security." I cannot do this horrific story justice, please read the chilling details in [714, Chapter 1].

This sort of thing is not supposed to happen in the U.S. or Canada (or any civilized common law country). First of all there is the right of *Habeas corpus* which has been part of civilized society since the Magna Carta of 1215 AD. A writ of habeas corpus is a court order addressed to a prison official ordering that a detainee be brought to court so it can be determined if that person is imprisoned lawfully or not, and if that person should be released or not. The right to petition for such a writ is the law, for anyone on our soil. Then there is the Constitution of the United States of America, which confers on all, citizens or not, certain fundamental rights. With the passage of the USA Patriot Act of 2001 and the Military Commissions Act of 2006 much of the Constitution of the United States has been suspended—even for citizens. What has been U.S. policy abroad for some time is becoming practiced policy in the homeland, cf., [351, 433, 529, 484, 273, 274, 502, 65, 41, 42, 43, 742, 743, 331, 335, 334, 302, 303, 592, 125, 256, 739, 428, 623, 647, 463] (See also any number of books by Noam Chomsky on U.S. foreign policy, which are too numerous to list all of them here. For two examples see [87, 86].) Basically, if you are accused of some crime you are supposed to get "your day in court" and thus a chance to see what evidence there is against you—but no longer is this guaranteed. Anyone can be "disappeared." (It might be a mistake, but ....)

Research the CIA kidnapping and torture case of German citizen, Khaled El Masri, [714, p. 186–7], cf., also the CBS "60 Minutes" archive for his story and interview in video. Research the story of U.S. citizen José Padilla, [259] and *The New York Times*, December 4, 2006, page A1, page A22. See http://ghostplane.net where 3,000 flight logs are published which expose the CIA rendition, i.e., kidnapping and torture, flights. See [259, 451, 513, 429, 582, 457, 275, 111], and *The New Yorker*, Oct. 30. 2006, pp. 33–37.

Even "ordinary" detention of migrants can be rough, cf., "Officials Obscured Truth Of Migrant Deaths in Jail: Evidence of Mistreatment Was Frequently Covered Up, Documents Show," *The New York Times*, by Nina Bernstein. Freedom of Information Act (FOIA) requests by the *The Times* and the ACLU revealed 107 deaths in detention counted by Immigration and Customs Enforcement (ICE) since October 2003, when ICE was created within the Department of Homeland Security. Video tapes obtained via FOIA offer confirmation. It was not easy to get this information!

U.S. citizens should consider the following quote: *"State-sponsored torture and abuse ... commonly thought to be practiced only outside this Nation's borders, are hardly unknown within this Nation's prisons."* U.S. Supreme Court Justice, Harry A. Blackmun, *Hudson v. McMillian*, February 25, 1992.

Exercise 23.3 Patterns in Torture and Assassination

(i) Research police and prison guard brutality in the U.S. and internationally, cf., [83], www.hrw.org, www.amnesty.org. For example, there is a rather (in)famous August 1997 case in New York of one Abner Louima, a 30 year-old Haitian immigrant who said he was tortured by New York police, 70th Precinct. More precisely he said he was sodomized with a bathroom plunger handle (It came out during the subsequent trial that actually a broken broom handle was used.) which was then shoved into his mouth, breaking his teeth. When Louima arrived at the hospital he was in critical condition with a torn colon and bladder. Were it not for a brave nurse, Magalie Laurent, this torture might have been successfully covered up, as other cases apparently have been. (Principal assaulting officer, Justin Volpe was sentenced to 30-year jail term.)

Another example, a completely innocent and unarmed Amadou Diallo was killed February 4, 1999, by four New York police officers. (The police were not fined or punished, but acquitted. A civil settlement was reached.)

(ii) Research the number of assassinations (accomplished, attempted and/or planned) in which the U.S. has been involved and/or implicated. Hints: A list of over three dozen prominent assassination targets is in [42, Chapter 3]. See also the other books by Blum, as well as the books by Perkins, e.g., *Confessions of an Economic Hit Man*, [529, 530]. See [484]. Also of interest is the work of Seymour Hersh, in particular, search on his name and JSOC, Joint Special Operations Command. Note that *The Baltimore Sun* (www.sunspot.net) on January 27 and 29, 1997 ran articles on CIA training manuals which detail methods of torture to be used against subversives. The *Sun* filed a Freedom of Information Act request for the documents on May 26, 1994.

(iii) Why did Abu Ghraib make the news in the first decade of the 21st century? See [303, 456, 672, 256].

(iv) Research the torture of Everardo, the husband of Jennifer Harbury, [273, 274, 275]. Research the torture of Sister Dianna Ortiz, [502]. What is the School of the Americas, renamed WHISC, Western Hemisphere Institute for Security Cooperation, run by the U.S.? Is WHISC linked to torture?

(v) I have a bookshelf of references on this topic which there is no room to discuss here. In particular, it is unfortunate that many people have died at the hands of others for standing up for environmental sanity and/or justice. Research this topic.

(vi) Give at least one very compelling reason, from the point of view of a soldier, to be fighting on the side of a nation known and assumed to be humane by opposing forces, i.e., a nation that does not torture.

(v) Research *Blackwater*, now named *Xe*, arguably the world's most powerful mercenary army, cf., [600]. What was Blackwater's role in the Katrina disaster in New Orleans?

Officially denied for years, how was it discovered that the U.S. government was doing things worthy of any number of despotic governments throughout history? The details can be found in *Torture Taxi: On the Trail of the CIA's Rendition Flights*, by Trevor Paglen and A. C Thompson, [513], and in *Ghost Plane: The True Story of the CIA Torture Program*, by Stephen Grey, [259]. A very interesting part of this detective story involves some amateur mathematics and a "sport" called planespotting to which I now turn.

23.3 Planespotting: A Self-Organizing Countermeasure the CIA Did Not Anticipate

Unknown to most of us, and also apparently at one point to the CIA, is a hobby called *planespotting*. There are various levels of participation ranging from sitting near an airport with a pencil, paper and binoculars; to sitting near an airport with digital cameras, telephoto lenses and recording devices; to sitting near an airport with all of the above and a laptop computer equipped with hacked software that can listen in to all telecommunications relevant to the sport, even those blocked and/or encrypted by the CIA. Any computer security measure is regarded by brilliant, suitably mathematically inclined puzzle solvers in or out of the NSA as a puzzle to be solved. These planespotters meticulously record the comings and goings of aircraft using such things as tail numbers (which can change), serial numbers (which do not), and logs of communications from the extremely mundane to the tantalizingly decrypted.

There is, perhaps, the highest level of planespotter; those who pursue the Freedom of Information Act to access data, examine multiple databases, do a little additional sleuthing, all in an attempt to see patterns and to figure out what all of the data gathered by planespotters from around the world means.

No place in the world where a plane can land (or fly), from the most remote dirt runways in deserts that shall remain nameless, to the giant hubs of the world's major metropolitan areas is guaranteed to be opaque to the collective community of planespotters. They meticulously record one bit of data after another, often with no other immediate goal than to be as precise as possible. They share data on the Web at sites such as Airliners.net and Planespotters.net. Most of the time they record in anonymity. On some occasions their lives are threatened.

The patterns revealed CIA "destinations" in Romania, Afghanistan, Poland. Bits of information became webs of reinforcing patterns of information, complete with photographs, corroborating otherwise unbelievable stories of kidnap and torture by a few actual victims. After several years of investigation, folks like Trevor Paglen, an expert on clandestine military installations, and Stephen Grey, British investigative journalist, and A.C. Thompson, award-winning American journalist brought these facts to the attention of the "general public," cf., [513, pp. 95–121] and [259]. Bits of the story have now appeared on CBS, eg., "60 Minutes," a Brian Ross report on ABC News, and in *The New York Times*, *Newsweek* and *The Guardian*.

There is more to be said and written, some of which reads like a real action-packed spy novel, except that it is all heavily documented as reality. I have provided enough leads so you can now explore this story and the following fundamental "countermeasures" principle as deeply as you wish.

A Fundamental Principle of "Games:" Countermeasures and Mathematics. No matter what your beliefs, priorities, social status or access to resources, if measures are taken which you believe to be against your self-interest, i.e., measures that do not "work" for you, *countermeasures* exist which you can take in defense of your self-interest. This topic might fall into a category called extremely applied game theory; and the countermeasures may not be obvious, may require imagination to create. Sometimes courage is required, sometimes not. Planespotting, intended or not, was a countermeasure that revealed the truth about the CIA torture program.

On December 18, 2009, an Associated Press article by Pauline Jelinek revealed that enemies of the U.S. in Iraq and Afghanistan were able to implement countermeasures against U.S. Predator drones using $26 software.

Using programs such as SkyGrabber, available for $25.95 on the internet, Shiite fighters in Iraq were able to regularly capture video feeds from Predator spy, surveillance, and intelligence drones. Similarly in Afghanistan. Although the Pentagon knew of this vulnerability for more than ten years, it assumed adversaries would not figure it out.

23.4 Bigger Pictures and the CIA

The following is one of the more interesting "getting a Bigger Picture" exercises that I have done. It led to some dark corners I am still investigating. It shows how just a little curiosity and insistence on original documentation can lead to a bit of adventure. It also turns out to be a lesson in *media literacy.*

On Thursday, October 24, 1996 some media sources carried a story about hearings before the Senate Select Committee on Intelligence held the previous day. As a typical example consider the following quote from an article by Thomas Farragher of Knight-Ridder Newspapers.

"No direct link connects the Central Intelligence Agency to a crack cocaine epidemic in Los Angeles, a former investigator into the labyrinthine Central American drug wars told a Senate panel Wednesday."

"The testimony came from Jack Blum, who was special counsel to a Senate subcommittee that investigated Contra[4] *drug operations in 1987–88."*

I was curious whether the public was getting the full story. Briefly, with the help of staff persons in the office of, my then Congressman David Skaggs, I obtained transcripts of the entire testimony of Jack Blum. Here is an excerpt.[5]

"We found no evidence to suggest that people at the highest level of the United States government adopted a policy of supporting the Contras by encouraging drug sales." .... *"There was, however, plenty of evidence that policy makers closed their eyes to the criminal*

---

[4]Contra here refers to (para)military forces supported by the Reagan administration in the United States in the early 1980s. The Contras worked from bases in Honduras and Nicaragua to overthrow the Sandinista government of Nicaragua. Such U.S. support for the Contras became illegal when Congress passed the Boland Amendment making such illegal, but the support continued. Millions of dollars in drug profits made from the selling of tons of cocaine to Los Angeles street gangs was funneled back to the CIA-backed Nicaraguan Contras. Documentation of these seemingly outrageous claims are in the book by Gary Webb, cf., [713]. Even a CIA inspector general's report corroborated about half of Webb's claims. Additional moneys were raised by selling missiles to Iran. The Reagan administration's direct knowing involvement can be verified by reading now declassified (2006) documents at the Web site of the National Security Archive. See also [361].

[5]I have also subsequently acquired from Pacifica Radio and/or www.democracynow.org audio of some of Jack Blum's testimony which are consistent with the written transcripts.

behavior of some of America's allies and supporters in the Contra war. The policy makers ignored their drug dealing, their stealing, their human rights violations."...

"In short, what you say about drugs and the Contras depends totally on the question. If the question is did the CIA sell crack in the inner city to support the contra war, the answer is a categorical no. If you ask whether the United States government ignored the drug problem and subverted law enforcement to prevent embarrassment and to reward our allies in the Contra war, the answer is yes."...

"I might add that the Justice Department did everything possible to block our investigation."...

Jack Blum arrived at the following conclusion:[6]

If you ask: In the process of fighting a war against the Sandinistas, did people connected with the US government open channels which allowed drug traffickers to move drugs to the United States, did they know the drug traffickers were doing it, and did they protect them from law enforcement? The answer to all those questions is yes.

Exercise 23.4 Jack Blum and a Bigger Picture

(i) Without worrying for the moment about whether Jack Blum is telling the truth, without worrying for the moment about the ethics of government agencies facilitating drug traffic, what are the chances that the megamedia reporting on this story would select the one sentence it did from several pages of Jack Blum's testimony to the exclusion of all the rest?

(ii) The November 30, 1998 issue of *The Nation* had an article entitled "CIA: Didn't Just Say No." The article begins: *It's official. During the Reagan years, the CIA worked with suspected drug dealers to bolster the* contras *in their war against Nicaragua's Sandinista government. Who says so? The CIA does.*

The article then points to CIA inspector general, Fred Hitz's report on some of the issues covered by Gary Webb. The report was posted on the CIA Web site www.odci.gov/cia/publications/cocaine2 ; and the report was "covered" in the mainstream press rather lightly, if at all. Hitz said to *The Washington Post*: "This is grist for more work, if anyone wants to do it." The mainstream press did not demand more work be done; they all but buried the story.

iii) The nice thing about this exercise is that you can go to a well stocked university library, or perhaps on-line resources, and research and verify this story for yourself. At least three reporters were key in breaking this CIA-crack cocaine story: Robert Parry, in 1986 was the first to my knowledge, cf., [519, 520];[7] see also Leslie Cockburn, [96]; Gary Webb was the last. Parry and Webb lost their jobs as reporters, Parry is alive, Webb is dead. All megamedia news condemned Webb and his work, from *The New York Times* to *The Los Angeles Times*. The megamedia was (is) determined that this story go away and be buried. Do you think there is a lot more to this story than what I have included here?

The CIA is into everything. Consider the following quote from the September 18, 2006 issue of *Forbes*, p. 105. *"Since then Cassatt has raised another $30 from investors that include In-Q-Tel, an investment arm created by the Central Intelligence*

[6] The following italicized paragraph was taken from Lusane, Clarence, "Cracking the CIA-Contra Drug Connection," *CovertAction Quarterly*, Number 59, Winter 1996–97, pp. 53–59.

[7] In [520] Parry discusses many issues surrounding the Iran-Contra scandal of the '80s, at least one of which is relevant today. He gives his answer to why the vaunted Watergate press corps of the '70s devolved into the Monica Lewinsky press corps of the '90s; and I might add the "weapons of mass destruction" press corps of 2001–2006 and beyond.

Agency." Also consider the following quote from William Colby, former director of the CIA, cf., [745, p. 141], *"The Central Intelligence Agency owns everyone of any significance in the major media."*

This last quote is consistent with the following fact. In 1975 the Senate Select Intelligence Committee found that the CIA owned outright more than 200 wire services, newspapers, magazines, and book publishing complexes and subsidized many more. About a decade later 50 further media outlets and over 12 book publishing houses had been discovered to be CIA related, [436, p. 186], [518].

Combining government institutions such as the military, FBI, CIA with corporate public relations and other interests that hire public relations firms to represent them to the public, there is a significant chance that any particular "news" item you read or see has been planted by such an entity for some particular purpose and not the result of a reporter's investigation. With cutbacks in the number of reporters and news budgets, the chances for such happening can only increase.

Exercise 23.5 Bigger Picture Exercises

(i) Pick a newspaper article and do the exercise I did with the "Jack Blum-CIA-crack cocaine" story. Is the article straight news? Was it placed, by whom, for what reasons? Can you find a bigger picture that the news article is a part of?

(ii) Find two people/classmates who disagree on some subject. (You can be one of the persons.) Analyze as best you can where each gets their information and from what set of assumptions each is reasoning. (If no one is reasoning, try again with some other people.)

(iii) Is there a bigger picture that contains the theory of evolution and creationism?

(iv) Why do I talk about "a bigger picture" and NOT "the biggest picture?"

(v) Discuss the following assertion. One of the greatest impediments to getting the big picture from television is that television breaks information up into short sound bites and television follows market demand as assessed by advertisers.

(vi) Discuss the following assertion. One of the impediments from getting the big picture in an educational institution is the fact that much of the information available therein is divided up into narrowly defined disciplines, and interdisciplinary interaction is not common.[8]

(vii) Should a prerequisite to leadership be having a big picture? Should leaders have to do more homework than most folks?

(viii) George Bernard Shaw wrote: *"The power of accurate observation is commonly called cynicism by those who have not got it."* Calling an accurate observer a cynic is what type of fallacy?

(ix) Discuss the following two quotes of Malcom X.

"Never accept images that have been created for you by someone else, it is always better to form the habit of learning how to see things for yourself."[100, pp. 212–3].

"I'm sorry to say that the subject I most disliked was mathematics. I have thought about it. I think the reason was that mathematics leaves no room for argument. If you made a mistake, that was all there was to it."[9]

---

[8] I was once told that the Chinese government once made interdisciplinary studies illegal—scholars worked only within their field of specialization. Why would they do this? Could this same end be achieved, perhaps inadvertently, in America by different means?

[9] This quote was taken from *The American Mathematical Monthly*, Vol. 103, No. 10, December 1996, p. 916.

(viii) Discuss the process of ordering your life's priorities and what that process might have to do with fuzzy logic and finding bigger pictures.

(ix) Discuss the following rule of thumb: When your mental model fails to work, it is time make a bigger picture.

(x) On July 19, 2010 the first of a series of articles in *The Washington Post* by Bill Arkin and Dana Priest investigating what amounts to a fourth branch of government called Top Secret America was published, cf., http://projects.washingtonpost.com/top-secret-america. This fourth branch of "government" consists of 1,931 private corporations (at the time of writing) and 1,271 government entities working in about 10,000 locations across the United States. The *Washington Post* Web site has an interactive map of the U.S. where you can investigate the activities of this intelligence apparatus in your state or local community. Over 850,000 people with top secret security clearance are embedded in this network, some of them doing the same job at a private firm that they were doing, say, for the CIA but for two or three times the salary. First created after 9/11/2001, this network has grown without much feedback, indeed without much if any oversight by representatives of the U.S. citizenry. It has been estimated that about 8 congresspersons are allowed to look at what is going on with this secret organization. As an exercise, estimate the budget of this Top Secret America at the time you read this. What are they doing? Is there any redundancy? What is the closest physical site of this operation to where you live?

We mention that Arkin and Priest worked on this story from 2008 to 2010, and that Tim Shorrock, in [633], perhaps "got the ball rolling" with his book "Spies for Hire: The Secret World of Intelligence Outsourcing." You can find interviews with Arkin, Priest, and Shorrock at www.democracynow.org, July 19, 2010.

Chapter 24

Voting in the 21st Century

It is enough that the people know there was an election. The people who cast the votes decide nothing. The people who count the votes decide everything.

Joseph Stalin

24.1 Stealing Elections Is a Time-Honored Tradition

Although one can find references to about fifty different techniques that have been used to steal elections in various political units through the years (Chicago comes to mind), I am going to focus on those new techniques made possible by the introduction of electronic voting machines in my own country, the U.S.A. Information on these techniques is not as well disseminated as it might be. In their 2005 and 2006 editions, for example, Project Censored rated among the most censored news stories ones related to election fraud in the U.S. Additionally they gave updates on election fraud in their 2007 edition as well as revisiting megamedia misrepresentations of fact in their 2009 edition, cf., "*New York Times* Perpetuates the Myth that George Bush Won the 2000 Election," [534, pp. 125–129]. I claim that the likely "that isn't true" reaction of many readers to the sentiments of the article just mentioned attests to said censorship. In the 2008 New Hampshire primary, glaring statistical anomalies associated with electronic voting machines arose in the Clinton-Obama race, cf, [535, Chapter 11]. Despite the fact that there is ample evidence recent elections in the United States have been tampered with, indeed, stolen, cf., [190, 191, 458, 459, 467, 205, 248, 264, 280, 514], and the video [179]; such is not the most important observation to be made. Whether you believe the data in these references or not, the main point I want to make to folks of all political persuasions is this: *electronic voting machines should immediately be replaced by paper ballots, hand counted in a transparent manner.*

Before giving the reasons for doing away with electronic voting machines, I wanted to visit some mathematical evidence that an assortment of the "fifty techniques" for stealing elections was used in recent national U.S. elections.

"Ohio, like the rest of the nation that day[1] *was the site of numerous statistical anomalies—so many that the number is itself statistically anomalous, as every single one of them took votes from Kerry."* I will let you investigate the specifics of these anomalies, bringing to your attention only a couple of the most obvious. First, in Gahann, Ohio, Ward 1-B, a total of 638 people voted. The total number of votes tallied for George W. Bush, however, was 4,258, cf., Associated Press, Nov. 5, 2004. In Miami County nearly 19,000 votes appeared in Bush's column *after* all precincts had reported; in Franklin County Bush similarly got an extra 4,000 votes from one computer. Throughout Perry County the number of Bush votes somehow exceeded the number of registered voters, leading to voter turnout rates as high as 124 percent, cf., [458, pp. 30–31].

Somehow, when a shortage of functioning voting machines occurred, it was in a Democratic leaning district. In the last century one of the most reliable predictors of election outcomes has been the media *exit poll*. Voters, after having voted, are sampled as to how they voted. The exit polls in the 2004 presidential election predicted a win for Kerry; but when the votes were *counted* Kerry's margin of victory was "flipped" and it went to George W. Bush.

Cyber security expert, Stephen Spoonamore, noted that if *one-tenth* the number of anomalies that occurred in the 2004 presidential election were to occur in the credit card business, an immediate fraud investigation would be launched, cf., Spoonamore's video interview.[2] It turns out that an investigation of sorts was attempted in that Michael Connell, GOP IT (Republican Information Technology) expert had been directed to appear in court to answer questions. Michael Connell died in a plane crash before he was forced to testify about his possible involvement in a "man in the middle" attack in the 2004 election. The cause of that crash has not yet been ascertained.[3]

Electronic Voting Machines: Unreliable and Unjustified. One of the justifications, if not the only one, for introducing electronic voting machines was the occurrence of "hanging chads," or more specifically the malfunction of paper ballots in the 2000 presidential election in Florida. Paper card technology was reliable enough to land Americans on the moon; and paper ballots worked throughout much of the United States in 2000—as paper ballots have continued to perform well ever since in Switzerland, Germany, England, Canada and so on. Though not well known, an investigation found that the problem with paper ballots was caused by the intentional changing of specifications for the paper used in the ballots in some precincts in the 2000 Florida presidential election.[4] Independent of intentions, however, any problem could have been

[1] The day of the national 2004 presidential election, cf., [458, p. 30].

[2] See http://www.velvetrevolution.us/prosecute_rove/images/SpoonIntvw3.wmv.

[3] The Web site http://www.velvetrevolution.us/ follows the Michael Connell story. "Man in the Middle" is a now classic hacker technique.

[4] See the report: "The Trouble with Touch Screens." In transcript form this report is available at http://election-reform.org/dan_rather.html, and in video form at http://election-

easily corrected by enforcing quality standards for paper ballots! Instead, the United States is in the untenable position of having some or all of its elections being carried out by private corporations (some with strongly held partisan positions) with citizens voting on electronic machines, many proven to be unreliable in at least two different ways, running proprietary (hence secret) software. The votes are then counted *secretly* and it is *impossible to independently audit* any intermediate or final counts since there is no tangible record of the vote available! Strangely, often the same manufacturer of voting machines also makes ATM machines for banks which routinely print out a paper receipt. Thus electing a president is evidently less in need of documentation than a routine $20 transaction at an ATM. Incidentally, voting electronically is much more expensive than voting with paper ballots. For example, in Denver in the 2006 election, the election process cost taxpayers $20 per vote—and the election was a disaster from my point of view. Thousands of voters were turned away due to a malfunctioning system. And Denver was unfortunately not unique.

So how are electronic machines unreliable? First of all they are plagued by technical malfunctions of every sort, the worst being those that are not detectable by voters. Second, electronic voting machines are notoriously open to cyber attack. We are in the position where the last best hacker, native or foreign, can determine the outcome of an election![5] It is essentially a mathematical theorem that no computer-based voting system can be made secure and reliable without voters giving up their privacy, i.e., their secret ballot. Electronic voting is intrinsically unreliable and unstable because one human can effectively interfere with an entire state's election process in 90 seconds (or less) and go undetected.

Taxation Without Representation? The percentage of eligible voters who turn out in a given election is related to a variety of things, and techniques exist to lower this percentage, [541]. A supreme court decision handed down in

---

reform.org/dan_rather.html (episode number 227).

[5]In 2006 a group of computer scientists at Princeton, including Edward Felton, director of the Center for Information Technology Policy, created demonstration vote-stealing software that can be installed within a minute on a common electronic voting machine. The software can fraudulently change vote counts without being detected. See http://itpolicy.princeton.edu/voting.)
This project is not unique. A preliminary report from the National Institute of Standards and Technology (NIST) posted (see http://vote.nist.gov/DraftWhitePaperOnSIinVVSG2007-20061120.pdf) Dec. 1. 2006, was referred to by *The Washington Post* as a condemnation of the security of electronic voting. (See "Security Of Electronic Voting Is Condemned, Paper Systems Should Be Included, Agency Says" by Cameron W. Barr, *The Washington Post*, Dec. 1, 2006. The article begins: *Paperless electronic voting machines used throughout the Washington region and much of the country "cannot be made secure," according to draft recommendations issued this week by a federal agency that advises the U.S. Election Assistance Commission. The assessment by the National Institute of Standards and Technology, one of the government's premier research centers, is the most sweeping condemnation of such voting systems by a federal agency.*

January, 2010, *Citizens United v. Federal Election Commission*, may increase the numbers of those who do not vote. In this decision five of the nine justices decided that corporations are people with free speech rights and that they can spend as much money as they wish, without regulation, influencing elections. As discussed in Chapter 9, absent a system of public funding for elections or absent a constitutional amendment that reserves constitutional rights for actual humans, our president, senate and congresspeople will not be able to win an election without partnering with corporations—and doing their corporate sponsors' bidding. What makes the supreme court decision a form of treason for some is that foreign corporations are allowed unlimited influence on U.S. elections. Of the 100 largest economic units on earth about half are corporations and half are countries. Thus "the people" will be paying taxes to a government that will be in the hands of whatever coalition of corporations happens to "win" a given election. Our "founding fathers" would undoubtedly have found this to be a form of *taxation without representation.*

The *Citizens United v. FEC* decision will likely have some obvious and not so obvious effects. As noted in Chapter 9, the campaigns by corporations via public relations (propaganda) firms will be designed to appear to be from "grassroots." We are likely to see continuous campaigning—or at least the threat of continuous campaigning. Every elected official will have in the back of his/her mind: "What kind of attack ads will be coming if I do so-and-so?" Just having the enormous corporate tool provided by *Citizens United v. FEC* available could get the results desired with only occasional actual implementation. Thus a constitutional amendment declaring corporations non-human artifacts of the law and public funding of elections will be essential and necessary (but perhaps not sufficient) for survival of any semblance of a free state.

Congress is Capable of Acting! A classic example, cf., [213, p. 90], is the Video Privacy Protection Act of 1988, cf., 18 USC 270. During the Senate hearings for Judge Bork's nomination to the Supreme Court, his record of video rentals was introduced. Apparently shocked by how easily the record of every video Bork had ever rented was made public, strict privacy provisions were enacted with respect to video rental information. Congress is also capable of voting for great health care for itself.

Amend the Constitution. The Constitution of the United States is very much like a list of axioms, basic rules of government. It is also not entirely dissimilar from the constitutions of other nations, such as Russia during this and much of the last century. But at any time a constitution of any country eventually means what the people who live in that country want it to mean. As Judge Learned Hand once said: *"Liberty lies in the hearts of men and women; when it dies there, no constitution, no law, no courts can even do much to help it. While it lies there, it needs no constitution, no law, no court to save it."* Of course, what is written has institutional momentum, but if sufficiently many people don't get behind liberty and push for it continuously, it is taken away. History repeatedly shows this; at least histories such as [742], as opposed to histories written from the vantage point of victors—political and economic.

Exercise 24.1 The Declaration of Independence: a Terrorist Manifesto?

(i) King George, monarch of the British Empire during the birth of the United States, regarded our "Founding Fathers" as, to use a current term, terrorists. Had King George had access to the signers of the Declaration of Independence, for example, he would have had them executed.

Review the definition of terrorist as defined by the U.S.A. Patriot Act of 2001. Then imagine the following paragraph being enuciated in a letter to the editor of your local paper, perhaps using modern language, but with identical intent.

"*. . . That to secure these rights, Governments are instituted among Men, deriving their just powers from the consent of the governed, That whenever any Form of Government becomes destructive of these ends, it is the Right of the People to alter or abolish it, and to institute new Government, laying its foundation on such principles and organizing its powers in such form, as to them shall seem most likely to effect their Safety and Happiness. . . .*"

Could such persons who signed such a letter be considered terrorists according to the definition of terrorist in the aforementioned 2001 act?

(ii) Search the media at the time you read this and for a decade or two prior and list all the people who have been referred to as terrorists, especially, eco-terrorists. For example, Judi Bari. Find and list as many professors and instructors who have been singled out as "unpatriotic" or similarly found wanting because they voiced disagreement with official government policies.[6]

(iii) From [710, 711] it is highly likely that any person A can be linked to any person B via a chain of acquaintances with no more than 6 people in it. From the definition of terrorist in the act of 2001, does this mean that everyone is a candidate for arrest and detention?

(iv) *"If we don't believe in freedom of expression for the people we despise, we don't believe in it at all."* Noam Chomsky. Comment.

(v) If corporations (even foreign corporations) have unlimited and unregulated influence on elections of people to political office in the U.S., who might such a system designate as terrorist? What conflicts do you foresee between environmentalists and the government if the government is completely run by corporations?

Data Bases and Elections. As noted in Chapter 22 there is an immense amount of information collected and stored about us. From our preferences in shopping (via cards issued by businesses and credit cards in general) to our political views, even our daily travels (via, for example, EZ pass transponders and GPS devices). Thus, besides direct attacks on the voting process, there is an immense amount of data available to those who would now try to manipulate or influence elections. Geographical Information System, GIS, databanks now include virtually every person/home/apartment in the United States. Combining this database with commercial data, credit data and so on,

[6]For example, shortly after Sept. 2001, Senator Lieberman and Lynne Cheney and ACTA put out a "black list" of professors/instructors in a publication entitled: "Defending Civilization: How Our Universities Are Failing America and What Can Be Done About It." Seriously questioning American foreign policy, or pondering what might motivate terrorists could get one on this list. The climate of the moment was not conducive to deep analysis, when, in fact, it was much needed. If you look for the publication, try to find the original from 2001 which actually names names of dissident academics.

predicts how a given person in a given precinct is likely to vote. This information is quite useful in either legal or illegal plans to manipulate an election. By the way, these databases are used by advertisers who want to know what foods/products persons will likely buy, by insurance companies who want to know one's medical history, or by anyone who wants to know your age, memberships, criminal record, list of associates, employers, computer sites visited, books/magazines/newpapers/music/television read, listened to or viewed. All of this data and more is likely available to persons with benign and/or nefarious intent.

24.2 A Simple Solution Exists

If the Swiss can hold elections on paper ballots, then so can the United States. If we insist on using computers, there needs to be a paper ballot, verified by each voter, and securely handled. These paper ballots need to be counted by an independent process to verify and cross check the electronic count. Nothing less is acceptable. Better still, make the paper ballots the only official vote; and count them mechanically.

Further references of relevance are [347, 264, 541, 540].

24.3 Two Modest Proposals

At the moment in the United States there are two viable political parties, the Democrats and the Republicans. Probably one reason that there are only two is mathematical. In any election the winner takes all with 50% plus 1 or more of the votes.

It sometimes happens that the sentiments of the majority are not reflected in the outcome of an election. For example, suppose Green Party candidate, A, is somewhat popular, candidate B is as well, with less pronounced but still pro-environmental leanings, while candidate C decidedly puts the environment low on his/her list of priorities. If candidates A and B collectively get 55% of the vote (but A gets 15% and B gets 40%) and C get 41% with other candidates, say D and E, getting 3% and 1%, respectively, the environment loses.

Modest Proposal Number One: IRV. One way to address this problem is with IRV, Instant Runoff Voting. This is how it might work in the above case. Everyone would list candidates A, B, C, D, E, say, in their order of preference. Then a sequence of counts would ensue. After the first vote, candidate E, say, would have the fewest votes as first choice. Candidate E would then be eliminated and all of his/her votes would then be transferred to the number two choice of those voters. To make life simple, let's say they all preferred C as their second choice. Then the votes would all be counted again, this time D would have the fewest number one votes and be eliminated. Again, for simplicity let's assume that all of D's votes would be transferred to C, being the next choice for all of those voters. At this point all of those who chose A, B, or C as their first choice would still have their votes going to A, B, or C respectively. Suppose 15% choose A as their first choice, 40% choose B as their first choice, and—after elimination of D and E—45% of the voters rank C highest. Then in the next round of counting A, with 15% first choice votes, would be eliminated and if, say 11% of those voters choose B as their next choice and 4% choose C as their next choice, we are left with B with 51% of the vote, C with 49% of the vote.

Thus in the one-round-winner-take-all system that we have now, C wins by 1%. In the IRV system, B wins by 1%. The advantage of this system is that everyone can give a higher priority to how they really feel about candidates and their positions. If there is a real Nazi running, voters with those inclinations can show their support; and it might be alarming. If there is someone who would like to declare half the country wildlife habitat and move the humans out, whomever felt that way could express that sentiment. These fringe candidates would not likely win. However, a greater variety of positions would get aired during elections, and greater voter participation might be encouraged.

All Voting Systems Have Drawbacks. For people who study the mathematics of voting there is a well-known result: Kenneth Arrow's Impossibility Theorem, [681, p. 248–251], see also [593, 594]. Without going into details, it says roughly that no matter what system of voting you employ (assuming you do not live in a dictatorship) there will exist scenarios where the "will of the people" will not be expressed by the outcome of the election.

So there must be a scenario where IRV leads to a mess. Here it is. Suppose there are three candidates: A, B and C. Suppose that one third of the voters list their preferences from first to third as follows: A, B, C. Suppose another third list their preferences as: C, A, B. And suppose a third list their preferences as B, C, and A.

Exercise 24.2 Instant Runoff Voting

(i) In the immediately preceding scenario, show that no matter how you count the votes, winner-take-all or with instant runoff voting, there is no clear winner. By this I mean a *transitive* outcome, with one candidate clearly ranked first, another clearly ranked second, another clearly ranked third.

(ii) Is there a scenario where IRV leads to a mess, winner-take-all leads to a clear victor?

(iii) Is there a scenario where winner-take-all leads to a mess, but IRV yields a clear victor?

(iv) In the political climate of the United States at the time you read this, which system of voting do you think would yield results that more closely reflect the will of the people: winner-take-all or IRV?

(v) For a project, you could start with a complete statement of Arrow's Impossibility Theorem and a reading of its proof. Then you could analyze various voting methods and decide which method(s) would be best for your group, city, state, and/or nation, cf., [315, 681, 593, 594]. Which methods might lead to more than two viable political parties?

A Second Modest Proposal. The definition of a *plutocracy* is one where the wealthy rule. One might call this type of government "one dollar one vote" government. We might call a democracy "one person one vote" government. Each election costs more and more money. Even midterm U.S. elections now cost in excess of two billion dollars, and new records are set with each new round of elections. Most candidates have to spend a great deal of their time raising money, just to run for office. There is ample data showing correlations between campaign contributions from X and votes of elected officials funneling profits back to X, cf., [668, 636, 395, 396, 397, 398].

In the states of Maine and Arizona, for example, a system of public funding for elections has been voted in by the electorate. Sometimes the publicly funded candidate wins, other times the privately funded candidate wins. My modest proposal would be to make elections a public, as opposed to a privatized, affair, by universalizing something similar to the Arizona or Maine model. Such a system might harness the energy of the wisdom of crowds and self-organizing systems. On the contrary, as an example, if too few people, with too narrow vision and economic interest, control the processes of government and society, "we" may choose to fuel the future with fossil fuels, and thus possibly bring on the end of complex civilization prematurely and unnecessarily. If those with access to the most money decide our governance, we most likely will have a WACU future, cf., Chapter 9. This is a problem that Nature presents to the entire human race, not just Americans. But what Americans choose to do may have global impacts.

Exercise 24.3 Plutocracy vs. Democracy

(i) On the fuzzy or measured logic scale from 0, for pure plutocracy, to 1, for pure democracy, estimate where the American system of government is at the time you read this. How do different people's estimations differ? What perceptions, what personal conditions color one's answer to this estimation?

(ii) Would either IRV or public financing of elections ultimately change your answer to part (i)?

(iii) Do you have any suggestions for structural changes that would move America closer to a democracy and further from a plutocracy?

(iv) When elections are funded "by the people," i.e., publicly funded, could "the people" actually save more money in other parts of the economy than is spent on elections? (For example, consider health care in the United States in 2010.) What has been the experience of populations that have adopted public funding of elections?

(v) What is the status of public funding of elections at the time you read this? For example, in Arizona it has been a constant struggle to maintain the public system in the face of repeated challenges to it in the courts. Hint: In 2010 the U.S. Supreme Court cut off matching funds for candidates in Arizona's governor's race, invalidating part of Arizona's clean elections law that had been in effect for over 10 years. The U.S. Supreme Court, however, allows unlimited funding from other sources, see page 207.

Part VII

Economics

Chapter 25

What Exactly Is Economics?

25.1 It Takes the Longest Time to Think of the Simplest Things

I will recount some of the history of economics, discuss some very fundamental notions and be guided along the way by the Bio-Copernican Axiom which in this context says that the human economic system, however it is defined, is a part of a larger system, namely Nature. If you do not accept this axiom you can skip ahead to the sections on the mathematics of debt and related subjects; from this mathematics no one can escape as long as there are such things as money, banks, laws, governments, police and prisons.

It follows from the Bio-Copernican axiom that whatever humans do, including whatever economics is, all actions of humans must obey the laws of Nature. For example, gravity exists, and no human system of thought or law, economic or otherwise, can create a zone where the law of gravity does not apply.[1] So far, I have gotten little argument about my claim that the law of gravity is inviolable. The same is not true for what I am about to discuss, however.

25.2 A Preview of Two Laws of Nature

Let me preview a subject I will discuss in more depth later on: the laws of thermodynamics. The first law is rather simple, it is the law of conservation of energy, which briefly says that in a closed box the total amount of energy and matter is constant. You can't make something out of nothing. The second law is a bit more subtle, and I can (and will) state it in several ways. For now, let me explain it by pointing out the difference in the heat energy in the air in a closed room with no wind, and the energy contained in a good,

[1]Scientists are trained never to say never. However, until someone exhibits a place where the law of gravity does not apply I will assume that this law is universal, at least on earth.

stiff breeze outside. Clearly we can and do build machines that can extract useful energy from wind. I have been advocating doing more of this most of my life. However, the second law of thermodynamics says that you cannot build a little car that converts the heat energy in a closed room[2] into directed mechanical motion of the car from some point A to another point B.

Scientists invented thermodynamics before the molecular theory of matter. The idea that all matter, like air, for example, is made up of molecules is an extremely powerful tool.[3] The scientists and engineers that first formulated the laws of thermodynamics did not have this molecular-matter-theory-tool to work with. They directly looked at Nature and observed some patterns empirically, and they recorded these patterns using mathematics. Today, however, we understand that the air outside, or in a closed room, is made up of molecules of various types, O_2, N_2 and so on. The difference between the heat energy in the air in a closed room with no wind, and the energy in the wind outside is in: *organization*, or *order*. The heat energy in the air in a room is really there, but it is the energy of motion of many molecules bouncing around and against each other with little order to it, i.e., randomly. Wind, on the other hand, is the motion of air molecules in some direction. This motion is more organized than the random motion of air molecules in our closed room; it exhibits a certain amount of order.

Now by historical accident, we do not have a good name for the amount of organization or order that a system of matter and energy has.[4] We do, however, have a name for the amount of disorganization, or disorder a certain system has. It is called *entropy*. A system with low entropy is more organized, has more order than a system with high entropy. Later I will give some entropy-related formulas that we can use to analyze "heat-engine" systems, like hurricanes, from the point of view of thermodynamics.

How do the laws of thermodynamics stack up when compared to, say, the law of gravity? Allow me to quote an authority on these subjects, Albert Einstein: *A theory is the more impressive, the greater the simplicity of its premises is, the more different kinds of things it relates, and the more extended is its area of applicability. Therefore the deep impression that classical thermodynamics made upon me. It is the only physical theory of universal content which I am convinced that, within the framework of applicability of its basic concepts, it will never be overthrown.*

Some Economists Do Not Agree. Now compare the above discussion to the following experience I had after I gave a lecture, based on this chapter of my book, at a conference wherein there were mathematicians, economists, biologists ... in the audience. At the end of my talk a well respected economist,

---

[2] I am assuming the temperature throughout the room is the same.

[3] Using this idea and a good deal of mathematics, we can get a much deeper understanding of where the laws of thermodynamics come from and what they mean. I will touch on this later on in this chapter.

[4] Some people have tried to introduce the concept of negative entropy or negentropy, but such terms have never garnered the recognition that the term entropy has historically.

head of economic research for a well respected organization headquartered in Washington D.C. came up to me and said: "The second law of thermodynamics applies in physics, but it does not apply to humans."

On another occasion I had a conversation with a professor of economics who told me the following. As an undergraduate he had studied what I will be calling ecological economics, a cornerstone of which is acknowledging that humans must obey the laws of Nature, the laws of thermodynamics in particular. He was eager to pursue this line of reasoning as a graduate student and was accepted into the economics department of what is considered to be one of the best universities in the world. One day early in his studies in (the successful) pursuit of a doctorate in economics he was told by his adviser: "If you mention entropy one more time, I am turning around and never speaking to you again."[5]

Exercise 25.1 Will We Insert Our Economics into Nature, or Will Nature Do It for Us?

(i) Economist Kenneth Boulding (1910–1993) said: *"Anyone who believes exponential growth can go on forever in a finite world is either a madman or an economist."* The absolute size of the human economy as measured by its impacts on Nature has been growing exponentially for some time. When do you think the size of the human economy will be great enough that our economic theories will be required to take into account the realities of these impacts? Has it already happened? Will it happen? When? Explain as best you can.

(ii) Physicist Max Planck (1858–1947), referring to the then newly emerging science of quantum mechanics, said: *"A new truth does not triumph by convincing its opponents and making them see the light but rather because its opponents eventually die and a new generation grows up that is familiar with it."* Do you think Planck's comments apply to other academic fields besides physics? Discuss.

25.3 Three Kinds of Economists

I will now give a simplified classification of economists that is based solely on the degree of acceptance of the Bio-Copernican Axiom. For the time being I will not be considering any other adjectives such as Marxist, Capitalist, Neoliberal, Keynesian, ..., for these are mostly descriptors of how humans relate to other humans.[6] I want to focus for now on how humans think they relate to all of Nature, which includes humans but much more. If you assume the Bio-Copernican Axiom then we humans have no choice but to obey the

[5] I count among my friends at least a couple "ecological economists." I am told, and it appears to be true, that it is difficult to be an ecological economist and get a job as an economist.

[6] The way humans treat other humans can vary a lot, depending on what various folks are willing to put up with. There is lots of room for choices here on how we organize our economic system—how we treat each other. I deal with this briefly, later.

laws of Nature, like gravity, thermodynamics, and Arrhenius' law, no matter what our mental models are.

Classical Economists. I define the *classical economist* as one who omits Nature, the ambient environment, from their mental model—sometimes in a most spectacular fashion. Classical economists can use the fanciest of mathematics to study the most important and complex of topics, but they do not acknowledge that Nature exists. For example, I have encountered economic theories with mathematical equations devoid of variables representing resources, the stuff humans use labor on to create capital and goods and to perform services. Classical economists do not necessarily accept the fact that humans need to obey the second law of thermodynamics. The subject is unlikely to come up with them, and if it does they are likely to declare such laws of Nature to be irrelevant to the study of economics.

This is understandable, since for most of human history resources were readily available, essentially an unbounded supply; and Nature processed our waste products without humans giving it much thought, essentially a limitless sink. Local exceptions cropped up from time to time, as when the Romans deforested what is now England or when people dropped dead from rat-flea borne plagues in garbage strewn cities. But by and large for most of human history Nature took care of business without our acknowledgment. For the classical economists the ambient environment can be ignored with impunity, human activities are center stage.

The diagram, Figure 25.1, of the "circular flow of value" appears in the first few pages of most introductory textbooks on economics. Now this circulatory

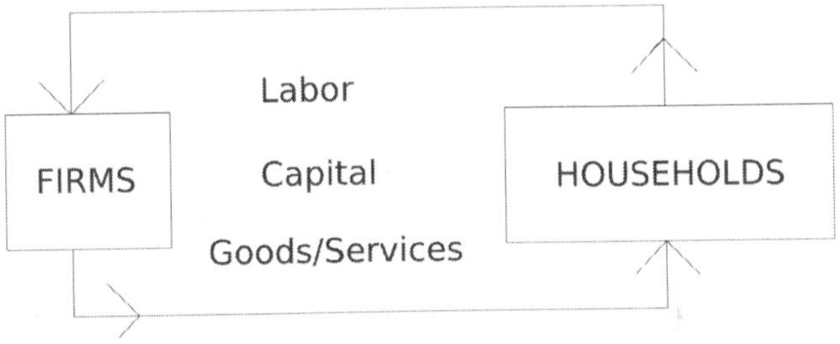

FIGURE 25.1: Circulatory System of the Human Economy

system does indeed exist, but something important is missing. Can you guess what it is?

Environmental Economists. Some economists realize that Nature plays a role in the human economy, but they can not give up totally on the implicit assumption that the human economy is the center of the economic universe. The way such a person deals with this is by introducing Nature as an *external-*

ity. Economists define an externality, roughly, as a party outside of the buyers and sellers who is directly affected by the transactions of the buyers and sellers. Nature, if viewed as an externality, is outside the human economy; but it is something to be included in the economic model—off to the side. I define an *environmental economist* as one who acknowledges that Nature exists but considers Nature to be an externality to the human economy,

Ecological Economists. What is missing from the classical circulatory system in Figure 25.1 is, briefly, a digestive tract, cf., Figure 25.2. The human economy takes *inputs* from the environment, called *resources*, uses them and produces *outputs* which include everything the economy creates, including *wastes*. Everything created by the human economy such as new humans,

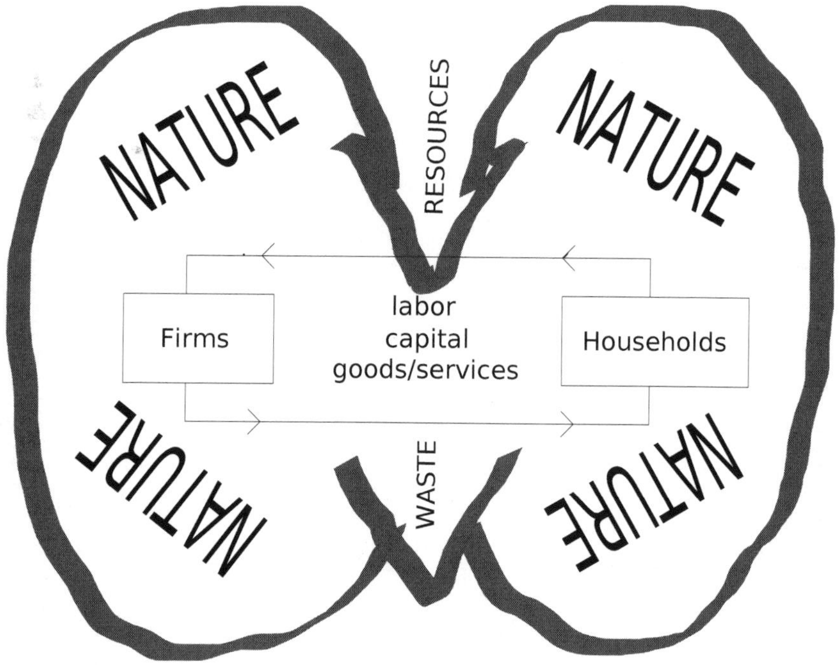

FIGURE 25.2: Human Economic System with a Digestive Tract

manufactured products, food, ...all eventually become wastes, along with the wastes created by the functioning of the economic processes, such as air and water pollution, waste heat, ... Nature, with or without the assistance of the human economy, recycles wastes back into resources of some sort. To do this requires energy, mostly in the form of solar energy.[7] Thus the human economy depends on two cycles, a circulatory cycle and a digestive cycle. Figure 25.2 is a fundamental part of an ecological economist's model of the economy.

25.4 The Human Economy Depends on Nature's Flows of Energy and Entropy

The important bit of science that I have so far not mentioned is that even though all things in Nature are resources for something, e.g., dung is a resource for dung beetles; resources for the human economy have lower entropy than the waste products of the human economy. And what is required to reduce the entropy of wastes so that they can become useful resources is energy and time. Thus the human economy lives not only on *energy flows*, it lives on (reverse) *entropy flows*.[8]

To emphasize the structure just described I make the following definitions.

Definitions: The human economy is a *super organism*, in that it ingests energy and matter resources (with lower entropy) and excretes wastes (energy and matter of higher entropy than that of the resources ingested). The flow of matter and energy through the human economy from a state of resources to a state of higher entropy wastes is called the *throughput* of the human economy.

Exercise 25.2 Low Entropy Resources

(i) An old growth tree is a considerable source of low entropy inputs for the human economy. Why are old growth forests so rare, i.e., why are most of them converted to products and money already?

(ii) Which source of energy is the most concentrated: fossil fuels (like oil, natural gas, coal) or solar energy? Which of these fuels can theoretically be used to generate electricity with the highest efficiency?

(iii) Can whales be viewed as a low entropy resource? What is the status of whales at the time you read this?

(iv) Are large schools of fish an example of a low entropy resource? Peruvian anchovy schools, California and Japanese sardine schools, Southwest African pilchard, North Sea

[7]Humans, of course, have been supplementing solar energy with fossilized solar energy, nuclear energy and some chemical energy—causing some problems in the interim, as has been noted in previous chapters.

[8]What I mean by using the adjective "reverse" is this: As resources flow through the human economic system, there is a net degradation of order, i.e., net entropy increases, and the eventual product is waste. In the long term, the entropy of this waste must be lowered, i.e., waste must be reorganized, so that it can become a resource once again. Sewage and compost become grass and corn, for example; energy and time are expended in this process, most of it by Nature, some of it by humans.

herring and cod off the New England coast of the U.S. are (were?) examples of low entropy resources. What is the status of these fisheries at the time you read this?

(v) Rich mineral deposits are examples of low entropy resources. What is the status of low entropy mineral deposits, such as gold nuggets and gold ores, concentrated iron ore deposits, and so on, at the time you read this?

(vi) If you have the time for a project, plot the entropy of various global resources with respect to time. For example, gold-ore deposits can be measured in terms of the density of gold per tonne of ore and/or the amount of energy needed to extract a gram of gold from said ore.

(vii) Do you think the following is a true statement? Humans tend to exploit the lowest entropy resources available first, then the next lowest entropy resources available and so on.

The above discussion of the role of entropy in the human economy and the associated digestive cycle is either obvious, if you are an ecological economist, or possibly irrelevant/absurd if you are not. This argument has been going on since at least the time of Frederick Soddy.[9] Soddy gives a more scientifically sophisticated analysis than, but in the tradition of, the Physiocrats.[10] Soddy said that any comprehensive theory of economics must have biophysical laws as its first principles since *"life derives the whole of its physical energy or power ... solely from the inanimate world."* Soddy particularly emphasized the centrality of solar energy in empowering the life process, cf., [374, p. 30].

Soddy's critics maintain that human technological innovations can offset any degradation in the quality of, i.e., increase in the entropy of, resources, and any problem in the treatment of wastes. These critics do have a point in so far as human creativity delivers the goods. For example, it is now possible to communicate over vast distances by means of electronic devices; the pony express is not needed to deliver a simple bit of information any more. However, it is discomforting to note that much of what is touted as human creativity since the time of Soddy is based in large measure on the burning of more fossil fuels.

Thus the critics of ecological economics make the fundamental assumption, e.g., [637, 408], that there are no limits to the positive effect that human innovation can have on the human economic process. More and more can be done with less and less. When one resource is degraded, new technology will either squeeze more utility out of the degraded resource or find a suitable substitute. The problem of the disposition of waste similarly will always find a technological solution. Thus there are no practical limits to growth of the

[9] Frederick Soddy (1877–1956) won the 1921 Nobel Prize in chemistry "for his contributions to our knowledge of the chemistry of radioactive substances, and his investigations into the origin and nature of isotopes." From 1903 on, Soddy maintained an energetic critique of classical economic theory.

[10] The French Physiocrats, were founded by François Quesnay (1694–1774). The Physiocrats believe(d) that the ultimate source of economic value resides in natural resources—in direct proportion to the degree of order stored in said resources and in inverse proportion to the physical cost of finding and extracting that order. Land and solar energy were sources of value for the Physiocrats along with labor. Of course, order can now be actually calculated using the concept of entropy.

human economy. Thus Figure 25.2 is unnecessary, irrelevant. However, as Herman Daly points out in [123, pp. 193–198],[11] if introductory economics texts were to take Nature seriously, all would need to be rewritten. Left fundamentally unanswered are the challenges to the "no limits" conclusion raised by [198, 449, 450, 277], for example.

When a hurricane destroys an entire urban area and its infrastructure the arduous and expensive process of rebuilding is counted on the plus side of the accounting ledger for some economists who measure an entity called the gross national product, GNP.[12] Economic theory and practice preoccupied with humans, neglecting the role of Nature, responds with difficulty if at all, when Nature crushes a whole state's economy for days, as did the "Blizzard of '06" in Colorado.[13] If an economic theory does not take Nature into account, only surprise can be manifested if Nature destroys the human enterprise here, there or perhaps everywhere—and for much longer time periods than that of a blizzard or hurricane. People are part of Nature too. If an economy ignores vast numbers of people, the social structure that makes a viable economy possible can disintegrate. When multitudes of people are economically marginalized, for whatever reason, to the point where they have absolutely nothing left to lose, instead of dying quietly they may attempt to take the food and shelter they need to survive and destroy any remaining semblance of "an economy" in the process.[14]

Exercise 25.3 Are There Limits to Growth?

(i) Given the spectacular growth of the human economy (and population) in the last century or two, is Julian Simon (1932–1998), [637], right; human innovation can overcome any obstacle to human population growth and economic growth?[15]

(ii) In 2006 about half the human population (over 3 billion people) lived on $2 a day or less. In the year 2010 the situation remains similar but with a larger population. How are

---

[11]See Herman Daly's review of, [637], in http://rpuchalsky.home.att.net/sci_env/simon.txt.

[12]The GNP is the total market value of all goods and services produced by a nation in a given interval of time. Thus the services that cleaned up the aftermath of destruction of the World Trade Center, as well as the economic activity generated by a person who hires a lawyer to handle a divorce, while undergoing expensive open heart surgery after totaling a car in a wreck—all count toward a positive GNP, just as would the invention of the electric car, had it been allowed to flourish.

[13]Sufficient snow was dumped on Colorado in the week before Christmas to shut down airports and highways and hence shopping! A rapid "dig out" depended on, guess what, many plows powered by fossil fuels.

[14]The Chinese government, for example, must always monitor the level of discontent of its population, for millions of people can quickly descend into the starvation zone where there is nothing left to lose. In 2006, for example, the Chinese government allowed its version of unionization of workers in such places as Wal-Mart, over the objections of American corporations, such as Nike, Dell, Ford, GE, Microsoft, etc., that complained ferociously to the Chinese government about its socialistic practices, cf., www.jimhightower.com, November 20, 2006. See also [657]. What other "revolutions" have occurred that you are aware of?

[15]Herman Daly responds in http://rpuchalsky.home.att.net/sci_env/simon.txt, and he poses some interesting mathematical challenges regarding the understanding of the concept of "infinite."

things in the year you read this? Is this a fuzzy limit to growth manifesting itself, or not? If not, why is half the world so poor?

(iii) Suppose we reach a population of 10 billion people and 8 billion are in abject poverty, would this be a fuzzy limit to growth manifesting itself or not? Would the extinction of *homo sapiens* indicate any limits to growth? Are there any facts, data, circumstances that could "prove" conclusively that humans are approaching or have passed the limits to growth?[16]

(iv) Does water present any practical limit to growth? How much water are you using per day? Don't forget your ecological footprint calculation, which says that every bite of food you take indirectly uses some nontrivial amount of water, and so on. Indirect use of water is just as relevant as direct uses, such as baths/showers; although I suppose we could give those up. Is there any lower limit on the amount of water a human requires to maintain life? A decent quality of life? Can any such lower limit be used to calculate an upper limit on human population? Are there any substitutes for water? Are there any other "essentials," such as water, necessary for life?

(v) Is there a difference between growth of the human economy in the sense of physical growth, i.e., increased energy and material throughput, and growth in the sense of development, i.e., increase in knowledge, culture, peace and other things which are not primarily made of energy and matter?

(vi) In December 2006 the U.S. mint announced that it costs 1.73 cents to produce a 1-cent coin and 8.34 cents to produce a 5-cent coin.[17] Can this be understood solely in terms of classical economics, cf., Figure 25.1? Can this be understood in terms of ecological economics, cf., Figure 25.2?

(vii) On the day before I first wrote this it was announced in *The New York Times*, that the baiji, a rare nearly blind white dolphin that survived for 20 million years, is effectively extinct. The baiji, i.e., the Yantze River dolphin, would be the first large aquatic mammal driven to extinction since hunting and overfishing killed off the Caribbean monk seal in the 1950s. What are the substitutes for the baiji, the Caribbean monk seal and other species that are going or have gone extinct? Is there any connection between this question and part (iv) above?[18] Is there a number of extinctions beyond which no practical substitutions can be found? For example, consider the decreasing number of species of food that industrial humans rely on, cf., [560]?

Exercise 25.4 Judge for Yourself

(i) Herman Daly quotes,[19] cf., [122, p 208], M.I.T. economist, Lester C. Thurow as follows:

In the context of zero economic growth and other countries, a fallacious "impossibility argument" is often made to demonstrate the need for zero economic growth. The argument starts with a question. How many tons of this or that non-renewable resource would the world need if everyone in the world now had the consumption standards enjoyed by those in the U.S.? The answer is designed to be a mind-boggling number in comparison with

---

[16] I would venture to say that for some people no such proof, in the sense of convincing argument, can be given. To the extent that such mental models "run human societies/behavior," and if there are in fact limits to human economic growth that Nature is enforcing (or will enforce), then much more misery is in store for much of humanity.

[17] The reason is the jump in the months previous in the cost of metal, particularly zinc and copper. The penny and the nickle are mostly zinc and copper. The nickel is 75% copper and 25% nickel. A single red cent is 97.5% zinc and 2.5% copper. It is now illegal to melt down pennies and nickels to get the metal they are made of.

[18] December 14, 2006. August Pfluger helped organize the search for the dolphins. Human impacts on the dolphin, from ship traffic and overfishing to pollution, are the causes given for the extinction.

[19] Thurow's quote is from his book: *The Zero-Sum Society*.

current supplies of such resources. The problem with both the question and the answer is that it assumes that the rest of the world is going to achieve the consumption standards of the average American without at the same time achieving the productivity standards of the average American. This is, of course, algebraically impossible. The world can consume only what it can produce. When the rest of the world has consumption standards equal to those of the U.S., it will be producing at the same rate and providing as much of an increment to the world-wide supplies of goods and services as it does to the demands for goods and services.

Based on this quote alone (which is not fair, of course) would you say that Professor Thurow is or is not an ecological economist? Discuss your reasoning fully.

ii) Briefly research the books of economist Manfred Max-Neef and decide if he is an ecological economist.

25.5 Nature's Services and Human Wealth: Important Calculations

No matter what type of economist one might be, or what type of human one might be for that matter, it should be of interest to actually look quantitatively at the bigger picture, Figure 25.2. What exactly are the contributions that Nature is making to the human economy? How are these contributions to be measured? Could humans dispense with Nature's contributions by performing Nature's services? Would humans really want to, even if it this were possible?

As pointed out in [121, 478, 92], Nature really is performing vital services for the human economy, and we will take a brief look at a few of them. A hardcore ecological economist would demand that these services be measured not only in terms of the current value of money, say the U.S. dollar, but also in more absolute biophysical terms such as quantities of energy and so on.

One Very Concrete Example. In the year 2000 it was reported in the mega-media that half of the American population buys bottled water for drinking. Interestingly enough, the city of New York protected its water supply from people by investing $600 million to preserve its watershed in the Catskills in good health. Calculations showed that it was cheaper for the city to let a wilder Nature provide it with water than to allow human development in its watershed and then rely more on science, technology and humans to clean up the water the city needs for drinking and other uses. The alternative water treatment plant would have cost $6 billion.[20]

In 2010 a proposal to drill in the Marcellus Shale for natural gas, which also entails injecting a cocktail of secret chemicals into the ground, was proposed. Despite industry claims to the contrary, such a project would put pristine water sources of New York City and others at risk. Does this sound like a

[20]According to *The New York Times*, May 20, 1997, B7.

temporary source of energy pitted against a permanent source of drinking water? At the time you read this, who won?

Other Examples. How much is clean water and air worth? In a *New York Times* January 3, 2003, editorial by Mike Dombeck, professor of environmental management at the University of Wisconsin-Stevens Point, and former chief of the United States Forest Service from 1997 to 2001, I read: *"Yet water from our national forests has an economic value of more than $3.7 billion a year, according to a Forest Service report issued in 2000."*

More generally, various dollar estimates of the services performed by the U.S. national forests in terms of cleansing our air and water and in terms of providing habitat for all forms of life are all very substantial. You can estimate these for yourself and see if you agree with other estimates that are around at the time you read this.

In [410, p. 183] James Lovelock, the creator of the Gaia hypothesis,[21] outlines a very interesting calculation. The conclusion is this: *"On this basis a reasonable estimate of the worth of the refrigeration system that is the whole of Amazonia is about $150 trillion."* This dollar estimate is in 1991 dollars, and you should compare this number to total world economic production in a typical year. What do you think that number is? In an age of a global warming, perhaps the value of the refrigeration provided by the world's forests is beyond practical measure. And, of course, the world's forests, tropical and otherwise, provide many other services besides refrigeration. Can you name, say, two others?

The Biggest Business of All. Former U.S. Senator from Colorado, Timothy Wirth[22] in April 2001, stated during a speech:

"The biggest business of all is global ecosystem services."

The interesting thing about this business[23] Wirth mentioned is that Nature clearly creates great wealth, but Nature is not paid for it. One of the greatest mistakes of most human political/economic systems is their failure to acknowledge—or even be aware of—the free services and wealth provided by Nature. The grander the scale of the Human Super-Organism the greater this mistake becomes. In America, the Clean Air Act, the Clean Water Act, the Endangered Species Act and other late 20th century legislative products are hard won attempts to acknowledge Nature's gifts—and to attempt to stop those who would take those gifts from us.

[21] Briefly, this is the hypothesis that the earth's biosphere behaves like a living organism, a super-super organism in our terminology. Although sometimes criticized as unscientific, the Gaia hypothesis has captured many an imagination and encouraged a great deal of useful thought.

[22] No one is "perfect" and I have criticized Senator Wirth upon occasion. However, I and the environment owe Senator Wirth a debt of gratitude for his role in preserving the old growth forest, Bowen Gulch, on the western boundary of Rocky Mountain National Park.

[23] I believe the next biggest businesses are "oil, weapons and drugs." Can you verify or debunk this claim?

This next exercise thus deals with perhaps the greatest contribution bigger picture ecological economics has to give, and economists in particular and humanity in general ignore the value of Nature's services at the peril of us all.

Exercise 25.5 Nature's Services: How Much Are They Worth?

A basic flaw in the accounting methods of our current political/economic system is its failure to acknowledge in any meaningful way the services Nature provides for us. This is due in part to our lack of awareness and the fact that until recently Nature has been able to carry on despite our lack of awareness. Accounting for Nature's services is not trivial, and often a true accounting shows that a human project can cause such a loss in services by Nature so as to render the project uneconomic, i.e., the costs are greater than the benefits. Since decision makers often benefit themselves at the expense of Nature and society at large, uneconomic projects often proceed nevertheless.

In this exercise I attempt to begin an honest, more comprehensive accounting of what Nature does for humanity. One useful reference is [121].

(i) Can you estimate the economic value of Nature's pollinators to humanity? Can humans replace the current pollination services of Nature with human pollination services? If so, what is the cost measured in dollars, energy, time (time in person-hours, for example)?

(ii) When the city of New York purchased a portion of its watershed to prevent other human developments, a cost/benefit calculation was done. This analysis presumably showed that it was cheaper to let Nature provide clean water than to let humans pollute it and then clean it up later. (Even though the human activities provided some benefits.) Can you estimate the value of water purification done by the undisturbed watershed?

(iii) Can you estimate the value of the services provided by large tracts of undisturbed forest ecosystems? Can humans replace these services with human services? At what cost measured in energy, dollars, time?

(iv) Can you estimate the value of services provided by wetlands? Coastal mangroves and marshes? Can humans replace these services with human services? At what cost measured in energy, dollars, time?

(v) Go through the details of Lovelock's analysis of the value of the refrigeration services of Amazonia.

(vi) What is the value of services provided by topsoil? Can humans replace these services? At what cost?

(vii) What is the genetic library of biodiversity and what is it worth?

(viii) List ten other services provided by Nature, whether or not you can estimate their value in dollars.

(ix) In the May 20, 1997 edition of *The New York Times* there was a report on an study by 13 ecologists, economists and geographers headed up by Dr. Robert Costanza, University of Maryland. This study, which appeared in the journal *Nature*, put an annual value of between \$16 trillion and \$54 trillion, with a most likely value of \$33 trillion on a set of 17 ecological services. In that year the gross national product of the entire world was \$18 trillion, i.e., the value of all the goods and services produced by all the people in the world. What is the status of these estimates when you read this?

(x) Are there any services provided by Nature that are priceless, i.e., there is no meaningful way to estimate a dollar value for them, they are not replaceable? Are there things that humans have the capability to do which should just not be done? Discuss examples.

(xi) How is the *radical (ecological) transparency* effort doing? This is a movement to get an assessment of a product's ecological impact/footprint stated right next to the price in \$, i.e., the ecological price next to the asking price for purchase. Who is doing the rating of these products (if it is being done)? For example, see [234].

Exercise 25.6 Natural Capitalism: Paying Nature for What We Get?

(i) Reference [289] outlines how humans might be able to create a sustainable economy by paying closer attention to how Nature operates. There are a great many innovations

FIGURE 25.3: Seeing Nature over the Dollar

both technological and philosophical discussed in this book—many with a mathematical component. Find at least ten concepts mentioned in this book for which mathematics assists in understanding.

(ii) Devise a general method to account for the contribution of Nature's services and human contributions/subtractions to the human political economy. (If you cannot do this try parts (iii) or (iv) which are next.) Can your method be used in place of the Gross Domestic Product?[24]

(iii) Currently humans are not incorporating the full, true costs of burning fossil fuels into the human economy. Can you think of a system that would build all of these true costs into the human economy? Would such a system hasten the replacement of fossil fuels with, say, renewable energy? Incidentally, what are some of the true costs of fossil fuel dependence/use that humanity is not paying for in terms of dollars (or other monetary unit)?[25]

(iv) Currently the world's fisheries are under stress and heading for potential collapse, cf., [736]. Is there a way to include all of the costs to the environment of fishing in the price of fish? Is there a system that will allow sustained fishing for the foreseeable future but not drive one species after another into economic and then biological extinction?

(v) True wilderness by definition is not controlled by, or part of, the human social, political, economic system. Is there a way to acknowledge the services wilderness provides humanity and still have it remain wilderness?

---

[24]The Gross Domestic Product is defined to be the total market value of all goods and services produced within the boundaries of a given nation in a specific time period, such as a year.

[25]Just one cost, often ignored, is mercury pollution. What is your exposure to mercury through food, for example, cf., http://gotmercury.org? What are the effects of mercury on people, life in general?

25.6 How We Treat Each Other: How We Treat Nature – The Tragedy of the Commons

Mutual Coercion Mutually Agreed Upon. One solution to the problems posed in Exercise 25.6 (ii), (iii), (iv) is *mutual coercion mutually agreed upon.* This can take at least two forms: pricing/taxes and/or regulations/laws. The system can work if there is a flow of information sufficient to the task. Thus if everyone realized the true costs of burning fossil fuels,[26] to personal health and to global climate, the majority might be willing to accept a carbon[27] tax—the proceeds of which could be used to make the transition to renewable energy, for example. We might elect people who would sign treaties with other nations to reduce dependence on fossil fuels. We are dealing with a dynamic system: Nature's processes, scientists study of same, flows of information pro and con to people at large, all followed by political action either to obey Nature's laws or not. If the social system does not obey Nature's laws it will pay a price of a different sort, via misery or possible extinction. The aboriginal tribes of Australia who did not invent kinship relations to avoid genetic deterioration, if there were such tribes, surely went extinct, cf., Part III. The challenges for any population may change, but the principle is the same: Humans must obey the laws of Nature, irrespective of their level of understanding of those laws. Our social systems appear to be more complex than, say, that of the Warlpiri. This level of size and complexity is on trial, and the verdict is not looking completely positive at this point.

Privatizing Profits: Commonizing Costs. Dee Hock invented the VISA card, he founded the U.S. VISA organization/VISA credit card company in 1968 and VISA international in 1974. He said: *"Institutions that operate so as to capitalize all gain in the interests of the few, while socializing all loss to the detriment of the many, are ethically, socially and operationally unsound. Yet that is precisely what far too many corporations demand and far too many societies tolerate. It must change."*[28]

---

[26]Of course, this brings us back Chapter [1], Section [1] of this book; how do you know what is true? In the case of climate change/global warming the overwhelming evidence of science says one thing. This information has been given weight equal to that from "dissenters" most of whom investigation has shown to have financial conflicts of interest which should be (have been) pointed out every time they speak (spoke) and write (wrote) on the issue—but it is (was) not.

[27]A carbon tax is one way to convey information during every purchase of fossil fuel, but it is not the only possible way. Rationing is more draconian, but it is a way. If everyone understood and believed the most pessimistic science, we could invent a myriad of social organizations to start building a fossil-fuel-free-future. We would not need taxes or rationing, but then I am talking about a form of Utopia.

[28]I had the privilege of talking to Stephen Viedemann in person at which time he shared with me a personal communication from Dee Hock to Stephen Viedemann of the Noyes Foundation, http://www.noyes.org.

In the economics game, if there are no laws enforced to prevent it, there is a tendency for people to appropriate for themselves those parts of Nature's goods and services that are available to all. In a classic 1968 article, cf., [276], *"The Tragedy of the Commons,"* Garett Hardin paints the following picture.

Imagine a common pasture shared by, say, a dozen herdsmen. This common pasture is currently at or near its capacity to sustain the grazing animals already there. Each herdsman must answer the following question: "What is the utility[29] *to me* of adding one more animal to my herd?"

Let's try to "count or measure" the total utility. From the point of view of the individual herdsman the initial (short-term) utility of adding one animal is +1 since all of the *profits*[30] from introducing the animal go to the private herdsman. However, the benefit of +1 is offset by the negative impact (negative utility) of −1 that the animal has on the pasture through, say, overgrazing. The main point is this: since the pasture is a commons the impact of −1 is shared by all dozen herdsmen and by the whole web of life of the commons—as long as the herdsman is able to and chooses to ignore his responsibilities to the other herdsmen and the web of life. Thus, the costs (to the social and natural environment) of the additional animal are commonized. Our individual herdsman probably sees[31] $\frac{-1}{12}$, more or less, as the cost to him of adding one animal to his herd. Thus, if each individual herdsman takes a narrow, short-term view, each will add an animal, and another, and another...
.

Of course, you being an alert reader have noted that if each of the herdsmen adds an animal, then each will contribute a $\frac{-1}{12}$ utility to each of the others. Thus, each herdsman will see a −1 utility if every herdsman adds one animal. So why do it if the net *long term gain* is 0 (or worse if the pasture collapses)? The key words here are long term. Obviously, each herdsman might say to himself: "I may as well add one animal and get the profits in the *short run* before the collapse (or decline) of the pasture (or before other herdsmen add animals!), because if I don't do it the others will anyway." In the short run the individual privatizes +1 in profit and commonizes −1 in costs. Before the social and/or environmental costs catch up with him, our herdsman will be hoping to eye a greener pasture elsewhere.

---

[29] In (neoclassical) economics utility is defined to be the benefit derived from consuming a good or service. See [270], for example.

[30] Some of my students only measure profit and wealth in terms of money. For example, a student once said: "You don't know that the herdsman made a profit on the animal until he sells it." Though I have not carefully defined profit, it is dangerous to use money as the only tool to measure wealth. A quote of biologist E.O.Wilson makes this point well: *"Every country can be said to have three forms of wealth: material, cultural and biological. The first two we understand very well, because they are the substance of our everyday lives. Biological wealth is taken much less seriously. This is a serious strategic error, one that will be increasingly regretted as time passes."*

[31] How the herdsman calculates this negative utility depends on his psychology, level of understanding and compassion, values and so forth. It is not a very precise figure.

The above argument is not new. As Hardin's article mentions, it was first proposed by a mathematical amateur, William Forster Lloyd, in 1833, as a rebuttal to economist Adam Smith's "invisible hand"[32] argument.

Exercise 25.7 Hidden Assumptions and Consequences

(i) In the above discussion I am assuming that each herdsman in question is selfish or believes that at least some of the other herdsmen will be selfish. I also assumed that the herdsmen are free to act. The tragedy of the commons does not happen with every commons. What alternative assumptions might hold which would avert the tragedy of the commons?

(ii) Is the above conclusion of Hardin changed if the commons is very large and the number of herdsmen is very small, i.e., the collective grazing capacity of all of the herds is not near the carrying capacity of the commons?

(iii) If there are strictly enforced laws that regulate the total number of animals that can be introduced to the commons, does that change the story?

(iv) Do you know of examples of commons where the "tragedy of the commons" is happening now, cf., [21, 609]? Can you give an example of a commons where the tragedy is not happening?

Garrett Hardin points out that disaster can be avoided if the herdsmen adopt a system of *mutual coercion mutually agreed upon.* That is they can pass laws or regulations that all must follow in order to stay within the carrying capacity[33] of the commons.

Of course, the herdsmen might even stop short of using all of the carrying capacity of the pasture for themselves; they could share their environment with the species that called the pasture home before they arrived. Should real herdsman not have such generous inclinations, however, the tragedy of the commons should be avoided at all costs by smart herdsmen any way—if for no other reason than that global human population growth has left us with few if any of those "greener pastures" elsewhere.

I cannot resist telling you an anecdote. I have read about Garrett Hardin's article in several places, I even enjoyed discussing Hardin's article (and other subjects) with Professor Hardin himself in the early 70s when the United States was experiencing an "environmental awakening." I had never, until preparing this section, actually read the original article, however. Since my goal is to write an "honest" book, and since I'm a great believer in direct experience/original sources, I thought that I should go to the library and

[32] Adam Smith (1723–1790), wrote in *Inquiry into the Nature and Causes of the Wealth of Nations* that an individual who "intends only his own gain," is, as it were, "led by an invisible hand to promote ... the public interest." See the next section also. Smith's idea is that individuals acting in their own self-interest will collectively benefit the system (society, earth) as a whole.

[33] The carrying capacity of a pasture in terms of, say, domestic animals, would be the maximum number of domestic animals that the pasture can support on a sustained basis without deterioration. The carrying capacity of the pasture in terms of several species is a little more complicated to define, but the idea put simply is "how much life can the pasture carry on a sustained basis without deterioration?" This number is not necessarily constant in time!

look up the article. When I got to the shelf and pulled out the appropriate volume, I found when I turned to the appropriate page that Hardin's article *had been stolen!* In fact, it had been stolen so many times that there were directions to go to the library desk and ask for "file folder 44 (which contains the article)"—after turning in an I.D. card for security.

The tragedy of the commons is alive and well in my university library. In the following exercises (some of them really research projects) you can examine if the principle of privatizing profits while commonizing costs is operating in the world today.

Exercise 25.8 Privatizing Profits, Commonizing Costs

(i) Do you think that if the herdsmen know each other, i.e., they are all from a community smaller than the Dunbar number, cf., Chapter 9, that it is more likely that the tragedy of the commons will be averted? Increasing the number of herdsmen from 147.8 to 1000, how might the psychology of the herdsmen change? How might their connection to the land change?

(ii) If I am a herdsman calculating the net negative utility of all herdsmen adding one more animal I could assume that the commons actually "pays" some fraction of the -1 utility due to each additional animal. If this is the case, then the $+1$ utility of adding one animal will be strictly greater than the sum of all the negative utilities passed on to me by the other herdsman (assuming each herdsman adds one animal). As animals are added, discuss qualitatively how the ability of the commons to pay this negative utility might change. Discuss how the group of herdsmen might be viewing the ability of the commons to pay as they add animals. Do you think that the herdsmen's mental models of the ability of the commons to pay will accurately reflect the actual ability of the commons to pay? Why?

(iii) G. Hardin says in his original article that population growth is an example of the tragedy of the commons. Each parent gets most of the benefits from having a child, while the "commons", i.e., the earth, absorbs the costs. Discuss.

(iv) Is the killing of whales for profit an example of the tragedy of the commons, cf., [64]?

(v) Is fishing in the oceans of the world an example of the tragedy of the commons, cf., [168]?

(vi) Is air pollution of an airshed (or the air around your home) an example of the tragedy of the commons? Is water pollution of an underground aquifer or river an example? Is the pumping of an underground aquifer (such as the "fossil water" of the Ogallala aquifer) a tragedy of the commons?

(vii) Are United States National Parks and Wilderness Areas examples of the tragedy of the commons? Discuss.

(viii) Is the treatment of native forests by ancient and modern civilizations an example of the tragedy of the commons, cf., [531, 726]?

(ix) Is the treatment in general of our public lands[34] in the United States an example of

[34]Most of the public lands in the United States are either under the management of the U.S. Forest Service (in the Department of Agriculture) or the Bureau of Land Management (in the Department of the Interior). These agencies were created in part to stem destructive exploitation of public lands. As with any regulatory body, there is a constant struggle, i.e., "game," for control of the agency between those who would honestly regulate and those who would exploit. Nature has "traditionally used" these, and all lands and waters in fact, as homes for many other species besides humans. Humans subsist, make a living, make a home, take a loss, make a profit, graze animals, grow food, kill wildlife, mine, recreate, cut down forests, thin forests and so forth on public (and private) lands. The quantitative amounts of these human and non-human activities and the quantitative effects

the tragedy of the commons? Are the treatments of some private lands an example of the tragedy of the commons or simply biological tragedies? See, for example, [609, 47, 692].

(x) On November 21, 2004, Congress replaced the Recreation Fee Demonstration Program (Fee Demo) with a Recreation Access Tax (RAT). The government is now charging people just to enter U.S. public lands, and a daily fee is usually more than the government charges for a cow and a calf to graze grass for a month. Some of the RAT money collected will be given to private corporations who will help the government "manage" and "improve" the public lands. Some critics have called this the Disneyfication of our public lands. What is the status of our public lands regarding this issue when you read this? Does privatization of public lands (even partially) change them in any way?[35]

(xi) Two different people can look at the same piece of art and one will see a masterpiece the other a piece of junk. Two different people can look at the same situation and one will see a biological tragedy the other might see a wonderful, healthy economy. Some people really believe that Adam Smith's principle of the invisible hand (that people will act rationally as individuals and will do what is best for themselves, and this will automatically benefit the whole) is the dominant principle that "explains economic prosperity." Some believe that the Lloyd-Hardin principle of the tragedy of the commons (privatizing profits while commonizing costs) is the dominant principle that explains the "unraveling of the web of life both locally and globally."

Using the data you have and your beliefs, evaluate each of the above examples of "tragedies." Are they tragedies? Are they examples of economic prosperity? Is each of the above situations in question sustainable? For each of the examples above discuss: Does either "the invisible hand principle" or "the tragedy of the commons principle" dominate? Does some mixture of the two principles operate? Are there other principles we haven't thought of yet at work? If you think that some of the above are tragedies, how do you suggest that the tragedies be avoided? Whatever your beliefs and actions regarding the above ten part exercise, there probably is a range of "correct" answers and Nature is doing the grading. What "grade" do you think you (your class, your civilization) will be given by Nature?

(xii) For a model wherein a commons is successfully managed by groups using it see the works of Elinor Ostrom, cf., page 482, [508, 509].

Exercise 25.9 How Are The Factory Versus Family Pig Farms Doing?

In March 1999 Jim Hightower stated: "In North Carolina, for example, corporate hog factories have caused almost a 300-percent increase in the number of hogs—and the state is now awash in 38 billion pounds of feces and urine from these porkers. That's more waste than the City of Charlotte generates in 58 years!" Look up the population of Charlotte and check to see if Hightower's claim is accurate.

Further on in this same program Hightower says: "The second mess being made is economic. The factory facilities are churning out such an oversupply of hogs that the price paid to independent farmers by the packing houses has plummeted to Depression-era levels. These farmers are getting $20 for an animal that cost more than $100 to produce—so thousands of them are being squeezed out of business. The factory farms, however, can survive because they are being paid a much higher price by the packing houses. Why? Because the packing houses often own a good-sized chunk of the corporate hog operation. Meanwhile, these monopolistic meat packers are able to keep charging us customers a high price for pork."

(i) Estimate the money the factory hog farms save by not treating their sewage.

---

of these activities are very important subjects that I hope you will investigate. Very often the amount of human activity in an area makes it impossible for Nature to carry out its services for all life, cf., Exercise 25.5, or for some species to live at all. The middle of a freeway is not very hospitable to many species (maybe some bacteria?).

[35]See http://www.wildwilderness.org/.

(ii) What percentage of pork in the United States is produced on factory farms when you read this? Are there any independent, family pig farmers left?

Exercise 25.10 A Hog Report on U.S. Public Forests

On November 1, 1999 Jim Hightower's radio commentary went like this:

"Time for another Hightower radio hog report. These hogs are the giant timber and paper companies. They're granted logging privileges in America's national forests. They're granted much more than the privilege to log, however. We tax payers also provide millions of dollars in annual subsidies by letting the corporations take our trees at way below the market price. Plus, we even build thousands of miles of logging roads in the forests to make it easier and cheaper for them to take the timber out.

"Now Mother Jones *magazine reports that the U.S. Forest Service is providing an additional, high tech, cosmetic subsidy for them. It amounts to a corporate P.R. scam paid for by you and me. It's a 3 million dollar state-of-the-art computer software program that helps the loggers hide their environment-destroying clearcutting practices from the public eye.*

"Basically it's a computerized game of hide and seek to reduce the public's outrage at coming to a national forest and seeing acres and acres of stumps where majestic thousand year old trees once stood. Computer programs with names like SMART FOREST and VIRTUAL FOREST allow the corporations to analyze the terrain so their clearcuts can be done behind ridges, down in ravines and on the back sides of mountains. These out of sight out of mind software programs are being put in all 800 Forest Service offices.

"The Forest Service also helps the companies create beauty strips along the tourist roads, a veneer of trees that look great as you drive along; but behind the strip and beyond your view is the distressing sight of our magnificent forests entirely stripped of their trees. All of this is the Forest Service's way of letting us have our forests while letting the timber companies eat them."[36]

(i) Discuss the above situation in light of the tragedy of the commons discussion.

(ii) How many dollars a year has the Forest Service of the U.S. used to subsidize the building of roads in our national forests so that timber cutting operations can be carried out?

(iii) Compare your answer to part (ii) with the amount of money brought in by charging Americans for access to public lands.

[36] A related book is: *Beyond the Beauty Strip: Saving What's Left of Our Forests*, [384], by Mitch Lansky. Lansky's home was burned down by people who disliked his activism on behalf of the forests.

Chapter 26

Mathematical Concepts and Economics

The late economist, Kenneth Boulding, was a professor at my university for many years. He was quite original and fundamental in his thinking; and he was sufficiently well-respected to be nominated for a Nobel Prize. He is the one who said it takes the longest time to think of the simplest things. One way to interpret this saying from the point of view of this text is as follows. Finding patterns in a complex situation requires stripping away all "nonessential" information, in order to get to the simple heart of the matter. Finding this "simple core pattern" takes much time and experience, as history records. Boulding was the originator of the term "throughput" and can be considered one of the ecological economists. His philosophy of economics was well connected with Nature, and he had a profound understanding of the interconnectedness of many disciplines.

Some might find it surprising that I agree wholeheartedly with another famous Boulding quote: *"Mathematics brought rigor to Economics. Unfortunately, it also brought mortis."*

26.1 Misapplied Mathematics

The mathematics that Boulding is referring to is primarily calculus, differential equations and similar mathematics which was created in large part to describe patterns of motion such as are found in the orbits of planets around the sun, and the trajectories of rockets. Economists took mathematics off the classical physics shelf, mathematics that described systems with, as Boulding referred to them, stable parameters. They then applied it to economics, instead of looking at economics and letting economics tell them what patterns, hence what mathematics, is best to describe their discipline. Boulding identified, correctly in my mind, the key difficulty in applying "classical physics math" to economics. Life, the economy, they are complex systems which in general do not have "stable parameters"; economics and life are most interesting when they are not in "equilibrium." Intuitively what this is saying is that the patterns observed in the economy and in life in general are not the patterns of planets going around the sun or the smooth trajectories of missiles. The patterns of life are much more interesting and complex than that!

The stock market, for example, can be in a state of equilibrium or near equilibrium. In situations like this, the "classical physics math" is suitable. However, the stock market is much more exhilarating if it is booming, or terrifying if it is crashing. In such nonequilibrium cases classical physics mathematics can be totally inadequate. An example which illustrates what I am saying is the story of the collapse of the trillion-dollar hedge fund, Long Term-Capital Management. A history of this example/story can be found in [414].

Very briefly, Black-Scholes Theory won the Nobel Prize in economics.[1] This was a mathematical theory which was being used quite successfully to manage large amounts of money, and make large amounts of money. But if you look carefully at the theory you will see phrases such as "continuous time model" and "differential equations" and so forth.

This is mathematics largely off the classical physics shelf. There are many situations in the stock market where the hypotheses of this theory are satisfied sufficiently well, so that the conclusions can be trusted. But then there are many situations where the hypotheses are not so satisfied, and the conclusions can fail.[2]

I bring this up so that you do not fall into the trap of believing that just because deep and fancy mathematics and/or computers are used to do some calculations, that the calculations will agree with reality. You must understand the logical structure at some level, and you must make sure that the hypotheses of the theorems used are actually satisfied by the real world situation in which you are using them.

It is much better to have a deep intuitive grasp of the patterns involved in a given situation, say in economics, than a superficial understanding expressed in terms of fancy mathematics. In such situations you are likely to be luckier with an experienced intuition than a blind calculator. Or as the sign said that hung over the door of the room where I took my sophomore physics exams:

Chance favors only the prepared mind.

Louis Pasteur[3]

While the following observation will surely not pass without controversy, allow me to quote the opinion of a successful, intuitive perceiver of patterns in finance, cf., [654, pp. 40–41]. The following quote of George Soros, one of

[1]Robert K. Merton and Myron S. Scholes won the Nobel Prize in Economics for the Black-Scholes Theory. Merton tackled a problem that had been partially solved earlier by Black and Scholes, and Merton solved it.

[2]It turns out that the mathematics of Section 19.8, if I were to have continued it for 50 more pages or so, would be closer to what is required than classical physics mathematics. A book which begins this discussion is [653].

[3]In the original French the quote is: "Dan les champs de l'observation, le hasard ne favorise que les esprits préparés."

the world's richest men, is about the very precise but misapplied mathematics just mentioned, i.e., Black-Scholes.

I have to confess that I am not familiar with the prevailing theories about efficient markets and rational expectations. I consider them irrelevant and I never bothered to study them because I seemed to get along quite well without them—which was perhaps just as well, judging by the recent collapse of Long-Term Capital Management, a hedge fund whose managers aimed to profit from the application of modern equilibrium theory and whose arbitrage strategies were inspired, in part, by the joint winners of the 1997 economics Nobel Prize—a prize they won for their theoretical work on options pricing. The fact that some successful participants in financial markets have found modern theories supposedly explaining how financial markets function completely useless may be considered a scathing criticism in itself but it does not quite amount to a formal demonstration of their inadequacy. The failure of Long-Term Capital Management is much more conclusive.

26.2 New Mathematical Patterns: Self-Organizing Systems

One can think of the mathematics of classical physics, the math of the orbits of planets[4] and such, as deterministic mathematics. If you know everything there is to know about an object, like the precise position and momentum of some object at some precise time, then you know precisely the future path of that object for all time, according to the classical theory. However, the mathematics that I got tantalizingly close to in Section 19.8, called chaos theory, says that even in apparently deterministic systems, the tiniest of changes in the initial position and momentum of some object will lead to arbitrarily large deviations from the previously predicted future path of the object. Since no one can measure the initial position and momentum of any object with perfect precision[5] we really cannot perfectly predict the future path of any object for the long term.[6]

There are two extremes: completely predictable, i.e., deterministic systems, and completely unpredictable, i.e., "random," systems. Boulding thought of economies as evolutionary systems, not deterministic solar systems. The mathematics necessary to describe evolutionary systems was not yet on the shelf; it is just now beginning to appear—making this an exciting time in

[4]Calculus, differential equations, analysis are names given to this type of mathematics.

[5]Even in classical mechanics this is not possible, because every measurement in real life will have at least a tiny bit of error. But things are worse than that, however, since quantum mechanics tells us that it actually is impossible—even in theory—to measure the position and momentum of some object with perfect precision.

[6]With huge objects like planets, however, you can do pretty well; hence the high degree of accuracy of astronomical calculations.

mathematics. It turns out that systems like life and economics are neither deterministic nor random; they are complex systems that occur in the fuzzy middle ground between these two types of systems.

Complex systems, while not completely understood, exhibit certain properties that are understood. The behavior of a complex system cannot be predicted by simply "adding up" the properties of the individual parts that make up the system. Complex systems are *nonlinear*.[7] The earth's biosphere is a complex system. Who could have predicted the diverse forms of life—an elephant, a human, a bee, a dinosaur, a flowering tree—during the Archean eon some 4 billion years ago? Another property of complex systems, already noted, is that small changes can develop over time into large changes, much larger than would have been predicted if the system were simple and linear. Finally, a complex system might be made up of individuals, like atoms in a crystal, with simple interactions. Such a system is predominantly deterministic. As noted, another extreme could be a system of individual atoms bouncing around randomly in a plasma or gas, quite nondeterministic, yet with individual atoms relating to each other rather simply. Complex interaction among individuals, i.e., "life," is most likely to occur between these two extremes, and it seems to develop in time by means of three principles or patterns.

Variation, Selection, Amplification. In [157] Nobelist, Christian de Duve, panoramically discusses three basic patterns that arise in trying to understand the system of life on earth: variation, selection and amplification. These three basic mathematical patterns, and possibly others, are associated with the development of complex systems, such as economic systems, political systems, and life itself. De Duve puts forth much evidence that life is not only mathematically possible, it is virtually a certain conclusion of the structure of the universe. In a similar vein, see [160, 340, 78]. For a contrary view, see [700].[8]

Though much remains to be discovered and imagined, there is compelling evidence that a process of self-organization can proceed from the smallest

[7] In a linear system if a certain stimulus gets a response of "X," then doubling the stimulus will double the response to "X+X." In a nonlinear system doubling the stimulus can evoke a response many times greater, or smaller than "X." For example, Mr. Latimer, page 73, took a medication for ulcers and sprayed a pesticide on his lawn. Neither of these alone would have altered Mr. Latimer's life significantly. Together, his life was nearly destroyed.

[8] As for the controversy between theories of evolution and religion, there is a simple answer from the view of this text. If your mental model works for you, then enjoy. From the point of view of mathematics, postulating that life was created by God, whether true or not, is not very interesting; because the complexity of life is explained by postulating the existence of an even more complex entity. Evolutionary theories, on the other hand, are mathematically interesting because they attempt to explain how complex systems can arise from simpler ones. Also, even for adherents to the "God did it" belief, the curious among them ask the further question: "How did God do it?" The fact that evolutionary theories are not perfect, have gaps and so on, make such theories very much like the rest of science: incomplete. Nature always has more exercises for the reader. Interestingly, various religions have themselves evolved over time.

scales on up, starting on the atomic scale, passing eventually to complex molecules including amino and nucleic acids, proteins, Ribonucleic acid (RNA), Dioxyribonucleic acid (DNA), then on to protocells, single cells, multicellular organisms, multicellular organisms with neural networks, including us. And beyond?

At each stage a part of Nature provides (provided) a variety of entities via processes which were neither purely random nor purely deterministic. A careful, complete (in-so-far as it can be complete) and rigorous analysis of the origins and evolution of life requires some simple yet subtle mathematics. The following is a key observation. If one assumes that Nature came up with variant life forms by way of a purely random process, the estimated age of the earth of around 4 billion years is not long enough to guarantee self-organization and life. The other extreme assumption, that the forms of life are the product of a purely deterministic process does not work for different reasons; it is inflexible and cannot adapt to a changing environment. The fuzzy middle ground between pure randomness and pure determinism miraculously allows life to self-organize in the time allotted, and to adapt to changing circumstances. The possible variety of entities (in this case, life forms) produced at each stage is neither completely determined nor completely random. And thus there is perhaps room for the free will of individuals?

At each stage of development of a complex system the diverse entities (the variation) interact with their environment, and some "do better than others" in the environment that exists at the time (the selection). The successful entities sustainably reproduce and/or grow in number (the amplification), the unsuccessful may simply hang around in relatively small numbers (waiting for an environment more favorable to them) or disappear.[9]

In the above paragraph I have used the word entities since the patterns of variation, selection and amplification are found not only in the evolution of life forms but in the evolution of our immune system, in the evolution of neural networks (of which our nervous system and brains are a special case) and in the evolution of economic, political, social systems.

Just as single cells somehow "got together" to form multicellular organisms, humans got (get) together to form social systems with many subsystems, of which the political economy is one. Our social systems, like our languages, are self-organizing. For example, English did not come into being by the action of a central authority who determined a complete version of English (together with the rules of English grammar) and who then ordered people to use the version of English presented to them. People presumably have been free to communicate with one another in any manner they wish to try (variation). Some methods worked better than others (selection). Various workable languages propagated (amplification), one of which was English.

[9] Or as Dr. Ian Malcolm, the mathematician in *Jurassic Park (1993)* says "... life ... finds a way." He also said, in the same movie, "Boy, do I hate being right all the time."

Exercise 26.1 Systems Self-Organize Via Variation, Selection, Amplification

(i) Simon Winchester's book *The Professor and the Madman: A Tale of Murder, Insanity, and the Making of the Oxford English Dictionary*, Harperperennial, 1999, tells the very interesting story of how the preeminent standard/authority on the English language came into being. Two observations: English was in use for centuries before the dictionary was written. The dictionary is continually revised to keep up with the development of the English language! Discuss how the dictionary which defines standard English can change.[10]

(ii) Consider the concept economists refer to as the market. Is the market a self-organizing system? Discuss.

(iii) There is much more to be said—and discovered—about how complexity can arise from simpler constituents. Investigate what is known, cf., the work of Stuart Kauffman for example, [340], and what has been discovered by the time you read this.

(iv) Elinor Ostrom received (half of) the Nobel Prize in Economics in 2009, although she is a political scientist. What are her "design principles" for how local communities should manage their "commons?" Does Ostrom's work provide an alternative (successful) model to Hardin's "Tragedy of the Commons," cf., page 471. How does she suggest that humans deal with climate change? How does her research relate to the mathematics of self-organizing systems? See, for example, [508, 509].

For further related reading from a variety of fields consider [340, 333, 740, 506, 650, 710, 621]. Also search under topics such as self-organization, emergence, chaos, complexity.

26.3 Finding a Niche: Habits and Habitats

Every living organism in Nature occupies some niche in the "web of life." Where and how it ingests and excretes defines the niche, or place of the organism in the larger web; its job, if you will. While the life of any human is necessarily supported by Nature, it is becoming increasingly rare—but not impossible—for an individual person, or small group of people, to extract a living directly from Nature, or to live autonomously, not subject to the will of a much larger central power.[11]

Habits and Habitats. Just about everyone finds themselves in a habitat seemingly defined almost entirely by other humans. We humans change our habitats, then our habitats change us. This is a fundamental feedback loop which may or may not be stable. Many have made this observation before:

[10] Hint: use the concept of feedback loop.

[11] The U'wa, Kayapo, Yanomami, Quichua, Achuar of South America, the Ogiek of Kenya, the San of southern Africa, the Penan of Borneo, the Amungme of West Papua, the Innuit, Gwich'in and Saami of the North, the traditional Hopi and Diné (Navajo) of the American Southwest, the Himba of Namibia and other indigenous peoples around the world are under increasing pressure to "assimilate" into "modern" civilization if they have not already. Usually the cultures of these indigenous people have gone largely unaltered until "modern" civilization discovers that said indigenous people are in the way of the extraction of resources that "modern" civilization wants. Chiapas is an example of an autonomous region in Mexico, cf., [586].

Winston Churchill (1874–1965), a famous Prime Minister of England, said: *"We shape our buildings and afterwards our buildings shape our world."*

Frederick Douglass (1817–95), a famous American antislavery leader said: *"A man is worked on by what he works on. He may carve out his circumstances, but his circumstances will carve him out as well."*

And our habitats create responses in each of us: our habits. Or as William James says in *The Principles of Psychology* (1890): *"When we look at living creatures ... one of the first things that strike us is that they are bundles of habits."*

We live in cultures which encapsulate our collective habits. Our created habitats may appear to have insulated us from Nature; but regardless of our political, economic or other theories, we are often moments away from an abrupt environmental reminder that we depend on Nature for our lives.[12] More dangerous, perhaps, are less obvious, less immediate cultural contradictions with Nature's rules. For example, we have built a habitat of suburbs

[12]A sequence of earthquakes and tsunamis around the world in the first decade of the twenty-first century, come to mind. Less dramatically many people have experienced situations where their habits have not responded quickly enough to a changing environment. When that first snow storm hits in Colorado after a long drought, and the streets go from dry to glare ice in a few minutes—some people crash because dry road habits do not work on ice.

I note with interest stories of people who venture into America's backcountry equipped with the latest electronic devices—cellular phones and/or GPS devices—but perhaps without a compass, a map or experience. Batteries go dead, technology falters, disaster awaits when centuries of simple common sense habits are replaced by television-induced fantasies in the minds of people more accustomed to walking around in malls than in wilderness. See, for example, a front page story on this subject in the October 8, 1996 *The Wall Street Journal*.

In 1977 I went on an expedition to eastern Canada with three friends. While in Nain, Labrador, we visited with some Inuit families. One night an elder villager recounted a winter tragedy that had recently befallen Nain's best group of hunters. The hunters had just switched from dog sleds to snowmobiles and from traditional dress to modern snowmobile outfits. Ninety miles out on frozen ocean ice their snowmobiles quit, most likely due to dirt in the fuel. The long walk back was not welcome; but surviving the ordeal would ordinarily have been well within the ability of these strong, young hunters. Our eyes were riveted on the elder as he recounted with hand motions and a sad voice the details of how the hunters died. A simple difference in design between the modern and traditional dress caused the hunters to lose the use of their hands to frostbite. The old skin suits had sleeves that extended below the hands and shed freezing rain and snow. The modern snowmobile suit has a gauntleted glove which captures snow, sleet and freezing rain while walking. Another complication: sled dogs can be eaten if there is no alternative, but snowmobiles are notoriously hard to chew. Habits that do not fit your habitat can kill you.

An outstanding habitat-habit feedback loop is exemplified in some current efforts to save certain endangered species. In 1997 I was privileged to have the information in this paragraph corroborated by a conversation with the founder of conservation biology, Michael Soulé. What has become clear is that any species is not itself if it is removed for some time from its ambient environment. Endangered species kept in zoos for a couple generations often lose natural environment connections necessary for their continued existence. The genes of an organism can be intact while the organism fails to survive when reintroduced into its once native habitat. It turns out that much behavior—for mammals, birds and so on—is learned from the parents and/or the environment. Thus, for example, a zoo animal often does not recognize its predators—with the obvious consequences.

and agriculture dependent on fossil fuels and the urgency to change our habits and infrastructure has yet to manifest itself.[13]

Exercise 26.2 Your Real Speed in a Car and on a Bicycle

(i) If you have a vehicle, compute the ratio $\frac{D}{T}$, your real average speed, in the following way. Let D equal the total number of miles (or kilometers) that you drove last year. Let T equal the total amount of time (measured in hours) in the last year that you devoted to your vehicle. Include the time you spent in the vehicle, driving, sitting in traffic jams, carpooling. Include the time you spent earning money to pay for the vehicle, pay for the insurance and fuel, plus the hours you spent repairing and maintaining the vehicle (or the hours spent earning money to pay someone else to repair and maintain it). We can call this speed your vehicle's real average speed. Is this real average speed greater or less than the speed you can maintain on a bicycle? If you do not have a vehicle, do the exercise for someone who does. Redo this exercise, if necessary, assuming that you make the minimum wage, which in 2010 was (about) \$7.25/hr.[14] Do the exercise once again assuming that you earn \$203 million a year (or about \$100,000 per hour).[15]

(ii) You can make this an open-ended problem by trying to calculate the total cost of your automobile to the environment as well as its total cost to you. Can you measure environmental costs entirely in terms of dollars?

Exercise 26.3 Habits and the Cost of Gas and Food

(i) Analyze as best you can how your local habitat has changed over the last one hundred years. How have these changes altered the habits of local inhabitants?

(ii) How would your habits change if the price of gasoline were ten times more expensive? One tenth as expensive? How would your habitat change in response to the changes in the price of gasoline?

(iii) How would your habits change if the price of food were ten times more expensive, respectively, one tenth as expensive. This can become an open-ended exercise as you vary more things in your habitat and as you try to do careful quantitative estimates of the effects of making these changes.

---

[13] Some people voluntarily change their lives and habits in order to live in a less hectic habitat. I remember the story of one time professor Scott Nearing, driven from academia. He moved his family to a farm in New England. As he did daily chores around his farm he formed the habit of picking up stones along the way and moving them to where they would be needed for future construction projects, thus avoiding special trips and saving a lot of personal energy.

[14] On October 1, 1996 the minimum wage—at its lowest purchasing power in 40 years—rose from \$4.25 to \$4.75. A second hike, to \$5.15, took effect Sept. 1, 1997. The latest minimum wage legislation included measures which helped compensate corporations for the effects of the raise in the minimum wage. The minimum wage can differ from \$7.25 depending on the state. In my state of Colorado the minimum wage was *lowered* in 2009.

[15] According to the magazine, *Business Week*, April 25, 1994, page 52, the top corporate officer of Walt Disney Corp., Michael Eisner, earned \$203 million in pay and stock grants in 1993. (That comes to about \$100,000 an hour!) On April 27, 1999 Jim Hightower's report disclosed that Mr. Eisner's compensation for the year previous was \$575 million (about \$287,500 per hour!). Mr. Eisner is now retired. The average chief executive officer's compensation in 1999 in the major corporations was about \$10,000,000. In 1999 the average CEO earned 419 times the average worker. The average worker's wage in 1999 was 12% lower than it was 25 years previously. How is this pattern holding up in the 21st century?

Exercise 26.4 What is Your Bioregional IQ?

(i) Compute the fraction $\frac{N}{C}$, where N is the number of native plants that you can name and recognize in the field and C is the number of corporations that you can name and whose corporate logos you can recognize.

(ii) Consider the following test of your bioregional knowledge:[16]

1. What soil series are you standing on?
2. When is the last time a fire burned your area?
3. Name five native edible plants in your region and their seasons of availability.
4. From what direction do winter storms generally come in your region?
5. Where does your garbage go?
6. How long is the growing season where you live?
7. Name five grasses in your area. Are any of them native?
8. Name five resident and five migratory birds in your area?
9. What primary geological event or processes influenced the land from where you live?
10. What species have become extinct in your area?
11. What are the major plant associations[17] in your region?

(iii) Exercise (ii) can be considered a first attempt to answer the question: What is my ecological address? Keep going, be as complete as you can be. For example, ask: What watershed do I live in?

(iv) In the area in which you live, what has been its history for the last 100 years? 1000 years? 10,000 years? one million years? one billion years?[18]

(v) Professor Tom Kelly of Tufts University while speaking at a conference at the University of Colorado in 1996 mentioned the following four states of consciousness relative to our environmental competence: unconscious incompetence, conscious incompetence, conscious competence, unconscious competence. If you interpret the state of unconscious incompetence as having "bad" habits, and the fourth state of unconscious competence as having "good" habits, give an example of a successful transition through these four states—starting with a bad habit, ending with a good habit. Have you ever successfully changed one of your habits, becoming more environmentally competent in the process?

(vi) Have you acquired the habits of attending class and doing homework regularly? What survival value might these habits have?

Exercise 26.5 A Personal Habit

(i) My office partner in graduate school was from a country which found it somewhat strange that we Americans used toilet paper instead of water to clean up after going to the bathroom. He referred to this habit of ours as "dry cleaning." Can you give at least a partial list of countries where dry cleaning is an established habit? Do the same for water cleaning?[19]

(ii) Can you estimate the environmental impact of having all Americans change from "dry cleaning" to water cleaning? For example, how much paper would be saved? Would

---

[16] From *Co-Evolution Quarterly* 32, Winter 1981–2, referenced in [500, p. 137].

[17] As in what grass species are found together in a tall grass prairie, for example; not as in what people associate to study plants.

[18] An example of an answer to this question is the book, Mitchell, John, *Ceremonial Time: Fifteen Thousand Years on One Square Mile*, Houghton Mifflin Company, Boston, 1984. The author charts the history of one square mile of Middlesex County, Massachusetts for the last 15,000 years.

[19] I lived for a short time in a country with the water cleaning habit; and, while at first put off by the idea, I eventually thought of it as more "natural" than dry cleaning.

extra water be used? Compare how much water is used in the manufacturing of a year's worth of toilet paper used in the United States with the amount of water used in the "direct application" method of cleaning up.

(iii) An almost universal habit of industrial civilizations is the mixing together in sewer systems human excretions and toxic wastes together with lots of water as the medium of transport. If someone told you that this is one of the most stupid habits imaginable, what would be your response? How might this habit be changed to a smarter one?

Get a Job: What is Your Socio-Economic Class? Most humans find themselves thoroughly enmeshed in a web of human activity, and in order to eat, a job must be created or found. Unavoidably, the mere act of staying alive involves politics, e.g., interacting with other people and gaining enough power and wealth to survive. Thus economics and politics are inextricably interconnected for all of us at the most basic level. And, of course, the whole human enterprise is embedded in Nature.

One of the most prevalent, recurring patterns in social and economic systems around the world and throughout history is the notion of *class*. One's niche in the web of humanity more than likely belongs to a class defined in part by the amount of wealth and income one has.[20] Wealth is the stock of resources in one's "niche box" and income is the flow into that box. Wealth takes many forms: material possessions, health, education, access to energy and social services and health care, political connections, friendships, family. Two forms of wealth, not easily measured, are infrastructure support provided by Nature in general and the social support of other humans, in particular. Few of us would have as much wealth as we do if we lived isolated lives without the benefits provided by the work done of many other people, seen or unseen by us. The complexity of society in and of itself is a priceless resource.

[20]For some people this is an unpleasant topic; for some it is a forbidden topic; for others it is the only topic. While I was a student in a sociology class a "sociogram" was constructed and I found that I was the lowest member of the class according to family wealth and education indicators of class. I happened to graduate with the highest GPA in my class; thus my concern that anyone who really wants to have an education should be get the opportunity. It is now acknowledged in the megamedia that class in America exists. For example, in the December 6, 2006 edition of *The New York Times*, p. C3, it is noted that 1% of Americans held 32% of the nation's wealth in 2001, and that the top 10% held 69.8% of the wealth. Worldwide the top 1% of the world's population held 40% of the world's total net worth in 2000. This is fleshed out a bit more in [109], where in the United States the top 1% holds 32%, those in the 95th through 98th percentiles claim another 25%. Thus the top 5% holds well over half the wealth of America, while the bottom 50% holds 2.8%. The *NY Times* article does not mention this last fact, but it does mention that the wealth of the average American is nearly $144,000. This is a time honored way of using mathematics. To illustrate this the story is told of nine workers in a bar, with modest means. The richest man in American walks into the bar, net assets well over twice 10 billion dollars. All of a sudden, on average, everyone in the bar is a billionaire!

Chapter 27

The Concept of Money

One of the ways of measuring wealth and income is with *money*. Understanding money takes one far beyond the standard definitions[1] of "medium of exchange," "store of value," "tool for organizing human endeavors." I will only be able to barely begin exploring the meaning of money and its implications for society.[2] What I can tell you is that the invention of money has opened up endless opportunities to play mathematical games.[3] Some of these games can profoundly effect social structure, sometimes negatively, sometimes positively, sometimes in unusual and community building ways. Just in case there is another Great Depression, cf., Chapter 2, the following should be kept in mind: community or local currencies and related projects supporting local self-sufficiency are possible.[4]

27.1 Financial Wealth and Real Wealth

Before the invention of money, wealth was primarily real, in the sense that a person's possessions were real objects, such as stores of grain/food, animals, access to clean water and air, and so on. Such real wealth is subject to decay, in the sense of increasing entropy. Money, on the other hand, can be stored for a much longer period than an apple can survive, or a horse may live. The concept of interest on a debt can also be introduced, about which ecological economist, Frederick Soddy, had this to say: *"You cannot permanently pit an absurd human convention, such as the spontaneous increment of debt [compound interest], against the natural law of the spontaneous decrement of wealth [entropy]."*

[1] See, for example, [712, 209, 324].

[2] For example, while researching for this chapter I discovered that as of March 23, 2006, the government/Federal Reserve stopped reporting on the M3 money supply. It is now a secret! Thus there is no way for me to find out how much U.S. money is being printed. One might ask why the secrecy? The answers are not reassuring.

[3] Ask any student studying "financial math," for example.

[4] For example, local communities can and have created their own currencies/money with positive effect, even in the United States, cf., [652], [401, 402], [148, 149], [253, 255, 254].

Thus somewhat detached from reality, wealth in the form of money, becomes a mathematical abstraction called financial wealth.[5] Computers have only aided this detachment from reality, for with these electronic communications tools trillions of dollars are traded in financial markets every day—far more than the real wealth supposedly represented. In Figure 27.1, I show a simple fundamental relationship between financial wealth, growing at about 10% per year for a period of time, vs. real wealth in the global human economy growing at about 4% per year during the same time interval.

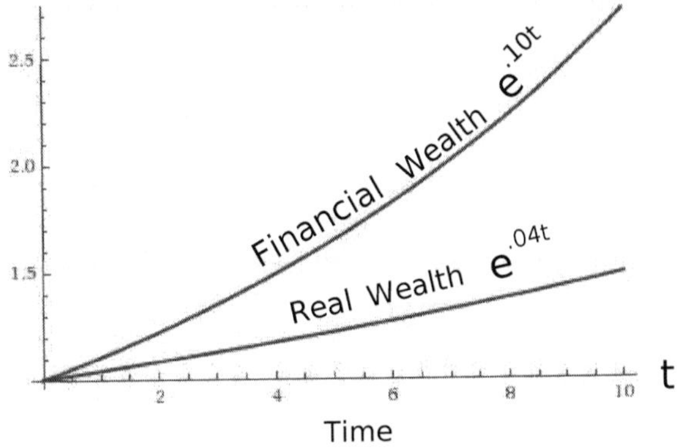

FIGURE 27.1: Financial Wealth and Real Wealth

[5]Money is just one form of financial wealth. There now many forms of financial instruments such as stocks, bonds, options, futures, derivatives, ... In fact, [301, p. 51] lists a whole page of financial innovations. It is important to observe that real wealth as measured by the gross domestic product of the world was estimated, in 1997, for example, to be $18 trillion. The financial instruments that supposedly can be converted to real wealth form a far larger number and have grown increasingly abstract and disassociated from real wealth as the following quote, cf., [548], about one form of financial instrument, derivatives, demonstrates:

Because of their notional quality and because of the secrecy in which they are typically traded, the volume of derivatives is difficult to measure; but taking currency trades, one of their most common forms, as an index, we can begin to glimpse their size. The International Bank of Settlements estimates that in 2001 the total value of derivatives contracts traded approached one hundred trillion dollars, which is approximately the value of the total global manufacturing production for the last millennium. In fact, one reason that derivatives trades have to be electronic instead of involving exchanges of capital is that the sums being circulated exceed the total quantity of the world's physical currencies.

Thus the abstraction of derivatives is even larger than the abstraction of currency. Financial wealth far exceeds the real wealth it represents, but (abstract) financial wealth has very real impacts on all of our lives. Compare with Ponzi schemes in Chapter 2.

Exercise 27.1 Inflation, Corrections and Collapse

In Figure 27.1 both financial and real wealth are growing exponentially, as unconstrained population growth did in Chapter 18. Financial worth tends to follow an exponential curve because investors of financial worth demand 6%, 10%, 20% and more return each year on their investments. The rate of growth demanded of financial capital is usually greater than the rate of growth of the real economy. Hence the graph above.

The important things to note about the graphs in Figure 27.1, however, are that both curves rise as time passes and financial wealth rises faster than real wealth. This is because the real economy is growing and humans have not yet figured out how to have an economy that follows a straight, horizontal line or some varying approximation to such a steady-state economy.

(i) If financial wealth continues to grow faster than real wealth, what is happening to the ratio of financial wealth to real wealth?

(ii) Assuming financial wealth grows faster than real wealth, as time passes how much real wealth does "one dollar," or other financial instrument, represent?

(iii) If financial wealth continues to grow faster than real wealth indefinitely, what will happen to the "real value" of a dollar far in the future, i.e., how much real stuff will you be able to buy in the far future?

(iv) You no doubt have heard of terms such as recession, the Great Depression, market corrections, financial collapse and so on. Graphically each of these economic phenomena would be represented by a decrease in the graph of financial wealth as time passes. In fact, the decrease can be quite abrupt. Can you graph a recession and a depression?

(v) Thus the two graphs in Figure 27.1 either diverge forever, or now and then they come together. This can happen as in part (iv) or it could happen by making the real economy grow faster to keep up with the financial economy. Are there any problems with having the real economy grow at 10% to 20% per year indefinitely?

(v) Is there any problem with real economic growth of 1% per year indefinitely?

(vi) Which is possible, which is not possible on earth: economic development for one hundred thousand years, economic growth for one hundred thousand years?

(vii) How would you redraw Figure 27.1 during a period of deflation?

27.2 Is Financial Collapse Possible Now?

Collapse of the American financial system (and others) did happen during the Great Depression following 1929. It is worth understanding as best we can how this collapse happened. For example, the Glass-Steagall Act of 1933 was a Great Depression era piece of legislation that legally separated the three financial functions: banking, insuring and stock trading. (Why did/does the mixing of these three lead to financial instability? For example, see Chapter 2.) In 1998 the Glass-Steagall Act was repealed and replaced, allowing "big banks" like Citicorps and "big insurance" like Travelers to legally merge.[6] Now such corporations can mix financial operations and share databases as well.

[6]Actually they merged while Glass-Steagall was still in effect, then lobbied to have the law changed.

At the time of the Great Depression investors were allowed to purchase stock by paying a small fraction of the actual worth of the stock. As a result it had long been the law that when you buy a stock you have to pay at least half the price of the stock, borrowing the rest. The so-called Commodity Futures Modernization Act of 2000 allows you to buy stock futures with almost nothing up front, borrowing more than 10 times your own investment.

The repeal of the "New Deal" legislation passed during the administration of President Franklin D. Roosevelt (1933–1945) has been the goal of some Americans ever since the end of World War II. For example, the privatization of Social Security[7] has been one of these goals.

Exercise 27.2 What Was the Great Depression Like? Why Did It Happen?

(i) Research the Glass-Steagall Act and why it was passed to increase financial stability. Research the practice of buying stocks on margin, and why this is a potentially destabilizing practice.

(ii) Research the role of the Robber Barons in the world and U.S. economies in the decades prior to the Great Depression, compare with the 1% of the Occupy Movement of 2011-12.

(iii) If you like the concept of a weekend and a 40 hour work week, find out what social forces brought these into being.

(iv) Research the Bonus Army, a mass of poor folks who marched on Washington D.C. in 1932. Why? Who put them down, cf., [139, 195]?

Exercise 27.3 Is Financial Collapse Possible Now?

(i) Research the Savings and Loan Crisis and/or Scandal of the mid 1980s at least to the extent that you can figure out how much you and/or your family have paid, are paying and are going to pay in taxes because of it. Also, find out who profited from the Savings and Loan Scandal of the mid 1980s.[8]

It sometimes helps to look in detail at one piece of the picture. To this end, research the role Silverado Savings of Denver, Colorado, played in the Savings and Loan Scandal. One

[7]I had a whole chapter on this at one point, since Social Security is perhaps the most successful antipoverty program of all time. The most efficient as well, since it spends less than one cent out of every tax dollar collected on administrative costs. Mathematically there is NO social security crisis. Very simple fixes can be put in place should shortfalls appear at some time in the future. Just about every statement made by would-be Social Security privatizers that had (has) mathematical content was (is) either misleading or outright wrong. See [21, 310, 7]. All Social Security privatization schemes I have seen amount simply to "theft." Also it is rarely mentioned that Social Security is not only a retirement program but also a very effective insurance program. For example, Social Security went quickly, quietly and efficiently to work for the families of the victims of the World Trade Center attacks of Sept. 11, 2001. There seems to be a never ending attack on Social Security, the latest as I write is one disguised as a form of "deficit reduction." A way to begin informing yourself on these various attacks and the abundance of misinformation about Social Security in the media is to go to www.fair.org and search "Social Security."

[8]To get you started I quote from [346, p. 128]: *"Americans are still paying for the $480.9 billion savings and loan debacle, which occurred largely because a Democratic-controlled Congress, inundated with $11.6 million in contributions from the (savings and loan) industry, freed savings and loan associations from traditional constraints in the 1980s."* See also [67]. Unfortunately, the Republicans were also part of the problem—perhaps that is why neither party was eager to bring attention to the scandal.

resource for this is [244, Chapter 8].[9] The Silverado Savings and Loan contributed at least one billion to the $480.9 billion tax burden mentioned in [346]. One important number that you should remember for later on[10] is this: federal regulators before the scandal required a 5 percent minimum capital-to-loan ratio, but Silverado had a ratio of 2.5 percent. Federal regulators subsequently reduced the requirement to 3 percent.

(ii) Research the collapse of the trillion-dollar hedge fund, Long-Term Capital Management in the 1990s, cf., [414].

(iii) Find and research the causes and consequences of a third financial failure. Avoid opportunistic bankruptcies as detailed, for example, in [338].

(iv) On Monday, December 3, 2001 *The New York Times* announced that Enron Corporation had filed for bankruptcy.[11] This is very interesting for many reasons, one of which is that Enron was a principal financial backer of George W. Bush's candidacy for both the Governorship of Texas and the Presidency. The second reason is that Enron was a prime mover for energy deregulation in Texas and nationally. George W. Bush in turn helped Enron reach its objectives while Governor of Texas and also while President. Those objectives are summed up in the words of consumer advocate Douglas Heller: *"Enron spearheaded a deregulation system that added expenses, but no consumer benefits."*[12]

---

[9]Note: DIA refers to Denver International Airport, an entity not unrelated to the Silverado role in the scandal. Answer the question: Why abandon a great airport, Stapleton Airport, relatively close to the city of Denver, for a remote airport in "Tornado Alley," more susceptible to blizzards, and causing millions of extra passenger-car driven miles and air pollution? Your answer will reveal much about modern American politics.

[10]See Fractional Reserve Banking, Section 27.6.

[11]The largest bankruptcy in U.S. history up to that time in the sense that its assets at the time of bankruptcy were $49.8 billion, counting only 13 subsidiaries. The second largest was the bankruptcy of Texaco on April 12, 1987, with assets of $35 billion. The fourth largest was that of Pacific Gas and Electric on April 6, 2001, with assets of $21 billion. This last case is also related to energy deregulation in California. Untold is the financial hardship exacted on the taxpayers and energy rate payers in places like California, cf., page 211.

[12]The feedback loop from Enron to the Bush administration and back allowed the deregulated Enron to take advantage of "market power" and buy low and sell high, e.g., to California. For example, from the December 4, 2001 edition of *The New York Times*, *"As the value of energy contracts soared into hundred of billions of dollars—at one point up to half of all electricity and natural gas transactions passed through Enron's trading operation, by some estimates—Enron scrambled to hold off Congress and regulators."* Campaign contributions did the trick. In the year 2000 Enron was the seventh largest American corporation in terms of revenues, its stock traded for about $90. In December, 2001, Enron stock traded at a fraction of a dollar. In fact, about a dozen shares would not net enough money to pay for a hot dog at Enron Field in Houston, a center of the energy trading business that Enron "pioneered," some say "got away with." You see, they traded energy, they did not produce it.

While Enron arranged for its privacy, it was cooking its books, i.e., not being honest about its financial transactions. In November, 2001, executives in the corporation sold their Enron holdings while the stock price was relatively high, and in the process Enron stock lost about $1.2 billion in market value. Enron employees who did not have access to "inside information" were left with virtually nothing in their Enron stock retirement accounts by the time they found out what was going on.

It is commonly assumed that the CEO of Enron, Kenneth L. Lay, met with Dick Cheney to formulate our national energy policy. It is not known for sure, since President Bush has kept secret all information about the meetings of his administration regarding the formulation of our national energy policy. (President Bush has kept a great deal of information concerning his administration secret. Can you name other things?) One question: Did Dick Cheney meet with any other corporation besides Enron? Has this information ever been revealed—

What is a municipally-owned utility, MUNI? (See Section 28.2.) Enron was an investor-owned utility. Why do you think that the megamedia made almost no mention of the fact the Los Angeles Department of Water and Power is a MUNI and had no problems during the energy crisis in California created by investor-owned utilities?

Very few news sources correctly identified the cause of the California energy crisis of 2000–2001 to be exertion of market power by investor-owned utilities. Some blamed environmentalists! If Enron had not gone bankrupt, perhaps they never would have correctly analyzed the problem. What news sources got the story right from the beginning?

(iv) In June of 2002 WorldCom suffered financial collapse after several billion dollars of "aggressive, Enron-style accounting" came to light. It set the record at the time for the largest bankruptcy. Why did WorldCom collapse? What structures could have prevented this collapse, had they been in place?

(v) List all of the banks (some of the largest in the world), accounting firms, stock brokerages and other corporations that either lied with numbers ("committed aggressive accounting") and collapsed or aided in the lying with numbers since, say, January 2002.

(vi) Go back to Chapter 2 and study the financial collapse of 2008 and following years. Do you see any patterns common to previous examples of collapse?

27.3 Follow the Money

I interrupt the flow of logic here to interject a principle which is mathematically appealing in its simplicity, quantitative objectivity and precision, and which is extremely useful in the analysis of economics and politics: *FOLLOW THE MONEY!* Carefully following the money, not always easy since some would like to keep such flows secret, gives insights into who does what for whom and why in any given situation—such as in the financial collapses discussed in Exercise 27.3.

Smedley Darlington Butler. As perhaps one of the most outstanding and instructive examples I offer Brigadier General Smedley Darlington Butler (1881–

say by the time you read this?

Enron had operations all over the world: in Asia, South America, Europe, Central America and the Caribbean and, of course, the United States. Research the incredibly lucrative deal Enron negotiated for itself with the government of Maharashtra, India. (Of course, the citizens pay.) A critical account of Enron's role in India can be found in the book [454].

Research the special favors in the form of laws passed by Congress or regulatory favors that Enron was able to obtain. For example, the Investment Company Act of 1940 and the Public Utility Holding Company Act of 1935 were passed to prevent certain corporate practices that contributed to the Great Depression. What were these practices? How did Enron get the government to grant Enron exemptions from these laws? You might start with January 23, 2002 edition of *The New York Times.*

Both major parties, the Democrats and the Republicans, were heavily involved in the Enron "situation" just as they were in the Savings and Loan Crisis/Scandal of part (i) of this exercise.

Research what happened to all those responsible for the Enron "situation." Some court cases were still active a decade after Enron went bankrupt and was exposed. Did some Enron executives "escape" with their portfolios intact? Compare the Enron scandal to the Tea Pot Dome scandal. See [35, 61, 117, 427, 362].

1940), winner of two Congressional Medals of Honor. General Butler was one of the most successful and decorated military men in American history, yet I have found that almost no students have ever heard of him. The fact that General Butler refused his first Medal of Honor, the fact that he was a whistle-blower on war profiteering and the fact that he took his oath to uphold the Constitution deadly seriously—well, that is probably why he is not in most history books, cf., [607].

For example, when the poor World War I veterans marched on Washington D.C. in the action called "the Bonus Army," General Butler took the side of the veterans while Douglas MacArthur and Dwight D. Eisenhower suppressed them. In his book [65] General Butler detailed who got rich off of the wars he spent his life fighting for America. Butler was famous for his direct, "no bull" speeches—to wit:

> *"War is a racket. Our stake in that racket has never been greater in all our peace-time history. It may seem odd for me, a military man, to adopt such a comparison. Truthfulness compels me to. I spent 33 years and 4 months in active service as a member of our country's most agile military force—the Marine Corps... .*
>
> *"I helped make Mexico and especially Tampico safe for American oil interests in 1914. I helped make Haiti and Cuba a decent place for the National City Bank boys to collect revenues in. I helped in the raping of half a dozen Central American republics for the benefit of Wall Street. The record of racketeering is long. I helped purify Nicaragua for the international banking house of Brown Brothers in 1909–12. I brought light to the Dominican Republic for American sugar interests in 1916. I helped make Honduras 'right' for American fruit companies in 1903...*
>
> *"Looking back on it, I feel I might have given Al Capone a few hints. The best *he* could do was to operate his racket in three city districts. We Marines operated on three *continents*."*[13]

Wars and war profiteering are an old and recurring pattern. Recent references dealing with more current wars are [719, 286, 595]. For an intimate view of war from the "on the ground" point of view of killing someone else, cf., [50].

Minds and Money on Autopilot. The recurring patterns of violence around the world and at many levels might lead one to question the logical organization of human societies in an effort to understand why. Since increasing disorder, i.e., increasing entropy, can be viewed as a common enemy of all humans, and since violence contributes to disorder, it is perhaps imperative to ask why violence, war, in particular, continually recurs. Derrick Jensen, just to name one, is an author who has thought about the logical premises upon which civilizations are based. In particular, in [329, p. ix], the third of twenty premises reads:

> *"Our way of living—industrial civilization—is based on, requires, and would collapse very quickly without persistent and widespread violence."*

[13] This quote is from *Common Sense* magazine, 1935.

Exercise 27.4 Are Economic and Political Decisions Resource Related?
(i) Regarding the logical structure of civilizations it is tempting to posit as an axiom the following: *Access to and/or acquisition of social resources*[14] *and/or natural resources are major (if not determinative) considerations in virtually all economic and political decisions.* Do you believe this or not? Comment. Is this axiom consistent with ecological economics?
(ii) Going to war is one of the major decisions governments make. Is the previously stated axiom verified in the situations surrounding the decisions to go to war that you are familiar with?

We will have to change some basic habits/assumptions and how we relate to each other and the environment in order to have a sustainable society, including a less violent world. Ultimately humanity must discover a new way of being, presumably by means of the principles of self-organization. As we have seen, however, the process of self-organization can be forced in one direction, at least for a period of time, by a minority.[15]

There are two non-technical books which I would like to mention, [432], on the madness of crowds, and [675], on the wisdom of crowds. How can crowds be both mad and wise? Crowds (sufficiently large) are complex systems, which certainly can be mad, as is well documented in [432]. In [675] examples of crowds behaving wisely are given; and these crowds have three properties: diversity, independence, and a particular kind of decentralization. The important thing to note is that a "crowd," i.e., a large group of humans, can behave wisely or not depending on the circumstances. I regard both the wisdom and madness of crowds as two possible outcomes of the process of self-organization.

Where particularly large concentrations of the wealth of society are found, we find also disproportionate political power, hence disproportionate responsibility. "Rules" such as laws passed by Congress, financial rules, rules governing information flow (in the media and on the internet, for example[16]) and other conventions of society are in large part formulated by the wealthy and powerful, at the moment. We must self-organize to accommodate new realities and this process is pushed in certain directions by these "rules." If the resulting social transformation is unsuccessful, total collapse is possible. Of course, megacorporations have concentrated much power and wealth. This is not a minority view. In a poll done by *Business Week* and published in their September 11, 2000 issue they found (emphasis in the original): *72% of Americans say business has TOO MUCH POWER over too many aspects of American life.*

---

[14]Social resources include labor, slavery and other economic activities of people.

[15]I am thinking at the moment of the destruction of the electric rail option for transportation in the United States during the 20th century, cf., Chapter 1.

[16]There is a critical issue at the time I write this called "internet neutrality." Roughly, the situation now is that everyone has equal access to the Web. If internet neutrality is lost, then the more money you have the better access you will have to the internet. In the extreme case, monied interests will be able to effectively block access to the Web of people they deem undesirable. Thus the Web as a democratic force will be greatly compromised, if not entirely lost. Also lost will be a great deal of the force for innovation, greatly needed at this time, that the Web represents, cf., www.savetheinternet.com.

Corporations[17] have not always had such a position in society. As Richard Grossman (and POCLAD) and Ralph Estes document,[18] long ago corporations were only allowed to exist, i.e., obtain a charter, if they clearly served a public purpose. Not only have most corporations evolved to serve a private purpose, they measure their performance quite narrowly with a mathematical measure called "return on investment," i.e., the bottom line. This is an instance of the autopilot which may be heading all of us to places we would rather not be, corporations included. Estes in his book and POCLAD develop broader methods of measurement of corporations and meaningful feedback (with clout) from society to corporations that would reform (or replace) corporations so as to profit all stakeholders, i.e., people who have a stake in what corporations do to or for them.

Jeff Gates, cf., [216], while extolling the virtues of capitalism notes that the current practice of same has led to extreme concentration of ownership as in the days of "robber baron" capitalism. Gates notes that capitalism as now practiced has not been good at creating capitalists (lots of people who share in the ownership of capital) and he states: *The reason for this is poorly understood: contemporary capitalism is not designed to create capitalists, but to finance capital.* Gates proposes ways for increasing the number of people who share in, "who own a part of," the economy. For one thing he proposes that ownership is an underutilized feedback loop. Gates also notes: *Financial capital in the United States today resides largely in two camps. While much of it is concentrated in the hands of an ownership aristocracy, the balance has no true owner because over the past two decades approximately $12 trillion in assets has come to be held by institutional investors; pension plans, mutual funds, insurance companies, banks, foundations and endowments of various sorts. Today's detached and disconnected capitalism is now largely "on automatic," with investment decisions based on a "by-the-numbers" process that is incapable of taking into account many longer-term concerns, including the impact those investments have on the social fabric, on the fiscal condition of the nation, and on the environment.* This is another form of autopilot, and organized shareholders/investors sometimes form feedback/activist groups to guide such things as pension funds in more socially and environmentally sensitive directions.

Exercise 27.5 Reorganize Civilization

How would you redesign the society in which you live to put it on a nonviolent, sustainable course? How would you effect such changes? Consider such tools as financial instruments (local currencies, etc.), information flow (community radio, internet etc.), rules for corporate charters (public purpose required again, three felony strikes and corporation is out, etc.), political and union organizing, organizing around common interests such as health, etc.

[17]For a concise history of corporations and corporate law see Richard Grossman, Frank Adams, *Taking Care of Business: Citizenship and the Charter of Incorporation*, POCLAD, 1993. POCLAD stands for Program on Corporations, Law and Democracy and is found on the Web at www.poclad.org.

[18]See www.poclad.org, [578], [180], for example.

Bits of history, such as the attitude of some U.S. presidents toward corporations, have been filtered out of most of our educations. For fun and to partially remedy this information gap consider the following exercise. It is more relevant today than ever, since 51 of the world's 100 largest economies are not countries but corporations, cf., [464].

Exercise 27.6 Who Said THAT?[19]

The following are quotes from well known persons in history. Jot down who you think is the author of each quote before you check the answers.

(i) "I hope that we shall crush in its birth the aristocracy of our monied corporations, which dare already to challenge our government to a trial of strength, and bid defiance to the laws of our country."

(ii) "I know of no safe depository of the ultimate powers of society but the people themselves; and if we think them not enlightened enough to exercise their control with a wholesome discretion, the remedy is not to take it from them, but to inform their discretion."

(iii) "Every man is equally entitled to protection by law; but when the laws undertake to add...artificial distinctions, to grant titles, gratuities, and exclusive privileges, to make the rich richer and the potent more powerful, the humble members of society—the farmers, mechanics, and laborers—who have neither the time nor the means of securing like favors to themselves, have a right to complain of the injustice of their government."

(iv) "We may congratulate ourselves that this cruel war is nearing its end. It has cost a vast amount of treasure and blood. It has indeed been a trying hour for the Republic; but I see in the near future a crisis approaching that unnerves me and causes me to tremble for the safety of my country. As a result of the war, corporations have been enthroned and an era of corruption in high places will follow, and the money power of the country will endeavor to prolong its reign by working upon the prejudices of the people until all wealth is aggregated in a few hands and the Republic is destroyed. I feel at this moment more anxiety for the safety of my country than ever before, even in the midst of war. God grant that my suspicions may prove groundless."

(v) "Shall the railroads govern the country or shall the people govern the railroads?...This is a government of the people, by the people, and for the people no longer. It is a government of the corporations, by the corporations, and for the corporations."

(vi) "Corporations, which should be the carefully restrained creatures of the law and the servants of the people, are fast becoming the people's masters."

(vii) "Our government, national and state, must be freed from the sinister influence or control of special interests. Exactly as the special interests of cotton and slavery threatened our political integrity before the Civil War, so now the great special business interests too often control and corrupt the men and methods of government for their own profit. We must drive the special interests out of politics. That is one of our tasks today... The citizens of the United States must effectively control the mighty commercial forces which they have themselves called into being. There can be no effective control of corporations while their political activity remains. To put an end to it will be neither a short nor an easy task, but it can be done."

(viii) "Great corporations exist only because they are created and safeguarded by our institutions; and it is therefore our right and our duty to see that they work in harmony with those institutions."

(ix) "Big business is not dangerous because it is big, but because its bigness is an unwholesome inflation created by privileges and exemptions which it ought not to enjoy."

(x) "Monopoly persists, monopoly will always sit at the helm of government. I do not expect monopoly to restrain itself. If there are men in this country big enough to own the government of the United States, they are going to own it."

[19]Why do you think that most of my students have never heard any of these quotes before?

(xi) "No business is above government; and government must be empowered to deal adequately with any business that tries to rise above government."

(xii) "Out of this modern civilization royalists carved new dynasties... The royalists of the economic order have conceded that political freedom was the business of the Government, but they have maintained that economic slavery was nobody's business."

(xiii) "Private enterprise is ceasing to be free enterprise. Private enterprise, indeed, became too private. It became privileged enterprise, not free enterprise."

(xiv) "In the councils of government, we must guard against the acquisition of unwarranted influence, whether sought or unsought, by the military-industrial complex. The potential for disastrous rise of misplaced power exists and will persist."

(xv) "We haven't done anything for business this week—but it is only Monday morning."

(xvi) The definition of a corporation: "An ingenious device for obtaining individual profit without individual responsibility."

(xvii) "Fascism should more properly be called the corporatism, since it is the merger of state and corporate power."[20]

(xviii) "The twentieth century has been characterized by three developments of great political importance: the growth of democracy, the growth of corporate power, and the growth of corporate propaganda as a means of protecting corporate power against democracy." Comment.

Self-organization can take many forms. If society's key decision-makers see Nature's message writ large and act accordingly, that is one path we might follow. Whether or not that happens one can always choose to minimize one's impact on the environment and others by living as simply as possible with regard to material possessions, cf., [172]. If it looks like our leaders are about to take us over a cliff, either socially or environmentally, then the process of self-organization might follow what I call: The Law of Thelma and Louise (and Frederick Douglass), to wit, *When dealing with humans, you get what you settle for.*

In the movie *Thelma and Louise*, at one point Louise tells Thelma: "Honey, you get what you settle for." Now there are various levels at which the above law applies. You can sometimes not settle for the rude behavior of another by walking away. Laborers in the past, in order to not settle for 7-day workweeks, organized unions and fought for the 40 hour week. In fact there was a worldwide labor strike for the 8 hour day. Democracies in mid century had to organize armies to not settle for the Nazis. Now while Nature does not negotiate with us as to which of Nature's laws we would like to follow, for

[20] From http://revthom.blogspot.com/2006/03/sermon-what-davidson-loehr-says-about.html we read the following excerpt (does any of it sound familiar?): According to political scientist Lawrence Britt, fascism is identifiable by fourteen characteristics that include, 1) powerful nationalism, 2) the emphasizing of a single enemy as a threat to be eliminated, 3) the supremacy of the military in abolishing that threat, 4) disdain for human rights, 5) obsession with national security, 6) controlled mass media, 7) religion and government intertwined, 8) the rise of corporate power, 9) suppression of labor power, 10) disdain for intellectuals and the arts, 11) obsession with crime and punishment, 12) fraudulent elections, 13) rampant cronyism and corruption, and, 14) insistence on male domination and control of sexuality. How does Britt's analysis compare with Naomi Wolf's "Ten Steps" in *The End of America*, [732]?

good or ill when dealing with humans there is a lot more flexibility—for a while, anyway.

A longer, more eloquent statement of the above "law of Thelma and Louise" was made by Frederick Douglass in 1849 in a letter to an abolitionist associate.

Let me give you a word on the philosophy of reform. The whole history of the progress of human liberty shows that all concessions yet made to her august claims have been born of earnest struggle. The conflict has been exciting, agitating, all absorbing, and for the time being putting all other tumults to silence. It must do this or it does nothing. If there is no struggle there is no progress. Those who profess to favor freedom, and yet depreciate agitation, are men who want crops without plowing up the ground. They want rain without thunder and lightning. They want the ocean without the awful roar of its many waters. This struggle may be a moral one; or it may be a physical one; or it may be both moral and physical; but it must be a struggle. Power concedes nothing without a demand. It never did and it never will. Find out just what people will submit to, and you have found the exact amount of injustice and wrong which will be imposed upon them; and these will continue until they are resisted with either words or blows, or with both. The limits of tyrants are prescribed by the endurance of those whom they oppress.

Which Is More Important: National Citizenship or Global Class? In a succinctly written article by Paul Krugman: "The End of Middle-Class America (and the Triumph of the Plutocrats)," cf., [375], a panoramic picture is painted of wealth distribution in America for the last century. Very roughly the "age of the American middle class" runs from the 1940s through the 1970s. Before and after this period are so-called Gilded Ages, times of large income (and wealth) inequality. For example, from Krugman's article: *"The 13,000 richest families in America now have almost as much income as the 20 million poorest. And those 13,000 families have incomes 300 times that of average families. As the rich get richer, they can buy a lot besides goods and services. Money buys political influence; used cleverly, it also buys intellectual influence."* The norms of the post World War II New Deal have unraveled, and with that has come the radical reduction of American equality and depopulation of the American middle class.

Notice that in Figure 27.2[21] the "valley" in the graph of the Gini Index corresponds to "age of the American middle class" discussed by Krugman. The "peaks" correspond to concentration of national income[22] in relatively fewer hands. See the next Exercise.

---

[21] Graph created by Left Business Observer, (LBO), (www.leftbusinessobserver.com). Data sources: 1913–1946, Eugene Smolensky and Robert Plotnick, "Inequality and Poverty in the United States: 1900 to 1990," University of California, Berkeley, Graduate School of Public Policy, Working Paper #193 (July 8, 1992). 1947–2008, U.S. Bureau of the Census. I wish to thank Doug Henwood at LBO for his hard work and permission to use this informative graph.

[22] It is easy to confuse "income" and "wealth." If you imagine a person's assets as a "box," income is the flow into the box and wealth is the total accumulation in the box, minus the flow out due to expenses and losses.

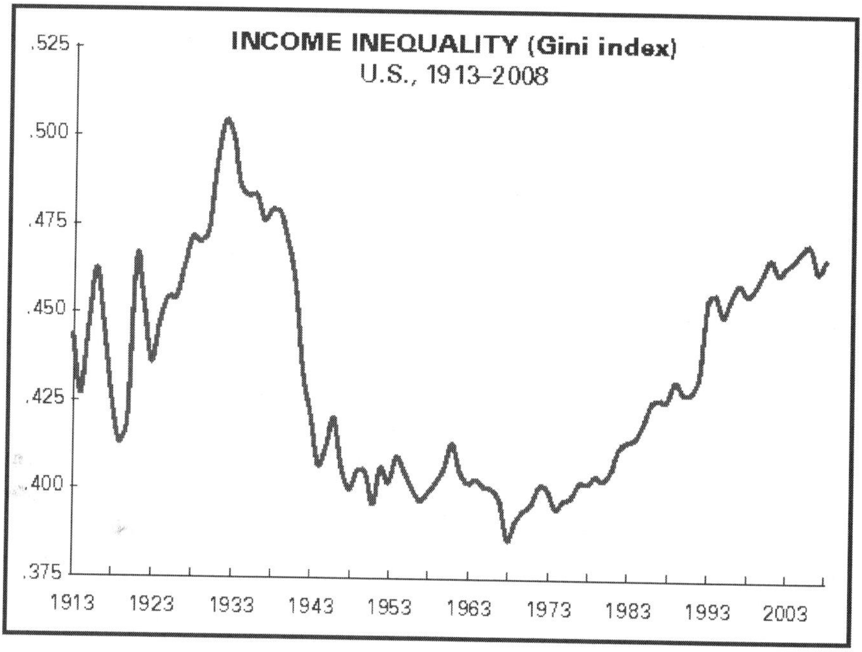

FIGURE 27.2: The Gini Index

Exercise 27.7 The Gini Index

One measure of income equality/inequality is the Gini Index. The Gini coefficient (or index) is a measurement of a population's income inequality on a scale of 0 to 1. A country with total equality of incomes would have a Gini coefficient of 0; a country with complete inequality (one person with everything and the rest having nothing) would have a Gini coefficient of 1.

(i) Do a search on the Web for the Gini Index and look at graphs of this index in America during the 20th century. Often the graph is in terms of a Lorenz curve. Can you see the connection of graphs in this form with graphs in the form of Figure 27.2? See if you can find the "Age of the Middle Class" in America with such a graph. Note that the Gini Index reaches a peak just before the Great Depression. It then goes down and comes back up. At the time of writing the Gini Index is roughly where it was before the Great Depression. What is the Gini Index in America at the time you read this? Comment.

(ii) On page 6 of [203] you will find a graph of the Gini Index in America from 1913 to 1998. Does this graph agree with other such graphs you can find on the Web, for example, [301, p.115] and Figure 27.2?

(iii) The Gini Index measures income inequality. It turns out that wealth inequality is much greater than income inequality. What is the difference between income and wealth? Can you find a measure of wealth inequality in the U.S.? Can you find a measure of wealth inequality in the world?

(iv) For want of a better place to bring this up I suggest that you read the following books and see what they have to do with income/wealth inequality in the United States. See [183, 733, 540, 541].

Although most people live in nation-states, it is making more and more sense for the person whose job has just been outsourced to a foreign country to identify more with the international class of workers with whom he/she competes and could co-operate, than with billionaires. Also it makes more sense for multi-billionaires to feel a greater sense of community with the "Dunbar number" of fellow multi-billionaires around the world, than with the working poor. Jeff Faux, in [183], clearly puts forth the case that international corporations are a global class not tied to any one particular country. What is good for these corporations, such as Exxon, Wal-Mart, General Motors, Goldman Sachs, or Microsoft, even if they are "American," is no longer necessarily good for Americans. If one follows the money, one may find that the likes of Wall-Mart, Dell, Fidelity Investments, Boeing and Cabella's get tax breaks in the U.S. (made up for by other tax paying Americans) while not delivering on promised job creation in the U.S., as documented in [392]. One of the most consistent chroniclers of class in America and the plight of working poor and the stressed middle class has been Barbara Ehrenreich, cf., [163, 164, 165, 166, 167]. See also [392], [150, 704, 611, 612, 626]. A few of the references where hard data on what amounts to growing economic class stratification can be found, for example, in [460, 108, 733, 107, 583][23]

Of course, it is possible to organize businesses, especially ones that employ over, say, 2000, so that meaningful feedback is taken from the workforce and the immediate community. Business organizations exist where workers serve on the board of directors, seriously participating in decision-making; and government laws prevent, for example, outsourcing, without meaningful input from affected citizens. Contemporary Germany is one such place, cf., [734].

27.4 Are You Paying More or Less than Your Fair Share of Taxes?

Before you answer this question you need to follow the money. First of all there are many taxes: federal, state, and municipal income taxes; sales taxes; taxes on food, fuel; luxury taxes; property taxes; and so on. When thinking of taxes it is important to consider the *total* taxes paid by a person as a percentage of their income from all sources. Then, where do the taxes paid go? There are two broad categories: interest on debt, and goods & services. Much of our debt is incurred by transferring public money to private, but well-connected hands. For just one example recall war profiteering, cf., [595,

[23]See also *A Decade of Executive Excess: The 1990s, Sixth Annual Executive Compensation Survey,* by United for a Fair Economy (Chuck Collins, Chris Hartman) and the Institute for Policy Studies (Sarah Anderson, John Cavanagh, Ralph Estes), Sept. 1, 1999, (www.ufenet.org). This Web site has a number of more recent, relevant publications.

690, 286, 542]. The United States became a net debtor nation to foreigners in 1987 during the presidency of Ronald Reagan. Prior to that the United States was a net creditor nation to foreigners. The national debt can be viewed as a classical case of a "tragedy of the commons," cf., page 471. Except in this case relatively a few benefit, in that the debt was incurred on their behalf, and all Americans must pay taxes to pay off the debt—most of whom did not profit much from the incurrence of the debt. (This pattern is global, not exclusive to the U.S.!) Also, if you are of the international billionaire class, your fortunes are probably not tied exclusively to the value of the U.S. dollar anyway.

Exercise 27.8 Taxes, National Debt, Economic Mobility

(i) Which of the following three income tax structures impacts your finances the least? Before you answer I recommend you take a look at [336] and [32, 397, 201].

(1) A steeply progressive tax.[24]

(2) A flat tax, i.e., everyone pays the same percentage of their income.

(3) The federal income tax is replaced by a national sales tax.

(ii) Go to www.ustreas.gov, the Web site of the U.S. Treasury Department and find out what your share of the National Debt is? How much of your taxes each year go to pay for this debt? To whom are you paying this debt? Does the debt effect you, cf., [418]?

(iii) Compared to other rich industrialized countries, is it easier or more difficult to work your way out of poverty in America? How difficult is it to work your way from lower middle class to upper middle class? From middle to upper class? In other words, what is the level of economic mobility in the United States compared to other comparable nations?

(iv) President Herbert Hoover said: *"Blessed are the young, for they shall inherit the national debt."* Comment.

(v) John K. Galbraith said: *"People of privilege will always risk their complete destruction rather than surrender any material part of their advantage."* Comment.

(vi) Are the International Monetary Fund, the World Bank and World Trade Organization benefiting your global class or not, cf., [141, 75, 25, 697, 564]? Has the United States surrendered its national sovereignty to any international trade organizations? If so, who controls these trade organizations, cf., [440]?

(vii) It is quite common for entire nations, cf., Greece's 2010 financial crisis and many other examples, to go into debt to "bond holders," banks, "investors," and the like, while simultaneously not seriously taxing the profits made by these lenders on the compound interest gained on the loans. (For example, the tax rate for ordinary workers can be at a higher rate.) Discuss how such economic structures can foster inequality. For example, the bank bailout money (see Chapter 2) financed by U.S. taxpayers went to banks for a fraction of 1%, say 0% to $\frac{1}{2}$ %, and the banks then loaned money back to the U.S. government (read U.S. taxpayers) at 3 % or more.[25] Coincidentally, *"... But Bank of America, Goldman Sachs and JP Morgan Chase & Company produced the equivalent of a trio of perfect* (baseball) *games during the first quarter.* (of 2010) *Each one finished the period without losing money for even one day."*[26] Given that through May 2010 there have

[24] During the (latest) Iraq war and occupation (2003–??) tax cuts were distributed disproportionately to the very rich. During World War II the marginal tax rate on the very rich was 90%, i.e., 90% of income above a certain point was taken in federal taxes to support the war effort. What is top marginal tax rate as you read this?

[25] "Wall Street's Simple Formula for Staying Rich," www.alternet.org, April 24, 2010; "Penny Saved, And Far From Earned," *The New York Times*, April 16, 2010.

[26] Article by Eric Dash, "3 Big Banks Score Perfect 61-day Run," *The New York Times*, May 12, 2010, p. B1.

been only 20 perfect baseball games in the history of the major league baseball, what are the chances of 3 perfect games in 1 year?

(viii) If income taxes are successfully avoided by one sector of the population other taxes tend to increase. If federal income taxes are not sufficient to pay for certain goods and services, either these things are foregone or their costs are pushed onto the states, which in turn may pass them on to cities and counties—resulting in higher local property taxes and/or higher local sales taxes, for example. Consider three cases: a person in the top 1% of income and/or wealth, a person in the 60th percentile of income and/or wealth, and a person in the 30th percentile of income and/or wealth. Given the tax structures currently in place when you read this, calculate the impact of the *total* taxes paid as percentage of income and/or wealth on each of the these three.

The following words of caution come from M.I.T. economist, Lester C. Thurow, in [683, p. 131]:

Many today would argue that international pressures to regulate less and tax less are good pressures, not bad pressures. Yet it is well to remember that most of our current system of business regulation arose in two real-world experiences—the "robber barons" era of the last half of the nineteenth century and the financial collapses and Great Depression (during which unemployment hit 27 percent) of the 1920s and 1930s. Those who were alive then saw something that needed to be regulated. Without regulations, perhaps we too will again see something that needs to be regulated.

But the era of national economic regulation is ending and the era of global economic regulation is not yet here. For at least a while capitalism is going to be tried with much less government regulation.

It would not be outrageous to predict that if the global economic system does not develop regulatory feedback mechanisms, it will collapse, just as it did before in the 1920s and 30s. It would be prudent to prepare for this, especially in light of the fact that we will not have abundant fossil fuels to assist a global recovery this time around. (I wrote this paragraph in 2000.)

27.5 Financial Growth vs. Fish Growth

In Figure 27.3 I have graphed the growth of money, or financial instruments, vs. the growth of a biological resource such as fish (it could have been trees, and so on). If the financial economy is growing at 10% a year and fish growth is 4% a year, it is clear that in the absence of regulations, a love for fish or some other ethical concern, it is to a fisherperson's advantage to transfer the fish from the fish curve to the money curve as fast as possible.

Even if the curves were reversed, there is incentive to transfer fish to the money curve, since money can be spent on anything that is for sale in the market place.

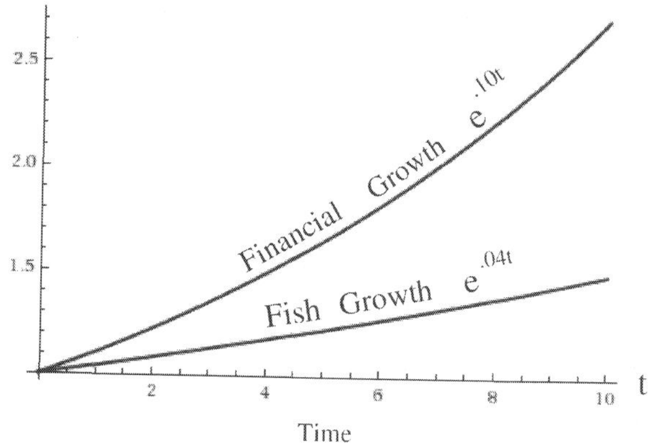

FIGURE 27.3: Fish Growth and Financial Growth

Exercise 27.9 Does Economic (or Actual) Extinction Have to Happen In the Absence of Regulations?

Assume in parts (i), (ii) and (iii) that the fish are valued *only* for the number of dollars they can bring at the market.

(i) Assuming that the Figure 27.3 holds, i.e., fish stocks are growing at 4% per year and financial wealth is growing at 10% per year (if you have money you can invest it and get a 10% per year return), which of the following strategies, (a) or (b), will make you the most money quickly? (Quickly here means that the next three months are most important to you, the next year or two is really as far into the future as you can imagine.)

(a) Converting the fish to money as fast as possible.

(b) Converting some of the fish to money each year, but in a sustainable fashion so that the fish stocks do not diminish.

(ii) If the fish become rare and the corresponding price they fetch in the market goes up, what would be the response of those catching the fish if what they value most is money?

(iii) Which of the two strategies in part (i) will make the most money over seven generations of fishing?

(iv) If you valued eating fish more than converting them into money how would you behave? Would you need to maintain a "navy" to protect the fish? Would this be economically viable? (See www.seashepherd.org).

(v) In today's economic climate do you see any way to avoid either actual or economic extinction of the fish? What must you do?

(vi) If you want to deeply study the mathematics involved in managing biological resources such as fish, I recommend starting with [90].

27.6 Fractional Reserve Banking: An Amazing Mathematical Trick

An ecological economist once told me that 90% of the money in America was in the form of debt to commercial banks. Such money is not in the form of bills/banknotes or coins, but rather in the form of "zeros and ones" in a computer. Finding this rather incredible I asked for an explanation. I was immediately taken back to the time of Frederick Soddy, chemist turned economist in the 1920s and was introduced to fractional reserve banking. Frederick Soddy regarded fractional reserve banking as one of the true sources of all economic and social problems. Doing away with fractional reserve banking would, indeed, have profound impacts on modern society. I want to do one calculation that you already have the tools to understand, so you can get a rough idea of what was bothering Professor Soddy. By the way, economists generally held (hold?) Soddy in low esteem. The feeling was mutual.

Suppose you have earned \$1000 and that you put it in a bank, which offers you 4% simple interest per year. (The actual interest rate on my savings as I write is $\frac{1}{2}$ %.) Depending on the laws in effect at the time, a bank can loan out a fraction of your \$1000 and keep the remaining fraction. Let me assume that the bank is required to keep 10% reserves.

Thus the bank can loan out \$900 and keep \$100. Let me assume that the bank will charge 10% simple interest per year on the \$900 loan.[27] If I stopped at this step and did no further analysis I might conclude that the bank pays out $.04 * 1000 = 40$ dollars in interest to you and collects $.10 * 900 = 90$ dollars in interest from the borrower of the \$900. Not bad for the bank you say, but it turns out that this so far is a very incomplete analysis.

Here comes the mathematical magic that the banking system loves. What really happens to the \$900? It is virtually impossible for the money not to be eventually deposited back in the banking system. The \$900 might become a deposit in a bank right away, financing whatever project the borrower pursues. The borrower could spend the money at some businesses, in which case the money would likely be deposited in the businesses' bank accounts. The borrower could buy some stocks, thus transferring the money to someone else who might put the money in a bank or transfer it again to someone else who does put the money in a bank. About the only way to keep the money out of the banking system is to bury it in the back yard, or keep it under a mattress. This is not likely, since interest is accumulating; and one borrows money for many purposes, none of which are back yard or mattress burial.

So I am going to make the reasonable assumption that the \$900 borrowed goes back rather quickly into the banking system as a deposit. Now the virtually perpetual fun begins. The bank is allowed to loan out 90% of the \$900, i.e., $.90 * 900 = 810$ dollars, while keeping 10%, or $.10 * 900 = 90$ dollars.

But the $.90 * 900$ dollars makes its way back into the banking system whereupon a bank can loan out 90% of it, i.e., $.9 * .9 * 900$ dollars, while keeping 10%, i.e., $.9 * 900 * .1$ dollars. The following exercise examines in detail what happens after n such steps.

[27] As an exercise the reader can redo this section under the assumption that banks pay less than 1% interest, one-half percent, for example, on savings accounts to depositors and collect interest of at least 14% (and up), say, on credit card debt and other loans. At the time of writing 30-year fixed-rate mortgage loans are 5%, but rates are predicted to rise.

Exercise 27.10 Fractional Reserve Banking Is Magical

(i) Verify that in our discussion above (involving 3 steps) the bank has kept in reserve a total (in dollars) of $1000 * .1 + (1 - .1) * 1000 * .1 + (1 - .1)^2 * 1000 * .1$.

(ii) Verify that in our discussion above (involving 3 steps) the banking system has loaned out (in dollars) a total of $(1 - .1) * 1000 + (1 - .1)^2 * 1000 + (1 - .1)^3 * 1000$.

(iii) Verify that in 5 steps the banking system will have kept in reserve (in dollars) $1000 * .1 + (1 - .1) * 1000 * .1 + (1 - .1)^2 * 1000 * .1 + (1 - .1)^3 * 1000 * .1 + (1 - .1)^4 * 1000 * .1$.

(iv) Verify that in 5 steps the banking system will have loaned out (in dollars) $(1 - .1) * 1000 + (1 - .1)^2 * 1000 + (1 - .1)^3 * 1000 + (1 - .1)^4 * 1000 + (1 - .1)^5 * 1000$.

Do you see a pattern here?

(v) Verify that in n steps the banking system will have kept in reserve $\sum_{k=0}^{n-1}(1 - .1)^k * 1000 * .1$ (in dollars).

(vi) Verify that in n steps the banking system will have loaned out $\sum_{k=1}^{n}(1 - .1)^k * 1000$ (in dollars).

(vii) Verify that the sum $\sum_{k=0}^{\infty}(1 - .1)^k * 1000 * .1$ equals $100 * \frac{1}{1-.9} = 1000$ (in dollars). (Hint: If you have trouble with this look back at the discussion of the linear multiplier effect in Section 20.2.)

(viii) Verify that the sum $\sum_{k=1}^{\infty}(1 - .1)^k * 1000$ equals $1000 * \frac{.9}{1-.9} = 9000$ (in dollars). (Hint: See hint in part (vii).)

Thus I conclude that after many steps the banking system will have kept almost \$1000 in reserve and it will have loaned out almost \$9000!

Thus, with my assumptions of 4% interest on savings and 10% interest on loans (rather conservative assumptions when I write this)[28] the banks at the end of one year will pay you \$40 in interest on your initial deposit and take in 10% of \$9000, or \$900 on loans! (This assumes that all transactions take place instantaneously, and then we all sit around for one year. A more complex analysis comes up with a similar result since many such processes are started throughout the year, and eventually the banking system will have made nearly \$9000, say, in loans. Thus even if my hypotheses are not entirely satisfied and hence my conclusions are not perfect, the approximations given are sufficiently accurate to give one pause for thought!)

The bank will pay 4% interest to all of the other savers as well (assuming no one loses interest due to early withdrawal), and you should verify that the total amount of savings interest paid (including yours) is almost $\sum_{k=0}^{\infty} .9^k * 1000 * .04 = 400$ dollars. This is because during the whole process a total of \$10,000 is deposited in banks (this includes your initial \$1000 deposit), \$9000 of which

[28]The 4% interest rate is at the time of writing not realistic. From a Reuters article by Rob Cox and Lauren Silva Laughlin, reprinted in *The New York Times*, March 29, 2010, we read: *"... in the fourth quarter of 2009, institutions* (banks) *with more than \$100 billion of assets paid an average of 0.77 percent annual interest on deposits, according to F.D.I.C. data. By comparison, institutions with less than \$10 billion of assets paid an average of 1.73 percent. That difference—nearly 1 percentage point—is one measure of the benefit that big banks enjoy from implicit government backing. ... The 10 largest banks hold about \$3.2 trillion of America's \$7.7 trillion of domestic deposits. Apply the differential in deposit interest rates, and those 10 appear (sic.) be saving nearly \$30 billion a year thanks to their size. ... Figures from the F.D.I.C. show that banks in all size categories paid 3.6 percent to 3.65 percent on deposits in the last quarter of 2006."*

is loaned out by the banks—leaving your $1000 as a reserve in the banking system. (Again this assumes that all transactions occur instantaneously, and then we all sit around for one year. And again, a more complex analysis comes up with a similar result, since many such processes are started throughout a given year. Thus eventually I can expect my conclusions to be approximately true.)

It is somewhat amazing that your $1000 deposit can give rise to so much economic activity. It is also amazing that the banking system ends up making (in dollars) almost $900 - 400 = 500$ in this somewhat idealized process. Now depending on what laws are in effect at the time you read this, the banks have to pay some taxes, fees and wages; in other words the banks have some costs in doing business. (The banks also have, besides interest, other sources of income from a variety of fees, penalties, and so on.) But the amount that the banks make in excess of those costs accrues to said banks simply because they are given the legal power to engage in fractional reserve banking as described above. This is what bothered Professor Soddy and many other people throughout history.

Exercise 27.11 How Much Do Banks Profit in Real Life From Fractional Reserve Loans?

(i) In the idealized example above what was the percentage income the banking system made on your 1000 dollar deposit? How much of that do you imagine is profit?

(ii) Do some searching in the literature and get an estimate of what profit banks make in the real world on the fractional reserve system. See if what you find is consistent with the following quote from James Robertson, economist and former civil servant (in England) with the treasury and the author of many books, including *Sane Alternatives*. The following quote came from an article (written several years ago) "Free Lunches, Yes; Free Markets, No" in *Resurgence*, Issue 204:[29]

At present in Britain less than 5% of new money is issued and put into circulation by the government and the Bank of England as cash (coins and banknotes). The remaining 95% is created by the commercial banks and put into circulation as non-cash money in our current accounts. As J.K. Galbraith has commented, "The process by which banks create money is so simple that the mind is repelled. Where something so important is involved, a deeper mystery seems only decent." The banks simply print the money out of thin air into the current accounts of their customers—as interest-bearing, profit-making loans.

Interest on these loans is estimated to give the UK banks supernormal, special profits of $21 billion a year. The annual loss of public revenue from allowing the banks to create non-cash money in that way is greater—about $45 billion. **Total banking profits from this source in the USA, UK, Eurozone countries and Japan are about $140 billion a year.** *With a free lunch on that scale, no wonder some of the cats are fat!* (bold face emphasis mine).

(iii) During the Savings and Loan Crisis, Silverado Savings and Loan had a reserve rate reported to be 2.5%, not the 10% used in the text preceding this exercise. Verify that with such a low reserve rate a $1000 deposit by you gives rise to $39,000 in loans, $40,000 in deposits and $1000 held in reserve by the system.

Some religions have (have had) rules against charging interest on money. President Andrew Jackson staked his presidency on the platform of killing

[29] Currently available on-line at http://resurgence.gn.apc.org

the U.S. central bank. Jackson did kill the U.S. central bank with his Bank Veto[30] of July 10, 1932, and it did not reappear until 1913 in the form of the (private) Federal Reserve Bank.[31]

The Populists of the late 1800s struggled—not for a central (private) bank like the Federal Reserve Bank—but for a democratically controlled currency. In a populist model, profits accrued through banking magic, if allowed, could go to the U.S. treasury and help offset the costs of government—that is, lower people's taxes. In most states it is still legal for communities of people to realize some of the objectives of the Populists and to create local currencies, local financial instruments and a local banking system that is community owned. Thus local communities can more directly benefit from banking magic and the magic of the multiplier effect discussed in Section 20.2. I have worked (with a little success, but more failure) with a group of citizens to implement a local community currency in Boulder County, Colorado. Whereas the Federal Reserve manages money for the benefit of financial capital, regarding full employment as something to be avoided, a community can manage its local currency with the goal of full local employment. There are many fascinating and positive properties of a community currency. One can easily envision fair trade among a system of local communities[32] in addition to having each local community interact with the global community.[33]

Exercise 27.12 Guess Who Didn't Like The Central Bank?

Try to identify the author of each quote below.

(i) "If the American people ever allow private banks to control the issue of their currency, first by inflation, then by deflation, the banks ... will deprive the people of all property until their children wakeup homeless on the continent their fathers conquered ... The issuing power should be taken from the banks and restored to the people to whom it properly belongs."

(ii) "History records that the money changers have used every form of abuse, intrigue, deceit, and violent means possible to maintain their control over governments by controlling money and its issuance."

---

[30] In his veto message President Jackson referred to the Bank of the United States as unconstitutional, a violation of the principle of equal rights and a giving of "a gratuity of many millions to the stockholders." He also said it would "make the rich richer and potent more powerful." You can read Jackson's entire veto message and much more in [707].

[31] For a detailed look at the Federal Reserve Bank see [258]. Note that the Federal Reserve Bank, with Alan Greenspan its chair, bailed out Long-Term Capital Management, a hedge fund with several former Federal Reserve Bank officials on its board. A much bigger bailout was to come, see Chapter 2.

[32] One can participate in community building with any currency, cf., the organization Co-op America, www.coopamereica.org which advocates putting your money where your values are.

[33] This certainly must sound strange to those who have not encountered the concept of local currencies before. I am not advocating leaving the United States to form "mini countries" or doing away with the U.S. dollar. I am, however, advocating that local communities take some small steps to regain some local self sufficiency, supplementing dependence on multinationals and the U.S. dollar. See [148], wherein it is argued that local energy generation sufficient to grow a substantial portion of locally needed food plus a local financial system are necessary for a self-sustainable, local community. See also [652, 322].

(iii) "The Government should create, issue, and circulate all the currency and credits needed to satisfy the spending power of the Government and the buying power of consumers. By the adoption of these principles, the taxpayers will be saved immense sums of interest. Money will cease to be master and become the servant of humanity."

(iv) "I am a most unhappy man. I have unwittingly ruined my country. A great industrial nation is controlled by its system of credit. Our system of credit is concentrated. The growth of the nation, therefore, and all our activities are in the hands of a few men. We have come to be one of the worst ruled, one of the most completely controlled and dominated Governments in the civilized world—no longer a Government by free opinion, no longer a Government by conviction and the vote of the majority, but the Government by the opinion and duress of small groups of dominated men."

(v) "The modern theory of the perpetuation of debt has drenched the earth with blood, and crushed its inhabitants under burdens ever accumulating."

(vi) (The following is not a quote but a sentiment.) It has been said that the Federal Reserve Bank provides "free money" to large banks and does not collect interest on behalf of the American taxpayer who provided the money. What does this mean, and can you find documented examples of this practice?

It is a very clever strategy to convert one's real wealth into a debt someone else owes you—thus making it possible to live on the interest. The steady growth of financial wealth via compound interest/return on investment (not possible with real material wealth) means that financial wealth must eventually be debased as time passes. As a mathematician I view the current state of affairs regarding compound interest and related topics merely as rules now in force in the human economic game. Some people play the game very well to their own advantage. Unfortunately some people who are truly creative and productive are not rewarded as well financially for their work and positive contributions as some who understand how, and are in a good position, to play the economic game.

I will give one example. The real economy of real goods and services, the global gross product if you will, for one full year at the end of the 20th century has been estimated to be about \$18 trillion dollars. In 1998 the global trade in goods and services was about \$6.5 trillion. *Every day* in 2000 it was estimated that \$1.5 trillion changed hands in global currency markets (made possible no doubt by the computer/electronic revolution), up from \$.08 trillion, i.e., \$80 billion, in 1980. Most of this \$1.5 trillion activity in the financial economy was not part of the real economy, since in about $4\frac{1}{3}$ days as much financial, global-trade economic activity occurred as occurred in the real, global-trade economy in one full year!

This casino economy has real effects on real people. For example, it is possible to wake up one morning and find that the value of money that you have carefully saved over time is worth a fraction of what it means to you in terms of the actual productive work you did to earn the money.[34] It has

[34]See [175]. Can you find the misprint on page 72 of this book (mathematics will help!)? Ellwood also points out that via the inexorable force of compound interest the World Bank and International Monetary Fund (supposedly, originally created in part to help poor countries) now extracts from poor countries far more in interest payments on debts than these countries spend on health care and education. This is not hard to believe when one

been suggested that this phenomenal amount of global financial activity be taxed.[35] The proceeds from this tax could help offset the damage done, and perhaps slow down the casino economy.

In so far as folks try to live on compound interest, they are relying on growth. And growth cannot continue forever.[36]

Exercise 27.13 Reserve Rates and Their Implications

Redo the Exercise 27.10, replacing the .1 reserve rate with r, an unspecified reserve rate.

(i) Find a formula for the answer to each part of that exercise. If the process starts with a \$1000 deposit, what does $\frac{1}{r}\$1000$ represent?

(ii) If you let $r = .025$, which was the actual reserve rate at one time of Silverado Savings and Loan, what numerical answers do you get for the previous exercise?

(iii) Find an example of a negative effect of the casino economy on a real person.

(iv) Find an example of negative effect on a real person arising from the fact that the country in which this person resides must pay extraordinary amounts of interest on debts.

(v) If you were to carefully define the concept of a casino economy, would such have any relation to the concept of a Ponzi scheme, cf., Chapter 2?

---

realizes that the poor countries have most people living on \$1 to \$2 a day. Such persons cannot afford to pay interest on large national debts and pay for health care and education, not to mention buy food or inputs to grow food. Another helpful reference is [12]. This book is a product of the Institute for Policy Studies.

[35] Some economists refer to this proposed tax as the Tobin tax, after the fellow who thought it up and brought it to the attention of many.

[36] An economist sent me an e-mail on July 5, 2001 telling me that Japan had 0% interest at that time. How interesting.

Chapter 28

Distributed vs. Centralized Control and Decision Making

In this chapter I revisit the fuzzy boundary between two models of decision making in society: in one case decisions are distributed for the most part among the many individuals regarding the structure of agriculture, energy production, media, banking, insurance, health care and pharmaceuticals, resource ownership, income and wealth in general; in the other case decision making in most matters of importance is concentrated in the hands of relatively few. Even the fundamental decision making process of voting can be controlled by very few, cf., Chapter 24.

28.1 Farms: To Be Run by Few or by Many?

The family farm, once (and perhaps someday again) a major and numerically numerous American institution, is held up by many as the paragon of private ownership—independence, freedom. However, these days enormous corporations sell patented seeds,[1] (even patented animals), chemicals and equipment to farmers at prices that are not really negotiable. Enormous corporations then dominate the market—and price—for farm products—thus reducing many farmers to tenant status even if they somehow manage to keep their land. Such a situation does not conjure up the traditional, nostalgic image of the family farmer. The family farmer and global corporate agriculture models are both private ownership models—but of quite different types. In one case controlling decisions are made by a diverse network of many family farmers in close touch with the earth and with innumerable locally determined details. In the other case most controlling decisions are made by a few persons, far removed from the actual process of farm work. Ultimately all of us together determine what models are followed, either by choice or default, cf., Chapter 5. We get what we settle for.

[1] The practice, thousands of years old, of saving seeds from one year to the next is thus not allowed and in some cases not possible.

28.2 Utilities: MUNI or Investor-Owned?

Consider the case of electrical energy and domestic water supplies. There are two basic models: the MUNI, i.e., municipally-owned utility, and investor-owned utility, IOU. Either can be managed well and perform, but there are basic structural differences, made abundantly clear by the case of Enron, for example, cf., pages 208, 211, 492. The investor-owned utility is motivated to obtain electricity or water at the lowest price and sell it at the highest price the market will bear, since the customers are not the owners. A municipally-owned utility is motivated to get electricity or water at the lowest possible price[2] and sell it with minimal markup, since the owners are the customers. In a MUNI control is distributed among the entire population served. Folks are hired to run the MUNI, and if the population served is not pleased the hired hands can and have been fired. In the IOU control is centralized, and true democratic processes are rare. Enron has been shown to have been a criminal corporation, but if it had not gone bankrupt—due to the inexorable march of mathematics—the false information about the California energy crisis of 2000–2001 might likely be the accepted history. Enron's friends in government blamed the crisis on environmentalists preventing the building of generating capacity, which was not true. There are those who say that the market corrected itself with the bankruptcy of Enron, cf., page 492. Such people should tell that to the thousands of Enron employees who now live in poverty after losing everything with the crash of Enron. Criminal and/or stupid behavior by our political and economic leaders can cause the crash of whole nations, and the untold and unnecessary misery is not mitigated by calling it a market correction.

Thus the LADWP, the Los Angeles Department of Water and Power, a MUNI outperformed Enron and all the other IOUs during the criminally contrived energy crisis of 2000–2001. The LADWP owned its own generating capacity, promoted conservation and renewables—all because that is the logical thing to do if the customers are the owners. MUNIs are structurally responsible to democratic institutions. Corporations are not, but they could be if appropriate laws were passed/enforced. In 1898 Samuel Insull, an "electric energy baron" called for state regulation of his and all IOUs in response to the rapid expansion of MUNIs.[3] He correctly saw the importance of electricity, and he knew without regulation the "Enrons" of his day would eventually drive everyone into the arms of a MUNI. MUNIs have always provided services more cheaply and reliably than IOUs, but you would not know that history

[2]Hence a MUNI is not necessarily any more environmentally sensitive than any other collection of humans.

[3]If regulated, the IOU has a chance of controlling the regulators and increasing profits; whereas this is not possible in the long term with a MUNI.

from reading the megamedia today. With the exception of Paul Krugman, *The New York Times* (and other megamedia) did not reliably inform the public in real time about the LADWP MUNI success vs. the Enron criminal failure. During the Enron induced crisis I informally asked hundreds of people (some within the service area of the LADWP!) if they knew that the LADWP had no problem. Not one person knew this. The megamedia did not inform the people about the most basic facts relevant to this criminal crisis as it was happening. See [455, 706, 469, 632, 514]. The same structure applies to domestic water supplies, and there will likely be epic battles over the privatization of water, IOU model, vs. the MUNI model, as water becomes more and more of a rare, irreplaceable resource for which more and more people compete. See for example Section 9.5, and [616]. Whether electricity or water we get what we settle for.

Central control is mathematically attractive, since the few doing the controlling can extract resources from the many. Nowhere is this truer than in medicine or computers.

28.3 Linux vs. Microsoft

Computer science had its birth as a mathematical discipline. Until recently most mathematicians that I have known would think up new results and freely share those results with anyone who might be interested. In part, this might be because mathematical results most often have no immediate monetary value—but not always. For example, how much was it worth to crack the secret codes of the Germans in World War II? (Quite a bit!)

Some of this "doing it for free, for love of the subject" ethic still exists in computer science as the next exercise shows. Although there are some modest costs that must be paid, a co-operative, public model for doing business in computer software is emerging as a viable alternative to predatory competition. Which operating system do you use? You get what you settle for.

Exercise 28.1 Linus Torvalds vs. Bill Gates: Public vs. Private; Distributed vs. Central Control in Computer Operating Systems.

In order to function, a computer needs what is called an "operating system." As I write one of the best known operating systems is Windows^{TM}, owned and sold by the Microsoft Corporation. Another operating system is called Linux, the creation of which was initiated by Linus Torvalds in 1991 while he was a student at Helsinki University, Finland. There are significant differences between these two operating systems. The Windows^{TM}operating system is copyrighted. Windows^{TM}is physically a computer program consisting of lines of code. Unless you are working on this system for Microsoft Corporation you cannot see any of the lines of code which make up Windows^{TM}. It is secret, protected by law. The Windows^{TM}operating system is proprietary.

Linux, on the other hand, is "copylefted." It is is distributed under what is called a GNU

Public License, cf., www.gnu.org.[4] Linux, unlike Windows^{TM}, is *open source.* This means that anyone can see the lines of code that make up Linux. Linux is available for free. Linus Torvalds manages thousands of volunteers via the internet in the development of Linux, while he and the volunteers simultaneously maintain paying jobs. At this writing there are several versions of Linux: Redhat, Debian, Slackware (the system I am using at this very minute), and several others. I purchased my copy of Linux Slackware on a CD for two dollars at a "citizen grassroots Linux meeting." I am not a computer scientist, but I have been told by more than one computer scientist that Linux is more reliable than Windows^{TM}. What is lacking at the moment are certain applications and ease of installation, but these things (as all things computer related) are changing rapidly.

(i) An article in the business section of the June 4, 2001 *The New York Times* by Laurie J. Flynn is titled, "Despite Microsoft's best efforts to kill it, the free-software movement shows no sign of quietly rolling over and dying." The article reports on Linus Torvalds' development of Linux and Richard M. Stallman's Free Software Foundation (founded by Stallman in 1984). The article mentions that IBM will spend $1 billion on GNU-Linux in 2001 to help make the software a standard. The article also mentions that GNU-Linux is now the fastest-growing operating system for network-server computers. In 1992 there were 1,000 GNU-Linux users, in 2001 there were 9 million.

Eben Moglen, a law professor at Columbia and general counsel for the Free Software Foundation, wrote: "Microsoft, which used to say all the time that the software business was ruthlessly competitive, is now matched against a competitor whose model of production and distribution is so much better that Microsoft stands no chance of prevailing in the long run. They're simply trying to scare people out of dealing with a competitor they can't buy, can't intimidate and can't stop."

Bruce Perens, an adviser to Hewlett-Packard, has said, "In contrast, the business model of open source is to reduce the cost of software development and maintenance by distributing it among many collaborators."

Linux Torvalds has said:

I'm a big believer in Darwinism. I believe that the better system will prevail in the end.

It appears to be a nontrivial problem to estimate the number of computers running Linux. The "Linux Counter" estimated 18 million worldwide in 2001 and 29 million in 2010. Of course, you can be considered at least a partial Linux user if you interact with a computer running Linux. Put this way it is difficult to estimate who is *not* a Linux user. Do your best to estimate the status of the Microsoft vs. Linux competition when you read this.

(ii) Which product do you think will be the most used in twenty years, Microsoft Operating Systems or Linux Operating Systems?

(iii) Hunt down a copy of the famous internal Microsoft "Halloween memo" which leaked out and which discusses how Microsoft intended to deal with the "Linux problem."

(iv) What are the comparative advantages (disadvantages) of open source software and what are the comparative advantages (disadvantages) of closed source, i.e., proprietary, software like Windows^{TM}?

Compare your comments with the "gnu manifesto" at www.gnu.org/gnu/manifesto.html. Does the Web site bring up any issues you did not think of? Do you agree or disagree with the various statements made on the Web site about GNU licensing?

(v) If you use (or have used) the Windows XP^{TM}, Vista^{TM}, Windows 7^{TM}, or subsequent MS operating system, how much information about you is owned by Microsoft Corporation?

(vi) What is active X technology (and its successors), and what does it have to do with your privacy?

---

[4]GNU stands for GNU's not Unix. Unix is another copyrighted operating system which has been popular at universities and in certain mainframe business applications. The fact that Linus Torvalds liked Unix but could not afford it was one reason that led to his development of Linux, I am told. If you read the Web site www.gnu.org/gnu/manifesto.html many questions about the nature of GNU licensing are answered.

28.4 Medicine for People or for Profit or Both?

On April 19, 2001 there appeared a United Press International article by Les Kjos titled "Pharmaceutical companies investigated for kickback scheme." From the article: *"The way the scheme allegedly works is that the drug makers inflate the wholesale price of the drugs they sell, which is the benchmark used by Medicare and Medicaid when they determine payment. Then the drug companies sell their products to doctors for less than the said wholesale price. The doctors can then bill Medicaid and Medicare the full wholesale price."*

At the time the article was published, offices of six state attorneys general and the Justice Department were investigating the above scheme. The Florida Assistant Attorney General, Mark Schlein said: *"It's bad for government, for taxpayers and for patients. If you're being treated by a doctor, you need to believe the drugs he is prescribing are the best for you. When you inject this element of greed, and the doctor says, 'When I prescribe a certain drug I get $150,' it has the potential of corrupting the doctor and patient relationship. I'm also suggesting some middlemen might bribe them by splitting the difference. I'm not suggesting that most doctors do that, but some do,"* Schlein said.

Bayer Corp. settled with the Justice Department on January 23, 2001 for $14 million. Bristol-Myers Squibb was being investigated by attorneys general in Florida, Nevada, Massachusetts, Georgia and Maine.

Zachary Bentley was the whistle blower in the Bayer case and received $1.5 million of the settlement. Bentley said he got involved when his Key West, Fla., pharmacy, Venacare, was asked to participate in a scheme Bentley thought defrauded Medicaid and Medicare. He said he was told if he didn't participate, the pharmaceutical companies would *"run him out of business." "They were true to their word. They ran me out of business,"* Bentley said.

If you are working full time but are sleeping on a heating grate near the subway because you cannot afford more luxurious accommodations, health care may not be your most immediate concern—probably eating comes first. But if the same standard of health care presently available (in the United States) could be made universally available for less cost if only it were organized more efficiently and honestly—we should not settle for less.[5]

It is possible to organize health care along several different models. Could public-minded, government supported scientists develop new drugs? They already do on National Institutes of Health grants, and often the results are handed over to private corporations—the public pays twice.

[5]It is reported (*The New Yorker*, November 6, 2006, page 45) that a Republican congressman became the pharmaceutical industry's top lobbyist (at a $2(10^6)$ dollar a year salary) after negotiating into law the bill that forbids the government to negotiate prices for prescription medicines. This provision remains in the 2010 "health care reform." So much for "free market" efficiencies.

Books have been written on how to make state-of-the-art medicine and pharmaceuticals more economically available to all Americans, but we get what we settle for, cf., [571, 114, 252, 311, 257, 245, 13, 626].

Exercise 28.2 Public and Private Models in Medicine

(i) Before the passage of "health care reform" in 2010, more money was spent denying medical (insurance) coverage than would have been required to insure the tens of millions of Americans then without health insurance. Has this situation changed at the time you read this? What is "single-payer" (or more precisely "single-risk-pool") health care insurance? What is "medicare for all?" What is socialized medicine, and is it different from single-payer (single-risk-pool) insurance or medicare? Read [571] and observe that there are many models of health insurance/health care operating successfully (define success!) and at less than the cost per person of the U.S. model. The model in Switzerland is built around (nonprofit) private insurance companies, for example. However, the U.S. is the only country which builds health insurance around for-profit insurance companies. Why do you think this is? What are the consequences?

(ii) Jonas Salk (1914–1995) discovered the polio vaccine in 1955 while he was the head of the Virus Research Laboratory at the University of Pittsburgh School of Medicine. Dr. Salk never made a penny from it. When asked who owned the vaccine, Salk answered: *"the people."* Dr. Salk was famous, not obscure. He was not poor. Should he have held out for a patent and a "piece of the action?" What could have motivated him? Did Salk have any help in discovering the polio vaccine? Was there any competition among scientists to find the polio vaccine?

(iii) In 1980 a new law in America, the Bayh-Dole Act, allowed medical academies to patent and license discoveries made on federally funded projects. Researchers are now allowed a cut of the royalties. One effect of this, as Marcia Angell, at one time the editor of the *New England Journal of Medicine*, said: *"Increasingly, academic medicine is merging its mission with that of the drug companies."* Why do you suppose this is happening? What return on investment goes to the taxpayer from which federal funds flow?

(iv) What change in research priorities might there be in going from "public medical research" a la Salk to joint academic/corporate research?

(v) Which model provides the greater sharing of information of research discoveries, the public, "Salk" model, or the corporate or corporate/academic joint venture model? Which is analogous to open source software? Which is proprietary?

(vi) When a university joins a joint business venture, does that affect who becomes a professor? Does that affect your confidence in what the professors say? Does the professor have a conflict of interest?

(vii) I read an article by Richard Gwyn in the March 7, 2001 issue of *The Toronto Star*, which pointed out that 31 multinational drug companies had gone to court in South Africa to get a law there declared unconstitutional. That law empowers the South African government to "prescribe conditions for the supply of more affordable medicines . . . so as to protect the health of the public."[6] GlaxoSmithKline, one of the 31, had just announced profits for the year 2000 to be 5.8 billion U.S. dollars. The entire Third World accounts for less than 2 percent of the drug company's total sales, according to Gwyn's article. What is 2 percent of 5.8 billion dollars?

Njoke Njehu, a social activist from Kenya, stated in a speech in Boulder, Colorado, on March 8, 2001, that a typical AIDS/HIV "cocktail" treatment costs between 10,000 and 15,000 U.S. dollars per patient per year. The World Health Organization has a special program which offers the same brand name drugs at a cost of 2,000 U.S. dollars per patient

[6] The BBC (British Broadcasting Corporation) announced on April 19, 2001 that the (eventually 39) pharmaceutical firms suing South Africa had dropped their law suit. It was conjectured that they did so since the lawsuit was generating considerable negative public relations.

per year. Ms. Njehu stated that generic versions of the "cocktail" manufactured in India cost between 239 and 250 U.S. dollars per patient per year. The article by Gwyn states: *"...the Indian generic drug company Cipla is offering South Africa a cocktail of drugs at $350 to $600 per patient, while Brazil has offered to supply generic drug-manufacturing equipment."*

Are Gwyn and Njehu figures on drug costs consistent with one another? Are there people in the so-called First World who cannot afford life-saving drugs? What is the annual income after taxes of a person on minimum wage in America? What is the annual income per person of the majority of people in South Africa? In Africa in general?

If you could make the decision in the South African court case mentioned above, what would you decide?

(viii) Gwyn states: *"In a public relations counter-strike, the drug companies have cut their prices to Third World countries by as much as 80 percent (for certain products under certain circumstances). And they've negotiated cheap drug agreements with three African countries, although these are mostly for show, since the agreement with Senegal provides these drugs to just 500 patients. At whatever cost, these drugs are merely palliatives. A real solution would require the development of an AIDS vaccine. But since there's now limited demand for it in developed countries, the drug companies are spending only $300 million on vaccine research. 'Patients before patents.' That's a war cry that's going to become louder and louder."*

At an 80 percent discount, what would the cost range be per patient per year for an HIV/AIDS "cocktail?" Comment on Gwyn's statement.

What is the status of HIV/AIDS drug distribution in the world at the time you read this? Who is paying for the drugs and how much?

(ix) What was the prevalence of polio before the Salk vaccine was introduced? What effect did polio have on people who contracted it? What is the incidence of polio when you read this? Do you know anyone who has had polio? What was the role of the human cells, unknowingly "donated" to our medical "commons" by Henrietta Lacks, to Salk's work? Who was Henrietta Lacks, cf., [639]?

(x) Is there any reason why a person in, say, America or Western Europe, should care about helping people in developing nations control disease? Consider HIV, tuberculosis, or malaria, for example.

(x) In the April 6, 2001 edition of *The Nation* author John le Carré has an article: "In Place of Nations." He refers to the multinational pharmaceutical world as Big Pharma, and the following are two quotes from that article. *"In a few years time, if Big Pharma continues unchecked on its present happy path, unbought medical opinion will be hard to find."* Also: *"Big Pharma did not invent these lifesaving drugs that they have patented and arbitrarily overpriced, incidentally. Anti-retrovirals were for the most part discovered by publicly funded US research projects into other diseases, and only later entrusted to pharmaceutical companies for marketing and exploitation. Once the pharmas had the patent, they charged whatever they thought an AIDS-desperate Western market would stand:* $12,000 *to* $15,000 *a year for compounds that cost a few hundred to run up. Thus a price tag was attached, and the Western world, by and large, fell for it. Nobody said it was a massive confidence trick. Nobody remarked that, while Africa has 80 percent of the world's AIDS patients, it comprises 1 percent of Big Pharma's market."* Comment on each of these quotes. In particular, do you believe them? Do you have any evidence to back up your beliefs? In the decade(s) since Carré made his comment about "unbought medical opinion," to what extent has his prediction come true?

(xi) The U.S. government funds basic health care research with many billions of dollars per year through such agencies as the N.I.H., the National Institutes of Health. Private pharmaceutical companies spend money bringing drugs to market. As best you can, find out how much public money is invested in innovation of new drugs vs. how much private money is so invested. Similarly find out how much public money is invested in bringing drugs to market (field testing, marketing etc.) vs. how much private money is so invested. The quantitative answers to these questions are key to the debate as to whether private pharmaceutical firms are setting the prices of drugs at a "fair" level or not. (Note: A reference from the consumer advocate point of view is *Rx R&D Myths: The Case Against the Drug Industry's R&D Scare Card,* by Public Citizen, 1600 20th Street, N.W., Washington, D.C. 20009-1001. Phone 1-800-289-3787.) How much does the pharmaceutical industry spend on advertising every year?

Sports: Community vs. Corporate. In sports you get what you settle for, as in everything else.

Exercise 28.3 Community Owned Football

The Green Bay Packers are a community owned football team, founded in 1923. It is the only such, since the "modern" football leagues made such community owned teams illegal.

(i) Why do you think they did this? Who are "they"?

(ii) Do you think professional sports would collapse if it followed the Green Bay Packers model? How might professional sports be different if the Green Bay Packers model were followed?

28.5 A Little History

I will let [124] bring to your attention the ecological economists' perspective on Adam Smith (1723–1790) and David Ricardo (1772–1823). I will only mention that current mainstream discussion of these and other classical economists definitely filters out a good deal of their world-view. As for the invisible hand, I mention that to the extent that it exists (and is not negated by the tragedy of the commons) it is an example of self-organization, (see also the work of Elinor Ostrom, [508, 509]).

Class Structure is Built Into the U.S. Constitution! I do want to point out a connection between economics and politics discovered by a particularly honest American historian, Charles Austin Beard (1874–1948).[7] I found two books by Beard: *The Economic Basis of Politics*, Alfred A. Knopf, New York, 1922, 1934, and *An Economic Interpretation of the Constitution of the United States,* Macmillan, New York, 1913, 1935. (There are more recent printings of both books.)

The first book is a grand tour of economics, politics and ethics starting with Aristotle and ending in what was then modern times. Beard states that his (Beard's) book is a confirmation of the thesis of James Madison ("Father of the United States Constitution"), to wit:

> *civilized societies are divided into economic groups of interests, according to different degrees and kinds of property-possessions and occupations, whether private or bureaucratic; and forms of government rest upon this social configuration, and politics is concerned with conflicts among interests.*

[7] He once said: *"You need only reflect that one of the best ways to get yourself a reputation as a dangerous citizen these days is to go about repeating the very phrases which our founding fathers used in the struggle for independence."*

Beard quotes the following excerpt of Madison's from *The Federalist*, Number 10:

> *The diversity in the faculties of men, from which the rights of property originate, is not less an insuperable obstacle to a uniformity of interests.* **The protection of these faculties is the first object of government.**[8] *From the protection of different and unequal faculties of acquiring property, the possession of different degrees and kinds of property immediately results; and from the influence of these on the sentiments and views of the respective proprietors, ensues a division of society into different interests and parties ... .*
>
> *The most common and durable source of factions has been the various and unequal distribution of property. Those who hold and those who are without property have ever formed distinct interests in society. Those who are creditors, and those who are debtors, fall under a like discrimination. A landed interest, a moneyed interest, with many lesser interests, grow up of necessity in civilized nations and divide them into different classes, actuated by different sentiments and views. The regulation of these various and interfering interests forms the principal task of modern legislation, and involves the spirit of party and faction in the necessary and ordinary operations of government.*

In Beard's preface to the 1935 edition of *An Economic Interpretation of the Constitution of the United States* I read the following passage which resonated with me for two reasons: he admonishes one to "check out the original sources" and after doing so he discovers that "I was never taught that in school!" Beard says:

> *One thing, however, my masters taught me, and that was to go behind the pages of history written by my contemporaries and read "the sources." In applying this method, I read the letters, papers, and documents pertaining to the Constitution written by the men who took part in framing and adopting it. And to my surprise I found that many Fathers of the Republic regarded the conflict over the Constitution as springing essentially out of conflicts of economic interests, which had a certain geographical or sectional distribution. This discovery, coming at a time when such conceptions of history were neglected by writers of history, gave me "the shock of my life."*

Although I have had bigger shocks in my life than the discovery that America's founding fathers had class structure on their mind, before preparing for this course I was at best vaguely aware that economic class considerations played such an enormous role in the minds of the drafters of the supreme law of my land. Economics and politics are indeed inseparable and have been since the time of Aristotle (at least), and they are indeed inseparable in the very foundation of my country.

Except for Benjamin Franklin, the founding fathers were OK with slavery, which is definitely a class structure.

It is a challenging exercise to calculate reparations to the descendants of slaves, cf., [580, 730, 38], but mathematical estimates can be made. The

[8] Emphasis mine. Thus the founding fathers wrote the Constitution, in part, to protect class structure.

politics of the situation is another thing altogether. Real slavery persisted as late as the 1950s in the United States. In fact, real slavery exists in many places around the world today.[9]

The existence of slavery today says something about humans in general. It is troubling to think that our fellow humans could enslave us for real, or for certain employ us under conditions approaching slavery such as in sweatshops.[10]

I was once challenged by an administrator, who was dealing with student protests against sweatshops, to prove that sweatshops exist. I did, cf., [185].[11]

---

[9] I read a story by Len Cooper in the June 16, 1996 edition of *The Washington Post* documenting the existence of real slavery in the United States as late as the mid 1950s, although slavery was declared illegal nearly a century before. I heard about this article on the July 3, 1996 radio program "Democracy Now." Mr. Cooper started his research for this article in 1984. I had to actually see this article to believe that it even existed!

Incidentally, I read an Associated Press article on Friday, March 17, 1995, which noted that the state House of Representatives of Mississippi ratified the 13th Amendment, which abolished slavery (in 1865). In 1865 the union comprised 36 states, and only Mississippi had not ratified the amendment.

A reference you may find interesting and disturbing is [22]. Desmond Tutu has this to say about the foregoing book: *"A well-researched, scholarly and deeply disturbing exposé of modern day slavery with well-thought-out strategies for what to do to combat this scourge."* Also Volume 29, No. 1, January/February 2000 Issue 192 of *Index On Censorship*, (www.indexoncensorship.org), is devoted to "The New Slavery." A general history of slavery is [452].

Unfortunately it is a fact that slavery is all too common globally. In the April 2001 edition of *Scientific American*, pp. 80–88, there is an article by Kevin Bales titled "The Social Psychology of Modern Slavery." From this article I quote: *"Contrary to conventional wisdom, slavery has not disappeared from the world. Social scientists are trying to explain its persistence."* What relationship, if any, can you find between the global economy and slavery.

[10] Sweatshops are places of employment characterized by long hours, low wages and generally unhealthy conditions.

[11] The starting point of my logic was the following axiom: any profitable action that is possible will be taken by some business unless all businesses are restrained from taking such action. Since sweatshops are certainly possible and are not prohibited (and not even discouraged), for example by the World Trade Organization, sweatshops will happen. (There is a variation of the traditional sweatshop sometimes called the "invisible assembly line." This refers to the outsourcing of sweatshop labor to "sweat-homes." There are no benefits, large quotas, long hours, easy firing but short commutes to work. Sweat-homes are hard to find by authorities or nosey human rights activists, and it is even harder to enforce a minimum wage and other standards.)

I once went to a panel discussion where several executives of corporations stated that they had to go abroad and seek cheap labor to be competitive. (They did not say they were using sweatshops, but ... .)

One of the organizations most persuasive in its documentation of sweatshops is the National Labor Committee (NLC); See www.nlcnet.org in particular, its Executive Director Charles Kernaghan has done some rather definitive sleuthing.

For example, while searching a garbage dump in the Dominican Republic, the NLC found Nike's SAM requirements, i.e., Standard Allotted Minutes. For every step of every item, Nike calculates (within .001 seconds!) how much time it allows for that step to be carried out. For example, the time allotted for the creation—and labeling and packaging—of a kiddie sweatshirt (starting with raw cloth) is 6.6 minutes. Wages in Dominican sweatshops

28.6 An Example of the Need for Fuzzy Logic: The Definition of Poverty

Let us look at a simple, but fundamental example that amply illustrates the need for and use of fuzzy logic.[12] Suppose we want to discuss wealth and poverty, concepts important in politics and economics. For example, I want

(at the time of writing) are about 70 cents per hour. The sweatshirt in question retails for $22.99. Question: What fraction of the final retail price is for labor by the workers who created the garment out of raw cloth? (Answer: $\frac{.077}{22.99} = .00335$. The NLC generously rounds up its calculations and gets .005.)

Kernaghan has documented wages in Cambodia of 19 cents per hour, in China 20 to 25 cents per hour. He has used "back engineering" to verify such numbers. By getting documents detailing, for example, declared values at points of entry into the United States, and then subtracting verifiable costs for materials and transportation (all known nonlabor costs), one can verify labor costs.

The NLC has connected Wal-Mart, J.C. Penney, Sears, Target and many others to sweatshops. Yet Kernaghan is cautious and does not call for boycotts, rather a flood of letters and e-mails to the appropriate CEOs.

More excuses are given for these labor arrangements than I have space for here, but I will mention a couple. Often the U.S. corporation says it is not responsible since it subcontracts jobs out. I regard this as a very lame answer.

Possibly more credible is the statement that "living wages" are low in the countries where sweatshops are located. It is not a difficult exercise to calculate what food costs in such countries, and determine whether on that basis alone the wages are appropriate.

Jim Keady was an assistant coach for the Saint Johns University men's soccer team, but he quit rather than be a walking advertisement for Nike whom he believed condoned sweatshop production of Nike goods. After being criticized again and again for never having been to a sweatshop, Keady decided to see for himself. In May of 1999, Jim Keady offered to work for six months in a Nike shoe factory in South-East Asia to dispel the myth that "these are great jobs for those people." Brad Figel of Nike's Labor Practices Department responded, "We are not interested in your offer." So, Keady and project assistant, Leslie Kretzu, went to Jakarta, Indonesia and adopted the lifestyle, diet, customs and culture of the factory workers; and they lived on the prevailing wages over approximately a two-month period. They found living on $1.20 a day quite challenging; search the internet for "Keady + Nike" for their full report.

The argument that "these jobs are better than anything else they can get" loses credibility every time native farmers are moved off their land (sometimes due to so-called structural adjustment requirements of the International Monetary Fund) so that export crops can be grown instead of food for the local population. More often than not, displaced farmers find themselves in cities looking for work—at a sweatshop. Beatings and bad air in some of these facilities have been documented.

Other key elements in the "proof," or "convincing argument," that sweatshops exist is that labor organizing is not allowed. My university administration's response to this was that collective bargaining by state employees is not allowed at my university either. Secrecy also raises suspicions. Always follow the money and see how much of the value of the labor and resources of a region goes back into education, education of women, in particular. This is one of the necessary conditions for population stabilization, and poor regions are often criticized for their fecundity. If those who rule, export the wealth and do not educate then .... I have far from exhausted the evidence in support of the existence and nature of sweatshops, but I must go on. You can continue, however.

[12] This example was taken from [597, p. 8].

to define what it means to be poor, or, equivalently, define the "set of poor people." This definition may vary as one travels around the world, but in the United States the federal government has a method of calculating what it calls the "poverty line."

In the mid-1960s, Mollie Orshansky, an employee of the Social Security Administration, was assigned the task of defining an official poverty line.[13] She assumed that food was one-third of a (poor) family's budget and then went about calculating how much money it would take to keep a family properly fed. From this Orshansky deduced the official poverty line for families of various sizes.

The official U.S. poverty line in 1995, adjusted for inflation, for a family of 3 was $12,158. For family of 2 it was $10,259. For a family of 1 it was $7,929. For a family of 9 it was $31,280.

Exercise 28.4 Problems With the Poverty Line

(i) A family (in 1995) of 3 with an income of $12,157 is poor. Is a family of 3 (in 1995) with an income of $12,159 not poor?

(ii) Do poor families still spend about one-third of their income on food? If not, what has become a much larger part of the poor family's budget than it was in the mid-1960s?

(iii) Does the "real" definition of poverty depend on where you are living?

(iv) Long ago it used to be the elderly who were disproportionately poor. Today, single mothers and children have that distinction. Why?

(v) Adjusting $7,929 for inflation for the year you are reading this, could you as a single person live on that amount of money for one year?

(vi) What is the definition of the poverty line in the U.S. at the time you read this? Has fuzzy logic been incorporated into any federal legislation?

I now want to tackle the problem of defining poverty in Denver, Colorado, in the year 2000,[14] and in doing so show why fuzzy logic is necessary.

Certainly anyone who has an annual income of $100 or less is in the set of poor people. Most people would also agree with the statement that: *If a person has an annual income of X dollars and is poor, then if that annual income is increased to $X + 1$ dollars that person is still poor.* If you are poor, getting one more dollar a year cannot get you out of poverty.

We are now in a rather absurd situation. We can start with a person who has an annual income of $100 and declare that person poor. Then certainly a person with an annual income of $101 is also poor, as is a person with an annual income of $102, of $103 and so on. By repeating the argument 1,000,000 times we end up with a person with an annual income of $1,000,100 being declared poor. It certainly is silly to declare a person with an annual income (in the year 2000) over $1,000,000 poor, you say—as would any government statistician.

To avoid this silly situation the U.S. government has by law defined a "poverty line" as discussed above. But now we have a new silliness. Whatever

[13]See Deborah A. Stone, "Making the Poor Count," *The American Prospect*, Spring 1994, pp.84–88.

[14]You should do this exercise for the city closest to you, for the year you are reading this.

that "poverty line" is, say it is Z dollars, then anyone making $Z + 1$ or more dollars a year is not poor. Thus one dollar does make a difference between the legal definition of being poor or not poor, contrary to our intuitive understanding of poverty. Unfortunately for some people, their annual income is just "over the poverty line;" and such people are denied food and/or health benefits, say, that they would have gotten had they been "just under the poverty line." As it actually happens, some people "just under the poverty line" get benefits that make their effective real annual income greater than someone "just over the poverty line."

Whatever your opinion might be concerning helping out a fellow citizen in poverty, you probably are thinking that there must be a better way to define poverty—and there is. Fuzzy logic to the rescue!

We can introduce a poverty (or wealth) function for a single person, as shown in the Figure 28.1.

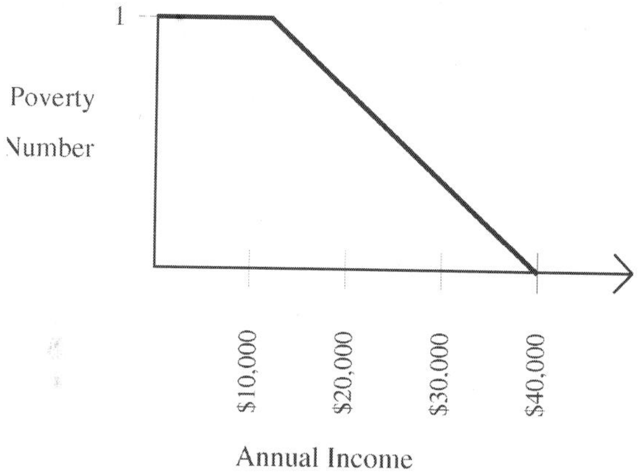

FIGURE 28.1: A Fuzzy Definition of Poverty

Thus according to the poverty function in Figure 28.1, if your annual income is less than $12,000 your "poverty index" or "poverty number" is 1. If your income is over $40,000 your "poverty index" is 0. If your annual income is between $12,000 and $40,000 then your "poverty index" is somewhere between 1 and 0.

Now I just made up the poverty function in Figure 28.1. If I really wanted to represent what the people in Denver, Colorado meant in the year 2000 by the word "poor," I would do a survey in that year and ask everyone what they meant. I could then assign a "poverty index" to a given annual income, using the results of the survey. This "poverty index" could then be used for any number of purposes. For example, assistance in buying needed food and shelter could be allocated in a manner which very closely reflected the current meaning of the word "poor" in Denver in the year 2000. There would be no dreaded "poverty line." Such a line is an artifact of "sharp Aristotelian logic" which is much less precise than fuzzy logic for representing the concept of poverty. Do you think that the government should actually use fuzzy logic to define poverty? If you do, tell your representatives in congress.

A technical point worth noting is that in the real world the above mentioned survey (to find out what people in Denver mean by the word "poor") would be done using rather sophisticated mathematics called statistical sampling. By cleverly selecting a small fraction of the population of Denver to participate in the survey we could come very close to what we would find if we included everyone, and at a fraction of the cost of asking everyone. It is also important to note that the graph of the "poverty index" vs. annual income in Figure 28.1 is not a probability function. The graph in Figure 28.1 illustrates the fuzziness in the meaning of the word "poor;" it does not represent any probability, such as the probability of having a given annual income.

Chapter 29

Energy and Thermodynamics

29.1 Energy and the First Law of Thermodynamics

What is thermodynamics, you ask? Thermodynamics determines and studies the relationships among the various properties of materials, without knowing the internal structure of the materials. Thermodynamics, in particular, deals with energy; its transmission from one body to another and its transformations from one form to another. The laws of thermodynamics were discovered empirically, i.e., through many observations and experiments, before humans understood molecules and atoms and the microscopic structure of matter.[1] As the prefix "thermo" suggests, thermodynamics started with studies of heat, steam engines to be precise. In fact, engineers started and led the early study of thermodynamics, not physicists or mathematicians. The *caloric theory* of heat was dominant in the beginning of the study of thermodynamics. This theory holds that heat is a material substance called caloric, and possession of caloric is what makes hot objects hot. Scientists of the day used the caloric theory to explain thermal expansion and why water boils when heated and many other phenomena. The unit of heat energy in name and concept, the Calorie, is a legacy of the caloric theory.

When James Joule began his serious study of energy in 1837, he had adopted the *kinetic theory* which holds that what makes a body hot is the kinetic energy (energy due to motion) of invisible molecules that make up the body.[2] Since the kinetic theory of heat is taken for granted today (after all, we now have tools that can "see" molecules for us) it is perhaps difficult for today's students to appreciate how genuinely revolutionary the kinetic theory was when it was introduced—and remained for some time afterward.

Work, Potential Energy and Kinetic Energy. Being well-versed in Newton's laws, Joule knew that it took work to set a body in motion and that the kinetic energy (energy due to motion) of the body was precisely equal to the work

[1]Sadi Carnot (1796–1832) and James Joule (1818–1889) are credited with discovering the Second and First Laws of Thermodynamics, respectively. The Second Law of Thermodynamics was discovered before the First!

[2]Joule's study of energy was motivated by his desire to find an inexpensive source of power for his family's brewery!

done in getting the body moving. The following exercise gives a little practice in understanding this last sentence.

Exercise 29.1 Work, Mechanical Energy and Kinetic Energy

Physicists define *work* as a force, like a newton, acting through a distance, such as a meter. In particular, the work, W, done by a force F acting through a distance D is given by: $W = FD$, where juxtaposition of the F and the D means multiply the two.

A *joule* of work is 1 newton-meter $= 1\ Nm$. (A *newton* is a metric unit denoting force, a *meter* is a metric unit of distance.) A *joule* of energy is also 1 newton-meter. It takes 1 J (one joule) of energy to do 1 J (one joule) of work. It is helpful to know[3] that $1\ newton\ = 1\ \frac{kg-m}{sec^2}$. It is also helpful to know that $1\ kg\ = 2.2\ pounds$, approximately. This last equation means one kilogram of mass equals 2.2 pounds of mass. The English units are confusing, since the term "pound" is also used to denote an entirely different quantity, namely, force. It takes one pound of force to lift one pound of mass (at the surface of the earth). One pound of force equals 4.45 newtons, i.e., $1\,lb\ (force) = 4.45\ N$.

(i) If a book weighs 5 pounds (thus it takes 5 pounds of force to lift it) and is lifted vertically through a distance of 4 feet, how many joules of energy does it take?[4]

(ii) If a person who weighs 220 pounds climbs 3,000 feet to the top of a small mountain, how much energy in joules does it take, neglecting horizontal distance traveled and friction, and so forth, only taking into account the vertical distance traveled?

Our mountain climber sitting on top of the mountain in part (ii) has increased his/her *potential* energy by the amount of work done. The formula for the *kinetic* energy, $E_{kinetic}$, of a mass m traveling at a speed v is $E_{kinetic} = \frac{1}{2}mv^2$.

(iii) If the energy climbing 3,000 feet expended by the mountain climber in part (ii) were completely converted into kinetic energy (as would happen in a frictionless—no air resistance—and otherwise unencumbered vertical free fall all the way down to the bottom of the 3000 foot cliff face, of the mountain), what would be the person's speed?

Further on, Exercise 29.16 will give you some more practice in doing energy calculations of the type above. (You could actually look at that exercise now if you wanted to see how much of a certain type of food you would have to eat to climb a given mountain.) In the above exercise, mechanical work, i.e., exerting a force through a distance, created either potential or kinetic energy depending how it was done. (Climbing the mountain and coming to rest at the top created potential energy. The force of gravity acting on a person free-falling back down the same height produces kinetic energy.) Thus an equivalence between work and kinetic (or potential) energy is seen. Joule saw this equivalence clearly.

Thermal Energy. Now suppose you have a tire pump which you use to compress air to blow up a tire. What happens to the pump? It gets warm, or even hot. Thus the total force times distance you exert (the total work you do) using the tire pump is a certain amount of mechanical energy. Some of this mechanical energy goes into the energy required to inflate the tire. Some of your work goes into heating the air in the pump and also into heating the pump itself.

[3]Note that $kg - m$ means kg times m.

[4]Hint: 1 foot-pound $= 1.356\ J$.

Joule carefully measured the energy into and out of such processes and found that the energy *in* equaled the energy *out*. But in the case of the tire pump, mechanical energy produced, at least in part, a different form of energy, namely heat or *thermal* energy. Joule reasoned correctly (from the current molecular theory of matter) that the heat energy was due to the kinetic energy or energy due to motion of the molecules in the air and the pump.

The next exercise deals with the conversion of joules of energy to "thermal" units of energy.

Exercise 29.2 joules and calories

Using modern equipment it is found that $1\ J = .2390\ cal$ (calories). Recall that $1\ cal$ is the amount of energy it takes to raise the temperature of 1 gram of pure water 1 degree Celsius. A *kilocal* or $1{,}000\ cal$ is denoted by $1\ Cal$.

(i) If the climber in parts (ii) and (iii) of the previous exercise converted all of his/her energy gained[5] into thermal energy, i.e, heat—how many calories would that be? How many Cal?

(ii) Suppose a one ounce lead bullet is traveling at twice the speed of sound and then hits an armored plate. Suppose all of the kinetic energy of the bullet is converted to thermal energy. How much thermal energy is that? It is useful to know that $16\ ounces = 453.6\ grams$. Also the speed of sound in air is $331.4\ meters/second$.

(iii) If the bullet in part (ii) is made from depleted uranium (pure U^{238}) and has the same volume, approximately how much thermal energy is generated? It is useful to know that the density of lead is $11\ g/cm^3$, and the density of uranium is $19\ g/cm^3$.

(iv) What forms of pollution do lead bullets cause? What forms of pollution do depleted uranium bullets cause?

Electric Energy. In Joule's time it was understood that if one rotated a coil of copper wire in a suitably situated magnetic field then an electric current was created in the wire. Such a device is called an electric generator. Conversely, if you take a similar device (a copper coil in a suitably situated magnetic field) and run an electric current through it, you can cause the coil to rotate and now you have an electric motor—capable of doing work. In fact, Joule experimented with such devices in his search for motors to run his family business.

One of the well known applications of this technology today is the hydroelectric dam. Water stored behind a dam has a lot of potential energy. The water was lifted into the air, most likely via solar energy, it then precipitated out and eventually flowed into the lake behind the dam. On occasion the water is allowed to fall through a pipe and turn a turbine which is connected to a massive coil of copper wires rotating in a suitably situated, massive magnetic field. Thus the potential energy of the water is converted to mechanical energy, turning of the turbine, which converts that mechanical energy into electrical energy. Not all of the water's energy is converted into electrical energy in a

[5] The climber has potential energy if he/she is sitting on top of the mountain. The climber has kinetic energy after free fall off a 3,000 foot cliff back to the bottom—where the sudden impact takes the organized kinetic energy of the climber's body and converts it, via "energetic disassembly," primarily into unorganized molecular motion, i.e., thermal energy.

real hydroelectric facility, some is converted to heat energy due to friction of the moving parts involved. It turns out that this lost energy is a constant companion when changing energy from one form to another in the real world. This will be discussed when I get to the Second Law of Thermodynamics.

At this moment I need to recall that there is a difference between *energy* and *power.* One joule is a unit of energy. If I want to use that joule of energy to do work, I can expend that joule and hence do the desired work in a short time or a long time. Please forgive the example, but if a martial artist slowly extends his/her foot into the abdomen of a sparring partner and slowly pushes that partner away it is said that this is a low power movement. If that same movement is executed in a fraction of a second, it is said to be a powerful, or high power, movement. The results are also different. Thus I need to distinguish such situations. Physicists define a joule of energy expended in one second to be a watt, W, i.e., $1W = 1\frac{J}{sec}$. A watt is a unit of *power.* A kilowatt, kW is a thousand watts, i.e., $10^3 W$. In the following exercise I investigate the numbers for converting to or from electrical energy and power.

Exercise 29.3 Electrical Energy Conversion

(i) Is a kilowatt-hour a unit of power or energy? How much (state your answer using joules and/or seconds if necessary)?

(ii) Suppose the climber in Exercise 29.1 is able to convert all of his/her potential energy into electrical energy by means of a perfect frictionless machine (say an ideal cord attached to the climber, going over a pully that turns a turbine as the climber jumps off the 3,000 foot cliff and descends to the bottom of the cliff). How much electrical energy is generated?

(iii) Would the climber in this Exercise part (ii) descend slower, the same, or faster than if he/she was in free-fall, not attached to the cord that runs the electrical generator?

The Maximum Power Principle. Since I have just defined power I would be remiss if I did not at least briefly discuss what is known as the *Maximum Power Principle.* Howard T. Odum, cf., [503, p. 101], has this to say about the Maximum Power Principle: *Maximization of useful power may be the most general design principle of self-organizing systems... . It may be time to recognize the maximum power principle as a fourth thermodynamic law as suggested by Lotka (1922) and use it more centrally in basic teaching of science and engineering.*

Since I have not yet stated the First, Second or Third Laws of thermodynamics it may seem premature to be exposing you to what some consider the Fourth Law. However, there is a beautiful example from economics which anyone can understand which richly illustrates the Maximum Power Principle.[6]

The Maximum Power Principle has not yet been elevated to Fourth Law status by the scientific community, but it is certainly worth understanding. The basic idea is that for a physical or biological system that processes energy for its own use there is a trade-off between the rate at which energy is used,

[6]This example is adapted from [270, pp. 63–66]. In this book Hall et al. give an interesting example of the Maximum Power Principle from biology which is a bit less easy to analyze mathematically than the example I discuss from economics.

i.e., magnitude of power consumption, and the magnitude of the efficiency with which energy is used. At an intermediate point useful power output is maximized, and according to proponents such as Dr. Odum, Nature selects such systems that maximize useful power. It is not agreed that such selections take place in all cases of self-organizing systems or that the principle is universally applicable to all self-organizing systems. Nevertheless, I want to take a look at the following example from economics.

Suppose that at the top of our 3,000 foot vertical mountain cliff an industrialist, Joe, builds a textile factory powered by coal, e.g. a steam engine as existed in England in the early nineteenth century. (This is a "thought experiment," I do not suggest actually doing this.) The industrialist, Joe, has a supply of rocks on the mountain top that he can use to lift coal, which is in a mine at the foot of the cliff. Joe uses a pulley with two buckets attached to opposite ends of a cable running over the pulley to lift the coal with the rocks, as shown in Figure 29.1.

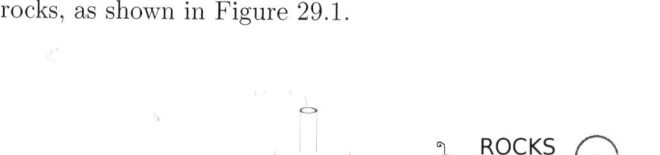

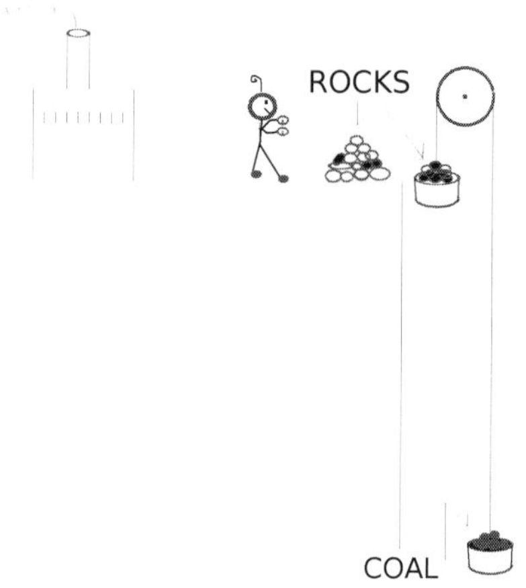

FIGURE 29.1: Maximum Power Principle

Now Joe is competing with two other industrialists, Sue and Moe, with identical factories sitting atop identical 3,000 foot cliffs, each with a coal mine at the base and a pulley machine with two buckets and an identical supply of rocks. Each of our three industrialists wants to capture as much market share as possible and maximize profit.

Now Joe thinks that the way to win this competition is to bring coal to the mountain top as quickly as possible, so Joe puts ten tonnes of rocks in one bucket to lift one tonne of coal in the other bucket. Sure enough, the bucket of rocks quickly descends, lifting the coal quickly to the cliff top (I could calculate quantitatively how fast, but I will restrain myself). Joe gets his coal quickly, but each load of rocks hits the ground with a terrific thud. Most of the potential energy that the rocks had at the top of the mountain is converted into thermal energy when the rocks hit the ground at the base of the cliff. Thus Joe converts only a small fraction of the potential energy of his rock supply to useful work, the running of the factory.

Now industrialist Moe is convinced that energy efficiency is the way to win the economic competition, so he puts 1 tonne of rocks—plus a one gram pebble—into the rock bucket to lift one tonne of coal. The extra pebble is enough to lift the coal (since Moe has an ideal, frictionless pulley machine); but the coal is lifted very, very slowly. This means that Moe's factory produces textiles very, very slowly. But the bucket of rocks contacts the ground losing almost no energy to impact and thermal energy generation. Moe is converting most of the potential energy of his rock supply to useful work, but the rate at which useful work is produced is very low.

Now Sue decides to use a strategy between those of Moe and Joe. She loads her bucket with two tonnes of rocks to lift one tonne of coal. Sue thus converts a much larger portion of the potential energy of her rock supply to useful work (than Joe does) and loses much less to waste heat (than Joe does) as her loads of rocks hit the ground at the base of the cliff. Sue's energy efficiency is greater than Joe's but less than Moe's. Sue is bringing coal to her cliff top much more quickly than Moe, but slower than Joe.

The important observation, however, is that Sue is doing something better than either Joe or Moe; but what, exactly, is she doing better? Answer: Sue is producing more (than Joe or Moe) useful work per unit of time (maximum useful power) per unit of resource (rocks) consumed. Sue's costs of production are correspondingly less, and she is more competitive in the market. The following exercise goes into some (but not all) of the quantitative details.

Exercise 29.4 The Maximum Power Principle

(i) Do you see that Joe is using 9 times the net upward force (in excess of gravity's force) on his bucket of coal as Sue is using?

It turns out that the time the bucket of coal spends in transit from the bottom of the cliff to the top varies inversely as the square root of the bucket's acceleration (in this calculation you can replace acceleration by net upward force). Thus Joe's machine runs 3 times faster than Sue's. For this problem I will define "one unit of Sue time" to be the time it takes Sue to raise one tonne of coal from the bottom to the top of the cliff with 2 tonnes of rocks.

(ii) What is the cost of production in tonnes of rocks per unit of textiles produced per unit of Sue time for Sue and Joe? (You can simplify by equating a unit of textiles produced with a tonne of coal raised to the top of the cliff.)

(iii) How many tonnes of coal are raised to the top of the cliff (equivalently, how many units of textiles are produced) per "unit of Sue time" per tonne of rocks used by Joe? Answer the same question for Sue.

(iv) You can define "useful power produced" to be units of textiles produced per unit of Sue time, or, equivalently, the number of tonnes of coal raised per unit of Sue time. What is the ratio of Sue's useful power produced per tonne of rocks to Joe's useful power produced per tonne of rocks?[7]

[7]Hint: In part (iii) you just computed "useful power produced per tonne of rocks used" for

(v) Even if Sue and Joe each have an infinite supply of rocks, who could sell their goods more cheaply? How much more cheaply? Assume that they both pay the same non-zero amount for each tonne of "rock potential energy."

(vi) If Sue and Joe have a finite supply of rocks who will run out of rocks first? How much faster will one run out of rocks than the other?

(vii) Assuming that Sue and Joe have the same finite-size rock supply, how many more units of textiles will Sue produce than Joe? (You can neglect time.)

(viii) Why will the very energy efficient Moe not be able to compete with Sue in the real-world textile market?

(ix) Is there any strategy that would yield a higher useful power (per unit of rock resource) than Sue's strategy? In other words, does Sue's strategy produce the maximum useful power per unit of (rock potential energy) resource used? This problem is not trivial. Let $X + 1$ be the number of tonnes of rocks put into the bucket to raise one tonne of coal in the other bucket. For Sue $X = 1$, for Joe $X = 9$ and for Moe $X = 10^{-6}$. Is there a value for X that yields a higher maximum useful power (per unit of rock-resource) than what Sue gets?[8] [9]

(x) If the cost of rocks to Joe, Moe and Sue is zero, who wins the economic competition for as long as the rocks last? Does something like this happen when a business totally neglects costs it imposes on Nature or is subsidized unduly by society at large? (Joe's economic strategy, followed by some in real life, was called "strength through exhaustion" by David Brower.)

29.2 The First Law of Thermodynamics

I now will state the first law of thermodynamics and discuss what it means.

First Law of Thermodynamics (The Law of Conservation of Energy): *Energy can neither be created nor destroyed. The total amount of energy in a closed system is constant.*

A closed system is one from which nothing can escape (to the outside of the system) and into which nothing can enter (from outside the system). Perfectly closed systems are easier to imagine than to build in the real world. Now a good question to ask is: What is energy? Actually no one knows precisely. What we do know is that energy takes many measurable forms as I have discussed in previous exercises: kinetic, potential, thermal, electric, chemical to name a few. In experiment after experiment, with increasing accuracy

both Sue and Joe.

[8]Hint 1: Consider the function $\frac{X^{\frac{1}{2}}}{X+1}$ and note that $X^{\frac{1}{2}}$ is the number of units of textiles produced (equivalently the number of tonnes of coal lifted to the top of the cliff) in "one unit of Sue time," i.e., the length of time it takes Sue to lift one tonne of coal to the top of the cliff. The function $\frac{X^{\frac{1}{2}}}{X+1}$ is what one gets after simplifying 1 unit of coal lifted per $\frac{1}{X^{\frac{1}{2}}}$ units of Sue time per $X + 1$ tonnes of rock.

[9]Hint 2: Try to graph this function and see where its maximum value is. If you do not know any calculus, which I am assuming is the case, just plot a lot of points and make an educated guess.

as better and better measuring devices are invented, when one measures the total energy in all of its forms at the beginning of an experiment and then measures the total energy in all of its forms at the end of an experiment one always gets the same answer.

This First Law is related to another law, viz.,

The Law of Conservation of Mass: *Matter (or mass) can neither be created nor destroyed. The total mass in a closed system is constant.*

I used this law whenever I discussed matter flowing in cycles among a set of "boxes." The tacit assumption there was that matter never disappears or magically appears out of nowhere; the total matter in all the boxes in a closed system at any time during a cycle is the same as the total amount of matter in all the boxes at any other time in a cycle.

Now the alert reader will know that sometimes matter can change into energy (and vice versa). For example, when the nuclear reactor Chernobyl spewed radioactive products around the world, radioactive iodine (I^{131}) was among the effluent. Given an amount of I^{131}, with half-life 8.1 days, one can expect that after 8.1 days half of the I^{131} initially present has undergone radioactive decay. When an atom of I^{131} decays it produces an atom of Xe^{131} (Xenon) which has a little less mass than I^{131}. The difference in mass did not just disappear, for if one measures carefully one finds that not only Xe^{131} is produced upon decay of I^{131}, but an electron (beta particle) and gamma ray (energy) are also emitted. If you use the famous formula of Einstein equating energy and mass, i.e., $E = mc^2$, where E is energy, m is mass and c is the speed of light, you find that the total amount of energy plus mass at the beginning always equals the total amount of energy plus mass at the end of any process or experiment (in a closed system).

When a nuclear weapon explodes the energy produced is just the right amount to account for the mass lost during the explosion. Although a nuclear weapon exploding nearby hopefully never happens, radioactive decay is quite common. For example, radon gas, present in the air in many homes, is radioactive with a half-life of 3.8 days. Though common, radon gas can present a danger to health. The Environmental Protection Agency has set the acceptable limit of radon in a home to be no more than 4.0 picocuries per liter of air. For a discussion of the politics and science involved in arriving at this value see [105].

Exercise 29.5 Radon Gas in the Home

(i) A direct way of measuring radioactivity is to count the number of radioactive decompositions/decays per second. One curie = 1 Ci = $3.7*10^{10}$ decays per second. If you have 4 picocuries per liter of radon gas activity (the EPA limit), how many decays are occurring in 1 cubic meter of air each second?

(ii) Suppose you spend 8 hours per 24 hour day sleeping in a room which has the EPA limit of radon gas. In one year how many radioactive decays of radon occur in your lungs, assuming that each of your lungs takes in about a half liter of fresh air with each breath?[10]

[10]For a more detailed discussion of radon gas see [283].

Thus the first law in modern textbooks reads as follows:

The First Law of Thermodynamics (revised): *The total amount of energy-matter in a closed system is constant. Energy-matter can neither be created nor destroyed.*

This law was arrived at empirically. Many scientists and engineers over a couple centuries have verified this law in the laboratory with experiment after experiment. No exceptions have ever been measured. It turns out that some scientists think that the total energy-matter of the universe is changing since the universe is expanding. I am going to ignore such cosmic considerations, since this theory does not affect the human political economy on earth, as far as I know. I cannot resist mentioning, however, that there remain mysteries in the First and Second Laws of Thermodynamics which require for their unraveling a deeper understanding of the universe, and in particular, more profound insights into the early history of the universe.

The First Law of Thermodynamics comes in handy in many engineering problems. Often a mechanism has a very complex internal structure that is beyond complete comprehension. But many times the complex inner workings of a device can be ignored and conclusions can be reached using only the First Law of Thermodynamics.

Exercise 29.6 Car Lights and Mileage

(i) Consider a car (with a gasoline engine) being driven from point A to point B along a path. Suppose the car is driven with the lights off, and then the car is driven from A to B with everything exactly the same except that this time the car lights are turned on. Is the gas mileage exactly the same in both cases, or is the gas mileage slightly greater in one case?

(ii) In Exercise 29.1 (iii) explain how the law of conservation of energy is tacitly used.

29.3 Entropy and the Second Law of Thermodynamics

The First Law of Thermodynamics is about energy, the Second Law of Thermodynamics in its most modern form is about entropy. So what is entropy? Any matter has properties. For example, a diamond has the property that it is hard. Water at 10 degrees Celsius has the property that it is a liquid, and so on. Any matter also has a property called entropy, but this property is not immediately obvious to our senses. In fact, the Second Law of Thermodynamics was first stated in various forms without mentioning entropy since most people had not explicitly realized that matter had such a property. I will follow an abbreviated version of the historical path to the discovery of entropy since it is the most intuitive for me and hopefully for you as well.

Heat Engines. Although the study of thermodynamics eventually produced profound laws of Nature, thermodynamics began with studies of the steam

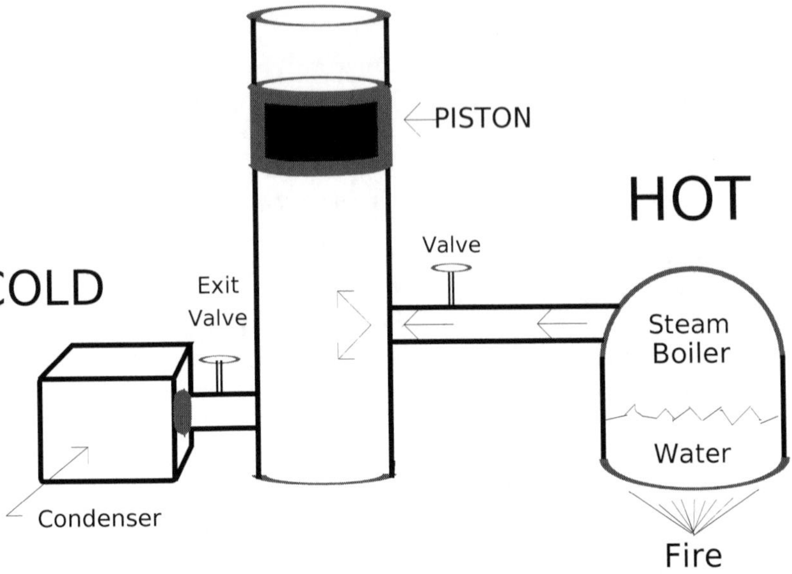

FIGURE 29.2: Basic Structure of a Steam Engine

engine. A sketch of a steam engine is found in the Figure 29.2. A steam engine is a special type of a more general type of engine called a *heat engine*.

What early steam engineers found was that if one boiled water in a chamber (called a boiler) and opened a valve allowing the steam into a second (otherwise closed) chamber with a piston at one end, the pressure of the steam would force the piston out. If the valve which allowed the steam in was then closed and an exit valve was then opened, connecting the piston's chamber with another chamber (called the condenser) which was cooled with cold water, the steam would condense causing a partial vacuum in the piston's chamber. Air pressure would then forcefully push the piston back in. By hooking various devices, like a loom for making textiles,[11] to the piston, the steam engine could do a lot of work. Trains with a steam engine driven locomotive were another well known application.

In all heat engines there is a HOT end (the source) and a COLD end (the sink), and useful work is extracted from the process of passing from the hot end to the cold end.

It was of great economic importance to run such steam engines at the highest possible efficiency. Efficiency, E_{ff}, here has a precise definition, viz., $E_{ff} = \frac{W}{Q}$, where W is the useful work done by the machine and Q is the heat

[11] Such power looms (coupled with enclosure acts which excluded people from public common lands) had profound social consequences, c.f., [596].

energy put into the machine. Engineers worked hard to figure out ways to run these engines as economically as possible, that is, get the most amount of work, W, out for a given amount of heat, Q, in.

Exercise 29.7 Efficiency Can Never Be Greater Than One

(i) Deduce from the First Law of Thermodynamics that $E_{ff} = \frac{W}{Q}$ must be less than or equal to 1, no matter what the internal structure of the steam engine might be.

(ii) Can you think of any reasons why $E_{ff} = \frac{W}{Q}$ would never equal 1 for a real steam engine?

29.4 Early Statements of the Second Law of Thermodynamics

Now a bright idea that has occurred to more than one person is this. How about inventing an engine powered by the heat energy of, say, the air, or ocean water (instead of a steam boiler, for example)? Recall that the molecules in the air are all moving, and that a molecule of mass m and speed v has kinetic energy associated with this motion of $\frac{1}{2}mv^2$. It turns out that a tonne of air at about 20 degrees Celsius has roughly 34 kilowatt-hours of this thermal energy, i.e., kinetic energy in the form of this molecular motion.

If all the molecules of air were moving in the same direction, the air could be used to push a sail or a turbine and thus a car could be propelled. But that is not what happens. Yes, wind exists and wind does power electric generating turbines as well as sailboats. But wind energy of air comes from a net organized flow of air molecules in some direction; the thermal energy of air is associated with random, or unorganized motion of air molecules. It is relatively easy to extract useful work from wind energy. It is not so easy to extract useful work from the thermal energy of air. But could I not invent a clever car engine that would extract about ten percent of the thermal energy of the air, say 3 kilowatt-hours per tonne of air, and thus solve a big energy problem?

Such an engine is not ruled out by the First Law of Thermodynamics, since if my car engine converted thermal energy from the air into kinetic energy of my automobile the total amount of energy could be constant during a drive into town. The Law of Conservation of Energy does insist that my car engine find some energy somewhere to convert into energy of motion of my car in some direction. If I claimed to have invented a car engine that required no input of energy but which propelled my car around, I would have what is called a perpetual motion machine of the first kind (since it would violate the First Law of Thermodynamics).

No one has yet invented a car engine whose sole effect in the universe is to extract thermal energy from the air and convert it into energy of motion of that car in some direction. In some sense this is not surprising, for I do not expect it to be possible to convert the unorganized motion of air molecules into the organized motion of my car without doing something else at the same time. This is the essence of a law of Nature first stated by Sadi Carnot (1824) and later stated using the energy concept by Lord Kelvin (1851) and Clausius (1850–1863).

Kelvin's Statement of the Second Law of Thermodynamics: *There exists no process whose* **sole** *effect is to extract a quantity of heat from a given heat reservoir and convert it entirely into work.*

A machine (like the car engine imagined above) which would extract heat (thermal energy) out of a heat reservoir (like the air or the ocean) and convert it to work without changing anything else in the universe is called a perpetual motion machine of the second kind (because it violates the Second Law of Thermodynamics). A formulation of the Second Law of Thermodynamics attributed to a scientist named Ostwald is: It is impossible to construct a perpetual motion machine of the second kind.

An equivalent form of the Second Law of Thermodynamics due to Clausius is:

Statement of Clausius of the Second Law of Thermodynamics: *There exists no process whose* **sole** *effect is to extract a quantity of heat from a colder reservoir and to deliver it to a hotter reservoir.*

The laws of thermodynamics are the encapsulated wisdom of about two centuries of careful observation and experiment. It is highly unlikely that anyone will be building a perpetual motion machine of the first or second kind any time soon.

Exercise 29.8 The Statements of Kelvin and Clausius are Equivalent

(i) Assume Kelvin's statement of the Second Law of Thermodynamics and give a convincing argument that the statement of Clausius follows. Hint: It may be easier to show that if the statement of Clausius is not true, then the statement of Kelvin is not true. "Statement A implies statement B" is logically equivalent to "Not statement B implies not statement A."

(ii) Assume the statement of Clausius and show that Kelvin's statement follows. Hint: It may be easier to assume that Kelvin's statement is not true and then show that the statement of Clausius is then not true.

29.5 Algebraic Statement of the Second Law of Thermodynamics

I now want to state *the entropy equation*, which has been called the "center of the universe" of thermodynamics.[12] Engineer, Sadi Carnot, performed a brilliant bit reasoning with what he called a reversible heat engine, now called a Carnot engine, in his honor. He proved (for a reversible heat engine) that if an amount of heat energy, Q_H, enters the hot end (source) of a heat engine at temperature, T_H, and an amount of heat, Q_L, exits the cold end (sink) of the heat engine at temperature, T_L, then the following relationship must hold:

$$\frac{Q_H}{T_H} = \frac{Q_L}{T_L}. \qquad \textbf{(The Entropy Equation)}$$

It is important to note that this equation only holds for an ideal, i.e., reversible, heat engine. For such an engine the (maximum) amount of useful work that can be done, according to the First Law of Thermodynamics, is $Q_H - Q_L$. A Carnot engine is the theoretically most efficient heat engine possible. All other, i.e., real, heat engines have an efficiency that is less. Another bit of science that is confusing for some is the difference between an amount of heat energy, say Q_H, and the temperature of that heat energy, say T_H. A swimming pool full of water at 10 degrees Celsius contains a lot more heat energy than a cubic centimeter of boiling water at 100 degrees Celsius. Heat energy is measured in calories, for example; while temperature is measured in degrees. Temperature is a measure of the average kinetic energy of the molecules in substance, while the heat energy of that same substance is the total kinetic energy of its molecules.

There is just one last tricky thing about the Entropy Equation: T_L and T_H are measured in degrees Kelvin. The formula[13] that relates temperature on the Centigrade scale to the Kelvin scale is this: Degrees Kelvin = Degrees Celsius + 273.15.

Assuming the entropy equation I can now prove the following.

The Algebraic Statement of the Second Law of Thermodynamics: *No heat engine taking heat Q_H from a source at constant temperature T_H and delivering heat Q_L to a sink at constant temperature T_L can do more work than a reversible engine, which can do work W given by:*

$$W = Q_H - Q_L = Q_H * (\frac{T_H - T_L}{T_H}),$$

[12] A complete derivation of this entropy equation is extremely fascinating and fun, and if you would like to see it I know of no better explanation than that of Professor Richard P. Feynman, cf., [187]. Feynman's proof uses just a wee bit of calculus. For a further discussion of thermodynamics without calculus, using only mathematics at the level of my book, see [233].

[13] There is also the Fahrenheit scale, where Degrees Fahrenheit = $\frac{9}{5}$Degrees Celsius + 32.

where temperatures are given in degrees Kelvin.

The efficiency, $E_{ff\ max}$, of a reversible heat engine taking heat from a source at constant temperature T_H and delivering heat to a sink at constant temperature T_L is

$$E_{ff\ max} = \frac{T_H - T_L}{T_H}.$$

This is the maximum efficiency that any heat engine, operating between temperatures T_H and T_L can have. Most real heat engines have efficiencies much less than the theoretical maximum.

Exercise 29.9 Equivalence of the Second Law and the Entropy Equation

Assume you have a reversible heat engine taking heat Q_H from a source at constant temperature T_H and delivering heat Q_L to a sink at constant temperature T_L.

(i) Assuming the entropy equation $\frac{Q_H}{T_H} = \frac{Q_L}{T_L}$, derive The Algebraic Statement of the Second Law of Thermodynamics. You can assume that $W = Q_H - Q_L$ since this follows from the First Law.

(ii) Assuming The Algebraic Statement of the Second Law of Thermodynamics derive the entropy equation $\frac{Q_H}{T_H} = \frac{Q_L}{T_L}$.

(iii) Is it true that the Algebraic Statement of the Second Law of Thermodynamics and the entropy equation (in some sense the "idea of entropy") are equivalent?

(iv) By definition the efficiency of this reversible heat engine is $E_{ff} = \frac{W}{Q_H}$. Can this efficiency ever equal 1?

29.6 So What Is Entropy and Can We Measure It?

The entropy equation is screaming something at us: in the (reversible) heat engine there is a mathematically measurable quantity that is conserved, i.e., is the same at the hot end as it is at the cold end of the (reversible) heat engine. That quantity is the ratio $\frac{Q}{T}$. So for a reversible heat engine the entropy S is given (at any point in the process) by: $S = \frac{Q}{T}$.

A reversible heat engine defines an equivalence between an amount of heat Q at (Kelvin) temperature T and an amount of heat Q' at temperature T' in the sense that if the heat engine goes through a reversible cycle and "picks up" $S = \frac{Q}{T}$ at one point in the cycle and "drops off" $S' = \frac{Q'}{T'}$ at another point in the cycle, then $S = S'$. While going around a reversible cycle, there is no net change in the entropy of a heat engine. The units of measurement for entropy are joules per degree Kelvin.

From our definition of entropy if I let my heat engine take in an amount of heat Q at temperature T and then deliver an amount of heat Q_S at 1 degree Kelvin, then $S = \frac{Q_S}{1deg}$.

You might ask what happens when $T = 0$ deg Kelvin. A scientist, Nernst, postulated what has become the Third Law of Thermodynamics:

The Third Law of Thermodynamics: *When the temperature of a substance is 0 degrees Kelvin, its entropy is 0 joules per degree Kelvin.*

Now I want to give a very concrete example of entropy so that you will believe it really exists physically and is not just a mathematical abstraction.

Before the Third Law was proposed, scientists worked as though only differences in entropy in the states of a system existed. The "zero point" for entropy was taken at any point convenient for experiments. It is in this spirit that I reproduce a table from [233] giving the *relative* entropy and energy of one kilogram of water at various temperatures given in degrees Celsius.

Temperature ($°C$)	E (joules)	S (joules per Kelvin)
0	0	0
10	41,840	151
20	83,680	297
30	125,500	435
40	167,400	573
50	209,200	703
60	251,000	833
70	292,900	954
80	334,700	1,075
90	376,600	1,192
100	418,400	1,305

The Relative Energy, E, and Entropy, S, of 1 kilogram of water

A Table of Entropy

In the above Table water at 0 deg Celsius is assigned entropy of 0 joules per Kelvin. Although the numerical value of the entropy of a kilogram of water at a given temperature is affected by this choice of zero point, the *change* in value of entropy in going from one temperature to another is not. The same comments apply to the energies listed in the Table of Entropy; they are relative energies not absolute energies.

Now water has a remarkable property which allows me to use it as an example of an approximately reversible thermal (heat) process. When water is slowly boiled (heat energy is added) at 100 deg Celsius and 1 atm pressure[14] it is converted from a liquid (at 100 deg Celsius and 1 atm pressure) to a gas, i.e., steam (at 100 deg Celsius and 1 atm pressure). One kilogram of water absorbs $2.258*10^6\ J$ of heat energy when going from a liquid state to a gaseous

[14]1 atm refers to atmospheric pressure, and 1 atmosphere of pressure is 14.7 pounds per square inch. It turns out that a column of mercury 76 centimeters high exerts the same pressure as 1 atmosphere.

(steam) state at 100 deg Celsius. This is called the latent heat of vaporization. (This process can be reversed by subtracting heat, condensing the steam into liquid water.)

The absolute temperature corresponding to 100 deg Celsius is 373.15 deg Kelvin. If I divide $2.258 * 10^6\, J$ by 373.15 deg Kelvin I get $\frac{2.258*10^6\, J}{373.15\, deg\, Kelvin} = 6051\, \frac{J}{deg\, Kelvin}$, for 1 kilogram of water passing from liquid to gas at 373.15 deg Kelvin.

The following is true: If an amount of heat ΔQ is added *reversibly* to a system at (absolute) temperature T, then the increase in entropy of the system is

$$\Delta S = \frac{\Delta Q}{T}.$$

Thus the $6051\, \frac{J}{deg\, Kelvin}$ is the increase in entropy of our kilogram of water, as it passes from the liquid to the gaseous state at 100 deg Celsius. If I condensed a kilogram of steam into water at 100 deg Celsius, then there would be a decrease in entropy of the same amount.

Using the Table of Entropy I want to calculate the change in entropy of a process that is not reversible. Now assuming that it takes 1 calorie of energy to raise the temperature of 1 gram of water 1 deg Celsius no matter what the temperature of the water is between 0 deg Celsius and 99 deg Celsius,[15] I want to mix 1 kilogram of water at 0 deg Celsius with one kilogram of water at 100 deg Celsius. By the First Law (Conservation of Energy) I expect that I will end up with two kilograms of water at 50 deg Celsius. (Imagine the 100 deg Celsius kilogram giving up 50 Calories, lowering its temperature to 50 deg Celsius; and the 0 deg Celsius kilogram of water absorbing 50 Calories, raising its temperature to 50 deg Celsius.) This process is not reversible![16]

The initial entropy of our two kilograms of water is the sum of the entropies of the 0 deg kilogram and the 100 deg kilogram, and from the Table of Entropy this is $(0 + 1,305)\frac{J}{deg\, Kelvin} = 1,305\, \frac{J}{deg\, Kelvin}$. The final entropy is twice the entropy of 1 kilogram of water at 50 deg Celsius, which from the Table of Entropy is $2 * 703\frac{J}{deg\, Kelvin} = 1,406\, \frac{J}{deg\, Kelvin}$. The entropy of the 100 deg kilogram decreased by $(1,305 - 703)\frac{J}{deg\, Kelvin} = 602\frac{J}{deg\, Kelvin}$. The entropy of the 0 deg kilogram increased by $703\frac{J}{deg\, Kelvin}$. The net change in the entropy of the entire irreversible process is an increase of $101\frac{J}{deg\, Kelvin}$.

This leads us to the following:

Qualitative Statement of the Second Law of Thermodynamics: *The total entropy of a closed system cannot decrease. If a system experiences an irreversible change, the total entropy of the system increases.*

---

[15] This is not exactly true, but it is close enough to the truth for our purposes here.

[16] If you have two kilograms of water at 50 deg Celsius, you would have to use energy to heat one of those kilograms to 100 deg Celsius; and you would have to use energy to refrigerate the other kilogram down to 0 deg Celsius.

A corollary of this statement would be that the entropy of the universe is increasing.

Exercise 29.10 Entropy Calculations
(i) Suppose you have 1 kilogram of water at 60 deg Celsius and 1 kilogram of water at 0 deg Celsius, what is the total entropy of the 2 kilograms of water before they are uniformly mixed and after? What is the net change in entropy?
(ii) Suppose you have 2 kilograms of water at 90 deg Celsius and 1 kilogram of water at 30 deg Celsius. What is the total entropy of the 3 kilograms of water before they are uniformly mixed and after? What is the net change in entropy?
(iii) Suppose you have 2 kilograms of water at 90 deg Celsius, 1 kilogram at 80 deg Celsius, 1 kilogram at 60 deg Celsius and 1 kilogram at 30 deg Celsius. What is the total entropy of the 5 kilograms of water before and after they are uniformly mixed? What is the net change in entropy?
(iv) Suppose you have 2 kilograms of steam at 100 deg Celsius and 1 atm of pressure. If the steam is reversibly condensed to 2 kilograms of liquid water at 100 deg Celsius, what is the change in entropy?

It is this inexorable increase of entropy with the passing of time that leads scientists to call entropy the "arrow of time." If you dissolve sugar in a cup of tea, the net entropy increases. If a gas expands isothermally its entropy increases. If you burn a log, entropy increases. If you cook an egg, entropy increases. If you build a house, net entropy increases. Living beings have the property that they can decrease their personal entropy, but this takes place at the expense of an increase in entropy somewhere else—and the result is a net increase in the entropy of the "organism plus environment system." This is because almost all the processes humans engage in (indeed all real processes in the political economy) are not reversible.

This is what led me to define the human economic system as a super organism, cf., page 462, and create Exercise 25.2. Allow me to recall one example of humans exploiting low and then higher and higher entropy resources: gold in Colorado and California, starting with the pioneers about 1850. It was not impossible for a very lucky miner to pick up a gold nugget. A piece of earth which had, say, about one ounce of gold per one ounce of ore. In 1999 I ran across the proposal for a gold mine called the Imperial Project in Southern California, where it was estimated that to extract one ounce of gold 422 tonnes of ore (and waste material disturbed) would have to be processed.[17] The gold nugget is an example of a low entropy gold resource. The entropy of the gold to be extracted in a typical cyanide-heap-leaching gold mine today is much, much higher. Humans have basically run out of low entropy gold-ore. It takes a lot of energy to extract one ounce of pure gold (low entropy gold)

[17]A 1,231 acre heap-leach gold mine in Colorado, the Summitville mine, was deserted by its owners in 1991. Years later the site continues to send water laced with acid and metals into tributaries of the Alamosa River—killing most life forms. From 1992 to 2001 the federal government spent $160 million "reclaiming" the mine site. The EPA estimates that it will take another 100 years and a total of $235 million to finish cleaning up the mine site. After a long court battle, the owners will contribute $30 million, the U.S. taxpayers $205 million. Such are some of the current costs of extracting gold from increasingly larger entropy ore.

from 422 tonnes of high entropy gold-ore. There are many more examples like this. Can you think of any?

29.7 Some Applications of The Second Law of Thermodynamics: Power Plants and Hurricanes

Let's apply our algebraic form of the second law to some real life situations. What the original thermodynamic thinkers like Carnot had in mind, presumably, when studying reversible heat engines, were idealized steam engines. It turns out that most "modern" electrical generating plants are not all that different from a classical steam engine. Thus, there is a source of heat, usually a fossil fuel like coal or natural gas or a nuclear reaction. This source of heat is used to boil water and produce steam under pressure which is used to turn a steam turbine.[18] This rotating turbine drives an electrical generator, just like the hydroelectric generators at the base of dams mentioned earlier. The following exercise is from [283, pp. 79–82].

Exercise 29.11 Efficiency of a Coal-Fired Power Plant
(i) If the pressurized steam coming out of the boilers of a coal-fired power plant is at 500 deg Celsius, what is this temperature in deg Kelvin?
(ii) This pressurized steam is directed at turbine blades, causing them to turn; and the steam is then condensed and cooled to room temperature (say 25 deg Celsius) as it exits the turbine. What is this "room" temperature in deg Kelvin?
(iii) Calculate the maximum possible theoretical efficiency of this coal-fired power plant.
(iv) Why did I ask parts (i) and (ii) of this exercise?
(v) Why do you think the steam in part (i) is under high pressure?

The actual efficiency of a modern coal-fired power plant is less than the ideal, maximum efficiency you just calculated. The real world efficiency is likely between 30% and 40%. The reason for this less than optimal efficiency comes from the fact that heat energy is lost up the smoke stack, energy is lost to internal friction and small thermal leaks and so on. In the next problem (also from [283]) I will examine the waste heat from a power plant.

Exercise 29.12 Analysis of a 1,000 Megawatt Coal-Fired Power Plant
(i) Assume the efficiency of the power plant in question is 40%. Can you write an equation which involves: "Energy out," i.e., electric power output; "Energy in," i.e., heat power supplied by burning coal; and the efficiency, i.e., 40%?

[18] After over 200 years of boiling water in various ways as a means to generate useful energy (like electricity) I hope our society soon invests in more innovative approaches both to do more with less and to capture what energy is needed from the sun with much higher efficiency than is now the norm. If one converts solar energy directly into electricity, for example, without using a heat engine, it is theoretically possible to attain much, much higher efficiencies than that of a heat engine whose efficiency is limited by the Second Law!

(ii) Why does the following equation hold for a power plant: $\frac{Energy\ Out}{Energy\ In} = \frac{Power\ Out}{Power\ In} = Efficiency$?

(iii) Suppose P_W is the wasted heat-power coming out of the plant. The equation $P_W = (Rate\ of\ heat\ input) - (Rate\ of\ electrical\ energy\ output)$ holds. Why? Use the equation from part (i) (or part (ii)) to express the right-hand of this equation, $P_W = (Rate\ of\ heat\ input) - (Rate\ of\ electrical\ energy\ output)$, in terms only of the efficiency and the rate of electrical energy output, which is the same as electrical power output.

(iv) If the electrical power output of the coal-fired power plant is 1 MW, what is P_W? Express your answer in MW (megawatts) and in joules per year.

(v) Assume that 15% of P_W goes up the smoke stack. The remaining 85% must be absorbed via some type of cooling device associated with the power plant. How much power must be absorbed by this cooling device for the plant in question?

A so-called wet-cooling tower is one method for absorbing the waste heat calculated in the previous exercise. Such towers are quite large and perform the cooling function by evaporating water from huge fins. This water is thus consumed in that this water goes into the air as a gas.

Exercise 29.13 Cooling a Coal-Fired Power Plant

(i) Express your answer from part (v) of the previous exercise in calories.

(ii) Suppose the water used to cool our power plant starts out at 17 deg Celsius. In changing water from a 17 deg liquid to water vapor suppose each kilogram of water absorbs $2.459 * 10^6 J$ of heat energy. How many calories is this?

(iii) How many kilograms of water will this cooling device consume in one year?

(iv) What is the volume of the water in part (iii), and at what rate is this water flowing?

(v) Describe qualitatively what is happening to the entropy of the coal and water used in the power plant.

When such plants are built in the arid Southwestern United States, not only are there visible environmental effects coming out of the smoke stack; but local water resources are impacted as well. A large Peabody Coal mine in a place called Big Mountain,[19] Arizona, impacts the water table because water is pumped out of the ground (lowering the water table) so as to create a coal slurry. That is, water and coal are mixed together and pumped to its final destination.

Exercise 29.14 Some Environmental Impacts

(i) If the precipitation in the area of Big Mountain is equivalent to 10 cm of water in one year, the precipitation over what land area is required to cool the plant?

(ii) Assuming that electric water pumps operate at 90% efficiency and that the water being pumped must be lifted 100 meters from below ground, how much energy is used to pump the cooling water in one year? What fraction of the output of the electrical power plant is used to pump this water?

(iii) Do part (ii) again, assuming the water is at a depth of 500 meters.

(iv) Do part (ii) again, assuming the water is at a depth of 1000 meters.

(v) At what depth of water table is all of the ouput of the power plant used in pumping cooling water?

Another example of a heat engine that converts heat energy into mechanical (wind) energy is a hurricane.

[19] This particular place is sacred to the traditional Native Americans living in the area.

Exercise 29.15 A Hurricane Heat Engine

(i) A hurricane is a heat engine that operates on the temperature difference between the surface of the ocean and the stratosphere. If the temperature of the seawater is 30 degrees Celsius and the temperature of the stratosphere is −80 degrees Celsius, what is the maximum theoretical efficiency of the hurricane according to the laws of thermodynamics?

(ii) A typical hurricane might have a diameter of 1,600 km. How many joules of heat energy are stored in the volume of the ocean under a hurricane[20] to a depth of 1 meter, per degree Celsius?

(iii) How does the answer from part (ii) compare to the energy in the form of petroleum consumed in the U.S. in one year?

(iv) Research the ecological impacts of hurricanes. For example, NASA has discovered that hurricanes foster phytoplankton blooms in their wake via their bringing of nutrients to the ocean's surface.

(v) Are there human activities that increase the negative impact of hurricanes on humans?

29.8 Hiking Up a Mountain

You may have noticed that processed food in the United States has a "Nutrition Facts" label.[21] I am at the moment of writing this line looking at such a label on a box of organic dry breakfast cereal. It states that one serving of this cereal is 1 cup (55 *grams*). It states that one serving of this cereal has 210 *Calories* (50 *Calories* coming from fat), and that this serving of cereal with $\frac{1}{2}$ cup of skim milk has 250 *Calories* (still 50 *Calories* from fat).

You might ask how much work with your skeletal muscles can you expect to do if you eat one serving of this cereal with (skim) milk. Obviously not all of the energy (*Calories*) can be converted to such work by your muscles, since much of the energy will be used by other parts of your body to carry on life's processes, and much energy will be "wasted" in that it will not be used by your body. From various sources I have read that anywhere from 15% to 30% of the energy in the food you eat is available for your skeletal muscles to do work.[22] For the sake of definiteness I will assume that 20% of the energy

[20]The model of a hurricane implied by this exercise is a good beginning, but it is not complete. For example, hurricanes cause mixing of ocean waters to a depth, say, of 100 meters! Hurricanes are, however, examples of heat engines!

[21]That these labels exist is a testament to the effectiveness of citizen action. Truth in labeling is routinely opposed by the relevant industries, and the excuse is usually cost. This rings hollow given the billions of dollars in advertising that are spent to "tell us" about food products. Currently the industry has successfully defeated attempts by citizens (and even states) to require labeling of irradiated food, genetically engineered food and food containing certain substances such as rBGH, a genetically engineered growth hormone.

[22]A colleague of mine in the Biology Department tells me that his study of various ethnic groups and indigenous people shows that different folks have different levels of efficiency. He postulates that some peoples evolved in situations that required higher efficiency in the utilization of food for survival.

in your food is available for your skeletal muscles to do work. If this is the case then $.20 * 250\ Calories = 50\ Calories$, of this one serving of cereal with skim milk is available for voluntary work on your part. From the exercises on energy conversion, it is possible to convert 50 *Calories* into kilowatt-hours or newton-meters or foot-pounds. A newton-meter is a joule, so I will first see how many joules 50 *Calories* is. Since $1\ Calorie = 10^3\ cal = \frac{10^3\ J}{.2390} = 4.184 * 10^3\ J$, $50\ Calories = 2.092 * 10^5\ J$. So you can do that many joules of work. Since the United States is one of the last places on earth where foot-pounds are used, I note that $1\ foot - pound = 1.356\ J$, so you will have $\frac{2.092*10^5\ J}{1} * \frac{1\ foot-pound}{1.356\ J} = 1.543 * 10^5\ foot - pounds$ of energy available for voluntary work.

What work do you want to do? Work done is force times the distance through which the force acts. If the force varies along the distance, you need calculus or some numerical techniques to do problems; so I will assume the force is constant. If you choose to climb up a mountain near the place where I am writing this, that force can be taken equal to your weight in pounds. If you weigh 150 *pounds*, how high on the mountain can you climb neglecting friction, energy lost in horizontal movement, energy lost in moving your jaw talking, and so on?

Well, if you have $1.543 * 10^5\ foot - pounds$ of energy available to exert through a vertical distance x a force of 150 *pounds* you get:

$$150\ pounds * x = 1.543 * 10^5\ foot - pounds.$$

Solving for x I get that you can climb a total vertical distance of $x = \frac{1.543*10^5\ foot-pounds}{1} * \frac{1}{150\ pounds} = 1,029\ feet.$

Of course, that answer is approximate for a number of reasons, and $1,029\ feet$ is not very far, considering you might like to hike starting at $10,000\ feet$, ending at $14,000\ feet$.

Exercise 29.16 Hiking Up a Mountain Propelled by Food

(i) How far is $1,029\ feet$ in *meters*?

(ii) If you eat 2 servings of cereal with skim milk (as described above) how high up the mountain can you climb if you weigh 150 *pounds* and are 20% efficient?

(iii) How many servings would the 150 *pound* person above have to eat to climb $4,000\ feet$?

(iv) Suppose you eat a can of blueberries with 110 *Calories*, none from fat. Suppose that your body is 30% efficient and that you weigh 200 *pounds*. How high on the mountain will the blueberries propel you?

(v) Using your real body weight and the information from the Nutrition Facts label on some package of processed food available to you, calculate how high up a mountain you can climb assuming you are 15% efficient and you eat everything in the package.

(vi) In what period of time does the "typical" American eat the amount of sugar that it took one year for our ancestors of 10,000 years ago to consume?

Recall the Maximum (Useful) Power Principle, cf., subsection 29.1. Since the energy available to any given species is limited, that species cannot ignore its level and rate of energy consumption. That is, such a species cannot always focus only on being time efficient and behave like Joe, in always using large quantities of energy quickly to save time and "live fast." Likewise, a species usually cannot focus solely on conserving energy, like Moe, and live "too slowly." Apparently depending somewhat on the environment, a species evolves a compromise between "living fast" and "living slow," and for humans that leads to an efficiency of voluntary muscles between 15% and 30%.

Body Mass Index: BMI. So how do you decide if you are the "right weight?" I actually hesitated to include the following discussion of the *Body Mass Index*[23] for at least three reasons. First, if someone thinks of his/her body as "fat," he/she might be tempted to act quickly—say take diet pills—with possibly fatal results, see a discussion of Fen-Phen, for example in [473]. Second, people are complex[24] and individual; and I am not qualified to be doling out "medical advice." Third, the BMI is but one number and it is easy to get overly concerned about the value of "your body mass index;" especially when the BMI gets discussed in places like the Style section of *The New York Times*.[25] The BMI is used, however, by many to define "overweight" and "underweight" and thus worth knowing. This index was one of the concepts introduced by Belgian mathematician Adolphe Quetelet[26] (1796–1874).

The following is the formula for the Body Mass Index, BMI, of someone with height H in *meters* and weight W in *kilograms*:

$$BMI = \frac{W}{H^2} \qquad \text{(Body Mass Index Formula)}$$

Exercise 29.17 The Body Mass Index

(i) Some books give the following formula for calculating the BMI: Take your weight in *pounds* and multiply by 705, then divide the result by your height in *inches*; then divide that result again by your height in *inches*. Explain why this should give the same numerical result as the formula above, where your weight in *kilograms* is divided by the square of your height in *meters*.

(ii) For someone five feet six inches tall, what is the healthy range of weight? (What is your definition of "healthy range"of weight?)

(iii) What is your BMI? (This is for your private information, do not hand it in.)

(iv) Could a body mass index be defined by taking a person's weight and dividing it by the cube of that person's height?

(v) For further reading on how the food industry influences what we eat and hence our health, cf., [489, 606].

---

[23] For a fuller discussion of the Body Mass Index, i.e., BMI, and related health issues see *The U.C. Berkeley Wellness Self-Care Handbook: The Everyday Guide to Prevention & Home Remedies* by John Edward Swartzberg, M.D., F.A.C.P. and Sheldon Margen, M.D., Rebus, New York, p. 102.

[24] Very muscular and athletic persons can have a BMI of over 30 and be extremely fit.

[25] See the Thursday, December 28, 2006, edition,

[26] Quetelet was not only a mathematician, he was an astronomer, statistician, and sociologist known for his application of statistics and the theory of probability to social phenomena. The Quetelet Index, now known as the Body Mass Index, was related to his concept of the homme moyen ("average man"). Quetelet measured various human traits and claimed that they were distributed according the normal distribution (see the next section). The normal distribution, or "bell shaped" curve, has a central or mean value. The Quetelet Index was one of his ways of measuring people. If one does not obsess about it, it can be useful.

29.9 Understanding Entropy with a Little Mathematics

A much deeper understanding of entropy can be had by using just a little bit of the mathematics that we already learned in Part IV on counting. Imagine an airtight box with a rigid airtight membrane dividing the box into two equal halves, as in Figure 29.3.

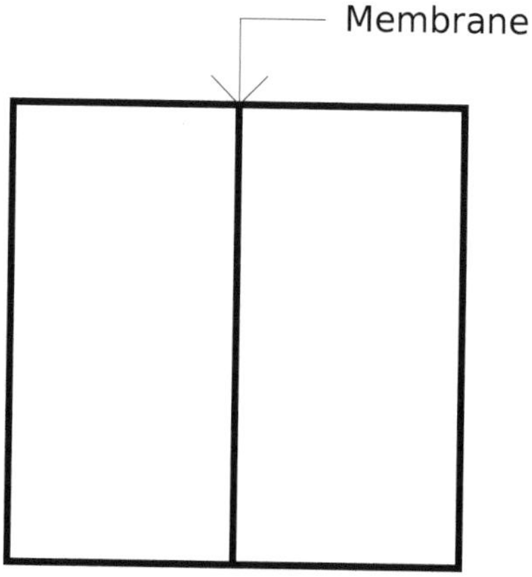

FIGURE 29.3: Two-Compartment Box

Now if there were a bunch of air molecules on the left side of the membrane, and none on the right side, the air molecules would be banging into the membrane with a net force to the right the size of which would depend on the number (and mass) of molecules and the speed with which the molecules hit the membrane.[27] Now when studying things as small as molecules it usually requires quantum mechanics to get the full picture. I am going to ignore quantum mechanics and pretend that my molecules are miniature marbles banging around, bouncing off the sides of the container and each other as though they were billiard balls. It turns out that the calculations I am about to do actually reflect reality, perhaps because I am going to be sticking to the very basic facts of counting. In any event, experiment is the ultimate arbiter; and experiments tell us that our calculations below agree with reality.

[27]The ideal gas law, $PV = NRT$, gives us the approximate, quantitative details (which I do not need to know in this particular discussion) where P is the pressure (force per unit area on the inside of the box containing air), V is the volume of the box containing air, N is the number of moles of air, R is a constant and T is the temperature (which keeps track of the average kinetic energy, i.e., $\frac{1}{2}mv^2$, m is the mass of a molecule and v is its speed).

If I made a small pinhole in the membrane separating the right half of my box (with no air initially) and the left half (with air initially) it is obvious what would happen. Air would leak through the pinhole until the pressure on the right and left sides of the membrane were equal. It would not surprise anyone if I asserted that air would move through the pinhole until there were approximately the same number of molecules of air on the left and right sides of the membrane. Now assuming the temperature of my air does not change, the entropy of the air will have increased.[28] So how can this be? Entropy is a property of matter, in this case the air. The only thing that changes about the air is that it takes up twice the volume as before—but that is enough. When the air was confined on one side of the box I had more information about the air, and I will say that the air was more ordered. When the air is free to roam both sides of the box I have less information about the air and it has less order—the air is more disordered. I am using the terms order, disorder and information intuitively here; I assure you I can define them more precisely, but probably less understandably. So let me use them intuitively.

Now if there are approximately an equal number of air molecules on the left and right of the membrane with a pinhole, no one would expect that in an interval of a million years all of the air molecules would at any time all rush over to the left (or right) side of the box, even if the membrane were not there. Why is that? Well it agrees with our experience, that's why. I can give a better argument than that, however; and in the process I can explain the concept of entropy more intuitively than I have done thus far.

Suppose that I have N molecules of air and I want to put them in the box with either a real or imaginary membrane dividing the box into two equal halves as above. Suppose also that each molecule has its individual identifying number (from 1 to N) printed on it, but you can only see this number if you look closely. If you do not look closely you cannot see the number printed on a given molecule, and then the molecules all look alike. I want to count how many ways I can put these molecules into the box, with only two choices being for a given molecule: Either, (i) I put the molecule in the left half of the box; or, (ii) I put the molecule in the right half of the box. That's all I want to count right now. How many ways can I do this?

From the Fundamental Principle of Counting, cf., page 309, I can think of the N distinct molecules as N "slots." Each slot I can fill two ways: put the molecule in the left half, or put the molecule in the right half. Each way of filling the N slots corresponds precisely to one way of distributing the molecules. The total number of ways of filling the slots, hence the total number of ways of distributing the molecules, is 2^N.

Most likely the number N will be really big, like Avagadro's number. Before confronting reality, however, I will look at smaller values for the number N

[28] I could prove this using what we already know, but the proof requires just a little bit of calculus; so trust me for the moment on this. I will show you how to understand this using "counting" shortly.

to get warmed up. Suppose $N = 1$. I can put the molecule either on the left or on the right, and that's it. (Note: $2^1 = 2$ is the number of ways I can distribute the molecule.) This corresponds to the first row in Figure 29.4.

Suppose $N = 2$, then I can put 0 molecules on the left (hence 2 on the right). I can put 1 molecule on the left (hence one on the right)—and I can do this *2 ways* since the molecules are all distinguishable (they each are imprinted with an individual identifying number from 1 to $N = 2$); or I can put 2 molecules on the left and 0 on the right. This corresponds to the second row in Figure 29.4. (Note: $2^2 = 4$ is the number of ways I can distribute the 2 molecules.)

Suppose $N = 3$. I can put 0 molecules on the left and 3 on the right (and I can do this only 1 way). I can put 1 molecule on the left and 2 on the right (and I can do this 3 ways). I can put 2 molecules on the left and 1 on the right (and I can do this 3 ways). I can put 3 molecules on the left and 0 on the right (and I can do this 1 way). (Note: $2^3 = 8$ is the number of ways I can distribute the 3 molecules.) This corresponds to the third row in Figure 29.4.

Exercise 29.18 Distributing Molecules to the Left or Right

(i) Verify that the fourth row in Figure 29.4 correctly lists the various ways of distributing 4 molecules between left and right halves of the box.

(ii) Verify that the fifth row in Figure 29.4 correctly lists the various ways of distributing 5 molecules between left and right halves of the box.

(iii) Verify that the sum of the numbers in the Nth row is 2^N.

(iv) Review the meaning of $\binom{n}{k}$ (if necessary review Exercises 17.5 and 17.17), i.e., $\binom{n}{k}$ is the number of ways of choosing k things from n things if the order of choosing does not matter.

(v) Verify that $\binom{n}{k} = \frac{n!}{(n-k)!k!}$, at least for $n = 1, 2, 3, 4, 5, 6$. This formula is true for any positive (or zero) whole number n and any value of k from 0 up to and including n. Recall that $0! = 1$ and that $n! = n * (n-1) * \cdots * 3 * 2 * 1$, i.e., the product of all the positive whole numbers from 1 up to n. So $1! = 1$, $2! = 2 * 1$, $3! = 3 * 2 * 1$, and so on.

	TOTAL
1 1	2
1 2 1	2^2
1 3 3 1	2^3
1 4 6 4 1	2^4
1 5 10 10 5 1	2^5
1 6 15 20 15 6 1	2^6

FIGURE 29.4: Binomial Distribution of Molecules

I now want to look at the case where I have N molecules. I can choose to put 0 molecules on the left (hence N molecules on the right) and I can do this 1 way. I can choose to put 1 molecule on the left (the rest on the right) and since I have N distinct molecules I can do this N ways. The number of ways

I can select 1 thing from N things can be written "N choose 1," or $\binom{N}{1}$. Note that $\binom{N}{1} = N$, and "N choose 0" is $\binom{N}{0} = 1$.

Next, I can select 2 molecules and put them on the left, putting the rest on the right. The number of ways I can do this is "N choose 2," and this is equal to $\binom{N}{2}$.

In general, I can put K molecules on the left, and I can select these K molecules from my N molecules in "N choose K" ways, i.e., in $\binom{N}{K}$ ways.

I now want to define two terms: *microstate* and *macrostate.* If I can see, for example, K molecules on the left and $N - K$ molecules on the right; but I cannot read the numbers on the individual molecules—this is a macrostate. If I look closely I can see the numbers that are printed on each of the molecules, and I can then see what microstate the system is in. There are $\binom{N}{K}$ microstates of my "air in the box system" corresponding to the one macrostate which has K molecules on the left.[29]

Now if I add up all the (microstate) ways I can distribute my N distinct molecules into the left and right compartments I get

$$\sum_{K=0}^{N} \binom{N}{K} = 2^N.$$

I know this is the answer from my use of the Fundamental Counting Principle above, or by use of the Binomial Theorem with $a = b = 1$, cf., Exercise 17.17.

Exercise 29.19 Why the Molecules Are Not All On the Left or Right

Suppose I have N molecules and I look at all the ways I can distribute these N molecules into the left and right halves of my box.

(i) How many (microstate) ways total can I distribute N molecules into the right and left halves?

(ii) How many ways can I distribute all the molecules into the left half of my box?

(iii) How many ways can I distribute all the molecules into the right half of my box?

(iv) What fraction of all the (microstate) distributions are such that all of the molecules are in the left half (respectively, right half) of the box?[30]

(v) It turns out that one mole of an ideal gas at standard temperature (i.e., 0 deg Celsius) and pressure (i.e., 1 atm) occupies 22.4 liters of volume. Suppose that our box has one mole of air molecules in it. What is N in this case?

(vi) Using the value for N in part (v) calculate the fraction of all distributions for which all the molecules are in the left half of the box.[31]

(vii) If one mole of air molecules are randomly distributed into the left and right halves of the box what are the "chances" that all of the molecules will be in the right half (or all of them will be in the left half) of the box?

At this point we have accomplished our main goal: understanding why from a purely mathematical point of view (plus an understanding that matter

[29] For those budding mathematics majors reading this, can you see that a macrostate can be defined to be an *equivalence class* of microstates? See Chapter 17.

[30] Hint: What is the answer to part (ii) divided by the answer to part (i)?

[31] Hint: Calculate the answer to part (iv) using this specific value for N.

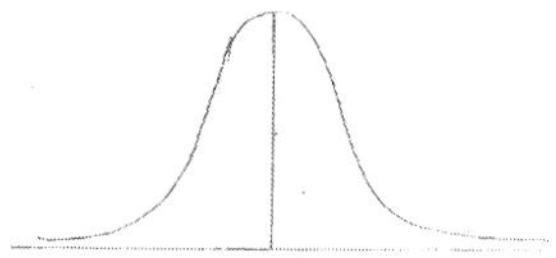

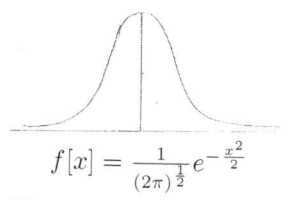

$$f[x] = \frac{1}{(2\pi)^{\frac{1}{2}}} e^{-\frac{x^2}{2}}$$

FIGURE 29.5: A Normal Distribution: The Bell-Shaped Curve

is made up of "many" molecules) it is astronomically unlikely that all the molecules of air in a box will end up in half of the box.

The number of ways of distributing a mole of molecules into the right and left halves of the box is truly beyond astronomical! For my purposes it is practically infinite. To get some idea of how big this number is, consider that the distance light travels in one year is called one light-year. The most distant visible stars are 10^{10} light-years away. Measured in centimeters this gives 10^{28} *cm*. Using the diameter of a hydrogen atom as the fundamental unit, viz., an angström, I would have to lay 10^{36} hydrogen atoms side by side to span the distance from here on earth to the most distant visible star. This number, 10^{36}, is practically "zero" compared to the number of ways of distributing a mole of molecules into the right and left halves of the box, viz., $2^{(6*10^{23})}$, or about $10^{(10^{23})}$. Can you, for example, even calculate the ratio on your calculator, viz., $\frac{10^{36}}{10^{(10^{23})}}$?

A mole of molecules is roughly 10^{23} molecules, and I could never write down the $10^{23^{rd}}$ row in Figure 29.4! This row would give us the answer to the question: what fraction of all of the distributions of molecules would have fraction x in the right half of the box? (Where x is some number between 0 and 1.) Fortunately I do not have to write down the $10^{23^{rd}}$ row in Figure 29.4; for there is a mathematical trick, or better said, a Theorem, that rescues us from that troublesome task.

The French mathematician,[32] Abraham De Moivre (1667–1754), discovered an example of what is claimed by many to be the most important single theorem in modern probability theory, viz., The Central Limit Theorem which you can read about in [186].

What he showed was that the distribution of the Nth row of the binomial distribution we have been looking at, when suitably "normalized," approaches a bell shaped curve when N grows large beyond all bounds. This bell shaped curve is drawn in Figure 29.5 and it is called the Normal Distribution.

[32]De Moivre was jailed in France (he was a Protestant when that was not popular in France), so he fled to England where he became a close friend of Sir Isaac Newton. De Moivre gained prominence as a mathematician and was a pioneer of probability theory; but he never got a permanent position—surviving as a tutor and a consultant to gambling and insurance interests.

The e in Figure 29.5 is approximately 2.718281828, the base of natural logarithms you have met before. The most likely distribution of molecules in our box is for about half of them to be in each half of the box. This is represented by the "peak" of the Bell-Shaped Curve.

Now I can do a more sophisticated analysis by looking at a box with many compartments.[33] I will stop and state the following mathematician's definition of entropy.

The Mathematical Definition of Entropy: *Given a physical system in a given macrostate, the entropy of that macrostate is the logarithm (to base e) of the number of microstates corresponding to the given macrostate.*

The alert reader will note an inconsistency with physics here, since the mathematician's definition of entropy is a pure number and the entropy we have calculated in previous exercises had units of $\frac{joules}{deg\ Kelvin} = \frac{J}{K}$. In order to relate the pure mathematics we have just done to the entropy equation of the Carnot engine requires some logical analysis and understanding of laboratory experiments. Ludwig Eduard Boltzmann (1844–1906), calculated the relationship between the mathematician's definition of entropy and that of the physicist. He came up with the following physicist's definition of entropy.

The Physicist's Definition of Entropy: *Let W_M be the number of microstates of a physical system corresponding to a given macrostate M of that system. The entropy, S_M, of macrostate M is:*

$$S_M = k\ ln\, W_M,$$

where $k = 1.38 * 10^{-23} \frac{J}{K}$; k is called Boltzmann's constant.

Why Did I Do This To You? I want you to know that entropy is as real as energy or matter, it just takes a little more mathematics than most folks know to get a grasp of it. The human economy is a Super-Organism that lives on energy, matter and entropy flows. Humans have been spending low entropy at an increasingly phenomenal rate: depleting fisheries, cutting old growth forests, mining ores, mining topsoil, polluting ground and surface waters and air sheds and on and on. Any low entropy resource that humans exploit needs to be given the chance (with or without human help) to replenish its store of low entropy—the current levels of exploitation cannot be sustained! And since many leading economists ignore—or hate—the concept of entropy, it will take concerned citizens like you to educate them.

[33] I can get even more sophisticated by looking at quantum states of molecules and doing what is called quantum statistical mechanics. But we have done enough for now.

Chapter 30

The Financial Mathematics of Loans, Debts, and Compound Interest

30.1 Simple and Compound Interest: A Review

We have already covered the concept of interest in Exercise 19.18, for example; but it will not hurt to review the results, if not the entire arguments leading up to those results.

If you borrow P dollars, where the P stands for what economists (and bankers as well as loan sharks) call the principal, and you pay an interest rate R compounded annually, then if you do not pay anything back for n years you will still owe the P dollars plus interest. How much is this? Recall that at the end of one year you will owe $P + R * P = P * (1 + R)$ dollars, where R is written as a decimal or a fraction.[1] Replacing P dollars of debt, that you started the first year with, by $P * (1 + R)$ dollars of debt that you begin year two with, hopefully you can see that after two years you would owe $P * (1+R) + R * P(1+R) = P * (1+R)^2$ dollars. After three years you would owe $P * (1 + R)^2 + R * P * (1 + R)^2 = P * (1 + R)^3$ dollars. And, if you see the pattern, after n years you would owe $P * (1+R)^n$ dollars. In the next two exercises assume that interest accumulates and you never make any payments.

Exercise 30.1 Calculating Simple Interest on a Loan Compounded Yearly
(i) If you borrow \$100 at 10% interest compounded yearly, how much do you owe after 5 years? How much interest do you owe?
(ii) If you borrow \$100 at 19% interest compounded yearly, how much do you owe after 5 years? How much interest do you owe? (How does your answer here compare with the answer in part (i)?)
(iii) If you borrow \$10,000 at 18% compounded yearly, how much do you owe after 3 years? How much interest do you owe?

If you borrow P dollars at an annual interest rate R, compounded daily, for one year, how much do you owe? Answer: recall from Exercise 19.18 and nearby material that you will owe $P * (1 + \frac{R}{365})^{365}$ dollars, unless it is a leap year.

[1] For example, if R is 12% we would write $R = .12$. If R is 5% we would write $R = .05$.

Exercise 30.2 Calculating Interest Compounded Daily

(i) If it is a leap year and you borrow P dollars, what do you owe after one year if you are paying an annual interest rate R compounded daily?

(ii) Ignoring leap years (assume all years have 365 days), if you borrow P dollars at an annual interest rate R compounded daily, what do you owe after 5 years?

(iii) What is the accumulated interest in part (ii) after the first month? (Depending on legal details of your loan, this will likely be the minimum payment you can make at the end of the first month. If you make this minimum payment you will have paid off the first month's interest, but you will still owe P dollars since you will not have paid off any of the principal.)

Reading the equation for simple interest "backwards" it is easy to understand one of the concepts in classical economics, viz., present discounted value, PDV. Economists define PDV to be the amount that future payments are worth in the present if they were to be received immediately, and could then be invested at the prevailing interest rate. Thus if FV is that future value (n years in the future), and R is the prevailing interest rate which is assumed not to change for n years, then PDV and FV are related by the equation: $PDV = \frac{FV}{(1+R)^n}$. According to this equation, a present dollar is worth more than a future dollar. Economists refer to this as the time value of money.

Exercise 30.3 Calculating Present Discounted Value

(i) What is the PDV of 100 dollars 10 years in the future, if $R = .07$?

(ii) If a resource can be converted into money, what considerations determine if you do the conversion now or in the future (or never)?

(iii) The tacit assumption being made in this discussion is that there is *inflation*, i.e., interest rates are positive. Is it possible to have *deflation*? Do a little checking of the U.S. economy during the Great Depression and during the last half of the 19th century and see if at any time the value of 1 dollar actually increased with time, i.e., it would buy more in the future than in the present.

(iv) Could deflation happen now? Investigate the subject as best you can, decide what circumstances are required for deflation to occur. (For example, see a relatively recent article on the possibility of deflation by Paul Krugman in the December 31, 2002 edition of *The New York Times*. In 2006 Krugman changed his mind.)

30.2 How Much Does a Debt Really Cost You? Buying on Time and/or Installment Plans. Amortization. The Four Important Numbers: P, R, r, n

When you buy a car or a house or something(s) on a credit card, you usually end up making a payment of some fixed size each month for a long time until your debt is paid off. If you can avoid it, I recommend never going into debt[2]

[2]But if you go into debt, do it only with someone or some institution you really trust with your financial life. The federal bankruptcy reform legislation, H.R. 2415, passed in 2000 makes life considerably more painful for those who cannot pay their debts. It is interesting

(unless you absolutely have to—for a lifesaving operation, an education, a home otherwise unaffordable and the like). There is an exception to this, if you can make more money (that is take more money in) than you pay out, by borrowing money, then it makes sense financially. I have heard rather wealthy people say that the best way to make money is by using other people's money.

Rule of thumb: Only borrow money if doing so allows you to be more productive, and eventually improves your financial situation. (Borrowing money for the latest model automobile when an older, cheaper model can do everything you need is an example of borrowing money which violates the rule of thumb.)

I am not going to get involved the least bit in a discussion of how to make money in any clever way, since the only way I actually make money is to earn it by working and saving some fraction of what I make. The money saved can be invested in any number of ways, of which socially responsible investing is supposed by some to be the best for us and the Earth.[3] Working for a living rarely leads to great wealth, but it often is part of a happy life.

If you have to borrow money, which is what you do when you use a credit card, for example, then it is a good idea to have a firm grasp of what you are getting into! I resisted getting a credit card for some time, and I never did completely figure out the fine print in any credit card contract—so I always pay off my credit card before the "no interest grace period" is up. One of the reasons it is hard to figure out the contract is that it does not always tell you in very simple terms how you are to be charged. (If you doubt me, obtain a credit card contract or statement with an "explanation of finance charges" on the back; and try to figure it out in less than one hour.) I have known people who thought they had a 19% annual interest rate, only to find one month that it had gone up to 35% after they were considered a high risk financially. In addition to interest charges there are various fees and other charges that you should get clear about—and in writing—before borrowing money. Later I will recount some horror stories for you from [319], so you will have some idea of what to avoid if you are ever forced to borrow money. Borrowing money can be a life threatening experience, and the poorer you are the greater the chance you will preyed upon and sucked dry. There is a lot of money to be made on poverty, but usually not by the poor.[4]

If you borrow money there are four important numbers that you should know, which we will label P, R, r, n. These four (or at least the last three)

to note that 45% of all bankruptcies in the U.S. involve a medical reason or a large medical debt, cf., [311].

[3]For some ideas on what is considered socially responsible investing see *Co-op America's National Green Pages: A directory of products and services for people and the planet*, 2001 Edition, Co-op America Foundation, Washington, D.C., www.coopamerica.org

[4]See, for example, an article by Howard Karger, "America's Fringe Economy: Financial services for the poor and credit-challenged are big business," *Dollars & Sense*, November/December 2006, pp. 16–21.

numbers are very tightly related mathematically, and I am now going to derive an equation that expresses this relationship. The number P represents the number of dollars you borrow, and P is called the principal. I will ignore transaction fees, hidden charges and so forth because there is no way for me to know what they may be—and they can be very well hidden and huge if the person or institution you are dealing with is sufficiently greedy. So watch out!

The second number you need to know is the annual percentage interest rate, which I will label R. If your annual percentage interest rate is 19%, for example, then $R = .19$. The third number you should know is what your monthly payments will be. We will assume that each month you will pay a fixed amount, $r * P = rP$, where r is some number between 0 and 1. For example, you might pay $\frac{1}{20}$ of P each month, i.e., $r = \frac{1}{20} = .05$. The fourth number you should know is, n, how many payments, i.e., how many months will I have to make monthly payments in the amount rP? Let's now derive a mathematical relationship among these four numbers. I will assume that each month I will be charged an interest rate of $\frac{R}{12}$. (Your payments may exceed the interest, but it is important to know the amount of interest you are charged.)

So after one month what do I owe? The answer is:

$$P - rP + \frac{R}{12}P = P * (1 + \frac{R}{12}) - rP.$$

We used the distributive law to combine the first and last terms on the left side of the equation to get the first term on the right side of the equation. The $-rP$ is what I paid, and the $\frac{R}{12}P$ is the interest I owe at the end of one month (this has a $+$ in front of it). The right side of the equation shows: what I owe at the end of one month is what I owed at the beginning of the first month, namely P, times $(1 + \frac{R}{12})$, minus rP. This can be written as $P * (1 + \frac{R}{12} - r)$. Do you see this?

I now want to make two observations about the relationship between R and r in any real borrowing situation. Assume the R is greater than zero.

Exercise 30.4 Some Relationships among R, r and n

(i) Do you see that it is necessary that r be larger than $\frac{R}{12}$ if you are ever going to pay off your loan? (Note that "r is larger than $\frac{R}{12}$" is written in mathese as $r > \frac{R}{12}$.) Give your reason(s).

(ii) If $r = \frac{R}{12}$ do you see that you will owe exactly P dollars forever? Give your reason(s). Thus if $r = \frac{R}{12}$ you are making what might be called a "minimum payment." In this case you just pay off the interest that has accumulated over the last month and you pay off none of the principal.

(iii) Do you see that r must be larger than $\frac{1}{n}$? Given your reason(s). Remember that n is the number of payments that you will be making. By definition of n you will not owe any more after you have made n payments.

(iv) Do you see that $r * n$ must be larger than 1? (In mathese we write $rn > 1$.) Give your reason(s).

So after two months what do I owe? The answer is:

$$[P*(1+\frac{R}{12}-r)]-rP+\frac{R}{12}*[P*(1+\frac{R}{12}-r)] = [P*(1+\frac{R}{12}-r)]*(1+\frac{R}{12})-rP.$$

Again I used the distributive law to combine the first and last terms on the left side of the equation to get the first (of two) terms on the right side. The left side of the equation says that I paid rP but that, in addition to the $P*(1+\frac{R}{12}-r)$ which I started the second month owing, I owe $\frac{R}{12}*P*(1+\frac{R}{12}-r)$ in additional interest.

The right side of this last equation says: what I owe at the end of the second month is what I owed at the beginning of the second month, namely $P*(1+\frac{R}{12}-r)$, times $(1+\frac{R}{12})$, minus rP.

I see a pattern here. What I owe at the end of a month is what I owe at the beginning of that month times $(1+\frac{R}{12})$, minus rP. I am led (see the next Exercise) to the following mathematical expression of what I owe after k months:

$$P*(1+\frac{R}{12})^k - r*P*\sum_{j=0}^{k-1}(1+\frac{R}{12})^j.$$

Exercise 30.5 Verifying a Pattern

Let's verify that the equation immediately above is what you get from the pattern I observed.

(i) Start with P dollars owed at 0 months, and multiply by $(1+\frac{R}{12})$ to get $P*(1+\frac{R}{12})$. Now subtract rP to get $P*(1+\frac{R}{12})-rP$ for what is owed at the end of 1 month.

(ii) To get what is owed at the end of 2 months multiply the result from part (i) by $(1+\frac{R}{12})$ and subtract rP. Do you get (something equivalent to) $P*(1+\frac{R}{12})^2 - r*P*[(1+\frac{R}{12})+1]$?

(iii) To get what is owed at the end of 3 months multiply the result from part (ii) by $(1+\frac{R}{12})$ and subtract rP. Do you get $P*(1+\frac{R}{12})^3 - r*P*[(1+\frac{R}{12})^2+(1+\frac{R}{12})+1]$?

(iv) To get what is owed at the end of 4 months multiply the result from part (iii) by $(1+\frac{R}{12})$ and subtract rP. Do you get $P*(1+\frac{R}{12})^4-r*P*[(1+\frac{R}{12})^3+(1+\frac{R}{12})^2+(1+\frac{R}{12})+1]$? Now recall:

$$(1+\frac{R}{12})^3+(1+\frac{R}{12})^2+(1+\frac{R}{12})+1 = \sum_{j=0}^{4-1}(1+\frac{R}{12})^j.$$

(v) Continue on in this way until you can "see" the formula for what is owed at the end of k months immediately preceding this exercise.

We can simplify $\sum_{j=0}^{k-1}(1+\frac{R}{12})^j$ and thus simplify the formula, or mathematical expression, for what is owed after k months. Recall from Exercise 20.2 that (if $x \neq 1$) $\sum_{j=0}^{k-1} x^j = \frac{1-x^k}{1-x}$. If we substitute $(1+\frac{R}{12})$ in for x we get:

$$\sum_{j=0}^{k-1}(1+\frac{R}{12})^j = \frac{1-(1+\frac{R}{12})^k}{1-(1+\frac{R}{12})}.$$

Simplifying the right hand side of this equation we get:

$$\sum_{j=0}^{k-1}(1+\frac{R}{12})^j = \frac{1-(1+\frac{R}{12})^k}{-\frac{R}{12}} = \frac{(1+\frac{R}{12})^k-1}{\frac{R}{12}}.$$

Substituting this last expression into the formula just before Exercise 30.5, we get that the amount owed after k months is:

$$P * (1 + \frac{R}{12})^k - r * P * \frac{(1 + \frac{R}{12})^k - 1}{\frac{R}{12}}.$$

We are nearing the end of our search for an equation that relates P, R, r, and n. Do you remember what n is? It is the number of months you make payments. Thus n is the number of months that makes the mathematical expression of what you owe equal to zero![5] Thus:

$$P * (1 + \frac{R}{12})^n - r * P * \frac{(1 + \frac{R}{12})^n - 1}{\frac{R}{12}} = 0. \tag{$*$}$$

If you did not understand all of the details in the derivation of $(*)$, all is not lost—although you should go back and review the derivation as many times as is necessary until you understand it. I will now use equation $(*)$ as a new starting point to answer three common questions you might ask when borrowing money or buying something on the "installment plan."

The first observation I make is that if you borrow an amount of money P that is not 0, then equation $(*)$ can be simplified to read:

$$(1 + \frac{R}{12})^n - r * \frac{(1 + \frac{R}{12})^n - 1}{\frac{R}{12}} = 0. \tag{$**$}$$

Exercise 30.6 The Fundamental Equation Relating R, r and n

(i) Can you derive $(**)$ from $(*)$?

(ii) Does $(**)$ say that the relationship among the three numbers R, r, n is independent of the amount of money you initially borrow?

Equation $(**)$ will show us some interesting things about mathematics as I answer three questions.

The first question is this. If I borrow some money and the interest rate I am charged on the money I owe in a given month is $\frac{R}{12}$, and if I want to pay off my loan in n monthly payments, what fraction r must I pay each month? In other words, if I know $\frac{R}{12}$ and n in equation $(**)$, can I solve for r? We answer this question in the following exercise.

[5] There is a slight refinement of this statement that I must tell you about for future reference. It does not always work out that your n^{th} payment, that is your last payment, is exactly $r * P$, it may be less. Thus, when solving for n in some circumstances, n may not come out to be a whole number. This does not change the fact that I define n to be that number that makes the formula for what you owe equal to zero, even if n comes out to be a whole number of months plus a fraction of one month.

Exercise 30.7 Solving for r, the Size of the Monthly Payment, Given the Interest Rate R and the Total Number of Payments, n

(i) Let's do a problem with concrete numbers in it first, to warm up. Suppose that $R = .12$ and we want to pay off our loan in exactly two years, i.e., $n = 24$. Can you see that equation (**) becomes

$$(1 + \frac{.12}{12})^{24} - r * \frac{(1 + \frac{.12}{12})^{24} - 1}{\frac{.12}{12}} = 0?$$

(ii) Solve (i) for r.

(iii) Assuming R and n are known (but you do not know actual numerical values for R and n) in equation (**), solve for r. How many algebraic steps do you need to make to do this problem? Describe them.

Assuming R and n are known, one way to solve (**) for r yields:

$$r = \frac{\frac{R}{12} * (1 + \frac{R}{12})^n}{(1 + \frac{R}{12})^n - 1}.$$

Thus, if R and n are known, we can find r in a straightforward manner. Try the following exercise to test your understanding of how to use the above formula for r in terms of R and n.

Exercise 30.8 Examples of Calculating r Given R and n

(i) Suppose $R = .12$ and $n = 24$, what is r according to the formula for r above?

(ii) Suppose $R = .19$ and $n = 36$ what is r according to the formula for r above?

(iii) Suppose a credit card company considers you a bad risk and charges you an annual percentage interest rate of $R = .38$ and you want to pay your loan off in two years, what fraction do you have to pay each month? If your initial loan is for \$2,000, what are your monthly payments.? How much money total will you have paid the credit card company at the end of two years? How much interest will you have paid at the end of two years?

The second of my three questions involves solving (**) for n, viz., suppose I know my annual percentage interest rate, R, and my monthly payments, determined by the number r, how many months will I have to pay? That is, what is n?

We can rewrite (**) (by inverting the denominator of the second term in (**), $\frac{R}{12}$, and multiplying this second term by $\frac{12}{R}$) as

$$(1 + \frac{R}{12})^n - r * \frac{12}{R} * (1 + \frac{R}{12})^n + r * \frac{12}{R} = 0.$$

This can be written

$$(1 + \frac{R}{12})^n * [1 - r * \frac{12}{R}] = -r * \frac{12}{R}.$$

Solving first for $(1 + \frac{R}{12})^n$ we get

$$(1 + \frac{R}{12})^n = \frac{-r * \frac{12}{R}}{[1 - r * \frac{12}{R}]} = \frac{r * \frac{12}{R}}{[r * \frac{12}{R} - 1]}.$$

Now I can solve for n, do you remember how? I will now take the logarithm (to base 10 or e) of both sides and get:

$$n * log\,(1+\frac{R}{12}) = log\,\frac{r*\frac{12}{R}}{[r*\frac{12}{R}-1]}.$$

Solving for n I get

$$n = \frac{log\,\frac{\frac{r*12}{R}}{(\frac{r*12}{R}-1)}}{log\,(1+\frac{R}{12})} = \frac{log\,\frac{r}{(r-\frac{R}{12})}}{log\,(1+\frac{R}{12})}.$$

The following exercise will give you some practice using this formula for n.

Exercise 30.9 Calculating Your Total Number of Payments, n, Given the Size of Your Monthly Payment, r and the Interest Rate, R

(i) Suppose that $R = .38$ and $r = .0601125$. What do you get for n? Compare this with Exercise 30.8 (iii) above. Explain why n is not exactly 24.

(ii) If $R = .19$ is the annual interest rate on a credit card, and you pay off $r = .05$ of your debt each month, how many months will you be making payments?

Now I ask the third, and hardest, of the three questions. Suppose I know the size of my monthly payment, i.e., r, and how long I will be paying, i.e., n, what is R, my annual rate of interest?

You might say, given r and n, just solve $(**)$ for R. But this cannot be done in general using algebraic steps like those we have learned (or by using algebraic steps of any kind!). This fact is interesting,[6] even though it makes our question harder to answer. The reason why this is an important question is evident from the following.

Exercise 30.10 Calculating the Interest Rate, R, When You Buy a Car (or Anything Else) on the Installment Plan

Suppose you want to buy a car for $15,000 and the car dealer has asked you to pay $400 per month for 4 years. Suppose your local credit union is offering to loan you $15,000 at an annual percentage interest rate of 6% compounded monthly. Which should you do: make the monthly payments of $400 to the car dealer or to your credit union?

(i) One solution is to figure out what the annual interest rate compounded monthly is that the car dealer is charging. See the discussion following this exercise to see if this interest rate is less than or greater than 6%.

(ii) A second solution is to figure the total cost of the loan from the credit union. What do you get? Is it more or less than what the car dealer is charging?

I can rewrite $(**)$ in the following form

$$r * [1-(1+\frac{R}{12})^{-n}] = \frac{R}{12}.$$

[6] The interested reader can follow up on this deep bit of mathematics by investigating the topics: "zeros of polynomial equations" and "Galois theory." Polynomial equations of degree 5 or higher cannot be solved in general using only standard algebraic steps.

I look at this equation assuming that I know r and n and am looking for R, or equivalently, $\frac{R}{12}$. Thus if I think of the left-hand side of the equation as a "machine" or "formula" or function which acts on $\frac{R}{12}$, if I plug the correct value in for $\frac{R}{12}$, this function produces $\frac{R}{12}$. That is, this function leaves the correct value of $\frac{R}{12}$ fixed. In mathese the answer we seek, viz., $\frac{R}{12}$, is a *fixed point* of the function $f[x] = r * [1 - (1 + x)^{-n}]$. By this I mean, if you plug $\frac{R}{12}$ in for x in the expression for f, f produces $\frac{R}{12}$ as the result.

Now in the above exercise I want to know if the R that the car dealer is charging is larger than or less than, or equal to, .06. I can test right away if $R = .06$ by plugging it into the equation and see if $\frac{.06}{12} = .005$ comes out.

Thus, since I am paying \$400 per month, my $r = \frac{400}{15000} = \frac{4}{150}$, and $n = 48$, so

$$r * [1 - (1 + \frac{R}{12})^{-n}] = \frac{4}{150} * [1 - (1 + \frac{.06}{12})^{-48}] = .0056774.$$

So our function did not produce .005 exactly when we put $\frac{R}{12} = .005$ in, it produced something a little larger. Thus $\frac{R}{12} = \frac{.06}{12} = .005$ does not use the correct value for R, since .005 is not left fixed by f.

I am now going to try a mathematical trick called *iteration* which is motivated by nothing more than the fact that I have tried such a trick before and it has led me to the answer I seek. The first step of this iteration is as follows. I am going to take my function $f[x] = r * [1 - (1 + x)^{-n}]$ and plug the result I got from plugging .005 in for x, namely, .0056774, back in for x. Thus I get

$$f[.0056774] = \frac{4}{150} * [1 - (1 + .0056774)^{-48}] = .0063453.$$

Now .0063453 is bigger than .0056774; the number that comes out is not the same as the number I plugged in. I will continue this process, I will now plug in .0063453 for x in the formula for $f[x]$ and see what I get.

Now I can continue this way with my calculator, or I can use a spreadsheet.[7] If I use a spreadsheet I get the following output:

0.0050000
0.0056774
0.0063453
0.0069828
0.0075721
0.0081008
0.0085625
0.0089561
0.0092846
0.0095542

[7] I only need the first column of a spreadsheet. In the top box, A1, I will put a number say, .005. In the next box down, A2, I will type =($\frac{4}{150}$)*(1-(1+ A1)^(-48)); and I will then "click and drag" on box A2 and get the resulting column shown in the text.

0.0097721
0.0099463
0.0100841
0.0101924
0.0102769
0.0103426
0.0103935
0.0104328
0.0104630
0.0104863
0.0105042
0.0105179
0.0105284
0.0105365
0.0105427
0.0105474
0.0105510
0.0105538
0.0105559
0.0105575
0.0105588
0.0105597
0.0105605
0.0105610
0.0105614
0.0105618
0.0105620
0.0105622
0.0105624
0.0105625
0.0105626
0.0105626
0.0105627
0.0105627
0.0105627
0.0105628
0.0105628

Note that I can stop when I start getting the same number repeated.[8] Thus $f[0.0105628] = 0.0105628$, I get the same number out that I put in! This means that $\frac{R}{12} = .0105628$ is a fixed number (or fixed point) for my function f and is (hopefully) the correct answer I seek. Thus $R = .12675$ (to

[8]If I had calculated numbers to greater accuracy, say 10 digits, then I would have had to go on a bit more.

at least 4 digit accuracy). This means that (assuming everything I have done is correct) the car dealer is charging me more than 12.67 % annual percentage interest, compounded monthly. Thus it is better to borrow the money from the credit union!

There are several questions that should occur to you: (1) Did I really get the correct answer? Is the car dealer really charging an interest rate of more than double the credit union's? (2) Will the type of calculation I did just above work in any other similar situation? (3) Will it work in *every* other similar situation? (4) Did I get *the only possible* answer, i.e., is there any other non-zero "fixed point" value that will work for $\frac{R}{12}$ besides .0105628?

The answers to all of the above questions are summarized by the following. There is only one possible answer, or non-zero fixed point, for problems of the type I just solved by using iteration. The method of iteration I used will work, i.e., it will find this unique answer, for any similar problem, assuming my initial guess is a positive number. (It may take time to find the answer using a hand calculator, but if you persist you will find it.) The beauty of the calculation that I just did, using the iteration method, is that it will always work on equation (**) to find R, and you can understand what is going on!

By the way, you will notice that each of the numbers produced during the iterative calculation is bigger than the one preceding it, until the last step. This sequence of numbers is called a *monotone increasing sequence of numbers*.

In order to understand what is going on in the above calculation, and in order to see that it will always work—and why—I want to start by graphing the function that I was using in the lengthy calculation above, namely,

$$f[x] = r * [1 - (1 + x)^{-n}]. \qquad (***)$$

In the Figure 30.1 I graph f with the values $r = \frac{4}{150}$ and $n = 48$, which is exactly the values I was using above. The horizontal dashed line is at height $\frac{4}{150}$.

I want to make some crucial observations about this graph that hold in general for any f as in $(***)$ when looking for an unknown R, given known values for r and n. It appears that this function f starts at (0,0) and increases as I go from left to right, but the rate of increase seems to diminish with the graph leveling off and approaching from below the value for $r = \frac{4}{150} \approx .026666667$. (The graph of $r = \frac{4}{150} \approx .026666667$, which is the dashed horizontal line, is called a horizontal asymptote of the graph of f.)

You can verify that f is indeed increasing by plotting lots of points with a computer, as I have done in Figure 30.1; or, more rigorously, you can use techniques from calculus that involve looking at what is called the first derivative of f. Hopefully this motivates you to study a little calculus, but I will not go into calculus here. It is true, however, that the function f is always increasing for positive x.

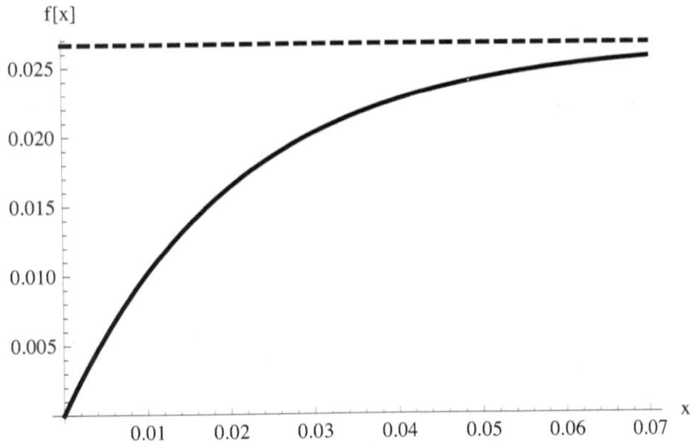

FIGURE 30.1: Graph of $f[x] = \frac{4}{150} * [1 - (1+x)^{-48}]$

Exercise 30.11 A Closer Look at f

Given function $f[x] = r * [1 - (1+x)^{-n}]$ as above, answer the following.

(i) Show that $f[0] = 0$.

(ii) Show that $f[x] < r$ for all positive x.

(iii) Show that as x gets larger and larger, $f[x]$ gets closer and closer to r.[9]

In Figure 30.2 I show with a "picture" some of the steps in the iteration calculation I did above.[10] I want to take a closer look at the graph of f.

It turns out that you can show (using calculus) that the slope of the tangent line to the graph of f at (0,0) is equal to $r * n$ and by Exercise 30.4 (iv), $r * n > 1$. This means that the graph of f as it leaves (0,0) will start out above the "$y = x$" line which has a slope of 1. Again, (using calculus) the slope of the tangent line to the graph of f at point $(x, f[x])$ is given by the formula $\frac{rn}{(1+x)^{n+1}}$ so as x increases the slope of f decreases. (The slope remains positive, but the slope gets closer and closer to 0 as you travel to the right. The graph of f "levels out" and approaches the horizontal asymptote, the horizontal line "$y = r$".) Thus as you follow the graph of f to the right you

[9]Hints for (ii) and (iii): The quantity $1 - (1+x)^{-n}$ is always less than 1 (for $x > 0$), and as x grows large, $(1+x)^{-n}$ gets as small as you like (assuming n is positive). In mathese it is said that $\lim_{x\to\infty}(1+x)^{-n} = 0$.

[10]The careful reader will observe that the value of $\frac{R}{12}$ in Figure 30.2 is about .024 (actually about .024234240) not the value .0105628 for $\frac{R}{12}$ that we calculated before. This is because the $f[x]$ graphed in Figure 30.2 is actually $f[x] = \frac{4}{150} * [1-(1+x)^{-100}]$, i.e., it is almost the same function, f, except that $n = 100$, not 48. Thus for this function $R \approx 12 * .024234240 \approx .2908$. Thus if you take 100 months to pay off the loan, the effective interest rate is about 29.08%, compared to 12.67% if you pay it off in 48 months. So why did I do this: change the 48 to 100? I needed a graph where there was enough space trapped between the graph of f and the "$y = x$" line to the left of the fixed point $\frac{R}{12}$ so I could show you in a picture (that you can actually see!) what is going on with our iteration procedure. If you will look at the corresponding area in Figure 30.3 you will see what I mean.

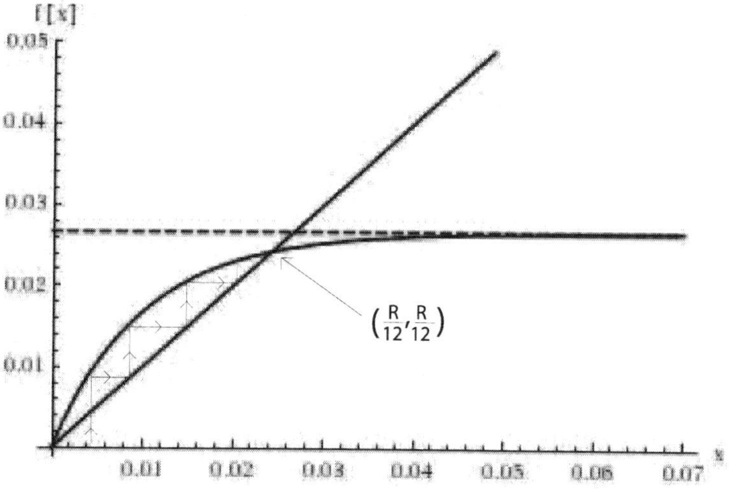

FIGURE 30.2: Visualization of Iteration 1

will see the graph of f cross the line "$y = x$" at some point. There will be one point, and no more than one point other than (0,0), where the graph of f crosses the line "$y = x$."

What is this point? It is precisely the fixed point where $f[x] = x$. (This is because this point is on the line "$y = x$," so the first and second co-ordinates of this point have the same numerical value. This point is also on the graph of the function f, so the point is of the form $(x, f[x])$. Thus $f[x] = x$.)

The co-ordinates of this fixed point for f are $(\frac{R}{12}, \frac{R}{12})$ as I noted (several pages) above. Note from Exercise 30.4 (i), that $\frac{R}{12} < r$, which is consistent with the picture under scrutiny.

So why does the iteration process "home in on" or converge to $\frac{R}{12}$ no matter where you pick your initial guess for $\frac{R}{12}$ (as long as your guess is a positive number)[11]? This can be seen from the graph of f and its relationship to the "$y = x$" line. If I guess the fixed point is .005, as in the long iteration calculation above, I then calculate $f[.005]$.[12] In Figure 30.2 this is pictorially represented by tracing your pencil first along the vertical line segment starting at (.005,0) and going up, stopping at point $(.005, f[.005])$, which is a point on the graph of f! Second, trace your pencil horizontally to the right along the line segment beginning at $(.005, f[.005])$ and ending at $(f[.005], f[.005])$ which is a point on the "$y = x$" line. Now starting at this point trace your pencil

[11]Recall that R is the number you are looking for.

[12]Actually in Figure 30.2 the iteration starts at a point close to .005, but it is actually closer to .004. Will starting at this point get us heading to the same fixed point as if we started exactly at .005?

up along the vertical line segment, beginning at ($f[.005]$,$f[.005]$) and ending at ($f[.005]$, $f[f[.005]]$). Continue this process and note that your pencil tip is trapped between the graph of f and the line "$y = x$" at all times, and your pencil tip is heading for the point where these two graphs intersect, which is the fixed point you are trying to find numerically!

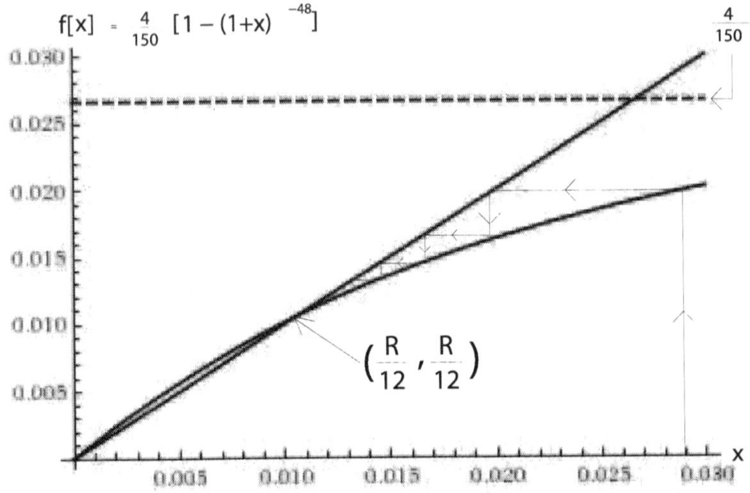

FIGURE 30.3: Visualization of Iteration 2

Exercise 30.12 A Picture of Iteration

Using the same function as above, viz., $f[x] = \frac{4}{150} * [1 - (1+x)^{-48}]$, try the two iteration procedures below. You know the fixed point is about .0105628282, but pretend you do not know this and see if you can rediscover it.

(i) Pick a "large" value for your initial guess of the fixed point, i.e., some number bigger than, say .011. (Such a choice starts your iteration to the right of the fixed point.) Do the steps of the iteration and follow each step graphically, as in Figure 30.3.

(ii) Pick a "small" value for your initial guess of the fixed point, i.e., some positive number less than, say, .01. (This choice starts your iteration to the left of the fixed point.) Do the steps of the iteration and follow each step graphically, as in Figure 30.2.

30.3 Examples of Individual Debt: Rent-to-Own, Credit Cards, and Loans

According to the chapter, " 'Rent-to-Own' The Slick Cousin of Paying On Time," in *Merchants of Misery*, [319], from the perspective of the person who owes the debt there is not a great deal of difference between "renting-to-own" and buying something on time. True, the sales pitches are a bit different and "renting-to-own" is legally different from buying on the installment plan. Hence many of the legal regulations on installment buying (some states limit the amount of interest that can be charged, for example) do not apply to "renting-to-own."

True, if you "rent-to-own" you are promised repairs for free, but similar options are available for purchasers on the installment plan. If you miss payments, or if you want to stop paying, your "rent-to-own" item will be taken back by the merchant. To me this does not seem a great deal different practically than having your car, TV or whatever repossessed for not making installment payments.

Take a look at the following example from *Merchants of Misery*, pp. 146–7. (The author of [319] has a more recent book, [320], also of interest.)

> *Rent-to-own customers routinely pay two, three, and four times what merchandise would cost if they could afford to pay cash. For example: A Rent-A-Center store in Roanoke, Virginia, recently offered a 20-inch Zenith TV for $14.99 a week for 74 weeks—or $1,109.26. Across town at Sears, the same TV was on sale for $329.99. Putting aside $15 a week, it would take just 22 weeks to save enough to buy it retail.*

Exercise 30.13 Some Mathematics of "Renting-to Own"

(i) Verify the mathematics in the above paragraph. Are the mathematical claims correct?

(ii) If you were to buy the TV mentioned above, what would it cost you to buy a service contract to guarantee repairs and/or replacement at no further cost?

(iii) What does the "Rent-to-Own" merchant offer in exchange for your payments that might not be available to you otherwise? Would it be worth it to you?

(iv) What is the interest rate (before weekly compounding) that the Rent-to-Own merchant is charging?[13]

(v) Would a consumer loan from a bank offer a lower or higher interest rate than the rate R found in part (iv)? If you bought the TV with a credit card what sort of interest rate might you be charged?

(vi) Why would a person not buy the TV with a consumer loan or a credit card?

(vii) Find a real "Rent-to-Own" business and, pick out an item and analyze the total costs were you to purchase that item on the rent-to-own plan offered. Find out how much you can buy that particular item for (elsewhere but close by) if you have cash. Calculate the interest charge were you to buy an item on the "rent-to-own" plan.

Exercise 30.14 How Much Did I End Up Paying?

Looking back at the general situation, if I start with a principal P and I pay rP, n times, then my total payments are rPn. The effective interest you pay is the total payments minus the principal, or $rnP - P$.

(i) Show that in the general situation the fraction of P that you in effect end up paying in *interest payments* is $rn - 1$.

---

[13]Hint: You are looking for R, the interest rate, and you are given $n = 74$ *weeks* (this is different from the unit of "monthly payment" used in the previous section), and you may take $r = \frac{14.99}{329.99} = .0454256189$. Assume that each week you are charged interest of $\frac{R}{52}$. Thus you can iterate the same "f" as in the previous section, but the fixed point you converge to is $\frac{R}{52}$.

(ii) In the Rent-to-Own situation in part (iv) of the previous exercise, calculate your final, total costs. Those total costs minus the "real" original price of the TV (which I am taking to be $P = \$329.99$ in this case) divided by the original price of the of the TV should be $rn - 1$. How much did you finally end up paying in interest for this TV? What percentage of the originally stated price of the TV is this amount? Note: Without doing some comparison shopping you do not know what the "real" original price of the TV is; and you hence cannot know r or R, and hence you do not know what the Rent-to-Own merchant is charging you in interest!

I want to do one exercise with a typical credit card bill since you are quite likely to get such a bill some day. It takes a little work to understand exactly what, for example, the APR or annual percentage rate means. I will not go into the fine print on late fees, finance charges and so on; but beware they loom in the background and increase the amount you can owe. Also, interest rates can be increased if the credit card company deems you unreliable.

Exercise 30.15 Can I (We) Understand a Credit Card Bill?

Suppose you get a credit card bill with the following information: You owe \$3, 228.47, the annual percentage rate of interest (APR) is stated to be 19.40% and your minimum payment is \$64.00.

(i) What rate of interest R compounded monthly would result in an annual interest rate of .1940? (Hint: Solve $(1 + \frac{R}{12})^{12} = 1.1940$ for R.) What would your monthly payment be if it were $\frac{R}{12} * 3228.47$ dollars?

(ii) According to the contract for the particular credit card in question, the interest charged on the credit card debt of \$3, 228.47 for one month is $\frac{.1940}{12} * 3228.47 = 52.19$. Is this more than you might have expected to pay considering part (i)?

(iii) Neglect any other fees or charges for the moment and assume that you let your credit card debt of \$3, 228.47 just sit for a year without charging anything additional to that card. At the end of the first month you would owe $\$3,228.47 + \$\frac{.1940}{12} * 3228.47 = \$(1+\frac{.1940}{12})*3228.47$. At the end of the second month you would owe $\$3,228.47*(1+\frac{.1940}{12})^2$ and so on until at the end of one year, i.e., 12 months, you would owe $\$3,228.47*(1+\frac{.1940}{12})^{12}$. What would be the actual annual percentage interest rate (how much interest would you owe and what percentage of \$3, 228.47 would it be)? Would it be less than, equal to or greater than 19.40%?

(iv) I called the credit card company in question and asked them how they computed the minimum payment of \$64.00. They said that one gets the minimum payment by taking 2% of the balance, in this case, of \$3, 228.47. Was this calculation done correctly in this case?

(v) If you make the minimum payment in this case, what is the maximum amount of principal that you will be paying off?

(vi) At the rate indicated by the first minimum payment, if you continue making minimum payments how long will it take to pay off the debt? (Assume no extra fees or penalties.) (Hint: Calculate r, $\frac{R}{12}$ and finally n.)

(vii) If you miss a payment do you think that charges in addition to interest will be owed? If this is the case then the "real" interest rate you calculated in part (iii) will be too low.

(viii) Find another credit card bill and (with the permission of the credit card holder) do an analysis like the one above.

Freelance journalist Eric Rorer wrote an article for the June 22, 1994 edition of *The San Francisco Bay Guardian*: "Shark Bait: How some consumer-finance companies make a killing off people who badly need money." This article is reproduced in *Merchants of Misery*. Rorer applied for loans at the local office of five consumer-finance companies. He did not identify himself as a freelance journalist, but all other personal and financial information he provided to the loan officers was accurate.

Rorer, a part-time word processor, made roughly \$1,000 per month; and he asked each company for a loan of \$1,500 for the purposes of buying new tires and a stereo for his car.

At Avco Financial Services he recounts that after much wrangling the Avco executive finally agreed to show him a copy of the loan contract. Rorer states:

It wasn't easy: First he wanted me to sign several documents, which he said were simply preliminary financial-disclosure papers that didn't obligate me to anything. But I had been warned by several attorneys from different consumer groups not to sign anything, period. So I insisted on seeing the actual contract before putting pen to paper.

The man from Avco already had a copy of my credit report, my addresses and phone numbers for home and work, my parents' addresses and phone numbers, a copy of my driver's license, and a copy of my last paycheck. He also told me that, because the loan was going to be secured by my car, he would have to take and keep the title before I got the loan, even though the low blue-book value of the car is \$7,000—more than four times the amount of the loan.

When Rorer looked at the contract he saw the following data: A \$1,500 loan at 24.5 percent interest with a monthly payment of \$80 for three years, resulting in a total payoff of \$2,880. Rorer asked to take a copy of the contract home (since he was suspicious of the numbers which did not seem right to him). He was told: "That's against company policy."

Rorer then asked if he could just copy down the interest rate, monthly payment and the number of payments from the contract. Again Rorer was told that that was against company policy. So Rorer memorized the numbers and went back to his office to do some calculations. He did a mathematical exercise (with some help), can you do it (with what you have learned above)?

Exercise 30.16 Catching a Company Lie with Mathematics

The data from Rorer's contract (which he had to wrangle to see): $P = \$1,500$, $r = \frac{80}{1500} = .053333333$, and $n = 36$. Rorer thought that the interest rate actually being charged was higher than the 24.5 percent claimed on the contract.

(i) Using the values given in the contract for r and and n, calculate (using the iteration technique) what the actual interest rate is and what R (before any compounding) is.

(ii) If the interest rate R really were 24.5 percent and $n = 36$ what would be the monthly payments for 3 years? (Hint: calculate r.)

It turns out that there were \$756 in charges in Rorer's contract for things that the loan officer never mentioned. Included was a "credit insurance" charge of \$281.69 and additional insurance for Rorer's car at a cost of \$220. It turns out that it is illegal to require "credit insurance" as a condition for a loan. After some period of time during which the loan officer was quite insistent that Rorer take the credit insurance, the loan officer relented and said he could have the loan without it (after checking with his manager). However, the loan officer insisted that Rorer buy the additional insurance for his car, without telling Rorer that he could just buy a supplement to his existing policy with his usual car insurer.

Exercise 30.17 Another Lie Caught with Mathematics

Rorer went to Beneficial of California, again asking for a loan of \$1,500. Beneficial offered Rorer the loan at (what they claimed was) 25.3 percent interest with monthly payments of \$106 for two years.

(i) What is the actual interest rate R if $P = 1500$, $r = \frac{106}{1500} = .0706666667$, and $n = 24$?

(ii) If the interest rate, R, really were 25.3 percent, what would be the monthly payments for two years?

The loan officer for Beneficial actually told Rorer that he was being charged for insurance, but the loan officer never informed Rorer that such charges were optional.

For me the lessons are clear. Besides bringing a programmable calculator or a lap top computer when shopping for a consumer loan, it wouldn't hurt to bring a finance lawyer along as well. Better yet, do without the consumer loan if at all possible. If you do go shopping for a consumer loan, be ready to read the fine print and understand it. Also, an audio or video tape of the transaction would be a good idea, just in case your name is forged onto papers you did not intend to sign. If such happened it would be your word against that of the company you are dealing with, which no doubt includes some company lawyers. I say this because on page 3 of *Merchants of Misery* I read:

Wilma Jean Henderson walked unsteadily out of Associates Financial Services[14] *and climbed behind the wheel of her car. "I went to start the ignition and my legs went to shaking so bad. And I took a deep breath and I turned my car back off and I put my head on my steering wheel and I started to cry."*

The mother of seven children and stepchildren had gone to an Associates office in Montgomery, Alabama, the month before and borrowed \$2,000 *to fix her '87 Blazer. It was a big company, one she thought she could trust. After all, it was owned by Ford Motor Company. When she sat down with the loan officer to close the deal, she recalled, "we really didn't talk about the loan. He was talking about he was having some trouble with his car—his car was one of those little foreign ones—at the same time saying, 'Sign this. Sign this.' " She testified later that the loan officer flipped through the papers so only the signature portion of the documents showed, and some of the numbers on one document had not been filled in until after she signed it. She didn't read anything, she said, because "I trusted him—to do right."*

But when Henderson went to make her first payment, she testified later, she learned that along with the \$2,000, *she owed another* \$1,200 *for "add-ons" she didn't know a thing about—three kinds of credit insurance and an auto club membership. And her interest rate was 33.99 percent.*

[14] According to Michael Hudson, in 1993 three-fifths of Ford Motor Company's earnings came from its financial services subsidiaries. The biggest earner among these is Ford Motor Credit Corp., with a large chunk of its profits coming from a lesser-known subsidiary, Associates Corp. of North America.

To make a long story short, Henderson sued, settled out of court for an undisclosed amount and Associates denied any wrong-doing. There are many more such stories like this in *Merchants of Misery.*

I had an interesting personal experience once when I went shopping for a used car. The sales people (there were three working on me) asked me for my driver's license so they could verify something. I was being pressured into signing some forms when I told the salespeople that I would not consider buying the car that looked interesting to me without first having a friend of mine who is a mechanic come check it out. Until that time I was not signing anything.

Time passed, I did not sign; and when I asked for my driver's license back it just did not appear. I got up to leave, asking them to mail my driver's license to the home address written on it. They said to me that I should not leave since it is not legal to drive without a license. I informed them that I was riding my bicycle and did not need the license that day. (When my mechanic eventually looked at the car that the salespeople wanted to sell me, he told me that the transmission was trashed—the car was junk.) This experience and several others have reinforced my belief: when doing business, *caveat emptor* or, "keep your left up."

Exercise 30.18 Monthly Payments: How Much Principal? How Much Interest?
This problem is not completely well posed in the sense that there is more than one "correct" answer. The following actually happened. In at least one part of the U.S. during 2002 interest rates for home mortgages dropped from 10% to 6%, and in response some homes that were listed at \$200,000 raised their listed price to, let me say, \$X. What value for X will give the same monthly payments at 6% interest that \$200,000 will give at 10% interest? Do you think this strategy always pays off?

Part VIII

Media Literacy

Chapter 31

Information Flow in the 21st Century

31.1 Investigative Journalism Requires Cash

I challenge anyone to demonstrate how democracy, in the United States or anywhere, can function without a citizenry that has ready access to information on virtually any subject, or without journalists who—rather than answer to the powerful—monitor them and tell us what they are doing. The originators of the United States were so clear on this that the only business guaranteed freedom, in fact the only business even mentioned, in the U.S. Constitution is the press.

In early America the government subsidy for "media," which in those days meant newspapers, was considerable. It was commensurate with what modern democracies, other than America, spend on subsidies today. From [424, pp. 124–4] we read:

"How extensive were these subsidies? 'Between 1792 and 1845,' Culver Smith, a scholar on the subject of subsidies, wrote, 'the minimum charge for a letter ranged from six to twenty-five cents, depending on the distance, but the maximum postage for a newspaper for any distance was one and one-half cents.' All the original research by the leading scholars on the subject finds that newspapers and pamphlets accounted for around 95 percent of the weight of the mail and less than 15 percent of the Post Office revenues."

Eventually the flow of information in the United States became one supported by advertising: ads in newspapers, magazines, and on television and radio. American media are now in what some refer to as a crisis, since the justification for placing an ad in a newspaper or even a television show is increasingly hard to make. If you want to sell a house, or some other possession, there are "lists" for that purpose available via the internet. If you want to reach a particular audience for your product, computers and the Web make available far more sophisticated and cost-effective tools than placing an ad "in the general media."

This "ad based" U.S. media developed along with the PR, public relations, industry. For example, in 1980 the number of public relations specialists and managers was a bit more than the numbers employed as journalists, i.e., as editors, reporters and announcers in newspaper, radio, and TV industries. In 2008 the ratio was nearly 4 to 1, PR persons to journalists, as I just broadly defined them, cf., [424, p. 49].

Newspapers are disappearing across the United States. Having been bought up by and become just another profit center for larger corporations many newspapers are being jettisoned as their circulation and ad revenues and hence their profits decline. If there is no available buyer, the newspaper dies. The traditional niche for investigative journalism, the newspaper, is disappearing. The big question is this. If serious journalism is to survive, how is it going to be paid for?

A great many people report getting their news from local TV news programs. (I do not count here "commentators" who are very long on opinion and very thin on "facts," selected narrowly to support those opinions. These folks are, of course, influential; but they are not professional journalists.) Such programs have shrinking budgets for serious journalism as is the case for print media. In fact, TV news programs, like most of the Web, get most of their real news from the remaining newspapers. It is ironic that those who would refer to professional journalism being subsidized by tax dollars as socialism, watch local TV "news" that is subsidized by tax dollars. How is that? Have you ever heard the phrase "if it bleeds, it leads?" The bulk of "local news" consists of crimes/disasters, weather, and sports. Crime and disaster reports are provided for free by local police and sheriff departments, jobs paid for by tax dollars. Weather news comes for free from tax supported government agencies such as, NCAR, National Center for Atmospheric Research, see www.rap.ucar.edu/weather/; and NOAA, National Oceanic and Atmospheric Administration, www.noaa.gov. Finally, sports, as entertaining as it is, is not as essential to a democracy as the many news items not covered.

Internet Web sites, thus far, are consolidators of news gathered by the few remaining print journalists—rather than sources of investigative journalism. This could change, but a method for paying for professional internet journalism remains to be found. Thus the following exercise.

Exercise 31.1 How Will Professional Journalism be Funded?

(i) Dean Baker, co-director with Mark Weisbrot of the Center for Economic and Policy Research in Washington, DC., has suggested an *artistic freedom voucher*, AFV, system to support future (current) professional journalism, as well as artists, musicians, writers and other creative workers. In his original proposal[1] Baker wrote: *"In exchange for receiving AFV support, creative workers would be ineligible for copyright protection for a significant period of time (e.g. five years) ... The AFV would create a vast amount of uncopyrighted material. A $100 per adult voucher would be sufficient to pay 500,000 writers, musicians, singers, actors, or other creative workers $40,000 a year. All of the material produced by these workers would be placed in the public domain where it could be freely reproduced."* Did Baker estimate the average salary correctly? The $100 is proposed to take the form of a refundable tax credit. Baker proposes a simple registration system for creative workers to prevent fraud. Give all the pro and con arguments you can think of for Baker's proposal. Is this a self-organizing or centrally controlled system? What do you predict is most likely to evolve under this proposal?

[1]The Artistic Freedom Voucher: An Internet Age Alternative to Copyrights, November 5, 2003; www.cepr.net

(ii) Should Americans just get all their news from foreign (subsidized) sources, like the BBC, British Broadcasting Company?

(iv) Glenn Greenwald said, *"If corporations that own media outlets engage in quid pro quos to prevent critical reporting about one another, then large corporations—which own the Congress and control regulatory agencies—have no checks imposed on them at all."* [424, p. 43]. Greenwald refers to deals such as one allegedly struck between General Electric's MSNBC and News Corp.'s Fox News, [424, p. 41]. Comment.

(v) What is *internet neutrality*? Why is it important for the free flow of information in society? Does it still exist when you read this?

(vi) How will the internet affect the evolution of the media? A place to start the discussion might be [224] and other books by Dan Gillmor.

(vii) In the spirit of Edison's Algorithm and its Corollary, cf., Section 1.3, is there a "list of rules" necessary for effective, independent journalism?

(viii) Can you come up with a system of journalism that minimizes its internal conflicts of interest and funds serious investigative reporting on the rich and powerful who are making decisions that affect us all? What does a country without independent journalists look like?

I now turn in part to a critique of the flow of information as it exists, largely in the United States, at the time of writing.

31.2 Thesis: The Range of Debate Is Too Narrow Now

My main thesis for this section is that the range of debate now in the megamedia is too narrow to inform the general public in a timely fashion about many matters of importance. The concept of range of debate can be defined and measured in various ways. By a timely fashion I mean in time to actually do something about it. The definition of megamedia is unfortunately rather easy, since the number of corporations controlling the majority of the American media has decreased from over 50 to 5 or 6 while I have been teaching.

Interestingly, many of those who would agree with the above thesis do not agree with each other, some calling the media "liberal" others calling the media "conservative." I do not find those two adjectives particularly helpful, but I do find the concept *deference to power* very useful in measuring the media. Recall that the only business given special attention in the U.S. Constitution is the press, understood more broadly today to include electronic news as well as print news. This special status was given to the press because it is the *sine qua non* of a vibrant society, not to mention a democracy; the press needs to be critical of power not a stenographer to power. Our Constitution gives the press the freedom to speak truth to power, but only the public can compel the press to do so. An argument can be made that the megamedia press has other priorities than being critical. For example, suppose a corporation spends much money lobbying for certain legislation or supporting a politician. Any media owned by that corporation would be in an awkward position if it were critical of that legislation or politician—to a degree that it mattered in real time.

Exercise 31.2 Who are the MegaMedia? Is Informing the Public Their Top Priority?

I define the *media* to include not only newspapers and magazines, but books and all print media, radio, television and internet connectivity and all electronic media, music, CDs, DVDs and movies.

(i) Go to http://www.mediachannel.org/ownership/chart.shtml and find the 5 or 6 corporations who own the majority of the media in the United States. Cross check this with, say, the Web site of the Columbia Journalism Review, http://www.cjr.org/tools/owners/. Do additional cross checking until you are satisfied you can name with some degree of confidence the 5 or 6 corporations that own the majority of the media. List these corporations by name. Owners of news media also own other businesses, what are these businesses?

(ii) Do the best job you can in following the amount of money allocated by owners of news media to investigative reporting. Has this money been going up, down or held steady over the past 40 years, say? How many foreign correspondents do the media support? Has this number been going up, down or held steady for the last few decades?[2]

(iii) Ben H. Bagdikian has chronicled the increasing concentration of the media for decades, cf., [20]. What are his credentials for doing so? Look up reviews of his earlier books going back to 1983 where he predicted the current state of ownership of the media. Did those reviews agree with him?

(iv) Has media consolidation affected the coverage of news? References you might want to take into consideration are [101, 423, 424, 420, 422, 421, 19, 39, 6, 664, 240, 603, 241, 49, 99, 27, 314, 425].

The Bagdikian Number. In honor of Ben H. Bagdikian I make the following definition.[3]

Definition of the Bagdikian Number: Define a *media monopoly* for a country, say the United States, to be a set of corporations which collectively own/control the majority of the media of said country. Look at the collection of all media monopolies of a given country. Pick a media monopoly which has the smallest number of corporations in it. The *Bagdikian number* for a country is equal to that smallest number.

You can generalize the above to the Bagdikian number of a region, such as a city or state or the world. What is the Bagdikian number of the United States? What is the Bagdikian number of the city you are living in? How many independent owners are there of television, radio and newspapers in your city? How can you find out?

---

[2] Consider the following example. When Time and Warner merged, the new chair of Time-Warner, Steve Ross, who came from the Warner side of the merger, told *Variety Magazine*: "Journalists cannot afford to be anywhere but part of a strong diverse company with global reach and responsibility. A diverse financially strong media company makes it possible for managers to attract and nurture talented journalists." Steve Ross's income the next year after the merger was estimated to be $78 million for the year. But Time-Warner laid off 600 magazine employees including 19 of Time's 75 correspondents. This is not a unique case study.

[3] Students of mathematics will recognize the relationship of this definition to the concepts of maxs and mins, or sups and infs.

31.3 Time Series Test and Multiple Source Test

The following test of a given news source is somewhat self-evident. The main problem is that it requires careful attention over a period of time; or a study of a body of work representing output over a period of time.

Time Series Truth Test: *Analyze a given news source such as an individual reporter, columnist, TV or radio commentator, book or magazine article author ... over a period of years. Go back and check the accuracy of their work from the standpoint of history. How often has the source been accurate, inaccurate? How often has the source completely missed an important event/story?*

Note that analyzing history is not without its own problems. A standing joke[4] about Russia or China is that the only thing harder to predict than the future is the past, since there is a tendency to doctor the official record. Given these realities, one must do the best one can in any given situation. To that end there is another principle that one can employ to test the accuracy of any data, current or historical.

Multiple Source Consistency Test: *Find as many* ***independent*** *sources for a given piece of data or information as you can. Examine for consistency. From the information available construct the biggest consistent picture you can.*

Exercise 31.3 Applying the Multiple Source and Time Series Tests

(i) I often ask my students how they could characterize a difference in my work and that of, say, Rush Limbaugh. The answer is that I document everything, fairly carefully I hope. However, copious documentation is not a guarantee of truth. You can find much documentation that the earth is flat.[5] References [637, 408] are well documented as is the present text. I invite the reader to apply the time series test and multiple source tests to these three works and decide who is most reliable. Not surprisingly, results may vary according to who is applying the tests. Thus those who would test should submit themselves to the same time series and multiple source tests!

(ii) Consider the data leading up to the U.S. invasion of Iraq in 2003. One of the main justifications of the invasion was to protect the United States from WMD, weapons of mass destruction, ostensibly in the possession of Iraq.[6] Now the United States had indeed sold WMD to Iraq in decades past, and had the receipts to prove it. However, in 2003 there were no WMD, and none were to be found. With regard to this specific issue, compare the performance of Rush Limbaugh, *The New York Times*, *The Washington Post*, www.democracynow.org, www.fair.org, www.jimhightower.com. Please feel free to expand

[4]The joke is potentially applicable to any country. Witness the absence of a history in the United States about the killing of electric rail transit, cf., Chapter 1.

[5]Finding a million references in support of the theory that the earth is flat is no longer a problem, given the electronic search tools that are available.

[6]Also there was the constant conflating (by the Bush administration and others) of Saddam Hussein and the attacks on the World Trade Center in 2001, although no connection has ever been established and the Bush administration has stated that there is no connection.

the list of media sources that you apply the time series and multiple source tests to in regard to the non existence of WMDs in Iraq in 2003.

(iii) During the California energy crisis of 2000–2001, it was nearly impossible to find the following information. The MUNIs, Municipally Owned Utilities, such as the LADWP and SMUD,[7] were not having a crisis. It was also difficult to find any media discussion of the difference in performance of the two models, IOUs, Investor Owned Utilities, such as Enron and MUNIs. Why do you think this is the case?

(iv) Apply the time series and multiple source tests to Matthew Wald and Gina Kolata of *The New York Times*.

(v) Pick some issue that is either more current or of more interest to you when you read this and apply the time series and multiple source tests.

31.4 Measuring the Range of Debate

There are some very simple methods for measuring the range of debate of any given piece of media. The following might be called constructing a histogram.

Exercise 31.4 Measuring the Range of Debate: a Histogram Method

(i) A story on the October 30, 2000 edition of CounterSpin[8] discussed a very simple but telling analysis done of the three presidential debates between Al Gore and George W. Bush by Jeff Milchen, founder of ReclaimDemocracy.org.

He took complete electronic transcripts of the three debates and counted the number of times various words were mentioned using a computer. Cumulative mentions of specific words/phrases by either candidate during the three CPD (presidential) debates in 2000:

Wealthiest 20, Poorest 1, Prosperity 16, Homeless 0, Middle Class 15, Working Class 0, Poverty 1,

Tax(es) 144, Social Security 67, Seniors 64, Medicare 58, Prevention 0,

Drug(s) (Prescription) 60, Drug War or War on Drugs 0, Slobodan Milosevic 17, Colombia 0, Crime 23, Corporate Crime 0, Police Brutality 0, Prison 0,

WTO[9] 0, NAFTA[10] 0, Corporation(s) 0, Anti-trust 0, Labor 1, Free Trade 0.

Do you think that the above word frequency analysis would have been different if third party candidates (such as Nader and Buchanan in 2000) were allowed in the debates? Can you draw any conclusions about the "range of debate" in these debates?

Pick a subject or time period in forms of the media that you follow and do a Milchen analysis. Can you reach any conclusions about the "range of debate" in the media you have analyzed?

(ii) Suppose someone says that the list of words ReclaimDemocracy.org chose is prejudiced in some manner. What is a way to answer this challenge?

Another method of analyzing the media is given by the following exercise.

[7] Los Angeles Department of Water and Power and Sacramento Municipal Utility District

[8] See the audio archive at www.fair.org.

[9] World Trade Organization, subject of biggest public demonstrations since the Vietnam War

[10] North American Free Trade Agreement

Exercise 31.5 A Chomsky Method of Analyzing the Media

(i) There is a very simple, if possibly time consuming, way to see if the output of a given news source is biased in its reporting of some issue or issues. I call this the *Chomsky method* after Noam Chomsky, an author and professor at the Massachusetts Institute of Technology. Pick a topic like the North American Free Trade Agreement (NAFTA),[11] or the Endangered Species Act or logging forests in the U.S.A. (or the world). Then identify two or more groups that have an intense interest in the subject. In the case of NAFTA[12] three obvious groups are business management, labor, and environmentalists. In the case of forests you might identify timber companies, timber workers, and environmentalists (and salmon fishermen if you are dealing with the Pacific Northwest) for starters. Then examine your news source as follows. If the media source you are looking at is a newspaper, count column inches and numbers of quotes from the various groups. How often is the subject covered in the paper(s) you are analyzing?[13] In addition to this quantitative analysis you can do more subjective assessments. For example, are most of the references from the labor group vaguely stated then immediately countered by specific authoritative government or business sources (or vice-versa)? When analyzing radio or TV programs you can count words or seconds devoted to topics by group.

[11] For a stark look at a slice of life that most Americans are insulated from read: Charles Bowden, "While You Were Sleeping: In Juárez, Photographers Expose the Violent Realities of Free Trade," *Harper's Magazine*, December 1996, pp. 44–52. I bring this situation which exists in a town/city on the U.S.-Mexico border to your attention for a number of reasons. One of the questions I would like us to contemplate is this. Is the horrendous situation in Juárez impacted positively, negatively, or not at all by such "abstract" things as the North American Free Trade Act? How are things in Juárez when you read this?

[12] NAFTA is not unique in being a "global economic agreement" that takes precedence over many local (city, county, state and national) laws. There is GATT, the General Agreement on Tariffs and Trade, and the MAI, Multilateral Agreement on Investments and presumably others. To find information on these topics, meaningful details of which I have not been able to find in the megamedia, see the Web sites: http://www.citizen.org and http://www.rtk.net. The first site belongs to Public Citizen and the second to the Right-To-Know Network.

[13] For example, in the case of NAFTA there were more than 10 pro-NAFTA quotes for every anti-NAFTA quote. A behind the scenes look by an investigative reporter at the process that brought us NAFTA can be found in John R. MacArthur Jr., *The Selling of Free Trade: NAFTA, Washington and the Subversion of American Democracy*, Hill and Wang Pub., 2000. Compare what the media said about NAFTA with what it has wrought. For example, out of 1,300 daily papers only 2 (of which I am aware) editorially opposed NAFTA. In 2000, since NAFTA's passage in 1993 the U.S. trade balance with Mexico had shifted from a $1.7 billion surplus in 1993 to a deficit of over $14.7 billion by 1998, which some equate to a loss of roughly a half-million factory jobs. Also, manufacturing wages in the U.S. have stagnated while those in Mexico have fallen sharply. The U.S.-Mexican border has seen a nearly 50% increase in the number of U.S.-owned maquiladora plants paying $.80 to $1.00 an hour while creating toxic air and water problems. What promises were made regarding labor and the environment while NAFTA was being passed? These promises have not been kept. This is in contrast to the fact that the vote in Congress was almost 50% for to 50% against—after an extremely intense pro-NAFTA lobbying effort by industry and a highly aggressive pro-NAFTA President Clinton who handed out much "pork" to buy pro-NAFTA votes. The point is that there was substantial anti-NAFTA sentiment in the country, felt by Congress, but almost totally ignored by the megamedia. This situation is not unique to the NAFTA debate. It is interesting to note, that just three years after the passage of NAFTA many of the fears of environmentalists and labor advocates regarding NAFTA were born out. For an update on the effects of NAFTA contact *Public Citizen*, 215 Pennsylvania Ave. SE, Washington, D.C. 20003; phone (202) 546–4996.

(ii) Find the same story in at least two different newspapers or on-line journals. Compare the stories for length, biases, omissions and so forth.

(iii) A more difficult exercise is to locate topics that are not reported on or which are "underreported." For this you need to do original research. Or you may need to examine sources which may not be readily available.[14] Of course, social networks, organizations with special interests, the internet generally, and sites like WikiLeaks in particular, are possible sources of important happenings not reported in the megamedia.

31.5 Distractions and Illusions

Always be on guard for tricks the mind might play on you. These tricks generally go by the name of *illusions*, and here are a couple of the many that are possible.

Consider the following optical illusions which should give you pause for thought about the "infallibility" of even direct observation.[15] It is interesting to note that the illusion that compares a <—> line with a >—< line, see Figure 31.1,[16] called the Müller-Lyer effect, is not an equally effective illusion

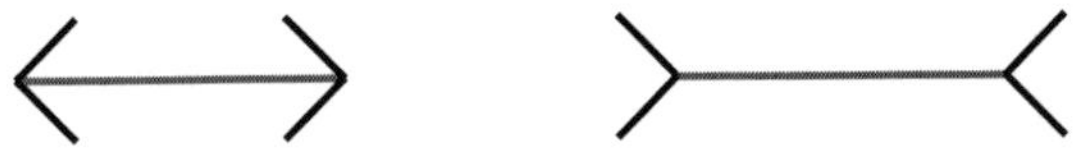

FIGURE 31.1: A Culturally Dependent Illusion

in all cultures. Indigenous people in the Torres Straight, for example, see at a glance that the lines are of equal length. The opposite is true for a second illusion, namely, given a horizontal line segment and a vertical line segment of equal length, the Torres Straight folks saw the vertical line as longer far

[14]Such as the latest edition (and back editions) of the annually produced "Censored: The News That Did Not Make the News and Why," [533], for one example. Or look at *EXTRA* a magazine published by FAIR (Fairness and Accuracy In Reporting), 130 West 25th Street, N.Y., N.Y., 10001, 800-847-3993.
A large fraction of the "undercovered" stories are environmental in nature. For example, there is the news that selenium poisoning is a growing problem in the western United States due in no small part to irrigation practices in agriculture. The problem may be larger in magnitude than the Exxon Valdez oil spill. I personally have seen this story in only one place, *High Country News*, a newspaper out of Paonia, Colorado. I was told that it has also appeared in one other place, *Sports Illustrated.*

[15]Through the wonder of the World Wide Web, much fancier illusions are available. For example, try http://samiam.colorado.edu/ mcclella/cips/graphs.html and look for "Illusion Works," a few pages in. You can also try to find Illusion Works by phone at 1-888-Il Works.

[16]The two lines being bracketed are of equal length, but the brackets make one look longer than the other.

more often than English or Americans. It is believed that the environment in which a person grows up has a profound effect on that person's perceptions of reality, see [46, pp. 114–118]. Some even see too many people growing up with a Nature-deficit disorder with negative consequences for us all, cf., [409].

Studies[17] have shown that if two cars travel the same distance but one starts before the other, young (preschool) children often say that the car which commenced traveling first traveled farthest. In a similar vein, Piaget[18] showed in a classical experiment that if (preschool) children are presented with two glasses, demonstrably having the same volume (one glass exactly full of water when poured into the other glass exactly fills it), one short and squat the other taller and thinner—the children invariably say that the tall glass holds more. This is the case even after the children see water poured from the (full) tall glass into the short glass exactly filling the latter.

Given the fact that we don't always see "the truth" even when it is simple and in front of our faces, we should be ever on our guard. *We often see what we want to see, we often believe what we want to be right about.*[19] Have we completely outgrown all "child-like" vulnerabilities? To complicate life there is an entire industry based on getting people like you and me to believe things that are not in our own best interest but in the interest of others, cf., [665]. Advertising/public relations is now a sophisticated discipline that knows how to exploit intrinsic human foibles. If we can be fooled as children and our eyes can be fooled as adults, you must ask: "How might I be fooled today, and by whom?"[20] For one small example see "How to Lie with Statistics", cf., [321]. See also [685, 686, 687]. Can you think of one example in your own life where you are sure, or you suspect, that you were tricked into believing something that is not true or is not in your own best interest?

Purposeful Illusions/Distractions. You should be aware that a sizeable number of items that pass for "news" have been placed there by someone or some organization that wants to sell you either something or some idea. You should research the prevalence, use, and status of VNRs, for example. These are "video news releases" that TV stations run without telling you where they came from. They may be nothing more than a clever advertisement disguised as a news story, and they are quite dishonest. Corporations and the government both use this handy tool of disinformation.

---

[17] Jean Piaget, *The Psychology of the Child*, New York: Basic Books, Inc., 1969.

[18] ibid.

[19] *"Science is an ongoing collision between our ideas and our experience in the real, material universe. Though it sounds cynical, we cannot expect to make progress in our understanding of the nature of things unless we can show that what we thought we knew is in some sense wrong, or at least incomplete. It is the job of all scientists to ask new questions, to be sure, but also to keep a weather eye on old answers to see if they still hold up in the light of new observations."* Niles Eldredge, *Time Frames: The Evolution of Punctuated Equilibria.*

[20] The answer here is often: "I fool myself—quite willingly."

In 1975 a Senate intelligence committee found that the CIA owned outright more that 200 wire services, newspapers, magazines and book publishing complexes and subsidized many more. Since then 50 further media outlets and over 12 book publishing houses have been discovered.[21] Can you find a copy of this Senate report? Can you find out how much of the media has been infiltrated by the CIA at the time you read this? Don't forget to look for the presence of the NSA (National Security Agency), various police, and the FBI while you are at it.

Exercise 31.6 Find the Source(s) of a News Article

(i) Pick a significant number of articles presented as news from a newspaper (if any still exist when you read this), or from some news source you follow. Examine as best you can the source(s) of the article. Was it placed there by a military interest, e.g., the Pentagon or a military contractor, a major corporation, a citizen group, and so on? With the shrinking financial support for reporters, newspapers may find it easier to base an article on a "news release" from some organization.

(ii) Do the same for TV news. Can you find any VNRs, i.e., video news releases?

(iii) Do the same for news Web sites.

(iv) Find "news" items in the megamedia that contain significant false statements and/or major omission(s) of relevant fact.

(v) The biography of journalist I.F. Stone by Myra MacPherson is titled: *"ALL Governments LIE!" The Life and Times of Rebel Journalist I.F. Stone.* This title contains one of Stone's more famous quotes. With the publication of confidential government documents by WikiLeaks, cf., Exercise 32.2, including documents from the U.S. Pentagon and the U.S. State Department, for example, how many examples of a government lying can you find? The leaks of cables from the U.S. State Department involve governments from around the world. How has WikiLeaks informed the subjects dealt with in Chapters 22, 23, and 9? What has WikiLeaks revealed about the role of the United States in international negotiations dealing with climate change, the subject of Chapter 1?

(vi) Have the world's powerful political and economic forces, e.g., the world's governments and corporations, successfully eliminated the ability of WikiLeaks or any similar organization to communicate? Yes? No? What does your answer say about the flow of information in the world? Compare Exercise 32.2.

(vii) For WikiLeaks in particular, and the class of whistleblowers in general, what is the focus of dicussion and media attention? Is that focus on what has been exposed? Or is the focus on the whistleblower? Is the focus on the message(s), or the messenger?

(viii) George Orwell once said/wrote: *"In a time of universal deceit, telling the truth is a revolutionay act."* Discuss.

[21] This comes from page 196 of McMurtry, John, *Unequal Freedoms: The Global Market as an Ethical System*, [436]. McMurtry quotes Parenti, Michael, *Inventing Reality: The Politics of the Mass Media*, [518].

Chapter 32

Media Literacy: Censorship and Propaganda

32.1 Filters and Censors

Filters. I define a *filter* to be anything that directly or indirectly influences information as it flows from its source, i.e., a *fact* in Nature, to you. Possible examples of filters are reporters, editors (of magazines, newspapers, radio and television programs, Web sites), advertisers, boards of directors of corporations, the government, professors, parents, public relations firms, friends, would-be internet censors, and so forth. The filter that managed to eliminate all but one line from Jack Blum's testimony, cf., page [443], on our government's relation to the importation of cocaine into the U.S. in the '80s is one of the more spectacular filters I have run across.

Think about all those facts in reality about which you need complete and accurate information. If you are not getting your information about these facts from direct experience, what do you know about the filters between you and the facts?

Exercise 32.1 Filters

(i) Since there is far too much going on in the world for any one person to experience directly, each of us must choose filters through which we receive much of our information. Do you choose your filters consciously or unconsciously? If consciously, what criteria do you use for selecting your filters?

(ii) Pick one of your filters and determine as best you can *when* you chose that particular filter.

(iii) The owners of the filters through which flows the bulk of the information you receive have self-interests. Do these owner's self-interests coincide or conflict with your self-interests?

(iv) Suppose a filter is owned by an entity whose self-interest largely conflicts with yours. Do you think it would be stupid for you to get most of your information through that filter?

(v) Who owns the various television filters? What are their self-interests?

(vi) General Electric owns NBC. It has been asserted by some that since GE is a giant in the nuclear industry, you will not find much information on NBC which is uncomplimentary to the nuclear industry. What has been your experience in this regard?

(vii) Politicians must raise ever larger amounts of money to run for office. It has been estimated (in 1997) that to run for the U.S. Senate from Colorado one must raise in excess of

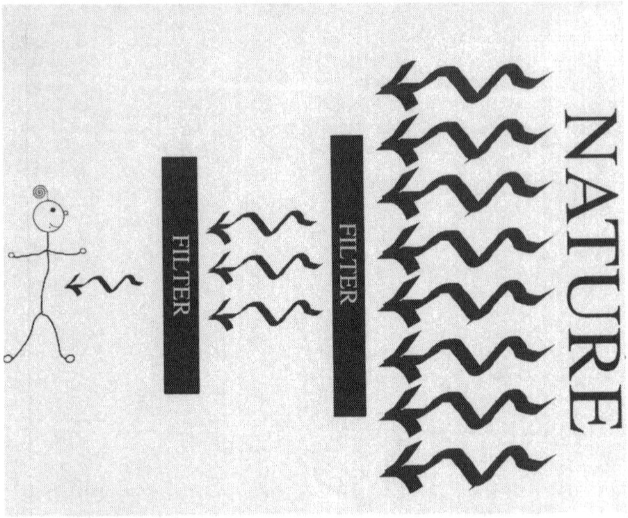

FIGURE 32.1: Filters between You and Nature

$12,000 a week for 6 years. Much of this money is provided by corporations; money much of which is then paid to media corporations for advertising. Analyze the cash flow and information flow of this system of picking our politicians. Is access/non-access to money a filter?[1]

(viii) Can you measure the reliability of the information you are getting through one of your filters? How?

(ix) Do you choose your filters so that the bulk of the information you get agrees with your current mental model of the world?

(x) If you get information that threatens your niche, or job, how do you react toward the filter through which that information came?

(xi) Do you ever change filters or just look for new ones out of curiosity?

(xii) Is there any information that you suspect is not making it through most of your filters? Give an example.

(xiii) Do you think that you might have to do some homework/searching to find filters whose owners' self-interest largely overlap with yours?

(xiv) Does TV news tend to emphasize stories with "good visuals"? Is this a form of filter?

[1] In 1981 while I was in Washington D.C. I was told by then chairman of the Republican National Committee, Richard Richards, *"You environmentalists are missing the boat. You have to get in way before the elections and play a role in picking the candidates who run."*

32.2 Censorship: External and Internal

Active censorship happens often enough for me to have hundreds of pages of documentation of the practice. The most effective censorship is internal self-censorship. When you realize that you will probably be fired for discussing, writing about or otherwise revealing certain information, the tendency is to not be a whistleblower, i.e., not speak truth to power. The exceptions are notable enough to have books and magazine articles written about them. External censorship happens as well, however, when the internal censorship does not work.

A documentary that was never shown on Public Television, nor on any network, concerned the first Gulf War. Jon Alpert, an Emmy Award winning journalist, risked his life to get into Iraq during the Gulf War. He managed to avoid both the Pentagon and Iraqi military censors and got film of the actual lethal effects on civilians of the U.S.'s so-called "surgical strikes." Upon getting back to his employer, NBC, producers gave him a hero's welcome and proposed a lead story on the national news. Before the piece ran the CEO of NBC, which was (is) owned by General Electric Corporation, quashed it and fired Alpert the same day. No longer employed by NBC, Alpert went to the news director of CBS who was so impressed with Alpert's footage that he was planning on running it the next day on the news. That CBS news director was relieved of his duties at 2 A.M., and Alpert's video footage was not shown. Public Television did not run this story either. Why?[2]

A particularly egregious consequence of the shortness of the list of media moguls was the summary cancellation by ABC (soon after Disney bought ABC in the Fall of 1995) of the radio show by populist Jim Hightower.[3] Hightower's show was new, rapidly growing in popularity and paying its own way. Had he not been censored from the megamedia he likely would be as well known as a number of today's popular commentators. Who is Jim Hightower and what did he have to say? Good question.[4] Just as interestingly I note that I have found only three articles on this cancellation.[5] It remains to be seen how

[2] A one hour video covering this and other stories entitled "Fear and Favor in the Newsroom" is available (at the time of writing) from California News Reel, 149 9th Street, San Francisco, California, 94103; phone (415) 621-6196. You can always do an internet search as well. However, even if you see this now, it should have been shown in 1991–2.

[3] Hightower had been critical of the then Disney CEO M. Eisner. Hightower was fired the same day Eisner took control of ABC.

[4] See www.jimhightower.com

[5] One was on page 410 of the October 16, 1995 issue of *The Nation.* The other was on page 4 of *EXTRA! Update*, the Bimonthly Newsletter of FAIR (Fairness and Accuracy in Reporting), December, 1995. The third was in *Public Citizen Magazine,* 1600 20th St. NW, Washington, D.C. 20009-1001, November/December 1995, p. 13. Note also the electronic mail address: publiccitizen@citizen.org. These three are not "megamedia" references. After searching for months I found a fourth reference in a story written by Al Brumley for the

much room remains on American radio for the point of view Jim Hightower represents.

I told the stories about Jon Alpert and Jim Hightower to prove one point. If one of the six CEOs who run one of the six major media owners does not want a particular story or point of view run, the chances are we will not see it in the megamedia in a timely fashion—if ever. Lesson: do your homework and develop other sources of information outside of the megamedia.

If you had hopes that Public Broadcasting (PBS) would reliably fill the void, think again. PBS is subject to many of the same pressures, and censors many programs. For a long list of programs censored by PBS see [664]. PBS once censored an Audubon special, "Rage Over Trees," an act which I found personally upsetting, see [196].

A pattern on NPR, National Public Radio, is particularly telling to some. From the March 2010 edition of *Extra!*, page 14, "NPR Puts Right-Wing Hate in Howard Zinn's Obit," by Jim Naureckas, (see also www.fair.org) I quote:

> When progressive historian Howard Zinn died on January 27, **NPR**'s **All Things Considered** (1/28/10) marked his passing with a declaration that his life's work was worthless.
>
> After quoting positive assessments from Noam Chomsky and Julian Bond, **NPR**'s Allison Keyes turned to far-right activist David Horowitz, a practitioner of what the **Nation** (11/12/07) calls the New McCarthyism, for a ritual denunciation. "There is absolutely nothing in Howard Zinn's intellectual output that is worthy of any kind of respect," Horowitz proclaimed. "Zinn represents a fringe mentality which has unfortunately seduced millions of people at this point in time. So he did certainly alter the consciousness of millions of younger people for the worse."

Howard Zinn authored a number of books, most notably *A People's History of the United States: 1492–Present*, [742]. Feel free to read it, as millions have, and then evaluate Horowitz's comments. We might note that Horowitz's statements have no substance. He is not pointing out any errors of fact, any omissions or fabrications, and so on. Horowitz is basically saying that he dislikes Zinn. Everyone who has done anything of note, as has Zinn, will have detractors. So far, there is no pattern.

However, *Extra!* goes on to point out that NPR has done complimentary obituaries of Robert Novak, Oral Roberts, and William F. Buckley, folks with political philosophies polar-opposite to that of Zinn. None of them contained negative remarks from critics. For example, NPR aired *6* segments commem-

---

Dallas Morning News, reprinted in *The Denver Post* on April 14, 1996. This is because Hightower is (was) a prominent politician in Texas. I once offered a reward to the first person who could bring in an article about this cancellation from any other media source besides *The Nation* or *EXTRA! Update*. No one answered the challenge. You can get some idea of Jim Hightower's point of view from his Website, www.jimhightower.com; his books, [307] and [306], the original Hightower report, unexpurgated, of the Agribusiness Accountability Project on the failure of America's land grant college complex and selected additional views of the problems and prospects of American agriculture in the late seventies; and several other more recent works.

orating William F. Buckley, and never mentioned were some interesting facts about him. From the same article in *Extra!* mentioned above we read:

> During his life, Buckley was an intensely controversial figure who supported, among other things, white supremacism in the U.S. South and South Africa, McCarthyism, nuclear war against China and the tattooing of AIDS patients (*Extra!* 5–6/08).

Now these are notable "outputs" of Buckley's intellect which some, no doubt, will find complimentary. However, NPR filtered them out. Why do you think the above pattern exists?

Book censorship happens also. One example of corporate book censorship is the suppression (by Warner publishing) of the first edition of Noam Chomsky, Edward S. Herman, *The Washington Connection and Third World Fascism,* South End Press, Boston, Mass., 1979. Fascinating details of this suppression are recounted in the prefatory note in the edition of that book eventually published at South End Press.

The book *In the Spirit of Crazy Horse*, by Peter Matthiessen, [446], was kept off the shelves for eight years because of one of the longest and most bitterly fought legal cases in publishing history. In this case the publisher, Viking, had to fight the government's effort at censorship.

As an exercise I ask you to investigate the following claims: (1) reporters Jane Akre and Steve Wilson were fired (http://www.foxBGHsuit.com) while trying to do a well researched news story on genetically Monsanto-engineered rBGH[6] in milk; (2) the Sept/Oct 1998 issue of the *Ecologist* magazine was shredded by the printer (it had an article subtitled "The Monsanto Files") (available, as I write, on the Web at http://www.gn.apc.org/ecologist/SeptOct); (3) thousands of copies of the first edition of [386] were shredded—why?

Oprah Winfrey was sued and had to spend millions defending herself and her program for having Howard Lyman on her show talking about BSE, or "Mad Cow Disease." See Chapter 5. In 13 states similar lawsuits can take place, for there are laws that prevent anyone saying "bad things" about food. So much for free speech. So don't expect anyone to talk openly in the media about BSE, except in books like [562] which could be in trouble if sufficiently widely read.

Wars have always produced censored news, now more than ever, cf., [10]. Journalists are often at risk, not only in times of war. Many are killed, cf., [487, 290].

If you have any doubts that wealthy corporations dominate the legal system read [479]. If you think you have the freedom to speak out and tell the truth as you know it read about "strategic lawsuits against public participation," i.e., SLAPP suits, in [554].

SHUT UP AND EAT! At no time does it become more clear who the powerful are in society as when you try to find out what is in your food. Allow me one example. In April 1994, the State of Vermont enacted a law requiring

[6] This is recombinant bovine growth hormone, sometimes called rBST.

products containing rBGH to be labeled. From the August 9, 1996 edition of the *Burlington Free Press*, we learn that the International Dairy Foods Association, the Milk Industry Foundation, the International Ice Cream Association, the National Cheese Institute, the Grocery Manufacturers of America, Inc., and the National Food Processors Association (all trade, lobbying, and promotional corporations) sued the Attorney General of Vermont to get the aforementioned law declared unconstitutional. For good measure, Monsanto Corporation joined the lawsuit as a friend of the court.

The Second Circuit, United States Court of Appeals, ruled 2 to 1 against Vermont.[7] Translating the legalese of the two judges in the majority, they said that Vermont cannot require corporations to tell the truth about their products; such a law causes the dairy manufacturers irreparable harm! Furthermore, the legal basis of this decision is the First Amendment to the Constitution, which guarantees the right not to speak, the two also said. (Real people do not have this right, when, for example, called before a grand jury.)

It gets more interesting. Oakhurst Dairy of Maine, which did its best to not use milk from rBGH-treated cows, started advertising its milk as rBGH free. In 2003 Monsanto sued Oakhurst dairy with the goal of stopping them from so labeling their milk.[8] Monsanto and Oakhurst settled out of court, with the result that Oakhurst has a label in the spirit of the one you can read on a carton of Ben and Jerry's ice cream. Namely, there must be included on the label the statement to the effect that the FDA has found no significant difference in milk from cows treated with rBGH (or rBST). Apparently the controversy is not over, since the struggle over labeling is now being carried out in various states.

Thus, not only are citizens[9] not told if they are eating genetically altered food (no labels required by the FDA); it is a struggle for those who do not sell genetically engineered food to say so.

The good news is that thus far, food certified and labeled "organic" must meet certain criteria, such as being free of genetically engineered ingredients and pesticides. But it is a periodic struggle to keep the official definition of "organic food" true to the meaning it has as I write.

Incidentally, as I write, the FDA is considering approval of its first genetically engineered animal product, salmon.[10] It also appears certain that the above story is about to be retold: there will be no labeling of genetically altered salmon, but there will be a (virtual) ban on the truthful labeling of salmon that are not genetically altered. In my humble opinion, there is something very wrong with a system that denies a seller of a food the right to

[7]See International Dairy Foods Association v. Amestoy 92 F.3d 67 (1996).

[8]See http://www.reclaimdemocracy.org/ for a more detailed account.

[9]Commonly referred to solely as consumers.

[10]See the article by Andrew Pollack, "Panel Leans in Favor of Engineered Salmon," *The New York Times*, Sept. 20, 2010.

speak the truth about his/her product—especially when a significant number of buyers would like to know what they are eating.

Exercise 32.2 The WikiLeaks Media Niche and SAFE Information

(i) I introduced Section 31.2 by stating that rather than trying to evaluate a source of information as conservative or liberal, I found it more meaningful to evaluate the extent that the source shows deference to power. I denote as SAFE information any that does not challenge power in any serious way. Seemingly unSAFE information often appears long after the time anyone can meaningfully act on that information, hence it is actually SAFE. Pick a media source you use and rate its deference to power. While you are at it, compare it to WikiLeaks, see part (ii) below.

(ii) As I write the Web site that is a candidate for showing the least deference to power is WikiLeaks.org. For example, see the article by Raffi Khatchadourian, "No Secrets," *The New Yorker*, June 7, 2010, pp. 40–51. A brave network of folks, including Assange, are implementing the vision of WikiLeaks founder, Julian Paul Assange, to make available on the Web significant secrets of governments and other institutions. WikiLeaks, out of necessity, must be a filter; but its performance thus far indicates that it is fearless. It also provides citizens of the world invaluable information, not available initially anywhere else. It no doubt is viewed as a criminal enterprise by many governments. It no doubt is the object of legal and physical attacks. Does WikiLeaks exist at the time you read this? Have persons involved been successfully legally attacked, and or imprisoned, or worse?

WikiLeaks occupies a media niche not filled by any other entity. Thus if it is successfully killed between the time I write and the time you read this exercise, has another similar entity arisen to fill this niche? Has any entity been able to breach the security wall of Wikileaks by time you read this?

List at least two spectacular secrets that WikiLeaks has presented to the world.

32.3 Conclusion and Epilog: Where Are the Adults?

I could have gone on at book length about obstacles to getting information that you need. Knowing that there are many filters out there trying to keep the public in the dark about many issues—many involving our food and our health—you are forewarned, hence forearmed. By all means, do not tolerate it; but in the meantime do what you need to do to connect with sources of information and work around the filters.[11] Always support your local libraries. The internet can be an invaluable tool for connectivity and information access. Do what you can to defend equal internet access, e.g., fight for what is called "internet neutrality." Support any local government initiatives to have an internet MUNI, i.e., a municipal wireless network freely accessible by anyone in your city.

I want to include in this conclusion one story that made the mainstream press, albeit half-a-century late.

[11] For example, as of January 1, 2007 the National Archives declassified millions of pages of documents that have been secret but now are over 25 years old. It should be fascinating to keep track of this unfolding story of things we should have probably learned long ago.

On Tuesday, September 30, 1997 in *The New York Times* appeared (well inside the paper) an article by Matthew L. Wald, "U.S. Warned Film Plants, Not Public, About Nuclear Fallout." Allow me to quote the first two paragraphs:

"WASHINGTON, Sept. 29 – Through most of the 1950s, while the Government reassured the public that there was no health threat from atmospheric nuclear tests, the Atomic Energy Commission regularly warned the Eastman Kodak Company and other film manufacturers about fallout that could damage their products, according to a review of Government literature by a private watchdog group here.

The warnings were confirmed today by people in the photographic industry."

At another place in the article we read:

"Senator Tom Harkin, an Iowa Democrat who is the ranking minority member of an Appropriations Committee subcommittee that will hold the hearing on Wednesday, said, 'It really is odd that the Government would warn Kodak about its film but it wouldn't warn the general public about the milk it was drinking.' Iodine-131 is absorbed by cows and incorporated into milk. In humans, it concentrates in the thyroid gland, where it can cause cancer." (Senator Harkin lost a close relative to thyroid cancer.)

More from the article. *"Kodak discovered in the early 1950s that some film was fogged before use, and it traced the problem to fallout from atmospheric nuclear tests, both American and Russian. The watchdog group, the Institute for Energy and Environmental Research, a nonprofit organization that specializes in nuclear weapons issues, said Kodak had threatened to sue the Atomic Energy Commission, which had then promised to warn Kodak about future tests."*

EPILOG

The human population has grown enormously in my lifetime. During this time humans overall have, for the most part, taken from Nature (including the less fortunate humans) what they wanted, when they wanted it, most often without asking or showing evidence of concern over the consequences. Only 4% of the world's oceans remain pristine, a single digit percentage of the the world's wilderness forests remain. Both these sources of life are shrinking as life forms are vanishing at record rates. Humans overall, cannot reliably follow known rules for the safety of workers or citizens in general, cannot reliably follow known rules necessary for a stable financial system, cannot stop polluting and killing the very supports for human life. Despite glowing rhetoric to the contrary, decision-makers fail to make decisions as if our life support systems really matter. Human political systems can ignore for a time—but cannot alter—the inexorable action of the laws of Nature that mathematics and science attempt to describe.

There will be change, of that I am certain. Whether our culture changes voluntarily or Nature—the ultimate "adult"—does it for us, that is the question. Where to begin? Begin where you are, in the middle of the muddle. Organize. Do your best.

Do not be fooled into acting against your own survival. Do not forget; you get what you settle for. And Nature bats last.

References

[1] Sasha Abramsky, *Breadline USA: The Hidden Scandal of American Hunger and How to Fix it*, Poli-PointPress, Sausalito, California (2009).

[2] Nafeez Mosaddeq Ahmed, *The War on Truth: 9/11, Disinformation, and the Anatomy of Terrorism*, Olive Branch Press, Northampton, Massachusetts (2005).

[3] Matthew M. Aid, *The Secret Sentry: The Untold History of the National Security Agency*, Bloomsbury Press, New York (2009).

[4] Katherine Albrecht, Liz McIntyre, *Spychips: How Major Corporations and Government Plan to Track Your Every Move with RFID*, Nelson Current, Nashville, Tennessee (2005).

[5] C. Fred Alford, *Whistleblowers: Broken Lives and Organizational Power*, Cornell University Press, Ithaca, New York (2001).

[6] Eric Alterman, *What Liberal Media? The Truth About BIAS and the News*, Basic Books, New York (2003).

[7] Nancy J. Altman, *The Battle for Social Security: From FDR's Vision to Bush's Gamble*, John Wiley & Sons, Inc., New Jersey (2005).

[8] L. W. Alvarez, W. Alvarez, F. Asaro, H. V. Michel, "Extraterrestrial cause for the Cretaceous-Tertiary extinction," *Science* 208, pp. 1095-1108, (1980).

[9] Luke Anderson, *Genetic Engineering, Food, and Our Environment*, Chelsea Green Publishing Co., White River Junction, Vermont (1999).

[10] Robin Anderson, *A Century of Media, A Century of War*, Peter Lang Publishing, New York (2006).

[11] Roy M. Anderson, Robert M. May, *Infectious Diseases of Humans*, Oxford University Press, Oxford, New York, Tokyo (1992).

[12] Sarah Anderson, John Cavanagh, with Thea Lee, *Field Guide to the Global Economy*, The New Press, New York (2000).

[13] Marcia Angell, M.D., *The Truth About the Drug Companies: How They Deceive Us and What to Do About It*, Random House, New York (2004).

[14] Harvey Arden, *Prison Writings: My Life is My Sundance–Leonard Peltier*, St. Martin's Griffin, New York (1999).

[15] Erik Assadourian, Project Director, *Vital Signs 2007–2008: The Trends That are Shaping Our Future*, Worldwatch Institute, W.W. Norton & Company, New York 2007.

[16] Svante August Arrhenius, "On the Influence of Carbonic Acid in the Air Upon the Temperature of the Ground," *Philosophical Magazine* 1896 (41): pp. 237–76.

[17] Marcia Ascher, *Ethnomathematics*, Chapman & Hall, New York (1991).

[18] Nicholas Ashford, Claudia Miller, *Chemical Exposures: Low Levels and High Stakes, Second Edition*, Van Nostrand Reinhold, New York (1998).

[19] Ben H. Bagdikian, *Double Vision: Reflections on My Heritage, Life, and Profession*, Beacon Press, Boston, Massachusetts (1995).

[20] Ben H. Bagdikian, *The New Media Monopoly*, Beacon Press, Boston (2004).

[21] Dean Baker, Mark Weisbrot, *Social Security: The Phony Crisis*, The University of Chicago Press, Chicago, Illinois (1999).

[22] Kevin Bales, *Disposable People: New Slavery In The Global Economy*, University of California Press, Berkeley, California (1999).

[23] James Bamford, *The Puzzle Palace: Inside the National Security Agency, America's Most Secret Intelligence Organization*, Penguin, New York (1983).

[24] James Bamford, *Body of Secrets: Anatomy of the Ultra-Secret National Security Agency*, Anchor Books, New York (2002).

[25] Doug Bandow, Ian Vásquez, *Perpetuating Poverty: The World Bank, the IMF, and the Developing World*, Cato Institute, Washington, D.C. (1994).

[26] Judi Bari, *Timber Wars*, Common Courage Press, Monroe, Maine (1994).

[27] David Barsamian, *The Decline and Fall of Public Broadcasting*, South End Press, Cambridge, Massachusetts (2001).

[28] Albert A. Bartlett, *The Essential Exponential (for the future of our planet)*, ISBN: 0975897306 (2004).

[29] Donald L. Barlett, James B. Steele, *America: What Went Wrong?*, Andrews and McMeel, Kansas City, Mo. (1992).

[30] Donald L. Barlett, James B. Steele, *America: Who Really Pays the Taxes?*, Touchstone, New York (1994).

[31] Donald L. Barlett, James B. Steele, *America: Who Stole the Dream?*, Andrews and McMeel, Kansas City, Mo. (1996).

[32] Donald L. Barlett, James B. Steele, *The Great American Tax Dodge: How Spiraling Fraud and Avoidance Are Killing Fairness, Destroying the Income Tax and Costing You*, Little Brown and Co., New York (2000).

[33] Maude Barlow, *Blue Gold: The Fight to Stop the Corporate Theft of the World's Water*, The New Press, New York (2002).

[34] Robert T. Bakker, *The Dinosaur Heresies*, Zebra Books, New York (1991).

[35] Sharon Beder, *Power Play: The Fight to Control the World's Electricity*, The New Press, New York (2003).

[36] Walden Bello, *The Food Wars*, Verso, New York (2009).

[37] Kermit C. Berger, *Sun, Soil & Survival*, University of Oklahoma Press, Norman (1965).

[38] Boris I. Bittker, *The Case for Black Reparations*, Beacon Press, Boston, Massachusetts (2003).

[39] Edwin Black, *Internal Combustion: How Corporations and Governments Addicted the World to Oil and Derailed the Alternatives*, St. Martin's Press, New York (2006).

[40] William Black, *The Best Way to Rob a Bank Is to Own One: How Corporate Executives and Politicians Looted the S&L Industry*, University of Texas Press, Texas (2005).

[41] William Blum, *Killing Hope: U.S. Military and CIA Interventions Since World War II*, Common Courage Press, Monroe, Maine (1995).

[42] William Blum, *Rogue State: A Guide to the World's Only Superpower*, Common Courage Press, Monroe, Maine (2000).

[43] William Blum, *Freeing the World to Death: Essays on the American Empire*, Common Courage Press, Monroe, Maine (2005).

[44] Eric Boehlert, *Lapdogs: How the Press Rolled Over for Bush*, Free Press, New York (2006).

[45] Derek Bok, *Universities in the Marketplace: The Commercialization of Higher Education*, Princeton University Press, Princeton, New Jersey (2003).

[46] Edmund Blair Bolles, *A Second Way of Knowing: The Riddle of Human Perception*, Prentice Hall Press, New York (1991).

[47] David Bollier, *Silent Theft: The Private Plunder of Our Common Wealth*, Routledge, New York (2002).

[48] Richard Bookstaber, *A Demon of Our Own Design: Markets, Hedge Funds, and the Perils of Financial Innovation*, John Wiley & Sons, Inc., Hoboken, New Jersey (2007).

[49] Kristina Borjesson, Editor, *Into the Buzzsaw: Leading Journalists Expose the Myth of a Free Press*, Prometheus Books, Amherst, New York (2002).

[50] Joanna Bourke, *An Intimate History of Killing: Face to Face Killing in 20th Century Warfare*, Basic Books, New York (1999).

[51] Mark Bowen, *Thin Ice: Unlocking the Secrets of Climate in the World's Highest Mountains*, Henry Holt, New York (2005).

[52] Mark Bowen, *Censoring Science: Inside the Political Attack on Dr. James Hansen and the Truth of Global Warming*, Dutton, New York (2008).

[53] Stephen R. Bown, *Scurvy: How a Surgeon, a Mariner, and a Gentleman Solved the Greatest Medical Mystery of the Age of Sail*, St. Martin's Press, New York (2003).

[54] Daniel G. Boyce, Marlon R. Lewis, Boris Worm, Global phytoplankton decline over the past century, *Nature*, 466, 591-596 (29 July 2010).

[55] Jules Boykoff, *Beyond Bullets: The Suppression of Dissent in the United States*, AK Press, Oakland, California (2007).

[56] Stewart Brand, *Whole Earth Discipline: An Ecopragmatist Manifesto*, Viking Adult (2009).

[57] Lester R. Brown, Gary Gardner, Brian Halweil, *Beyond Malthus: Sixteen Dimensions of the Population Problem*, WorldWatch Paper 143, Worldwatch Institute, Washington D.C. (1998).

[58] Lester Brown, *Plan B: Rescuing a Planet under Stress and a Civilization in Trouble*, W.W. Norton & Company, New York (2003).

[59] Lester Brown, *Plan B 2.0: Rescuing a Planet under Stress and a Civilization in Trouble*, W.W. Norton & Company, New York (2006), available for free download at www.earthpolicy.org.

[60] Lester Brown, *Plan B 4.0: Mobilizing to Save Civilizations*, W. W. Norton & Company, New York (2009), available for free download from www.earthpolicy.org.

[61] Robert Bryce, *Pipe Dreams: Greed, Ego, and the Death of Enron*, Public Affairs, New York (2002).

[62] Christopher Bryson, *The Fluoride Deception*, Seven Stories Press, New York (2004).

[63] Daniel Burton-Rose, Editor, *The Celling of America: An Inside Look at the U.S. Prison Industry*, Common Courage Press, Monroe, Maine (1998).

[64] Robert Burton, *The Life and Death of Whales, second edition*, Universe Books (1980).

[65] Smedley D. Butler, *War Is a Racket*, Feral House, Los Angeles (1935, 2003).

[66] Seth Cagin, Philip Dray, *Between Earth and Sky: How CFCs Changed Our World and Endangered the Ozone Layer*, Pantheon Books, New York (1993).

[67] Kitty Calavita, Henry N. Pontell, Robert H. Tillman, *Big Money Crime: Fraud and Politics in the Savings and Loan Crisis*, University of California Press, Berkeley, California (1997).

[68] Ernest Callenbach, *Ecology: A Pocket Guide*, University of California Press, Berkeley (1998).

[69] Colin J. Campbell, J.H. Laherrere, "The End of Cheap Oil," *Scientific American*, pp. 78–83, (March 1998).

[70] C. J. Campbell, *The Coming Oil Crisis*, Multi-Science Publishing Company, Ltd., Essex, England (2004).

[71] Fritjof Capra, *The Hidden Connections: A Science for Sustainable Living*, Anchor Books, New York (2004).

[72] David Carrasco, Eduardo M. Moctezuma, *Moctezuma's Mexico, Visions of the Aztec World*, University Press of Colorado, Niwot, Colorado (1992).

[73] Timothy J. Cartwright, *Modeling the World in a Spreadsheet: Environmental Simulation on a Microcomputer*, The Johns Hopkins University Press, Baltimore (1993).

[74] William R. Catton, Jr., *Overshoot: The Ecological Basis of Revolutionary Change*, University of Illinois Press, Chicago, Illinois (1982).

[75] Catherine Caufield, *Masters of Illusion: The World Bank and the Poverty of Nations*, Henry Holt, New York (1996).

[76] Jerry Cederblom, David W. Paulsen, *Critical Reasoning–Understanding and Criticizing Arguments and Theories*, Wadsworth Publishing Company, (1991).

[77] Christopher Cerf, Victor Navasky, *The Experts Speak*, Villard, New York, (1998).

[78] Eric J. Chaisson, *Cosmic Evolution: The Rise of Complexity in Nature*, Harvard University Press, Cambridge, Massachusetts (2001).

[79] Edward Chancellor, *Devil Take the Hindmost: A History of Financial Speculation*, Plume, Penguin Group, New York (2000).

[80] Nancy Chang, *Silencing Political Dissent: How Post-September 11 Anti-Terrorism Measures Threaten Our Civil Liberties*, Seven Stories Press, New York (2002).

[81] Daniel Charles, *Master Mind: The Rise and Fall of Fritz Haber, the Nobel Laureate Who Launched the Age of Chemical Warfare*, Harper-Collins, New York (2005).

[82] Ira Chernus, *Eisenhower's Atoms for Peace*, TAMU Press, (2002).

[83] Paul Chevigny, *Edge of the Knife: Police Violence in the Americas*, The New Press, New York (1995).

[84] Dan Chiras, Dave Wann, *Superbia! 31 Ways to Create Sustainable Neighborhoods*, New Society Publishers, Gabriola Island, BC, Canada (2003).

[85] Dan Chiras, *The Homeowner's Guide to Renewable Energy: Achieving Energy Independence through Solar, Wind, Biomass and Hydropower*, New Society Publishers, A Mother Earth News Book for Wiser Living, (2006)

[86] Noam Chomsky, *Objectivity and Liberal Scholarship*, The New Press, New York (2003).

[87] Noam Chomsky, *Failed States: The Abuse of Power and the Assault on Democracy*, Metropolitan Books, New York (2006).

[88] Ward Churchill, Jim Vander Wall, *The Cointelpro Papers: Documents from the FBI's Secret Wars Against Dissent in the United States*, South End Press, Boston, Massachusetts (1990).

[89] Ward Churchill, Jim Vander Wall, *Agents of Repression: The FBI's Secret Wars Against the Black Panther Party and the American Indian Movement*, South End Press Classics, South End Press, Cambridge, Massachusetts (2002).

[90] Colin W. Clark, *Mathematical Bioeconomics: The Optimal Management of Renewable Resources, Second Editon*, Wiley Interscience, New York (1990).

[91] Robin Clarke, Jannet King, *The Water Atlas*, The New Press, New York (2004).

[92] Cutler J. Cleveland, David I. Stern, Robert Costanza, *The Economics of Nature and the Nature of Economics (Advances in Ecological Economics Series)*, Edward Elgar Publishing (2002).

[93] Liane Clorfene-Casten, *Breast Cancer: Poisons, Profits, and Prevention*, Common Courage Press, Monroe, Me (1996).

[94] Charles Clover, *The End of the Line: How Overfishing Is Changing the World and What We Eat*, New Press, New York (2006).

[95] Andrew Cockburn, *Rumsfeld: His Rise, Fall, and Catastrophic Legacy*, Scribner, New York (2007).

[96] Leslie Cockburn, *Out of Control: The Story of the Reagan Administration's Secret War in Nicaragua, the Illegal Arms Pipeline, and the Contra Drug Connection*, Atlantic Monthly Press, New York (1987).

[97] Robert Coenraads, Chief Consultant, *Natural Disasters and How We Cope: The world's greatest natural disasters-Tsunamis-Fires-Hurricanes-Floods-Droughts-Diseases-Avalanches*, Millenium House Pty Ltd, New South Wales, Australia (2006).

[98] Steve Coffel, *But Not a Drop to Drink! The Lifesaving Guide to Good Water*, Rawson Associates, New York (1989).

[99] Elliot D. Cohen, PhD, Editor, *New Incorporated: Corporate Media Ownership and Its Threat to Democracy*, Prometheus Books, Amherst, New York (2005).

[100] Jeff Cohen, Norman Solomon, *Adventures in Medialand*, Common Courage Press, Monroe, Maine (1993).

[101] Jeff Cohen, *Cable News Confidential: My Misadventures in Corporate Media*, PoliPointPress, Sausalito, California (2006).

[102] Joel E. Cohen, *How Many People Can the Earth Support?*, W.W. Norton and Company, New York (1995).

[103] Theo Colborn, Dianne Dumanoski, John Peterson Myers, *Our Stolen Future: Are We Threatening our Fertility, Intelligence, and Survival? – A Scientific Detective Story*, Dutton, New York (1996).

[104] David Cole, *No Equal Justice: Race and Class in the American Criminal Justice System*, The New Press, New York (1999).

[105] Leonard A. Cole, *Element of Risk : The Politics of Radon*, Oxford University Press, New York (1994).

[106] Paul A. Colinvaux, *Why Big Fierce Animals are Rare*, Princeton University Press, Princeton, New Jersey (1978).

[107] Chuck Collins, Betsy Leondar-Wright, Holly Sklar, *Shifting Fortunes: The Perils of the Growing American Wealth Gap*, United for a Fair Economy (1999).

[108] Chuck Collins, Felice Yeskel, *Economic Apartheid in America: A Primer on Economic Inequality & Insecurity*, The New Press, New York (2000).

[109] Chuck Collins, Amy Gluckman, Meizhu Lui, Betsy Leondar-Wright, Amy Offner, Adria Scharf, Editors, *The Wealth Inequality Reader*, Dollars & Sense Economic Affairs Bureau, Cambridge, Massachusetts (2004).

[110] John M. Connor, *Archer Daniels Midland: Price-Fixer to the World, Fourth Edition*, Dept. of Agricultural Economics, Purdue University, West Lafayette, Indiana, Staff Paper # 00-11 (2000).

[111] John Conroy, *Unspeakable Acts, Ordinary People: The Dynamics of Torture, An Examination of the Practice of Torture in Three Democracies*, University of California Press, Berkeley, California (2000).

[112] John H. Conway, Richard K. Guy, *The Book of Numbers*, Springer-Verlag, New York (1996).

[113] Christopher D. Cook, *Diet for a Dead Planet: Big Business and the Coming Food Crisis*, The New Press, New York (2007).

[114] Jamie Court, Francis Smith, *Making a Killing: HMOs and the Threat to Your Health*, Common Courage Press, Monroe, Maine (1999).

[115] Jacques-Yves Cousteau, *The Cousteau Almanac: An Inventory of Life on Our Water Planet*, Doubleday, Garden City, N.Y. (1981),

[116] Thomas Crump, *The Anthropology of Numbers*, Cambridge University Press, (1992).

[117] Brian Cruver, *Anatomy of Greed: The Unshredded Truth from an Enron Insider*, Carroll & Graf Publishers, New York (2002).

[118] Francisco Ramirez Cuellar, *The Profits of Extermination: How U.S. Corporate Power Is Destroying Columbia*, Common Courage Press, Monroe, Maine (2005).

[119] Debra Lynn Dadd, *Nontoxic & Natural: How to Avoid Dangerous Everyday Products and Buy or Make Safe Ones*, Tarcher/St. Martin's Press, New York (1986).

[120] Debra Lynn Dadd, *The Nontoxic Home & Office: Protecting Yourself and Your Family from Everyday Toxics and Health Hazards*, Tarcher/Putnam, New York (1992).

[121] Gretchen C. Daily, *Nature's Services: Societal Dependence on Natural Ecosystems*, Island Press, Covelo, California (1997).

[122] Herman E. Daly, *Steady-State Economics, Second Edition*, Island Press, Covelo, California (1991).

[123] Herman E. Daly, *Beyond Growth*, Beacon Press, Boston, Massachusetts (1996).

[124] Herman E. Daly, John B. Cobb, Jr., *For the Common Good: Redirecting the Economy Toward Community, the Environment, and a Sustainable Future*, Beacon Press, Boston, Massachusetts (1998).

[125] Mark Danner, *Torture and Truth: America, Abu Ghraib, and the War on Terror*, New York Review Books, New York (2004).

[126] Tobias Dantzig, *Number: The Language of Science*, Fourth edition, Doubleday & Company, Garden City, New York (1954).

[127] Julian Darley, *High Noon for Natural Gas: The New Energy Crisis*, Chelsea Green, White River Junction, Vermont (2004).

[128] Mike Davis, *The Monster at Our Door: The Global Threat of Avian Flu*, The New Press, New York (2005).

[129] Mike Davis, *Planet of Slums*, Verso, New York (2007).

[130] Kenneth S. Deffeyes, *Hubbert's Peak: The Impending World Oil Shortage*, Princeton University Press, Princeton (2001).

[131] Kenneth S. Deffeyes, *Beyond Oil: The View From Hubbert's Peak*, Hill and Wang, New York (2005).

[132] Quentin Dempster, *Whistleblowers: People with the Courage to Put the Truth First*, ABC Books, GPO Box 9994, Sydney, NSW, Australia (1997).

[133] Mark Denny, Steven Gaines, *Chance in Biology: Using Probability to Explore Nature*, Princeton University Press, Princeton, New Jersey (2000).

[134] Suraje Dessai, Martin Walter, " Self-Organized Criticality and the Atmospheric Sciences: Selected Review, New Findings and Future Directions" *XE Extreme Events: Developing a Research Agenda for the 21st Century*, National Center for Atmospheric Research, August 2000, pp. 34–44.

[135] Robert L. Devaney, *An Introduction to Chaotic Dynamical Systems, 2nd Edition*, Westview Press (2003).

[136] Keith Devlin, *Mathematics, The Science of Patterns: The search for order in life, mind, and the universe*, Scientific American Library, New York (1994).

[137] Jared Diamond, *Collapse: How Societies Choose to Fail or Succeed*, Penguin Group, New York (2005).

[138] John Dicker, *The United States of Wal-Mart*, Penguin, New York (2005).

[139] Paul Dickson, Thomas B. Allen, *The Bonus Army: An American Epic*, Walker & Company, New York (2004).

[140] Douglas W. Dockery, C. Arden Pope, Xiping Xu, John D. Spengler, James H. Ware, Martha E. Fay, Benjamin G. Ferris, Frank E. Speizer, "An Association between Air Pollution and Mortality in Six U.S. Cities," *New England Journal of Medicine*, Volume 329:1753–1759, Number 24 (December 9, 1993).

[141] Kevin Danaher, *50 Years Is Enough: The Case Against the World Bank and the International Monetary Fund*, South End Press, Boston, Massachusetts (1994).

[142] Jack Doyle, *Taken for a Ride: Detroit's Big Three and the Politics of Pollution*, Four Walls Eight Windows, New York (2000).

[143] Jack Doyle, *Trespass Against Us: Dow Chemical & The Toxic Century*, Common Courage Press, Monroe, Maine (2004).

[144] William Holland Drury, Jr., *Chance and Change: Ecology for Conservationists*, University of California Press, Berkeley, California (1998).

[145] Bill Devall, editor, *Clearcut: The Tragedy of Industrial Forestry*, Earth Island Press, San Francisco, California (1993).

[146] Scott C. Doney, "The Dangers of Ocean Acidification," *Scientific American*, pp. 58–65, (March 2006).

[147] Beshara Doumani, *Academic Freedom after September 11*, Zone Books, Brooklyn, New York (2006).

[148] Richard Douthwaite, *Short Circuit: Strengthening Local Economies for Security in an Unstable World*, The Lilliput Press, Ltd., Dublin, Ireland (1996).

[149] Richard Douthwaite, *The Ecology of Money*, Green Books, Ltd., The Schumacher Society, Bristol, England (1999).

[150] Tamara Draut, *Strapped: Why America's 20- and 30- Somethings Can't Get Ahead*, Doubleday, New York (2006).

[151] Madeline Drexler, *Secret Agents: The Menace of Emerging Infections*, Penguin, New York (2002).

[152] Karl Drlica, *Understanding DNA and Gene Cloning: A Guide for the Curious*, John Wiley & Sons, Inc., New Jersey (2004).

[153] Robin Dunbar, *Grooming, Gossip, and the Evolution of Language*, Harvard University Press, Cambridge (1996).

[154] David Dunbar, Brad Reagan, *Debunking 9/11 Myths: Why Conspiracy Theories Can't Stand Up to the Facts*, PopularMechanics, Hearst Books, Sterling Publishing Co., Inc., New York (2006).

[155] Seth Dunn, *Hydrogen Futures: Toward a Sustainable Energy System*, Worldwatch Paper 157, Worldwatch Institute, Washington, D.C. (August, 2001).

[156] Alan Durning, *How Much Is Enough?*, W.W. Norton & Company/Worldwatch Books, New York (1992).

[157] Christian de Duve, *Vital Dust: The Origin and Evolution of Life on Earth (Life as a Cosmic Imperative)*, Basic Books, New York (1995).

[158] Joel Dyer, *The Perpetual Prisoner Machine: How America Profits from Crime*, Westview Press, Boulder, Colorado (2000).

[159] Joel Dyer, *Harvest of Rage: Why Oklahoma City is Only the Beginning*, Westview Press, Boulder, Colorado (2001).

[160] Freeman Dyson, *Origins of Life, Second Edition*, Cambridge University Press, New York (1999).

[161] Silvia A. Earle, *The World is Blue: How our Fate and the Ocean's are One*, National Geographic Society, Washington D.C. (2009).

[162] Leah Edelstein-Keshet, *Mathematical Models in Biology*, Society for Industrial and Applied Mathematics (SIAM) (2005).

[163] Barbara Ehrenreich, *Fear of Falling: The Inner Life of the Middle Class*, HarperPerennial, New York (1989).

[164] Barbara Ehrenreich, *The Snarling Citizen*, HarperPerennial, New York (1995).

[165] Barbara Ehrenreich, *Blood Rites: Origins and History of the Passions of War*, Owl Books, New York (1997).

[166] Barbara Ehrenreich, *Nickel and Dimed: On (Not) Getting By in America*, Metropolitan Books, New York (2001).

[167] Barbara Ehrenreich, *Bait and Switch: The (Futile) Pursuit of the American Dream*, Metropolitan Books, New York (2005).

[168] Paul R. Ehrlich, Anne H. Ehrlich, *Population, Resources, Environment: Issues in Human Ecology*, W. H. Freeman and Company, San Francisco (1970).

[169] Paul R. Ehrlich, *Human Natures: Genes, Cultures, and the Human Prospect*, Penguin Books, New York (2002).

[170] Kurt Eichenwald, *The Informant: The FBI Was Ready to Take Down America's Most Politically Powerful Corporation. But There Was One Thing They Didn't Count On.*, Broadway Books, New York (2000).

[171] Manfred Eigen, Ruthild Winkler, *Laws of the Game: How the Principles of Nature Govern Chance*, Princeton University Press, Princeton, New Jersey (1993).

[172] Duane Elgin, *Voluntary Simplicity: Toward a Way of Life That Is Outwardly Simple, Inwardly Rich*, Quill, William Morrow, New York (1993).

[173] S. Ellis, A. Mellor, *Soils and Environment*, Routledge, London (1995).

[174] Charles S. Elton, *The Ecology of Invasions by Animals and Plants*, Chapman and Hall, London, England (1977).

[175] Wayne Ellwood, *The No-Nonsense Guide to Globalization*, Verso, London, England (2001).

[176] Daniel Ellsberg, *Secrets: A Memoir of Vietnam and the Pentagon Papers*, Penguin (2003).

[177] Daniel Ellsberg, *Papers on the War*, Simon & Schuster, (2009).

[178] Samuel S. Epstein, M.D., *The Politics of Cancer Revisited*, East Ridge Press, New York (1998).

[179] David Earnhardt, *Uncounted: The New Math of American Elections*, (a film), Earnhardt Pirkle, Inc. (2007).

[180] Ralph Estes, *Tyranny of the Bottom Line: Why Corporations Make Good People Do Bad Things*, Berrett-Koehler, San Francisco (1996).

[181] Stuart Ewen, *PR! A Social History of Spin*, Basic Books, New York (1996).

[182] Dan Fagin, Marianne Lavelle, the Center for Public Integrity, *Toxic Deception: How the Chemical Industry Manipulates Science, Bends the Law, and Endangers Your Health*, Birch Lane Press, New Jersey (1996).

[183] Jeff Faux, *The Global Class War: How America's Bipartisan Elite Lost Our Future–and What It Will Take to Win It Back*, John Wiley & Sons, New York (2006).

[184] W. Ward Fearnside, William B. Holther, *Fallacy – The Counterfeit of Argument*, Prentice-Hall, Inc., New York (1959).

[185] Liza Featherstone, United Students Against Sweatshops, *Students Against Sweatshops*, Verso, New York (2002).

[186] William Feller, *An Introduction to Probability Theory and Its Applications,* John Wiley & Sons, New York (1968).

[187] Richard P. Feynman, Robert B. Leighton, Matthew Sands, *The Feynman Lectures on Physics, Vol. 1*, Addison-Wesley, Reading, Massachusetts (1963).

[188] Richard P. Feynman, *"What Do You Care What Other People Think?"*, W.W. Norton, New York (1988).

[189] John Firor, Judith Jacobsen, *The Crowded Greenhouse: Population, Climate Change and Creating a Sustainable World*, Yale University Press, New Haven (2002).

[190] Robert J. Fitrakis, Steven Rosenfeld, Harvey Wasserman, *What Happened in Ohio? A Documentary Record of Theft and Fraud in the 2004 Election*, The New Press, New York (2006).

[191] Bob Fitrakis, Harvey Wasserman, *How the GOP Stole America's 2004 Election & Is Rigging 2008*, www.freepress.org, CICJ Books, www.harveywasserman.com, Bexley, Ohio (2005).

[192] Tim Flannery, *The Future Eaters*, Reed Books, 35 Cotham Road, Kew, Victoria 3101, Australia (1997).

[193] Anthony Flint, *This Land: The Battle over Sprawl and the Future of America*, The Johns Hopkins University Press, Baltimore, Maryland (2006).

[194] Stephen Flynn, *The Edge of Disaster*, Random House, New York (2007).

[195] Franklin Folsom, *Impatient Armies of the Poor: The Story of Collective Action of the Unemployed 1808–1942*, University Press of Colorado (1991).

[196] Dave Foreman, *Confessions of an Eco–Warrior*, Harmony Books (1991).

[197] Dave Foreman, *Ecodefense: A Field Guide to Monkeywrenching, Third Edition*, Abbzug Press, Chico, California (1993).

[198] Jay W. Forrester, *World Dynamics*, Wright-Allen Press, Inc., Cambridge, Massachusetts (1971).

[199] J. B. Joseph Fourier, "Remarques Générales Sur Les Températures Du Globe Terrestre Et Des Espaces Planétaires," *Annales de Chemie et de Physique* (1824) Vol. 27. pp. 136–67.

[200] J. B. Joseph Fourier, "Mémoire Sur Les Températures Du Globe Terrestre Et Des Espaces Planétaires," *Mémoires de l'Académie Royale des Sciences* (1827) Vol. 7. pp. 569–604

[201] John O. Fox, *If Americans Really Understood the Income Tax*, Westview, Boulder, Colorado (2001).

[202] Nicols Fox, *Spoiled: The Dangerous Truth About a Food Chain Gone Haywire*, BasicBooks, HarperCollins, New York (1997).

[203] Thomas Frank, *One Market Under God: Extreme Capitalism, Market Populism, and the End of Economic Democracy*, Doubleday, New York (2000).

[204] H. Bruce Franklin, *The Most Important Fish in the Sea: Menhaden and America*, Island Press/Shearwater Books, Washington, D.C. (2007).

[205] Steven F. Freeman, Joel Bleifuss, *Was the 2004 Presidential Election Stolen? Exit Polls, Election Fraud, and the Official Count*, Seven Stories Press, New York (2006).

[206] Samuel Fromartz, *Organic, Inc.*, Harcourt, Inc., Orlando, Florida (2006).

[207] Annette Fuentes, "Autism in a Needle?" *In These Times*, (November 11, 2003).

[208] John Kenneth Galbraith, *The Great Crash 1929*, Houghton Mifflin Company, Boston, Massachusetts (1961).

[209] John Kenneth Galbraith, *Money: Whence It Came, Where It Went*, Bantam Books, New York (1980).

[210] James K. Galbraith, Naomi Prins, "December Surprise," *Mother Jones*, July/August 2008, pp 38–40.

[211] Gary Gardner, Brian Halweil, *Underfed and Overfed: the Global Epidemic of Malnutrition*, WorldWatch Paper 150, Worldwatch Institute, Washington, D.C. (2000).

[212] Lloyd C. Gardner, Marilyn B. Young, editors, *Iraq and the Lessons of Vietnam, or How NOT to Learn from the Past*, The New Press, New York (2007).

[213] Simson Garfinkel, *Database Nation: The Death of Privacy in the 21st Century*, O'Reilly, Sebastopol, California (January 2001).

[214] Laurie Garrett, *The Coming Plague: Newly Emerging Diseases in a World Out of Balance*, Penguin Books, New York (1994).

[215] Laurie Garrett, *Betrayal of Trust: The Collapse of Global Public Health*, Hyperion, New York (2000).

[216] Jeff Gates, *The Ownership Solution: Toward a Shared Capitalism for the 21st Century*, Perseus Books, New York (1998).

[217] Ross Gelbspan, *The Heat Is on: The Climate Crisis, the Cover-Up, the Prescription*, Perseus Publishing, New York (1998).

[218] Ross Gelbspan, *Boiling Point: How Politicians, Big Oil and Coal, Journalists, and Activists Have Fueled the Climate Crisis–and What We Can Do to Avert Disaster*, Basic Books, New York (2004).

[219] Thomas Geoghegan, "Infinite Debt: How unlimited interest rates destroyed the economy," *Harper's Magazine*, April 2009.

[220] Allan Gerson, Jerry Adler, *The Price of Terror: Lessons of Lockerbie for a World on the Brink*, Harper Collins, New York (2001).

[221] John Gever, Robert Kaufmann, David Skole, Charles Vörösmarty, *Beyond Oil: The Threat to Food and Fuel in the Coming Decades*, University Press of Colorado, (1991).

[222] Lois Marie Gibbs, *Love Canal: My Story*, Grove Press, (1982)

[223] Lois Marie Gibbs, *Love Canal: The Story Continues*, New Society Publishers, P.O. Box 189, Gabriola Island, BC VOR 1XO, Canada (1998).

[224] Dan Gillmor, *We the Media: Grassroots Journalism by the People, for the People*, O'Reilly Media, Inc., Sebastopol, California (2006).

[225] Malcolm Gladwell, *The Tipping Point: How Little Things Can Make a Big Difference*, Little, Brown and Company, New York (2002).

[226] James Glanz, *Saving Our Soil: Solutions for Sustaining Earth's Vital Resource*, Johnson Books, Boulder, Colorado (1995).

[227] Myron Peretz Glazer, Penina Migdal Glazer, *The Whistleblowers: Exposing Corruption in Government and Industry*, BasicBooks, New York (1989).

[228] James Gleick, *Chaos: Making a New Science*, Penguin, New York (1988).

[229] James Gleick, *Genius: The Life and Science of Richard Feynman*, Vintage Books, New York, (1992).

[230] Peter H. Gleick, editor, *Water in Crisis: A Guide to the World's Fresh Water Resources*,

[231] Brian Glick, *War At Home: Covert Action Against U.S. Activists and What We Can Do About It*, South End Press, Boston, Massachusetts (1989). Oxford University Press, New York (1993).

[232] Benjamin A. Goldman, *The Truth About Where You Live: An Atlas for Action on Toxins and Mortality*, Random House, New York (1991).

[233] Martin Goldstein, Inge F. Goldstein, *The Refrigerator and the Universe: Understanding the Laws of Energy*, Harvard University Press, Cambridge, Massachusetts (1995).

[234] Daniel Goleman, *Ecological Intelligence: How Knowing the Hidden Impacts of What We Buy Can Change Everything*, Broadway Books, New York (2009).

[235] Adam Leith Gollner, *The Fruit Hunters: A Story of Nature, Adventure, Commerce and Obsession*, Scribner, New York (2008).

[236] Larry Gonick, Alice Outwater, *The Cartoon Guide to the Environment*, HarperPerennial, New York (1996).

[237] Larry Gonick, Woollcott Smith, *The Cartoon Guide to Statistics*, HarperPerennial, New York (1993).

[238] Juan González, *Fallout: The Environmental Consequences of the World Trade Center Collapse*, The New Press, New York (2002).

[239] Jeff Goodell, *Big Coal: The Dirty Secret Behind America's Energy Future*, Houghton Mifflin, New York (2006).

[240] Amy Goodman, David Goodman, *The Exception to the Rulers: Exposing Oily Politicians, War Profiteers, and the Media that Love Them*, Hyperion, New York (2004).

[241] Amy Goodman, David Goodman, *Static: Government Liars, Media Cheerleaders, and the People Who Fight Back*, Hyperion, New York (2006).

[242] Amy Goodman, David Goodman, *Standing Up to the Madness: Ordinary Heroes in Extraordinary Times*, Hyperion, New York (2008).

[243] David Goodstein, *Out of Gas: The End of the Age of Oil*, W. W. Norton, New York (2004).

[244] Phil Goodstein, *DIA and Other Scams: A People's History of the Modern Mile High City*, New Social Publications, Denver, Colorado (2000).

[245] Merrill Goozner, *The $800 Million Pill: The Truth Behind the Cost of New Drugs*, University of California Press, Berkeley and Los Angeles, California (2004).

[246] Al Gore, *An Inconvenient Truth: The Planetary Emergency of Global Warming and What We Can Do About It*, Rodale Press, Emmaus, Pa. (2006).

[247] Al Gore, *Our Choice: A Plan to Solve the Climate Crisis*, Rodale Press, Emmaus, Pa. (2009).

[248] Andrew Goudie, *The Human Impact on the Natural Environment: Past, Present, and Future*, Blackwell, Oxford, UK (2005).

[249] K. A. Gourlay, *World of Waste: Dilemmas of Industrial Development*, Zed Books, New Jersey (1992).

[250] Serhiy Grabarchuk, *The New Puzzle Classics: Ingenious Twists on Timeless Favorites*, Sterling Publishing Co., New York (2005).

[251] Joe Graedon, Teresa Graedon, Ph.D., *The People's Guide to Deadly Drug Interactions: How to Protect Yourself from Life-Threatening Drug/Drug, Drug/Food, Drug/Vitamin Combinations*, St. Martin's Press, New York (1995).

[252] Mike Gray, *Drug Crazy: How We Got Into this Mess and How We Can Get Out*, Routledge, New York (2000).

[253] Thomas H. Greco, Jr., *Money and Debt: A Solution to the Global Crisis, Second Edition*, Thomas H. Greco, Jr., Publisher, P.O. Box 42663, Tucson, Arizona (1990).

[254] Thomas H. Greco, Jr., *New Money for Healthy Communities*, Thomas H. Greco, Jr., Publisher, P.O. Box 42663, Tucson, Arizona (1994).

[255] Thomas H. Greco, Jr., *Money: Understanding and Creating Alternatives to Legal Tender*, Chelsea Green Publishing Company, White River Junction, Vermont (2001).

[256] Karen J. Greenberg, Joshua L. Dratel, Editors, *The Torture Papers: The Road to Abu Ghraib*, Cambridge University Press, New York (2005).

[257] Katharine Greider, *The Big Fix: How the Pharmaceutical Industry Rips Off American Consumers*, Public Affairs, New York (2003).

[258] William Greider, *Secrets of the Temple: How the Federal Reserve Runs the Country*, Simon & Schuster, New York (1989).

[259] Stephen Grey, *Ghost Plane: The True Story of the CIA Torture Program*, St. Martin's Press, New York (2006).

[260] David Ray Griffin, *The New Pearl Harbor: Disturbing Questions about the Bush Administration and 9/11*, Olive Branch Press, Northampton, Massachusetts (2004).

[261] David Ray Griffin, *The 9/11 Commission Report: Omissions and Distortions*, Olive Branch Press, Northampton, Massachusetts (2005).

[262] David Ray Griffin, *Debunking 9/11 Debunking: An Answer to Popular Mechanics and Other Defenders of the Offical Conspiracy Theory*, Olive Branch Books, Northampton, Massachusetts (2007).

[263] Trauger Groh, Steven McFadden, *Farms of Tomorrow Revisited: Community Supported Farms – Farm Supported Communities*, Biodynamic Farming and Gardening Association, Kimberton, Pa. (1997).

[264] Andrew Gumbel, *Steal This Vote*, Nation Books, New York (2006).

[265] Doug Gurian-Sherman, *Failure to Yield*, http://www.ucsusa.org/food_and_agriculture/science_and_impacts/science/failure-to-yield.html

[266] Owen D. Gutfreund, *20th-Century Sprawl: Highways and the Reshaping of the American Landscape*, Oxford University Press, New York (2004).

[267] Nicky Hager, "Exposing the Global Surveillance System," *Covert Action Quarterly*, Number 59, Washington, DC (Winter 1996/97).

[268] Nicky Hager, *Secret Power: New Zealand's Role in the International Spy Network*, Craig Potton Publishing, Box 555, Nelson, New Zealand (1996).

[269] Nicky Hager, *Secrets and Lies: The Anatomy of an Anti-Environmental PR Campaign*, Common Courage Press, Monroe, Maine (1999).

[270] Charles A. S. Hall, Cutler J. Cleveland, Robert Kaufmann, *Energy and Resource Quality: The Ecology of the Economic Process*, University Press of Colorado (1992).

[271] C. L. Hamblin, *Fallacies,* Methuen & Co. LTD, (1970).

[272] James Hansen, *Storms of My Grandchildren: The Truth About the Coming Climate Catastrophe and Our Last Chance to Save Humanity*, Bloomsbury, U.S.A. (2009).

[273] Jennifer K. Harbury, *Bridge of Courage: Life Stories of the Guatemalan Compañeros and Compañeras* Common Courage Press, Monroe, Maine (1995).

[274] Jennifer K. Harbury, *Searching for Everardo: A Story of Love, War, and the CIA in Guatemala*, Warner Books, New York (1997).

[275] Jennifer K. Harbury, *Truth, Torture, and the American Way: The History and Consequences of U.S. Involvement in Torture*, Beacon Press, Boston, Massachusetts (2005).

[276] Garrett Hardin, "The Tragedy of the Commons," *Science*, Vol 162, pp. 1243–1248, (December 13, 1968).

[277] Garrett Hardin, *Living Within Limits: Ecology, Economics, and Population Taboos*, Oxford University Press, New York (1993).

[278] Vincent Harding, *Martin Luther King: The Inconvenient Hero*, Orbis Books, Maryknoll, New York (2008).

[279] Vincent Harding, *Hope and History: Why We Must Share the Story of the Movement*, Orbis Books, Maryknoll, New York (2009).

[280] Bev Harris, *Black Box Voting: Ballot-tampering in the 21st Century*, Plan Nine Publishing, 1237 Elon Place, High Point, NC 27263 (2003).

[281] David A. Harris, *Good Cops: The Case for Preventive Policing*, The New Press, New York (2005).

[282] Marvin Harris, *Cannibals and Kings: Origins of Cultures*, Vintage, (1991).

[283] John Harte, *Consider a Spherical Cow: A Course in Environmental Problem Solving*, University Science Books, Sausalito, California (1988).

[284] John Harte, Cheryl Holdren, Richard Schneider, Christine Shirley, *Toxics A to Z: A Guide to Everyday Pollution Hazards*, University of California Press, Berkeley (1991).

[285] Thom Hartmann, *The Last Hours of Ancient Sunlight: The Fate of the World and What We Can Do Before It's Too Late*, Three Rivers Press, New York (2004).

[286] William D. Hartung, *How Much Are You Making on the War, Daddy? A Quick and Dirty Guide to War Profiteering in the Bush Administration*, Nation Books, New York (2003).

[287] Jeffrey Haas, *The Assassination of Fred Hampton: How the FBI and the Chicago Police Murdered a Black Panther*, Lawrence Hill Books, Chicago, Illinois (2010).

[288] Shane Harris, *The Watchers: The Rise of America's Surveillance State*, The Penguin Press, (2010).

[289] Paul Hawken, Amory Lovins, L. Hunter Lovins, *Natural Capitalism: Creating the Next Industrial Revolution*, Little, Brown and Company, New York (1999).

[290] Chris Hedges, *War Is a Force That Gives Us Meaning*, Public Affairs, New York (2002).

[291] William Heffernan, *Consolidation in the Food and Agriculture System*, Report to the National Farmer's Union, February 5, 1999, (available at http://home.hiwaay.net/~becraft/NFUFarmCrisis.htm).

[292] William Heffernan, Mary Hendrickson, *Multinational Concentrated Food Processing and Marketing Systems and the Farm Crisis*, presented to the American Association for the Advancement of Science, February 19, 2002 (available at http://www.agribusinessaccountability.org/page/148/).

[293] Richard Heinberg, *The Party's Over: Oil, War and the Fate of Industrial Societies*, New Society Publishers, P.O. Box 189, Gabriola Island, BC VOR 1XO, Canada (2005).

[294] Richard Heinberg, *Peak Everything: Waking Up to the Century of Declines*, New Society Publishers, P.O. Box 189, Gabriola Island, BC VOR 1XO, Canada (2007).

[295] David Helvarg, *The War Against the Greens: The "Wise-Use" Movement, the New Right, and Anti-Environmental Violence*, University of California Press, Berkeley (1994).

[296] David Helvarg, *Blue Frontier: Saving America's Living Seas*, Owl Books, Henry Holt, New York (2001).

[297] David Helvarg, *Saved by the Sea: A Love Story with Fish*, Thomas Dunne Books, St. Martin's Press, New York (2010).

[298] David Helvarg, *Blue Frontier: Dispatches from America's Ocean Wilderness, 2nd edition*, Sierra Club Books, (2006).

[299] Michael G. Henle, "Forget Not the Lowly Spreadsheet," *The College Mathematics Journal*, 26, No. 4 (September 1995).

[300] Robert Henson, *The Rough Guide to Climate Change: The Symptoms, The Science, The Solutions*, Rough Guides, New York (2006).

[301] Doug Henwood, *Wall Street*, Verso, New York (1997).

[302] Seymour M. Hersh, *Against All Enemies: Gulf War Syndrome: The War Between America's Ailing Veterans and Their Government*, The Ballantine Publishing Group, New York (1998).

[303] Seymour M. Hersh, *Chain of Command: The Road from 9/11 to Abu Ghraib*, Harper Collins, New York (2004).

[304] James F. Hettinger, Stanley D. Tooley, *Small Town, Giant Corporation: Japanese Manufacturing Investment and Community Economic Development in the United States*, University Press of America, Lanham, Maryland (1994).

[305] Philip B. Heymann, *Terrorism and America: A Commonsense Strategy for a Democratic Society*, The MIT Press, Cambridge, Massachusetts (1998).

[306] Jim Hightower, *Hard Tomatoes, Hard Times*, forewords by Harry M. Scoble and James Abourezk, Cambridge, Mass., Schenkman Pub. Co. (1978).

[307] Jim Hightower, *There's Nothing in the Middle of the Road but Yellow Stripes and Dead Armadillos : A Work of Political Subversion*, HarperCollins, New York (1997).

[308] Daniel J. Hillel, *Out of the Earth: Civilization and the Life of the Soil*, The Free Press, A Division of Macmillan, Inc., New York (1991).

[309] Philip J. Hilts, *Protecting America's Health: The FDA, Business, and One Hundred Years of Regulation*, Alfred A. Knopf, New York (2003).

[310] Michael A. Hiltzik, *The Plot Against Social Security: How the Bush Plan is Endangering Our Financial Future*, Harper Collins, New York (2005).

[311] David Himmelstein, M.D., Steffie Woolhandler, M.D., M.P.H., Ida Hellander, M.D., *Bleeding the Patient: The Consequences of Corporate Health Care*, Common Courage Press, Monroe, Maine (2001).

[312] Robert E. Hinshaw, *Living with Nature's Extremes: The Life of Gilbert Fowler White*, Johnson Books, Boulder, Colorado (2006).

[313] Dr. Mae-Wan Ho, *Genetic Engineering Dreams or Nightmares? The Brave New World of Bad Science and Big Business*, Research Foundation for Science, Technology and Ecology/Third World Network, New Delhi, India (1997).

[314] Errol Hodge, *Radio Wars: Truth, Propaganda, and the Struggle for Radio Australia*, Cambridge University Press, Melbourne, Australia (1995).

[315] Jonathan K. Hodge, Richard E. Klima, *The Mathematics of Voting and Elections: A Hands-On Approach*, American Mathematical Society, Mathematical World, Volume 22, Providence, Rhode Island (2005).

[316] Richard Hofrichter, editor, *Toxic Struggles: The Theory & Practice of Environmental Justice*, The University of Utah Press, Salt Lake City, Utah (2002).

[317] James Hoggan, Richard Littlemore (contributor), *Climate Cover-up: The Crusade to Deny Global Warming*, Greystone Books, (2009).

[318] Allen M. Hornblum, *Acres of SKIN: Human Experiments at Holmesburg Prison*, Routledge, New York (1998).

[319] Michael Hudson, *Merchants of Misery: How Corporate America Profits from Poverty*, Common Courage Press, Monroe, Maine (1996).

[320] Michael Hudson, *The Monster: How a Gang of Predatory Lenders and Wall Street Bankers Fleeced America–and Spawned a Global Crisis*, Times Books, (2010).

[321] Darrel Huff, *How to Lie with Statistics*, W. W. Norton & Company, New York (1954).

[322] Francis Hutchinson, Mary Mellor, Wendy Olsen, *The Politics of Money: Towards Sustainability and Economic Democracy*, Pluto Press, Sterling, Virginia (2002).

[323] Daniel Imhoff, *Food Fight: The Citizen's Guide to a Food and Farm Bill*, University of California Press (2007).

[324] Kevin Jackson, Editor, *The Oxford Book of Money*, Oxford University Press, New York (1995).

[325] Lynn Jacobs, *Waste of the West: Public Lands Ranching*, Lynn Jacobs, P.O. Box 5784, Tucson, Arizona 85703, (602) 578-3173, (1992).

[326] Mark Z. Jacobson, Mark A. Delucchi, "A Path to Sustainable Energy by 2030," *Scientific American*, November 2009, pp. 58–65, (www.ScientificAmerican.com) (2009).

[327] Joy James, Editor, *The Angela Y. Davis Reader*, Blackwell Publishing, Malden, Massachusetts (2004).

[328] Derrick Jensen, George Draffan, *Strangely Like War: The Global Assault on Forests*, Chelsea Green Publishing Company, White River Junction, Vermont (2003).

[329] Derrick Jensen, *Endgame: Volume 1: The Problem of Civilization*, Seven Stories Press, New York (2006).

[330] Derrick Jensen, *Endgame: Volume 2: Resistance*, Seven Stories Press, New York (2006).

[331] Robert Jensen, *Citizens of Empire: The Struggle to Claim Our Humanity*, City Lights Books, San Francisco, California (2004).

[332] Fred Jerome, *The Einstein File: J. Edgar Hoover's War Against the World's Greatest Scientist*, St. Martin's Press, New York (2002).

[333] Robert Jervis, *System Effects: Complexity in Political and Social Life*, Princeton University Press, Princeton (1997).

[334] Chalmers Johnson, *Blowback: The Costs and Consequences of American Empire*, Owl Books, New York (2000).

[335] Chalmers Johnson, *The Sorrows of Empire: Militarism, Secrecy, and the End of the Republic*, Metropolitan Books, New York (2004).

[336] David Cay Johnston, *Perfectly Legal: The Covert Campaign to Rig our Tax System to Benefit the Super Rich – and Cheat Everybody Else*, Portfolio, New York (2003).

[337] Antonia Juhasz, *The Bush Agenda: Invading the World, One Economy at a Time*, Regan Books, Harper Collins, New York (2006).

[338] Laurence H. Kallen, *Corporate Welfare: The Mega Bankruptcies of the 80s and 90s*, Carol Publishing Group, New Jersey (1991).

[339] Donald Kaufman, Cecillia Franz, *Biosphere 2000: Protecting Our Global Environment*, Kendall/Hunt (1999).

[340] Stuart A. Kauffman, *The Origins of Order: Self-Organization and Selection in Evolution*, Oxford University Press, New York (1993).

[341] Jane Holtz Kay, *Asphalt Nation: How the Automobile Took Over America and How We Can Take It Back*, University of California Press, Los Angeles, California (1997).

[342] Patrick Radden Keefe, *Chatter: Dispatches from the Secret World of Global Evesdropping*, Random House, New York (2005).

[343] Charles M. Kelly, *The Great Limbaugh Con: And Other Right-Wing Assaults on Common Sense*, Fithian Press, Santa Barbara, California (1994).

[344] John F. Kelly, Phillip K. Wearne, *Tainting Evidence: Inside the Scandals at the FBI Crime Lab*, The Free Press, New York (1998).

[345] Patricia Clark Kenschaft, *Mathematics for Human Survival*, Whittier Publications, Island Park, New York (2002).

[346] Ronald Kessler, *Inside Congress: the Shocking Scandals, Corruption, and Abuse of Power Behind the Scenes on Capitol Hill*, Pocket Books, New York (1997).

[347] Alexander Keyssar, *The Right to Vote: The Contested History of Democracy in the United States*, Basic Books, New York (2000).

[348] Andrew Kimbrell, editor, *The Fatal Harvest Reader: The Tragedy of Industrial Agriculture*, Island Press, Washington, Covelo, London (2002).

[349] Andrew Kimbrell, editor, *Fatal Harvest: The Tragedy of Industrial Agriculture*, (illustrated), Island Press, Washington, Covelo, London (2002).

[350] Charles P. Kindleberger, *Manias, Panics, and Crashes: A History of Financial Crises*, fifth edition, John Wiley & Sons, Inc., Hoboken, New Jersey (2005).

[351] Stephen Kinzer, *Overthrow: America's Century of Regime Change from Hawaii to Iraq*, Times Books, New York (2006).

[352] David Kirby, *Animal Factory: The Looming Threat of Industrial Pig, Dairy, and Poultry Farms to Humans and the Environment*, St. Martin's Press, New York (2010).

[353] Lorne T. Kirby, *DNA Fingerprinting: An Introduction*, Oxford University Press, New York (1992).

[354] Michael T. Klare, *Resource Wars: The New Landscape of Global Conflict*, Metropolitan Books, Henry Holt, New York (2001).

[355] Michael T. Klare, *Blood and Oil: The Dangers and Consequences of America's Growing Dependency on Imported Petroleum*, Metropolitan Books, Henry Holt, New York (2004).

[356] Mark Klein, *Wiring Up The Big Brother Machine...And Fighting It*, BookSurge Publishing, (2009).

[357] Morris Kline, *Mathematics, the Loss of Certainty*, Oxford University Press, New York (1980).

[358] George J. Klir and Tina A. Folger, *Fuzzy Sets, Uncertainty, And Information*, Prentice Hall, New Jersey (1998).

[359] Elizabeth Kolbert, *Field Notes from a Catastrophe: Man, Nature, and Climate Change*, Bloomsbury USA (2006).

[360] Melvin Konner, *The Tangled Wing: Biological Constraints on the Human Spirit*, Henry Holt and Company, New York (1982).

[361] Peter Kornbluh, Malcolm Byrne, *The Iran-Contra Scandal: The Declassified History–a National Security Archives Documents Reader*, The New Press, New York (1993).

[362] Peter Kornbluh, *The Pinochet File: A Declassified Dossier on Atrocity and Accountability–a National Security Archive Book*, The New Press, New York (2003).

[363] Alan Charles Kors, Harvey A. Silvergate, *The Shadow University: The Betrayal of Liberty on America's Campuses*, HarperCollins, New York (1998).

[364] David C. Korten, *The Great Turning: From Empire to Earth Community*, Berrett-Koehler Publishers, Inc., San Francisco California (2006).

[365] Bart Kosko, *Fuzzy Thinking: The New Science of Fuzzy Logic*, Hyperion, New York (1993).

[366] Jonathan Kozol, *Death at an Early Age*, Penguin Books, New York (1985).

[367] Jonathan Kozol, *Savage Inequalities: Children in America's Schools*, Crown Publishers, New York (1991).

[368] Jonathan Kozol, *Amazing Grace: The Lives of Children and the Conscience of a Nation*, Perennial, New York (2000).

[369] Jonathan Kozol, *Ordinary Resurrections: Children in the Years of Hope*, Perennial, New York (2001).

[370] Jonathan Kozol, *The Shame of the Nation: The Restoration of Apartheid Schooling in America*, Crown Publishers, New York (2005).

[371] Chris Kraft, *Flight : My Life in Mission Control*, E.P. Dutton, New York (2001).

[372] A. V. Krebs, *The Corporate Reapers: The Book of Agribusiness*, Essential Books, P.O. Box 19405, Washington D.C. 20036 (1992).

[373] Albert Krebs, *The Agribusiness Examiner*, (available at http://www.ea1.com/CARP/).

[374] Rajaram Krishnan, Jonathan M. Harris, Neva R. Goodwin, Editors, *A Survey of Ecological Economics*, Island Press, Covelo, California (1995).

[375] Paul Krugman, "The End of Middle-Class America (and the Triumph of the Plutocrats)," *The New York Times Magazine*, Sunday (October 20, 2002).

[376] Paul Krugman, *The Great Unraveling: Losing Our Way in the New Century*, W. W. Norton & Company, New York (2003).

[377] James Howard Kunstler, *The Geography of Nowhere: The Rise and Decline of America's Man-made Landscape*, Simon & Schuster, New York (1993).

[378] James Howard Kunstler, *The Long Emergency: Surviving the Converging Catastrophes of the Twenty-First Century*, Atlantic Monthly Press, New York (2005).

[379] George Lakoff, Rafael E. Núñez, *Where Mathematics Comes From: How the Embodied Mind Brings Mathematics into Being*, Basic Books, New York (2000).

[380] George Lakoff, *Don't Think of an Elephant! Know Your Values and Frame the Debate*, Chelsea Green Publishing, White River Junction, Vermont (2004).

[381] George Lakoff, *Whose Freedom? The Battle Over America's Most Important Idea*, Farrar, Straus and Giroux, New York (2006).

[382] Philip J. Landrigan, M.D.; Herbert L. Needleman, M.D.; Mary Landrigan, M.P.A., *Raising Healthy Children in a Toxic World: 101 Smart Solutions for Every Family*, Rodale Press, Emmaus, P.A. (2001).

[383] Greg Langkamp, Joseph Hull, *Quantitative Reasoning and the Environment: Mathematical Modeling in Context*, Pearson, Prentice Hall, Upper Saddle River, New Jersey (2006).

[384] Mitch Lansky, *Beyond the Beauty Strip: Saving What's Left of Our Forests*, Tilbury House Publishers, Gardiner, Maine (1992).

[385] Marc Lappé, *Chemical Deception: The Toxic Threat to Health and the Environment*, Sierra Club Books, San Francisco (1991).

[386] Marc Lappé, Britt Bailey, *Against the Grain: Biotechnology and the Corporate Takeover of Your Food*, Common Courage Press, Monroe, Me (1998).

[387] Peter Latz, *Bushfires & Bushtucker*, IAD Press, P.O. Box 2531, Alice Springs, Northern Territory 0871, Australia (1996).

[388] Richard Leakey, Roger Lewin, *The Sixth Extinction: Biodiversity and its Survival*, Doubleday, New York (1995).

[389] Johannes Lehmann, Stephen Joseph, *Biochar for Environmental Management: Science and Technology*, Earthscan Publications Ltd., (2009).

[390] Richard C. Leone, Greg Anrig, Jr., Editors, *The War on Our Freedoms: Civil Liberties in an Age of Terrorism*, A Century Foundation Book, Public Affairs, New York (2003).

[391] Les Leopold, *The Man Who Hated Work and Loved Labor: The Life and Times of Tony Mazzocchi*, Chelsea Green Publishing Company, White River Junction, Vermont (2007).

[392] Greg LeRoy, *The Great American Jobs Scam: Corporate Tax Dodging and the Myth of Job Creation*, Berrett Koehler, San Francisco (2005).

[393] Lawrence Lessig, *Free Culture: How Big Media Uses Technology and the Law to Lock Down Culture and Control Creativity*, The Penguin Press, (2004).

[394] Steven Levy, *Crypto: How the Code Rebels Beat the Government–Saving Privacy in the Digital Age*, Penguin, New York (2001).

[395] Charles Lewis and the Center for Public Integrity, *The Buying of the Congress: How Special Interests Have Stolen Your Right to Life, Liberty, and the Pursuit of Happiness*, Avon Books, New York (1998).

[396] Charles Lewis and the Center for Public Integrity, *The Buying of the President 2000: The Authoritative Guide to the Big-Money Interests Behind This Year's Presidential Candidates*, Avon Books, New York (2000).

[397] Charles Lewis and the Center for Public Integrity, *The Cheating of America: How Tax Avoidance and Evasion by the Super Rich Are Costing the Country Billions – and What You Can Do About It*, William Morrow, New York (2001).

[398] Charles Lewis and the Center for Public Integrity, *The Buying of the President 2004: Who's Really Bankrolling Bush and His Democratic Challengers – and What They Expect in Return*, Perennial, New York (2004).

[399] Eric Lichtblau, *Bush's Law: The Remaking of American Justice*, Pantheon Books, New York (2008).

[400] James B. Lieber, *Rats in the Grain: The Dirty Tricks and Trials of Archer Daniels Midland The Supermarket to the World*, Four Walls Eight Windows, New York (2000).

[401] Bernard Lietaer, *The Future of Money: Creating New Wealth, Work and a Wiser World*, Century, London, England (2001).

[402] Bernard A. Lietaer, Stephen M. Belgin, *Of Human Wealth: Beyond Greed & Scarcity*, Human Wealth Books, Boulder, Colorado (2003).

[403] Robert J. Lifton, *The Nazi Doctors: Medical Killing and the Psychology of Genocide*, Basic Books, New York (1986).

[404] Rush Limbaugh, *The Way Things Ought to Be*, Pocket Books, New York (1992).

[405] Dave Lindorff, *This Can't Be Happening*, Common Courage Press, Monroe, Maine (2004).

[406] Thomas Linzey with Anneke Campbell, *BE THE Change: How to Get What you Want in Your Community*, Gibbs Smith, Layton, Utah (2009).

[407] Andrew Lipscomb, Albert Ellergy Bergh, editors, *The Writings of Thomas Jefferson: Memorial Edition*, 20 Vols., Thomas Jefferson Memorial Association, Washington, D.C. (1903–04).

[408] Bjørn Lomberg, *Skeptical Environmentalist: Measuring the Real State of the World*, Cambridge University Press, (2001).

[409] Richard Louv, *Last Child in the Woods: Saving Our Children from Nature-Deficit Disorder*, Algonquin Books of Chapel Hill, Chapel Hill, NC (2005).

[410] James Lovelock, *Healing Gaia: Practical Medicine for the Planet*, Harmony Books, New York (1991).

[411] Amory B. Lovins, L. Hunter Lovins, Florentin Krause, Wilfrid Bach, *Least-Cost Energy: Solving the CO_2 Problem*, Rocky Mountain Institute, Snowmass, Colorado (1981), reprint of the original Brick House edition with additional preface.

[412] Amory B. Lovins, E. Kyle Datta, Odd-Even Bustnes, Jonathan G. Koomey, Nathan J. Glasgow, *Winning the Oil Endgame: Innovation for Profits, Jobs, and Security*, Rocky Mountain Institute, Snowmass, Colorado (2004), available for individual use for free download at www.oilendgame.com.

[413] Amory B. Lovins, *Four Nuclear Myths: A commentary on Stewart Brand's "Whole Earth Discipline" and on similar writings*, downloadable from www.rmi.org the web site of the Rocky Mountain Institute, Snowmass, Colorado (October 13, 2009).

[414] Roger Lowenstein, *When Genius Failed: The Rise and Fall of Long-Term Capital Management*, Random House, New York (2000).

[415] Roger Lowenstein, *The End of Wall Street*, Penguin Press, New York (2010).

[416] Stan Luger, *Corporate Power, American Democracy, and the Automobile Industry*, Cambridge University Press, (2000).

[417] Howard F. Lyman, Glen Merzer, *Mad Cowboy: Plain Truth from the Cattle Rancher Who Won't Eat Meat*, Scribner, (2001).

[418] Arthur MacEwan, *Debt & Disorder: International Economic Instability & U.S. Imperial Decline*, Monthly Review Press, New York (1990).

[419] Wil McCarthy, *Hacking Matter: Levitating Chairs, Quantum Mirages, and the Infinite Weirdness of Programmable Atoms*, Basic Books, New York (2003).

[420] Robert W. McChesney, *Telecommunications, Mass Media, & Democracy: The Battle for the Control of U.S. Broadcasting, 1928–1935*, Oxford University Press, New York (1993).

[421] Robert W. McChesney, *Corporate Media and the Threat to Democracy*, Seven Stories Press, New York (1997).

[422] Robert W. McChesney, *Rich Media, Poor Democracy: Communication Politics in Dubious Times*, University of Illinois Press, Chicago, Illinois (1999).

[423] Robert W. McChesney, John Nichols, *Our Media Not Theirs: The Democratic Struggle Against Corporate Media*, Seven Stories Press, New York (2002).

[424] Robert W. McChesney, John Nichols, *The Death and Life of American Journalism: The Media Revolution that Will Begin the World Again*, Nation Books, Philadelphia, Pennsylvania (2010).

[425] Richard McCord, *The Chain Gang: One Newspaper versus the Gannett Empire*, University of Missouri Press, Columbia, Missouri (1996).

[426] Alice McKeown, Project Director, *Vital Signs 2009*, Worldwatch Institute, Washington D.C. (2009).

[427] Betsy McLean, Peter Elkind, *The Smartest Guys in the Room: The Amazing Rise and Scandalous Fall of Enron*, Portfolio, New York (2003).

[428] Alfred McCoy, *The Politics of Heroin in Southeast Asia*, Harper & Row, New York (1972).

[429] Alfred W. McCoy, *A Question of Torture: CIA Interrogation, from the Cold War to the War on Terror*, Metropolitan Books, New York (2006).

[430] Jack McCroskey, *Light Rail and Heavy Politics: How Denver Set About Reviving Public Transportation*, Tenlie Publishing, Denver, Colorado (2003).

[431] Ian L. McHarg, *A Quest for Life: An Autobiography*, John Wiley & Sons, New York (2006).

[432] Charles Mackay, LL.D., *Extraordinary Popular Delusions and the Madness of Crowds*, Farrar, Straus and Giroux, New York (1932), reprint of 1841 and 1852 editions.

[433] Angus Mackenzie, *Secrets: The CIA's War at Home*, University of California Press, Berkeley, California (1997).

[434] Bill McKibben, *The End of Nature*, Random House, New York (1989).

[435] Bill McKibben, *Hope, Human and Wild: True Stories of Living Lightly on the Earth*, Little, Brown and Company, New York (1995).

[436] John McMurtry, *Unequal Freedoms: The Global Market as an Ethical System*, Kumarian Press, West Hartford, Connecticut (1998).

[437] Arjun Makhijani, Howard Hu, Katherine Yih, editors, *Nuclear Wastelands*, The MIT Press, Cambridge, Massachusetts (1995).

[438] Arjun Makhijani, *Carbon-Free and Nuclear-Free: A Roadmap for U.S. Energy Policy*, available and downloadable from www.ieer.org, the web site of the Institute for Energy and Environmental Research, Takoma Park, MD (2008).

[439] Stacy Malkan, *Not Just a Pretty Face: The Ugly Side of the Beauty Industry*, New Society Publishers, (2007).

[440] Jerry Mander, Edward Goldsmith, *The Case Against the Global Economy and for a Turn Toward the Local*, Sierra Club Books, San Francisco, California (1996).

[441] Jerry Mander, *In the Absence of the Sacred*, Sierra Club Books, San Francisco, Calif. (1992).

[442] Lynn Margulis, *Symbiotic Planet: A New Look at Evolution*, Basic Books, New York (1998).

[443] Gerald Markowitz, David Rosner, *Deceit and Denial: The Deadly Politics of Industrial Pollution*, University of California Press, Berkeley (2003).

[444] Jim Marrs, *Rule by Secrecy*, Perennial, New York (2001).

[445] Chris Maser, *Forest Primeval: The Natural History of an Ancient Forest*, Sierra Club Books, San Francisco, California (1989).

[446] Peter Matthiessen, *In the Spirit of Crazy Horse*, Viking-Penguin, New York (1991).

[447] Philip Mattera, Anna Purinton, *Shopping for Subsidies: How Wal-Mart Uses Taxpayer Money to Finance Its Never-Ending Growth*, Good Jobs First, Washington D.C. (2004).

[448] Robert M. May, "Simple mathematical models with very complicated dynamics," *Nature* 261 (1976).

[449] Donella H. Meadows, Dennis L. Meadows, Jørgen Randers, William W. Behrens III, *The Limits to Growth: A Report for The Club of Rome's Project on the Predicament of Mankind*, Universe Books, New York (1972).

[450] Donella H. Meadows, Dennis L. Meadows, Jørgen Randers, *Beyond the Limits: Confronting Global Collapse, Envisioning a Sustainable Future*, Chelsea Green Publishing Company, Post Hills, Vermont (1992).

[451] Rachel Meeropol, *America's Disappeared: Secret Imprisonment, Detainees, and the "War on Terror,"* Seven Stories Press, New York (2005).

[452] Milton Meltzer, *Slavery: A World History*, Da Capo Press, Chicago, Illinois (1993).

[453] Gavin Menzies, *1421 : The Year China Discovered America*, Perennial, New York (2004).

[454] Abhay Mehta, *Power Play: A Study of the Enron Project*, Orient Longman, Mumbai, India (1999).

[455] U.S. Senator Lee Metcalf, Vic Reinemer, *Overcharge: How Electric Utilities Exploit and Mislead the Public, and What You Can Do About It*, David McKay Company, Inc., New York (1967).

[456] S. G. Mestrovic, *The Trials of Abu Ghraib: An Expert Witness Account of Shame and Honor*, Paradigm Publishers, Boulder, Colorado, U.S.A. (2007).

[457] Steven H. Miles, M.D., *Oath Betrayed: Torture, Medical Complicity, and the War on Terror*, Random House, New York (2006).

[458] Mark Crispin Miller, *Fooled Again: How the Right Stole the 2004 Election & Why They'll Steal the Next One Too (Unless We Stop Them)*, Basic Books, New York (2005).

[459] Mark Crispin Miller, *Loser Take All: Election Fraud and the Subversion of Democracy, 2000–2008*, Ig Publishing, Brooklyn, New York (2008).

[460] Lawrence Mishel, Jared Bernstein, John Schmitt, *The State of Working America: 2000/2001*, Cornell University Press, New York (2001).

[461] Stacy Mitchell, *Big-Box Swindle: The True Cost of Mega-Retailers and the Fight for America's Independent Businesses*, Beacon Press, Boston (2006).

[462] Kevin D. Mitnick, *The Art of Deception: Controlling the Human Element of Security*, Wiley Publishing, Inc., Indianapolis, Indiana (2002).

[463] Louis R. Mizell, Jr., *Invasion of Privacy: The Complete Home Security Guide to Personal Computers, Phone Calls, Tax Records, Voice Mail, Medical Records, Credit Cards, E-mail, Private Property*, A Berkeley Book, New York (1998).

[464] Russel Mokhiber, Robert Weissman, *Corporate Predators: The Hunt for Mega-Profits and the Attack on Democracy*, Common Courage Press, Monroe, Maine (1999).

[465] Mark Monmonier, *Spying with Maps*, University of Chicago Press, Chicago, Illinois (2002).

[466] Chris Mooney, *The Republican War on Science*, Basic Books, New York (2005).

[467] David W. Moore, *How to Steal an Election: The Inside Story of How George W. Bush's Brother and Fox Network Miscalled the 2000 Election and Changed the Course of History*, Nation Books, New York (2006).

[468] David B. Morris, *Earth Warrior: Overboard with Paul Watson and the Sea Shepherd Conservation Society*, Fulcrum Publishing, Golden, Colorado (1995).

[469] David Morris, *Seeing the Light: Regaining Control of Our Electricity System*, Institute for Local Self-Reliance, Minneapolis, Minnesota (2001).

[470] Douglas E. Morris, *It's a Sprawl World After All: The Human Cost of Unplanned Growth – and Visions of a Better Future*, New Society Publishers, Gabriola Island, BC, Cananda (2005).

[471] Marion Moses, *Designer Poisons: How to Protect Your Health and Home from Toxic Pesticides*, Pesticide Education Center, San Francisco, California (1995).

[472] Jim Motavalli, Sally Deneen, Ross Gelbspan, David Helvarg, Mark Hertsgaard, Orna Izakson, Kieran Mulvaney, Dick Russell, Colin Woodard, Gary Braasch, *Feeling the Heat: Dispatches from the Frontlines of Climate Change*, Routledge, New York (2004).

[473] Alicia Mundy, *Dispensing With the Truth: The Victims, the Drug Companies, and the Dramatic Story Behind the Battle over Fen-Phen*, St. Martins Press, New York (2001).

[474] Barry Muslow with Yemi Katerere, Adriaan Ferf, Phil O'Keefe, *The Fuelwood Trap: A Study of the SADCC Region*, Earthscan Publications, International Institute for Environment and Development, London (1988).

[475] Norman Myers, *The Sinking Ark*, Pergamon Press, Oxford, England (1980).

[476] Gunnar Myhre, Eleanor J. Highwood and Keith P. Shine, Frode Stordal, "New estimates of radiative forcing due to well mixed greenhouse gases," *Geophysical Research Letters*, Vol. 25, No. 14, pp. 2715–18, July 15, 1998.

[477] Ted Nace, *Gangs of America: The Rise of Corporate Power and the Disabling of Democracy*, Berrett-Koehler Publishers, Inc., San Francisco, California (2003).

[478] Robert Nadeau, *The Wealth of Nature*, Columbian University Press (2003).

[479] Ralph Nader, Wesley J. Smith, *No Contest: Corporate Lawyers and the Perversion of Justice in America*, Random House, New York (1996).

[480] National Commission on Terrorist Attacks Upon the United States, *The 9/11 Commission Report*, W.W. Norton & Company, Inc., New York (2004).

[481] National Research Council, *Pesticides in the Diets of Infants and Children*, National Academy Press, Washington D.C. (1993).

[482] Loretta Napoleoni, *Terror Incorporated: Tracing the Dollars Behind the Terror Networks*, Seven Stories Press, New York (2005).

[483] Scott Nearing, *The Making of a Radical: A Political Autobiography*, Social Science Institute, Harborside, Maine (1972).

[484] Jack Nelson-Pallmeyer, *School of Assassins: Guns, Greed, and Globalization*, Orbis Books, Maryknoll, New York (2001).

[485] David Nelson, George Gheverghese Joseph, Julian Williams, *Multicultural Mathematics, Teaching Mathematics from a Global Perspective*, Oxford University Press (1993).

[486] Don A. Nelson, *NASA New Millenium Problems and Solutions: A Shocking Disclosure of NASA's Space Transportation Problems*, Xlibris Corporation, www.Xlibris.com (2001).

[487] John Nerone, *Violence Against the Press*, Oxford University Press, New York (1994).

[488] Marion Nestle, *Safe Food, Bacteria, Biotechnology, and Bioterrorism*, University of California Press, Berkeley (2003).

[489] Marion Nestle, *Food Politics: How the Food Industry Influences Nutrition and Health*, Univ. of California Press, Berkeley (2002).

[490] Marion Nestle, *What to Eat: An Aisle-by-Aisle Guide to Savvy Food Choices and Good Eating*, North Point Press, Farrar, Straus and Giroux, New York (2006).

[491] John Von Neumann, Oskar Morgenstern, *Theory of Games and Economic Behavior*, Princeton University Press, Princeton, New Jersey (1953).

[492] Peter Newman, Jeffrey Kenworthy, *Sustainability and Cities: Overcoming Automobile Dependence*, Island Press, Covelo, California (1999).

[493] Ronald E. Ney, Jr., *Where Did that Chemical Go? A Practical Guide to Chemical Fate and Transport in the Environment*,

[494] Andrew Nikiforuk, *Tar Sands: Dirty Oil and the Future of a Continent*, Greystone Books, Vancouver, Canada (2008).

[495] Helena Norberg-Hodge, Todd Merrifeld, Steven Goerlick, *Bringing the Food Economy Home: Local Alternatives to Global Agribusiness*, Zed Books, London; Kumarian Press, Inc., Bloomfield, Ct.; Fernwood Publishing Co., Ltd, Halifax, Canada (2002).

[496] Dom Nozzi, *Road to Ruin: An Introduction to Sprawl and How to Cure It*, Praeger, Westport, Connecticut (2003).

[497] Robyn O'Brien, *The Unhealthy Truth: How Our Food Is Making Us Sick – and What We Can Do About It*, Broadway Books, New York (2009).

[498] Peter C. Ordeshook, *Game Theory and Political Theory: An Introduction*, Cambridge University Press, (1986). Van Nostrand Reinhold (Thomson Learning), New York (1990).

[499] Leonard Orland, *Prisons: Houses of Darkness*, The Free Press, New York (1975).

[500] David W. Orr, *Ecological Literacy*, State University of New York Press, (1992).

[501] Bob Ortega, *In Sam We Trust: The Untold Story of Sam Walton and How Wal-Mart Is Devouring America*, Random House, New York (1998).

[502] Sister Dianna Ortiz, *The Blindfold's Eyes: My Journey from Torture to Truth*, Orbis Books, Maryknoll, New York (2002).

[503] Howard T. Odum, *Ecological and General Systems: An Introduction to Systems Ecology Revised Edition*, The University Press of Colorado, (1994)

[504] Jack Olsen, *Last Man Standing: The Tragedy and Triumph of Geronimo Pratt*, Doubleday, New York (2000).

[505] Barbara Olshansky, *Secret Trials and Executions: Military Tribunals and the Threat to Democracy*, Seven Stories Press, New York (2002).

[506] Paul Ormerod, *Butterfly Economics: A New General Theory of Social and Economic Behavior*, Pantheon Books, New York (2000).

[507] Michael T. Osterhold, John Schwartz, *Living Terrors: What America Needs to Know to Survive the Coming Bioterrorist Catastrophe*, Dekacorte Press, New York (2000).

[508] Elinor Ostrom, *Governing the Commons: The Evolution of Institutions for Collective Action*, Cambridge University Press, (1990).

[509] Elinor Ostrom and Charlotte Hess, editors, *Understanding Knowledge as a Commons: From Theory to Practice*, MIT Press, (2006).

[510] Riki Ott, *Sound Truth and Corporate Myth$: The Legacy of the Exxon Valdez Oil Spill*, Dragonfly Sisters Press, Cordova, Alaska (2005).

[511] Riki Ott, *Not One Drop: Betrayal and Courage in the Wake of the Exxon Valdez Oil Spill*, Chelsea Green Publishing, White River Junction, VT (2008).

[512] Ursula Owen, Editor in Chief, *Index on Censorship: The Privacy Issue*, Volume 29, No. 3, Issue 194 (May/June 2000).

[513] Trevor Paglen, A.C. Thompson, *Torture Taxi: On the Trail of the CIA's Rendition Flights*, Melville House Publishing, Hoboken, New Jersey (2006).

[514] Greg Palast, *The Best Democracy Money Can Buy: The Truth About Corporate Cons, Globalization, and High-Finance Fraudsters*, Plume, Penguin Putnam, Inc., New York (2003).

[515] Keith Coates Palgrave, *Trees of Southern Africa*, Struik Publishers, Cape Town, South Africa (1993).

[516] Christian Parenti, *Lockdown America: Police and Prisons in the Age of Crisis*, Verso, New York (2000).

[517] Christian Parenti, *The Soft Cage: Surveillance in America*, Basic Books, New York (2003).

[518] Michael Parenti, *Inventing Reality: The Politics of the Mass Media*, St. Martin's Press, New York (1986).

[519] Robert Parry, *Trick or Treason: The October Surprise Mystery*, Sheridan Square Press, New York (1993).

[520] Robert Parry, *Lost History: Contras, Cocaine, The Press and Project Truth*, The Media Consortium, Arlington, Virginia (1999).

[521] Raj Patel, *Stuffed & Starved: The Hidden Battle for the World Food System*, Melville House Publishing, Brooklyn, New York (2007).

[522] John Allen Paulos, *Innumeracy: Mathematical Illiteracy and Its Consequences*, Hill and Wang, New York (1998).

[523] Andrea Peacock, *Libby, Montana: Asbestos & the Deadly Silence of an American Corporation*, Johnson Books, Boulder, Colorado (2003).

[524] Doug Peacock, *The Grizzly Years: In Search of the American Wilderness*, Henry Holt & Co., New York (1990).

[525] Adam L. Penenberg, *Tragic Indifference: One Man's Battle with the Auto Industry over the Dangers of SUVs*, HarperBusiness, New York (2003).

[526] William F. Pepper, *Orders to Kill: The Truth Behind the Murder of Martin Luther King*, Carroll & Graf Publishers, Inc., New York (1995).

[527] William F. Pepper, *An Act of State: The Execution of Martin Luther King*, Verso, New York (2003).

[528] William F. Pepper, *An Act of State: The Execution of Martin Luther King, New and Updated Edition*, Verso, New York (2008).

[529] John Perkins, *Confessions of an Economic Hit Man*, Penguin, New York (2004).

[530] John Perkins, *The Secret History of the American Empire*, Dutton, New York (2007).

[531] John Perlin, *A Forest Journey: The Role of Wood in the Development of Civilization*, W. W. Norton, New York (1989).

[532] Kevin Phillips, *Bad Money: Reckless Finance, Failed Politics, and the Global Crisis of American Capitalism*, Viking/Penguin, New York, 2008.

[533] Peter Phillips, Project Censored, *Censored 1999: The News that Didn't Make the News – The Year's Top 25 Censored Stories*, Seven Stories Press, New York (1999).

[534] Peter Philips, Andrew Roth, *Censored 2009: The Top 25 Censored Stories of 2007–08*, Seven Stories Press, New York (2008).

[535] Peter Phillips, Mickey Huff, *Censored 2010*, Seven Stories Press, New York (2010).

[536] Gary M. Pierzynski, Thomas J. Sims, George F. Vance, *Soils and Environmental Quality*, Lewis Publishers, Boca Raton, Florida (1994).

[537] David Pimentel, Marcia Pimentel, editors, *Food, Energy, and Society, Revised Edition*, University Press of Colorado (1996).

[538] Stuart Pimm, Jeff Harvey, "No need to worry about the future," *Nature*, vol. 414, (November 8, 2001).

[539] Madsen Pirie, *The Book of the Fallacy–A Training Manual for Intellectual Subversives*, Routledge & Kegan Paul, (1985).

[540] Francis Fox Piven, Richard A. Cloward, *The Breaking of the American Social Compact*, The New Press, New York (1997).

[541] Francis Fox Piven, Richard A. Cloward, *Why Americans Still Don't Vote and Why Politicians Want It That Way*, Beacon Press, Boston, Massachusetts (2000).

[542] Frances Fox Piven, *The War at Home: The Domestic Costs of Bush's Militarism*, The New Press, New York (2004).

[543] Michael Pollan, *The Omnivore's Dilema: A Natural History of Four Meals*, The Penguin Press, New York (2006).

[544] Michael Pollan, *In Defense of Food: An Eater's Manifesto*, The Penguin Press, New York (2008).

[545] Norman Polmar, Thomas B. Allen, *Spy Book: The Encyclopedia of Espionage*, Random House, New York (1997).

[546] Clive Ponting, *A Green History of the World: The Environment and the Collapse of Great Civilizations*, Penguin Books, New York (1993).

[547] Clive Ponting, *A New Green History of the World: The Environment and the Collapse of Great Civilizations*, Penguin Books, New York (2007).

[548] Mary Poovey, "Can Numbers Ensure Honesty? Unrealistic Expectations and the U.S. Accounting Scandal," *Notices of the American Mathematical Society*, Vol. 50, Number 1, (January 2003).

[549] Wendell Potter, *Deadly Spin: An Insurance Company Insider Speaks Out on How Corporate PR is Killing Health Care and Deceiving Americans*, Bloomsbury Press, New York (2010).

[550] Richard Gid Powers, *Broken: The Troubled Past and Uncertain Future of the FBI*, The Free Press, New York (2004).

[551] Marc Pratarelli, *Niche Bandits: Why BIG Brains Consumed an Ecosystem*, Medici Publishing, Inc., Pueblo, Colorado (2003).

[552] Richard Preston, *The Hot Zone*, Random House, (1994).

[553] David H. Price, *Threatening Anthropology: McCarthyism and the FBI's Surveillance of Activist Anthropologists*, Duke University Press, Durham, North Carolina (2004).

[554] George W. Pring, Penelope Canan, *SLAPPs: Getting Sued for Speaking Out*, Temple University Press, Philadelphia, Pennsylvania (1996).

[555] Nomi Prins, *It Takes a Pillage: Behind the Bailouts, Bonuses, and Backroom Deals from Washington to Wall Street*, Wiley, (2009).

[556] Nomi Prins, Christopher Hayes, "Meet the Hazzards: If banks were people, here's what the full $17.5 trillion bailout would look like," *The Nation*, Oct. 12, 2009, pp.16–21.

[557] Christopher H. Pyle, *Military Surveillance of Civilian Politics [1967–1970]*, Garland Publishing, Inc., New York (1986).

[558] Stephen J. Pyne, *Fire In America: A Cultural History of Wildland and Rural Fire*, University of Washington Press, Seattle, Washington (1999).

[559] Bill Quinn, *How Wal-Mart is Destroying America: And What You Can Do About It*, Ten Speed Press, Berkeley, California (1998).

[560] Paul Raeburn, *The Last Harvest: The Genetic Gamble that Threatens to Destroy American Agriculture*, Simon & Schuster, New York (1995).

[561] Sheldon Rampton, John Stauber, *Toxic Sludge is Good for You! Lies, Damn Lies and the Public Relations Industry*, Common Courage Press, Monroe, Maine (1995).

[562] Sheldon Rampton, John Stauber, *Mad Cow U.S.A.: Could the Nightmare Happen Here?*, Common Courage Press, Monroe, Maine (1997).

[563] Sheldon Rampton, John Stauber, *Trust Us, We're Experts! How Industry Manipulates Science and Gambles with Your Future*, Tarcher/Putnam, New York (2001).

[564] David Ransom, *The No-Nonsense Guide to Fair Trade*, Verso, London, United Kingdom (2001).

[565] Doris J. Rapp, M.D., *Is This Your Child?: Discovering and Treating Unrecognized Allergies in Children and Adults*, Quill, William Morrow, New York (1991).

[566] Doris J. Rapp, M.D., *Is This Your Child's World?*, Bantam Books, New York (1996).

[567] Doris J. Rapp, M.D., *Our Toxic World: A Wake Up Call*, Environmental Medical Research Foundation, Buffalo, New York (2004).

[568] William Rathje, Cullen Murphy, *Rubbish! The Archaeology of Garbage: What Our Garbage Tells Us about Ourselves*, HarperCollins, New York (1992).

[569] Jim Redden, *Snitch Culture: How Citizens Are Turned into the Eyes and Ears of the State*, Feral House, Los Angeles, California (2000).

[570] Andy Rees, *Genetically Modified Food: A Short Guide for the Confused*, Pluto Press, Ann Arbor, Michigan (2006).

[571] T. R. Reid, *The Healing of America: A Global Quest for Better, Cheaper, and Fairer Health Care*, The Penguin Press, New York (2009).

[572] Steve Rendall, Jim Naureckas, Jeff Cohen, *The Way Things Aren't: Rush Limbaugh's Reign of Error: Over 100 Outrageously False and Foolish Statements from America's Most Powerful Radio and TV Commentator*, The New Press, New York (1995).

[573] Michael Renner, *The Anatomy of Resource Wars*, Worldwatch Paper 162, Worldwatch Institute, Washington, D.C. (October, 2002).

[574] Susan Reverby, *Examining Tuskegee: The Infamous Syphilis Study and Its Legacy*, The University of North Carolina Press, Chapel Hill, North Carolina (2009).

[575] John F. Richards, Richard P. Tucker, *World Deforestation in the Twentieth Century*, Duke University Press, Durham (1988).

[576] Jeremy Rifkin, *The Hydrogen Economy: The Creation of the Worldwide Energy Web and the Redistribution of Power on Earth*, Tarcher/Penguin Books, New York (2003).

[577] James Risen, *State of War: The Secret History of the CIA and the Bush Administration*, Free Press, Simon & Schuster, New York (2006).

[578] Dean Ritz, Editor, *Defying Corporations, Defining Democracy: A Book of History & Strategy*, The Apex Press, New York (2001).

[579] Sharon L. Roan, *Ozone Crisis: The 15 Year Evolution of a Sudden Global Emergency*, John Wiley & Sons, New York (1990).

[580] Randall Robinson, *The Debt: What America Owes to Blacks*, Dutton, New York (2000).

[581] Heather Rogers, *Gone Tomorrow: The Hidden Life of Garbage*, The New Press, New York (2005).

[582] David Rose, *Guantánamo: The War on Human Rights*, The New Press, New York (2004).

[583] Stephen J. Rose, *Social Stratification in the United States*, The New Press, New York (2000).

[584] Stephen M. Rosoff, Henry N. Pontell, Rofert H. Tillman, *Profit Without Honor: White-Collar Crime and the Looting of America*, Prentice Hall, New Jersey (2002).

[585] Stephen M. Rosoff, Henry N. Pontell, Robert H. Tillman, *Profit Without Honor: White-Collar Crime and the Looting of America, 5th edition*, Prentice Hall, New Jersey (2009).

[586] John Ross, *El Monstruo: Dread and Redemption in Mexico City*, Nation Books, New York (2009).

[587] Jeffrey Rothfeder, *Every Drop for Sale: Our Desperate Battle Over Water in a World About to Run Out*, Tarcher/Putnam, New York (2001).

[588] Andrew Rowell, *Don't Worry [It's Safe to Eat]: The True Story of GM Food, BSE and Foot and Mouth*, Earthscan Publications, Inc., Sterling, Virginia (2004).

[589] Elizabeth Royte, *Garbage Land: On the Secret Trail of Trash*, Back Bay Books, New York (2005).

[590] Elizabeth Royte, *Bottlemania: How Water Went on Sale and Why We Bought It*, Bloomsbury, New York (2008).

[591] Michael C. Ruppert, *Crossing the Rubicon: The Decline of the American Empire at the End of the Age of Oil*, New Society Publishers, Gabriola Island, BC, Canada (2004).

[592] Erik Saar, Viveca Novak, *Inside the Wire: A Military Intelligence Soldier's Eyewitness Account of Life at Guantánamo*, Penguin Press, New York (2005).

[593] Donald G. Saari, *Chaotic Elections! A Mathematician Looks at Voting*, American Mathematical Society (2001).

[594] Donald G. Saari, *Decisions and Elections: Explaining the Unexpected*, Cambridge University Press, New York (2001).

[595] Jeffrey St. Clair, *Grand Theft Pentagon: Tales of Corruption and Profiteering in the War on Terror*, Common Courage Press, Monroe, Maine (2005).

[596] Kirkpatrick Sale, *Rebels Against the Future: The Luddites and Their War on the Industrial Revolution: Lessons for the Computer Age*, Addison-Wesley, New York (1995).

[597] Arturo Sangalli, *The Importance of Being Fuzzy and Other Insights from the Border Between Math and Computers*, Princeton University Press, Princeton, New Jersey (1998).

[598] Janet L. Sawin, *Mainstreaming Renewable Energy in the 21st Century*, WorldWatch Paper 169, May (2004).

[599] Janet L. Sawin, Ph.D., Project Director and Senior Author, *American Energy: The Renewable Path to Energy Security*, Worldwatch Institute, Washington, D.C. (2006).

[600] Jeremy Scahill, *Blackwater: The Rise of the World's Most Powerful Mercenary Army*, Nation Books, New York (2007).

[601] Mark Schapiro, David Weir, *Circle of Poison: Pesticides and People in a Hungry World*, Food First Books, (1981).

[602] Mark Schapiro, *Exposed: The Toxic Chemistry of Everyday Products and What's at Stake for American Power*, Chelsea Green Publishing, White River Junction, Vermont (2007).

[603] Danny Schechter, *The More You Watch the Less You Know*, Seven Stories Press, New York (1997).

[604] Ted Schettler, M.D., Gina Solomon, M.D., Maria Valenti, Annette Huddle, *Generations at Risk: Reproductive Health and the Environment*, The M.I.T. Press, Cambridge, MA (2000).

[605] Arthur M. Schlesinger, Jr., *The Cycles of American History*, Mariner Books, Houghton Mifflin, New York (1999).

[606] Eric Schlosser, *Fast Food Nation: The Dark Side of the All-American Meal*, Houghton Mifflin, Boston, New York (2001).

[607] Hans Schmidt, *Maverick Marine: General Smedley D. Butler and the Contradictions of American Military History*, University Press of Kentucky, Lexington (1987).

[608] Andrew Schneider, David McCumber, *An Air That Kills: How the Asbestos Poisoning of Libby, Montana, Uncovered a National Scandal*, G.P. Putnam's Sons, New York (2004).

[609] Judith Scherff, Editor, *The Piracy of America: Profiteering in the Public Domain*, Clarity Press, Inc., Atlanta, Georgia (1999).

[610] Stephen Schneider, John P. Holdren, John Bongaarts, Thomas Lovejoy, "Misleading Math about the Earth," *Scientific American*, (January 2002).

[611] Juliet B. Schor, *The Overworked American: The Unexpected Decline of Leisure*, Basic Books, New York (1992).

[612] Juliet B. Schor, *The Overspent American: Why We Want What We Don't Need*, HarperPerennial, New York (1998).

[613] Ellen W. Schrecker, *No Ivory Tower: McCarthyism in the Universities*, Oxford University Press, New York (1986).

[614] Manfred Schroeder, *Fractals, Chaos, Power Law: Minutes from an Infinite Paradise*, W. H. Freeman, New York (1991).

[615] Jim Shultz, *The Democracy Owner's Manual: A Practical Guide to Changing the World*, Rutgers University Press, New Jersey (2003).

[616] Jim Shultz, Melissa Crane Draper, editors, *Dignity and Defiance: Stories from Bolivia's Challenge to Globalization*, University of California Press, Berkeley, California (2008).

[617] Robert Schultz, Ruth Schultz, *It Did Happen Here: Recollections of Political Repression in America*, University of California Press, Berkeley, California (1989).

[618] Maxime Schwartz, Tranlated by Edward Schneider, *How the Cows Turned Mad*, University of California Press, Berkeley and Los Angeles, California (2003).

[619] Richard H. Schwartz, "A Simple Mathematical Model for Population Growth," *Journal of Environmental Education*, Vol 12, No. 2. Winter (1980–81).

[620] Richard H. Schwartz, *Mathematics and Global Survival, Fourth Edition*, Ginn Press (1998).

[621] Frank Schweitzer, editor, *Self-Organization of Complex Structures: From Individual to Collective Dynamics*, Gordon and Breach Science Publishers, Amsterdam (1997).

[622] Elliott D. Sclar, *You Don't Always Get What You Pay For: The Economics of Privatization*, Cornell University Press, Ithaca, New York (2001).

[623] Peter Dale Scott, Jonathan Marshall, *Cocaine Politics; Drugs, Armies, and the CIA in Central America*, University of California Press, Berkeley, California (1991).

[624] Peter Dale Scott, *The Road to 9/11: Wealth, Empire, and the Future of America*, University of California Press, Berkeley, California (2007).

[625] Paul Seabright, *The Company of Strangers : A Natural History of Economic Life*, Princeton University Press, Princeton (2004).

[626] Susan Starr Sered, Rushika Fernandopulle, *Uninsured in America: Life & Death in the Land of Opportunity*, University of California Press, Berkeley, California (2005).

[627] Philip Shabecoff, Alice Shabecoff, *Poisoned Profits: The Toxic Assault on Our Children*, Random House, New York (2008).

[628] Judith Shapiro, *Mao's War Against Nature: Politics and the Environment in Revolutionary China*, Cambridge University Press, New York (2001).

[629] Robert Sherrill, *First Amendment Felon: The Story of Frank Wilkinson, His 132,000-Page FBI File, and His Epic Fight for Civil Rights and Liberties*, Nation Books, New York (2005).

[630] David K. Shipler, *The Working Poor: Invisible in America*, Vintage Books, New York (2005).

[631] Vandana Shiva, *Stolen Harvest: The Hijacking of the Global Food Supply*, South End Press, New York (2000).

[632] Vandana Shiva, *Water Wars: Privatization, Pollution, and Profit*, South End Press, Cambridge, MA (2002).

[633] Tim Shorrock, *Spies for Hire: The Secret World of Intelligence Outsourcing*, Simon & Schuster, (2001).

[634] Michael H. Shuman, *Going Local: Creating Self-Reliant Communities in a Global Age*, Routledge, (2000).

[635] Michael H. Shuman, *The Small-Mart Revolution: How Local Businesses Are Beating the Global Competition, 2nd edition*, Berrett-Koehler Publishers, (2007).

[636] Ken Silverstein, *Washington on $10 Million a Day: How Lobbyists Plunder the Nation*, Common Courage Press, Monroe, Me (1998).

[637] Julian Lincoln Simon, *The Ultimate Resource 2*, Princeton University Press, Princeton, New Jersey (1998).

[638] Charles Singer, *A Short History of Scientific Ideas to 1900*, Oxford University Press, London (1959).

[639] Rebecca Skloot, *The Immortal Life of Henrietta Lacks*, Crown Publishers, New York (2010).

[640] Stephen Sloan, *Ocean Bankruptcy: World Fisheries on the Brink of Disaster*, The Lyons Press, Guilford, CT (2003).

[641] Vaclav Smil, *Energies: An Illustrated Guide to the Biosphere and Civilization*, The MIT Press, Cambridge, Massachusetts (1999).

[642] Bruce Smith, *Insurmountable Risks: The Dangers of Using Nuclear Power to Combat Global Climate Change*, IEER Press, Takoma Park, Maryland, and RDR Press, Muskegon, Michigan (2006).

[643] D.E. Smith, *History of Mathematics, Volume 1*, Dover Publications, Inc., New York (1958).

[644] Jeffrey M. Smith, *Seeds of Deception: Exposing Industry and Government Lies About the Safety of the Genetically Engineered Foods You're Eating*, Yes! Books, Fairfield, Iowa (2003).

[645] Jeffrey M. Smith, *Genetic Roulette: The Documented Health Risks of Genetically Engineered Foods*, Yes! Books, Fairfield, Iowa (2007).

[646] Bradford Snell, *American Ground Transport: A Proposal for Restructuring the Automobile, Truck, Bus and Rail Industries*, a report submitted to the U. S. Congress, Senate Judiciary Committee, Subcommittee on Antitrust and Monopoly (February 26, 1974).

[647] Frank Snepp, *Irreparable Harm: A Firsthand Account of How One Agent Took on the CIA in an Epic Battle Over Secrecy and Free Speech*, Random House, New York (1999).

[648] Nancy Snow, *Propaganda, Inc.*, Seven Stories Press, New York (1998).

[649] Nancy Snow, *Information War: American Propaganda, Free Speech and Opinion Control Since 9-11*, Seven Stories Press, New York (2003).

[650] Ricard Solé, Brian Goodwin, *Signs of Life: How Complexity Pervades Biology*, Basic Books, New York (2000).

[651] Lawrence C. Soley, *Leasing the Ivory Tower: The Corporate Takeover of Academia*, South End Press, Boston, Massachusetts (1995).

[652] Lewis D. Solomon, *Rethinking Our Centralized Monetary System: A Case for a System of Local Currencies*, Praeger, Westport, Connecticut (1996).

[653] Didier Sornette, *Why Stock Markets Crash: Critical Events in Complex Financial Systems*, Princeton University Press, Princeton, New Jersey (2003).

[654] George Soros, *The Crisis of Global Capitalism*, Perseus Books, New York (1998).

[655] Michael Soulé, Editor, *Conservation Biology: the Science of Scarcity and Diversity*, Sinauer Associates, Inc., Sunderland, Massachusetts (1986).

[656] Wole Soyinka, *Climate of Fear*, Random House, New York (2005).

[657] Jonathan D. Spence, *The Search for Modern China*, W. W. Norton & Company, New York (1990).

[658] Vivien Spitz, *Doctors from Hell: The Horrific Account of Nazi Experiments on Humans*, Sentient Publications, LLC, Boulder, Colorado (2005).

[659] Greg Spotts, *Wal-Mart: The High Cost of Low Price*, The Disinformation Company, www.disinfo.com, New York (2005).

[660] A. M. Starfield, Karl A. Smith, A. L. Bleloch, *How to Model It: Problem Solving for the Computer Age*, McGraw Hill, New York (1990).

[661] A. M. Starfield, A. L. Bleloch, *Building Models for Conservation and Wildlife Management*, Distributed by Burgess International Group, Inc., Edina, Minnesota (1991).

[662] Linda Starke, editor, *Vital Signs 2006-2007: The Trends That Are Shaping Our Future*, The Worldwatch Institute, W.W. Norton and Company, New York, (2006).

[663] Linda Starke, editor, *2009 State of the World*, The Worldwatch Institute, W.W. Norton and Company, New York (2009).

[664] Jerold M. Starr, *Air Wars: The Fight to Reclaim Public Broadcasting*, Beacon Press, Boston, Massachusetts (2000).

[665] John Stauber, Sheldon Rampton, *Toxic Sludge Is Good for You!: Lies, Damn Lies and the Public Relations Industry*, Common Courage Press, Monroe, Me (1995).

[666] Rick Steiner, Kurt Byers, *Lessons of the Exxon Valdez*, Alaska Sea Grant College Program, University of Alaska Fairbanks, Fairbanks, Alaska (1990).

[667] Sandra Steingraber, *Living Downstream: An Ecologist Looks at Cancer and the Environment*, Addison Wesley, New York (1997).

[668] Philip M. Stern, *Still the Best Congress Money Can Buy*, Regnery Gateway, Washington D.C. (1992).

[669] Joseph E. Stiglitz, *Freefall: America, Free Markets, and the Sinking of the World Economy*, W. W. Norton & Company, New York (2010)

[670] Robert B. Stinnett, *Day of Deceit: The Truth About FDR and Pearl Harbor*, Touchstone, Simon & Schuster, Inc., New York (2000).

[671] Philip Straffin, *Game Theory and Strategy*, Mathematical Association of America, Washington, D.C. (1995).

[672] Steven Strasser, *The Abu Ghraib Investigations: The Official Reports of the Independent Panel and the Pentagon on the Shocking Prisoner Abuse in Iraq*, Public Affairs, New York (2004).

[673] Carol Van Strum, *A Bitter Fog: Herbicides and Human Rights*, Sierra Club Books, San Francisco (1983).

[674] Tristram Stuart, *Waste: Uncovering the Global Food Scandal*, W. W. Norton & Company, New York (2009).

[675] James Surowiecki, *The Wisdom of Crowds: Why the Many Are Smarter than the Few and How Collective Wisdom Shapes Business, Economies, Societies, and Nations*, Doubleday, New York (2004).

[676] Ron Suskind, *The Price of Loyalty: George W. Bush, the White House, and the Education of Paul O'Neill*, Simon & Schuster, New York (2004).

[677] M. Wesley Swearingen, *FBI Secrets: An Agent's Exposé*, South End Press, Boston, Massachusetts (1998).

[678] Joseph Tainter, *The Collapse of Complex Societies*, Cambridge University Press, Cambridge, U.K. (1988).

[679] Pete Takeda, *An Eye at the Top of the World: The Terrifying Legacy of the Cold War's Most Daring CIA Operation*, Thunder's Mouth Press, New York (2006).

[680] Terry Tamminen, *Lives Per Gallon: The True Cost of Our Oil Addiction*, Island Press, Covelo, California (2006).

[681] Alan D. Taylor, *Mathematics and Politics: Strategy, Voting, Power and Proof*, Springer-Verlag, New York (1995).

[682] Edward Tenner, *Why Things Bite Back: Technology and the Revenge of Unintended Consequences*, Alfred A. Knopf, New York (1996).

[683] Lester C. Thurow, *The Future of Capitalism*, Penguin Group, New York (1996).

[684] Will Toor, Spenser W. Havlick, *Transportation & Sustainable Campus Communities: Issues, Examples, Solutions*, Island Press, Covelo, California (2004).

[685] Edward R. Tufte, *The Visual Display of Quantitative Information*, Graphics Press, Cheshire, Connecticut (1983).

[686] Edward R. Tufte, *Envisioning Information*, Graphics Press, Cheshire, Connecticut (1990).

[687] Edward R. Tufte, *Visual Explanations: Images and Quantities, Evidence and Narrative*, Graphics Press, Cheshire, Connecticut (1997).

[688] B.L.Turner II, William C. Clark, Robert W. Kates, John F. Richards, Jessica T. Mathews, and William B. Meyer, *The Earth as Transformed by Human Action: Global and Regional Changes in the Biosphere over the Past 300 Years*, Cambridge University Press, New York (1993).

[689] Christy G. Turner II, Jacqueline Turner, *Man Corn: Cannibalism and Violence in the Prehistoric American Southwest*, University of Utah Press (1998).

[690] Nick Turse, *The Complex: How the Military Invades our Everday Lives*, Metropolitan Books, Henry Holt and Company, New York (2008).

[691] John Vandermeer, Ivette Perfecto, *Breakfast of Biodiversity: The Political Ecology of Rain Forest Destruction*, Food First Books, Oakland, California (2005).

[692] Stewart L. Udall, *The Quiet Crisis*, Avon Books (1964).

[693] Vladimir I. Vernadsky, *The Biosphere*, Copernicus, Springer-Verlag, New York (1997).

[694] Marq de Villiers, *Water*, Stoddart Publishing Co. Limited, Toronto, Canada (1999).

[695] Peter M. Vitousek, Paul R. Ehrlich, Anne H. Ehrlich, Pamela A. Matson, "Human Appropriation of the Products of Photosynthesis," *BioScience*, Vol. 36, No. 6, pp. 368–373, JSTOR, (June 1986).

[696] Mathis Wackernagel, William Rees, *Our Ecological Footprint: Reducing Human Impact on the Earth*, New Society Publishers, Gabriola Island, British Columbia, Canada (1996).

[697] Lori Wallach, Michelle Sforza, *Whose Trade Organization: Corporate Globalization and the Erosion of Democracy*, Public Citizen, (www.citizen.org), Washington D.C. (1999).

[698] Martin E. Walter, "Earthquakes and Weatherquakes: Mathematics and Climate Change," *Notices of the American Mathematical Society*, Volume 57, Number 10, pp. 1278–1284, November, (2010).

[699] Diane Raines Ward, *Water Wars: Drought, Flood, Folly, and the Politics of Thirst*, Riverhead Books, New York (2002).

[700] Peter D. Ward, Donald Brownlee, *Rare Earth: Why Complex Life Is Uncommon in the Universe*, Copernicus, New York (2000).

[701] Peter D. Ward, *Gorgon: Paleontology, Obsession, and the Greatest Catastrophe in Earth's History*, Viking, Penguin Group, New York (2004).

[702] John Wargo, *Our Children's Toxic Legacy: How Science and Law Fail to Protect Us from Pesticides*, Yale University Press, New Haven (1996).

[703] Christian Warren, *Brush with Death: A Social History of Lead Poisoning*, Johns Hopkins University Press, Baltimore (2000).

[704] Elizabeth Warren, Amelia Warren Tyagi, *The Two-Income Trap: Why Middle-Class Mothers & Fathers Are Going Broke*, Basic Books, New York (2003).

[705] Jennifer Washburn, *University Inc.: The Corporate Corruption of Higher Education*, Basic Books, Cambridge, Massachusetts (2005).

[706] Harvey Wasserman, *The Last Energy War: The Battle Over Utility Deregulation*, Seven Stories, New York (1999).

[707] Harry L. Watson, *Andrew Jackson vs. Henry Clay: Democracy and Development in Antebellum America*, Bedford/St. Martin's, Boston, Massachusetts (1998).

[708] Captain Paul Watson, *Earthforce! An Earth Warrior's Guide to Strategy*, Chaco Press, Los Angeles, California (1993).

[709] Captain Paul Watson, *Ocean Warrior: My Battle to End the Illegal Slaughter on the High Seas*, Key Porter Books Limited, Toronto, Ontario, Canada (1994).

[710] Duncan J. Watts, *Small Worlds: The Dynamics of Networks between Order and Randomness*, Princeton University Press, Princeton (1999).

[711] Duncan J Watts, *Six Degrees: The Science of a Connected Age*, W.W. Norton & Co., New York (2003).

[712] Jack Weatherford, *The History of Money*, Crown Publishers, Inc., New York (1997).

[713] Gary Webb, *Dark Alliance: The CIA, the Contras, and the Crack Cocaine Explosion*, Seven Stories Press, New York (1998).

[714] Maureen Webb, *Illusions of Security: Global Surveillance and Democracy in the Post-9/11 World,* City Lights, San Francisco (2007).

[715] Robert Weissman, et al., *Multinational Monitor*, Essential Information, Inc., Washington, D.C. (May/June 2006).

[716] Robert Weissman, et al. http://www.wallstreetwatch.org/reports/sold_out.pdf

[717] Robert Weissman, et al., "Wall Street Self Destructs," *Multinational Monitor*, January/February 2009.

[718] Eileen Welsome, *The Plutonium Files: America's Secret Medical Experiments in the Cold War*, Delta, Random House, New York (1999).

[719] Winslow T. Wheeler, *The Wastrels of Defense: How Congress Sabotages U.S. Security*, Naval Institute Press, Annapolis, Maryland (2004).

[720] Reg Whitaker, *The End of Privacy: How Total Surveillance Is Becomming a Reality*, The New Press, New York (1999).

[721] Geoffry D. White, with Flannery C. Hauck, *Campus Inc.: Corporate Power in the Ivory Tower*, Prometheus Books, New York (2000).

[722] Jon Wiener, *Professors, Politics and Pop*, Verso, New York (1991).

[723] Jon Wiener, *Gimme Some Truth: The John Lennon FBI Files*, University of California Press, Los Angeles, California (1999).

[724] Fred A. Wilcox, *Waiting for an Army to Die: The Tragedy of Agent Orange*, Seven Locks Press, Santa Ana, Ca (1989).

[725] Todd Wilkinson, *Science Under Siege: The Politician's War on Nature and Truth*, Johnson Books, Boulder, Colorado (1998).

[726] Michael Williams, *Deforesting the Earth: From Prehistory to Global Crisis*, The University of Chicago Press, Chicago (2006).

[727] Howard G. Wilshire, Jane E. Nielson, Richard W. Hazlett, *The American West at Risk: Science, Myths, and Politics of Land Abuse and Recovery*, Oxford University Press, New York (2008).

[728] Duff Wilson, *Fateful Harvest: The True Story of a Small Town, a Global Industry, and a Toxic Secret*, Perennial, New York (2002).

[729] Richard Wilson, John Spengler, editors, *Particles in Our Air: Concentrations and Health Effects*, Harvard University Press, Mass. (1996).

[730] Raymond A. Winbush, Ph.D., *Should America Pay? Slavery and the Raging Debate on Reparations*, Amistad, New York (2003).

[731] Carl K. Winter, James N. Seiber, Carole F. Nuckton, *Chemicals in the Human Food Chain*, Thomson Learning, New York (1990).

[732] Naomi Wolf, *The End of America: Letter of Warning to a Young Patriot*, Chelsea Green Publishing Company, Vermont (2007).

[733] Edward N. Wolff, *Top Heavy: The Increasing Inequality of Wealth in America and What Can Be Done About It*, The New Press, New York (1996).

[734] Richard D. Wolff, *Capitalism Hits the Fan: The Global Economic Meltdown and What to Do About It,* Olive Branch Press, (2009).

[735] Christine Woodside, *The Homeowner's Guide to Energy Independence: Alternative Power Sources for the Average American*, The Lyons Press, Guilford, Ct (2006).

[736] Boris Worm, Edward B. Barbier, Nicola Beaumont, J. Emmett Duffy, Carl Folke, Benjamin S. Halpern, Jeremy B. C. Jackson, Heike K. Lotze, Fiorenza Micheli, Stephen R. Palumbi, Enric Sala, Kimberley A. Selkoe, John J. Stachowicz, Reg Watson, "Impacts of Biodiversity Loss on Ocean Ecosystem Services," *Science*, pp. 770–772 (November 3, 2006).

[737] George Wuerthner, Mollie Matteson, *Welfare Ranching: The Subsidized Destruction of the American West*, Foundations for Deep Ecology 2, (2002).

[738] George Wuerthner, Editor, *The Wild Fire Reader: A Century of Failed Forest Policy*, Island Press, Covelo, California (2006).

[739] James Yee, *For God and Country: Faith and Patriotism Under Fire*, Public Affairs, New York (2005).

[740] H. Peyton Young, *Individual Strategy and Social Structure: An Evolutionary Theory of Institutions*, Princeton University Press, Princeton (1998).

[741] H. Peyton Young, *Equity: In Theory and Practice*, Princeton University Press, Princeton (1994).

[742] Howard Zinn, *A People's History of the United States: 1492–Present*, HarperCollins, New York (1995).

[743] Howard Zinn, Anthony Arnove, *Voices of a People's History of the United States*, Seven Stories Press, New York (2004).

[744] Robert Zubrin, *Energy Victory: Winning the War on Terror by Breaking Free of Oil*, Prometheus Books, (2009).

[745] Barrie Zwicker, *Towers of Deception: The Media Cover-Up of 9/11*, New Society Publishers, Gabriola Island, British Columbia, Canada (2006).

Index

9781439834725

An environmentally friendly book printed and bound in England by www.printondemand-worldwide.com

PEFC Certified

This product is from sustainably managed forests and controlled sources

PEFC™

PEFC/16-33-415

www.pefc.org

FSC®

www.fsc.org

MIX

Paper from responsible sources

FSC® C004959

This book is made entirely of sustainable materials; FSC paper for the cover and PEFC paper for the text pages.

Reprint of # - CO - 234/156/36 - CB - Lamination Gloss - Printed on 23-Sep-15 10:27